Annotated Instructor's Edition

Basic College Mathematics

FIFTH EDITION

Margaret L. Lial
American River College

Stanley A. Salzman
American River College

Diana L. Hestwood
Minneapolis Community and Technical College

Charles D. Miller

▲ ADDISON-WESLEY

An imprint of Addison Wesley Longman, Inc.

Reading, Massachusetts • Menlo Park, California • New York • Harlow, England
Don Mills, Ontario • Sydney • Mexico City • Madrid • Amsterdam

SEP 0 9 2002

Publisher: Jason Jordan

Acquisition Editor: Rita Ferrandino

Assistant Editor: K. B. Mello

Developmental Editor: Sandi Goldstein

Managing Editor: Ron Hampton

Production Services: Elm Street Publishing Services, Inc.

Art Development: Meredith Nightingale

Marketing Manager: Liz O'Neil

Senior Manufacturing Manager: Roy Logan

Manufacturing Manager: Ralph Mattivello

Text and Cover Designer: Susan Carsten

Cover Photograph: ©1997 Gary Price/Picture Perfect

Basic College Mathematics, Fifth Edition

Copyright © 1998 by Addison Wesley Longman. All rights reserved.

No part of this work may be reproduced or transmitted in any form or by any means, electronic or mechanical, including photocopying and recording, or by any information storage or retrieval system without the prior written permission of Addison Wesley Longman unless such copying is expressly permitted by federal copyright law. Address inquiries to Addison Wesley Longman, One Jacob Way, Reading, MA.

Printed in the U.S.A.

Library of Congress Cataloging-in-Publication Data

Miller, Charles David.
    Basic college mathematics/Charles D. Miller, Stanley A. Salzman,
Diana L. Hestwood.—5th ed.
        p.   cm.
    Includes index.
    ISBN 0-321-01265-8 (Student Edition)
    ISBN 0-321-40343-6 (Annotated Instructor's Edition)
    1. Mathematics.   I. Salzman, Stanley A.   II. Hestwood, Diana.
III. Title
QA37.2.M55   1997
513'.1—DC21   96-52938
                                                    CIP

123456789-DOW-01 00 99 98

Contents

Preface

This new edition of *Basic College Mathematics* is designed to help students achieve success in a developmental mathematics program. It is comprehensive, providing the necessary background and review in whole numbers, fractions, decimals, ratio and proportion, percent, and measurement, as well as an introduction to algebra and geometry and a preview of statistics.

We have addressed many of the concerns of the **American Mathematical Association of Two-Year Colleges** and the **National Council of Teachers of Mathematics** by providing the following:

▶ Exercises that build conceptual understanding;

▶ Opportunities for students to write about mathematics;

▶ Emphasis on estimation;

▶ Appropriate use of technology;

▶ Data presented in the form of graphs, charts, and tables; and

▶ Applications that emphasize the use of mathematics in real-world situations.

The following pages describe some of the key features of this text.

New Features

▶ *Numbers in the Real World: Collaborative Investigations* pages show students how mathematics is used in everyday life. Students read and interpret articles and graphs from newspapers, tables of data from food packaging, labels on consumer goods, and more. These investigations may be completed by individual students but are well-suited as collaborative assignments for pairs or small groups of students, or for open-ended discussion by an entire class.

▶ *Calculator Tips* are included at appropriate points to guide students in using their scientific calculators appropriately and effectively. The calculator is used to emphasize such concepts as order of operations, commutativity of addition and multiplication, rounding of decimal numbers, division by zero as undefined, and more. Appropriate calculator exercises are designated in selected sections. (If desired, the Calculator Tips and exercises may be omitted, with the exception of finding non-perfect square roots in Section 8.8, Pythagorean Theorem.)

▶ **Estimation** has been added to the fractions and percent chapters, building on the estimation skills already found in the whole numbers and decimals chapters.

▶ **Additional conceptual and writing exercises** have been added throughout the book. Students are asked to look for patterns, devise their own rules, use estimation to identify reasonable and unreasonable solutions, and extend their thinking to cover new possibilities.

▶ **Application exercises** appear more frequently, with up-to-date data drawn from a variety of real-world sources. The data is usually presented in the form of a table, chart, or graph.

▶ **Phonetic spellings** help students pronounce key terms correctly. This feature will be especially useful for students who speak English as their second language. A separate *Spanish Glossary* is also available.

▶ Instructors and students will have access to a World Wide Web site (**www.mathnotes.com),** where additional support material is available.

Continuing Features

We have retained the popular features of the previous editions of the text. Some of these features are:

▶ **Ample and varied exercise sets** Students of basic mathematics require a large number and variety of exercises. This text meets that need: approximately 7100 exercises.

▶ **Learning objectives** Each section begins with clearly stated, numbered objectives, and material in the section is keyed to these objectives. In this way, students and instructors know exactly what is being covered in each section.

▶ **Margin problems** Margin problems are found in every section, with answers immediately available at the bottom of the page. These problems allow the student to practice the material covered in the section in preparation for the exercise set that follows.

▶ **Opportunity for review** A group of exercises called "Review and Prepare" appear at the end of each section. These exercises review skills from earlier chapters, particularly ones that will be needed in the upcoming section. Each chapter concludes with a concise *chapter summary*, which includes key terms and a quick review, *chapter review exercises* with exercises keyed to individual sections, mixed review exercises, and a *chapter test*. Furthermore, following every chapter after Chapter 1, there is a set of cumulative review exercises that covers material going back to the first chapter. Students always have an opportunity to review material that appears earlier in the text, and this provides an excellent way to prepare for the final examination in the course.

▶ **Answers and solutions** Answers to all margin problems are provided at the bottom of the page on which the problem appears. The answer section at the back of the text provides answers to odd-numbered exercises in numbered sections and answers to *all* chapter test exercises, review exercises, and cumulative review exercises. The solutions section at the back of the text gives students complete, worked-out solutions to every fourth exercise in numbered sections. In this edition, we have also included sample answers to writing exercises.

All-New Supplements Package

Our extensive new supplements package includes an Annotated Instructor's Edition, testing materials, solutions, software, and videotapes.

FOR THE INSTRUCTOR

Annotated Instructor's Edition

This edition provides instructors with immediate access to the answers to every exercise in the text. Each answer is printed in color next to the corresponding text exercise. Answers with the ≈ symbol have been rounded according to exercise instructions. This will help instructors identify students who can solve the exercise but have trouble with rounding. Symbols are also used to identify the writing (✐) and conceptual (◉) exercises to assist in making homework assignments. Challenging (▲) exercises are also marked for the instructor for this purpose.

Instructor's Resource Guide

The *Instructor's Resource Guide* includes suggestions for using the textbook in a mathematics laboratory; short-answer and multiple-choice versions of a placement test; eight forms of chapter tests for each chapter, including six open-response and two multiple-choice forms; short-answer and multiple-choice forms of a final examination; and an extensive set of additional exercises, providing 10 to 20 exercises for each textbook objective, which instructors can use as another source of questions for tests, quizzes, or student review of difficult topics. In addition, a section containing teaching tips is included for the instructor's convenience.

Instructor's Solutions Manual

This manual includes solutions to all of the even-numbered section exercises.

Answer Book

The *Answer Book* includes answers to all exercises.

TestGen EQ with QuizMaster EQ

This test generation software is available in Windows and Macintosh versions and is fully networkable. TestGen EQ's friendly, graphical interface enables instructors to easily view, edit, and add questions; transfer questions to tests; and print tests in a variety of fonts and forms. Search and sort features allow instructors to quickly locate questions and arrange them in a preferred order. Six question formats are available, including short-answer, true-false, multiple-choice, essay, matching, and bimodal formats. A built-in question editor gives users power to create graphs, import graphics, insert mathematical symbols and templates, and insert variable numbers or text. Computerized testbanks include algorithmically defined problems organized according to each textbook.

QuizMaster EQ enables instructors to create and save tests using TestGen EQ so students can take them for practice or a grade on a computer network. Instructors can set preferences for how and when tests are administered. Quiz-Master EQ automatically grades the exams, stores results on disk, and allows instructors to view or print a variety of reports for individual students, classes, or courses.

InterAct Mathematics Plus—Management System

InterAct Math Plus combines course management and online testing with the features of the basic InterAct Math tutorial software to create an invaluable teaching resource. InterAct Math is available in either Windows or Macintosh versions; consult your Addison Wesley Longman representative for details.

FOR THE STUDENT

STUDENT'S SOLUTIONS MANUAL

This book contains solutions to every other odd-numbered section exercise (those not included at the back of the textbook) as well as solutions to all margin problems, chapter review exercises, chapter tests, and cumulative review exercises. (ISBN 0-321-01317-4)

InterAct Mathematics Tutorial Software

InterAct Math tutorial software has been developed and designed by professional software engineers working closely with a team of experienced developmental math educators.

InterAct Math tutorial software includes exercises that are linked with every objective in the textbook and require the same computational and problem-solving skills as their companion exercises in the text. Each exercise has an example and an interactive guided solution that are designed to involve students in the solution process and to help them identify precisely where they are having trouble. In addition, the software recognizes common student errors and provides students with appropriate customized feedback.

With its sophisticated answer-recognition capabilities, InterAct Math tutorial software recognizes appropriate forms of the same answer for any kind of input. It also tracks student activity and scores for each section, which can then be printed out. Available for Windows and Macintosh computers, the software is free to qualifying adopters or can be bundled with books for sale to students. (Macintosh: ISBN 0-321-00935-5, Windows: ISBN 0-321-00934-7)

Videotapes

A videotape series, "Real to Reel," has been developed to accompany *Basic College Mathematics,* Fifth Edition. In a separate lesson for each section in the book, the series covers all objectives, topics, and problem-solving techniques discussed within the text. The video series is free to qualifying adopters. (ISBN 0-321-01320-4)

Spanish Glossary

A separate *Spanish Glossary* is now being offered as part of the supplements package for this textbook series. This book contains the key terms from each of the four texts in the series and their Spanish translations. (ISBN 0-321-01647-5)

Acknowledgments

For a textbook to succeed through five editions, it is necessary for the authors to rely on comments and suggestions from users, non-users, instructors, and students. We are grateful for the many responses that we have received over the years. We wish to thank the following individuals who reviewed this edition of the text:

Susan Anderson, *University of Minnesota*

Jackie Back, *Ashland Community College*

Arlene Bakner, *Collin County Community College*

Jo Cummins, *South Texas Community College*

Sue Fader, *Del Technical and Community College*

Scott Higinbotham, *Middlesex Community College*

Jeff Koleno, *Lorain County Community College*

Barbara Kurt, *St. Louis Community College at Meramec*

Pat O'Brien, *Mesa Community College*

Bettie Truitt, *Blackhawk College*

No author can write a text of this magnitude without the help of many other individuals. Our sincere thanks goes to Rita Ferrandino of Addison Wesley Longman who coordinates the package of texts of which this book is a part. We are appreciative of the support given to us by Greg Tobin, also of Addison Wesley Longman, who helped make the transition from our former publisher go as smoothly as one could ever imagine. We wish to thank Sandi Goldstein for her efforts in coordinating the reviews and working with us in the early stages of preparation for production. Becky Troutman assisted in preparing the solutions that appear in the back of the book. Carmen Eldersveld, Scott Higinbotham, and Barbara Kurt helped in checking for the accuracy of the answers. Cathy Wacaser of Elm Street Publishing Services provided her usual excellent production work. She is indeed one of the best in the business. As usual, Paul Van Erden created an accurate, useful index. We are also grateful to Tommy Thompson who made suggestions for the feature "To the Student: Success in Mathematics."

Our sincere thanks also goes to Paul Eldersveld of the College of DuPage who has coordinated our print supplements for many years. The importance of his job cannot be overestimated, and we want to thank him for his work over the years.

In appreciation of your lasting support and never-ending enthusiasm: family, colleagues, and more than a generation of motivated students.

Stan Salzman

This book is dedicated to my dad, who always told me when I was young that girls could learn math, and to my students at Minneapolis Community and Technical College, who keep me in touch with the real world.

Diana L. Hestwood

Features Walk-Through

A new design updates and refreshes the overall flow of the text. New, full-color situational art throughout the text enhances student comprehension of the material.

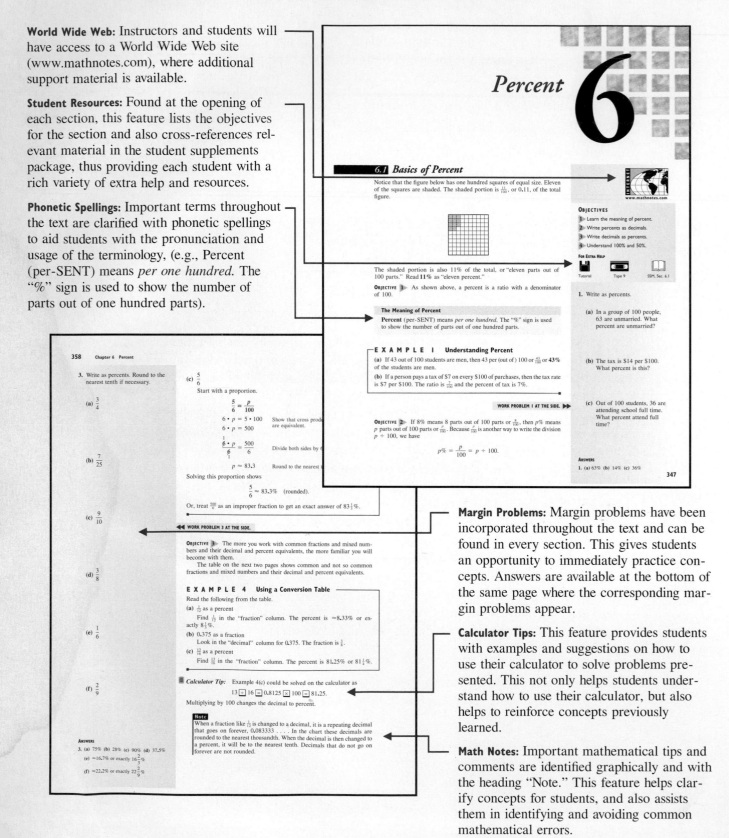

World Wide Web: Instructors and students will have access to a World Wide Web site (www.mathnotes.com), where additional support material is available.

Student Resources: Found at the opening of each section, this feature lists the objectives for the section and also cross-references relevant material in the student supplements package, thus providing each student with a rich variety of extra help and resources.

Phonetic Spellings: Important terms throughout the text are clarified with phonetic spellings to aid students with the pronunciation and usage of the terminology, (e.g., Percent (per-SENT) means *per one hundred*. The "%" sign is used to show the number of parts out of one hundred parts).

Margin Problems: Margin problems have been incorporated throughout the text and can be found in every section. This gives students an opportunity to immediately practice concepts. Answers are available at the bottom of the same page where the corresponding margin problems appear.

Calculator Tips: This feature provides students with examples and suggestions on how to use their calculator to solve problems presented. This not only helps students understand how to use their calculator, but also helps to reinforce concepts previously learned.

Math Notes: Important mathematical tips and comments are identified graphically and with the heading "Note." This feature helps clarify concepts for students, and also assists them in identifying and avoiding common mathematical errors.

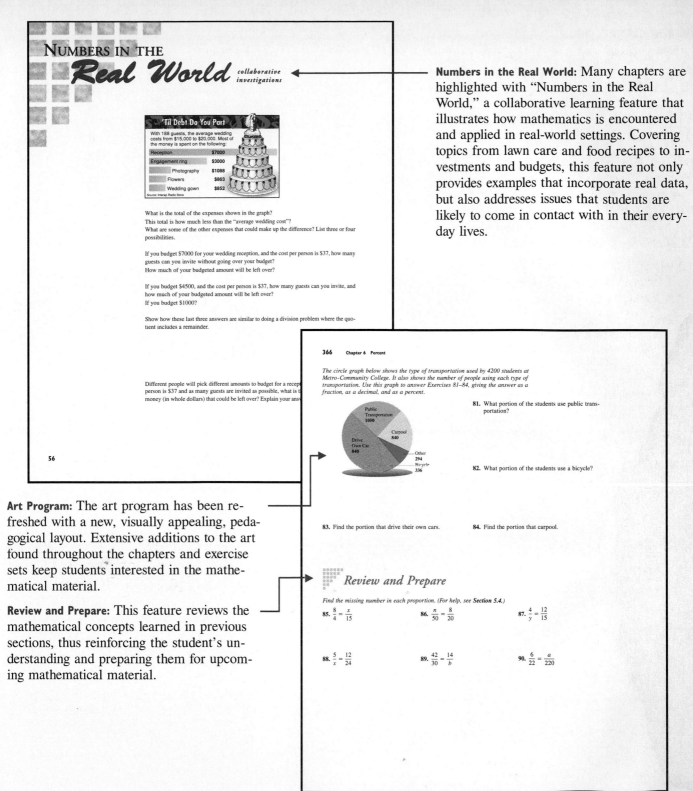

NUMBERS IN THE
Real World *collaborative investigations*

'Til Debt Do You Part

With 188 guests, the average wedding costs from $15,000 to $20,000. Most of the money is spent on the following:

Reception	$7000
Engagement ring	$3000
Photography	$1088
Flowers	$863
Wedding gown	$852

Source: Interep Radio Store

What is the total of the expenses shown in the graph?
This total is how much less than the "average wedding cost"?
What are some of the other expenses that could make up the difference? List three or four possibilities.

If you budget $7000 for your wedding reception, and the cost per person is $37, how many guests can you invite without going over your budget?
How much of your budgeted amount will be left over?

If you budget $4500, and the cost per person is $37, how many guests can you invite, and how much of your budgeted amount will be left over?
If you budget $1000?

Show how these last three answers are similar to doing a division problem where the quotient includes a remainder.

Different people will pick different amounts to budget for a recept___
person is $37 and as many guests are invited as possible, what is t___
money (in whole dollars) that could be left over? Explain your ans___

56

Numbers in the Real World: Many chapters are highlighted with "Numbers in the Real World," a collaborative learning feature that illustrates how mathematics is encountered and applied in real-world settings. Covering topics from lawn care and food recipes to investments and budgets, this feature not only provides examples that incorporate real data, but also addresses issues that students are likely to come in contact with in their everyday lives.

366 Chapter 6 Percent

The circle graph below shows the type of transportation used by 4200 students at Metro-Community College. It also shows the number of people using each type of transportation. Use this graph to answer Exercises 81–84, giving the answer as a fraction, as a decimal, and as a percent.

Public Transportation 1050
Carpool 840
Drive Own Car 840
Other 294
Bicycle 336

81. What portion of the students use public transportation?

82. What portion of the students use a bicycle?

83. Find the portion that drive their own cars.

84. Find the portion that carpool.

Review and Prepare

Find the missing number in each proportion. (For help, see **Section 5.4**.)

85. $\frac{8}{4} = \frac{x}{15}$

86. $\frac{n}{50} = \frac{8}{20}$

87. $\frac{4}{y} = \frac{12}{15}$

88. $\frac{5}{x} = \frac{12}{24}$

89. $\frac{42}{30} = \frac{14}{b}$

90. $\frac{6}{22} = \frac{a}{220}$

Art Program: The art program has been refreshed with a new, visually appealing, pedagogical layout. Extensive additions to the art found throughout the chapters and exercise sets keep students interested in the mathematical material.

Review and Prepare: This feature reviews the mathematical concepts learned in previous sections, thus reinforcing the student's understanding and preparing them for upcoming mathematical material.

Exercise Sets: Ample exercise sets provide the carefully designed practice students need. Also included are four special types of problems: Writing (✎), Conceptual (◉), Challenging (▲), and Estimation (≈). Each of these problem types is graphically represented in the Annotated Instructor's Edition by an icon to help guide instructors through the exercise sets in assigning appropriate problems. Additionally, nearly 50% of the exercises throughout the text have been enhanced with real-world applications and data.

Find the amount or rate of discount and the amount paid after the discount. Round money answers to the nearest cent if necessary.

Example: A Casio QV-30 Digital Camera is normally priced at $898. If it is discounted 20%, find the amount of discount and the sale price.

Solution: discount = rate of discount · original price
= 20% · $898
= 0.2 · $898
= $179.60   Amount of discount

The amount of discount is **$179.60** and the sale price is **$718.40** ($898 − $179.60).

	Original Price	Rate of Discount	Amount of Discount	Sale Price
17.	$100	15%	_____	_____
18.	$200	20%	_____	_____
19.	$180	_____	$54	_____
20.	$38	_____	$9.50	_____
21.	$17.50	25%	_____	_____
22.	$76	60%	_____	_____
23.	$37.50	10%	_____	_____
24.	$49.90	40%	_____	_____

✎ ◉ **25.** You are trying to decide between Company A paying a 10% commission and Company B ~~paying an 8%~~ commission. For which com~~pany would yo~~u prefer to work? Are there ~~factors~~ other than commission rate that ~~are impo~~rtant to you? What would

✎ ◉ **26.** Give four examples of where you might use the percent of increase or the percent of decrease in your own personal activities. Think in terms of work, school, home, hobbies, and sports. Write an increase or a decrease problem about one of these four examples, then show how to solve it.

CHAPTER 6 SUMMARY   423

KEY TERMS

6.1	percent	Percent means per one hundred. A percent is a ratio with a denominator of 100.
6.3	percent proportion	The proportion $\frac{\text{amount}}{\text{base}} = \frac{\text{percent}}{100}$ or $\frac{a}{b} = \frac{p}{100}$ is used to solve percent problems.
6.4	base	The base in a percent problem is the entire quantity, the total, or the whole.
	amount	The amount in a percent problem is the part being compared with the whole.
6.6	percent equation	The percent equation is amount = percent · base. It is another way to solve percent problems.
6.7	sales tax	Sales tax is a percent of the total sales charged as a tax.
	commission	Commission is a percent of the dollar value of total sales paid to a salesperson.
	discount	Discount is often expressed as a percent of the original price; it is then deducted from the original price, resulting in the sale price.
	percent of increase or decrease	Percent of increase or decrease is the amount of increase or decrease expressed as a percent of the original amount.
6.8	interest	Interest is a fee paid or a charge for lending or borrowing money.
	interest formula	The interest formula is used to calculate interest. It is interest = principal · rate · time or $I = p \cdot r \cdot t$.
	simple interest	Interest that is computed on the original principal.
	principal	Principal is the amount of money on which interest is earned.
	rate of interest	Often referred to as "rate," it is the charge for interest and is given as a percent.
6.9	compound interest	Compound interest is interest paid on past interest as well as on principal.
	compound amount	The total amount in an account including compound interest and the original principal.
	compounding	Interest that is **compounded** once each year is compounded **annually.**

QUICK REVIEW

Concepts	Examples
6.1 Basics of Percent	
Writing a Percent as a Decimal To write a percent as a decimal, move the decimal point two places to the left and drop the % sign.	50% (.50%) = 0.50 or just 0.5 3% (.03%) = 0.03
Writing a Decimal as a Percent To write a decimal as a percent, move the decimal point two places to the right and attach a % sign.	0.75 (0.75) = 75% 3.6 (3.60) = 360%
6.2 Writing a Fraction as a Percent Use a proportion and solve for p to change a fraction to percent.	$\frac{2}{5} = \frac{p}{100}$ Proportion $5 \cdot p = 2 \cdot 100$ Cross products $5 \cdot p = 200$ $\frac{5 \cdot p}{5} = \frac{200}{5}$ Divide both sides by 5. $p = 40$ $\frac{2}{5} = 40\%$ Attach % sign.

End-of-Chapter Material: In order to reinforce concepts learned and assist students with their retention, the end-of-chapter material includes a Chapter Summary of key terms, a Quick Review of concepts learned, Review Exercises, and a Chapter Test.

To the Student: Success in Mathematics

The main reason students have difficulty with mathematics is that they don't know how to study it. Studying mathematics *is* different from studying subjects like English or history. The key to success is regular practice. This should not be surprising. After all, can you learn to ski or play a musical instrument without a lot of regular practice? The same thing is true for learning mathematics. Working problems nearly every day is the key to becoming successful. Here are suggestions to help you succeed in studying mathematics.

1. Pay attention in class to what your instructor says and does, and make careful notes. Note the problems the instructor works on the board and copy the complete solutions. Keep these notes separate from your homework to avoid confusion when you read them later.

2. Feel free to ask questions in class. It is not a sign of weakness, but of strength. There are always other students with the same question who are too shy to ask.

3. Determine whether tutoring is available and know how to get help when needed. Use the instructor's office hours and contact the instructor for suggestions and direction if necessary.

4. Before you start on your homework assignment, rework the problems the instructor worked in class. This will reinforce what you have learned. Many students say, "I understand it perfectly when you do it, but I get stuck when I try to work the problem myself."

5. *Read your text carefully*. Many students read only enough to get by, usually only the examples. Reading the complete section will help you be successful with the homework problems. As a bonus you will be able to do the problems more quickly if you have read the text first. As you read the text, work the example problems and check the answers. This will test your understanding of what you have read. Pay special attention to highlighted statements and those labeled "Note."

6. Do your homework assignment only after reading the text and reviewing your notes from class. Estimate the answer before you begin working the problem in the worktext. Check your work before checking with the answers in the back of the book. If you get a problem wrong and are unable to understand why, mark that problem and ask your instructor about it.

7. Work as neatly as you can using a *pencil* and organize your work carefully. Write your symbols clearly, and make sure the problems are clearly separated from each other. Working neatly will help you to think clearly and also make it easier to review the homework before a test.

8. After you have completed a homework assignment, look over the text again. Try to decide what the main ideas are in the lesson. Often they are clearly highlighted or boxed in the text.

9. Keep any quizzes and tests that are returned to you for studying for future tests and the final exam. These quizzes and tests indicate what your instructor considers most important. Be sure to correct any problems on these tests that you missed, so you will have the corrected work to study. Write all quiz and test scores on the front page of your notebook.

10. Don't worry if you do not understand a new topic right away. As you read more about it and work through the problems, you will gain understanding. Each time you review a topic you will understand it a little better. No one understands each topic completely right from the start.

Whole Numbers

1

1.1 Reading and Writing Whole Numbers

OBJECTIVE 1 The **decimal system** of writing numbers uses the ten digits

$$0, 1, 2, 3, 4, 5, 6, 7, 8, 9$$

to write any number. For example, these digits can be used to write the **whole numbers:**

$$0, 1, 2, 3, 4, 5, 6, 7, 8, 9, 10, 11, 12, 13$$

and so on.

OBJECTIVE 2 Each digit in a whole number has a **place value.** The following place value chart shows the names of the different places used most often.

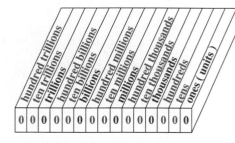

E X A M P L E I   Identifying Whole Numbers

(a) In the whole number 63, the 3 is in the ones place and has a value of 3 *ones.*

(b) In 37, the 3 is in the tens place and has a value of 3 *tens.*

(c) In 381, the 3 is in the hundreds place and has a value of 3 *hundreds.*
   The value of 3 in each number is different, depending on its location (place) in the number.

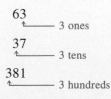

WORK PROBLEM I AT THE SIDE. ▶▶

INTERNET
www.mathnotes.com

OBJECTIVES

1 ▶ Identify whole numbers.

2 ▶ Give the place value of a digit.

3 ▶ Write a number in words or digits.

FOR EXTRA HELP

Tutorial   Tape I   SSM, Sec. 1.1

1. Identify the place value of the 4 in each whole number.

   (a) 341

   (b) 714

   (c) 479

ANSWERS
1. (a) tens (b) ones (c) hundreds

1

2. Identify the place value of each digit.

   (a) 24,386

   (b) 371,942

3. In the number 3,251,609,328 identify the digits in each of the following periods (groups).

   (a) billions period

   (b) millions period

   (c) thousands period

   (d) ones period

E X A M P L E 2   Identifying Place Values

Find the place value of each digit in the number 725,283.

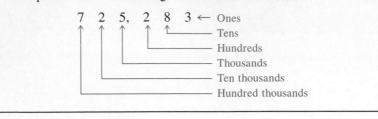

◀◀ **WORK PROBLEM 2 AT THE SIDE.**

Notice the comma between the hundreds and thousands position in the number 725,283 above.

Using Commas

Commas are used to separate each group of three digits, starting from the right. This makes numbers easier to read. (An exception: commas are frequently omitted in four-digit numbers such as 9748 or 1329.) Each three-digit group is called a **period.** Some instructors prefer to just call them **groups.**

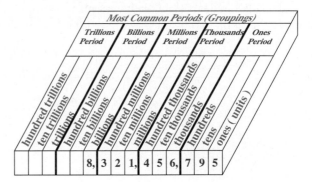

E X A M P L E 3   Knowing the Period or Group Names

Write the digits in each period of 8,321,456,795.

                          8,321,456,795

      8 billions ←
   321 millions ←
 456 thousands ←
    795 ones ←

◀◀ **WORK PROBLEM 3 AT THE SIDE.**

Use the following rule to read a number with more than three digits.

Reading Numbers

Start at the left when reading a number. Read the digits in each period (group), followed by the name of the period, except for the period name "ones," which is *not* used.

ANSWERS

2. (a) 2: ten thousands
     4: thousands
     3: hundreds
     8: tens
     6: ones
 (b) 3: hundred thousands
     7: ten thousands
     1: thousands
     9: hundreds
     4: tens
     2: ones

3. (a) 3 **(b)** 251 **(c)** 609 **(d)** 328

OBJECTIVE 3▷ The following examples show how to write names for whole numbers.

┌─
E X A M P L E 4 **Writing Numbers in Words**

Write each number in words.

(a) 57

This number means 5 tens and 7 ones, or 50 ones and 7 ones. Write the number as

fifty-seven.

(b) 94

ninety-four

(c) 874

eight hundred seventy-four

(d) 601

six hundred one
└────────────────────────■

Note
The word "and" should never be used when writing whole numbers. You will often hear someone say "five hundred and twenty-two," but the use of "and" is not correct since "522" is a whole number. When you work with decimal numbers, the word "and" is used to show the position of the decimal point. For example, 98.6 is read as "ninety-eight *and* six tenths." Practice with decimal numbers is the topic of **Section 4.1.**

WORK PROBLEM 4 AT THE SIDE. ▶▶

┌─
E X A M P L E 5 **Writing Numbers in Words by Using Period Names**

Write each number in words.

(a) 725,283

seven hundred twenty-five **thousand,** two hundred eighty-three

Number in period   Name of period   Number in period (not necessary to write "ones")

(b) 7835

seven **thousand,** eight hundred thirty-five

Name of period   No period name needed

(c) 111,356,075

one hundred eleven **million,** three hundred fifty-six **thousand,** seventy-five

(d) 17,000,017,000

seventeen **billion,** seventeen **thousand**
└────────────────────────■

WORK PROBLEM 5 AT THE SIDE. ▶▶

4. Write each number in words.

(a) 46

(b) 68

(c) 293

(d) 902

5. Write each number in words.

(a) 7309

(b) 95,372

(c) 100,075,002

(d) 11,022,040,000

ANSWERS

4. **(a)** forty-six **(b)** sixty-eight
   (c) two hundred ninety-three
   (d) nine hundred two
5. **(a)** seven thousand, three hundred nine
   (b) ninety-five thousand, three hundred seventy-two
   (c) one hundred million, seventy-five thousand, two
   (d) eleven billion, twenty-two mill forty thousand

6. Rewrite each of the following numbers using digits.

  (a) one thousand, four hundred thirty-seven

EXAMPLE 6 · **Writing Numbers in Digits**

Rewrite each of the following numbers using digits.

(a) seven thousand, eighty-five

$$7085$$

(b) two hundred fifty-six thousand, six hundred twelve

$$256,612$$

(c) nine million, five hundred fifty-nine

$$9,000,559$$

Zeros indicate
there are no thousands

◀◀ **WORK PROBLEM 6 AT THE SIDE.**

▦ *Calculator Tip:*   Does your calculator show a comma between each group of three digits? Probably not, but try entering a long number such as 34,629,075. Notice that there is no key with a comma on it, so you do not enter commas. A few calculators may show the position of the commas *above* the digits, like this.

34′629′075

Most of the time you will have to write in the commas.

(b) nine hundred seventy-one thousand, six

(c) eighty-two million, three hundred twenty-five

1.1 Exercises

*Fill in the digit for the given **place value** in each of the following whole numbers.*

Example:		Solution:
782	hundreds	7
	ones	2

1. 5031
thousands **5**
tens **3**

2. 7428
thousands **7**
ones **8**

3. 18,015
ten thousands **1**
hundreds **0**

4. 75,229
ten thousands **7**
ones **9**

5. 7,628,592,183
millions **8**
thousands **2**

6. 1,700,225,016
billions **1**
millions **0**

*Fill in the number for the given **period** (group) in each of the following whole numbers.*

Example:		Solution:
58,618	thousands	**58**
	ones	**618**

7. 7,536,175
millions **7**
thousands **536**
ones **175**

8. 28,785,203
millions **28**
thousands **785**
ones **203**

9. 60,000,502,109
billions **60**
millions **0**
thousands **502**
ones **109**

10. 100,258,100,006
billions **100**
millions **258**
thousands **100**
ones **6**

11. Do you think the fact that humans have four fingers and a thumb on each hand explains why we use a number system based on ten digits? Explain.

Evidence suggests that this is true. It is common to count using fingers.

12. The decimal system uses ten digits. Fingers and toes are often referred to as digits. In your opinion, is there a relationship here? Explain.

No doubt there is a relationship here. One answer might be that people could count using their fingers and toes and, therefore, thought of them as numbers or digits.

Rewrite the following numbers in words.

Example: 1,630,254	Solution: one million, six hundred thirty thousand, two hundred fifty-four

13. 64,215   **sixty-four thousand, two hundred fifteen**

14. 37,886   **thirty-seven thousand, eight hundred eighty-six**

15. 725,009   **seven hundred twenty-five thousand, nine**

16. 218,033   **two hundred eighteen thousand, thirty-three**

17. 25,756,665   **twenty-five million, seven hundred fifty-six thousand, six hundred sixty-five**

18. 999,993,000   **nine hundred ninety-nine million, nine hundred ninety-three thousand**

Writing   Conceptual   ▲ Challenging   ≈ Estimation

Rewrite each of the following numbers using digits.

> **Example:** three thousand, four hundred twenty   **Solution: 3420**

19. thirty-two thousand, five hundred twenty-six   **32,526**

20. ninety-five thousand, one hundred eleven   **95,111**

21. ten million, two hundred twenty-three   **10,000,223**

22. one hundred million, two hundred   **100,000,200**

Rewrite the numbers from the following sentences using digits.

> **Example:** Her income tax refund was one thousand, five hundred twelve dollars.   **Solution:** In digits, the number is **$1512.**

23. Java City sold four thousand, twenty cups of coffee in a two-day period.

4020

24. Twentieth Century Fox hoped video sales of the hit film *Independence Day* would exceed four hundred fifty million dollars.

$450,000,000

25. The Binney & Smith Company in Pennsylvania makes about two billion Crayola Crayons each year.

2,000,000,000

26. Yosemite National Park had three million, four hundred eighty-five thousand, five hundred visitors last year.

3,485,500

27. The United States Postal Service set a record of two hundred eighty million, four hundred eighty-nine thousand postmarked pieces of mail on a single day.

280,489,000

28. The total Aid to Families with Dependent Children (AFDC) in the United States last year was eighteen billion, six hundred thirty million, six hundred four thousand, seven hundred thirty-three dollars.

$18,630,604,733

▲ **29.** Rewrite eight hundred trillion, six hundred twenty-one million, twenty thousand, two hundred fifteen by using digits.

800,000,621,020,215

▲ **30.** Rewrite 70,306,735,002,102 in words.

seventy trillion, three hundred six billion, seven hundred thirty-five million, two thousand, one hundred two

1.2 *Addition of Whole Numbers*

There are 4 triangles at the left and 2 at the right. In all, there are 6 triangles.

The process of finding the total is called **addition.** Here 4 and 2 were added to get 6. Addition is written with a + sign, so that

$$4 + 2 = 6.$$

OBJECTIVE 1 In addition, the numbers being added are called **addends,** (AD-ends), and the resulting answer is called the **sum** or **total.**

$$
\begin{array}{r}
4 \leftarrow \text{Addend} \\
+\ 2 \leftarrow \text{Addend} \\
\hline
6 \leftarrow \text{Sum (answer)}
\end{array}
$$

Addition problems can also be written horizontally as follows.

$$
\begin{array}{ccccc}
4 & + & 2 & = & 6 \\
\uparrow & & \uparrow & & \uparrow \\
\text{Addend} & & \text{Addend} & & \text{Sum}
\end{array}
$$

Commutative Property of Addition

To change the order of the numbers in an addition problem we use the **commutative** (cuh-MUE-tuh-tiv) **property of addition.** This does not change the sum.

For example, the sum of $4 + 2$ is the same as the sum of $2 + 4$. This allows the addition of the same numbers in a different order.

EXAMPLE 1 Adding Two Single-Digit Numbers

Add the following and then change the order of numbers to write another addition problem.

(a) $6 + 2 = 8$ and $2 + 6 = 8$

(b) $5 + 9 = 14$ and $9 + 5 = 14$

(c) $8 + 3 = 11$ and $3 + 8 = 11$

(d) $8 + 8 = 16$

WORK PROBLEM 1 AT THE SIDE. ▶▶

Associative Property of Addition

By the **associative** (uh-SOH-shuh-tiv) **property of addition,** grouping the addition of numbers in any order does not change the sum.

For example, the sum of $3 + 5 + 6$ may be found as follows.

$$(3 + 5) + 6 = 8 + 6 = 14 \qquad \text{Parentheses tell what to do first.}$$

Another way to add the same numbers is shown below.

$$3 + (5 + 6) = 3 + 11 = 14$$

Either method gives the answer 14.

OBJECTIVES

1 ▶ Add two single-digit numbers.
2 ▶ Add more than two numbers.
3 ▶ Add when carrying is not required.
4 ▶ Add with carrying.
5 ▶ Solve application problems with carrying.
6 ▶ Check the answer in addition.

FOR EXTRA HELP

Tutorial   Tape 1   SSM, Sec. 1.2

1. Add and then change the order of numbers to write another addition problem.

   (a) $3 + 4$

   (b) $9 + 9$

   (c) $7 + 8$

   (d) $6 + 9$

ANSWERS

1. **(a)** 7; $4 + 3$ **(b)** 18 **(c)** 15; $8 + 7$
   (d) 15; $9 + 6$

2. Add the following columns of numbers.

(a)
```
   5
   4
   6
   9
+  2
```

(b)
```
   7
   5
   1
   2
+  6
```

(c)
```
   9
   2
   1
   3
+  4
```

(d)
```
   3
   8
   6
   4
+  8
```

OBJECTIVE 2   To add several numbers, first write them in a column. Add the first number to the second. Add this sum to the third digit; continue until all the digits are used.

E X A M P L E   2   Adding More Than Two Numbers

Add 2, 5, 6, 1, and 4.

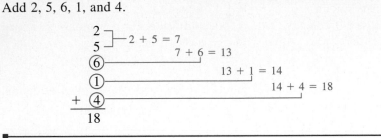

Note

By the commutative and associative properties of addition, numbers may also be added starting at the bottom of a column. Adding from the top or adding from the bottom will give the same answer.

◄◄ **WORK PROBLEM 2 AT THE SIDE.**

OBJECTIVE 3   If numbers have two or more digits, first you must arrange the numbers in columns so that the ones digits are in the same column, tens are in the same column, hundreds are in the same column, and so on. Next, you add column by column starting at the right.

E X A M P L E   3   Adding Without Carrying

Add 511 + 23 + 154 + 10.

First line up the numbers in columns, with the ones column at the right.

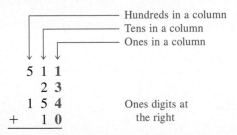

Now start at the right and add the ones digits. Add the tens digits next, and finally, the hundreds digits.

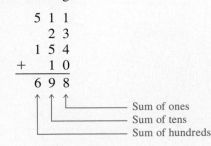

The sum of the four numbers is 698.

WORK PROBLEM 3 AT THE SIDE. ▶▶

OBJECTIVE ▶ 4 If the sum of the digits in a column is more than 9, use carrying.

E X A M P L E 4   Adding with Carrying

Add 47 and 29.

Add ones.

$$
\begin{array}{r}
47 \\
+\ 29 \\
\hline
\end{array}
$$
⌐—— Sum of ones is 16.

Because 16 is 1 ten plus 6 ones, place 6 in the ones column and carry 1 to the tens column.

$$
\begin{array}{r}
1 \\
47 \\
+\ 29 \\
\hline
6
\end{array}
$$
$7 + 9 = 16$

Add the tens column.

$$
\begin{array}{r}
1 \\
47 \\
+\ 29 \\
\hline
76
\end{array}
$$
⌐—— Sum of digits in tens column

WORK PROBLEM 4 AT THE SIDE. ▶▶

E X A M P L E 5   Adding with Carrying

Add $324 + 7855 + 23 + 7 + 86$.

Step 1 Add the digits in the ones column.

$$
\begin{array}{r}
2 \\
324 \\
7855 \\
23 \\
7 \\
+\quad 86 \\
\hline
5
\end{array}
$$

Carry 2 to the tens column.

Sum of the ones column is 25.

Write 5 in the ones column.

In 25, the 5 represents 5 ones and is written in the ones column, while 2 represents 2 tens and is carried to the tens column.

Step 2 Now add the digits in the tens column, including the carried 2.

$$
\begin{array}{r}
12 \\
324 \\
7855 \\
23 \\
7 \\
+\quad 86 \\
\hline
95
\end{array}
$$

Carry 1 to the hundreds column.

Sum of the tens column is 19.

Write 9 in the tens column.

CONTINUED ON NEXT PAGE

3. Add.

(a)
$$
\begin{array}{r}
25 \\
+\ 73 \\
\hline
\end{array}
$$

(b)
$$
\begin{array}{r}
364 \\
+\ 532 \\
\hline
\end{array}
$$

(c)
$$
\begin{array}{r}
42,305 \\
+\ 11,563 \\
\hline
\end{array}
$$

4. Add by using carrying.

(a)
$$
\begin{array}{r}
69 \\
+\ 26 \\
\hline
\end{array}
$$

(b)
$$
\begin{array}{r}
76 \\
+\ 18 \\
\hline
\end{array}
$$

(c)
$$
\begin{array}{r}
56 \\
+\ 37 \\
\hline
\end{array}
$$

(d)
$$
\begin{array}{r}
34 \\
+\ 49 \\
\hline
\end{array}
$$

ANSWERS
3. (a) 98 (b) 896 (c) 53,868
4. (a) 95 (b) 94 (c) 93 (d) 83

5. Add by carrying as necessary.

    (a)   481
               79
               38
       + 395

    (b)   4271
             372
           8976
       + 162

    (c)   57
                 4
           392
           804
            51
       + 27

    (d)   7821
             435
             72
            305
       + 1693

6. Add with mental carrying.

    (a)   278
            825
             14
              3
              7
       + 9275

    (b)   3305
            650
            708
             29
             40
              6
       +   3

    (c)   15,829
            765
             78
             15
              9
              7
       + 13,179

ANSWERS

5. (a) 993 **(b)** 13,781 **(c)** 1335
    (d) 10,326

6. (a) 10,402 **(b)** 4741 **(c)** 29,882

Step 3 Add the hundreds column, including the carried 1.

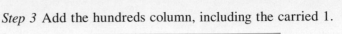

$$
\begin{array}{r}
112 \\
324 \\
7855 \\
23 \\
7 \\
+ \quad 86 \\
\hline
295
\end{array}
$$

Carry 1 to the thousands column.

Sum of the hundreds column is 12.

Write 2 in the hundreds column.

Step 4 Add the thousands column, including the carried 1.

$$
\begin{array}{r}
112 \\
324 \\
7855 \\
23 \\
7 \\
+ \quad 86 \\
\hline
\mathbf{8}295
\end{array}
$$

Sum of the thousands column is 8.

Finally, 324 + 7855 + 23 + 7 + 86 = 8295.

◀◀ **WORK PROBLEM 5 AT THE SIDE.**

Note

For additional speed, try to carry mentally. Do not write the number carried, but just carry the number mentally to the top of the next column being added. Try this method. If it works for you, use it.

◀◀ **WORK PROBLEM 6 AT THE SIDE.**

OBJECTIVE 5 In Section 1.9 we will describe how to solve application problems in more detail. The next two examples have application problems that require adding.

E X A M P L E 6   Applying Addition Skills

On this map, the distance in miles from one location to another is written alongside the road. Find the shortest distance from Altamonte Springs to Clear Lake.

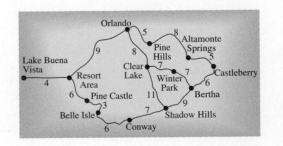

CONTINUED ON NEXT PAGE

Approach Add the mileage along various routes to determine the distances from Altamonte Springs to Clear Lake. Then select the shortest route.

Solution One way from Altamonte Springs to Clear Lake is through Orlando. Add the mileage numbers along this route.

$$
\begin{array}{rl}
8 & \text{Altamonte Springs to Pine Hills} \\
5 & \text{Pine Hills to Orlando} \\
+\ 8 & \text{Orlando to Clear Lake} \\
\hline
21 & \longrightarrow \text{miles from Altamonte Springs to} \\
& \text{Clear Lake, going through Orlando}
\end{array}
$$

Another way is through Bertha and Winter Park. Add the mileage numbers along this route.

$$
\begin{array}{rl}
5 & \text{Altamonte Springs to Castleberry} \\
6 & \text{Castleberry to Bertha} \\
7 & \text{Bertha to Winter Park} \\
+\ 7 & \text{Winter Park to Clear Lake} \\
\hline
25 & \longrightarrow \text{miles from Altamonte Springs to Clear Lake} \\
& \text{through Bertha and Winter Park}
\end{array}
$$

The shortest way from Altamonte Springs to Clear Lake is through Orlando.

WORK PROBLEM 7 AT THE SIDE. ▶▶

E X A M P L E 7   Finding a Total

Find the total distance from Shadow Hills to Castleberry to Orlando and back to Shadow Hills.

Approach Add the mileage from Shadow Hills to Castleberry to Orlando and back to Shadow Hills to find the total distance.

Solution Use the numbers from the map.

$$
\begin{array}{rl}
9 & \text{Shadow Hills to Bertha} \\
6 & \text{Bertha to Castleberry} \\
5 & \text{Castleberry to Altamonte Springs} \\
8 & \text{Altamonte Springs to Pine Hills} \\
5 & \text{Pine Hills to Orlando} \\
8 & \text{Orlando to Clear Lake} \\
+\ 11 & \text{Clear Lake to Shadow Hills} \\
\hline
52 & \longrightarrow \text{miles from Shadow Hills to Castleberry} \\
& \text{to Orlando and back to Shadow Hills}
\end{array}
$$

WORK PROBLEM 8 AT THE SIDE. ▶▶

OBJECTIVE 6 ▶ Checking the answer is an important part of problem solving. A common method for checking addition is to re-add from bottom to top. This is an application of the commutative and associative properties of addition.

---

7. Use the map to find the shortest distance from Lake Buena Vista to Conway.

8. The road is closed between Orlando and Clear Lake, so this route cannot be used. Use the map to find the next shortest distance from Orlando to Clear Lake.

ANSWERS

7. 19 miles

8.
$$
\begin{array}{lr}
\text{Orlando to Pine Hills} & 5 \\
\text{Pine Hills to Altamonte Springs} & 8 \\
\text{Altamonte Springs to Castleberry} & 5 \\
\text{Castleberry to Bertha} & 6 \\
\text{Bertha to Winter Park} & 7 \\
\text{Winter Park to Clear Lake} & +\ 7 \\
\hline
& 38 \\
& \text{miles}
\end{array}
$$

9. Check the following additions. If an answer is incorrect, give the correct answer.

(a)
```
    32
     8
     5
+   14
    59
```

(b)
```
   872
   539
    46
+  152
  1609
```

(c)
```
    79
   218
     7
+  639
   953
```

(d)
```
   21,892
   11,746
+  43,925
   79,563
```

EXAMPLE 8   Checking Addition

Check the following addition.

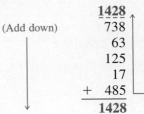

```
                    1428
(Add down)           738      Adding down and
                      63      adding up should give
                     125      the same answer.
                      17      (Add up)
                  +  485      Check
                    1428
```

Here the answers agree, so the sum is probably correct.

EXAMPLE 9   Checking Addition

Check the following additions. Are they correct?

```
                 1033       Correct, because both
(a)     785       785       answers are the same
         63        63       (Add up)
       + 185     + 185      Check
        1033      1033
```

```
                 2454       Error, because answers
(b)     635       635       are different
         73        73
        831       831       (Add up)
       + 915     + 915      Check
        2444      2444
```

Re-add to find that the correct answer is 2454.

◀◀ **WORK PROBLEM 9 AT THE SIDE.**

ANSWERS

9. (a) correct (b) correct
   (c) incorrect, should be 943
   (d) incorrect, should be 77,563

1.2 Exercises

Add.

Examples: 57
+ 42   23 + 721 + 834

Solutions: Line up the numbers in columns.

```
  57                23
+ 42               721
  99             + 834
                  1578
```
— Ones added
— Tens added

1. 26
+ 33
59

2. 18
+ 11
29

3. 44
+ 53
97

4. 83
+ 15
98

5. 317
+ 572
889

6. 651
+ 228
879

7. 258
421
+ 320
999

8. 135
253
+ 410
798

9. 6310
252
+ 1223
7785

10. 121
5705
+ 3163
8989

11. 274 + 302 + 421

997

12. 517 + 131 + 250

898

13. 1251 + 4311 + 2114

7676

14. 3241 + 1513 + 2014

6768

15. 12,142 + 43,201 + 23,103

78,446

16. 41,124 + 12,302 + 23,500

76,926

17. 3213 + 5715

8928

18. 6344 + 1655

7999

19. 38,204 + 21,020

59,224

20. 63,251 + 36,305

99,556

✍ Writing   ◉ Conceptual   ▲ Challenging   ≈ Estimation

Add the following numbers by carrying as necessary.

Example:	**Solution:**
185 + 769	11 185 + 769 **954**

21. 67
 + 83
 150

22. 78
 + 36
 114

23. 58
 + 96
 154

24. 37
 + 85
 122

25. 47
 + 74
 121

26. 68
 + 68
 136

27. 73
 + 89
 162

28. 96
 + 47
 143

29. 73
 + 29
 102

30. 68
 + 37
 105

31. 746
 + 905
 1651

32. 621
 + 359
 980

33. 306
 + 848
 1154

34. 798
 + 206
 1004

35. 278
 + 135
 413

36. 172
 + 156
 328

37. 928
 + 843
 1771

38. 686
 + 726
 1412

39. 526
 + 884
 1410

40. 116
 + 897
 1013

41. 7968
 + 1285
 9253

42. 1768
 + 8275
 10,043

43. 7896
 + 3728
 11,624

44. 9382
 + 7586
 16,968

45. 9625
 + 7986
 17,611

46. 6829
 6076
 + 8218
 21,123

47. 9056
 78
 6089
 + 731
 15,954

48. 4022
 709
 8621
 + 37
 13,389

49. 18
 708
 9286
 + 636
 10,648

50. 1708
 321
 61
 + 8926
 11,016

51. 218
 7022
 335
 + 9283
 16,858

52. 6505
 173
 7044
 + 168
 13,890

53. 321
 9603
 8
 21
 + 1604
 11,557

54. 7631
 5983
 7
 36
 + 505
 14,162

55. 2109
 63
 16
 3
 + 9887
 12,078

56.	57.	58.	59.	60.
244	553	3187	413	576
67	97	810	85	7934
7076	2772	527	9919	60
13	437	76	602	781
618	63	2665	31	5968
+ 3005	+ 328	+ 317	+ 1218	+ 371
11,023	4250	7582	12,268	15,690

Check the following additions. If an answer is incorrect, give the correct answer.

Example:

```
  ----
   835
   278
 + 422
  1535
```

Solution:

```
  1535  ↑ Correct
   835
   278
 + 422  ⌐ Check
  1535
```

61.	62.	63.	64.	65.
537	391	179	17	4713
382	873	214	296	28
+ 215	+ 684	+ 376	713	615
1134	1948	759	+ 94	+ 64
			1220	5420
correct	correct	incorrect; should be 769	incorrect; should be 1120	correct

66.	67.	68.	69.	70.
6 215	678	516	4 714	6 715
744	7 952	8 760	27	283
36	56	24	77	9 617
+ 4 284	718	189	8 878	13
11,279	+ 2 173	+ 1 723	+ 636	+ 81
	11,377	11,212	14,332	16,719
correct	incorrect; should be 11,577	correct	correct	incorrect; should be 16,709

71. Explain the commutative property of addition in your own words. How is this used when checking an addition problem?

Changing the order in which numbers are added does not change the sum. You can add from bottom to top when checking addition.

72. Explain in your own words the associative property of addition. How can this be used when adding columns of numbers?

Grouping the addition of numbers in any order does not change the sum. You can add numbers in any order. For example, you can add pairs of numbers which add to 10.

Using the map below find the shortest distance between the following cities.

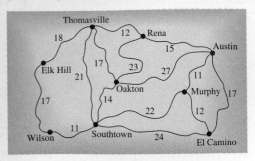

73. Southtown and Rena   **37 miles**

74. Elk Hill and Oakton   **35 miles**

75. Thomasville and Murphy   **38 miles**

76. Murphy and Thomasville   **38 miles**

Solve the following application problems.

77. A tune-up costs $65 and a tire rotation is $14. Find the total cost for both services.

$79

78. Jane Lim ordered 68 large sprinkler valves and 47 small valves for her hardware store. How many valves did she order altogether?

115 valves

79. There are 413 women and 286 men on the sales staff. How many people are on the sales staff?

699 people

80. One department in an office building has 283 employees while another department has 218 employees. How many employees are in the two departments?

501 employees

81. At a charity bazaar, a library has a total of 9792 books for sale, while a book dealer has 3259 books for sale. How many books are for sale?

13,051 books

82. A plane is flying at an altitude of 5924 feet. It then increases its altitude by 7284 feet. Find its new altitude.

13,208 feet

Find the perimeter or total distance around each of the following figures.

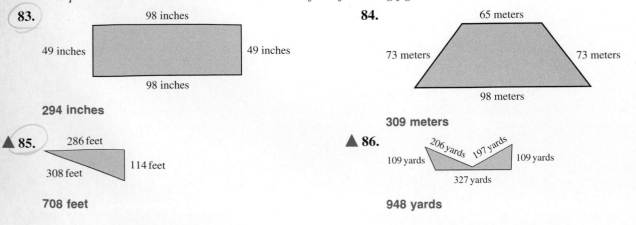

83.

98 inches

49 inches   49 inches

98 inches

294 inches

84.

65 meters

73 meters   73 meters

98 meters

309 meters

▲ **85.**

286 feet

114 feet

308 feet

708 feet

▲ **86.**

206 yards   197 yards

109 yards   109 yards

327 yards

948 yards

1.3 Subtraction of Whole Numbers

Suppose you have $8, and you spend $5 for gasoline. You then have $3 left. There are two different ways of looking at these numbers.

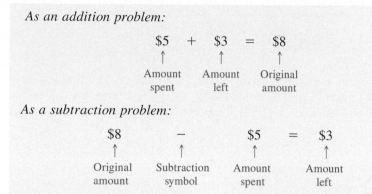

As an addition problem:

$$\$5 \;\; + \;\; \$3 \;\; = \;\; \$8$$

Amount spent   Amount left   Original amount

As a subtraction problem:

$$\$8 \;\; - \;\; \$5 \;\; = \;\; \$3$$

Original amount   Subtraction symbol   Amount spent   Amount left

OBJECTIVE 1 As this example shows, an addition problem can be changed to a subtraction problem and a subtraction problem can be changed to an addition problem.

E X A M P L E 1 **Changing Addition Problems to Subtraction**

Change each addition problem to a subtraction problem.

(a) $4 + 1 = 5$

Two subtraction problems are possible:

$$5 - 1 = 4 \quad \text{or} \quad 5 - 4 = 1$$

These figures show each subtraction problem.

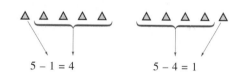

$5 - 1 = 4$ $5 - 4 = 1$

(b) $8 + 7 = 15$

$$15 - 7 = 8 \quad \text{or} \quad 15 - 8 = 7$$

WORK PROBLEM 1 AT THE SIDE. ▶▶

E X A M P L E 2 **Changing Subtraction Problems to Addition**

Change each subtraction problem to an addition problem.

(a) $8 - 3 = 5$

$$8 = 3 + 5$$

It is also correct to write $8 = 5 + 3$.

— CONTINUED ON NEXT PAGE

— CONTINUED ON NEXT PAGE

OBJECTIVES

1. Change addition problems to subtraction and subtraction problems to addition.
2. Identify the minuend, subtrahend, and difference.
3. Subtract when no borrowing is needed.
4. Check answers.
5. Subtract by borrowing.
6. Solve application problems with subtraction.

FOR EXTRA HELP

Tutorial   Tape 1   SSM, Sec. 1.3

1. Write two subtraction problems for each addition problem.

(a) $4 + 3 = 7$

(b) $6 + 5 = 11$

(c) $15 + 22 = 37$

(d) $23 + 55 = 78$

ANSWERS

1. **(a)** $7 - 3 = 4$ or $7 - 4 = 3$
   (b) $11 - 5 = 6$ or $11 - 6 = 5$
   (c) $37 - 22 = 15$ or $37 - 15 = 22$
   (d) $78 - 55 = 23$ or $78 - 23 = 55$

2. Write an addition problem for each subtraction problem.

(a) $5 - 3 = 2$

(b) $8 - 3 = 5$

(c) $21 - 15 = 6$

(d) $58 - 42 = 16$

3. Subtract.

(a) $\begin{array}{r} 56 \\ -\ 31 \\ \hline \end{array}$

(b) $\begin{array}{r} 38 \\ -\ 14 \\ \hline \end{array}$

(c) $\begin{array}{r} 378 \\ -\ 235 \\ \hline \end{array}$

(d) $\begin{array}{r} 3927 \\ -\ 2614 \\ \hline \end{array}$

(e) $\begin{array}{r} 5464 \\ -\ 324 \\ \hline \end{array}$

(b) $\begin{aligned} 19 - 14 &= 5 \\ 19 &= 14 + 5 \end{aligned}$

(c) $\begin{aligned} 29 - 13 &= 16 \\ 29 &= 13 + 16 \end{aligned}$

◀◀ **WORK PROBLEM 2 AT THE SIDE.**

OBJECTIVE 2 In subtraction, as in addition, the numbers in a problem have names. For example, in the problem, $8 - 5 = 3$, the number 8 is the **minuend** (MIN-yoo-end), 5 is the **subtrahend** (SUB-truh-hend), and 3 is the **difference** or answer.

$$\underset{\substack{\uparrow \\ \text{Minuend}}}{8} \quad - \quad \underset{\substack{\uparrow \\ \text{Subtrahend}}}{5} \quad = 3 \leftarrow \text{Difference (answer)}$$

$$\begin{array}{r} 8 \leftarrow \text{Minuend} \\ -\ 5 \leftarrow \text{Subtrahend} \\ \hline 3 \leftarrow \text{Difference} \end{array}$$

OBJECTIVE 3 Subtract two numbers by lining up the numbers in columns so the digits in the ones place are in the same column. Next subtract by columns, starting at the right with the ones column.

E X A M P L E 3   Subtracting Two Numbers

Subtract.

┌─── Ones digits are lined up in the same column.

(a) $\begin{array}{r} 53 \\ -\ 21 \\ \hline 32 \end{array}$

$3 - 1 = 2$
$5 - 2 = 3$

┌─── Ones digits are lined up.

(b) $\begin{array}{r} 385 \\ -\ 161 \\ \hline 224 \end{array}$ ← $5 - 1 = 4$

$8 - 6 = 2$
$3 - 1 = 2$

(c) $\begin{array}{r} 9431 \\ -\ 210 \\ \hline 9221 \end{array}$ ← $1 - 0 = 1$

$3 - 1 = 2$
$4 - 2 = 2$
$9 - 0 = 9$

◀◀ **WORK PROBLEM 3 AT THE SIDE.**

OBJECTIVE 4 Use addition to check your answer to a subtraction problem. For example, check $8 - 3 = 5$ by *adding* 3 and 5:

$$3 + 5 = 8, \quad \text{so} \quad 8 - 3 = 5 \quad \text{is correct.}$$

ANSWERS

2. (a) $5 = 3 + 2$ (b) $8 = 3 + 5$
(c) $21 = 15 + 6$ (d) $58 = 42 + 16$
3. (a) 25 (b) 24 (c) 143
(d) 1313 (e) 5140

E X A M P L E 4   Checking Subtraction

Check each answer.

(a)   89
       − 47
       ——
         42

Rewrite as an addition problem, as shown in Example 2.

Subtraction problem { 89 − 47 —— 42 —— 89 }   Addition problem   **47 + 42 —— 89**

Because 47 + 42 = 89, the subtraction was done correctly.

(b) 72 − 41 = 21

Rewrite as an addition problem.

$$72 = 41 + 21$$

But, 41 + 21 = 62, not 72, so the subtraction was done incorrectly. We rework the original subtraction to get the correct answer, 31.

(c)   374 ←—— Match ——┐
       − 141
       ——
         233   141 + 233 = 374

The answer checks.

WORK PROBLEM 4 AT THE SIDE. ▶▶

OBJECTIVE 5 If a digit in the minuend is less than the one directly below it we cannot subtract, so **borrowing** will be necessary.

E X A M P L E 5   Subtracting with Borrowing

Subtract 19 from 57.

Write the problem.

        57
      − 19

In the ones column, 7 is less than 9, so, in order to subtract, we must borrow a 10 from the 5 (which represents 5 tens, or 50).

50 − 10 = 40 ——▶ 4 17 ←— 10 + 7 = 17
                  5̸ 7̸
                − 1 9

Now we can subtract 17 − 9 in the ones column and then 4 − 1 in the tens column,

        4 17
        5̸ 7̸
      − 1 9
      ——
        3 8   Difference

Finally, 57 − 19 = 38. Check by adding 19 and 38. You should get 57.

WORK PROBLEM 5 AT THE SIDE. ▶▶

4. Decide whether these answers are correct. If incorrect, what should they be?

(a)   65
       − 23
       ——
         42

(b)   46
       − 32
       ——
         24

(c)   374
       − 251
       ——
         113

(d)   7531
       − 4301
       ——
        3230

5. Subtract.

(a)   67
       − 38

(b)   97
       − 29

(c)   31
       − 17

(d)   863
       − 47

(e)   762
       − 157

ANSWERS
4. (a) correct (b) incorrect, should be 14
   (c) incorrect, should be 123
   (d) correct
5. (a) 29 (b) 68 (c) 14 (d) 816 (e) 60.

6. Subtract.

(a)   354
         − 82

(b)   457
         − 68

(c)   874
         − 486

(d)   1437
         − 988

(e)   8739
         − 3892

ANSWERS
6. **(a)** 272 **(b)** 389 **(c)** 388
   (d) 449 **(e)** 4847

E X A M P L E 6 **Subtracting with Borrowing**

Subtract by borrowing as necessary.

(a)   7856
         − 137

There is no need to borrow, as
4 is greater than 3.

$$10 + 6 = 16$$
4 16

7 8 5̷ 6̷
−   1 3 7
7 7 1 9   Difference

(b)   635
         − 546

$100 + 20 = 120$ (12 tens = 120)
$600 − 100 = 500$ $10 + 5 = 15$

5 12 15

6̷ 3̷ 5̷
− 5 4 6
8 9   Difference

(c)   412
         − 225

3 10 12
4̷ 1̷ 2̷
− 2 2 5
1 8 7

◄◄ **WORK PROBLEM 6 AT THE SIDE.**

Sometimes a minuend has zeros in some of the positions. In such cases, borrowing may be a little more complicated than what we have shown so far.

E X A M P L E 7 **Borrowing with Zeros**

Subtract.

4607
− 3168

It is not possible to borrow from the tens position. Instead we must first borrow from the hundreds position.

$600 − 100 = 500$ $100 + 0 = 100$
5 10

4 6̷ 0̷ 7
− 3 1 6 8

Now we may borrow from the tens position.

9 ←—— $100 − 10 = 90$
5 10 17 ←—— $10 + 7 = 17$

4 6̷ 0̷ 7
− 3 1 6 8
9

CONTINUED ON NEXT PAGE

Complete the problem.

$$
\begin{array}{r}
9 \\
5\ \cancel{10}\ 17 \\
4\ \cancel{6}\ \cancel{0}\ 7 \\
-\ 3\ 1\ 6\ 8 \\
\hline
1\ 4\ 3\ 9 \quad \text{Difference}
\end{array}
$$

As above, check by adding 1439 and 3168; you should get 4607.

WORK PROBLEM 7 AT THE SIDE. ▶▶

EXAMPLE 8 Borrowing with Zeros

Subtract.

(a)
$$
\begin{array}{r}
708 \\
-\ 149 \\
\end{array}
$$

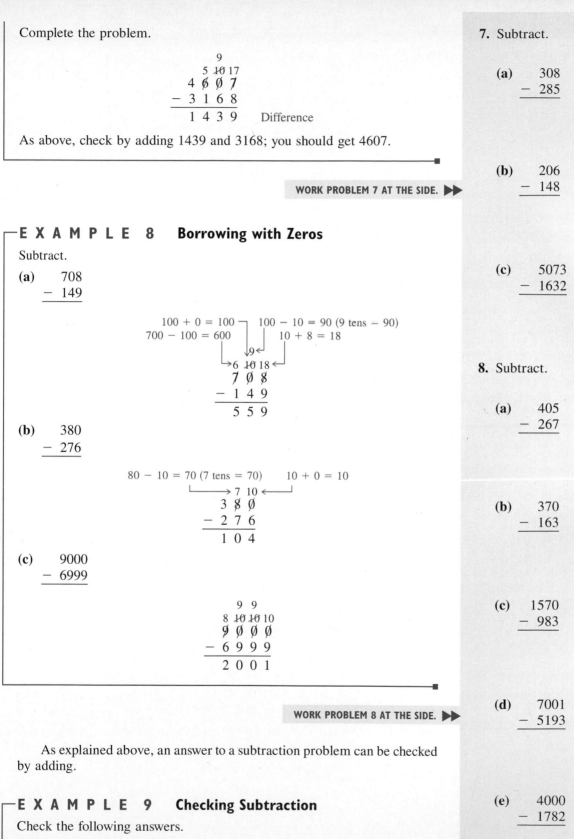

$$
\begin{array}{r}
100 + 0 = 100 \quad 100 - 10 = 90\ (9\ \text{tens} - 90) \\
700 - 100 = 600 \qquad 10 + 8 = 18 \\
9 \\
6\ \cancel{10}\ 18 \\
7\ \cancel{0}\ \cancel{8} \\
-\ 1\ 4\ 9 \\
\hline
5\ 5\ 9
\end{array}
$$

(b)
$$
\begin{array}{r}
380 \\
-\ 276 \\
\end{array}
$$

$$
\begin{array}{r}
80 - 10 = 70\ (7\ \text{tens} = 70) \qquad 10 + 0 = 10 \\
7\ 10 \\
3\ \cancel{8}\ \cancel{0} \\
-\ 2\ 7\ 6 \\
\hline
1\ 0\ 4
\end{array}
$$

(c)
$$
\begin{array}{r}
9000 \\
-\ 6999 \\
\end{array}
$$

$$
\begin{array}{r}
9\ \ 9 \\
8\ \cancel{10}\ \cancel{10}\ 10 \\
\cancel{9}\ \cancel{0}\ \cancel{0}\ 0 \\
-\ 6\ 9\ 9\ 9 \\
\hline
2\ 0\ 0\ 1
\end{array}
$$

WORK PROBLEM 8 AT THE SIDE. ▶▶

As explained above, an answer to a subtraction problem can be checked by adding.

EXAMPLE 9 Checking Subtraction

Check the following answers.

(a)
$$
\begin{array}{r}
613 \\
-\ 275 \\
\hline
338
\end{array}
\qquad
\begin{array}{r}
\text{Check} \\
275 \\
+\ 338 \\
\hline
613 \quad \text{Correct}
\end{array}
$$

Match

CONTINUED ON NEXT PAGE

---

7. Subtract.

(a)
$$
\begin{array}{r}
308 \\
-\ 285 \\
\end{array}
$$

(b)
$$
\begin{array}{r}
206 \\
-\ 148 \\
\end{array}
$$

(c)
$$
\begin{array}{r}
5073 \\
-\ 1632 \\
\end{array}
$$

8. Subtract.

(a)
$$
\begin{array}{r}
405 \\
-\ 267 \\
\end{array}
$$

(b)
$$
\begin{array}{r}
370 \\
-\ 163 \\
\end{array}
$$

(c)
$$
\begin{array}{r}
1570 \\
-\ 983 \\
\end{array}
$$

(d)
$$
\begin{array}{r}
7001 \\
-\ 5193 \\
\end{array}
$$

(e)
$$
\begin{array}{r}
4000 \\
-\ 1782 \\
\end{array}
$$

ANSWERS

7. (a) 23 **(b)** 58 **(c)** 3441

8. (a) 138 **(b)** 207 **(c)** 587
(d) 1808 **(e)** 2218

9. Check the answers in the following problems. If the answer is incorrect, give the correct answer.

(a)
$$
\begin{array}{r}
425 \\
- \ 368 \\
\hline
57
\end{array}
$$

(b)
$$
\begin{array}{r}
670 \\
- \ 439 \\
\hline
241
\end{array}
$$

(c)
$$
\begin{array}{r}
14{,}726 \\
- \ 8\ 839 \\
\hline
5\ 887
\end{array}
$$

10. Using the table from Example 10, how many more deliveries did Lopez make on

(a) Friday than on Tuesday?

(b) Tuesday than on Wednesday?

(b)
$$
\begin{array}{r}
1915 \\
- \ 1635 \\
\hline
280
\end{array}
$$
Check
$$
\begin{array}{r}
1635 \\
+ \ 280 \\
\hline
1915
\end{array}
$$
Match   Correct

(c)
$$
\begin{array}{r}
15{,}803 \\
- \ 7\ 325 \\
\hline
8\ 578
\end{array}
$$
Check
$$
\begin{array}{r}
7\ 325 \\
+ \ 8\ 578 \\
\hline
15{,}903
\end{array}
$$
No match   Error

Rework the original problem to get the correct answer, 8478.

◀◀ **WORK PROBLEM 9 AT THE SIDE.**

OBJECTIVE 6 As shown in the next example, subtraction can be used to solve an application problem.

E X A M P L E   10   Applying Subtraction Skills

Diana Lopez drives a United Parcel Service delivery truck. Using the table below, decide how many more deliveries were made by Lopez on Monday than on Thursday.

PACKAGE DELIVERY (Lopez)

Day	Number of Deliveries
Sunday	0
Monday	137
Tuesday	126
Wednesday	119
Thursday	89
Friday	147
Saturday	0

Lopez made 137 deliveries on Monday, but had only 89 deliveries on Thursday. Find how many more deliveries were made on Monday than on Thursday by subtracting 89 from 137.

$$
\begin{array}{rl}
137 & \text{Deliveries on Monday} \\
- \ \ 89 & \text{Deliveries on Thursday} \\
\hline
48 & \text{More deliveries on Monday}
\end{array}
$$

Lopez made 48 more deliveries on Monday than she made on Thursday.

◀◀ **WORK PROBLEM 10 AT THE SIDE.**

ANSWERS

9. (a) correct
   (b) incorrect, should be 231
   (c) correct
10. (a) 21 **(b)** 7

1.3 Exercises

Solve the following subtraction problems. Check each answer.

Example:	3722	**Solution:**	3722 ←	**Check:**	1610
	$-\ 1610$		$-\ 1610$		$+\ 2112$
			2112 *Match*		**3722**

The answer, 2112, checks.

1. 46
$-\ 24$
22

2. 18
$-\ 13$
5

3. 97
$-\ 64$
33

4. 59
$-\ 27$
32

5. 77
$-\ 60$
17

6. 87
$-\ 63$
24

7. 335
$-\ 122$
213

8. 602
$-\ 301$
301

9. 552
$-\ 451$
101

10. 888
$-\ 215$
673

11. 6821
$-\ 610$
6211

12. 4420
$-\ 310$
4110

13. 5546
$-\ 2134$
3412

14. 1875
$-\ 1362$
513

15. 6259
$-\ 4148$
2111

16. 8732
$-\ 1621$
7111

17. 24,392
$-\ 11,232$
13,160

18. 57,921
$-\ 34,801$
23,120

19. 46,253
$-\ 5\ 143$
41,110

20. 75,904
$-\ 3\ 702$
72,202

Check the following subtractions. If an answer is not correct, give the correct answer.

Example:	725	Add. $413 + 212 = 625$, which does not match 725.
	$-\ 413$	The answer does not check. Rework the subtraction to
	212	get the correct answer of **312**.

21. 37
$-\ 25$
12
correct

22. 69
$-\ 32$
37
correct

23. 89
$-\ 27$
63
**incorrect;
should be 62**

24. 47
$-\ 35$
13
**incorrect;
should be 12**

25. 382
$-\ 261$
131
**incorrect;
should
be 121**

26. 838
$-\ 516$
322
correct

27. 3767
$-\ 2456$
1311
correct

28. 5217
$-\ 4105$
1132
**incorrect;
should be 1112**

29. 8643
$-\ 1421$
7212
**incorrect;
should be 7222**

30. 9428
$-\ 3124$
6324
**incorrect;
should be
6304**

☑ Writing   ◉ Conceptual   ▲ Challenging   ≈ Estimation

Subtract by borrowing as necessary.

Example: 63
          − 47

Solution:
$$
\begin{array}{r}
\overset{5}{\cancel{6}}\,\overset{13}{\cancel{3}} \\
-\ 4\ 7 \\
\hline
1\ 6
\end{array}
$$

31. 36
       − 28
       ____
         8

32. 97
       − 39
       ____
        58

33. 83
       − 58
       ____
        25

34. 65
       − 28
       ____
        37

35. 45
       − 29
       ____
        16

36. 93
       − 37
       ____
        56

37. 719
       − 658
       ____
         61

38. 916
       − 618
       ____
        298

39. 771
       − 252
       ____
        519

40. 973
       − 788
       ____
        185

41. 9861
       − 684
       ____
       9177

42. 6171
       − 1182
       ____
       4989

43. 9988
       − 2399
       ____
       7589

44. 3576
       − 1658
       ____
       1918

45. 38,335
       − 29,476
       ____
        8859

46. 61,278
       − 3 559
       ____
       57,719

47. 40
       − 37
       ____
         3

48. 80
       − 73
       ____
         7

49. 60
       − 37
       ____
        23

50. 70
       − 27
       ____
        43

51. 407
       − 399
       ____
         8

52. 400
       − 399
       ____
         1

53. 4041
       − 1208
       ____
       2833

54. 4602
       − 2063
       ____
       2539

55. 9305
       − 1530
       ____
       7775

56. 7120
       − 6033
       ____
       1087

57. 1580
       − 1077
       ____
        503

58. 3068
       − 2105
       ____
        963

59. 2006
       − 1850
       ____
        156

60. 8203
       − 5365
       ____
       2838

61. 6020
       − 4078
       ____
       1942

62. 7050
       − 6045
       ____
       1005

63. 8503
       − 2816
       ____
       5687

64. 16,004
       − 5 087
       ____
       10,917

65. 80,705
       − 61,667
       ____
       19,038

66. 72,000
       − 44,234
       ____
       27,766

67. 66,000
       − 34,444
       ____
       31,556

68. 77,000
       − 65,308
       ____
       11,692

69. 20,080
       − 13,496
       ____
       6584

70. 80,056
       − 23,869
       ____
       56,187

Check the following subtractions. If an answer is incorrect, give the correct answer.

Example:	3084	**Solution:**	1278
	− 1278		+ 1806
	1806	Match	3084 Correct

71.   4791
    − 2853
     1938

   correct

72.   1671
    − 1325
     1346

   incorrect;
   should be 346

73.   1439
    − 1169
     270

   correct

74.   5274
    − 1130
     4144

   correct

75.   65,318
    − 23,429
     41,889

   correct

76.   82,357
    − 14,396
     68,961

   incorrect;
   should be 67,961

77.   27,689
    − 22,306
     5 383

   correct

78.   34,821
    − 17,735
     17,735

   Incorrect;
   should be 17,086

⊙ **79.** An addition problem can be changed to a subtraction problem and a subtraction problem can be changed to an addition problem. Give two examples of each to demonstrate this.

Possible answers are:
3 + 2 = 5 could be changed to 5 − 2 = 3 or
5 − 3 = 2
6 − 4 = 2 could be changed to 2 + 4 = 6 or
4 + 2 = 6

80. Can you use the commutative and the associative properties in subtraction? Explain.

No, you cannot. Numbers must be subtracted in the order given. The difference found in subtraction is the result of subtracting the subtrahend from the minuend.

Solve the following application problems.

81. A man burns 103 calories during 30 minutes of bowling while a woman burns 88 calories during 30 minutes of bowling. How many fewer calories did the woman burn than the man in 30 minutes of bowling?

82. Lynn Couch has $553 in her checking account. She writes a check for $308 for school fees. How much is left in her account?

$245

15 calories

83. An airplane is carrying 254 passengers. When it lands in Atlanta 133 passengers get off the plane. How many passengers are left on the plane?

121 passengers

84. The fastest animal in the world, the peregrine falcon, dives at 217 miles per hour while a Boeing 747 cruises at 580 miles per hour. How much faster does the plane fly than the falcon?

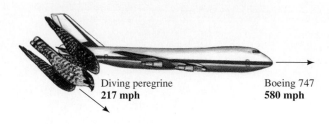

Diving peregrine   Boeing 747
217 mph   **580 mph**

363 miles per hour

The table below shows the average annual earnings of various types of medical doctors. Refer to the table to answer Exercises 85 and 86.

Type	Yearly Income
Anesthesiologist	$228,500
Radiologist	$253,300
Surgeon	$244,600
Family Doctor	$111,800
Pediatrician	$121,700

85. How much more does a radiologist earn than a surgeon?

$8700

86. How much more does an anesthesiologist earn than a pediatrician?

$106,800

87. Downtown Toronto's skyline is dominated by the CN Tower which rises 1821 feet. The Sears Tower in Chicago is 1454 feet high. Find the difference in height between the two structures.

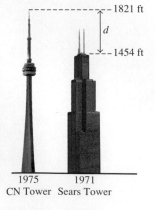

1821 ft
d
1454 ft

1975   1971
CN Tower   Sears Tower

367 feet

88. On Tuesday, 5822 people went to a soccer game, and on Friday, 7994 people went to a soccer game. How many more people went to the game on Friday?

2172 people

89. The average balance per credit card today is $1780 while ten years ago the average balance was $840. Find the increase in the average credit card balance.

$940

90. The 1964 Ford Mustang (first year) sold for $2500. In 1997 the Ford Mustang sold for $23,300. Find the increase in price.

$23,300 $2500

$20,800

91. Last fall 12,625 students enrolled in classes. In the spring semester, 11,296 students enrolled. How many more students enrolled in classes in the fall semester than in the spring?

1329 students

92. One bid for painting a house was $1954. A second bid was $1742. How much would be saved using the second bid?

$212

93. Patriot Flag Company manufactured 14,608 flags and sold 5069. How many flags remain?

9539 flags

94. Eye exams have been given to 14,679 children in the school district. If there are 23,156 students in the school district, find the number of children who have not received eye exams.

8477 children

▲ *Solve each of the following application problems. Add or subtract as necessary.*

95. A survey of large hotels found that the average salary for a general manager of a deluxe spa and tennis resort is one hundred one thousand, five hundred dollars per year, while spa and tennis directors earn $44,000. How much more does a general manager earn than a spa and tennis director?

$57,500

96. There are 24 million business enterprises in the United States. If only 7000 of these businesses are large businesses having 500 or more employees, while the rest are small and midsize businesses, find the number of small and midsize businesses.

23,993,000 small and midsize businesses

97. The Jordanos now pay rent of $650 per month. If they buy a house, their housing expense will be $913 per month. Find the increase in their monthly housing expense.

$263

98. A retired couple used to receive a social security payment of $1479 per month. Recently, benefits were increased and the couple now receives $1568 per month. Find the amount of the monthly increase.

$89

99. On Monday, 11,594 people visited Arcade Amusement Park, and 12,352 people visited the park on Tuesday. How many more people visited the park on Tuesday?

758 people

100. Last month Alice Blake earned $2382 while this month she earned $2671. How much more did she earn this month than last month?

$289

1.4 Multiplication of Whole Numbers

OBJECTIVES

1. Know the parts of a multiplication problem.
2. Do chain multiplications.
3. Multiply by single-digit numbers.
4. Multiply quickly by numbers ending in zeros.
5. Multiply by numbers having more than one digit.
6. Solve application problems with multiplication.

FOR EXTRA HELP

Tutorial   Tape 1   SSM, Sec. 1.4

Adding the number 3 a total of 4 times gives 12.

$$3 + 3 + 3 + 3 = 12$$

This result can also be shown with a figure.

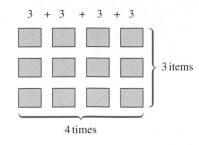

3 + 3 + 3 + 3

3 items

4 times

OBJECTIVE 1 Multiplication is a shortcut for this repeated addition. The numbers being multiplied are called **factors.** The answer is called the **product.** For example, the product of 3 and 4 can be written with the symbol ×, a raised dot, or parentheses, as follows.

$$3 \leftarrow \text{Factor (also called } multiplicand\text{)}$$
$$\underline{\times\ 4} \leftarrow \text{Factor (also called } multiplier\text{)}$$
$$12 \leftarrow \text{Product (answer)}$$

$$3 \times 4 = 12 \qquad 3 \cdot 4 = 12 \qquad (3)(4) = 12$$

▶▶ **WORK PROBLEM 1 AT THE SIDE.** ▶▶

Commutative Property of Multiplication

By the **commutative property of multiplication,** the answer or product remains the same when the order of the factors is changed. For example,

$$3 \times 5 = 15 \quad \text{and} \quad 5 \times 3 = 15$$

Note
Recall that addition also has a commutative property. Remember that $4 + 2$ is the same as $2 + 4$. Subtraction, however, is *not* commutative.

EXAMPLE 1   Multiplying Two Numbers

Multiply. (Remember that a raised dot means to multiply.)

(a) $3 \times 4 = 12$

(b) $6 \cdot 0 = 0$ (The product of any number and 0 is 0; if you give no money to each of 6 relatives, you give no money.)

(c) $(4)(8) = 32$

▶▶ **WORK PROBLEM 2 AT THE SIDE.** ▶▶

1. Identify the factors and the product in each multiplication problem.

(a) $3 \times 6 = 18$

(b) $8 \times 4 = 32$

(c) $5 \cdot 7 = 35$

(d) $(3)(9) = 27$

2. Multiply.

(a) 4×7

(b) 0×9

(c) $8 \cdot 6$

(d) $5 \cdot 5$

(e) $(3)(8)$

ANSWERS

1. (a) factors: 3, 6; product: 18
   (b) factors: 8, 4; product: 32
   (c) factors: 5, 7; product: 35
   (d) factors: 3, 9; product: 27
2. (a) 28 (b) 0 (c) 48 (d) 25 (e) 24

3. Multiply.

(a) $2 \times 3 \times 4$

OBJECTIVE ▶ Some multiplications contain more than two factors.

Associative Property of Multiplication

By the associative property of multiplication, the grouping of numbers in any order gives the same product.

E X A M P L E 2   Multiplying Three Numbers

Multiply: $2 \times 3 \times 5$.

$$(2 \times 3) \times 5 \qquad \text{Parentheses tell what to do first.}$$

$$6 \quad \times 5 = 30$$

Also,

$$2 \times (3 \times 5)$$

$$2 \times \quad 15 \qquad = 30$$

Either grouping results in the same answer.

(b) $6 \cdot 1 \cdot 5$

 *Calculator Tip:*   The calculator approach to Example 2 uses chain calculations.

$$2 \boxed{\times} 3 \boxed{\times} 5 \boxed{=} 30$$

A problem with more than two factors, such as the one in Example 2, is called a **chain multiplication.**

◀◀ WORK PROBLEM 3 AT THE SIDE.

OBJECTIVE ▶ Carrying may be needed in multiplication problems with larger factors.

E X A M P L E 3   Carrying with Multiplication

Multiply.

(c) $(8)(3)(0)$

(a) 53
   $\underline{\times \quad 4}$

Start by multiplying in the ones column.

$$\overset{1}{53}$$
$$\underline{\times \quad 4} \qquad 4 \times 3 = 12$$
$$2$$

Carry the 1 to the tens column. Write 2 in the ones column.

Next, multiply 4 ones and 5 tens.

$$\overset{1}{53}$$
$$\underline{\times \quad 4} \qquad 4 \times 5 = \mathbf{20} \text{ tens}$$
$$2$$

Add the 1 that was carried to the tens column.

$$\overset{1}{53}$$
$$\underline{\times \quad 4}$$
$$212 \qquad 20 + 1 = \mathbf{21} \text{ tens}$$

CONTINUED ON NEXT PAGE ─┘

(b) 724
     × 5

Work as shown.

        12
        724
      ×   5
        3620 ← 5 × 4 = **20** ones; write 0 ones and carry 2 tens.

                    5 × 2 = **10** tens; add the 2 tens to get 12 tens;
                    write 2 tens and carry 1 hundred.
                    5 × 7 = **35** hundreds; add the 1
                    hundred to get 36 hundreds

WORK PROBLEM 4 AT THE SIDE. ▶▶

OBJECTIVE 4 The product of two whole-number factors is also called a **multiple** of either factor. For example, since 4 • 2 = 8, the whole number 8 is a multiple of both 4 and 2. *Multiples of 10 are very useful when multiplying.* A **multiple of 10** is a whole number that ends in zero, such as 10, 20, or 30; 100, 200, or 300; 1000, 2000, or 3000. There is a short way to multiply by these multiples of 10. Look at the following examples.

$$26 \times 1 = 26$$
$$26 \times 10 = 260$$
$$26 \times 100 = 2600$$
$$26 \times 1000 = 26,000$$

Do you see a pattern in the multiplications using multiples of 10? These examples suggest the following rule.

> **Multiplying by Multiples of 10**
>
> Multiply a whole number by 10, 100, or 1000, by attaching one, two, or three zeros to the right of the whole number.

E X A M P L E 4   Using Multiples of 10 to Multiply

Multiply.

(a) 59 × 10 = 590
                    Attach 0.

(b) 74 × 100 = 7400
                    Attach 00.

(c) 803 × 1000 = 803,000 ← Attach 000.

WORK PROBLEM 5 AT THE SIDE. ▶▶

You can also find the product of other multiples of ten by attaching zeros.

4. Multiply.

(a)   52
        × 5

(b)   79
        × 0

(c)   862
        × 9

(d)   2831
        × 7

(e)   4714
        × 8

5. Multiply.

(a) 45 × 10

(b) 102 × 100

(c) 571 × 1000

ANSWERS

4. (a) 260 **(b)** 0 **(c)** 7758 **(d)** 19,817
(e) 37,712
5. (a) 450 **(b)** 10,200 **(c)** 571,000

6. Multiply.

(a) 14×50

(b) 68×400

(c) $\begin{array}{r} 180 \\ \times \quad 30 \\ \hline \end{array}$

(d) $\begin{array}{r} 6100 \\ \times \quad 90 \\ \hline \end{array}$

(e) $\begin{array}{r} 800 \\ \times \quad 200 \\ \hline \end{array}$

7. Complete each multiplication.

(a) $\begin{array}{r} 35 \\ \times \quad 54 \\ \hline 140 \\ 175 \\ \hline \end{array}$

(b) $\begin{array}{r} 76 \\ \times \quad 49 \\ \hline 684 \\ 304 \\ \hline \end{array}$

ANSWERS

6. (a) 700 **(b)** 27,200 **(c)** 5400
   (d) 549,000 **(e)** 160,000
7. (a) 1890 **(b)** 3724

E X A M P L E 5   Multiplying by Using Other Multiples of Ten

Multiply.

(a) 75×3000

Multiply 75 by 3 and attach 3 zeros.

$$\begin{array}{r} 75 \\ \times \quad 3 \\ \hline 225 \end{array} \qquad 75 \times 3000 = 225{,}000$$

— Attach 000.

(b) 150×70

Multiply 15 by 7, and then attach 2 zeros.

$$\begin{array}{r} 15 \\ \times \quad 7 \\ \hline 105 \end{array} \qquad 150 \times 70 = 10{,}500 \leftarrow \text{Attach 00.}$$

◀◀ **WORK PROBLEM 6 AT THE SIDE.**

OBJECTIVE 5 The next example shows multiplication when both factors have more than one digit.

E X A M P L E 6   Multiplying with More Than One Digit

Multiply 46 and 23.

First multiply 46 by 3.

$$\begin{array}{r} \overset{1}{} \\ 46 \\ \times \quad 3 \\ \hline 138 \end{array} \leftarrow 46 \times 3 = 138$$

Now multiply 46 by 20.

$$\begin{array}{r} \overset{1}{} \\ 46 \\ \times \quad 20 \\ \hline 920 \end{array} \leftarrow 46 \times 20 = 920$$

Add the results.

$$\begin{array}{r} 46 \\ \times \quad 23 \\ \hline 138 \\ +\ 920 \\ \hline 1058 \end{array} \begin{array}{l} \leftarrow 46 \times 3 \\ \leftarrow 46 \times 20 \end{array}$$

— Add.

Both 138 and 920 are called **partial products.** To save time, the zero in 920 is usually not written.

$$\begin{array}{r} 46 \\ \times \quad 23 \\ \hline 138 \\ 92 \\ \hline 1058 \end{array} \leftarrow \begin{array}{l} \text{0 not written. Be very careful to} \\ \text{place the 2 in the tens column.} \end{array}$$

◀◀ **WORK PROBLEM 7 AT THE SIDE.**

E X A M P L E 7   Using Partial Products

Multiply.

(a)
```
      2 3 3
   ×  1 3 2
   ─────────
      4 6 6
      6 9 9    (Tens lined up)
      2 3 3    (Hundreds lined up)
   ─────────
   3 0,7 5 6   Product
```

(b)
```
      5 3 8
   ×    4 6
```

First multiply by 6.
```
         2 4
         5 3 8
      ×    4 6     Carrying is
      ──────────   needed here.
         3 2 2 8
```

Now multiply by 4, being careful to line up the tens.

```
          1 3
          2 4
          5 3 8
       ×    4 6
       ──────────
          3 2 2 8  ⎤
          2 1 5 2  ⎦─ Finally, add the results.
       ──────────
          2 4,7 4 8
```

WORK PROBLEM 8 AT THE SIDE. ▶▶

When zero appears in the multiplier, be sure to move the partial products to the left to account for the position held by the zero.

E X A M P L E 8   Multiplication with Zeros

Multiply.

(a)
```
      1 3 7
   ×  3 0 6
   ─────────
      8 2 2
      0 0 0    (Tens lined up)
      4 1 1    (Hundreds lined up)
   ─────────
   4 1,9 2 2
```

(b)
```
         1 4 0 6                    1 4 0 6
      ×  2 0 0 1                 ×  2 0 0 1
      ──────────                 ──────────
         1 4 0 6                    1 4 0 6
      0 0 0 0   ← (0 to line up tens) ─────┐
      0 0 0 0   ← (0 to line up hundreds)  2 8 1 2 0 0   ← Zeros are
      2 8 1 2                              ──────────     written so this
      ──────────                           2,8 1 3,4 0 6  partial product
      2,8 1 3,4 0 6                                        starts in the
                                                           thousands
                                                           column.
```

Note

In Example 8(b) in the solution on the right, zeros were inserted so that thousands were placed in the thousands column.

---

8. Multiply.

(a)
```
      3 8
   ×  1 5
```

(b)
```
      3 1
   ×  4 3
```

(c)
```
      6 7
   ×  5 9
```

(d)
```
     2 3 4
   ×   7 3
```

(e)
```
     8 3 5
   ×  1 8 9
```

ANSWERS

8. (a) 570 **(b)** 1333 **(c)** 3953
(d) 17,082 **(e)** 157,815

9. Multiply.

(a)   28
    × 60

(b)   817
    ×   30

(c)   481
    × 206

(d)   3526
    × 6002

◀◀ WORK PROBLEM 9 AT THE SIDE.

OBJECTIVE 6▶ The next example shows how multiplication can be used to solve an application problem.

E X A M P L E 9   Applying Multiplication Skills

Find the total cost of 24 cordless telephones that cost $54 each.

Approach   To find the cost of all the telephones, multiply the number of telephones (24) by the cost of one telephone ($54).

Solution   Multiply 24 by 54.

$$
\begin{array}{r}
24 \\
\times\ 54 \\
\hline
96 \\
120 \\
\hline
1296
\end{array}
$$

The total cost of the cordless telephones is $1296.

⊞ *Calculator Tip:*   If you are using a calculator for Example 9 you will do this calculation

$$24\ \boxed{\times}\ 54\ \boxed{=}\ 1296.$$

◀◀ WORK PROBLEM 10 AT THE SIDE.

10. Find the total cost of the following items.

(a) 289 redwood planters at $12 per planter

(b) 180 cordless drills at $42 per drill

(c) 15 forklifts at $8218 per forklift

ANSWERS

9. (a) 1680 **(b)** 24,510 **(c)** 99,086
    (d) 21,163,052
10. (a) $3468 **(b)** $7560 **(c)** $123,270

1.4 Exercises

Work each of the following chain multiplications.

Example: $3 \times 2 \times 9$ **Solution:** $(3 \times 2) \times 9$ Or, $3 \times (2 \times 9)$

$6 \times 9 = 54$ $3 \times 18 = 54$

1. $3 \times 1 \times 3$

9

2. $2 \times 8 \times 2$

32

3. $9 \times 1 \times 7$

63

4. $2 \times 4 \times 5$

40

5. $9 \cdot 5 \cdot 0$

0

6. $6 \cdot 0 \cdot 8$

0

7. $4 \cdot 1 \cdot 6$

24

8. $1 \cdot 5 \cdot 7$

35

9. $(2)(3)(6)$

36

10. $(4)(1)(9)$

36

11. $(3)(0)(7)$

0

12. $(0)(9)(4)$

0

13. Explain in your own words the commutative property of multiplication. How do the commutative properties of addition and multiplication compare to each other?

Factors may be multiplied in any order to get the same answer. They are the same; you may add or multiply numbers in any order.

14. Explain in your own words the associative property of multiplication. How do the associative properties of addition and multiplication compare to each other?

You may group the multiplication of numbers in any order. Just as in addition, the different grouping results in the same answer.

Multiply.

Example: **Solution:**

$$\begin{array}{r} 24 \\ \times \ 2 \\ \hline 48 \end{array} \leftarrow 2 \times 4 = 8$$
$\uparrow \!\!-\!\!-\!\!-\ 2 \times 2 = 4$

Example: **Solution:**

$$\begin{array}{r} 5 \\ 37 \\ \times \ 8 \\ \hline 296 \end{array} \leftarrow \begin{array}{l} 8 \times 7 = 56; \text{ write} \\ 6 \text{ and carry 5.} \end{array}$$
$\uparrow \!\!-\!\!-\ 8 \times 3 = 24,$
$24 + 5 = 29$

15.
$$\begin{array}{r} 35 \\ \times \ 7 \\ \hline 245 \end{array}$$

16.
$$\begin{array}{r} 76 \\ \times \ 9 \\ \hline 684 \end{array}$$

17.
$$\begin{array}{r} 28 \\ \times \ 6 \\ \hline 168 \end{array}$$

18.
$$\begin{array}{r} 83 \\ \times \ 5 \\ \hline 415 \end{array}$$

19.
$$\begin{array}{r} 512 \\ \times \ 4 \\ \hline 2048 \end{array}$$

20.
$$\begin{array}{r} 472 \\ \times \ 4 \\ \hline 1888 \end{array}$$

21.
$$\begin{array}{r} 624 \\ \times \ 3 \\ \hline 1872 \end{array}$$

22.
$$\begin{array}{r} 852 \\ \times \ 7 \\ \hline 5964 \end{array}$$

23.
$$\begin{array}{r} 2153 \\ \times \ 4 \\ \hline 8612 \end{array}$$

24.
$$\begin{array}{r} 1137 \\ \times \ 3 \\ \hline 3411 \end{array}$$

25.
$$\begin{array}{r} 2521 \\ \times \ 4 \\ \hline 10{,}084 \end{array}$$

26.
$$\begin{array}{r} 2544 \\ \times \ 3 \\ \hline 7632 \end{array}$$

27.
$$\begin{array}{r} 3182 \\ \times \ 6 \\ \hline 19{,}092 \end{array}$$

28.
$$\begin{array}{r} 7326 \\ \times \ 5 \\ \hline 36{,}630 \end{array}$$

29.
$$\begin{array}{r} 36{,}921 \\ \times \ 7 \\ \hline 258{,}447 \end{array}$$

30.
$$\begin{array}{r} 28{,}116 \\ \times \ 4 \\ \hline 112{,}464 \end{array}$$

Writing   ◉ Conceptual   ▲ Challenging   ≈ Estimation

Multiply.

Example:

$$\begin{array}{r} 110 \\ \times\ 50 \\ \hline \end{array}$$

Solution:

First

$$\begin{array}{r} 11 \\ \times\ 5 \\ \hline 55 \end{array}$$

$110 \times 50 = 5500 \leftarrow$ Attach 00.

31.
$$\begin{array}{r} 20 \\ \times\ 6 \\ \hline 120 \end{array}$$

32.
$$\begin{array}{r} 50 \\ \times\ 7 \\ \hline 350 \end{array}$$

33.
$$\begin{array}{r} 30 \\ \times\ 5 \\ \hline 150 \end{array}$$

34.
$$\begin{array}{r} 90 \\ \times\ 7 \\ \hline 630 \end{array}$$

35.
$$\begin{array}{r} 740 \\ \times\ 3 \\ \hline 2220 \end{array}$$

36.
$$\begin{array}{r} 300 \\ \times\ 8 \\ \hline 2400 \end{array}$$

37.
$$\begin{array}{r} 500 \\ \times\ 4 \\ \hline 2000 \end{array}$$

38.
$$\begin{array}{r} 86 \\ \times\ 7 \\ \hline 602 \end{array}$$

39.
$$\begin{array}{r} 125 \\ \times\ 30 \\ \hline 3750 \end{array}$$

40.
$$\begin{array}{r} 246 \\ \times\ 50 \\ \hline 12{,}300 \end{array}$$

41.
$$\begin{array}{r} 1485 \\ \times\ 30 \\ \hline 44{,}550 \end{array}$$

42.
$$\begin{array}{r} 8522 \\ \times\ 50 \\ \hline 426{,}100 \end{array}$$

43.
$$\begin{array}{r} 900 \\ \times\ 300 \\ \hline 270{,}000 \end{array}$$

44.
$$\begin{array}{r} 400 \\ \times\ 700 \\ \hline 280{,}000 \end{array}$$

45.
$$\begin{array}{r} 43{,}000 \\ \times\ 2\ 000 \\ \hline 86{,}000{,}000 \end{array}$$

46.
$$\begin{array}{r} 11{,}000 \\ \times\ 9\ 000 \\ \hline 99{,}000{,}000 \end{array}$$

47. $970 \cdot 50$

48,500

48. $730 \cdot 40$

29,200

49. $400 \cdot 800$

320,000

50. $720 \cdot 600$

432,000

51. $9700 \cdot 200$

1,940,000

52. $10{,}050 \cdot 300$

3,015,000

Multiply.

Example:
$$\begin{array}{r} 63 \\ \times\ 28 \\ \hline \end{array}$$

Solution:
$$\begin{array}{r} 63 \\ \times\ 28 \\ \hline 504 \\ 126\ \ \\ \hline 1764 \end{array}$$

53.
$$\begin{array}{r} 27 \\ \times\ 15 \\ \hline 405 \end{array}$$

54.
$$\begin{array}{r} 14 \\ \times\ 54 \\ \hline 756 \end{array}$$

55.
$$\begin{array}{r} 68 \\ \times\ 22 \\ \hline 1496 \end{array}$$

56.
$$\begin{array}{r} 82 \\ \times\ 32 \\ \hline 2624 \end{array}$$

57.
$$\begin{array}{r} 83 \\ \times\ 45 \\ \hline 3735 \end{array}$$

58. $(43)(27)$

1161

59. $(58)(41)$

2378

60. $(82)(67)$

5494

61. $(67)(92)$

6164

62. $(26)(33)$

858

63. (32)(475)

15,200

64. (67)(218)

14,606

65. (729)(45)

32,805

66. (681)(47)

32,007

67. (44)(331)

14,564

68.   332
   × 772
256,304

69.   735
   × 112
82,320

70.   621
   × 415
257,715

71.   538
   × 342
183,996

72.   3228
   × 751
2,424,228

73.   8162
   × 198
1,616,076

74.   528
   × 106
55,968

75.   215
   × 307
66,005

76.   218
   × 106
23,108

77.   428
   × 201
86,028

78.   3706
   × 208
770,848

79.   6310
   × 3078
19,422,180

80.   3533
   × 5001
17,668,533

81.   2195
   × 1038
2,278,410

82.   1502
   × 2009
3,017,518

83. A classmate of yours is not clear on how to multiply a whole number by 10, by 100, or by 1000. Write a short note explaining how this can be done.

To multiply by 10, 100, or by 1000 just add the number of zeros to the number you are multiplying and that's your answer.

84. Show two ways to multiply when a zero is in the factor that is multiplying. Use the problem 291 × 307 to show this.

```
   291          291
 × 307        × 307
  2037         2037
   000         8730
  873         89,337
89,337
```

Solve the following application problems.

85. An encyclopedia has 30 volumes. Each volume has 800 pages. What is the total number of pages in the encyclopedia?

24,000 pages

86. A hospital has 20 bottles of thyroid medication, with each bottle containing 2500 tablets. How many of these tablets does the hospital have in all?

50,000 tablets

87. There are 12 tomato plants to a flat. If there are 18 flats, find the total number of tomato plants.

216 plants

88. A hummingbird's wings beat about 65 times per second. How many times do the hummingbird's wings beat in 30 seconds?

89. A new Saturn automobile gets 38 miles per gallon on the highway. How many miles can it go on 11 gallons of gas?

418 miles

90. Squid are being hauled out of the Santa Barbara Channel by the ton. They are then processed, renamed calamari, and exported. Last night 27 fishing boats each hauled out 40 tons of squid. Find the total catch for the night.

1080 tons

65 wingbeats per second

1950 times

Find the total cost of each of the following.

> **Example:** 54 hammers at $8 per hammer
>
> **Solution:** Multiply.
>
> $$\begin{array}{r} 54 \\ \times\ 8 \\ \hline 432 \end{array}$$
>
> The total cost is **$432.**

91. 16 gallons of paint at $18 per gallon

$288

92. 38 employees at $64 per day

$2432

93. 65 rebuilt alternators at $24 per alternator

$1560

94. 76 flats of flowers at $22 per flat

$1672

95. 108 sets of wrenches at $37 per set

$3996

96. 305 compact discs at $12 per disc

$3660

Multiply.

97. $21 \cdot 43 \cdot 56$ 50,568

98. (600)(8)(75)(40)   14,400,000

Use addition, subtraction, or multiplication, as needed, to solve each of the following.

99. Alison Leow counted 53 joggers one day and 122 joggers the next day. How many joggers did she count altogether during the two days?

175 joggers

100. The largest living land mammal is the African elephant, and the largest mammal of all time is the blue whale. If an African elephant weighs 15,225 pounds and a blue whale weighs 28 times that amount, find the weight of the blue whale.

426,300 pounds

101. A large meal contains 1406 calories, while a small meal contains 348 calories. How many more calories are in the large meal than the small one?

1058 calories

102. Last year the Chrysler Corporation gave each of its 79,000 union-represented employees a payment of $8000 as profit sharing. Find the total amount distributed as profit sharing.

$632,000,000

▲ 103. Dannie Sanchez bought 4 tires at $110 each, 2 seat covers at $49 each, and 6 socket wrenches at $3 each. Find the total amount that he spent.

$556

▲ 104. The distance from Reno, Nevada, to the Atlantic Ocean is 2695 miles, while the distance from Reno to the Pacific Ocean is 255 miles. How much farther is it to the Atlantic Ocean than it is to the Pacific Ocean?

2440 miles

---

* Color exercise numbers are used to indicate exercises designed for calculator use.

1.5 *Division of Whole Numbers*

Suppose $12 is to be divided into 3 equal parts. Each part would be $4, as shown here.

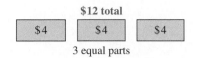

$12 total

| $4 | $4 | $4 |

3 equal parts

OBJECTIVE 1 Just as $3 \cdot 4$, 3×4, and $(3)(4)$ are different ways of indicating the multiplication of 3 and 4, there are several ways to write 12 divided by 3.

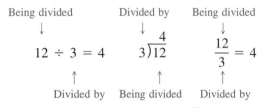

Being divided ↓   Divided by ↓   Being divided ↓

$$12 \div 3 = 4 \qquad 3\overline{)12} \,{}^{4} \qquad \frac{12}{3} = 4$$

↑ Divided by   ↑ Being divided   ↑ Divided by

We will use all three division symbols, $\div$, $\overline{)}\,$, and —. In courses such as algebra, a slash symbol, /, or a bar, —, is most often used.

EXAMPLE 1 Using Division Symbols

Write each division by using two other symbols.

(a) $12 \div 4 = 3$

This division can also be written as

$$4\overline{)12}\,{}^{3} \quad \text{or} \quad \frac{12}{4} = 3.$$

(b) $\frac{15}{5} = 3$

$$15 \div 5 = 3 \quad \text{or} \quad 5\overline{)15}\,{}^{3}$$

(c) $5\overline{)20}\,{}^{4}$

$$20 \div 5 = 4 \quad \text{or} \quad \frac{20}{5} = 4$$

WORK PROBLEM 1 AT THE SIDE. ▶▶

OBJECTIVE 2 In division, the number being divided is the **dividend** (DIV-uh-dend), the number divided by is the **divisor** (div-EYE-zer), and the answer is the **quotient** (KWOH-shunt).

$$\text{dividend} \div \text{divisor} = \text{quotient}$$

$$\text{divisor}\overline{)\text{dividend}}\,{}^{\text{quotient}} \qquad \frac{\text{dividend}}{\text{divisor}} = \text{quotient}$$

EXAMPLE 2 Identifying the Parts in a Division Problem

Identify the dividend, divisor, and quotient.

(a) $35 \div 7 = 5$

$$35 \div 7 = 5 \leftarrow \text{Quotient}$$

Dividend ↗   ↖ Divisor

--- **CONTINUED ON NEXT PAGE**

CONTINUED ON NEXT PAGE

OBJECTIVES

1 ▶ Write division problems in three ways.

2 ▶ Identify the parts of a division problem.

3 ▶ Divide zero by a number.

4 ▶ Recognize that a number cannot be divided by zero.

5 ▶ Divide a number by itself.

6 ▶ Use short division.

7 ▶ Check the answer to a division problem.

8 ▶ Use tests for divisibility.

FOR EXTRA HELP

Tutorial   Tape 1   SSM, Sec. 1.5

1. Write each division problem using two other symbols.

(a) $48 \div 6 = 8$

(b) $24 \div 6 = 4$

(c) $9\overline{)36}\,{}^{4}$

(d) $\frac{42}{6} = 7$

ANSWERS

1. (a) $6\overline{)48}\,{}^{8}$ and $\frac{48}{6} = 8$

(b) $6\overline{)24}\,{}^{4}$ and $\frac{24}{6} = 4$

(c) $36 \div 9 = 4$ and $\frac{36}{9} = 4$

(d) $6\overline{)42}\,{}^{7}$ and $42 \div 6 = 7$

2. Identify the dividend, divisor, and quotient.

(a) $10 \div 2 = 5$

(b) $30 \div 5 = 6$

(c) $\dfrac{28}{7} = 4$

(d) $2\overline{)36}$ with quotient 18

(b) $\dfrac{100}{20} = 5$

$$\underset{\uparrow}{\underset{\text{Divisor}}{\dfrac{\overset{\text{Dividend}}{\overset{\downarrow}{100}}}{20}}} = 5 \leftarrow \text{Quotient}$$

(c) $12\overline{)72}$ with quotient 6

$$\begin{array}{r} 6 \leftarrow \text{Quotient} \\ 12\overline{)72} \leftarrow \text{Dividend} \\ \uparrow \\ \text{Divisor} \end{array}$$

◄◄ **WORK PROBLEM 2 AT THE SIDE.**

OBJECTIVE 3 If no money, or $0, is divided equally among five people, each person gets $0. There is a general rule for dividing zero.

> **Dividing Zero**
>
> **Zero** divided by any nonzero number is **zero.**

EXAMPLE 3   Dividing Zero by a Number

Divide.

(a) $0 \div 12 = 0$

(b) $0 \div 1728 = 0$

(c) $\dfrac{0}{375} = 0$

(d) $129\overline{)0}$ with quotient 0

3. Divide.

(a) $0 \div 9$

(b) $\dfrac{0}{8}$

(c) $\dfrac{0}{36}$

(d) $57\overline{)0}$

◄◄ **WORK PROBLEM 3 AT THE SIDE.**

Just as a subtraction such as $8 - 3 = 5$ can be written as the addition $8 = 3 + 5$, any division can be written as a multiplication. For example, $12 \div 3 = 4$ can be written as

$$3 \times 4 = 12 \quad \text{or} \quad 4 \times 3 = 12$$

EXAMPLE 4   Converting Division to Multiplication

Convert each division to a multiplication.

(a) $\dfrac{20}{4} = 5$ becomes $4 \cdot 5 = 20$

(b) $8\overline{)48}$ with quotient 6   becomes $8 \cdot 6 = 48$

(c) $72 \div 9 = 8$ becomes $9 \cdot 8 = 72$

ANSWERS

2. (a) dividend: 10; divisor: 2; quotient: 5
   (b) dividend: 30; divisor: 5; quotient: 6
   (c) dividend: 28; divisor: 7; quotient: 4
   (d) dividend: 36; divisor: 2; quotient: 18
3. all 0

WORK PROBLEM 4 AT THE SIDE. ▶▶

OBJECTIVE 4 Division by zero cannot be done. To see why, try to find

$$9 \div 0.$$

As we have just seen, all division problems can be converted to a multiplication problem so that

$$\text{divisor} \cdot \text{quotient} = \text{dividend}.$$

If you convert the problem $9 \div 0 = ?$ to its multiplication counterpart, it reads

$$0 \cdot ? = 9.$$

You already know that zero times any number must always equal zero. Try any number you like to replace the "?" and you'll always get 0 instead of 9. Therefore, the division problem $9 \div 0$ cannot be done. Mathematicians say it is *undefined* and have agreed never to divide by zero. However, $0 \div 9$ *can* be done. Check by rewriting it as a multiplication problem.

$$0 \div 9 = 0 \quad \text{because} \quad 0 \cdot 9 = 0 \text{ is true.}$$

Dividing by Zero

Since dividing by zero cannot be done, we say that division by **zero** is undefined. It is impossible to compute an answer.

E X A M P L E 5   Dividing by Zero Is Undefined

All the following are undefined.

(a) $\dfrac{6}{0}$ is undefined

(b) $0\overline{)8}$ is undefined

(c) $18 \div 0$ is undefined

(d) $\dfrac{0}{0}$ is undefined

Division involving 0 is summarized below.

$$\frac{0}{\text{nonzero number}} = 0 \qquad \frac{\text{number}}{0} \text{ is undefined}$$

Note
When "0" is the divisor in a problem you write undefined. Never divide by zero.

WORK PROBLEM 5 AT THE SIDE. ▶▶

▦ *Calculator Tip:* Try these two problems on your calculator. Jot down your answers.

$$9 \boxed{\div} 0 \boxed{=} \underline{\qquad} \qquad 0 \boxed{\div} 9 = \underline{\qquad}$$

When you try to divide by zero, the calculator cannot do it, so it shows the word "Error" in the display, or the letter "E" (for "error").

4. Write each division problem as a multiplication problem.

(a) $6\overline{)18}$ with quotient 3

(b) $\dfrac{28}{4} = 7$

(c) $48 \div 8 = 6$

5. Work the following problems whenever possible.

(a) $\dfrac{8}{0}$

(b) $\dfrac{0}{8}$

(c) $0\overline{)32}$

(d) $32\overline{)0}$

(e) $100 \div 0$

(f) $0 \div 100$

ANSWERS
4. (a) $6 \cdot 3 = 18$ (b) $4 \cdot 7 = 28$
(c) $8 \cdot 6 = 48$
5. (a) undefined (b) 0 (c) undefined
(d) 0 (e) undefined (f) 0

42   Chapter 1   Whole Numbers

6. Divide.

  (a) $5 \div 5$

  (b) $14\overline{)14}$

  (c) $\dfrac{37}{37}$

7. Divide.

  (a) $2\overline{)18}$

  (b) $3\overline{)39}$

  (c) $4\overline{)88}$

  (d) $2\overline{)462}$

OBJECTIVE 5 What happens when a number is divided by itself? For example, $4 \div 4$ or $97 \div 97$?

> **Dividing a Number by Itself**
>
> Any nonzero number divided by itself is **one.**

E X A M P L E   6   **Dividing a Nonzero Number by Itself**

Divide.

(a) $16 \div 16 = 1$

(b) $32\overline{)\overset{1}{32}}$

(c) $\dfrac{57}{57} = 1$

◄◄ **WORK PROBLEM 6 AT THE SIDE.**

OBJECTIVE 6 **Short division** is a method of dividing a number by a one-digit divisor.

E X A M P L E   7   **Using Short Division**

Divide: $3\overline{)96}$.

First, divide 9 by 3.

$$3\overline{)96}^{\,3} \quad \leftarrow \frac{9}{3} = 3$$

Next, divide 6 by 3.

$$3\overline{)96}^{\,32} \quad \leftarrow \frac{6}{3} = 2$$

◄◄ **WORK PROBLEM 7 AT THE SIDE.**

When two numbers do not divide exactly, the leftover portion is called the **remainder.**

E X A M P L E   8   **Using Short Division with a Remainder**

Divide 147 by 4.

Write the problem.

$$4\overline{)147}$$

Because 1 cannot be divided by 4, divide 14 by 4.

$$4\overline{)14^2 7}^{\,3} \qquad \frac{14}{4} = 3 \text{ with 2 left over}$$

CONTINUED ON NEXT PAGE

6. all 1

7. (a) 9 **(b)** 13 **(c)** 22 **(d)** 231

Next, divide 27 by 4. The final number left over is the remainder. Write the remainder to the side. "R" stands for remainder.

$$\begin{array}{r} 3\ 6\ \textbf{R3} \\ 4\overline{)14^27} \end{array} \qquad \frac{27}{4} = 6 \text{ with 3 left over}$$

WORK PROBLEM 8 AT THE SIDE. ▶▶

E X A M P L E   9   Dividing with a Remainder

Divide 1809 by 7.

Divide 7 into 18.

$$\begin{array}{r} 2 \leftarrow \\ 7\overline{)18^409} \end{array} \qquad \frac{18}{7} = 2 \text{ with 4 left over}$$

Divide 7 into 40.

$$\begin{array}{r} 2\ 5 \\ 7\overline{)18^40^59} \end{array} \qquad \frac{40}{7} = 5 \text{ with 5 left over}$$

Divide 7 into 59.

$$\begin{array}{r} 2\ 5\ 8\ \textbf{R3} \\ 7\overline{)18^40^59} \end{array} \qquad \frac{59}{7} = 8 \text{ with 3 left over}$$

WORK PROBLEM 9 AT THE SIDE. ▶▶

Note

Short division takes practice but is useful in many situations.

OBJECTIVE 7 Check the answer to a division problem as follows.

Checking Division

$$\text{divisor} \times \text{quotient} + \text{remainder} = \text{dividend}$$

E X A M P L E   10   Checking Division by Using Multiplication

Check each answer.

(a) $5\overline{)458}$ with quotient 91 **R3**

$$\text{divisor} \times \text{quotient} + \text{remainder} = \text{dividend}$$
$$\downarrow \qquad \downarrow \qquad \qquad \downarrow$$
$$5 \quad \times \quad 91 \quad + \quad 3$$
$$\underbrace{\qquad\qquad\qquad} \qquad \downarrow$$
$$455 \quad + \quad 3 \quad = 458$$
$$\uparrow$$

Matches original dividend
so the division was done correctly.

— **CONTINUED ON NEXT PAGE**

8. Divide.

(a) $2\overline{)225}$

(b) $3\overline{)275}$

(c) $4\overline{)538}$

(d) $\dfrac{819}{5}$

9. Divide.

(a) $5\overline{)937}$

(b) $\dfrac{675}{7}$

(c) $3\overline{)1885}$

(d) $8\overline{)1135}$

ANSWERS

8. (a) 112 R1 **(b)** 91 R2
(c) 134 R2 **(d)** 163 R4

9. (a) 187 R2 **(b)** 96 R3
(c) 628 R1 **(d)** 141 R7

10. Check each division. If an answer is incorrect, give the correct answer.

$$\overset{38 \text{ R1}}{3\overline{)115}}$$

(a)

$$\overset{92 \text{ R2}}{8\overline{)739}}$$

(b)

$$\overset{328}{4\overline{)1312}}$$

(c)

$$\overset{476 \text{ R3}}{5\overline{)2383}}$$

(d)

(b)

$$\overset{239 \text{ R4}}{6\overline{)1437}}$$

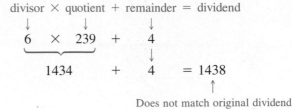

divisor × quotient + remainder = dividend

$$6 \times 239 + 4$$

$$1434 + 4 = 1438$$

Does not match original dividend

The answer does not check. Rework the original problem to get the correct answer, 239 **R3**.

Note
A common error when checking division is forgetting to add the remainder. Be sure to add any remainder when checking a division problem.

◄◄ **WORK PROBLEM 10 AT THE SIDE.**

OBJECTIVE 8 ▶ It is often important to know whether a number is divisible by another number. You will find this useful in Chapter 2 when writing fractions in lowest terms.

Divisibility

One whole number is **divisible** by another if the remainder is zero.

Decide whether one number is exactly divisible by another by using the following tests for divisibility.

Tests for Divisibility

A number is divisible by

2	if it ends in 0, 2, 4, 6, or 8.
3	if the sum of its digits is divisible by 3.
4	if the last two digits make a number that is divisible by 4.
5	if it ends in 0 or 5.
6	if it is divisible by both 2 and 3.
8	if the last three digits make a number that is divisible by 8.
9	if the sum of its digits is divisible by 9.
10	if it ends in 0.

The most commonly used tests are those for 2, 3, 5, and 10.

Divisibility by 2

A number is divisible by **2** if the number ends in 0, 2, 4, 6, or 8.

ANSWERS

10. (a) correct
  (b) incorrect, should be 92 R3
  (c) correct **(d)** correct

┌─ **E X A M P L E   11   Testing for Divisibility by 2**

Are the following numbers divisible by 2?

(a) 986
　　↑—— Ends in 6

　　Because the number ends in 6, which is in the list on the previous page, the number 986 is divisible by 2.

(b) 3255 is not divisible by 2.
　　↳ Ends in 5, and not in 0, 2, 4, 6, or 8
└──────────────────────────────■

WORK PROBLEM 11 AT THE SIDE. ▶▶

Divisibility by 3

A number is divisible by **3** if the sum of its digits is divisible by **3**.

┌─ **E X A M P L E   12   Testing for Divisibility by 3**

Are the following numbers divisible by 3?

(a) 4251

　　Add the digits.

$$4\ 2\ 5\ 1$$
$$4 + 2 + 5 + 1 = 12$$

Because 12 is divisible by 3, the number 4251 is divisible by 3.

(b) 29,806

　　Add the digits.

$$2\ 9\ 8\ 0\ 6$$
$$2 + 9 + 8 + 0 + 6 = 25$$

Because 25 is not divisible by 3, the number 29,806 is not divisible by 3.
└──────────────────────────────■

Note
Be careful when testing for divisibility by adding the digits. This method works only for the numbers 3 and 9.

WORK PROBLEM 12 AT THE SIDE. ▶▶

11. Which are divisible by 2?

　(a) 612

　(b) 315

　(c) 2714

　(d) 36,000

12. Which are divisible by 3?

　(a) 836

　(b) 7545

　(c) 242,913

　(d) 102,484

ANSWERS
11. all but (b)
12. (b) and (c)

13. Which are divisible by 5?

(a) 160

(b) 635

(c) 3381

(d) 108,605

14. Which are divisible by 10?

(a) 290

(b) 218

(c) 2020

(d) 11,670

Divisibility by 5 and by 10

A number is divisible by **5** if it ends in 0 or 5.
A number is divisible by **10** if it ends in 0.

E X A M P L E 13   Determining Divisibility by 5

Are the following numbers divisible by 5?

(a) 12,900 ends in 0 and is divisible by 5.

(b) 4325 ends in 5 and is divisible by 5.

(c) 392 ends in 2 and is not divisible by 5.

◄◄ WORK PROBLEM 13 AT THE SIDE.

E X A M P L E 14   Determining Divisibility by 10

Are the following numbers divisible by 10?

(a) 80, 700, and 9140 end in 0 and are divisible by 10.

(b) 29, 355, and 18,743 do not end in 0 and are not divisible by 10.

◄◄ WORK PROBLEM 14 AT THE SIDE.

ANSWERS
13. all but (c)
14. all but (b)

1.5 Exercises

Write each division problem by using two other symbols.

Example:

$$14 \div 2 = 7$$

Solution:

$$\frac{14}{2} = 7 \qquad 2\overline{)14}^{\,7}$$

1. $12 \div 4 = 3$

$$4\overline{)12}^{\,3} \qquad \frac{12}{4} = 3$$

2. $36 \div 6 = 6$

$$6\overline{)36}^{\,6} \qquad \frac{36}{6} = 6$$

3. $\frac{45}{9} = 5$

$$9\overline{)45}^{\,5} \qquad 45 \div 9 = 5$$

4. $\frac{56}{8} = 7$

$$56 \div 8 = 7$$
$$8\overline{)56}^{\,7}$$

5. $2\overline{)16}^{\,8}$

$$16 \div 2 = 8$$
$$\frac{16}{2} = 8$$

6. $8\overline{)48}^{\,6}$

$$48 \div 8 = 6$$
$$\frac{48}{8} = 6$$

Divide.

Examples:

$$21 \div 3 \qquad \frac{0}{5} \qquad 18 \div 0$$

Solutions:

$$21 \div 3 = \mathbf{7} \qquad \frac{0}{5} = \mathbf{0}$$
$$18 \div 0 \text{ is } \mathbf{undefined}$$

7. $8 \div 8$

1

8. $35 \div 7$

5

9. $\frac{12}{2}$

6

10. $\frac{9}{0}$

undefined

11. $24 \div 0$

undefined

12. $4 \div 4$

1

13. $\frac{0}{4}$

0

14. $\frac{24}{8}$

3

15. $12\overline{)0}$

0

16. $\frac{0}{7}$

0

17. $0\overline{)21}$

undefined

18. $\frac{2}{0}$

undefined

19. $\frac{0}{3}$

0

20. $\frac{0}{0}$

undefined

21. $\frac{8}{1}$

8

22. $\frac{0}{5}$

0

✎ Writing   ⊙ Conceptual   ▲ Challenging   ≈ Estimation

Divide by using short division. Check each answer.

Examples:	**Solutions:**	**Check:**
(a) $8\overline{)376}$	$\begin{array}{r} 4\ 7 \\ 8\overline{)37^5 6} \end{array}$	$8 \times 47 = \mathbf{376}$
(b) $6\overline{)1487}$	$\begin{array}{r} 2\ 4\ 7\ \textbf{R5} \\ 6\overline{)14^2 8^4 7} \end{array}$	$6 \times 247 + 5 = 1482 + 5 = \mathbf{1487}$

23. $\begin{array}{r} 27 \\ 4\overline{)108} \end{array}$

24. $\begin{array}{r} 27 \\ 5\overline{)135} \end{array}$

25. $\begin{array}{r} 36 \\ 9\overline{)324} \end{array}$

26. $\begin{array}{r} 22 \\ 8\overline{)176} \end{array}$

27. $\begin{array}{r} 608 \\ 4\overline{)2432} \end{array}$

28. $\begin{array}{r} 621 \\ 5\overline{)3105} \end{array}$

29. $\begin{array}{r} 627\ \text{R1} \\ 4\overline{)2509} \end{array}$

30. $\begin{array}{r} 166\ \text{R7} \\ 8\overline{)1335} \end{array}$

31. $\begin{array}{r} 1522\ \text{R5} \\ 6\overline{)9137} \end{array}$

32. $\begin{array}{r} 930\ \text{R1} \\ 9\overline{)8371} \end{array}$

33. $\begin{array}{r} 309 \\ 6\overline{)1854} \end{array}$

34. $\begin{array}{r} 107 \\ 8\overline{)856} \end{array}$

35. $4024 \div 4$

1006

36. $16,024 \div 8$

2003

37. $15,018 \div 3$

5006

38. $32,008 \div 8$

4001

39. $4867 \div 6$

811 R1

40. $5993 \div 7$

856 R1

41. $12,947 \div 5$

2589 R2

42. $33,285 \div 9$

3698 R3

43. $29,298 \div 4$

7324 R2

44. $17,937 \div 6$

2989 R3

45. $12,630 \div 4$

3157 R2

46. $46,560 \div 7$

6651 R3

47. $\dfrac{26,684}{4}$

6671

48. $\dfrac{16,398}{9}$

1822

49. $\dfrac{74,751}{6}$

12,458 R3

50. $\dfrac{72,543}{5}$

14,508 R3

51. $\dfrac{71,776}{7}$

10,253 R5

52. $\dfrac{77,621}{3}$

25,873 R2

53. $\dfrac{128,645}{7}$

18,377 R6

54. $\dfrac{172,255}{4}$

43,063 R3

Check each answer. If an answer is incorrect, give the correct answer.

Example: $9\overline{)1609}$ with **178 R7**

Solution:

$$\text{divisor} \times \text{quotient} + \text{remainder} = \text{dividend}$$
$$9 \times 178 + 7$$
$$1602 + 7 = \mathbf{1609}$$

The answer checks.

55. $4\overline{)218}$ **54 R2**

correct

56. $6\overline{)194}$ **32 R2**

correct

57. $3\overline{)5725}$ **1908 R2**

incorrect;
should be 1908 R1

58. $5\overline{)2158}$ **432 R3**

incorrect;
should be 431 R3

59. $7\overline{)4692}$ **650 R2**

incorrect;
should be 670 R2

60. $9\overline{)5974}$ **663 R5**

incorrect;
should be 663 R7

61. $6\overline{)21,409}$ **3 568 R2**

incorrect;
should be 3568 R1

62. $4\overline{)103,516}$ **25,879**

correct

63. $6\overline{)18,023}$ **3 003 R5**

correct

64. $8\overline{)33,664}$ **4 208**

correct

65. $6\overline{)69,140}$ **11,523 R2**

correct

66. $3\overline{)82,598}$ **27,532 R1**

incorrect;
should be 27,532 R2

67. $9\overline{)86,655}$ **9 628 R7**

incorrect;
should be 9628 R3

68. $7\overline{)50,809}$ **7 258 R4**

incorrect;
should be 7258 R3

69. $8\overline{)222,576}$ **27,822**

correct

70. $4\overline{)311,216}$ **77,804**

correct

71. Explain in your own words how to check a division problem using multiplication. Be sure to include what must be done if the quotient includes a remainder.

   Multiply the quotient by the divisor and add any remainder. The result should be the dividend.

72. Describe the three divisibility rules that you feel might be most useful to you and tell why.

   **Three choices might be:
   A number is divisible by 2 if it ends in a 0, 2, 4, 6, or 8.
   A number is divisible by 5 if it ends in 0 or 5.
   A number is divisible by 10 if it ends in 0.**

Solve each application problem.

73. A carton of antifreeze holds 4 one-gallon jugs. Find the number of cartons needed to package 624 one-gallon jugs.

   156 cartons

74. Eight people invested a total of $244,224 to buy a condominium. Each person invested the same amount of money. How much did each person invest?

   $30,528

75. If 6 identical service vans cost a total of $99,600, find the cost of each service van.

$16,600

76. One gallon of beverage will serve 9 people. How many gallons are needed for 3483 people?

387 gallons

77. An estate of $127,400 is divided equally among 7 family members. Find the amount received by each family member.

$18,200

78. How many 5-pound bags of rice can be filled from 8750 pounds of rice?

1750 bags

79. If 36 gallons of fertilizer are needed for each acre of land, find the number of acres that can be fertilized with 7380 gallons of fertilizer.

205 acres

80. A roofing contractor has purchased 2268 squares (10 feet by 10 feet) of roofing material. If each home needs 21 squares of material, find the number of homes that can be roofed.

108 homes

81. The Super Lotto payout of $8,100,000 will be divided equally by 36 people who purchased the winning ticket. Find the amount received by each person.

$225,000

82. Oprah Winfrey reportedly earned $171,000,000 over the past two years. Find her monthly income for that time period.

$7,125,000 per month

Put a ✓ mark in the blank if the number at the left is divisible by the number at the top.
Put an X in the blank if the number is not divisible by the number at the top.

Example: 40	Solution:				
The number 40 ends in zero so it can be divided by 2, 5, and 10. But $4 + 0 = 4$, which is not divisible by 3.		**2**	**3**	**5**	**10**
	40	✓	X	✓	✓

		2	3	5	10			2	3	5	10
83.	30	✓	✓	✓	✓	**84.**	25	X	X	✓	X
85.	184	✓	X	X	X	**86.**	192	✓	✓	X	X
87.	445	X	X	✓	X	**88.**	897	X	✓	X	X
89.	903	X	✓	X	X	**90.**	500	✓	X	✓	✓
91.	5166	✓	✓	X	X	**92.**	8302	✓	X	X	X
93.	21,763	X	X	X	X	**94.**	32,472	✓	✓	X	X

▲95. Kaci Salmon, a supervisor at Albany Electric, earns $36,540 per year. Find the amount of her earnings in a three month period.

$9135

▲96. A worker assembles 168 light diffusers in an 8-hour shift. Find the number assembled in 3 hours.

63 light diffusers

1.6 Long Division

Long division is used to divide by a number with more than one digit.

OBJECTIVE 1 In long division, estimate the various numbers by using a **trial divisor,** which is used to get a **trial quotient.**

┌─ **E X A M P L E 1**   **Using a Trial Divisor and a Trial Quotient**

Divide: $42\overline{)3066}$.

Because 42 is closer to 40 than to 50, use the first digit of the divisor as a trial divisor.

$$\underset{\uparrow\underline{\hspace{1cm}}\text{Trial divisor}}{42}$$

Try to divide the first digit of the dividend by 4. Since 3 cannot be divided by 4, use the first *two* digits, 30.

$$\frac{30}{4} = 7 \text{ with remainder } 2$$

$$\underset{42\overline{)3066}}{7} \leftarrow \text{Trial quotient}$$

7 goes over the 6, because $\frac{306}{42}$ is about 7.

Multiply 7 and 42 to get 294; next, subtract 294 from 306.

$$\begin{array}{r} 7 \\ 42\overline{)3066} \\ \underline{294} \leftarrow 7 \times 42 \\ 12 \leftarrow 306 - 294 \end{array}$$

Bring down the 6 at the right.

$$\begin{array}{r} 7 \\ 42\overline{)3066} \\ \underline{294\downarrow} \\ 126 \leftarrow 6 \text{ brought down} \end{array}$$

Use the trial divisor, 4.

First two digits of 126 → $\frac{12}{4} = 3$

$$\begin{array}{r} 73 \\ 42\overline{)3066} \\ \underline{294} \\ 126 \\ \underline{126} \leftarrow 3 \times 42 = 126 \\ 0 \end{array}$$

Check the answer by multiplying 42 and 73. The product should be 3066.

Note
The first digit on the left of the answer in long division must be placed in the proper position over the dividend.

WORK PROBLEM 1 AT THE SIDE. ▶▶

OBJECTIVES

1 Do long division.

2 Divide numbers ending in zero by numbers ending in zero.

3 Check answers.

FOR EXTRA HELP

Tutorial   Tape 2   SSM, Sec. 1.6

1. Divide.

(a) $25\overline{)1775}$

(b) $26\overline{)2132}$

(c) $51\overline{)2295}$

(d) $\dfrac{6552}{84}$

ANSWERS

1. (a) 71 **(b)** 82 **(c)** 45 **(d)** 78

2. Divide.

(a) 56$\overline{)2352}$

(b) 38$\overline{)1599}$

(c) 65$\overline{)5416}$

(d) 89$\overline{)6649}$

E X A M P L E 2 **Dividing to Find a Trial Quotient**

Divide: 58$\overline{)2730}$.

Use 6 as a trial divisor, since 58 is closer to 60 than to 50.

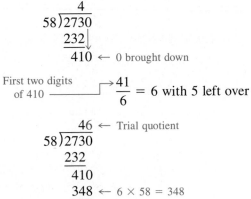

First two digits
of dividend $\longrightarrow \dfrac{27}{6} = 4$ with 3 left over

$$
\begin{array}{r}
4 \leftarrow \text{Trial quotient} \\
58\overline{)2730} \\
232 \leftarrow 4 \times 58 = 232 \\
\hline
41 \leftarrow 273 - 232 = 41 \text{ (smaller than 58,} \\
\text{the divisor)}
\end{array}
$$

Bring down the 0.

$$
\begin{array}{r}
4 \\
58\overline{)2730} \\
232\downarrow \\
\hline
410 \leftarrow 0 \text{ brought down}
\end{array}
$$

First two digits
of 410 $\longrightarrow \dfrac{41}{6} = 6$ with 5 left over

$$
\begin{array}{r}
46 \leftarrow \text{Trial quotient} \\
58\overline{)2730} \\
232 \\
\hline
410 \\
348 \leftarrow 6 \times 58 = 348 \\
\hline
62 \leftarrow \text{Greater than 58}
\end{array}
$$

The remainder, 62, is greater than the divisor, 58, so 7 should be used instead of 6.

$$
\begin{array}{r}
47 \text{ } \mathbf{R4} \leftarrow \\
58\overline{)2730} \\
232 \\
\hline
410 \\
406 \leftarrow 7 \times 58 = 406 \\
\hline
4 \leftarrow 410 - 406
\end{array}
$$

◄◄ **WORK PROBLEM 2 AT THE SIDE.**

Sometimes it is necessary to insert a zero in the quotient.

E X A M P L E 3 **Inserting Zeros in the Quotient**

Divide: 42$\overline{)8734}$.

Start as above.

$$
\begin{array}{r}
2 \\
42\overline{)8734} \\
84 \leftarrow 2 \times 42 = 84 \\
\hline
3 \leftarrow 87 - 84 = 3
\end{array}
$$

Bring down the 3.

$$
\begin{array}{r}
2 \\
42\overline{)8734} \\
84\downarrow \\
\hline
33 \leftarrow 3 \text{ brought down}
\end{array}
$$

ANSWERS

2. (a) 42 **(b)** 42 R3 **(c)** 83 R21
(d) 74 R63

CONTINUED ON NEXT PAGE

Since 33 cannot be divided by 42, place a 0 in the quotient as a placeholder.

$$\begin{array}{r} 20 \quad \leftarrow \text{0 in quotient} \\ 42\overline{)8734} \\ \underline{84} \\ 33 \end{array}$$

Bring down the final digit, the 4.

$$\begin{array}{r} 20 \\ 42\overline{)8734} \\ \underline{84}\downarrow \\ 334 \quad \leftarrow \text{4 brought down} \end{array}$$

Complete the problem.

$$\begin{array}{r} 207 \ \textbf{R}40 \\ 42\overline{)8734} \\ \underline{84} \\ 334 \\ \underline{294} \\ 40 \end{array}$$

The answer is 207 **R**40.

Note
There must be a digit in the quotient (answer) above every digit in the dividend once the answer has begun. Notice that in Example 3 a zero was used to assure an answer digit above every digit in the dividend.

WORK PROBLEM 3 AT THE SIDE. ▶▶

OBJECTIVE 2 When the divisor and dividend both contain zeros at the far right, recall that these numbers are multiples of 10. There is a short way to divide these multiples of 10. Look at the following examples.

$$26,000 \div 1 = 26,000$$

$$26,000 \div 10 = 2600$$

$$26,000 \div 100 = 260$$

$$26,000 \div 1000 = 26$$

Do you see a pattern in these divisions using multiples of 10? These examples suggest the following rule.

Dividing a Whole Number by 10, 100, or 1000

Divide a whole number by 10, 100, or 1000 by dropping the appropriate number of zeros from the whole number.

E X A M P L E 4   Dividing by Multiples of 10
Divide.

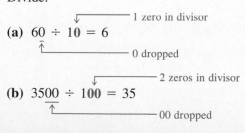

(a) $60 \div 10 = 6$
— 1 zero in divisor
— 0 dropped

(b) $3500 \div 100 = 35$
— 2 zeros in divisor
— 00 dropped

CONTINUED ON NEXT PAGE

3. Divide.

(a) $24\overline{)3127}$

(b) $52\overline{)10,660}$

(c) $39\overline{)15,933}$

(d) $78\overline{)23,462}$

ANSWERS

3. (a) 130 R7 **(b)** 205 **(c)** 408 R21
  (d) 300 R62

4. Divide.

(a) $50 \div 10$

(b) $1800 \div 100$

(c) $305,000 \div 1000$

5. Divide.

(a) $60\overline{)7200}$

(b) $130\overline{)131,040}$

(c) $2600\overline{)195,000}$

3 zeros in divisor

(c) $915,\underline{000} \div 1\mathbf{000} = 915$

000 dropped

◄◄ **WORK PROBLEM 4 AT THE SIDE.**

Find the quotient for other multiples of 10 by dropping zeros.

E X A M P L E   5   Dividing by Multiples of 10

Divide.

(a) $40\overline{)11,000}$ Drop 1 zero from the divisor and the dividend.

$$
\begin{array}{r}
275 \\
4\overline{)1100} \\
\underline{8} \\
30 \\
\underline{28} \\
20 \\
\underline{20} \\
0
\end{array}
$$

(b) $3500\overline{)31,500}$ 2 zeros dropped from the divisor and the dividend

$$
\begin{array}{r}
9 \\
35\overline{)315} \\
\underline{315} \\
0
\end{array}
$$

Note

Dropping zeros when dividing by multiples of 10 does not change the answer (quotient).

◄◄ **WORK PROBLEM 5 AT THE SIDE.**

OBJECTIVE 3 ▶ Answers in long division can be checked just as answers in short division were checked.

E X A M P L E   6   Checking Division

Check each answer.

$$
\begin{array}{r}
114 \ \mathbf{R}43 \\
(a) \ 48\overline{)5324}
\end{array}
$$

$$
\begin{array}{r}
114 \\
\times \quad 48 \quad \leftarrow \text{Multiply the quotient and the divisor.} \\
\hline
912 \\
456 \\
\hline
5472 \\
+ \quad 43 \quad \leftarrow \text{Add the remainder.} \\
\hline
5515 \quad \leftarrow \text{Result does not match dividend.}
\end{array}
$$

The answer does not check. Rework the original problem to get $110\ \mathbf{R}44$.

ANSWERS
4. (a) 5 (b) 18 (c) 305
5. (a) 120 (b) 1008 (c) 75

CONTINUED ON NEXT PAGE

(b) $716\overline{)26{,}492}$ $\dfrac{37}{}$

$$716$$
$$\times\ 37$$
$$\overline{5\ 012}$$
$$21\ 48$$
$$\overline{26{,}492}$$ Correct

$$\begin{array}{r} 37 \\ 716\overline{)26{,}492} \\ 21\ 48 \\ \hline 5\ 012 \\ 5\ 012 \\ \hline 0 \end{array}$$

Calculator Tip: To check the answer to Example 6(a) be certain to add the remainder.

$$48\ \boxed{\times}\ 110\ \boxed{+}\ 44\ \boxed{=}\ 5324$$

Note

When checking a division problem, first multiply the quotient and the divisor. Then be sure to add any remainder before checking it against the original dividend.

WORK PROBLEM 6 AT THE SIDE. ▶▶

6. Decide whether the following divisions are correct. If the answer is incorrect, find the correct answer.

(a) $\begin{array}{r} 43 \\ 18\overline{)774} \\ 72 \\ \hline 54 \\ 54 \\ \hline 0 \end{array}$

(b) $\begin{array}{r} 42\ \text{R}178 \\ 426\overline{)19{,}170} \\ 17\ 04 \\ \hline 1\ 130 \\ 952 \\ \hline 178 \end{array}$

(c) $\begin{array}{r} 57\ \text{R}18 \\ 514\overline{)29{,}316} \\ 25\ 70 \\ \hline 3\ 616 \\ 3\ 598 \\ \hline 18 \end{array}$

ANSWERS

6. (a) correct   **(b)** incorrect; should be 45
   (c) correct

NUMBERS IN THE
Real World collaborative investigations

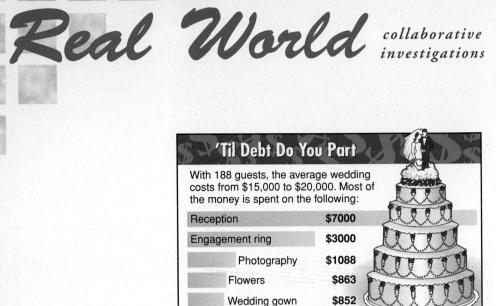

'Til Debt Do You Part

With 188 guests, the average wedding costs from $15,000 to $20,000. Most of the money is spent on the following:

Reception	$7000
Engagement ring	$3000
Photography	$1088
Flowers	$863
Wedding gown	$852

Source: Interep Radio Store

What is the total of the expenses shown in the graph? **$12,803**

This total is how much less than the "average wedding cost"? **$2197 to $7197**

What are some of the other expenses that could make up the difference? List three or four possibilities. **wedding rings, limousine rental, organist, gifts for bridal party, etc.**

If you budget $7000 for your wedding reception, and the cost per person is $37, how many guests can you invite without going over your budget? **189 guests**

How much of your budgeted amount will be left over? **$7 left over**

If you budget $4500, and the cost per person is $37, how many guests can you invite, and how much of your budgeted amount will be left over? **121 guests, $23 left over**

If you budget $1000? **27 guests, $1 left over**

Show how these last three answers are similar to doing a division problem where the quotient includes a remainder.

$$
\begin{array}{r}
189 \\
37{\overline{\smash{\big)}\,7000}} \\
\underline{6993} \\
7
\end{array}
\quad\quad
\begin{array}{r}
121 \\
37{\overline{\smash{\big)}\,4500}} \\
\underline{-4477} \\
23
\end{array}
\quad\quad
\begin{array}{r}
27 \\
37{\overline{\smash{\big)}\,1000}} \\
\underline{-999} \\
1
\end{array}
$$

← guests

← money left over

Different people will pick different amounts to budget for a reception. But if the price per person is $37 and as many guests are invited as possible, what is the maximum amount of money in whole dollars that could be left over? Explain your answer. **$36; If $37 or more were left over, you'd have enough to invite another guest.**

1.6 Exercises

Without doing the actual division, circle the correct answer from the three choices given.

Example: $42\overline{)7560}$ 18   180   1800   **Solution:** $42\overline{)7560}^{\,1}$ 18   (180)   1800

└ 1 goes over the 5, because $\frac{75}{42}$ is about 1.
The answer must then be a three-digit
number or 180.

---

1. $24\overline{)768}$

   3   (32)   320

2. $35\overline{)805}$

   2   (23)   230

3. $18\overline{)4500}$

   2   25   (250)

4. $28\overline{)3500}$

   12   (125)   1250

5. $86\overline{)10,327}$

   12   (120 R7)   1200

6. $46\overline{)24,026}$

   5   52   (522 R14)

7. $52\overline{)68,025}$

   13   130 R1   (1308 R9)

8. $12\overline{)116,953}$

   974 R2   (9746 R1)   97,460

9. $21\overline{)149,826}$

   71   713   (7134 R12)

10. $32\overline{)247,892}$

   77 R1   (7746 R20)   77,460

11. $523\overline{)470,800}$

   9 R100   90 R100   (900 R100)

12. $230\overline{)253,230}$

   11   110   (1101)

Divide by using long division. Check each answer.

Example: $41\overline{)2388}$ **Solution:**

$$\begin{array}{r} 58\text{ R10} \\ 41\overline{)2388} \\ \underline{205} \\ 338 \\ \underline{328} \\ 10 \end{array}$$

Check:

$$\begin{array}{r} 58 \\ \times\ 41 \\ \hline 58 \\ 232 \\ \hline 2378 \\ +\ \ 10 \leftarrow \text{Remainder (added)} \\ \hline \mathbf{2388} \leftarrow \text{Matches dividend} \end{array}$$

---

13. $\overset{207\text{ R5}}{42\overline{)8699}}$

14. $\overset{38}{58\overline{)2204}}$

15. $\overset{236\text{ R29}}{47\overline{)11,121}}$

16. $\overset{478\text{ R18}}{83\overline{)39,692}}$

17. $\overset{2407\text{ R1}}{26\overline{)62,583}}$

18. $\overset{3008\text{ R25}}{28\overline{)84,249}}$

19. $\overset{1239\text{ R15}}{63\overline{)78,072}}$

20. $\overset{785\text{ R118}}{238\overline{)186,948}}$

21. $\overset{3331\text{ R82}}{153\overline{)509,725}}$

22. $\overset{73}{402\overline{)29,346}}$

23. $\overset{850}{420\overline{)357,000}}$

24. $\overset{170}{900\overline{)153,000}}$

✎ Writing   ◉ Conceptual   ▲ Challenging   ≈ Estimation

Check each answer. If an answer is incorrect, give the correct answer.

25. $\overset{106\ R17}{56\overline{)5943}}$ 26. $\overset{37\ R37}{87\overline{)3254}}$ 27. $\overset{658\ R9}{28\overline{)18,424}}$

incorrect; should be 106 R7   incorrect; should be 37 R35   incorrect; should be 658

28. $\overset{463\ R171}{191\overline{)88,604}}$ 29. $\overset{62\ R3}{614\overline{)38,068}}$ 30. $\overset{174\ R368}{557\overline{)97,286}}$

correct   incorrect; should be 62   correct

31. Describe in your own words how you can divide a multiple of 10 by 10, by 100, or by 1000 by dropping zeros from the whole number. Write an example problem and solve it.

When dividing by 10, 100, or 1000 drop the same number of zeros from the dividend to get the quotient. One example is
2500 ÷ 100 = 25.

32. Suppose you have a division problem with a remainder in the answer. Explain how to check your answer by using an example problem that has a remainder.

Multiply the quotient and the divisor and add any remainder. One example is
18 ÷ 5 = 3 R3   check: 3 × 5 = 15 + 3 = 18

Solve each application problem by using addition, subtraction, multiplication, or division as needed.

33. A car travels 1350 miles at 54 miles per hour. How many hours did it travel?

25 hours

34. The Government Printing Office uses 255,000 pounds of ink each year. If they do an equal amount of printing on each of 200 work days in a year, find the weight of the ink used each day.

1275 pounds

35. There were 1838 medals made for the 1996 Olympic Games. The ancient Olympic stadium is shown on one side, the pictogram of the sport on the other side. Of the medals made, 604 were gold, 604 were silver and the remainder were bronze. Find the number of bronze medals.

630 bronze medals

36. Two separated parents each share some of the education costs of their child which amount to $3718. If one parent pays $1880, find the amount paid by the other parent.

$1838

37. Judy Martinez owes $3888 on a loan. Find her monthly payment if the loan is to be paid off in 36 months.

$108

38. A consultant charged $13,050 for studying a school's compliance with the Americans with Disabilities Act. If the consultant worked 225 hours, find the rate charged per hour.

$58

39. Clarence Hanks can assemble 42 circuits in 1 hour. How many circuits can he assemble in a 5-day workweek of 8 hours per day?

1680 circuits

40. There are two conveyer lines in a factory each of which packages 240 sacks of salt per hour. If the lines operate for 8 hours, find the total number of sacks of salt packaged by the two lines.

3840 sacks

41. A youth soccer association raised $7588 in fund-raising projects. There were expenses of $838 that had to be paid first, with the balance of the money divided evenly among the 18 teams. How much did each team receive?

$375

42. Feather Farms Egg Ranch collects 3545 eggs in the morning and 2575 eggs in the afternoon. If the eggs are packed in flats containing 30 eggs each, find the number of flats needed for packing.

204 flats

1.7 Rounding Whole Numbers

One way to get a rough check on an answer is to *round* the numbers in the problem. **Rounding** a number means finding a number that is close to the original number, but easier to work with.

For example, a superintendent of schools in a large city might be discussing the need to build new schools. In making her point, it probably would not be necessary to say that the school district has 152,807 students—it would probably be sufficient to say there are 153,000 students, or even 150,000 students.

OBJECTIVE 1 The first step in rounding a number is to locate the *place to which the number is to be rounded.*

─ E X A M P L E 1 **Finding the Place to Which a Number Is to Be Rounded**

Locate and draw a line under the place to which each number is to be rounded.

(a) Round 83 to the nearest ten. Is 83 closer to 8̲0 or to 9̲0?

83 is closer to 80.

Tens place ────┘

(b) Round 54,702 to the nearest thousand. Is it closer to 54̲,000 or to 55̲,000?

54,702 is closer to 55,000.

Thousands place ────┘

(c) Round 2,806,124 to the nearest hundred thousand. Is it closer to 2,8̲00,000 or to 2,9̲00,000?

2,806,124 is closer to 2,800,000.

└─Hundred thousands place

WORK PROBLEM 1 AT THE SIDE. ▶▶

OBJECTIVE 2 Use the following rules for rounding whole numbers.

Step 1 Locate the **place** to which the number is to be rounded. Draw a line under that place.

Step 2A Look only at the next digit to the right of the one you underlined. If it is 5 or more, increase the underlined digit by 1.

Step 2B If the next digit to the right is 4 or less, do not change the digit in the underlined place.

Step 3 **Change** all digits to the right of the underlined place to zeros.

─ E X A M P L E 2 **Using Rounding Rules for 4 or Less**

Round 349 to the nearest hundred.

Step 1 Locate the place to which the number is being rounded. Draw a line under that place.

3̲49

└─ Hundreds place

── **CONTINUED ON NEXT PAGE**

OBJECTIVES

1 Locate the place to which a number is to be rounded.

2 Round numbers.

3 Round numbers to estimate an answer.

4 Use front end rounding to estimate an answer.

FOR EXTRA HELP

Tutorial   Tape 2   SSM, Sec. 1.7

1. Locate and draw a line under the place to which the number is to be rounded. Then answer the question.

(a) 746 (nearest ten)

Is it closer to 740 or to 750? _____

(b) 2412 (nearest thousand)

Is it closer to 2000 or to 3000? _____

(c) 89,512 (nearest hundred)

Is it closer to 89,500 or 89,600? _____

(d) 546,325 (nearest ten thousand)

Is it closer to 540,000 or 550,000? _____

ANSWERS

1. **(a)** 74̲6 is closer to 75̲0
   (b) 2̲412 is closer to 2̲000
   (c) 89,5̲12 is closer to 89,5̲00
   (d) 54̲6,325 is closer to 55̲0,000

59

2. Round to the nearest ten.

(a) 34

(b) 71

(c) 143

(d) 5732

3. Round to the nearest thousand.

(a) 1725

(b) 6511

(c) 56,899

(d) 82,608

Step 2 Because the next digit to the right of the underlined place is 4, which is 4 or less, do *not* change the digit in the underlined place.

Next digit is 4 or less.

3̲49

3 remains 3.

Step 3 Change all digits to the right of the underlined place to zeros.

3̲49 rounded to the nearest hundred is 300.

In other words, 349 is closer to 300 than to 400.

◀◀ **WORK PROBLEM 2 AT THE SIDE.**

E X A M P L E 3   Rounding Rules for 5 or More

Round 36,833 to the nearest thousand.

Step 1 Find the place to which the number is to be rounded and draw a line under that place.

36,833

Thousands

Step 2 Because the next digit to the right of the underlined place is 8, which is 5 or more, add 1 to the underlined place.

Next digit is 5 or more.

36,833

Change 6 to 7.

Step 3 All digits to the right of the underlined place are changed to zeros.

Change to 0.

36,833 rounded to the nearest thousand is 37,000.

Change 6 to 7.

In other words, 36,833 is closer to 37,000 than to 36,000.

◀◀ **WORK PROBLEM 3 AT THE SIDE.**

E X A M P L E 4   Using Rules for Rounding

(a) Round 2382 to the nearest ten.

Step 1 2382

Tens place

Step 2 The next digit to the right is 2, which is 4 or less.

Next digit is 4 or less.

2382

Leave 8 as 8.

Step 3 2382   Change to 0.

2382 rounded to the nearest ten is 2380.

CONTINUED ON NEXT PAGE

ANSWERS

2. (a) 30 **(b)** 70 **(c)** 140 **(d)** 5730
3. (a) 2000 **(b)** 7000 **(c)** 57,000
   (d) 83,000

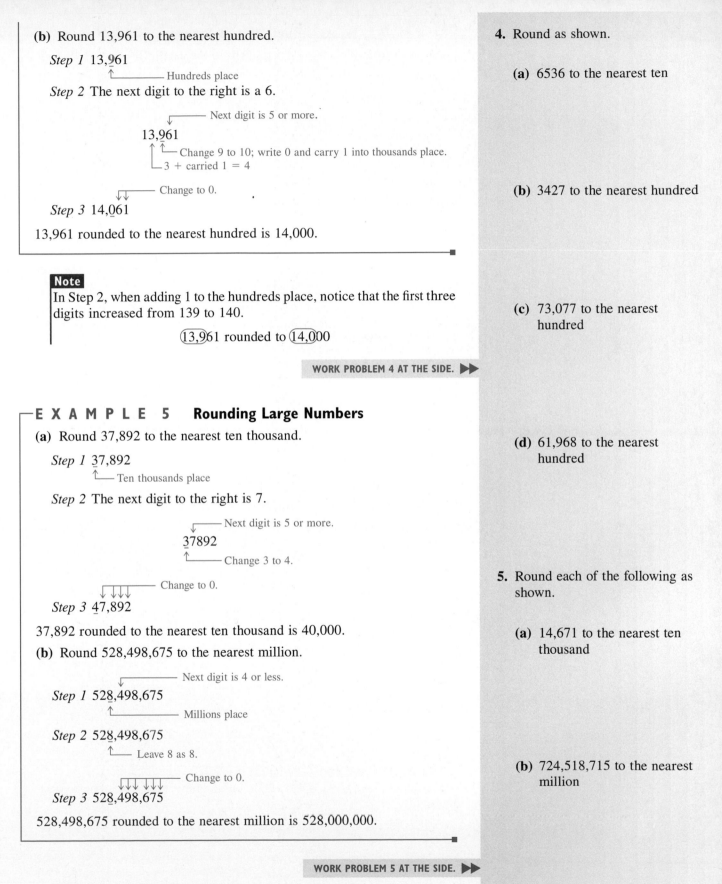

(b) Round 13,961 to the nearest hundred.

Step 1 13,961
↑——————— Hundreds place

Step 2 The next digit to the right is a 6.

——— Next digit is 5 or more.
↓
13,961
↑↑— Change 9 to 10; write 0 and carry 1 into thousands place.
└ 3 + carried 1 = 4

——— Change to 0.
↓↓
Step 3 14,061

13,961 rounded to the nearest hundred is 14,000.

Note
In Step 2, when adding 1 to the hundreds place, notice that the first three digits increased from 139 to 140.

⟨13,9⟩61 rounded to ⟨14,0⟩00

WORK PROBLEM 4 AT THE SIDE. ▶▶

E X A M P L E 5   Rounding Large Numbers
(a) Round 37,892 to the nearest ten thousand.

Step 1 37,892
↑—— Ten thousands place

Step 2 The next digit to the right is 7.

——— Next digit is 5 or more.
↓
37892
↑——————— Change 3 to 4.

——— Change to 0.
↓↓↓↓
Step 3 47,892

37,892 rounded to the nearest ten thousand is 40,000.

(b) Round 528,498,675 to the nearest million.

——— Next digit is 4 or less.
↓
Step 1 528,498,675
↑——————— Millions place

Step 2 528,498,675
↑—— Leave 8 as 8.

——— Change to 0.
↓↓↓ ↓↓↓
Step 3 528,498,675

528,498,675 rounded to the nearest million is 528,000,000.

WORK PROBLEM 5 AT THE SIDE. ▶▶

Sometimes a number must be rounded to different places.

4. Round as shown.

(a) 6536 to the nearest ten

(b) 3427 to the nearest hundred

(c) 73,077 to the nearest hundred

(d) 61,968 to the nearest hundred

5. Round each of the following as shown.

(a) 14,671 to the nearest ten thousand

(b) 724,518,715 to the nearest million

ANSWERS
4. (a) 6540 **(b)** 3400 **(c)** 73,100
(d) 62,000
5. (a) 10,000 **(b)** 725,000,000

6. Round each of the following numbers to the nearest ten and then to the nearest hundred.

(a) 156

(b) 649

(c) 9809

E X A M P L E 6   Rounding to Different Places

Round 648 **(a)** to the nearest ten and **(b)** to the nearest hundred.

(a) to the nearest ten

Next digit is 5 or more.

648

Tens place (4 + 1 = 5)

648 to the nearest ten is 650.

(b) to the nearest hundred

Next digit is 4 or less.

648

Hundreds place stays the same.

648 to the nearest hundred is 600.

If 648 is rounded to the nearest ten as 650, and then 650 is rounded to the nearest hundred, the result is 700. If, however, 648 is rounded to the nearest hundred, the result is 600.

Note
Always go back to the *original* number when rounding to a different place.

◀◀ **WORK PROBLEM 6 AT THE SIDE.**

E X A M P L E 7   Applying Rounding Rules

Round each of the following to the nearest ten, nearest hundred, and nearest thousand.

(a) 4358

First round 4358 to the nearest ten.

Next digit is 5 or more.

4358

Tens place (5 + 1 = 6)

4358 rounded to the nearest ten is 4360.

Now go back to the *original* number to round to the nearest hundred.

Next digit is 5 or more.

4358

Hundreds place (3 + 1 = 4)

4358 rounded to the nearest hundred is 4400.

Again, go back to the *original* number to round, this time to the nearest thousand.

Next digit is 4 or less.

4358

Thousands place stays the same.

4358 rounded to the nearest thousand is 4000.

ANSWERS

6. (a) 160; 200
(b) 650; 600
(c) 9810; 9800

CONTINUED ON NEXT PAGE

(b) 680,914

First, round to the nearest ten.

Next digit is 4 or less.

680,9_14_

Tens place stays the same.

680,914 rounded to the nearest ten is 680,910.

Go back to the *original* number to round to the nearest hundred.

Next digit is 4 or less.

680,_9_14

Hundreds place stays the same.

680,914 rounded to the nearest hundred is 680,900.

Go back to the *original* number to round to the nearest thousand.

Next digit is 5 or more.

68_0_,914

Thousands place (0 + 1 = 1)

680,914 rounded to the nearest thousand is 681,000.

WORK PROBLEM 7 AT THE SIDE. ▶▶

OBJECTIVE 3 Numbers are rounded to estimate an answer. An estimated answer is one that is close to the exact answer and may be used as a check when the exact answer is found. The "≈" sign is often used to show that an answer has been rounded or estimated and is almost equal to the exact answer. It means "approximately equal to."

┌ **E X A M P L E   8   Using Rounding to Estimate an Answer**

Estimate the following answers by rounding to the nearest ten.

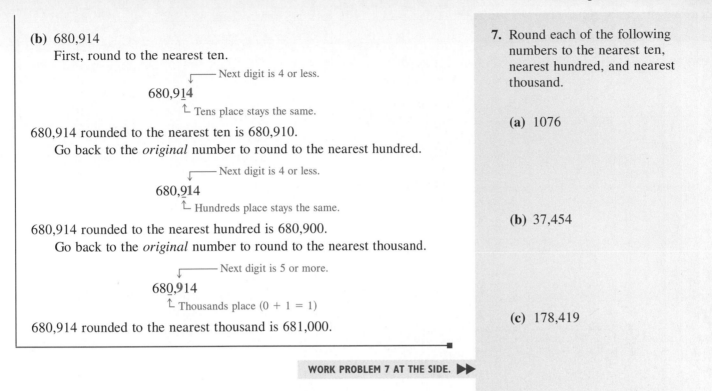

(a)

76	80 ⎫	
53	50 ⎬	Rounded to the nearest ten
38	40 ⎪	
+ 91	+ 90 ⎭	
	260	Estimated answer

(b)

27	30 ⎫	Rounded to the nearest ten
− 14	− 10 ⎭	
	20	Estimated answer

(c)

16	20 ⎫	Rounded to the nearest ten
× 21	× 20 ⎭	
	400	Estimated answer

WORK PROBLEM 8 AT THE SIDE. ▶▶

┌ **E X A M P L E   9   Using Rounding to Estimate an Answer**

Estimate the following answers by rounding to the nearest hundred.

(a)

152	200 ⎫	
749	700 ⎬	Rounded to the nearest hundred
576	600 ⎪	
+ 819	+ 800 ⎭	
	2300	Estimated answer

└ **CONTINUED ON NEXT PAGE**

---

7. Round each of the following numbers to the nearest ten, nearest hundred, and nearest thousand.

(a) 1076

(b) 37,454

(c) 178,419

8. ≈ Estimate the following answers by rounding to the nearest ten.

(a)
18
73
57
+ 34

(b)
44
− 18

(c)
37
× 84

ANSWERS

7. (a) 1080; 1100; 1000
    (b) 37,450; 37,500; 37,000
    (c) 178,420; 178,400; 178,000
8. (a) 20 + 70 + 60 + 30 = 180
    (b) 40 − 20 = 20 **(c)** 80 × 40 = 3200

9. ≈ Estimate the following answers by rounding to the nearest hundred.

(a)
```
   175
   618
   739
 + 865
```

(b)
```
   739
 - 361
```

(c)
```
   723
 × 478
```

10. ≈ Use front end rounding to estimate each answer.

(a)
```
     36
   3852
    749
 + 5474
```

(b)
```
   2583
 -  765
```

(c)
```
   639
 ×  55
```

(b)
```
   780        800 ⎫
 - 536      - 500 ⎬  Rounded to the nearest hundred
 ─────      ───── ⎭
              300    Estimated answer
```

(c)
```
   664        700 ⎫
 × 843      × 800 ⎬  Rounded to the nearest hundred
 ─────      ───── ⎭
          560,000    Estimated answer
```

◀◀ **WORK PROBLEM 9 AT THE SIDE.**

OBJECTIVE 4 **Front end rounding** is used to estimate an answer. With front end rounding, each number is rounded so that all the digits are changed to zero except the first digit, which is rounded. Only one non-zero digit remains.

EXAMPLE 10   Using Front End Rounding to Estimate an Answer

Estimate the following answers by using front end rounding.

(a)
```
   3825        4000 ⎫
     72          70 ⎪  All digits changed to
    565         600 ⎬    zero except first
 + 2389      + 2000 ⎪    digit, which is rounded
 ──────      ────── ⎭
               6670    Estimated answer
```

(b)
```
   6712        7000 ⎫  First digit rounded and
 -  825      -  800 ⎬    all others changed to zero
 ─────       ───── ⎭
               6200    Estimated answer
```

(c)
```
   725         700 ⎫  Only one non-zero digit
 ×  86       ×  90 ⎬    remains
 ─────       ───── ⎭
             63,000    Estimated answer
```

> **Note**
> When using front end rounding, only 1 non-zero digit (first digit) remains. All digits to the right are zeros.

◀◀ **WORK PROBLEM 10 AT THE SIDE.**

ANSWERS

9. (a) 200 + 600 + 700 + 900 = 2400
   (b) 700 − 400 = 300
   (c) 500 × 700 = 350,000
10. (a) 40 + 4000 + 700 + 5000 = 9740
    (b) 3000 − 800 = 2200
    (c) 60 × 600 = 36,000

1.7 Exercises

Round as shown.

> **Example:** Round 4336 to the nearest ten.   **Solution:**
>
> 43$\underline{3}$6 — Next digit is 5 or more.
>
> Tens place changes (3 + 1 = 4). All digits to the right of the underlined place change to 0.
>
> $\approx$**4340**

1. 623 to the nearest ten

$\approx$**620**

2. 206 to the nearest ten

$\approx$**210**

3. 1085 to the nearest ten

$\approx$**1090**

4. 2439 to the nearest ten

$\approx$**2440**

5. 7862 to the nearest hundred

$\approx$**7900**

6. 6746 to the nearest hundred

$\approx$**6700**

7. 86,813 to the nearest hundred

$\approx$**86,800**

8. 17,211 to the nearest hundred

$\approx$**17,200**

9. 42,495 to the nearest hundred

$\approx$**42,500**

10. 18,273 to the nearest hundred

$\approx$**18,300**

11. 5996 to the nearest hundred

$\approx$**6000**

12. 8451 to the nearest hundred

$\approx$**8500**

13. 15,758 to the nearest hundred

$\approx$**15,800**

14. 28,065 to the nearest hundred

$\approx$**28,100**

15. 78,499 to the nearest thousand

$\approx$**78,000**

16. 14,314 to the nearest thousand

$\approx$**14,000**

17. 5847 to the nearest thousand

$\approx$**6000**

18. 49,706 to the nearest thousand

$\approx$**50,000**

19. 53,182 to the nearest thousand

$\approx$**53,000**

20. 13,124 to the nearest thousand

$\approx$**13,000**

21. 595,008 to the nearest ten thousand

$\approx$**600,000**

22. 725,182 to the nearest ten thousand

$\approx$**730,000**

23. 8,906,422 to the nearest million

$\approx$**9,000,000**

24. 13,713,409 to the nearest million

$\approx$**14,000,000**

✏ Writing   ⊙ Conceptual   ▲ Challenging   ≈ Estimation

Round each of the following to the nearest ten, nearest hundred, and nearest thousand.

Example: 6475 —— **Solution:** Always go back to the original number.

	Ten	Hundred	Thousand
	5 or more	5 or more	4 or less
	6475	6475	6475
	Change to 0. Add 1 to 7.	Change to 0. Add 1 to 4.	Change to 0. Leave 6 as 6.
	≈**6480**	≈**6500**	≈**6000**

Remember to round from the original number.

		Ten	*Hundred*	*Thousand*			*Ten*	*Hundred*	*Thousand*
25.	1476	≈1480	≈1500	≈1000	**26.**	3471	≈3470	≈3500	≈3000
27.	4483	≈4480	≈4500	≈4000	**28.**	8624	≈8620	≈8600	≈9000
29.	5049	≈5050	≈5000	≈5000	**30.**	7065	≈7070	≈7100	≈7000
31.	3132	≈3130	≈3100	≈3000	**32.**	7456	≈7460	≈7500	≈7000
33.	19,539	≈19,540	≈19,500	≈20,000	**34.**	59,806	≈59,810	≈59,800	≈60,000
35.	26,292	≈26,290	≈26,300	≈26,000	**36.**	78,519	≈78,520	≈78,500	≈79,000
37.	23,502	≈23,500	≈23,500	≈24,000	**38.**	84,639	≈84,640	≈84,600	≈85,000

39. Write in your own words the three steps that you will use to round a number when the digit to the right of the place to which you are rounding is 5 or more.

1. Locate the place to be rounded and underline it.
2. Look only at the next digit to the right. If this digit is 5 or more, increase the underlined digit by 1.
3. Change all digits to the right of the underlined place to zeros.

40. Write in your own words the three steps that you will use to round a number when the digit to the right of the place to which you are rounding is 4 or less.

1. Locate the place to be rounded and underline it.
2. Look only at the next digit to the right. If this digit is 4 or less, do not change the underlined digit.
3. Change all digits to the right of the underlined place to zeros.

*≈ *Estimate the following answers by rounding to the nearest ten. Then find the exact answers.*

41. *estimate*   *exact*

$$
\begin{array}{r}
70 \\
40 \\
90 \\
+\ 90 \\
\hline
290
\end{array}
\xleftarrow{\text{rounds to}}
\begin{array}{r}
66 \\
43 \\
89 \\
+\ 94 \\
\hline
292
\end{array}
$$

42. *estimate*   *exact*

$$
\begin{array}{r}
60 \\
20 \\
90 \\
+\ 70 \\
\hline
240
\end{array}
\qquad
\begin{array}{r}
56 \\
24 \\
85 \\
+\ 71 \\
\hline
236
\end{array}
$$

43.
$$
\begin{array}{r}
100 \\
-\ 30 \\
\hline
70
\end{array}
\qquad
\begin{array}{r}
97 \\
-\ 26 \\
\hline
71
\end{array}
$$

44.
$$
\begin{array}{r}
60 \\
-\ 20 \\
\hline
40
\end{array}
\qquad
\begin{array}{r}
57 \\
-\ 24 \\
\hline
33
\end{array}
$$

45.
$$
\begin{array}{r}
80 \\
\times\ 20 \\
\hline
1600
\end{array}
\qquad
\begin{array}{r}
76 \\
\times\ 22 \\
\hline
1672
\end{array}
$$

46.
$$
\begin{array}{r}
50 \\
\times\ 80 \\
\hline
4000
\end{array}
\qquad
\begin{array}{r}
53 \\
\times\ 75 \\
\hline
3975
\end{array}
$$

≈ *Estimate the following answers by rounding to the nearest hundred. Then find the exact answers.*

47. *estimate*   *exact*

$$
\begin{array}{r}
800 \\
800 \\
300 \\
+\ 700 \\
\hline
2600
\end{array}
\xleftarrow{\text{rounds to}}
\begin{array}{r}
786 \\
823 \\
342 \\
+\ 684 \\
\hline
2635
\end{array}
$$

48. *estimate*   *exact*

$$
\begin{array}{r}
600 \\
400 \\
200 \\
+\ 700 \\
\hline
1900
\end{array}
\qquad
\begin{array}{r}
623 \\
362 \\
189 \\
+\ 736 \\
\hline
1910
\end{array}
$$

49.
$$
\begin{array}{r}
900 \\
-\ 500 \\
\hline
400
\end{array}
\qquad
\begin{array}{r}
874 \\
-\ 458 \\
\hline
416
\end{array}
$$

50.
$$
\begin{array}{r}
600 \\
-\ 300 \\
\hline
300
\end{array}
\qquad
\begin{array}{r}
614 \\
-\ 276 \\
\hline
338
\end{array}
$$

51.
$$
\begin{array}{r}
400 \\
\times\ 400 \\
\hline
160{,}000
\end{array}
\qquad
\begin{array}{r}
368 \\
\times\ 436 \\
\hline
160{,}448
\end{array}
$$

52.
$$
\begin{array}{r}
700 \\
\times\ 500 \\
\hline
350{,}000
\end{array}
\qquad
\begin{array}{r}
739 \\
\times\ 487 \\
\hline
359{,}893
\end{array}
$$

≈ *Estimate the following answers by using front end rounding. Then find the exact answers.*

53. *estimate*   *exact*

$$
\begin{array}{r}
8000 \\
60 \\
700 \\
+\ 4000 \\
\hline
12{,}760
\end{array}
\xleftarrow{\text{rounds to}}
\begin{array}{r}
8215 \\
56 \\
729 \\
+\ 3605 \\
\hline
12{,}605
\end{array}
$$

54. *estimate*   *exact*

$$
\begin{array}{r}
3000 \\
70 \\
600 \\
+\ 7000 \\
\hline
10{,}670
\end{array}
\qquad
\begin{array}{r}
2685 \\
73 \\
592 \\
+\ 7183 \\
\hline
10{,}533
\end{array}
$$

*This symbol is used to indicate exercises for which you should estimate your answer.

55.

estimate	*exact*
700	681
− 300	− 316
400	365

56.

estimate	*exact*
900	942
− 300	− 286
600	656

57.

800	841
× 40	× 38
32,000	31,958

58.

900	864
× 70	× 74
63,000	63,936

59. The number 648 rounded to the nearest ten is 650, and 650 rounded to the nearest hundred is 700. But when 648 is rounded to the nearest hundred it becomes 600. Why is this true? Explain.

Perhaps the best explanation is that 648 is closer to 650 than 640, but 648 is closer to 600 than to 700.

60. The use of rounding is helpful when estimating the answer to a problem. Why is this true? Give an example using either addition, subtraction, multiplication, or division to show how this works.

Rounding numbers usually allows for faster calculation and results in an estimated answer prior to getting an exact answer. One example is

	400	432
estimate: −	200	*exact:* − 209
	200	223

61. Mexico City Bank has total assets of 3,025,935,000 pesos. Round this amount to the nearest ten thousand, nearest million, and nearest billion pesos.

≈3,025,940,000 pesos; ≈3,026,000,000 pesos; ≈3,000,000,000 pesos

62. Round 621,999,652 to the nearest thousand, nearest ten thousand, and nearest hundred thousand.

≈622,000,000; ≈622,000,000; ≈622,000,000

63. The gross national product for the United States (sum of all goods and services sold) was $5,465,485,362,159. Round this amount to the nearest hundred thousand, nearest hundred million, and nearest billion.

≈$5,465,485,400,000; ≈$5,465,500,000,000; ≈$5,465,000,000,000

64. The total Aid to Families with Dependent Children (AFDC) in the United States last year was $18,630,604,733. Round this amount to the nearest hundred thousand, nearest hundred million, and nearest ten billion.

≈$18,630,600,000; ≈$18,600,000,000; ≈$20,000,000,000

1.8 Roots and Order of Operations

OBJECTIVES

1. Identify an exponent and a base.
2. Find the square root of a number.
3. Use the order of operations.

FOR EXTRA HELP

Tutorial   Tape 2   SSM, Sec. 1.8

OBJECTIVE 1 The product $3 \cdot 3$ can be written as 3^2 (read as "3 squared"). The small raised number 2, called an **exponent** (EX-poh-nent) says to use 2 factors of 3. The number 3 is called the **base.** Writing 3^2 as 9 is called *simplifying the expression.*

E X A M P L E   1   Simplifying an Expression

Identify the exponent and the base. Simplify each expression.

(a) 4^3

$$\text{Base} \rightarrow 4^3 \leftarrow \text{Exponent} \qquad 4^3 = 4 \times 4 \times 4 = 64$$

(b) $2^5 = 2 \times 2 \times 2 \times 2 \times 2 = 32$

The base is 2 and the exponent is 5.

WORK PROBLEM 1 AT THE SIDE. ▶▶

1. Identify the exponent and the base. Simplify each expression.

   (a) 3^2

   (b) 6^3

   (c) 2^4

   (d) 3^4

OBJECTIVE 2 Because $3^2 = 9$, the number 3 is called the **square root** of 9. The square root of a number is one of two identical factors of that number. Square roots of numbers are written with the symbol $\sqrt{}$, so

$$\sqrt{9} = 3.$$

By definition,

Square Root

$$\sqrt{\text{number} \cdot \text{number}} = \sqrt{\text{number}^2} = \text{number}.$$

For example: $\sqrt{36} = \sqrt{6 \cdot 6} = \sqrt{6^2} = 6$

To find the square root of 64 ask, "What number can be multiplied by itself (that is, *squared*) to give 64?" The answer is 8, so

$$\sqrt{64} = \sqrt{8 \cdot 8} = \sqrt{8^2} = 8.$$

A **perfect square** is a number that is the square of a whole number. The first few perfect squares are listed here.

Perfect Squares Table

$0 = 0^2$	$16 = 4^2$	$64 = 8^2$	$144 = 12^2$
$1 = 1^2$	$25 = 5^2$	$81 = 9^2$	$169 = 13^2$
$4 = 2^2$	$36 = 6^2$	$100 = 10^2$	$196 = 14^2$
$9 = 3^2$	$49 = 7^2$	$121 = 11^2$	$225 = 15^2$

E X A M P L E   2   Using Perfect Squares

Find each square root.

(a) $\sqrt{16}$ Because $4^2 = 16$, $\sqrt{16} = 4$.   (b) $\sqrt{49} = 7$

(c) $\sqrt{0} = 0$ (d) $\sqrt{169} = 13$

2. Find each square root.

   (a) $\sqrt{4}$

   (b) $\sqrt{36}$

   (c) $\sqrt{81}$

   (d) $\sqrt{225}$

   (e) $\sqrt{1}$

WORK PROBLEM 2 AT THE SIDE. ▶▶

OBJECTIVE 3 Frequently problems may have parentheses, exponents, and square roots, and may involve more than one operation. Work these problems with the following order of operations.

ANSWERS

1. (a) 2; 3; 9 (b) 3; 6; 216
   (c) 4; 2; 16 (d) 4; 3; 81
2. (a) 2 (b) 6 (c) 9 (d) 15 (e) 1

3. Work each problem.

(a) $3 + 8 + 2^2$

(b) $3^2 + 2^3$

(c) $5 \cdot 8 \div 20 - 1$

(d) $40 \div 5 \div 2$

(e) $8 + (14 \div 2) \cdot 6$

4. Work each problem.

(a) $8 - 3 + 4^2$

(b) $2^2 + 3^2 - (5 \cdot 2)$

(c) $2 \cdot \sqrt{81} - 6 \cdot 2$

(d) $20 \div 2 + (7 - 5)$

(e) $15 \cdot \sqrt{9} - 8 \cdot \sqrt{4}$

ANSWERS

3. (a) 15 **(b)** 17 **(c)** 1 **(d)** 4 **(e)** 50
4. (a) 21 **(b)** 3 **(c)** 6 **(d)** 12 **(e)** 29

Order of Operations

1. Do all operations inside **parentheses.**
2. Simplify any expressions with **exponents** and find any **square roots.**
3. **Multiply** or **divide,** proceeding from left to right.
4. **Add** or **subtract,** proceeding from left to right.

E X A M P L E 3   **Understanding Order of Operations**

Work each problem.

(a) $8^2 + 5 + 2$

$$8^2 + 5 + 2$$
$$\underline{8 \cdot 8} + 5 + 2 \qquad \text{Evaluate exponent first; } 8^2 \text{ is } 8 \cdot 8.$$
$$\underline{64 + 5} + 2 \qquad \text{Add from left to right.}$$
$$69 \quad + 2 = 71$$

(b) $\underline{35 \div 5} \cdot 6 \qquad$ Divide first (start at left).
$$\quad 7 \quad \cdot 6 = 42 \qquad \text{Multiply.}$$

(c) $9 + \underline{(20 - 4)} \cdot 3 \qquad$ Work inside parentheses first.
$$9 + \quad \underline{16 \cdot 3} \qquad \text{Then multiply.}$$
$$9 + \qquad 48 \quad = 57 \qquad \text{Add last.}$$

(d) $12 \cdot \underline{\sqrt{16}} - 8 \cdot 4 \qquad$ Find square root first.
$$\underline{12 \cdot \quad 4} \quad - \underline{8 \cdot 4} \qquad \text{Multiply from left to right.}$$
$$48 \qquad - \qquad 32 \quad = 16 \qquad \text{Subtract last.}$$

◀◀ WORK PROBLEM 3 AT THE SIDE.

E X A M P L E 4   **Using Order of Operations**

Work each problem.

(a) $\underline{15 - 4} + 2 \qquad$ Subtract first (start at left).
$$11 \quad + 2 = 13 \qquad \text{Add.}$$

(b) $8 + \underline{(7 - 3)} \div 2 \qquad$ Work inside parentheses first.
$$8 + \quad \underline{4 \quad \div 2} \qquad \text{Divide.}$$
$$8 + \qquad 2 \quad = 10 \qquad \text{Add last.}$$

(c) $4^2 \cdot 2^2 + \underline{(7 + 3)} \cdot 2 \qquad$ Parentheses first.
$$\underline{4^2 \cdot 2^2} + \quad 10 \quad \cdot 2 \qquad \text{Evaluate exponents.}$$
$$\underline{16 \cdot 4} + \quad \underline{10 \quad \cdot 2} \qquad \text{Multiply from left to right.}$$
$$64 \quad + \qquad 20 \quad = 84 \qquad \text{Add last.}$$

(d) $4 \cdot \underline{\sqrt{25}} - 7 \cdot 2 \qquad$ Find square root first.
$$\underline{4 \cdot \quad 5} \quad - \underline{7 \cdot 2} \qquad \text{Multiply from left to right.}$$
$$20 \quad - \quad 14 = 6 \qquad \text{Subtract last.}$$

Note
Getting a correct answer always depends on using the order of operations.

◀◀ WORK PROBLEM 4 AT THE SIDE.

1.8 Exercises

Use the Perfect Squares Table on page 69 to find each square root.

> **Example:** $\sqrt{169}$ **Solution:** From the table, $13^2 = 169$, so $\sqrt{169} = $ **13.**

1. $\sqrt{9}$ 3

2. $\sqrt{36}$ 6

3. $\sqrt{16}$ 4

4. $\sqrt{25}$ 5

5. $\sqrt{144}$ 12

6. $\sqrt{100}$ 10

7. $\sqrt{121}$ 11

8. $\sqrt{225}$ 15

Identify the exponent and the base. Simplify each expression.

> **Example:** 4^2 **Solution:** $4^2 \leftarrow$ Exponent
> $\quad\quad\quad\quad\quad\quad\quad\;\; \llcorner\!\!\longrightarrow$ Base
> Simplify. $4^2 = 4 \times 4 = $ **16**

9. 5^2 2; 5; 25

10. 2^3 3; 2; 8

11. 6^2 2; 6; 36

12. 3^2 2; 3; 9

13. 12^2 2; 12; 144

14. 10^3 3; 10; 1000

15. 15^2 2; 15; 225

16. 11^3 3; 11; 1331

Complete each blank.

> **Example:** $23^2 = $ ___ so $\sqrt{} = 23$ **Solution:** $23^2 = $ **529,** so $\sqrt{529} = 23$

17. $10^2 = $ **100**   so $\sqrt{100} = 10$

18. $20^2 = $ **400**   so $\sqrt{400} = 20$

19. $15^2 = $ **225**   so $\sqrt{225} = 15$

20. $30^2 = $ **900**   so $\sqrt{900} = 30$

21. $35^2 = $ **1225**   so $\sqrt{1225} = 35$

22. $38^2 = $ **1444**   so $\sqrt{1444} = 38$

✍ Writing   ◉ Conceptual   ▲ Challenging   ≈ Estimation

23. $40^2 =$ 1600   so $\sqrt{1600} = 40$ **24.** $50^2 =$ 2500   so $\sqrt{2500} = 50$

25. $54^2 =$ 2916   so $\sqrt{2916} = 54$ **26.** $60^2 =$ 3600   so $\sqrt{3600} = 60$

27. Describe in your own words a perfect square. Of the two numbers 25 and 50, which is a perfect square and why.

A perfect square is the square of a whole number.
The number 25 is the square of 5 because $5 \cdot 5 = 25$. The number 50 is not a perfect square. There is no whole number that can be squared to get 50.

28. Use the following list of words and terms to write the four steps in the order of operations.

     add   square root
     exponents   subtract
     multiply   divide
     parentheses

1. **Do all operations inside parentheses.**
2. **Simplify expressions with exponents and square roots.**
3. **Multiply or divide from left to right.**
4. **Add or subtract from left to right.**

Work each problem by using the order of operations.

Examples: $15 + \underline{9 \cdot 4}$ Multiply first. $28 \div 7 + 3^2$ Evaluate exponent first.
Solutions: $15 + \underline{36} = \mathbf{51}$ Add last. $\underline{28 \div 7} + 9$ Then divide.
                                                  $4 + 9 = \mathbf{13}$ Add last.

29. $5^2 + 8 - 2$ **31**   **30.** $3^2 + 8 - 5$ **12**

31. $6 \cdot 5 - 4$ **26**   **32.** $3 \cdot 9 - 7$ **20**

33. $20 \cdot 3 \div 5$ **12**   **34.** $6 \cdot 8 \div 8$ **6**

35. $25 \div 5(8 - 4)$ **20**   **36.** $36 \div 18(7 - 3)$ **8**

37. $6 \cdot 2^2 + \dfrac{0}{6}$ **24**   **38.** $8 \cdot 3^2 - \dfrac{10}{2}$ **67**

39. $4 \cdot 1 + 8(9 - 2) + 3$ **63**   **40.** $3 \cdot 2 + 7(3 + 1) + 5$ **39**

41. $3^3 \cdot 2^2 + (10 - 5) \cdot 2$ **118**   **42.** $4^2 \cdot 5^2 + (20 - 9) \cdot 3$ **433**

43. $5 \cdot \sqrt{100} - 7 \cdot 2$ 36

44. $7 \cdot \sqrt{81} - 4 \cdot 5$ 43

45. $6 \cdot 4 + 5 \cdot 7 - 10$ 49

46. $9 \cdot 2 + 8 \cdot 3 + 15$ 57

47. $2^3 \cdot 3^2 + (14 - 4) \cdot 3$ 102

48. $3^2 \cdot 4^2 + (15 - 6) \cdot 2$ 162

49. $8 + 10 \div 5 + \dfrac{0}{3}$ 10

50. $6 + 8 \div 2 + \dfrac{0}{8}$ 10

51. $3^2 + 6^2 + (30 - 21) \cdot 2$ 63

52. $4^2 + 5^2 + (25 - 9) \cdot 3$ 89

53. $7 \cdot \sqrt{81} - 5 \cdot 6$ 33

54. $5 \cdot \sqrt{144} - 5 \cdot 7$ 25

55. $7 \cdot 2 + 8(2 \cdot 3) - 4$ 58

56. $5 \cdot 2 + 3(5 + 3) - 6$ 28

57. $4 \cdot \sqrt{49} - 7(5 - 2)$ 7

58. $3 \cdot \sqrt{25} - 6(3 - 1)$ 3

59. $6 \cdot (5 - 1) + \sqrt{4}$ 26

60. $5 \cdot (4 - 3) + \sqrt{9}$ 8

61. $6^2 + 2^2 - 6 + 2$ 36

62. $3^2 - 2^2 + 3 - 2$ 6

63. $5^2 \cdot 2^2 + (8 - 4) \cdot 2$ 108

64. $5^2 \cdot 3^2 + (30 - 20) \cdot 2$ 245

65. $7 + 6 \div 3 + 5 \cdot 2$ 19

66. $8 + 3 \div 3 + 6 \cdot 3$ 27

67. $5 \cdot \sqrt{36} - 7(7 - 4)$ 9

68. $9 \cdot \sqrt{64} - 5(4 + 2)$ 42

69. $4^2 - 2^3 + 5 \cdot 3$ **23**

70. $5^2 + 6^2 - 9 \cdot 4$ **25**

71. $8 + 5 \div 5 + 7 + \dfrac{0}{3}$ **16**

72. $2 + 12 \div 6 + 5 + \dfrac{0}{5}$ **9**

73. $4 \cdot \sqrt{25} - 6 \cdot 2$ **8**

74. $8 \cdot \sqrt{36} - 4 \cdot 6$ **24**

75. $3 \cdot \sqrt{25} - 4 \cdot \sqrt{9}$ **3**

76. $8 \cdot \sqrt{100} - 6 \cdot \sqrt{36}$ **44**

77. $7 \div 1 \cdot 8 \cdot 2 \div (21 - 5)$ **7**

78. $12 \div 4 \cdot 5 \cdot 4 \div (15 - 13)$ **30**

79. $15 \div 3 \cdot 2 \cdot 6 \div (14 - 11)$ **20**

80. $9 \div 1 \cdot 4 \cdot 2 \div (11 - 5)$ **12**

81. $4 \cdot \sqrt{16} - 3 \cdot \sqrt{9}$ **7**

82. $10 \cdot \sqrt{49} - 4 \cdot \sqrt{64}$ **38**

83. $5 \div 1 \cdot 10 \cdot 4 \div (17 - 9)$ **25**

84. $15 \div 3 \cdot 8 \cdot 9 \div (12 - 8)$ **90**

▲**85.** $8 \cdot 9 \div \sqrt{36} - 4 \div 2 + (14 - 8)$ **16**

▲**86.** $3 - 2 + 5 \cdot 4 \cdot \sqrt{144} \div \sqrt{36}$ **41**

▲**87.** $1 + 3 - 2 \cdot \sqrt{1} + 3 \cdot \sqrt{121} - 5 \cdot 3$ **20**

▲**88.** $6 - 4 + 2 \cdot 9 - 3 \cdot \sqrt{225} \div \sqrt{25}$ **11**

▲**89.** $6 \cdot \sqrt{25} \cdot \sqrt{100} \div 3 \cdot \sqrt{4} + 9$ **209**

▲**90.** $9 \cdot \sqrt{36} \cdot \sqrt{81} \div 2 + 6 - 3 - 5$ **241**

1.9 Solving Application Problems

Most problems involving applications of mathematics are written out in sentence form. You need to read the problem carefully to decide how to solve it.

OBJECTIVE 1 You have to look for **indicator words** in the application problem—words that indicate the necessary operations—either addition, subtraction, multiplication, or division. Some of these word indicators appear below.

OBJECTIVES

1▸ Find indicator words in application problems.

2▸ Solve application problems.

3▸ Estimate an answer.

FOR EXTRA HELP

Tutorial

Tape 2

SSM, Sec. 1.9

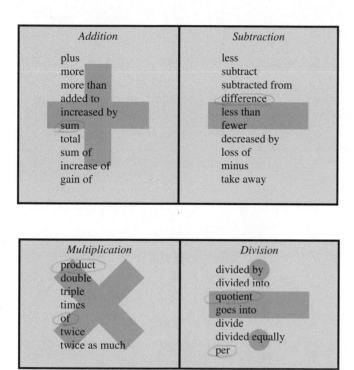

Addition	Subtraction
plus	less
more	subtract
more than	subtracted from
added to	difference
increased by	less than
sum	fewer
total	decreased by
sum of	loss of
increase of	minus
gain of	take away

Multiplication	Division
product	divided by
double	divided into
triple	quotient
times	goes into
of	divide
twice	divided equally
twice as much	per

Equals
is
the same as
equals
equal to
yields
results in
are

Note

The word "and" does not indicate addition and does not appear as an indicator word above. Notice how the "and" shows the location of an operation sign.

The sum of 6 *and* 2 is 6 + 2

The difference of 6 *and* 2 is 6 − 2

The product of 6 *and* 2 is 6 • 2

The quotient of 6 *and* 2 is 6 ÷ 2

1. ≈ Pick the most reasonable answer for each problem.

    (a) a clerk's hourly wage $2; $7; $60

    (b) a score on a 100-point test 6; 20; 74; 109

    (c) the cost of heart bypass surgery $500; $50,000; $5,000,000

OBJECTIVE 2 Solve application problems by using the following steps.

> **Steps for Solving Application Problems**
>
> *Step 1* Read the problem carefully and be certain you *understand* what the problem is asking. It may be necessary to read the problem several times.
>
> *Step 2* Before doing any calculations, work out a *plan* and try to visualize the problem. Know which facts are given and which must be found. Use *indicators* to help decide on the *plan*.
>
> *Step 3* Estimate a *reasonable answer* by using rounding.
>
> *Step 4* *Solve* the problem by using the facts given and your plan. If the answer is reasonable, *check* your work. If the answer is not reasonable, begin again by rereading the problem.

OBJECTIVE 3 These steps give a systematic approach for solving word problems. Each of the steps is important, but special emphasis should be placed on Step 3, estimating a *reasonable answer*. Many times an "answer" just does not fit the problem.

    What is a reasonable answer? Read the problem and try to determine the approximate size of the answer. Should the answer be part of a dollar, a few dollars, hundreds, thousands, or even millions of dollars? For example, if a problem asks for the cost of a man's shirt, would an answer of $20 be reasonable? $1000? $0.65? $65?

    Always make an estimate of a reasonable answer; then check the answer you get to see if it is close to your estimate.

◀◀ **WORK PROBLEM 1 AT THE SIDE.**

2. (a) On a recent geology field trip, 84 fossils were collected. If the fossils are divided equally among John, Sean, Jenn, and Kara, how many fossils will each receive?

(b) A company advertising campaign generates 264 sales leads. If there are twelve sales people who will divide these leads equally, how many will each receive?

E X A M P L E 1   Applying Division

At a recent garage sale, the total sales were $584. If the money was divided equally among Paul, Rachel, Maryangela, and Jose, how much did each person get? *expect an answer smaller than 584*

Approach To find the amount received by each person, divide the total amount of sales by the number of people.

Solution

    Step 1 A reading of the problem shows that the four members in the group divided $584 equally.

    Step 2 The word indicators, ***divided equally,*** show that the amount each received can be found by dividing $584 by 4.

    Step 3 A reasonable answer would be a little less than $150 each, since $600 ÷ 4 = $150 ($584 rounded to $600).

    Step 4 Find the actual answer by dividing $584 by 4.

$$\frac{146}{4)584}\qquad \text{Each person should get \$146.}$$

The answer $146 is reasonable, as $146 is close to the estimated answer of $150.

    Is the answer $146 correct? Check the work.

$146	Amount received by each person
× 4	Number of people
$584	Total sales

◀◀ **WORK PROBLEM 2 AT THE SIDE.**

ANSWERS

1. (a) $7 **(b)** 74 **(c)** $50,000

2. (a) 21 fossils **(b)** 22 leads

EXAMPLE 2 Applying Addition

Matt Owens earns $82 on Monday, $72 on Tuesday, $90 on Wednesday, $94 on Thursday, and $64 on Friday. Find his total earnings for the week.

Approach To find the total for the week, add the earnings for each day.

Solution

Step 1 In this problem, the earnings for each day are given and the total earnings for the week must be found.

Step 2 Add the daily earnings to arrive at the weekly total.

Step 3 Because the earnings were about $80 per day for a week of 5 days, a reasonable estimate would be around $400 (5 × $80 = $400).

Step 4 Find the actual answer by adding the earnings for the 5 days.

$$\begin{array}{r} \underline{\$402} \quad \text{Check by adding up} \\ \$\ 82 \\ 72 \\ 90 \\ 94 \\ +\quad 64 \\ \hline \$402 \quad \text{Earnings for the week} \end{array}$$

This answer is reasonable and correct.

Calculator Tip: The calculator solution to Example 2 uses chain calculations to get

82 + 72 + 90 + 94 + 64 = 402.

WORK PROBLEM 3 AT THE SIDE. ▶▶

EXAMPLE 3 Determining Whether Subtraction Is Necessary

The number of students enrolled in Chabot College this year is 4084 fewer than the number enrolled last year. Enrollment last year was 21,382. Find the enrollment this year. *expect answer smaller than 21382*

Approach To find the number of students enrolled this year, the enrollment decrease (fewer students) must be subtracted from last year.

Solution

Step 1 In this problem, the enrollment has decreased from last year to this year. The enrollment last year and the decrease in enrollment are given. This year's enrollment must be found.

Step 2 The word indicator, *fewer,* shows that subtraction must be used to find the number of students enrolled this year.

Step 3 Because the enrollment was about 21,000 students, and the decrease in enrollment is about 4000 students, a reasonable estimate would be 17,000 students (21,000 − 4000 = 17,000).

Step 4 Find the actual answer by subtracting 4084 from 21,382.

$$\begin{array}{r} 21,382 \\ -\quad 4,084 \\ \hline 17,298 \end{array}$$

CONTINUED ON NEXT PAGE

3. (a) During the semester, Cindy receives the following number of points on examinations and quizzes: 92, 81, 83, 98, 15, 14, 15, and 12. Find her total points for the semester.

(b) Stephanie Dixon works at the telephone order desk of a catalog sales company. One week she has the following number of customer contacts: Monday 78; Tuesday 64; Wednesday 118; Thursday 102; and Friday 196. Find her total number of customer contacts for the week.

ANSWERS
3. (a) 410 points
(b) 558 customer contacts

4. (a) One home occupies 1450 square feet, while an apartment occupies 980 square feet. Find the difference in the number of square feet of the two living units.

The enrollment this year is 17,298. The answer 17,298 is reasonable, as it is close to the estimate of 17,000. Check by adding.

$$\begin{array}{r} 17{,}298 \\ +\ 4{,}084 \\ \hline 21{,}382 \end{array}$$
Enrollment this year
Decrease in enrollment
Enrollment last year

◄◄ **WORK PROBLEM 4 AT THE SIDE.**

(b) Aim Electronics had 19,805 employees. After a layoff of 3980 employees, how many employees remain?

E X A M P L E   4   Solving a Two-Step Problem

A landlord receives $680 from each of five tenants. After paying $1880 in expenses, how much rent money does the landlord have left?
How much does he begin with?

Approach   To find the amount remaining, first find the total rent received. Next, subtract the expenses paid to find the amount remaining.

Solution

Step 1 There are five tenants and each pays the same rent.

Step 2 The wording *from each of five tenants* indicates that the five rents must be totaled. Since the rents are all the same, use multiplication to find the total rent received. Finally, subtract expenses.

Step 3 The amount of rent is about $700, making the total rent received about $3500 ($700 × 5). The expenses are about $2000. A reasonable estimate of the amount remaining is $1500 ($3500 − $2000).

5. (a) Gwen is paid $315 for each car that she sells. If she sells five cars and has $280 in sales expense, find the amount remaining after the expenses are deducted.

Step 4 Find the exact amount by first multiplying $680 by 5 (the number of tenants).

$$\begin{array}{r} \$680 \\ \times\ \ \ 5 \\ \hline \$3400 \end{array}$$

Finally, subtract the $1880 in expenses from $3400.

$$\$3400 - \$1880 = \$1520$$

The amount remaining is $1520.

The answer $1520 is reasonable, since it is close to the estimated answer of $1600. Check the amount by adding the expenses and then dividing by 5.

$$\$1520 + \$1880 = \$3400$$
$$5\overline{)3400} = \$680$$

(b) During a 4-hour period, 125 cars enter a parking lot each hour. In the same time period, 271 cars leave the lot. Find the number of cars remaining in the lot.

◄◄ **WORK PROBLEM 5 AT THE SIDE.**

ANSWERS

4. (a) 470 square feet
(b) 15,825 employees
5. (a) $1295 **(b)** 229 cars

1.9 Exercises

≈ *Solve the following application problems. First use front end rounding to estimate the answer. Then find the exact answer.*

1. A "riding type" lawn mower costs $500 more than a self-propelled lawn mower. If a self-propelled lawn mower costs $380, find the cost of a "riding type" mower.

estimate: **$400 + $500 = $900**
exact: **$880**

2. The price of the least expensive rear bagging lawn mower used in a recent test was $175. If this was $475 less than the most expensive model used in the test, find the price of the most expensive mower tested.

estimate: **$200 + $500 = $700**
exact: **$650**

3. John and Will Kellogg invented the first cold flakes cereal in 1894. Of the total 200 different types of cold cereals produced today, a large supermarket decides to sell all but 62 types. How many types of cereal does the supermarket sell?

estimate: **200 − 60 = 140 types**
exact: **138 types**

4. A truck weighs 9250 pounds when empty. After being loaded with firewood, it weighs 21,375 pounds. What is the weight of the firewood?

estimate: **20,000 − 9000 = 11,000 pounds**
exact: **12,125 pounds**

5. A packing machine can package 236 first-aid kits each hour. At this rate, find the number of first-aid kits packaged in 24 hours.

estimate: **200 × 20 = 4000 kits**
exact: **5664 kits**

6. If 450 admission tickets to the Classic Car Show are sold each day, how many tickets are sold in a 12-day period?

estimate: **500 × 10 = 5000 tickets**
exact: **5400 tickets**

7. Ted Slauson, coordinator of Toys for Tots, has collected 2628 toys. If his group can give the same number of toys to each of 657 children, how many toys will each child receive?

estimate: **3000 ÷ 700 ≈ 4 toys**
exact: **4 toys**

8. If profits of $680,000 are divided evenly among a firm's 1000 employees, how much money will each employee receive?

estimate: **$700,000 ÷ 1000 = $700**
exact: **$680**

9. The number of boaters and campers at the lake was 8392 on Friday. If this was 4218 more than the number of people at the lake on Wednesday, find the number of people at the lake on Wednesday.

estimate: **8000 − 4000 = 4000 people**
exact: **4174 people**

10. The community has raised $52,882 for the homeless shelter. If the total amount needed for the shelter is $75,650, find the additional amount needed to be collected.

estimate: **$80,000 − $50,000 = $30,000**
exact: **$22,768**

📝 Writing   ◉ Conceptual   ▲ Challenging   ≈ Estimation

11. Turn down the thermostat in the winter and you can save money and energy. In the upper Midwest, setting back the thermostat from 68° to 55° at night can save $14 per month on fuel. Find the amount of money saved in five months.

estimate: $10 × 5 = $50
exact: $70

12. The cost of tuition and fees is $785 per quarter. If Gale Klein has five quarters of college remaining, find the total amount that she will need for tuition and fees.

estimate: $800 × 5 = $4000
exact: $3925

13. The total number of miles covered on a cross-country bicycle trip was 3150. If the trip took 18 days and the same number of miles was traveled each day, how many miles were traveled per day?

estimate: 3000 ÷ 20 = 150 miles
exact: 175 miles

14. Erich Means completed a 2146-mile trip on his motorcycle and used 37 gallons of gasoline. How many miles did he travel on each gallon?

estimate: 2000 ÷ 40 = 50 miles
exact: 58 miles

15. Dorene Cox decides to establish a budget. She will spend $450 for rent, $325 for food, $320 for child care, $182 for transportation, $150 for other expenses, and she will put the remainder in savings. If her monthly take-home pay is $1620, find her monthly savings.

estimate: $2000 − $500 − $300 − $300 − $200 − $200 = $500
exact: $193

16. Jared Ueda had $2874 in his checking account. He wrote checks for $308 for auto repairs, $580 for child support and $778 for an insurance payment. Find the amount remaining in his account.

estimate: $3000 − $300 − $600 − $800 = $1300
exact: $1208

17. There are 43,560 square feet in one acre. How many square feet are there in 138 acres?

estimate: 40,000 × 100 = 4,000,000 square feet
exact: 6,011,280 square feet

18. The number of gallons of water polluted each day in an industrial area is 209,670. How many gallons of water are polluted each year? (Use a 365-day year.)

estimate: 200,000 × 400 = 80,000,000 gallons
exact: 76,529,550 gallons

An automobile manufacturer lists the following optional features and the price of each feature as shown below. Use this information to solve Exercises 19–22.

Safety and Security Items		Convenience and Comfort Items	
Item	**Cost**	**Item**	**Cost**
Air Bags	$475	Central Locking System	$245
Antilock Brakes	$780	Power Windows	$310
Rear Window Defroster	$130	Adjustable Steering Column	$185
Rear Window Wiper/Washer	$115	Air-conditioning	$975
Integrated Child Seat	$135	Mirror Options	$165

19. Find the total cost of all Safety and Security Items listed.

estimate: **$500 + $800 + $100 + $100 + $100 = $1600**
exact: **$1635**

20. Find the total cost of all Convenience and Comfort Items listed.

estimate: **$200 + $300 + $200 + $1000 + $200 = $1900**
exact: **$1880**

21. A manufacturer offers an option value package which includes air bags, antilock brakes, a central locking system, and air-conditioning at a cost of $1750. If the customer buys the value pack of options instead of paying for each item separately, how much can be saved?

estimate: **$500 + $800 + $200 + $1000 = $2500**
$2500 − $1800 = $700
exact: **$725**

22. A new car dealer offers a group of automobile options including air bags, antilock brakes, integrated child seat, adjustable steering column, air-conditioning, and mirror options for $1995. Find the amount saved if the group of options is purchased instead of paying for each item separately.

estimate: **$500 + $800 + $100 + $200 + $1000 + $200 = $2800**
$2800 − $2000 = $800
exact: **$720**

23. The Enabling Supply House purchases 6 wheelchairs at $1256 each and 15 speech compression recorder-players at $895 each. Find the total cost.

estimate: **($1000 × 6) + ($900 × 20) = $24,000**
exact: **$20,961**

24. Find the total cost if a college bookstore buys 17 computers at $506 each and 13 printers at $482 each.

estimate: **($500 × 20) + ($500 × 10) = $15,000**
exact: **$14,868**

25. Being able to identify indicator words is helpful in determining how to solve an application problem. Write three indicator words for each of these operations: add, subtract, multiply, and divide. Write two indicator words that mean equals.

Possible answers are:
Addition: more; total; gain of
Subtraction: less; loss of; decreased by
Multiplication: twice; of; product
Division: divided by; goes into; per
Equals: is; are

26. Identify and explain the four steps used to solve an application problem. You may refer to the text if you need help, but use your own words.

1. **Read the problem carefully.**
2. **Work out a plan.**
3. **Estimate a reasonable answer.**
4. **Solve the problem being certain to check your work.**

27. Write in your own words why it is important to estimate a reasonable answer. Give three examples of what might be a reasonable answer to a math problem in your daily activities.

Estimating the answer can help you avoid careless mistakes like decimal errors and calculation errors.
Examples of reasonable answers in daily life might be a $20 bag of groceries, $15 to fill the gas tank, or $45 for a phone bill.

28. First estimate by rounding to thousands, then find the exact answer to the following problem.

$$7438 + 6493 + 2380$$

Do the two answers vary by more than 1000? Why? Will estimated answers always vary from exact answers?

estimate: 7000 + 6000 + 2000 = 15,000
exact: 16,311
Yes, the answers vary by more than 1000 as a result of the rounding. However, the estimated answer and the exact answer are close enough to give some assurance that the answer is reasonable.

Solve the following application problems.

29. A package of 3 undershirts costs $12, and a package of 6 pairs of socks costs $15. Find the total cost of 30 undershirts and 18 pairs of socks.

$165

30. In one week, Brian earned $8 per hour for 38 hours. Maria earned $9 per hour for working 39 hours. Find their total combined income.

$655

31. A car weighs 2425 pounds. If its 582-pound engine is removed and replaced with a 634-pound engine, find the weight of the car after the engine change.

2477 pounds

32. Barbara has $2324 in her preschool operating account. After spending $734 from this account the class parents raise $568 in a rummage sale. Find the balance in the account after depositing the money from the rummage sale.

$2158

33. In a recent survey of high-priced hotels, the least expensive was Harrah's at a cost of $65 per night while several of the hotels in the group were $90 per night. Find the amount saved on a five-night stay at the least expensive hotel instead of staying at one of the more expensive hotels.

$125

34. The most expensive hotel room in a recent study was the Ritz-Carlton at $150 per night, while the least expensive was Motel 6 at $32 per night. Find the amount saved in a four-night stay at Motel 6 instead of staying at the Ritz-Carlton.

$472

35. Jim Peppa's vending machine company had 325 machines on hand at the beginning of the month. At different times during the month, machines were distributed to new locations; 35 machines were taken at one time, then 23 machines, and then 76 machines. During the same month additional machines were returned; 15 machines were returned at one time, then 38 machines, and then 108 machines. How many machines were on hand at the end of the month?

352 machines

36. Mike Fitzgerald owns 70 acres of land that he leases to an alfalfa farmer for $150 per acre per year. If property taxes are $28 per acre per year, find the total amount of yearly lease income he has left after taxes are paid.

$8540

37. A theater owner wants to provide enough seating for 1250 people. The main floor has 30 rows of 25 seats in each row. If the balcony has 25 rows, how many seats must be in each row to satisfy the owner's seating requirements?

20 seats

38. Jennie makes 24 grapevine wreaths per week to sell to gift shops. She works 40 weeks a year and packages six wreaths per box. If she ships equal quantities to each of five shops, find the number of boxes each store will receive.

32 boxes

1.1	**whole numbers**	The whole numbers are 0, 1, 2, 3, 4, 5, 6, 7, 8, and so on.
1.2	**addition**	The process of finding the total.
	addends	The numbers being added in an addition problem.
	sum (total)	The answer in an addition problem is called the sum (total).
	commutative property of addition	The commutative property of addition states that the order of numbers in an addition problem can be changed without changing the sum.
	associative property of addition	The associative property of addition states that grouping the addition of numbers in any order does not change the sum.
	carrying	The process of carrying is used in an addition problem when the sum of the digits in a column is greater than 9.
1.3	**minuend**	The number from which another number (the subtrahend) is subtracted in a subtraction problem.
	subtrahend	The number being subtracted or taken away in a subtraction problem.
	difference	The answer in a subtraction problem is called the difference.
	borrowing	The method of borrowing is used in subtraction if a digit is less than the one directly below.
1.4	**factors**	The numbers being multiplied are called factors. For example, in $3 \times 4 = 12$, both 3 and 4 are factors.
	product	The answer in a multiplication problem is called the product.
	commutative property of multiplication	The commutative property of multiplication states that the product in a multiplication problem remains the same when the order of the factors is changed.
	associative property of multiplication	The associative property of multiplication states that the grouping of numbers in any order gives the same product.
	multiple	The product of two whole-number factors is a multiple of those numbers.
1.5	**dividend**	The number being divided by another number in a division problem.
	divisor	The number doing the dividing in a division problem.
	quotient	The answer in a division problem is called the quotient.
	short division	A method of dividing a number by a one-digit divisor.
	remainder	The remainder is the number left over when two numbers do not divide exactly.
1.6	**long division**	The process of long division is used to divide by a number with more than one digit.
1.7	**rounding**	To find a number that is close to the original number, but easier to work with, we use rounding. Use the $\approx$ sign, which means "approximately equal to."
	front end rounding	Rounding a number so that only one nonzero digit remains. The front digit (left hand digit) is rounded and all other digits become zeros.
1.8	**square root**	The square root of a whole number is the number that can be multiplied by itself to produce the given (larger) number.
	perfect square	A number that is the square of a whole number is a perfect square.
1.9	**indicator words**	Words in a problem that indicate the necessary operations—either addition, subtraction, multiplication, or division.

QUICK REVIEW

Concepts	Examples
1.1 Reading and Writing Whole Numbers Do not use the word "and" with a whole number. Commas help divide the periods or groups for ones, thousands, millions, and billions. A comma is not needed with a number having four digits or fewer.	795 is written *seven hundred ninety-five.* 9,768,002 is written *nine million, seven hundred sixty-eight thousand, two.*

1.2 Addition of Whole Numbers

Add from top to bottom, starting with the ones column and working left. To check, add from bottom to top.

$$
\begin{array}{r}
1\;1\;4\;0 \\
6\;8\;7 \\
2\;6 \\
9 \\
+\;4\;1\;8 \\
\hline
1\;1\;4\;0
\end{array}
$$

(Add up to check) } Addends Sum

1.2 Commutative Property of Addition

The order of numbers in an addition problem can be changed without changing the sum.

$$2 + 4 = 6$$
$$4 + 2 = 6$$

By the commutative property, the sum is the same.

1.2 Associative Property of Addition

Grouping the addition of numbers in any order gives the same sum.

$$(2 + 3) + 4 = 9$$
$$2 + (3 + 4) = 9$$

By the associative property, the sum is the same.

1.3 Subtraction of Whole Numbers

Subtract the subtrahend from minuend to get the difference by borrowing when necessary. To check, add the difference to the subtrahend to get the minuend.

Problem

$$
\begin{array}{r}
6\;12\;18 \\
4\;7\;3\;8 \quad\leftarrow \text{Minuend}\\
-\;\;6\;4\;9 \quad\quad \text{Subtrahend}\\
\hline
4\;0\;8\;9 \quad\quad \text{Difference}
\end{array}
$$

Check

$$
\begin{array}{r}
4\;0\;8\;9 \\
+\;\;6\;4\;9 \\
\hline
4\;7\;3\;8
\end{array}
$$

1.4 Multiplication of Whole Numbers

The numbers being multiplied are called *factors.* The multiplicand is being multiplied by the multiplier, giving the product. When the multiplier has more than one digit, partial products must be used and added to find the product.

$$
\begin{array}{r}
78 \\
\times\;24 \\
\hline
312 \\
156 \\
\hline
1872
\end{array}
$$

Multiplicand } Factors
Multiplier
Partial product
Partial product (one position left)
Product

1.4 Commutative Property of Multiplication

The answer or product in multiplication remains the same when the order of the factors is changed.

$$3 \times 4 = 12$$
$$4 \times 3 = 12$$

By the commutative property, the product is the same.

Concepts	Examples
1.4 Associative Property of Multiplication The grouping of numbers in any order gives the same product.	$(2 \times 3) \times 4 = 24$ $2 \times (3 \times 4) = 24$ By the associative property, the product is the same.

1.5 Division of Whole Numbers

$\div$ and $\overline{)}$ mean divide.
Also a —, as in $\frac{25}{5}$, means to divide the top number (dividend) by the bottom number (divisor).

$$\begin{array}{r} 22 \leftarrow \text{Quotient} \\ \text{Divisor} \rightarrow 4\overline{)88} \leftarrow \text{Dividend} \\ \underline{88} \\ 0 \end{array}$$

$$88 \div 4 = 22$$
Dividend | Quotient
Divisor

$$\frac{88}{4} = 22 \quad \text{Quotient}$$

1.7 Rounding Whole Numbers

Rules for rounding:

1. Identify the position to be rounded, and draw a line under it.
2. If the next digit to the right is 5 or more, increase the underlined digit by 1. If the next digit is 4 or less, do not change the underlined digit.
3. Change all digits to the right of the underlined place to zeros.

Round 726 to the nearest ten.

Next digit is 5 or more.
$72\underline{6}$
Tens place increases by 1 ($2 + 1 = 3$).

726 rounds to 730.

Round 1,498,586 to the nearest million.

Next digit is 4 or less.
$1,498, 586$
Millions place does not change.

1,498,586 rounds to 1,000,000.

1.7 Front End Rounding

Front end rounding leaves only the first digit as a nonzero digit. All other digits are changed to zero.

Round each of the following using front end rounding.

76 rounds to 80
348 rounds to 300
6512 rounds to 7000
23,751 rounds to 20,000
652,179 rounds to 700,000

1.8 Order of Operations

Problems may have several operations. Work these problems with the following order of operations.

1. Do all operations inside parentheses.
2. Simplify any expressions with exponents and find any square roots.
3. Multiply or divide from left to right.
4. Add or subtract from left to right.

Solve, using the order of operations.

$7 \cdot \sqrt{9} - 4 \cdot 5$ Find square root.

$7 \cdot 3 - 4 \cdot 5$ Multiply from left to right.

$21 - 20 = 1$ Subtract.

Concepts	*Examples*

1.9 Application Problems
Follow these steps.
 	
1. Read the problem carefully, perhaps several times.
2. Work out a plan before starting.
3. Estimate a reasonable answer.
4. Solve the problem. If the answer is reasonable, check; if not, start over.
	Manuel earns \$118 on Sunday, \$87 on Monday, and \$63 on Tuesday. Find total earnings for the three days. *Total* means to add.
	
	1. The earnings for each day are given, and the total for the 3 days must be found.
	2. Add the daily earnings to find the total.
	3. Since the earnings were about \$100 + \$90 + \$60 = \$250, a reasonable estimate would be approximately \$250.
	4. \$268 Check by adding up
	     \$118
	       87
	  + 63
	    \$268 Total earnings
	
	Manuel's total earnings are \$268. The answer is reasonable because it is close to the estimate of \$250.

CHAPTER 1 REVIEW EXERCISES

If you need help with any of these review exercises, look in the section indicated in brackets.

[1.1] *Fill in the digits for the given period or group in each of the following numbers.*

1. 4621
   thousands   4
   ones   621

2. 87,328
   thousands   87
   ones   328

3. 105,724
   thousands   105
   ones   724

4. 1,768,710,618
   billions   1
   millions   768
   thousands   710
   ones   618

Rewrite the following numbers in words.

5. 725   seven hundred twenty-five

6. 12,412   twelve thousand, four hundred twelve

7. 319,215   three hundred nineteen thousand, two hundred fifteen

8. 62,500,005   sixty-two million, five hundred thousand, five

Rewrite each of the following numbers in digits.

9. four thousand, four   4004

10. two hundred million, four hundred fifty-five   200,000,455

[1.2] *Add.*

11.
```
  74
+ 18
────
  92
```

12.
```
  35
+ 78
────
 113
```

13.
```
   807
  4606
+   51
─────
  5464
```

14.
```
  8215
     9
+ 7433
──────
15,657
```

15.
```
  1108
   566
  7201
+  304
─────
  9179
```

16.
```
   187
  5543
   246
+ 1003
─────
  6979
```

17.
```
   5 732
  11,069
      37
   1 595
+ 22,169
───────
  40,602
```

18.
```
   3 451
  12,286
      43
   1 291
+ 32,784
───────
  49,855
```

✎ Writing   ◎ Conceptual   ▲ Challenging   ≈ Estimation

[1.3] *Subtract.*

19.
$$\begin{array}{r} 34 \\ -\ 12 \\ \hline 22 \end{array}$$

20.
$$\begin{array}{r} 56 \\ -\ 35 \\ \hline 21 \end{array}$$

21.
$$\begin{array}{r} 238 \\ -\ 199 \\ \hline 39 \end{array}$$

22.
$$\begin{array}{r} 573 \\ -\ 389 \\ \hline 184 \end{array}$$

23.
$$\begin{array}{r} 4380 \\ -\ 577 \\ \hline 3803 \end{array}$$

24.
$$\begin{array}{r} 5210 \\ -\ 883 \\ \hline 4327 \end{array}$$

25.
$$\begin{array}{r} 2210 \\ -\ 1986 \\ \hline 224 \end{array}$$

26.
$$\begin{array}{r} 99{,}704 \\ -\ 73{,}838 \\ \hline 25{,}866 \end{array}$$

[1.4] *Multiply.*

27.
$$\begin{array}{r} 5 \\ \times\ 5 \\ \hline 25 \end{array}$$

28.
$$\begin{array}{r} 8 \\ \times\ 0 \\ \hline 0 \end{array}$$

29. 7×3

21

30. 8×8

64

31. $(6)(7)$

42

32. $(4)(9)$

36

33. $7 \cdot 8$

56

34. $9 \cdot 9$

81

Work the following chain multiplications.

35. $2 \times 4 \times 6$

48

36. $9 \times 1 \times 5$

45

37. $4 \times 4 \times 3$

48

38. $2 \times 2 \times 2$

8

39. $(8)(0)(6)$

0

40. $(8)(8)(1)$

64

41. $6 \cdot 1 \cdot 8$

48

42. $7 \cdot 7 \cdot 0$

0

Multiply.

43.
$$\begin{array}{r} 43 \\ \times\ 4 \\ \hline 172 \end{array}$$

44.
$$\begin{array}{r} 62 \\ \times\ 7 \\ \hline 434 \end{array}$$

45.
$$\begin{array}{r} 58 \\ \times\ 9 \\ \hline 522 \end{array}$$

46.
$$\begin{array}{r} 98 \\ \times\ 1 \\ \hline 98 \end{array}$$

47.
$$\begin{array}{r} 639 \\ \times\ 6 \\ \hline 3834 \end{array}$$

48.
$$\begin{array}{r} 781 \\ \times\ 7 \\ \hline 5467 \end{array}$$

49.
$$\begin{array}{r} 1349 \\ \times\ 4 \\ \hline 5396 \end{array}$$

50.
$$\begin{array}{r} 9163 \\ \times\ 5 \\ \hline 45{,}815 \end{array}$$

51.
$$\begin{array}{r} 7259 \\ \times\ 2 \\ \hline 14{,}518 \end{array}$$

52.
$$\begin{array}{r} 5440 \\ \times\ 6 \\ \hline 32{,}640 \end{array}$$

53.
$$\begin{array}{r} 93{,}105 \\ \times\ 5 \\ \hline 465{,}525 \end{array}$$

54.
$$\begin{array}{r} 21{,}873 \\ \times\ 8 \\ \hline 174{,}984 \end{array}$$

55. 34
× 18
612

56. 52
× 36
1872

57. 98
× 12
1176

58. 68
× 75
5100

59. 655
× 21
13,755

60. 392
× 77
30,184

61. 4051
× 219
887,169

62. 1527
× 328
500,856

Find the total cost of each of the following.

63. 20 CD's at $15 per CD

$300

64. 48 T-shirts at $14 per shirt

$672

65. 278 batteries at $48 per battery

$13,344

66. 168 welders masks at $9 per mask

$1512

Multiply by using multiples of ten.

67. 320
× 60
19,200

68. 280
× 90
25,200

69. 517
× 400
206,800

70. 752
× 400
300,800

71. 16,000
× 8 000
128,000,000

72. 43,000
× 2 100
90,300,000

[1.5] *Divide whenever possible.*

73. 12 ÷ 4

3

74. 36 ÷ 6

6

75. 42 ÷ 7

6

76. 18 ÷ 9

2

77. $\frac{72}{8}$ 9

78. $\frac{36}{9}$ 4

79. $\frac{54}{6}$ 9

80. $\frac{0}{6}$ 0

81. $\frac{125}{0}$ undefined

82. $\frac{0}{35}$ 0

83. $\frac{64}{8}$ 8

84. $\frac{81}{9}$ 9

[1.5–1.6] *Divide.*

85. $4\overline{)432}$ $\dfrac{108}{}$

86. $9\overline{)216}$ $\dfrac{24}{}$

87. $9\overline{)56,259}$ $\dfrac{6251}{}$

88. $76\overline{)26,752}$ $\dfrac{352}{}$

89. $2704 \div 18$

     150 R4

90. $15,525 \div 125$

     124 R25

[1.7] *Round as shown.*

91. 318 to the nearest ten

   ≈320

92. 14,309 to the nearest hundred

   ≈14,300

93. 19,721 to the nearest thousand

   ≈20,000

94. 67,485 to the nearest ten thousand

   ≈70,000

Round each of the following to the nearest ten, nearest hundred, and nearest thousand. Remember to round from the original number.

	Ten	*Hundred*	*Thousand*
95. 2397	≈2400	≈2400	≈2000
96. 20,065	≈20,070	≈20,100	≈20,000
97. 98,201	≈98,200	≈98,200	≈98,000
98. 352,118	≈352,120	≈352,100	≈352,000

[1.8] *Find each square root by using the Perfect Squares Table on page 69.*

99. $\sqrt{36}$ 6   100. $\sqrt{49}$ 7   101. $\sqrt{144}$ 12   102. $\sqrt{196}$ 14

Identify the exponent and the base. Simplify each expression.

103. 3^2 2; 3; 9   **104.** 2^3 3; 2; 8   **105.** 5^3 3; 5; 125   **106.** 4^5 5; 4; 1024

Work each problem by using the order of operations.

107. $9^2 - 9$ 72   **108.** $3^2 - 5$ 4   **109.** $2 \cdot 3^2 \div 2$ 9

110. $9 \div 1 \cdot 2 \cdot 2 \div (11 - 2)$ 4   **111.** $\sqrt{9} + 2 \cdot 3$ 9   **112.** $6 \cdot \sqrt{16} - 6 \cdot \sqrt{9}$ 6

[1.9] $\approx$ *Solve each of the following application problems. First use front end rounding to estimate the answer. Then find the exact answer.*

113. Find the cost of 48 shovels at $11 per shovel.

estimate: 50 × $10 = $500
exact: $528

114. A pulley on an evaporative cooler turns 1400 revolutions per minute. How many revolutions will the pulley turn in 60 minutes?

estimate: 1000 × 60 = 60,000 revolutions
exact: 84,000 revolutions

115. A banquet size coffee pot makes 120 cups of coffee. Find the number of cups in 6 pots.

estimate: 100 × 6 = 600 cups
exact: 720 cups

116. A drum contains 6000 brackets. How many brackets are in 30 drums?

estimate: 6000 × 30 = 180,000 brackets
exact: 180,000 brackets

117. It takes 2000 hours of work to build 1 home. How many hours of work are needed to build 12 homes?

estimate: 2000 × 10 = 20,000 hours
exact: 24,000 hours

118. A Japanese bullet train travels 80 miles in 1 hour. Find the number of miles traveled in 5 hours.

estimate: 80 × 5 = 400 miles
exact: 400 miles

119. The Houston Space Center charges $15 for each adult admission and $12 for each child. Find the total cost to admit a group of 18 adults and 26 children.

estimate: ($20 × 20) + ($10 × 30) = $700
exact: $582

120. A newspaper carrier has 56 customers who take the paper daily and 23 customers who take the paper on weekends only. A daily customer pays $15 per month and a weekend-only customer pays $8 per month. Find the total monthly collections.

estimate: (60 × $20) + (20 × $10) = $1400
exact: $1024

121. In a recent test of side bagging lawn mowers, the most expensive model sold for $350 while the least expensive model sold for $170. Find the difference in price between the most expensive and the least expensive mower tested.

estimate: $400 − $200 = $200
exact: **$180**

122. Susan Hessney has $382 in her checking account. She writes a check for $135. How much does she have left in her account?

estimate: $400 − $100 = $300
exact: **$247**

123. A food canner uses 1 pound of pork for every 175 cans of pork and beans. How many pounds of pork are needed for 8750 cans of pork and beans?

estimate: 9000 ÷ 200 = 45 pounds
exact: **50 pounds**

124. A stamping machine produces 986 license plates each hour. How long will it take to produce 32,538 license plates?

estimate: 30,000 ÷ 1000 ≈ 30 hours
exact: **33 hours**

125. If an acre needs 250 pounds of fertilizer, how many acres can be fertilized with 5750 pounds of fertilizer?

estimate: 6000 ÷ 300 = 20 acres
exact: **23 acres**

126. Each home in a subdivision requires 180 feet of fencing. Find the number of homes that can be fenced with 5760 feet of fencing material.

estimate: 6000 ÷ 200 = 30 homes
exact: **32 homes**

MIXED REVIEW EXERCISES

Solve each of the following as indicated.

127.
$$\begin{array}{r} 47 \\ \times\ 6 \\ \hline 282 \end{array}$$

128.
$$\begin{array}{r} 78 \\ \times\ 7 \\ \hline 546 \end{array}$$

129.
$$\begin{array}{r} 182 \\ -\ 75 \\ \hline 107 \end{array}$$

130.
$$\begin{array}{r} 716 \\ -\ 153 \\ \hline 563 \end{array}$$

131.
$$\begin{array}{r} 662 \\ +\ 379 \\ \hline 1041 \end{array}$$

132.
$$\begin{array}{r} 352 \\ +\ 678 \\ \hline 1030 \end{array}$$

133.
$$\begin{array}{r} 38,140 \\ -\ 6\,078 \\ \hline 32,062 \end{array}$$

134.
$$\begin{array}{r} 29,156 \\ -\ 4\,209 \\ \hline 24,947 \end{array}$$

135. $21 \div 7$ 3

136. $\dfrac{42}{6}$ 7

137.
$$\begin{array}{r} 7\,218 \\ 3 \\ 18 \\ 1\,791 \\ 82,623 \\ +\ 1\,982 \\ \hline 93,635 \end{array}$$

138.
$$\begin{array}{r} 3\,812 \\ 5 \\ 22 \\ 1\,836 \\ 75,134 \\ +\ 2\,369 \\ \hline 83,178 \end{array}$$

139. $\dfrac{8}{0}$

undefined

140. $\dfrac{6}{1}$ 6

141. $55,200 \div 4$

13,800

142. $18,440 \div 8$

2305

143.   8430
$\times$ 128
1,079,040

144.   38,571
$\times$ 3
115,713

145. $34\overline{)3672}$ (108)

146. $68\overline{)14,076}$ (207)

147. Rewrite 286,753 in words.

two hundred eighty-six thousand, seven hundred fifty-three

148. Rewrite 108,210 in words.

one hundred eight thousand, two hundred ten

149. Round 3349 to the nearest hundred.

$\approx$3300

150. Round 200,498 to the nearest thousand.

$\approx$200,000

Find each square root.

151. $\sqrt{25}$ 5

152. $\sqrt{100}$ 10

Find the total cost of each of the following.

153. 36 pairs of in-line skates at $165 per pair

$5940

154. 65 refrigerators at $520 per refrigerator

$33,800

155. 185 team shirts at $12 per shirt

$2220

156. 607 boxes of avocados at $26 per box

$15,782

Solve each of the following application problems.

157. There are 52 cards in a deck. How many cards are there in 9 decks?

468 cards

158. Your college bookstore receives textbooks packed 12 books per carton. How many textbooks are received in a delivery of 238 cartons?

2856 textbooks

159. "Push type" gasoline-powered lawn mowers cost $100 less than self-propelled mowers that you walk behind. If a self-propelled mower costs $380, find the cost of a "push-type" mower.

$280

160. The Village School wants to raise $115,280 for a new library. If $87,340 has already been raised, how much more must be raised to reach the goal?

$27,940

American River Raft Rentals lists the following daily raft rental fees. Notice that there is an additional $2 launch fee payable to the park system for each raft rented. Use this information to solve Exercises 161–162.

AMERICAN RIVER RAFT RENTALS

Size	Rental Fee	Launch Fee
4-person	$28	$2
6-person	$38	$2
10-person	$70	$2
12-person	$75	$2
16-person	$85	$2

161. On a recent Tuesday the following rafts were rented: 6 4-person; 15 6-person; 10 10-person; 3 12-person; and 2 16-person. Find the total receipts including the $2 per raft launch fee.

$1905

162. On the 4th of July the following rafts were rented: 38 4-person; 73 6-person; 58 10-person; 34 12-person; and 18 16-person. Find the total receipts including the $2 per raft launch fee.

$12,420

Write the following numbers in words.

1. 8208

2. 75,065

3. Use digits to write "one hundred thirty-eight thousand, eight."

Add the following.

4.   984
            65
          4283
        + 7561

5.   17,063
                7
               12
            1 505
           93,710
         +   333

Subtract.

6.   7002
       − 3954

7.   5062
       − 1978

Multiply.

8. 5 × 7 × 4

9. 57 • 3000

10. (85)(21)

11.   7381
        × 603

Divide whenever possible.

12. 16)‾123,952‾

13. $\frac{791}{0}$

14. 38,472 ÷ 84

15. 280)‾44,800‾

Round as shown.

16. 4756 to the nearest ten

17. 67,509 to the nearest thousand

1. eight thousand, two hundred eight

2. seventy-five thousand, sixty-five

3. 138,008

4. 12,893

5. 112,630

6. 3048

7. 3084

8. 140

9. 171,000

10. 1785

11. 4,450,743

12. 7747

13. undefined

14. 458

15. 160

16. ≈4760

17. ≈68,000

18. __35__

19. __28__

estimate: $500 + $500 +
$500 + 400 −
$800 = $1100
20. *exact:* $1140

estimate: 60,000 ÷ 500 =
120 days
21. *exact:* 118 days

estimate: $1000 − $700 −
$200 − $70 = $30
22. *exact:* $165

estimate: (100 × 4) +
(100 × 4) =
800 ovens
23. *exact:* 1028 ovens

24. _____

25. _____

Work each problem.

18. $5^2 + 2 \times 5$

19. $7 \cdot \sqrt{64} - 14 \cdot 2$

≈ *Solve each of the following application problems. First use front end rounding to estimate the answer. Then find the exact answer.*

20. The monthly rents collected from the four units in an apartment building are $485, $500, $515, and $425. After expenses of $785 are paid, find the amount that remains.

21. Packard Bell assembles 472 personal computers each day. Find the number of work days it would take to assemble 55,696 personal computers.

22. Kenée Shadbourne paid $690 for tuition, $185 on books, and $68 on supplies. If this money was withdrawn from her checking account, which had a balance of $1108, find her new balance.

23. An appliance manufacturer assembles 118 self-cleaning ovens each hour for 4 hours and 139 standard ovens each hour for the next 4 hours. Find the total number of ovens assembled in the 8-hour period.

✐ 24. Explain in your own words the rules for rounding numbers. Give an example of rounding a number to the nearest ten thousand.
   1. **Locate the place to be rounded and underline it.**
   2. **Look only at the next digit to the right. If this digit is a 4 or less, do not change the underlined digit. If the digit is a 5 or more, increase the underlined digit by 1.**
   3. **Change all digits to the right of the underlined place to zeros. Each person's example will vary.**

✐ 25. List and describe the four steps for solving word problems. Be sure to include estimating a reasonable answer and checking your work in your description.
   1. **Read the problem carefully.**
   2. **Work out a plan.**
   3. **Estimate a reasonable answer.**
   4. **Solve the problem being certain to check your work.**

Multiplying and Dividing Fractions

Chapter 1 discussed whole numbers. Many times, however, we find that parts of whole numbers are considered. One way to write parts of a whole is with **fractions.** Another way is with decimals, which is discussed in Chapter 4.

2.1 Basics of Fractions

OBJECTIVE 1 The number $\frac{1}{8}$ is a fraction that represents 1 of 8 equal parts. Read $\frac{1}{8}$ as "one eighth."

E X A M P L E 1 · Identifying Fractions

Use fractions to represent the shaded portions and the unshaded portions.

(a) The figure on the left has 3 equal parts. The 2 shaded parts are represented by the fraction $\frac{2}{3}$. The *un*shaded part is $\frac{1}{3}$.

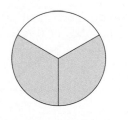

(b) The 4 shaded parts of the 7-part figure on the right are represented by the fraction $\frac{4}{7}$. The unshaded part is $\frac{3}{7}$.

WORK PROBLEM 1 AT THE SIDE. ▶▶

Fractions can be used to show more than one whole object.

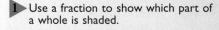
www.mathnotes.com

OBJECTIVES

1. ▶ Use a fraction to show which part of a whole is shaded.
2. ▶ Identify the numerator and denominator.
3. ▶ Identify proper and improper fractions.

FOR EXTRA HELP

Tutorial   Tape 2   SSM, Sec. 2.1

1. Write fractions for the shaded portions and the unshaded portions.

(a)

(b)

(c)

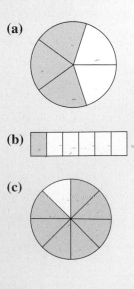

ANSWERS

1. (a) $\frac{3}{5}; \frac{2}{5}$ **(b)** $\frac{1}{6}; \frac{5}{6}$ **(c)** $\frac{7}{8}; \frac{1}{8}$

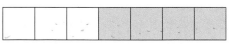

2. Write fractions for the shaded portions.

(a)

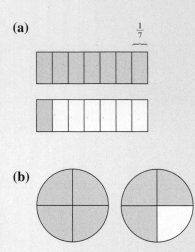

(b)

EXAMPLE 2   **Representing Fractions Greater Than One**

Use a fraction to represent the shaded part.

(a)

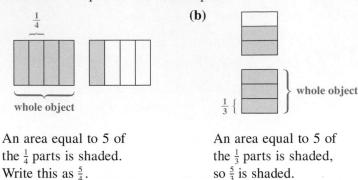

(b)

An area equal to 5 of the $\frac{1}{4}$ parts is shaded. Write this as $\frac{5}{4}$.

An area equal to 5 of the $\frac{1}{3}$ parts is shaded, so $\frac{5}{3}$ is shaded.

◀◀ **WORK PROBLEM 2 AT THE SIDE.**

OBJECTIVE 2 In the fraction $\frac{2}{3}$, the number 2 is the **numerator** (NOOM-er-ay-ter), and 3 is the **denominator** (di-NAHM-in-ay-ter). The bar between the numerator and the denominator is the *fraction bar*.

$$\text{Fraction bar} \rightarrow \frac{2}{3} \begin{array}{l} \leftarrow \text{Numerator} \\ \leftarrow \text{Denominator} \end{array}$$

3. Identify the numerator and the denominator. Draw a picture with shaded parts to show each fraction. Your drawings may vary, but they should have the correct number of shaded parts.

(a) $\dfrac{2}{3}$

(b) $\dfrac{1}{4}$

(c) $\dfrac{8}{5}$

(d) $\dfrac{5}{2}$

The Numerator and Denominator

The denominator of a fraction shows the number of equivalent parts in the whole, and the numerator shows how many parts are being considered.

Note

Remembering that a bar, —, is one of the division symbols, and that division by 0 is undefined, a fraction with a denominator of 0 is also undefined.

EXAMPLE 3   **Identifying Numerator and Denominator**

Identify the numerator and denominator in each fraction.

(a) $\dfrac{5}{9}$ **(b)** $\dfrac{11}{7}$

$$\frac{5}{9} \begin{array}{l} \leftarrow \text{Numerator} \\ \leftarrow \text{Denominator} \end{array} \qquad \frac{11}{7} \begin{array}{l} \leftarrow \text{Numerator} \\ \leftarrow \text{Denominator} \end{array}$$

◀◀ **WORK PROBLEM 3 AT THE SIDE.**

ANSWERS

2. (a) $\dfrac{8}{7}$ (b) $\dfrac{7}{4}$

3. (a) N: 2; D: 3

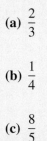

(b) N: 1; D: 4

(c) N: 8; D: 5

(d) N: 5; D: 2

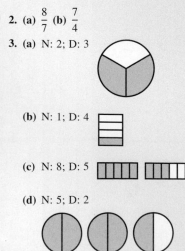

OBJECTIVE 3▶ Fractions are sometimes called *proper* or *improper* fractions.

Proper and Improper Fractions

If the numerator of a fraction is *smaller* than the denominator, the fraction is a **proper fraction.**

If the numerator is *greater than or equal to* the denominator, the fraction is an **improper fraction.**

Proper Fractions	*Improper Fractions*
$\dfrac{1}{2}, \dfrac{5}{11}, \dfrac{35}{36}$	$\dfrac{9}{7}, \dfrac{126}{125}, \dfrac{7}{7}$

E X A M P L E 4   Classifying Types of Fractions

(a) Identify all proper fractions in this list.

$$\frac{3}{4}, \frac{5}{9}, \frac{17}{5}, \frac{9}{7}, \frac{3}{3}, \frac{12}{25}, \frac{1}{9}, \frac{5}{3}$$

Proper fractions have a numerator that is smaller than the denominator. The proper fractions are:

$\dfrac{3}{4}$ ← 3 is smaller than 4 $\dfrac{5}{9}$ $\dfrac{12}{25}$ and $\dfrac{1}{9}$

(b) Identify all improper fractions in the list above.

Improper fractions have a numerator that is equal to or greater than the denominator. The improper fractions are:

$\dfrac{17}{5}$ ← 17 is greater than 5 $\dfrac{9}{7}$ $\dfrac{3}{3}$ and $\dfrac{5}{3}$.

WORK PROBLEM 4 AT THE SIDE. ▶▶

4. From the following group of fractions:

$$\frac{3}{4}, \frac{8}{7}, \frac{5}{7}, \frac{6}{6}, \frac{1}{2}, \frac{2}{1}$$

(a) list all proper fractions

(b) list all improper fractions.

ANSWERS

4. (a) $\dfrac{3}{4}, \dfrac{5}{7}, \dfrac{1}{2}$

  (b) $\dfrac{8}{7}, \dfrac{6}{6}, \dfrac{2}{1}$

NUMBERS IN THE
Real World *collaborative investigations*

People who make quilts often use designs based on a block cut into 4, 9, 16, or 25 squares. Then, they select the colors for the various pieces in the block. Finally, they calculate how much fabric of each color to buy, based on the fractional part of the block in that color. For each of these quilt designs, figure out what fraction of the block is red, what fraction is blue, and so on.

$\frac{4}{9}$ blue; $\frac{5}{9}$ red

$\frac{4}{9}$ green; $\frac{2}{9}$ blue;

$\frac{3}{9}$ or $\frac{1}{3}$ purple

$\frac{2}{8}$ or $\frac{1}{4}$ of each color

$\frac{4}{16}$ or $\frac{1}{4}$ blue; $\frac{6}{16}$ or $\frac{3}{8}$

each red and yellow

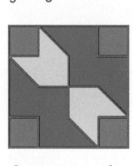

$\frac{8}{16}$ or $\frac{1}{2}$ blue; $\frac{4}{16}$ or $\frac{1}{4}$

each yellow and green

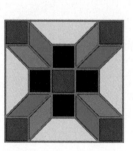

$\frac{8}{25}$ yellow; $\frac{5}{25}$ or $\frac{1}{5}$ red;

$\frac{4}{25}$ each black, green, blue

Use the blocks below to create and color your own quilt patterns. Then indicate what fractional part of the block is in each color.

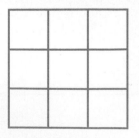

Find the next two numbers in this pattern: 4, 9, 16, 25, **36, 49**
Explain how the pattern works. **The pattern is 2², 3², 4², 5², 6², 7².**

2.1 Exercises

Write the fractions that represent the shaded and unshaded areas.

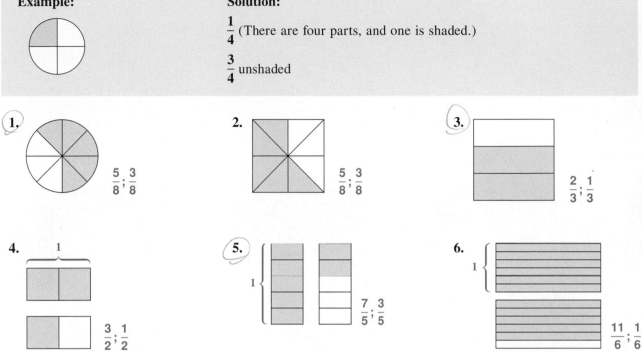

Example:

Solution:

$\frac{1}{4}$ (There are four parts, and one is shaded.)

$\frac{3}{4}$ unshaded

1.

$\frac{5}{8}; \frac{3}{8}$

2.

$\frac{5}{8}; \frac{3}{8}$

3.

$\frac{2}{3}; \frac{1}{3}$

4. 1

$\frac{3}{2}; \frac{1}{2}$

5. 1

$\frac{7}{5}; \frac{3}{5}$

6. 1

$\frac{11}{6}; \frac{1}{6}$

7. What fraction of these 11 coins are dimes?

$\frac{2}{11}$

8. What fraction of these 6 recording stars are men?

$\frac{5}{6}$

9. In an American Sign Language (A.S.L.) class of 25 students, 8 are hearing impaired. What fraction of the students are hearing impaired?

$\frac{8}{25}$

10. Of 35 motor cycles in the parking lot, 17 are Harley Davidsons. What fraction of the motor-cycles are *not* Harley Davidsons?

$\frac{18}{35}$

☑ Writing   ◉ Conceptual   ▲ Challenging   ≈ Estimation

11. Of 71 cars making up a freight train, 58 are boxcars. What fraction of the cars are *not* boxcars?

$$\frac{13}{71}$$

12. A college cheerleading squad has 12 members. If 5 of the cheerleaders are sophomores and the rest are freshmen, find the fraction of the members that are freshmen.

$$\frac{7}{12}$$

Identify the numerator and denominator.

Example: $\frac{9}{11}$ **Solution:** $\frac{9}{11}$ ← *Numerator* (on top)
← *Denominator* (on bottom)

	Numerator	*Denominator*		*Numerator*	*Denominator*
13. $\frac{3}{4}$	3	4	**14.** $\frac{5}{8}$	5	8
15. $\frac{12}{7}$	12	7	**16.** $\frac{8}{3}$	8	3

List the proper and improper fractions in each of the following groups.

Example:

$$\frac{3}{8}, \frac{7}{4}, \frac{5}{6}, \frac{2}{3}, \frac{9}{4}$$

Solution: *Proper*   *Improper*

$$\frac{3}{8}, \frac{5}{6}, \frac{2}{3} \qquad \frac{7}{4}, \frac{9}{4}$$

17. $\frac{8}{5}, \frac{1}{3}, \frac{5}{8}, \frac{6}{6}, \frac{12}{2}, \frac{7}{16}$

Proper
$$\frac{1}{3}, \frac{5}{8}, \frac{7}{16}$$

Improper
$$\frac{8}{5}, \frac{6}{6}, \frac{12}{2}$$

	Proper	*Improper*
18. $\dfrac{1}{6}, \dfrac{5}{8}, \dfrac{15}{14}, \dfrac{11}{9}, \dfrac{7}{7}, \dfrac{3}{4}$	$\dfrac{1}{6}, \dfrac{5}{8}, \dfrac{3}{4}$	$\dfrac{15}{14}, \dfrac{11}{9}, \dfrac{7}{7}$

19. $\dfrac{3}{4}, \dfrac{3}{2}, \dfrac{5}{5}, \dfrac{9}{11}, \dfrac{7}{15}, \dfrac{19}{18}$	$\dfrac{3}{4}, \dfrac{9}{11}, \dfrac{7}{15}$	$\dfrac{3}{2}, \dfrac{5}{5}, \dfrac{19}{18}$

20. $\dfrac{12}{12}, \dfrac{15}{11}, \dfrac{13}{12}, \dfrac{11}{8}, \dfrac{17}{17}, \dfrac{19}{12}$	none	$\dfrac{12}{12}, \dfrac{15}{11}, \dfrac{13}{12}, \dfrac{11}{8}, \dfrac{17}{17}, \dfrac{19}{12}$

21. Write a fraction of your own choice. Label the parts of the fraction and write a sentence describing what each part represents. Draw a picture with shaded parts showing your fraction.

One possibility is

$\dfrac{3}{4}$ ← Numerator
← Denominator

The denominator shows the number of equal parts in the whole and the numerator shows how many of the parts are being considered.

22. Give one example of a proper fraction and one example of an improper fraction. What determines whether a fraction is proper or improper? Draw pictures with shaded parts showing these fractions.

An example is $\dfrac{1}{2}$ as a proper fraction and $\dfrac{3}{2}$ as an improper fraction. A proper fraction has a numerator smaller than the denominator. An improper fraction has a numerator that is equal to or greater than the denominator.

$\dfrac{1}{2}$ $\dfrac{3}{2}$

Proper fraction   Improper fraction

Complete the following sentences.

23. The fraction $\frac{9}{16}$ represents ___9___ of the ___16___ equal parts into which a whole is divided.

24. The fraction $\frac{23}{24}$ represents ___23___ of the ___24___ equal parts into which a whole is divided.

Review and Prepare

Almost every exercise set in the rest of the book ends with a brief set of preview exercises. These exercises are designed to help you review ideas needed for the next few sections in the chapter. If you need help with these preview exercises, look in the chapter sections indicated.

*Multiply each of the following. (For help, see **Section 1.4**.)*

25. $4 \times 5 \times 5$ **100**

26. $3 \times 5 \times 2$ **30**

27. $7 \cdot 2 \cdot 0$ **0**

28. $2 \cdot 4 \cdot 24$ **192**

*Divide each of the following. (For help, see **Section 1.5**.)*

29. $32 \div 8$ **4**

30. $56 \div 8$ **7**

31. $209 \div 11$ **19**

32. $115 \div 23$ **5**

2.2 Mixed Numbers

OBJECTIVES
1. Identify mixed numbers.
2. Write mixed numbers as improper fractions.
3. Write improper fractions as mixed numbers.

FOR EXTRA HELP

Tutorial   Tape 3   SSM, Sec. 2.2

OBJECTIVE 1 When a fraction and a whole number are written together the result is a **mixed number.** For example, the mixed number

$$3\frac{1}{2} \quad \text{represents} \quad 3 + \frac{1}{2},$$

or 3 wholes and $\frac{1}{2}$ of a whole. Read $3\frac{1}{2}$ as "three and one half." As this figure shows, the mixed number $3\frac{1}{2}$ is equal to the improper fraction $\frac{7}{2}$.

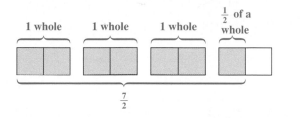

WORK PROBLEM 1 AT THE SIDE. ▶▶

OBJECTIVE 2 Use the following steps to write $3\frac{1}{2}$ as an improper fraction without drawing a figure.

Step 1 Multiply 3 and 2.

$$3\frac{1}{2} \quad 3 \cdot 2 = 6$$

Step 2 Add 1 to the product.

$$3\frac{1}{2} \qquad = 6 + 1 = 7$$

Step 3 Use 7, from Step 2, as the numerator and 2 as the denominator.

$$3\frac{1}{2} = \frac{7}{2}$$

Same denominator

In summary, use the following steps to *write a mixed number as an improper fraction.*

Write a Mixed Number as an Improper Fraction

Step 1 **Multiply** the denominator of the fraction and the whole number.

Step 2 **Add** to this product the numerator of the fraction.

Step 3 Write the result of Step 2 as the **numerator** and the original denominator as the **denominator.**

┌ **E X A M P L E 1** **Changing Mixed Numbers to Improper Fractions**

Write $7\frac{2}{3}$ as an improper fraction (numerator greater than denominator).

Step 1 $7\frac{2}{3}$ $7 \cdot 3 = 21$ Multiply 7 and 3.

└ **CONTINUED ON NEXT PAGE**

1. (a) Use these diagrams to write $1\frac{2}{3}$ as an improper fraction.

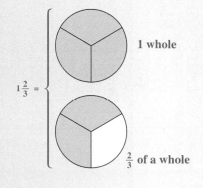

$1\frac{2}{3} =$ 1 whole

$\frac{2}{3}$ of a whole

(b) Use these diagrams to write $2\frac{1}{4}$ as an improper fraction.

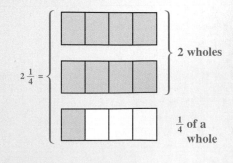

$2\frac{1}{4} =$ 2 wholes

$\frac{1}{4}$ of a whole

ANSWERS

1. (a) $\frac{5}{3}$ (b) $\frac{9}{4}$

105

2. Write as improper fractions.

(a) $3\frac{2}{3}$

Step 2 $7\frac{2}{3} = 21 + 2 = 23$ Add 2.

Step 3 $7\frac{2}{3} = \frac{23}{3}$ Same denominator

◀◀ **WORK PROBLEM 2 AT THE SIDE.**

(b) $4\frac{3}{8}$

OBJECTIVE 3 ▶ Write an improper fraction as a mixed number as follows.

> **Write an Improper Fraction as a Mixed Number**
>
> Write an **improper fraction** as a mixed number by dividing the numerator by the denominator. The quotient is the whole number (of the mixed number), the remainder is the numerator of the fraction part, and the denominator remains unchanged.

(c) $5\frac{3}{4}$

E X A M P L E 2   Changing Improper Fractions to Mixed Numbers

Write as mixed numbers.

(a) $\frac{17}{5}$

Divide 17 by 5.

$$\begin{array}{r} 3 \\ 5\overline{)17} \\ 15 \\ \hline 2 \end{array}$$ ← Whole number part

← Remainder

(d) $8\frac{5}{6}$

The quotient **3** is the whole number part of the mixed number. The remainder **2** is the numerator of the fraction and the denominator remains as **5.**

$$\frac{17}{5} = 3\frac{2}{5}$$ ← Remainder

Same denominator

3. Write as mixed numbers.

(a) $\frac{5}{2}$

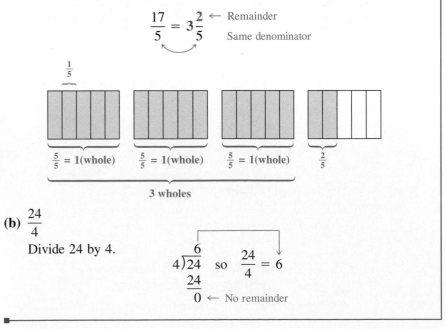

$\frac{5}{5} = 1$(whole) $\frac{5}{5} = 1$(whole) $\frac{5}{5} = 1$(whole) $\frac{2}{5}$

3 wholes

(b) $\frac{15}{4}$

(b) $\frac{24}{4}$

Divide 24 by 4.

$$\begin{array}{r} 6 \\ 4\overline{)24} \\ 24 \\ \hline 0 \end{array}$$ so $\frac{24}{4} = 6$

← No remainder

(c) $\frac{35}{5}$

(d) $\frac{75}{8}$

> **Note**
>
> A proper fraction has a value that is smaller than 1, while an improper fraction has a value that is 1 or greater.

ANSWERS

2. (a) $\frac{11}{3}$ **(b)** $\frac{35}{8}$ **(c)** $\frac{23}{4}$ **(d)** $\frac{53}{6}$

3. (a) $2\frac{1}{2}$ **(b)** $3\frac{3}{4}$ **(c)** 7 **(d)** $9\frac{3}{8}$

◀◀ **WORK PROBLEM 3 AT THE SIDE.**

2.2 Exercises

Write each mixed number as an improper fraction.

Example:	**Solution:**
$6\dfrac{3}{7}$	$6 \cdot 7 = 42 \qquad 42 + 3 = 45 \qquad 6\dfrac{3}{7} = \dfrac{45}{7}$ Same denominator

1. $1\dfrac{2}{3}$ $\dfrac{5}{3}$

2. $3\dfrac{1}{2}$ $\dfrac{7}{2}$

3. $2\dfrac{3}{4}$ $\dfrac{11}{4}$

4. $5\dfrac{1}{2}$ $\dfrac{11}{2}$

5. $4\dfrac{3}{4}$ $\dfrac{19}{4}$

6. $6\dfrac{1}{5}$ $\dfrac{31}{5}$

7. $6\dfrac{3}{4}$ $\dfrac{27}{4}$

8. $8\dfrac{1}{2}$ $\dfrac{17}{2}$

9. $1\dfrac{7}{11}$ $\dfrac{18}{11}$

10. $5\dfrac{4}{7}$ $\dfrac{39}{7}$

11. $6\dfrac{1}{3}$ $\dfrac{19}{3}$

12. $8\dfrac{2}{3}$ $\dfrac{26}{3}$

13. $11\dfrac{1}{3}$ $\dfrac{34}{3}$

14. $12\dfrac{2}{3}$ $\dfrac{38}{3}$

15. $10\dfrac{3}{4}$ $\dfrac{43}{4}$

16. $6\dfrac{1}{6}$ $\dfrac{37}{6}$

17. $3\dfrac{3}{8}$ $\dfrac{27}{8}$

18. $2\dfrac{8}{9}$ $\dfrac{26}{9}$

19. $8\dfrac{4}{5}$ $\dfrac{44}{5}$

20. $3\dfrac{4}{7}$ $\dfrac{25}{7}$

21. $4\dfrac{10}{11}$ $\dfrac{54}{11}$

✍ Writing   ◉ Conceptual   ▲ Challenging   ≈ Estimation

22. $13\frac{5}{9}$ $\frac{122}{9}$

23. $22\frac{7}{8}$ $\frac{183}{8}$

24. $12\frac{9}{10}$ $\frac{129}{10}$

25. $17\frac{12}{13}$ $\frac{233}{13}$

26. $19\frac{8}{11}$ $\frac{217}{11}$

27. $17\frac{14}{15}$ $\frac{269}{15}$

28. $8\frac{17}{24}$ $\frac{209}{24}$

29. $6\frac{7}{18}$ $\frac{115}{18}$

30. $9\frac{7}{12}$ $\frac{115}{12}$

Write each improper fraction as a mixed number.

Example: $\frac{13}{5}$ **Solution:** $5\overline{)13}$ $\frac{13}{5} = 2\frac{3}{5}$

31. $\frac{8}{3}$ $2\frac{2}{3}$

32. $\frac{9}{4}$ $2\frac{1}{4}$

33. $\frac{9}{5}$ $1\frac{4}{5}$

34. $\frac{11}{2}$ $5\frac{1}{2}$

35. $\frac{60}{12}$ 5

36. $\frac{56}{7}$ 8

37. $\frac{27}{8}$ $3\frac{3}{8}$

38. $\frac{27}{7}$ $3\frac{6}{7}$

39. $\frac{19}{4}$ $4\frac{3}{4}$

40. $\frac{40}{9}$ $4\frac{4}{9}$

41. $\frac{27}{3}$ 9

42. $\frac{78}{6}$ 13

43. $\frac{58}{5}$ $11\frac{3}{5}$

44. $\frac{19}{5}$ $3\frac{4}{5}$

45. $\frac{47}{9}$ $5\frac{2}{9}$

46. $\frac{65}{9}$ $7\frac{2}{9}$

47. $\dfrac{50}{7}$ $7\dfrac{1}{7}$

48. $\dfrac{30}{7}$ $4\dfrac{2}{7}$

49. $\dfrac{84}{5}$ $16\dfrac{4}{5}$

50. $\dfrac{92}{3}$ $30\dfrac{2}{3}$

51. $\dfrac{123}{4}$ $30\dfrac{3}{4}$

52. $\dfrac{118}{5}$ $23\dfrac{3}{5}$

53. $\dfrac{183}{7}$ $26\dfrac{1}{7}$

54. $\dfrac{212}{11}$ $19\dfrac{3}{11}$

Stop

55. Your classmate asks you how to change a mixed number to an improper fraction. Write a couple of sentences and give an example showing how this is done.

Multiply the denominator by the whole number and add the numerator. The result becomes the new numerator which is placed over the original denominator.

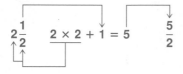

56. Explain in a sentence or two how to change an improper fraction to a mixed number. Give an example showing how this is done.

Divide the numerator by the denominator. The quotient is the whole number of the mixed number and the remainder is the numerator of the fraction part. The denominator is unchanged.

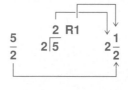

▲ *Write each mixed number as an improper fraction.*

57. $101\dfrac{1}{2}$ $\dfrac{203}{2}$

58. $218\dfrac{3}{5}$ $\dfrac{1093}{5}$

59. $333\dfrac{1}{3}$ $\dfrac{1000}{3}$

▲ *Write each improper fraction as a mixed number.*

60. $\dfrac{837}{8}$ $104\dfrac{5}{8}$

61. $\dfrac{2565}{15}$ 171

62. $\dfrac{2915}{16}$ $182\dfrac{3}{16}$

Review and Prepare

*Work each of the following problems. (For help, see **Section 1.8**.)*

63. $4^2 + 2^2$

20

64. $4^2 \cdot 2^2$

64

65. $15 \cdot 4 + 6$

66

66. $5 \cdot 6 + 8$

38

67. $6 \cdot 3^2 - 5$

49

68. $5 \cdot 5^2 - 9$

116

2.3 Factors

OBJECTIVE 1 You will recall that numbers multiplied to give a product are called **factors.** Because 2 • 5 = 10, both 2 and 5 are factors of 10. The numbers 1 and 10 are also factors of 10, because

$$1 \cdot 10 = 10$$

The various tests for divisibility show 1, 2, 5, and 10 are the only whole-number factors of 10. The products 2 • 5 and 1 • 10 are called **factorizations** (FAK-ter-uh-ZAY-shuns) of 10.

> **Note**
> You might want to review the tests for divisibility in Section 1.5. The ones that you will want to remember are those for 2, 3, 5, and 10.

EXAMPLE 1   Using Factors

Find all possible two-number factorizations of each number.

(a) 12

$$1 \cdot 12 \qquad 2 \cdot 6 \qquad 3 \cdot 4$$

The factors of 12 are 1, 2, 3, 4, 6, and 12.

(b) 60

$$1 \cdot 60 \qquad 2 \cdot 30$$
$$3 \cdot 20 \qquad 4 \cdot 15$$
$$5 \cdot 12 \qquad 6 \cdot 10$$

The factors of 60 are 1, 2, 3, 4, 5, 6, 10, 12, 15, 20, 30, and 60.

WORK PROBLEM 1 AT THE SIDE. ▶▶

Composite Numbers

A number with a factor other than itself or 1 is called a **composite** (kahm-PAHZ-it) **number.**

EXAMPLE 2   Identifying Composite Numbers

Which of the following numbers is (are) composite?

(a) 6

Because 6 has factors of 2 and 3, numbers other than 6 or 1, the number 6 is composite.

(b) 17

The number 17 has only two factors—17 and 1. It is not composite.

(c) 25

A factor of 25 is 5, so 25 is composite.

WORK PROBLEM 2 AT THE SIDE. ▶▶

OBJECTIVE 2 Whole numbers that are not composite are called **prime numbers,** except 0 and 1, which are neither prime nor composite.

Prime Numbers

A prime number is a whole number that has exactly *two different* factors, itself and 1.

OBJECTIVES

1 ▶ Find factors of a number.
2 ▶ Identify primes.
3 ▶ Find prime factorizations.

FOR EXTRA HELP

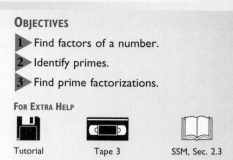

Tutorial   Tape 3   SSM, Sec. 2.3

1. Find all the factors of the following numbers.

   (a) 9

   (b) 18

   (c) 36

   (d) 80

2. Which of these numbers is (are) composite?

   2, 4, 5, 6, 8, 10, 11, 13, 19, 21, 27, 28, 33, 36, 42

ANSWERS

1. **(a)** 1, 3, 9 **(b)** 1, 2, 3, 6, 9, 18
   (c) 1, 2, 3, 4, 6, 9, 12, 18, 36
   (d) 1, 2, 4, 5, 8, 10, 16, 20, 40, 80
2. 4, 6, 8, 10, 21, 27, 28, 33, 36, 42

3. Which of the following are prime?

2, 3, 4, 7, 9, 13, 19, 29

The number 3 is a prime number, since it can be divided evenly only by itself and 1. The number 8 is not a prime number (it is composite), since 8 can be divided evenly by 2 and 4, as well as by itself and 1.

> **Note**
> A prime number has *only two* different factors, itself and 1. The number 1 is not a prime number because it does not have *two different* factors; the only factor of 1 is 1.

E X A M P L E   3   Finding Prime Numbers

Which of the following numbers are prime?

2   5   8   11   15

The number 8 can be divided by 4 and 2, so it is not prime. Also, because 15 can be divided by 5 and 3, 15 is not prime. All the other numbers in the list are divisible by only themselves and 1, and are prime.

◀◀ **WORK PROBLEM 3 AT THE SIDE.**

OBJECTIVE ▶ For reference, here are the primes smaller than 100.

$$2, \quad 3, \quad 5, \quad 7, \quad 11,$$
$$13, \quad 17, \quad 19, \quad 23, \quad 29,$$
$$31, \quad 37, \quad 41, \quad 43, \quad 47,$$
$$53, \quad 59, \quad 61, \quad 67, \quad 71,$$
$$73, \quad 79, \quad 83, \quad 89, \quad 97$$

> **Note**
> All prime numbers are odd numbers except the number 2. Be careful though, because *all odd numbers are not prime numbers*. For example, 9, 15, and 21 are odd numbers but are *not* prime numbers.

The **prime factorization** of a number can be especially useful when working with fractions.

> **Prime Factorization**
>
> A **prime factorization** of a number is a factorization in which every factor is a prime number.

E X A M P L E   4   Determining the Prime Factorization

Find the prime factorization of 12.

Try to divide 12 by the first prime, 2.

$$12 \div 2 = 6,$$

↑—— First prime

so

$$12 = 2 \cdot 6.$$

Try to divide 6 by the prime, 2.

$$6 \div 2 = 3,$$

CONTINUED ON NEXT PAGE ——

so

$$12 = 2 \cdot 2 \cdot 3.$$

Factorization of 6

Because all factors are prime, the prime factorization of 12 is

$$2 \cdot 2 \cdot 3.$$

WORK PROBLEM 4 AT THE SIDE. ▶▶

EXAMPLE 5   Factoring by Using the Division Method

Find the prime factorization of 48.

All prime factors
$$
\begin{array}{r}
2\,\overline{)\,48} \\
2\,\overline{)\,24} \\
2\,\overline{)\,12} \\
2\,\overline{)\,6} \\
3\,\overline{)\,3} \\
1
\end{array}
$$

Divide 48 by 2 (first prime).
Divide 24 by 2.
Divide 12 by 2.
Divide 6 by 2.
Divide 3 by 3.
Continue to divide until the quotient is 1.

Because all factors (divisors) are prime, the prime factorization of 48 is

$$2 \cdot 2 \cdot 2 \cdot 2 \cdot 3.$$

In Chapter 1 we wrote $2 \cdot 2 = 2^2$, so the prime factorization of 48 can be written, using exponents, as

$$2 \cdot 2 \cdot 2 \cdot 2 \cdot 3 = 2^4 \cdot 3.$$

WORK PROBLEM 5 AT THE SIDE. ▶▶

Note
When using the division method of factoring, the last quotient found is 1. The "1" is never used as a prime factor because 1 is neither prime nor composite. Besides, 1 times any number is the number itself.

EXAMPLE 6   Using Exponents with Prime Factorization

Find the prime factorization of 225.

All prime factors
$$
\begin{array}{r}
3\,\overline{)\,225} \\
3\,\overline{)\,75} \\
5\,\overline{)\,25} \\
5\,\overline{)\,5} \\
1
\end{array}
$$

225 is not divisible by 2; use 3.
Divide 75 by 3.
25 is not divisible by 3; use 5.
Divide by 5.

CONTINUED ON NEXT PAGE

---

4. Find the prime factorization of each number.

(a) 8

(b) 14

(c) 18

(d) 30

5. Find the prime factorization of each number. Write the factorizations with exponents.

(a) 18

(b) 36

(c) 60

(d) 126

ANSWERS
4. (a) $2 \cdot 2 \cdot 2$ **(b)** $2 \cdot 7$ **(c)** $2 \cdot 3 \cdot 3$
   (d) $2 \cdot 3 \cdot 5$
5. (a) $2 \cdot 3^2$ **(b)** $2^2 \cdot 3^2$ **(c)** $2^2 \cdot 3 \cdot 5$
   (d) $2 \cdot 3^2 \cdot 7$

6. Write the prime factorization of each number by using exponents.

(a) 50

(b) 88

(c) 90

(d) 150

(e) 280

Write the prime factorization,

$$3 \cdot 3 \cdot 5 \cdot 5$$

with exponents, as

$$3^2 \cdot 5^2.$$

◄◄ **WORK PROBLEM 6 AT THE SIDE.**

Another method of factoring uses what is called a factor tree.

E X A M P L E   7   Factoring by Using a Factor Tree

Find the prime factorization of each number.

(a) 30

Try to divide by the first prime, 2. Write the factors under the 30. Circle the 2, since it is a prime.

$$
\begin{array}{c}
30 \\
\swarrow \searrow \\
② \quad 15
\end{array}
$$

Since 15 cannot be divided evenly by 2, try the next prime, 3.

$$
\begin{array}{c}
30 \\
\swarrow \searrow \\
② \quad 15 \\
\quad \swarrow \searrow \\
\quad ③ \quad ⑤ \;\leftarrow \text{Circle, because they are primes.}
\end{array}
$$

No uncircled factors remain, so the prime factorization (the circled factors) has been found.

$$30 = 2 \cdot 3 \cdot 5$$

(b) 24

Divide by 2.

$$
\begin{array}{c}
24 \\
\swarrow \searrow \\
② \quad 12 \quad \leftarrow \text{Divide by 2, again.} \\
\quad \swarrow \searrow \\
\quad ② \quad 6 \quad \leftarrow \text{Divide by 2, a third time.} \\
\quad\quad \swarrow \searrow \\
\quad\quad ② \quad ③
\end{array}
$$

$24 = 2 \cdot 2 \cdot 2 \cdot 3$ or, using exponents, $24 = 2^3 \cdot 3$.

(c) 45

Because 45 cannot be divided by 2, try 3.

$$
\begin{array}{c}
45 \\
\swarrow \searrow \\
③ \quad 15 \quad \leftarrow \text{Divide by 3, again.} \\
\quad \swarrow \searrow \\
\quad ③ \quad ⑤
\end{array}
$$

$45 = 3 \cdot 3 \cdot 5$ or, using exponents, $45 = 3^2 \cdot 5$.

Note
The diagrams used in Example 7 look like tree branches, and that is why this method is referred to as using a factor tree.

◄◄ **WORK PROBLEM 7 AT THE SIDE.**

7. Complete each factor tree and give the prime factorization.

(a)
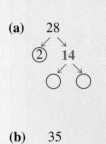
$$
\begin{array}{c}
28 \\
\swarrow \searrow \\
② \quad 14 \\
\quad \swarrow \searrow \\
\quad ○ \quad ○
\end{array}
$$

(b)
$$
\begin{array}{c}
35 \\
\swarrow \searrow \\
⑤
\end{array}
$$

(c)

$$
\begin{array}{c}
90 \\
\swarrow \searrow
\end{array}
$$

ANSWERS

6. (a) $2 \cdot 5^2$ (b) $2^3 \cdot 11$ (c) $2 \cdot 3^2 \cdot 5$
   (d) $2 \cdot 3 \cdot 5^2$ (e) $2^3 \cdot 5 \cdot 7$

7. (a)
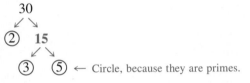
$$
\begin{array}{c}
28 \\
\swarrow \searrow \\
② \quad 14 \\
\quad \swarrow \searrow \\
\quad ② \quad ⑦
\end{array}
$$
$28 = 2 \cdot 2 \cdot 7 = 2^2 \cdot 7$

   (b)
$$
\begin{array}{c}
35 \\
\swarrow \searrow \\
⑤ \quad ⑦
\end{array}
$$
$35 = 5 \cdot 7$

   (c)
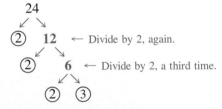
$$
\begin{array}{c}
90 \\
\swarrow \searrow \\
② \quad 45 \\
\quad \swarrow \searrow \\
\quad ③ \quad 15 \\
\quad\quad \swarrow \searrow \\
\quad\quad ③ \quad ⑤
\end{array}
$$
$90 = 2 \cdot 3 \cdot 3 \cdot 5 = 2 \cdot 3^2 \cdot 5$

2.3 Exercises

Find all the factors of each number.

Example: 18	**Solution:** Write all factorizations of 18 that have two factors.

$$1 \cdot 18 \qquad 2 \cdot 9 \qquad 3 \cdot 6$$

The factors of 18 are **1, 2, 3, 6, 9,** and **18.**

1. 8

1, 2, 4, 8

2. 12

1, 2, 3, 4, 6, 12

3. 6

1, 2, 3, 6

4. 18

1, 2, 3, 6, 9, 18

5. 25

1, 5, 25

6. 30

1, 2, 3, 5, 6, 10, 15, 30

7. 18

1, 2, 3, 6, 9, 18

8. 20

1, 2, 4, 5, 10, 20

9. 40

1, 2, 4, 5, 8, 10, 20, 40

10. 60

1, 2, 3, 4, 5, 6, 10, 12, 15, 20, 30, 60

11. 64

1, 2, 4, 8, 16, 32, 64

12. 80

1, 2, 4, 5, 8, 10, 16, 20, 40, 80

Which numbers are prime and which are composite?

Example: 5, 12	**Solution: 5 is prime** because it can be divided by only itself and 1. **12 is composite** because it can be divided by 2 and 3 (numbers *other than* 12 or 1).

13. 9

composite

14. 8

composite

15. 2

prime

16. 3

prime

✐ Writing   ◉ Conceptual   ▲ Challenging   ≈ Estimation

17. 10

composite

18. 12

composite

19. 11

prime

20. 15

composite

21. 19

prime

22. 23

prime

23. 25

composite

24. 26

composite

25. 34

composite

26. 43

prime

27. 45

composite

28. 47

prime

Find the prime factorization of the following numbers. Write answers with exponents when repeated factors appear.

Example:

40

Solution:

Division method

$2\overline{)40}$
  $2\overline{)20}$
    $2\overline{)10}$
      $5\overline{)5}$
        1

$40 = 2 \cdot 2 \cdot 2 \cdot 5 = \mathbf{2^3 \cdot 5}$

Factor tree

40 ← Divide by 2; circle 2, since it is prime.

② 20 ← Divide by 2, again.

② 10 ← Divide by 2, a third time.

② ⑤

29. 6

$2 \cdot 3$

30. 12

$2^2 \cdot 3$

31. 20

$2^2 \cdot 5$

32. 30

$2 \cdot 3 \cdot 5$

33. 25

5^2

34. 18

$2 \cdot 3^2$

35. 36

$2^2 \cdot 3^2$

36. 56

$2^3 \cdot 7$

37. 44

$2^2 \cdot 11$

38. 68

$2^2 \cdot 17$

39. 88

$2^3 \cdot 11$

40. 64

2^6

41. 75

$3 \cdot 5^2$

42. 80

$2^4 \cdot 5$

43. 100

$2^2 \cdot 5^2$

44. 108

$2^2 \cdot 3^3$

45. 120

$2^3 \cdot 3 \cdot 5$

46. 180

$2^2 \cdot 3^2 \cdot 5$

47. 225

$3^2 \cdot 5^2$

48. 300

$2^2 \cdot 3 \cdot 5^2$

49. 320

$2^6 \cdot 5$

50. 340

$2^2 \cdot 5 \cdot 17$

51. 360

$2^3 \cdot 3^2 \cdot 5$

52. 400

$2^4 \cdot 5^2$

53. Give a definition in your own words of both a composite number and a prime number. Give three examples of each.

A composite number has a factor(s) other than itself or 1. Examples include 4, 6, 8, 9, 10. A prime number is a whole number that has exactly two *different* factors, itself and 1. Examples include 2, 3, 5, 7, 11.

54. With the exception of the number 2, all prime numbers are odd numbers. Nevertheless, all odd numbers are not prime numbers. Explain why these statements are true.

No even number other than 2 is prime because all even numbers have 2 as a factor. Many odd numbers are multiples of prime numbers and are no longer prime. For example, 9, 21, 33, and 45 are all multiples of 3.

As a review, solve each of the following.

Examples: 2^4, $2^2 \cdot 3^4$

$2^4 = 2 \cdot 2 \cdot 2 \cdot 2 \leftarrow$ 4 factors of 2 $2^2 \cdot 3^4 = 2 \cdot 2 \cdot 3 \cdot 3 \cdot 3 \cdot 3$

$\qquad = 16$ $\qquad\qquad\qquad\qquad\qquad = 4 \cdot 81$

$\qquad\qquad\qquad\qquad\qquad\qquad\qquad\qquad = 324$

55. 2^3

8

56. 2^5

32

57. 5^3

125

58. 6^3

216

59. 3^4

81

60. 4^4

256

61. $2^2 \cdot 3^3$

108

62. $2^4 \cdot 3^2$

144

63. $5^3 \cdot 3^2$

1125

64. $2^4 \cdot 5^2$

400

65. $3^5 \cdot 2^2$

972

66. $5^2 \cdot 4^3$

1600

▲ *Find the prime factorization of each number. Write answers by using exponents.*

67. 280 $2^3 \cdot 5 \cdot 7$

68. 520 $2^3 \cdot 5 \cdot 13$

69. 960 $2^6 \cdot 3 \cdot 5$

70. 1125 $3^2 \cdot 5^3$

71. 1600 $2^6 \cdot 5^2$

72. 1575 $3^2 \cdot 5^2 \cdot 7$

Review and Prepare

*Multiply each of the following. (For help, see **Section 1.4**.)*

73. $4 \cdot 2 \cdot 3$ 24

74. $5 \cdot 3 \cdot 4$ 60

75. $6 \cdot 1 \cdot 8$ 48

76. $5 \cdot 3 \cdot 9$ 135

*Divide each of the following. (For help, see **Section 1.5**.)*

77. $36 \div 4$ 9

78. $28 \div 4$ 7

79. $135 \div 5$ 27

80. $75 \div 15$ 5

2.4 *Writing a Fraction in Lowest Terms*

OBJECTIVES

1. ▶ Tell whether a fraction is written in lowest terms.

2. ▶ Write a fraction in lowest terms using common factors.

3. ▶ Write a fraction in lowest terms using prime factors.

4. ▶ Tell whether two fractions are equivalent.

FOR EXTRA HELP

Tutorial   Tape 3   SSM, Sec. 2.4

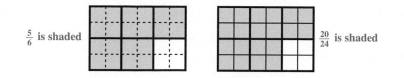

$\frac{5}{6}$ is shaded $\frac{20}{24}$ is shaded

The figures show areas that are $\frac{5}{6}$ shaded and $\frac{20}{24}$ shaded. Because the shaded areas are equivalent, the fractions $\frac{5}{6}$ and $\frac{20}{24}$ are **equivalent** (ee-KWIV-uh-lent) **fractions.**

$$\frac{5}{6} = \frac{20}{24}$$

Because the numbers 20 and 24 both have 4 as a factor, 4 is called a **common factor** of the numbers. Other common factors of 20 and 24 are 1 and 2.

WORK PROBLEM 1 AT THE SIDE. ▶▶

1. Decide whether the given factor is a common factor of both numbers.

(a) 14, 20; 2

OBJECTIVE 1 ▶ The fraction $\frac{5}{6}$ is in lowest terms because the numerator and denominator have no common factor other than 1; however, the fraction $\frac{20}{24}$ is *not* in lowest terms because its numerator and denominator have a common factor of 4.

(b) 32, 48; 16

Writing a Fraction in Lowest Terms

A fraction is written in *lowest terms* when the numerator and the denominator have no common factor other than 1.

(c) 24, 36; 8

E X A M P L E 1   Understanding Lowest Terms

Are the following fractions in lowest terms?

(a) $\frac{3}{8}$

The numerator and denominator have no common factor other than 1, so the fraction is in lowest terms.

(b) $\frac{21}{36}$

The numerator and denominator have a common factor of 3, so the fraction is not in lowest terms.

(d) 56, 73; 1

WORK PROBLEM 2 AT THE SIDE. ▶▶

2. Are the following fractions in lowest terms?

(a) $\frac{2}{3}$ (b) $\frac{4}{16}$

OBJECTIVE 2 ▶ There are two common methods for writing a fraction in lowest terms. These methods are shown in the next examples. The first method works best when the numerator and denominator are small numbers.

(c) $\frac{9}{11}$ (d) $\frac{15}{51}$

ANSWERS

1. (a) yes (b) yes (c) no (d) yes
2. (a) yes (b) no (c) yes (d) no

119

3. Write in lowest terms.

(a) $\dfrac{5}{10}$

(b) $\dfrac{9}{12}$

(c) $\dfrac{24}{30}$

(d) $\dfrac{15}{40}$

(e) $\dfrac{32}{80}$

E X A M P L E 2   Changing to Lowest Terms

Write each fraction in lowest terms.

(a) $\dfrac{20}{24}$

The largest common factor of 20 and 24 is 4. Divide both numerator and denominator by **4.**

$$\dfrac{20}{24} = \dfrac{20 \div 4}{24 \div 4} = \dfrac{5}{6}$$

(b) $\dfrac{30}{50} = \dfrac{30 \div 10}{50 \div 10} = \dfrac{3}{5}$ Divide both numerator and denominator by 10.

(c) $\dfrac{24}{42} = \dfrac{24 \div 6}{42 \div 6} = \dfrac{4}{7}$

(d) $\dfrac{60}{72}$

Suppose we made an error and thought 4 was the largest common factor of 60 and 72. Dividing by 4 would give

$$\dfrac{60}{72} = \dfrac{60 \div 4}{72 \div 4} = \dfrac{15}{18}.$$

But $\frac{15}{18}$ is not in lowest terms, because 15 and 18 have a common factor of 3. Divide by 3.

$$\dfrac{15}{18} = \dfrac{15 \div 3}{18 \div 3} = \dfrac{5}{6}$$

The fraction $\frac{60}{72}$ could have been written in lowest terms in one step by dividing by 12, the largest common factor of 60 and 72.

$$\dfrac{60}{72} = \dfrac{60 \div 12}{72 \div 12} = \dfrac{5}{6}$$

> **Note**
> Dividing the numerator and denminator by the same number results in an equivalent fraction.

This method of writing a fraction in lowest terms by dividing by a common factor is summarized in the following steps.

> **The Method of Dividing by a Common Factor**
>
> *Step 1* Find the largest number that will divide evenly into both the numerator and denominator. This number is a **common factor.**
>
> *Step 2* **Divide** both numerator and denominator by the common factor.
>
> *Step 3* **Check** to see if the new fraction has any common factors (besides 1). If it does, repeat Steps 2 and 3. If the only common factor is 1, the fraction is in lowest terms.

◀◀ **WORK PROBLEM 3 AT THE SIDE.**

ANSWERS

3. (a) $\dfrac{1}{2}$ (b) $\dfrac{3}{4}$ (c) $\dfrac{4}{5}$ (d) $\dfrac{3}{8}$ (e) $\dfrac{2}{5}$

OBJECTIVE 3 The method of writing a fraction in lowest terms by division works well for fractions with small numerators and denominators. For larger numbers, it is common to use the method of **prime factors,** which is shown in the next example.

E X A M P L E 3   Using Prime Factors

Write each of the following in lowest terms.

(a) $\dfrac{24}{42}$

Write the prime factorization of both numerator and denominator. See **Section 2.3** for help.

$$\frac{24}{42} = \frac{2 \cdot 2 \cdot 2 \cdot 3}{2 \cdot 3 \cdot 7}$$

Just as with the other method, divide numerator and denominator by any common factors. Use a shortcut called **cancellation** (kan-suh-LAY-shun) to show this division. Place a **1** by each factor that is canceled.

$$\frac{24}{42} = \frac{\overset{1}{\cancel{2}} \cdot 2 \cdot 2 \cdot \overset{1}{\cancel{3}}}{\underset{1}{\cancel{2}} \cdot \underset{1}{\cancel{3}} \cdot 7}$$

Multiply the remaining factors in both numerator and denominator.

$$\frac{24}{42} = \frac{1 \cdot 2 \cdot 2 \cdot 1}{1 \cdot 1 \cdot 7} = \frac{4}{7}$$

Finally, $\frac{24}{42}$, written in lowest terms, is $\frac{4}{7}$.

(b) $\dfrac{162}{54}$

Write the prime factorization of both numerator and denominator.

$$\frac{162}{54} = \frac{2 \cdot 3 \cdot 3 \cdot 3 \cdot 3}{2 \cdot 3 \cdot 3 \cdot 3}$$

Now cancel the common factors. **Do not forget the 1's.**

$$\frac{162}{54} = \frac{\overset{1}{\cancel{2}} \cdot \overset{1}{\cancel{3}} \cdot \overset{1}{\cancel{3}} \cdot \overset{1}{\cancel{3}} \cdot 3}{\underset{1}{\cancel{2}} \cdot \underset{1}{\cancel{3}} \cdot \underset{1}{\cancel{3}} \cdot \underset{1}{\cancel{3}}}$$

$$= \frac{1 \cdot 1 \cdot 1 \cdot 1 \cdot 3}{1 \cdot 1 \cdot 1 \cdot 1} = \frac{3}{1} = 3$$

(c) $\dfrac{18}{90}$

$$\frac{18}{90} = \frac{\overset{1}{\cancel{2}} \cdot \overset{1}{\cancel{3}} \cdot \overset{1}{\cancel{3}}}{\underset{1}{\cancel{2}} \cdot \underset{1}{\cancel{3}} \cdot \underset{1}{\cancel{3}} \cdot 5} = \frac{1 \cdot 1 \cdot 1}{1 \cdot 1 \cdot 1 \cdot 5} = \frac{1}{5}$$

Note

In Example 3(c), all factors of the numerator cancelled. But $1 \cdot 1 \cdot 1$ is still 1, so the final answer is $\frac{1}{5}$ (not 5).

4. Use the method of prime factors to write each fraction in lowest terms.

(a) $\dfrac{16}{48}$

(b) $\dfrac{28}{60}$

(c) $\dfrac{74}{111}$

(d) $\dfrac{124}{340}$

5. Are the following fractions equivalent?

(a) $\dfrac{1}{2}$ and $\dfrac{2}{4}$

(b) $\dfrac{3}{4}$ and $\dfrac{2}{3}$

(c) $\dfrac{6}{50}$ and $\dfrac{9}{75}$

(d) $\dfrac{12}{22}$ and $\dfrac{18}{32}$

This method of writing a fraction in lowest terms is summarized below.

The Method of Prime Factors

Step 1 Write the **prime factorization** of both numerator and denominator.

Step 2 Use **cancellation** to divide numerator and denominator by any common factors.

Step 3 **Multiply** the remaining factors in numerator and denominator.

◀◀ **WORK PROBLEM 4 AT THE SIDE.**

OBJECTIVE 4 The next example shows how to use the *equivalency test* to tell whether two fractions are equivalent.

EXAMPLE 4 Using the Equivalency Test ("Cross Multiplication")

Are the following fractions equivalent?

(a) $\dfrac{3}{4}$ and $\dfrac{6}{8}$

Find each *cross product*.

$$\dfrac{3}{4} \,\, \dfrac{6}{8} \qquad \begin{array}{l} 4 \cdot 6 = 24 \\ 3 \cdot 8 = 24 \end{array} \quad \text{Equivalent } (24 = 24)$$

Since the cross products are equal ($=$), the fractions are equivalent.

The cross products test is based on rewriting both fractions with the common denominator of 4×8 or 32. Multiplying the numerator and denominator by the same number results in an equivalent fraction.

$$\dfrac{3 \times 8}{4 \times 8} = \dfrac{24}{32} \qquad \dfrac{6 \times 4}{8 \times 4} = \dfrac{24}{32}$$

$\frac{3}{4}$ and $\frac{6}{8}$ are equivalent because they both equal $\frac{24}{32}$.

The cross products test uses a shortcut by comparing only the two numerators ($24 = 24$).

(b) $\dfrac{8}{12}$ and $\dfrac{21}{30}$

Find the cross products.

$$\dfrac{8}{12} \,\, \dfrac{21}{30} \qquad \begin{array}{l} 12 \cdot 21 = 252 \\ 8 \cdot 30 = 240 \end{array} \quad \text{Not equivalent } (252 \neq 240)$$

The cross products are *not equal* ($\neq$), so the fractions are *not equivalent*.

Note

Use cross multiplication to determine whether two fractions are *equivalent*. You will use this later to determine whether two ratios are *equivalent* (Section 5.3). This is **not the same** as *multiplying fractions*, which you will learn about in the next section.

◀◀ **WORK PROBLEM 5 AT THE SIDE.**

ANSWERS

4. (a) $\dfrac{1}{3}$ (b) $\dfrac{7}{15}$ (c) $\dfrac{2}{3}$ (d) $\dfrac{31}{85}$

5. (a) equivalent (b) not equivalent (c) equivalent (d) not equivalent

2.4 Exercises

Write each fraction in lowest terms.

Example: $\dfrac{12}{15}$ **Solution:** $\dfrac{12}{15} = \dfrac{12 \div 3}{15 \div 3} = \dfrac{4}{5}$

1. $\dfrac{8}{16}$ $\dfrac{1}{2}$

2. $\dfrac{6}{8}$ $\dfrac{3}{4}$

3. $\dfrac{32}{48}$ $\dfrac{2}{3}$

4. $\dfrac{9}{27}$ $\dfrac{1}{3}$

5. $\dfrac{20}{32}$ $\dfrac{5}{8}$

6. $\dfrac{14}{21}$ $\dfrac{2}{3}$

7. $\dfrac{36}{42}$ $\dfrac{6}{7}$

8. $\dfrac{22}{33}$ $\dfrac{2}{3}$

9. $\dfrac{63}{70}$ $\dfrac{9}{10}$

10. $\dfrac{27}{45}$ $\dfrac{3}{5}$

11. $\dfrac{180}{210}$ $\dfrac{6}{7}$

12. $\dfrac{72}{80}$ $\dfrac{9}{10}$

13. $\dfrac{36}{63}$ $\dfrac{4}{7}$

14. $\dfrac{73}{146}$ $\dfrac{1}{2}$

15. $\dfrac{12}{600}$ $\dfrac{1}{50}$

16. $\dfrac{54}{90}$ $\dfrac{3}{5}$

17. $\dfrac{96}{132}$ $\dfrac{8}{11}$

18. $\dfrac{165}{180}$ $\dfrac{11}{12}$

19. $\dfrac{60}{108}$ $\dfrac{5}{9}$

20. $\dfrac{112}{128}$ $\dfrac{7}{8}$

Write the numerator and denominator of each fraction as a product of prime factors. Then write the fraction in lowest terms.

Example: $\dfrac{24}{36}$ **Solution:** $\dfrac{24}{36} = \dfrac{\overset{1}{\cancel{2}} \cdot \overset{1}{\cancel{2}} \cdot 2 \cdot \overset{1}{\cancel{3}}}{\underset{1}{\cancel{2}} \cdot \underset{1}{\cancel{2}} \cdot \underset{1}{\cancel{3}} \cdot 3} = \dfrac{1 \cdot 1 \cdot 2 \cdot 1}{1 \cdot 1 \cdot 1 \cdot 3} = \dfrac{2}{3}$

21. $\dfrac{12}{18}$ $\dfrac{\overset{1}{\cancel{2}} \cdot 2 \cdot \overset{1}{\cancel{3}}}{\underset{1}{\cancel{2}} \cdot \underset{1}{\cancel{3}} \cdot 3} = \dfrac{2}{3}$

22. $\dfrac{20}{32}$ $\dfrac{\overset{1}{\cancel{2}} \cdot \overset{1}{\cancel{2}} \cdot 5}{\underset{1}{\cancel{2}} \cdot \underset{1}{\cancel{2}} \cdot 2 \cdot 2 \cdot 2} = \dfrac{5}{8}$

23. $\dfrac{35}{40}$ $\dfrac{\overset{1}{\cancel{5}} \cdot 7}{2 \cdot 2 \cdot 2 \cdot \underset{1}{\cancel{5}}} = \dfrac{7}{8}$

24. $\dfrac{36}{48}$ $\dfrac{\overset{1}{\cancel{2}} \cdot \overset{1}{\cancel{2}} \cdot \overset{1}{\cancel{3}} \cdot 3}{\underset{1}{\cancel{2}} \cdot \underset{1}{\cancel{2}} \cdot 2 \cdot 2 \cdot \underset{1}{\cancel{3}}} = \dfrac{3}{4}$

25. $\dfrac{90}{180}$ $\dfrac{\overset{1}{\cancel{2}} \cdot \overset{1}{\cancel{3}} \cdot \overset{1}{\cancel{3}} \cdot \overset{1}{\cancel{5}}}{\underset{1}{\cancel{2}} \cdot 2 \cdot \underset{1}{\cancel{3}} \cdot \underset{1}{\cancel{3}} \cdot \underset{1}{\cancel{5}}} = \dfrac{1}{2}$

26. $\dfrac{16}{64}$ $\dfrac{\overset{1}{\cancel{2}} \cdot \overset{1}{\cancel{2}} \cdot \overset{1}{\cancel{2}} \cdot \overset{1}{\cancel{2}}}{\underset{1}{\cancel{2}} \cdot \underset{1}{\cancel{2}} \cdot \underset{1}{\cancel{2}} \cdot \underset{1}{\cancel{2}} \cdot 2 \cdot 2} = \dfrac{1}{4}$

27. $\dfrac{36}{12}$ $\dfrac{\overset{1}{\cancel{2}} \cdot \overset{1}{\cancel{2}} \cdot \overset{1}{\cancel{3}} \cdot 3}{\underset{1}{\cancel{2}} \cdot \underset{1}{\cancel{2}} \cdot \underset{1}{\cancel{3}}} = 3$

28. $\dfrac{192}{48}$ $\dfrac{\overset{1}{\cancel{2}} \cdot \overset{1}{\cancel{2}} \cdot \overset{1}{\cancel{2}} \cdot \overset{1}{\cancel{2}} \cdot 2 \cdot 2 \cdot \overset{1}{\cancel{3}}}{\underset{1}{\cancel{2}} \cdot \underset{1}{\cancel{2}} \cdot \underset{1}{\cancel{2}} \cdot \underset{1}{\cancel{2}} \cdot \underset{1}{\cancel{3}}} = 4$

29. $\dfrac{77}{264}$ $\dfrac{7 \cdot \overset{1}{\cancel{11}}}{2 \cdot 2 \cdot 2 \cdot 3 \cdot \underset{1}{\cancel{11}}} = \dfrac{7}{24}$

30. $\dfrac{65}{234}$ $\dfrac{5 \cdot \overset{1}{\cancel{13}}}{2 \cdot 3 \cdot 3 \cdot \underset{1}{\cancel{13}}} = \dfrac{5}{18}$

✎ Writing   ◉ Conceptual   ▲ Challenging   ≈ Estimation

Decide whether the following pairs of fractions are equivalent or not equivalent.

Example:

$\frac{16}{20}$ and $\frac{36}{45}$

Solution: Find cross products.

$20 \cdot 36 = 720$

$\frac{16}{20} \bowtie \frac{36}{45}$

$16 \cdot 45 = 720$

Equivalent because $720 = 720$

The fractions are equivalent.

31. $\frac{1}{2}$ and $\frac{17}{34}$

equivalent

32. $\frac{2}{5}$ and $\frac{12}{30}$

equivalent

33. $\frac{5}{12}$ and $\frac{12}{30}$

not equivalent

34. $\frac{11}{16}$ and $\frac{32}{48}$

not equivalent

35. $\frac{15}{24}$ and $\frac{35}{52}$

not equivalent

36. $\frac{7}{11}$ and $\frac{9}{12}$

not equivalent

37. $\frac{14}{16}$ and $\frac{35}{40}$

equivalent

38. $\frac{9}{30}$ and $\frac{12}{40}$

equivalent

39. $\frac{7}{52}$ and $\frac{9}{40}$

not equivalent

40. $\frac{21}{28}$ and $\frac{54}{72}$

equivalent

41. $\frac{25}{30}$ and $\frac{65}{78}$

equivalent

42. $\frac{24}{72}$ and $\frac{30}{90}$

equivalent

43. What does it mean when a fraction is expressed in lowest terms? Give 3 examples.

A fraction is in lowest terms when the numerator and the denominator have no common factors other than 1.

Some examples: $\frac{1}{2}, \frac{3}{8}, \frac{2}{3}$

44. Explain what equivalent fractions are and give an example of a pair of equivalent fractions. Show that they are equivalent.

Two fractions are equivalent when they are of equal value. The fractions $\frac{2}{3}$ and $\frac{8}{12}$ are equivalent.

$\frac{2}{3} \bowtie \frac{8}{12}$ $3 \cdot 8 = 24$ $24 = 24$
$2 \cdot 12 = 24$

▲ *Write each fraction in lowest terms.*

45. $\frac{224}{256}$ $\frac{7}{8}$

46. $\frac{363}{528}$ $\frac{11}{16}$

47. $\frac{356}{178}$ $\frac{2}{1} = 2$

48. $\frac{175}{35}$ $\frac{5}{1} = 5$

Review and Prepare

*Find all the factors of each number. (For help, see **Section 2.3**.)*

49. 15   1, 3, 5, 15

50. 24   1, 2, 3, 4, 6, 8, 12, 24

51. 64   1, 2, 4, 8, 16, 32, 64

52. 55   1, 5, 11, 55

2.5 Multiplication of Fractions

OBJECTIVE ▶ Multiply the fractions $\frac{2}{3}$ and $\frac{1}{3}$.

$$\frac{2}{3} \cdot \frac{1}{3}$$

To multiply $\frac{2}{3}$ and $\frac{1}{3}$, we will find $\frac{2}{3}$ of $\frac{1}{3}$. Start with a figure showing $\frac{1}{3}$.

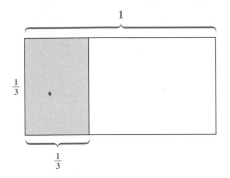

Next, take $\frac{2}{3}$ of the shaded area. (Here we are dividing $\frac{1}{3}$ into 3 equal parts and shading two of them.)

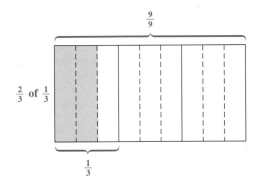

The shaded area in this second figure is equal to $\frac{2}{9}$ of the entire figure, so that

$$\frac{2}{3} \cdot \frac{1}{3} = \frac{2}{9}.$$

WORK PROBLEM 1 AT THE SIDE. ▶▶

The rule for multiplying fractions follows.

Multiplying Fractions

Multiply two fractions by multiplying the numerators and multiplying the denominators.

Use this rule to find the product of $\frac{2}{3}$ and $\frac{1}{3}$ $\left(\frac{2}{3} \text{ of } \frac{1}{3}\right)$.

$$\frac{2}{3} \cdot \frac{1}{3} = \frac{2 \cdot 1}{3 \cdot 3} \qquad \text{Multiply numerators.}$$
$$\text{Multiply denominators.}$$
$$= \frac{2}{9}$$

You will see that $\frac{2}{9}$ is in lowest terms.

$$\frac{2}{3} \cdot \frac{1}{3} = \frac{2 \cdot 1}{3 \cdot 3} = \frac{2}{9} \qquad \begin{array}{l} \leftarrow 2 \cdot 1 = 2 \\ \leftarrow 3 \cdot 3 = 9 \end{array}$$

OBJECTIVES

1 ▶ Multiply fractions.
2 ▶ Use cancellation.
3 ▶ Multiply a fraction and a whole number.
4 ▶ Find the area of a rectangle.

FOR EXTRA HELP

Tutorial   Tape 3   SSM, Sec. 2.5

1. Use these figures to find $\frac{1}{4}$ of $\frac{1}{2}$.

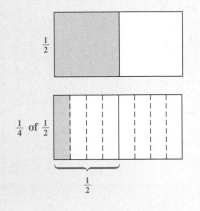

ANSWERS

1. $\frac{1}{8}$

125

2. Multiply. Write answers in lowest terms.

(a) $\dfrac{3}{4} \cdot \dfrac{1}{2}$

(b) $\dfrac{2}{5} \cdot \dfrac{2}{3}$

(c) $\dfrac{1}{4} \cdot \dfrac{5}{9} \cdot \dfrac{1}{2}$

(d) $\dfrac{1}{2} \cdot \dfrac{3}{4} \cdot \dfrac{3}{8}$

E X A M P L E 1   Multiplying Fractions

Multiply.

(a) $\dfrac{5}{8} \cdot \dfrac{3}{4}$

Multiply the numerators and multiply the denominators.

$$\frac{5}{8} \cdot \frac{3}{4} = \frac{5 \cdot 3}{8 \cdot 4} = \frac{15}{32}$$

Notice that there are no common factors other than 1 for 15 and 32, so the answer is in lowest terms.

(b) $\dfrac{4}{7} \cdot \dfrac{2}{5}$

$$\frac{4}{7} \cdot \frac{2}{5} = \frac{4 \cdot 2}{7 \cdot 5} = \frac{8}{35}$$

(c) $\dfrac{5}{8} \cdot \dfrac{3}{4} \cdot \dfrac{1}{2}$

$$\frac{5}{8} \cdot \frac{3}{4} \cdot \frac{1}{2} = \frac{5 \cdot 3 \cdot 1}{8 \cdot 4 \cdot 2} = \frac{15}{64}$$

 WORK PROBLEM 2 AT THE SIDE.

OBJECTIVE **2**   It is often easier to cancel before multiplying, as shown in Example 2.

E X A M P L E 2   Understanding Cancellation

Multiply $\frac{5}{6}$ and $\frac{9}{10}$.

$$\frac{5}{6} \cdot \frac{9}{10} = \frac{5 \cdot 9}{6 \cdot 10} = \frac{45}{60} \quad \text{(Not in lowest terms)}$$

The numerator and denominator have a common factor other than 1, so write the prime factorization of each number.

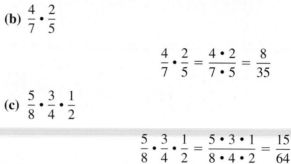

$$\frac{5}{6} \cdot \frac{9}{10} = \frac{5 \cdot 9}{6 \cdot 10} = \frac{5 \cdot 3 \cdot 3}{2 \cdot 3 \cdot 2 \cdot 5}$$

Next, cancel common factors.

$$\frac{5}{6} \cdot \frac{9}{10} = \frac{5 \cdot 9}{6 \cdot 10} = \frac{\overset{1}{\cancel{5}} \cdot \overset{1}{\cancel{3}} \cdot 3}{2 \cdot \underset{1}{\cancel{3}} \cdot 2 \cdot \underset{1}{\cancel{5}}}$$

Finally, multiply the remaining factors in the numerator and in the denominator.

$$\frac{5}{6} \cdot \frac{9}{10} = \frac{1 \cdot 1 \cdot 3}{2 \cdot 1 \cdot 2 \cdot 1} = \frac{3}{4} \quad \text{(Lowest terms)}$$

ANSWERS

2. (a) $\dfrac{3}{8}$ (b) $\dfrac{4}{15}$ (c) $\dfrac{5}{72}$ (d) $\dfrac{9}{64}$

As a shortcut, instead of writing the prime factorization of each number, find the product of $\frac{5}{6}$ and $\frac{9}{10}$ as follows.

First, divide both 5 and 10 by 5.
$$\frac{\overset{1}{\cancel{5}}}{6} \cdot \frac{9}{\underset{2}{\cancel{10}}}$$

Next, divide both 6 and 9 by 3.
$$\frac{\overset{1}{\cancel{5}}}{\underset{2}{\cancel{6}}} \cdot \frac{\overset{3}{\cancel{9}}}{\underset{2}{\cancel{10}}}$$

Finally, multiply.
$$\frac{1 \cdot 3}{2 \cdot 2} = \frac{3}{4}$$

> **Note**
> During **cancellation** you are dividing a numerator and a denominator. Be certain that you divide a numerator and a denominator by the same number. If you are able to do all possible cancellations your answer will be in lowest terms.

EXAMPLE 3 Using Cancellation

Use cancellation to multiply. Write in lowest terms or as mixed numbers.

(a) $\frac{6}{11} \cdot \frac{7}{8}$

Divide both 6 and 8 by 2. Next, multiply.

$$\frac{\overset{3}{\cancel{6}}}{11} \cdot \frac{7}{\underset{4}{\cancel{8}}} = \frac{3 \cdot 7}{11 \cdot 4} = \frac{21}{44} \qquad \text{Lowest terms}$$

(b) $\frac{7}{10} \cdot \frac{20}{21}$

Divide 7 and 21 by 7, and divide 10 and 20 by 10.

$$\frac{\overset{1}{\cancel{7}}}{\underset{1}{\cancel{10}}} \cdot \frac{\overset{2}{\cancel{20}}}{\underset{3}{\cancel{21}}} = \frac{1 \cdot 2}{1 \cdot 3} = \frac{2}{3} \qquad \text{Lowest terms}$$

(c) $\frac{35}{12} \cdot \frac{32}{25}$

$$\frac{\overset{7}{\cancel{35}}}{\underset{3}{\cancel{12}}} \cdot \frac{\overset{8}{\cancel{32}}}{\underset{5}{\cancel{25}}} = \frac{7 \cdot 8}{3 \cdot 5} = \frac{56}{15} \quad \text{or} \quad 3\frac{11}{15} \qquad \text{Mixed number}$$

(d) $\frac{2}{3} \cdot \frac{8}{15} \cdot \frac{3}{4}$

$$\frac{\overset{1}{\cancel{2}}}{\underset{1}{\cancel{3}}} \cdot \frac{\overset{4}{\cancel{8}}}{15} \cdot \frac{\overset{1}{\cancel{3}}}{\underset{\underset{1}{2}}{\cancel{4}}} = \frac{1 \cdot 4 \cdot 1}{1 \cdot 15 \cdot 1} = \frac{4}{15}$$

Cancellation is especially helpful when the fractions involve large numbers.

3. Use cancellation to find each of the following.

(a) $\dfrac{3}{4} \cdot \dfrac{2}{3}$

(b) $\dfrac{6}{11} \cdot \dfrac{33}{21}$

(c) $\dfrac{20}{4} \cdot \dfrac{3}{40} \cdot \dfrac{1}{3}$

(d) $\dfrac{18}{17} \cdot \dfrac{1}{36} \cdot \dfrac{2}{3}$

4. Multiply. Write all answers in lowest terms.

(a) $8 \cdot \dfrac{1}{8}$

(b) $12 \cdot \dfrac{3}{4} \cdot \dfrac{5}{3}$

(c) $\dfrac{7}{10} \cdot 50$

(d) $\dfrac{5}{11} \cdot 99 \cdot \dfrac{3}{25}$

ANSWERS

3. **(a)** $\dfrac{\overset{1}{\cancel{3}}}{\cancel{4}} \cdot \dfrac{\overset{1}{\cancel{2}}}{\cancel{3}} = \dfrac{1}{2}$ **(b)** $\dfrac{\overset{2}{\cancel{6}}}{\cancel{11}} \cdot \dfrac{\overset{3}{\cancel{33}}}{\cancel{21}} = \dfrac{6}{7}$

(c) $\dfrac{\overset{1}{\cancel{20}}}{\cancel{4}} \cdot \dfrac{\overset{1}{\cancel{3}}}{\cancel{40}} \cdot \dfrac{1}{\cancel{3}} = \dfrac{1}{8}$

(d) $\dfrac{\overset{1}{\cancel{18}}}{17} \cdot \dfrac{1}{\cancel{36}} \cdot \dfrac{\cancel{2}}{\cancel{3}} = \dfrac{1}{51}$

4. **(a)** 1 **(b)** 15 **(c)** 35 **(d)** $\dfrac{27}{5}$ or $5\dfrac{2}{5}$

Note
There is no specific order that must be used in cancellation as long as both a numerator and a denominator are divided by the same number.

◀◀ **WORK PROBLEM 3 AT THE SIDE.**

OBJECTIVE ▶ 3 The rule for multiplying a fraction and a whole number follows.

Multiplying a Whole Number and a Fraction

Multiply a whole number and a fraction by writing the whole number as a fraction with a denominator of 1.

For example, write the whole numbers 8, 10, and 25 as follows.

$$8 = \dfrac{8}{1}, \quad 10 = \dfrac{10}{1}, \quad \text{and} \quad 25 = \dfrac{25}{1}$$

EXAMPLE 4   Multiplying by a Whole Number

Multiply. Write all answers in lowest terms or whole numbers where possible.

(a) $8 \cdot \dfrac{3}{4}$

Write 8 as $\dfrac{8}{1}$ and multiply.

$$8 \cdot \dfrac{3}{4} = \dfrac{\overset{2}{\cancel{8}}}{1} \cdot \dfrac{3}{\cancel{4}} = \dfrac{2 \cdot 3}{1 \cdot 1} = \dfrac{6}{1} = 6$$

(b) $12 \cdot \dfrac{5}{6}$

$$12 \cdot \dfrac{5}{6} = \dfrac{\overset{2}{\cancel{12}}}{1} \cdot \dfrac{5}{\cancel{6}} = \dfrac{2 \cdot 5}{1 \cdot 1} = \dfrac{10}{1} = 10$$

◀◀ **WORK PROBLEM 4 AT THE SIDE.**

OBJECTIVE ▶ 4 To find the area of a rectangle (the amount of space inside the rectangle), use the following formula.

The Area of a Rectangle

The area of a rectangle is equal to the length multiplied by the width.

$$\textbf{area} = \textbf{length} \cdot \textbf{width}$$

For example, the rectangle shown here has an area of 12 square feet.

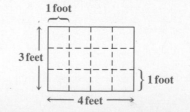

area = length • width
area = 4 feet • 3 feet
area = 12 square feet

─ E X A M P L E 5   **Applying Fraction Skills**

Find the area of each rectangle.

(a) [shaded rectangle] $\frac{3}{4}$ foot

$\frac{11}{12}$ foot

$$\text{area} = \text{length} \cdot \text{width}$$

$$\text{area} = \frac{11}{12} \cdot \frac{3}{4}$$

$$= \frac{11}{\underset{4}{\cancel{12}}} \cdot \frac{\overset{1}{\cancel{3}}}{4} \qquad \text{Cancel.}$$

$$= \frac{11}{16} \text{ square foot}$$

(b) a rectangle, $\frac{7}{9}$ inch by $\frac{3}{14}$ inch

Multiply the length and width.

$$\text{area} = \frac{7}{9} \cdot \frac{3}{14}$$

$$= \frac{\overset{1}{\cancel{7}}}{\underset{3}{\cancel{9}}} \cdot \frac{\overset{1}{\cancel{3}}}{\underset{2}{\cancel{14}}} \qquad \text{Cancel.}$$

$$= \frac{1}{6} \text{ square inch}$$

WORK PROBLEM 5 AT THE SIDE. ▶▶

5. Find the area of each rectangle.

(a) [rectangle] $\frac{1}{3}$ **yard**

$\frac{3}{4}$ **yard**

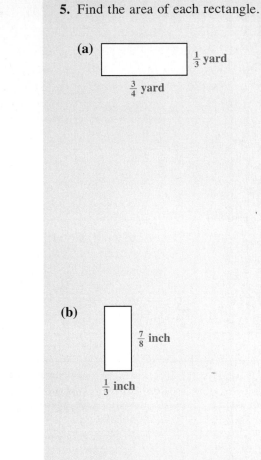

(b) [rectangle] $\frac{7}{8}$ inch

$\frac{1}{3}$ inch

(c) a rectangle, $\frac{7}{5}$ mile by $\frac{5}{8}$ mile

ANSWERS

5. (a) $\frac{1}{4}$ square yard

(b) $\frac{7}{24}$ square inch

(c) $\frac{7}{8}$ square mile
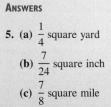

NUMBERS IN THE
Real World *collaborative investigations*

Math teachers attending conferences in Dallas, Texas and Tampa Bay, Florida received the following information about hotel rates:

	Distance from Convention Center	Single	Double/ Twin	Triple	Quad
Dallas, Texas February 12–14					
Wyndham Anatole	3 blocks	$135	$145	$155	$165
Wilson World Market Center	$3\frac{1}{2}$ blocks	$105	$105	$105	$105
Tampa Bay, Florida March 5–7					
Wyndham Harbor Island	3 blocks	$153	$168	$183	$198
Hyatt Regency	2 blocks	$132	$142	$152	$162
Club Hotel	$2\frac{1}{2}$ blocks	$99	$99	$109	$119

The double/twin rate is for 2 people sharing a room. So each person pays $\frac{1}{2}$ of the cost. Multiply the double/twin rate by $\frac{1}{2}$ to find each person's cost at each hotel.
Wyndham Anatole 72\frac{1}{2}$; Wilson 52\frac{1}{2}$; Wyndham Harbor $84; Hyatt $71; Club 49\frac{1}{2}$

Which hotel has the cheapest double/twin rate? **Club Hotel in Tampa**

Do you see a shortcut for finding $\frac{1}{2}$ of a number? Describe the shortcut. **Divide by 2**

Why does multiplying by $\frac{1}{2}$ give the same result as dividing by 2?
The 2 in $\frac{1}{2}$ means that the whole $1\frac{1}{2}$ has been cut into 2 equal parts. The 1 in $\frac{1}{2}$ means you are interested in only one of the parts. So dividing by 2 gives the same result.

The triple rate is for 3 people. What fraction of the room cost does each person pay? $\frac{1}{3}$

The quad rate is for 4 people. What fraction of the room cost does each person pay? $\frac{1}{4}$

Multiply each triple rate by $\frac{1}{3}$ and each quad rate by $\frac{1}{4}$ to find each person's cost.
Triple: Wyndham Anatole, 51\frac{2}{3}$; Wilson $35; Wyndham Harbor $61; Hyatt 50\frac{2}{3}$; Club 36\frac{1}{3}$

Quad: Wyndham Anatole 41\frac{1}{4}$; Wilson 26\frac{1}{4}$; Wyndham Harbor 49\frac{1}{2}$; Hyatt 40\frac{1}{2}$;

Club 29\frac{3}{4}$

Which hotel has the cheapest triple rate? **Wilson** The cheapest quad rate? **Wilson**

What shortcut could you use for finding $\frac{1}{3}$ of a number? For finding $\frac{1}{4}$ of a number?
Divide by 3; divide by 4

Name a city you would like to visit. How could you get information on room rates at various hotels in that city? **Some options are: call a travel agent; check out a travel guidebook; write or call the tourist information bureau or Chamber of Commerce in that city; look at hotel Web sites on the Internet.**

2.5 Exercises

Multiply. Write all answers in lowest terms.

Example: $\dfrac{9}{16} \cdot \dfrac{8}{27} \cdot \dfrac{9}{10}$ **Solution:** $\dfrac{\overset{1}{\cancel{9}}}{\underset{2}{\cancel{16}}} \cdot \dfrac{\overset{1}{\cancel{8}}}{\underset{3}{\cancel{27}}} \cdot \dfrac{\overset{3}{\cancel{9}}}{10} = \dfrac{1 \cdot 1 \cdot 3}{2 \cdot 1 \cdot 10} = \dfrac{3}{20}$

1. $\dfrac{3}{8} \times \dfrac{1}{2}$ $\dfrac{3}{16}$

2. $\dfrac{2}{3} \times \dfrac{5}{8}$ $\dfrac{5}{12}$

3. $\dfrac{2}{5} \times \dfrac{2}{3}$ $\dfrac{4}{15}$

4. $\dfrac{1}{3} \times \dfrac{2}{5}$ $\dfrac{2}{15}$

5. $\dfrac{3}{8} \cdot \dfrac{12}{5}$ $\dfrac{9}{10}$

6. $\dfrac{4}{9} \cdot \dfrac{12}{7}$ $\dfrac{16}{21}$

7. $\dfrac{5}{6} \cdot \dfrac{12}{25} \cdot \dfrac{3}{4}$ $\dfrac{3}{10}$

8. $\dfrac{7}{8} \cdot \dfrac{16}{21} \cdot \dfrac{1}{2}$ $\dfrac{1}{3}$

9. $\dfrac{3}{4} \cdot \dfrac{5}{6} \cdot \dfrac{2}{3}$ $\dfrac{5}{12}$

10. $\dfrac{2}{5} \cdot \dfrac{3}{8} \cdot \dfrac{2}{3}$ $\dfrac{1}{10}$

11. $\dfrac{9}{22} \cdot \dfrac{11}{16}$ $\dfrac{9}{32}$

12. $\dfrac{5}{12} \cdot \dfrac{7}{10}$ $\dfrac{7}{24}$

13. $\dfrac{21}{30} \cdot \dfrac{5}{7}$ $\dfrac{1}{2}$

14. $\dfrac{6}{11} \cdot \dfrac{22}{15}$ $\dfrac{4}{5}$

15. $\dfrac{14}{25} \cdot \dfrac{65}{48} \cdot \dfrac{15}{28}$ $\dfrac{13}{32}$

16. $\dfrac{35}{64} \cdot \dfrac{32}{15} \cdot \dfrac{27}{72}$ $\dfrac{7}{16}$

17. $\dfrac{16}{25} \cdot \dfrac{35}{32} \cdot \dfrac{15}{64}$ $\dfrac{21}{128}$

18. $\dfrac{39}{42} \cdot \dfrac{7}{13} \cdot \dfrac{7}{24}$ $\dfrac{7}{48}$

Multiply. Write all answers in lowest terms; change answers to whole or mixed numbers where possible.

Example: $27 \cdot \dfrac{5}{9}$ **Solution:** $27 \cdot \dfrac{5}{9} = \dfrac{\overset{3}{27}}{1} \cdot \dfrac{5}{\underset{1}{9}} = \dfrac{3 \cdot 5}{1 \cdot 1} = \dfrac{15}{1} = \mathbf{15}$

19. $5 \cdot \dfrac{3}{5}$ 3

20. $20 \cdot \dfrac{3}{4}$ 15

21. $\dfrac{4}{9} \cdot 81$ 36

22. $\dfrac{2}{3} \cdot 48$ 32

23. $32 \cdot \dfrac{3}{8}$ 12

24. $30 \cdot \dfrac{3}{10}$ 9

25. $42 \cdot \dfrac{7}{10} \cdot \dfrac{5}{7}$ 21

26. $35 \cdot \dfrac{3}{5} \cdot \dfrac{1}{2}$ $10\dfrac{1}{2}$

27. $100 \cdot \dfrac{21}{50} \cdot \dfrac{3}{4}$ $31\dfrac{1}{2}$

28. $300 \cdot \dfrac{5}{6}$ 250

29. $\dfrac{3}{5} \cdot 400$ 240

30. $\dfrac{5}{9} \cdot 360$ 200

31. $\dfrac{3}{4} \cdot 363$ $272\dfrac{1}{4}$

32. $\dfrac{12}{25} \cdot 430$ $206\dfrac{2}{5}$

33. $\dfrac{28}{21} \cdot 640 \cdot \dfrac{15}{32}$ 400

34. $\dfrac{21}{13} \cdot 520 \cdot \dfrac{7}{20}$ 294

35. $\dfrac{54}{38} \cdot 684 \cdot \dfrac{5}{6}$ 810

36. $\dfrac{76}{43} \cdot 473 \cdot \dfrac{5}{19}$ 220

Find the area of each rectangle.

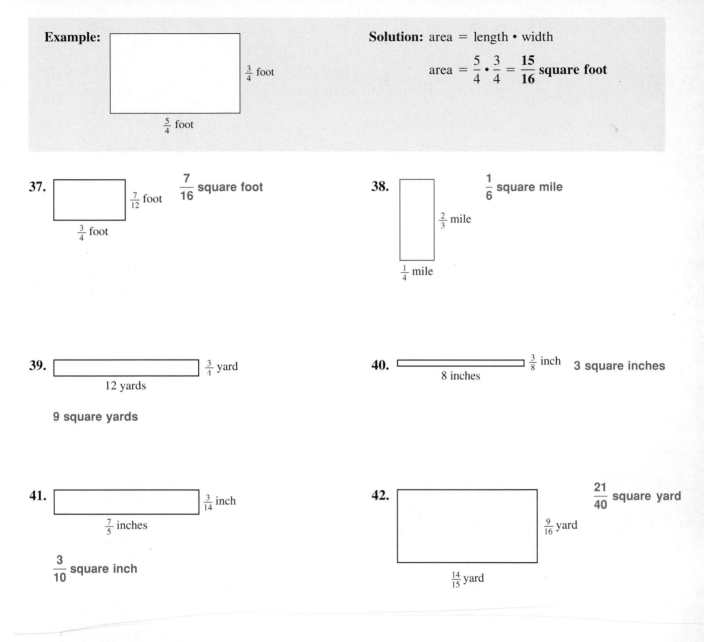

Example:

$\frac{3}{4}$ foot

$\frac{5}{4}$ foot

Solution: area = length • width

$$\text{area} = \frac{5}{4} \cdot \frac{3}{4} = \frac{15}{16} \text{ square foot}$$

37. $\frac{7}{16}$ **square foot**

$\frac{7}{12}$ foot

$\frac{3}{4}$ foot

38. $\frac{1}{6}$ **square mile**

$\frac{2}{3}$ mile

$\frac{1}{4}$ mile

39. $\frac{3}{4}$ yard

12 yards

9 square yards

40. $\frac{3}{8}$ inch **3 square inches**

8 inches

41. $\frac{3}{14}$ inch

$\frac{7}{5}$ inches

$\frac{3}{10}$ **square inch**

42. $\frac{21}{40}$ **square yard**

$\frac{9}{16}$ yard

$\frac{14}{15}$ yard

43. Write in your own words the rule for multiplying fractions. Make up an example problem of your own showing how this works.

Multiply the numerators and multiply the denominators. An example is

$$\frac{3}{4} \cdot \frac{1}{2} = \frac{3 \cdot 1}{4 \cdot 2} = \frac{3}{8}.$$

44. A useful shortcut when multiplying fractions is called cancellation. Describe how this works. Give an example to show how this works.

You must divide a numerator and a denominator by the same number. If you do all possible cancellations your answer will be in lowest terms. One example is

$$\frac{3}{4} \cdot \frac{2}{3} = \frac{3 \cdot 2}{4 \cdot 3} = \frac{\overset{1}{\cancel{3}} \cdot \overset{1}{\cancel{2}}}{\underset{2}{\cancel{4}} \cdot \underset{1}{\cancel{3}}} = \frac{1}{2}.$$

▲ *Solve each of the following application problems. Write answers in lowest terms or as mixed numbers.*

45. Find the floor area of a rabbit cage having a length of 2 yards and a width of $\frac{2}{3}$ yard.

$1\frac{1}{3}$ **square yards**

46. Find the area of the top of a bookcase having a length of 3 feet and a width of $\frac{11}{12}$ foot.

$2\frac{3}{4}$ **square feet**

47. A parcel of land measures $\frac{1}{2}$ mile by 2 miles. Find the total area of the parcel.

1 square mile

48. A motorcycle race course is $\frac{3}{4}$ mile wide by 6 miles long. Find the area of the race course.

$4\frac{1}{2}$ **square miles**

49. Parking lot A is $\frac{1}{4}$ mile long and $\frac{3}{16}$ mile wide while parking lot B is $\frac{3}{8}$ mile long and $\frac{1}{8}$ mile wide. Which parking lot has the larger area?

They are both the same size: $\frac{3}{64}$ **square mile**

50. The Rocking Horse Ranch is $\frac{3}{4}$ mile long and $\frac{1}{2}$ mile wide. The Silver Spur Ranch is $\frac{5}{8}$ mile long and $\frac{4}{5}$ mile wide. Which ranch has the larger area?

Rocking Horse Ranch is $\frac{3}{8}$ **square mile.**

Silver Spur Ranch is $\frac{1}{2}$ **square mile.**

Silver Spur Ranch has the larger area.

Review and Prepare

*Solve each of the following application problems. (For help, see **Section 1.9.**)*

51. If 795 cars enter a parking garage each day, how many cars enter the garage in a 365-day year?

290,175 cars

52. In 1986, the propellor plane *Voyager* made aviation history by flying nonstop around the world on a single tank of fuel. *Voyager* cruised at a speed of 122 miles per hour. Could *Voyager* have traveled across the country from Boston to San Francisco, a distance of 2603 miles, in a 24-hour period? How many miles would *Voyager* have flown in 24 hours?

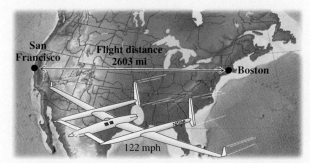

yes; 2928 miles

2.6 Applications of Multiplication

OBJECTIVE ▶ Many application problems are solved by multiplying fractions. Use the following indicator words for multiplication.

product
double
triple
times
of
twice
twice as much

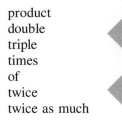

Look for these indicator words in the following examples.

OBJECTIVE

▶ Solve application problems using multiplication.

FOR EXTRA HELP

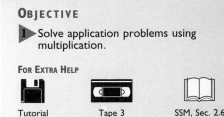

Tutorial   Tape 3   SSM, Sec. 2.6

EXAMPLE 1   Applying Indicator Words

Lois Stevens gives $\frac{1}{10}$ of her income to her church. One month she earned $1980. How much did she give to the church that month?

Approach   To find the amount given to the church, the fraction $\frac{1}{10}$ must be multiplied by the monthly earnings ($1980).

Solution   The indicator word is *of:* Stevens gave $\frac{1}{10}$ *of* her income. The word *of* indicates multiplication, so find the amount given to the church by multiplying $\frac{1}{10}$ and $1980.

$$\text{amount} = \frac{1}{\overset{}{\underset{1}{10}}} \cdot \frac{\overset{198}{\cancel{1980}}}{1} = \frac{198}{1} = 198$$

Stevens gave $198 to the church that month.

WORK PROBLEM 1 AT THE SIDE. ▶▶

EXAMPLE 2   Solving a Fraction Application Problem

Of the 42 students in a biology class, $\frac{2}{3}$ went on a field trip. How many went on the trip?

Approach   Find the number of students who went on the field trip by multiplying the fraction $\frac{2}{3}$ by the number of students in the class (42).

Solution   Reword the problem to read

$$\frac{2}{3} \text{ of the students went.}$$
$$\uparrow$$
Indicator word

Find the number who went by multiplying $\frac{2}{3}$ and 42.

$$\text{number who went} = \frac{2}{3} \cdot 42$$

$$= \frac{2}{\underset{1}{3}} \cdot \overset{14}{\cancel{42}} = \frac{28}{1} = 28$$

28 students went on the trip.

1. **(a)** At Frink Chevrolet $\frac{1}{3}$ of the new car buyers purchase the extended warranty. If the dealership sold 8397 new cars last year, find the number of extended warranties sold.

(b) A retiring police officer will receive $\frac{5}{8}$ of her highest annual salary as retirement income. If her highest annual salary is $48,000, how much will she receive as retirement income?

ANSWERS

1. **(a)** 2799 warranties **(b)** $30,000

2. At one pharmacy, $\frac{3}{16}$ of the prescriptions are paid by a third party (insurance company paid). If 2816 prescriptions are filled, find the number paid by a third party.

◀◀ **WORK PROBLEM 2 AT THE SIDE.**

EXAMPLE 3   Finding a Fraction of a Fraction

In her will, a woman divides her estate into 6 equal parts. 5 of the 6 parts are given to relatives. Of the sixth part, $\frac{1}{3}$ goes to the Salvation Army. What fraction of her total estate goes to the Salvation Army?

Approach   To find the fraction of the estate going to the Salvation Army, the part not going to relatives $\left(\frac{1}{6}\right)$ is multiplied by the fractional part going to the Salvation Army $\left(\frac{1}{3}\right)$.

Solution   The Salvation Army gets $\frac{1}{3}$ **of** $\frac{1}{6}$.
↑
Indicator word

To find the fraction that the Salvation Army is to receive, multiply $\frac{1}{3}$ and $\frac{1}{6}$.

$$\text{fraction to Salvation Army} = \frac{1}{3} \cdot \frac{1}{6}$$

$$= \frac{1}{18}$$

The Salvation Army gets $\frac{1}{18}$ of the total estate.

◀◀ **WORK PROBLEM 3 AT THE SIDE.**

3. In a certain community $\frac{1}{3}$ of the residents speak a foreign language. Of those speaking a foreign language, $\frac{3}{4}$ speak Spanish. What fraction of the residents speak Spanish?

2.6 Exercises

Solve each of the following application problems. Look for indicator words.

Example: Of the 96 units in an apartment building, $\frac{5}{8}$ of the units have two bedrooms. How many units have two bedrooms?

Approach: To find the number of two-bedroom units, multiply the total number of units (96) by the fraction that are two-bedroom units $\left(\frac{5}{8}\right)$. The indicator word for multiplication is *of* in $\frac{5}{8}$ *of* the units.

Solution: number of two-bedroom units $= \frac{5}{8} \cdot 96$

$$= \frac{5}{\underset{1}{8}} \cdot \frac{\overset{12}{96}}{1}$$

$$= \frac{60}{1} = 60$$

There are **60 two-bedroom units.**

---

1. A file cabinet top is $\frac{3}{4}$ yard by $\frac{2}{3}$ yard. Find its area.

$\frac{1}{2}$ **square yard**

2. A dog bed is $\frac{7}{8}$ yard by $\frac{10}{9}$ yards. Find its area.

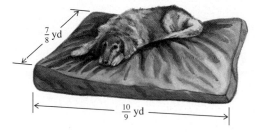

$\frac{35}{36}$ **square yard**

3. A laboratory tray is $\frac{4}{3}$ feet by $\frac{1}{2}$ foot. Find the area.

$\frac{2}{3}$ **square foot**

4. Al is helping Tim make a mahogony lamp table for Jill's birthday. Find the area of the top of the table if it is $\frac{4}{5}$ yard long by $\frac{3}{8}$ yard wide.

$\frac{3}{10}$ **square yard**

✍ Writing   ◉ Conceptual   ▲ Challenging   ≈ Estimation

5. According to *Forbes* magazine's listing of the 400 wealthiest Americans, 25 New England residents made the list. Of these 25 wealthy New Englanders, $\frac{2}{5}$ had over $700 million. How many New Englanders had over $700 million?

10 of them

6. A mini-market sells 2500 items, of which $\frac{3}{25}$ are classified as junk food. How many of the items are junk food?

300 items

7. Erica Green needs $2800 to go to Europe next summer. She earns $\frac{5}{8}$ of this amount during the summer. How much money does she earn during the summer?

$1750

8. Seth Torres paid $485 to have his transmission repaired. Of this amount, the transmission shop owner kept $\frac{1}{5}$ as profit. How much money did the owner keep?

$97

9. A school gives scholarships to $\frac{5}{24}$ of its 1800 freshmen. How many freshman students received scholarships?

375 students

10. Jason Todd estimates that it will cost him $8400, including living expenses, to attend college full time for one year. If he must earn $\frac{3}{4}$ of the cost and borrow the balance, find the amount that he must earn.

$6300

11. At the Garlic Festival Fun Run, $\frac{5}{12}$ of the runners are women. If there are 780 runners, how many are women?

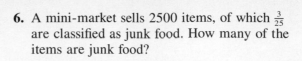

325 women

12. A hotel has 408 rooms. Of these rooms, $\frac{9}{17}$ are for non-smokers. How many rooms are for non-smokers?

216 rooms

The following table shows the earnings for the Gomes family last year and the circle graph shows how they spent their earnings. Use this information to solve Exercises 13–18.

Month	*Earnings*	*Month*	*Earnings*
January	$3050	July	$3160
February	$2875	August	$2355
March	$3325	September	$2780
April	$3020	October	$3675
May	$2880	November	$3310
June	$3265	December	$4305

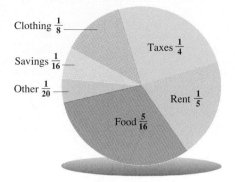

13. Find their total income for the year.

$38,000

14. Find the amount of their annual earnings that went to taxes.

$9500

15. Find the amount of their rent for the year.

$7600

16. How much did they spend for food during the year?

$11,875

17. Find their annual savings.

$2375

18. How much of their income was spent on clothing?

$4750

19. Here is how one student solved this multiplication problem. Find the error and solve the problem correctly.

$$\frac{9}{10} \times \frac{20}{21} = \frac{\overset{3}{\cancel{9}}}{\underset{1}{\cancel{10}}} \times \frac{\overset{2}{\cancel{20}}}{\underset{3}{\cancel{21}}} = \frac{6}{3} = 2$$

Solution is

$$\frac{9}{10} \times \frac{20}{21} = \frac{\overset{3}{\cancel{9}}}{\underset{1}{\cancel{10}}} \times \frac{\overset{2}{\cancel{20}}}{\underset{7}{\cancel{21}}} = \frac{6}{7}$$

20. When two whole numbers are multiplied, the product is always larger than the numbers being multiplied. When two common fractions are multiplied the answer is always smaller than the numbers being multiplied. Are these statements true? Why or why not?

Yes, the statements are true. Since whole numbers are 1 or greater, when you multiply, the product will always be greater than either of the numbers multiplied. But, when you multiply two common fractions, you are finding a fraction of a fraction, and the product will be smaller than either of the two common fractions.

▲ *Solve each of the following application problems.*

21. Pamela Denny is a waitress and earned $112 in one 8-hour day. How much money did she earn in 3 hours?

$42

22. Rita Ferrandino ran 20 miles in 4 hours. How far did she run in 3 hours?

15 miles

23. LaDonna Washington is running for city council. She needs to get $\frac{2}{3}$ of her votes from senior citizens. Ms. Washington will need 27,000 votes in all to win. How many votes does she need from voters other than the senior citizens?

9000 votes

24. The start-up cost of a Subs and Sandwich Shop is $32,000. If the bank will loan $\frac{9}{16}$ of the start up and the balance must be paid by the business owner, how much must be paid by the business owner?

$14,000

25. A will states that $\frac{7}{8}$ of the estate is to be divided among relatives. Of the remaining estate $\frac{1}{4}$ goes to the American Cancer Society. What fraction of the estate goes to the American Cancer Society?

$\dfrac{1}{32}$ **of the estate**

26. A couple has invested $\frac{1}{5}$ of their total investment in stocks. Of the remaining investment, $\frac{1}{8}$ is invested in bonds. What fraction of the total investment is invested in bonds?

$\dfrac{1}{10}$ **of the total investment**

Review and Prepare

▦ *Solve each of the following application problems. (For help, see Section 1.9.)*

27. There are 18 cholesterol test kits per carton. Find the number of cartons needed to supply 1332 test kits.

74 cartons

28. Your Mitsubishi Eclipse boasts 31 miles to the gallon on the highway. If you are planning a summer road trip of 2294 miles, how many gallons of gasoline will you use?

74 gallons

2.7 *Dividing Fractions*

As shown in Chapter 1, the division problem $12 \div 3$ asks how many 3's are in 12. In the same way, the divison problem $\frac{2}{3} \div \frac{1}{6}$ asks how many $\frac{1}{6}$'s are in $\frac{2}{3}$. Look at the figure.

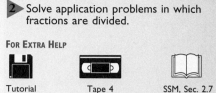
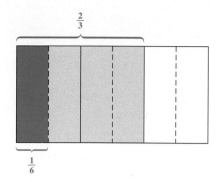

The figure shows that there are four of the $\frac{1}{6}$'s in $\frac{2}{3}$, or

$$\frac{2}{3} \div \frac{1}{6} = 4.$$

OBJECTIVE 1 ▶ Compare: $\quad \frac{2}{3} \div \frac{1}{6} = 4 \quad$ and $\quad \frac{2}{3} \cdot \frac{6}{1} = \frac{\overset{2}{\cancel{2}}}{\underset{1}{\cancel{3}}} \cdot \frac{\overset{2}{\cancel{6}}}{1} = \frac{4}{1} = 4$

Invert $\frac{1}{6}$ to get $\frac{6}{1}$.

Dividing Fractions

Divide two fractions by inverting the second fraction (divisor) and multiplying.

E X A M P L E   1   Dividing One Fraction by Another

Divide. Write answers in lowest terms.

(a) $\dfrac{7}{8} \div \dfrac{15}{16}$

Multiply.

$$\frac{7}{8} \div \frac{15}{16} = \frac{7}{8} \cdot \frac{16}{15} \quad \leftarrow \text{ Invert } \frac{15}{16} \text{ to get } \frac{16}{15}.$$

$$= \frac{7}{\underset{1}{\cancel{8}}} \cdot \frac{\overset{2}{\cancel{16}}}{15} \qquad \text{Cancel by dividing a numerator and a denominator by 8.}$$

$$= \frac{7 \cdot 2}{1 \cdot 15} \qquad \text{Multiply.}$$

$$= \frac{14}{15}$$

CONTINUED ON NEXT PAGE

1. Divide. Write all answers in lowest terms.

(a) $\dfrac{1}{4} \div \dfrac{3}{4}$

(b) $\dfrac{3}{8} \div \dfrac{5}{8}$

(c) $\dfrac{\frac{9}{10}}{\frac{3}{5}}$

(d) $\dfrac{\frac{5}{6}}{\frac{25}{24}}$

(b) $\dfrac{\frac{4}{5}}{\frac{3}{10}}$

$$\frac{\frac{4}{5}}{\frac{3}{10}} = \frac{4}{5} \div \frac{3}{10} \qquad \text{Rewrite by using the} \div \text{symbol for division.}$$

$$= \frac{4}{\underset{1}{5}} \cdot \frac{\overset{2}{10}}{3} \qquad \text{Invert, change "}\div\text{" to "•", and cancel.}$$

$$= \frac{4 \cdot 2}{1 \cdot 3} \qquad \text{Multiply.}$$

$$= \frac{8}{3} = 2\frac{2}{3} \qquad \text{Mixed number}$$

Note
Be certain that the divisor fraction is inverted before doing any cancelling.

◀◀ **WORK PROBLEM 1 AT THE SIDE.**

EXAMPLE 2 Dividing with a Whole Number

Divide. Write all answers in lowest terms.

(a) $5 \div \dfrac{1}{4}$

Write 5 as $\frac{5}{1}$. Next, invert $\frac{1}{4}$ and multiply.

$$5 \div \frac{1}{4} = \frac{5}{1} \cdot \frac{4}{1} \qquad \text{Invert } \tfrac{1}{4} \text{ to } \tfrac{4}{1}.$$

$$= \frac{5 \cdot 4}{1 \cdot 1} \qquad \text{Multiply.}$$

$$= \frac{20}{1} = 20 \qquad \text{Whole number}$$

(b) $\dfrac{2}{3} \div 6$

Write 6 as $\frac{6}{1}$. Next, invert $\frac{6}{1}$ and multiply.

$$\frac{2}{3} \div \frac{6}{1} = \frac{2}{3} \cdot \frac{1}{6}$$

$$\frac{\overset{1}{2}}{3} \cdot \frac{1}{\underset{3}{6}} = \frac{1 \cdot 1}{3 \cdot 3} \qquad \text{Cancel and multiply.}$$

$$= \frac{1}{9}$$

ANSWERS

1. (a) $\frac{1}{3}$ (b) $\frac{3}{5}$ (c) $1\frac{1}{2}$ (d) $\frac{4}{5}$

WORK PROBLEM 2 AT THE SIDE. ▶▶

OBJECTIVE **2**▶ Many application problems require division of fractions. Recall typical indicator words for division such as *goes into, per, divide, divided by, divided equally,* or *divided into.*

┌─
E X A M P L E 3   **Applying Fraction Skills**

Mary must fill a 12-gallon barrel with a chemical. She has only a $\frac{2}{3}$-gallon container to use. How many times must she fill the $\frac{2}{3}$-gallon container and empty it into the 12-gallon barrel?

Approach   To find the number of times Mary needs to fill the container, we need to divide the size of the barrel (12 gallons) by the size of the container $\left(\frac{2}{3} \text{ gallon}\right)$.

Solution   This problem can be solved by finding the number of times 12 can be divided by $\frac{2}{3}$.

$$12 \div \frac{2}{3} = \frac{12}{1} \cdot \frac{3}{2} \qquad \text{Invert } \frac{2}{3} \text{ to } \frac{3}{2} \text{ and change "} \div \text{" to "} \cdot \text{".}$$

$$= \frac{\overset{6}{\cancel{12}}}{1} \cdot \frac{3}{\underset{1}{\cancel{2}}} \qquad \text{Cancel and multiply.}$$

$$= \frac{18}{1} = 18$$

The container must be filled 18 times.
─┘

WORK PROBLEM 3 AT THE SIDE. ▶▶

┌─
E X A M P L E 4   **Applying Fraction Skills**

At the Happi-Time Day Care Center, $\frac{6}{7}$ of the total operating fund goes to classroom operation. If there are 18 classrooms, what fraction of the classroom operating amount does each classroom receive?

Approach   Since $\frac{6}{7}$ of the total operating fund must be split into 18 parts, we must find the fraction of the classroom operating amount received by each classroom. To do this we will divide the fraction of the total operating funds going to classroom operation $\left(\frac{6}{7}\right)$ by the number of classrooms (18).

└─ **CONTINUED ON NEXT PAGE**

2. Divide. Write all answers in lowest terms.

(a) $6 \div \dfrac{2}{3}$

(b) $9 \div \dfrac{3}{4}$

(c) $\dfrac{7}{8} \div 3$

(d) $\dfrac{7}{10} \div 3$

3. (a) How many times must a $\frac{2}{3}$-quart spray bottle be filled in order to use up 18 quarts of window cleaner?

(b) How many $\frac{3}{4}$-quart iced tea glasses may be filled from 24 quarts of iced tea?

ANSWERS

2. (a) 9 **(b)** 12 **(c)** $\dfrac{7}{24}$ **(d)** $\dfrac{7}{30}$

3. (a) 27 times **(b)** 32 glasses

4. (a) The Sweepstakes Lottery pays out $\frac{7}{8}$ of the total revenue to 14 top winners. What fraction of the total revenue does each winner receive?

Solution   This problem can be solved by dividing $\frac{6}{7}$ by 18.

$$\frac{6}{7} \div 18 = \frac{6}{7} \div \frac{18}{1}$$

$$= \frac{\overset{1}{\cancel{6}}}{7} \cdot \frac{1}{\underset{3}{\cancel{18}}} \qquad \text{Invert, change "}\div\text{" to "}\bullet\text{", and cancel.}$$

$$= \frac{1}{21} \qquad \text{Multiply.}$$

Each classroom receives $\frac{1}{21}$ of the total operating funds.

◀◀ **WORK PROBLEM 4 AT THE SIDE.**

(b) The 4 top-performing students at Tulsa Community College will divide $\frac{1}{3}$ of the scholarship money awarded to students. What fraction of the scholarship money will each of these top students receive?

ANSWERS

4. (a) $\frac{1}{16}$ of total revenue

(b) $\frac{1}{12}$ of the scholarship money

2.7 Exercises

Divide. Write all answers in lowest terms; change answers to whole or mixed numbers where possible.

Example: $\dfrac{3}{4} \div \dfrac{1}{2}$ **Solution:** $\dfrac{3}{\underset{2}{4}} \cdot \dfrac{\overset{1}{2}}{1} = \dfrac{3}{2} = 1\dfrac{1}{2}$

1. $\dfrac{1}{6} \div \dfrac{1}{3}$ $\dfrac{1}{2}$

2. $\dfrac{1}{2} \div \dfrac{2}{3}$ $\dfrac{3}{4}$

3. $\dfrac{7}{8} \div \dfrac{1}{3}$ $2\dfrac{5}{8}$

4. $\dfrac{3}{4} \div \dfrac{5}{8}$ $1\dfrac{1}{5}$

5. $\dfrac{3}{4} \div \dfrac{5}{3}$ $\dfrac{9}{20}$

6. $\dfrac{4}{5} \div \dfrac{9}{4}$ $\dfrac{16}{45}$

7. $\dfrac{7}{12} \div \dfrac{14}{15}$ $\dfrac{5}{8}$

8. $\dfrac{13}{20} \div \dfrac{4}{5}$ $\dfrac{13}{16}$

9. $\dfrac{\frac{7}{9}}{\frac{7}{36}}$ 4

10. $\dfrac{\frac{15}{32}}{\frac{5}{64}}$ 6

11. $\dfrac{\frac{36}{35}}{\frac{15}{14}}$ $\dfrac{24}{25}$

12. $\dfrac{\frac{28}{15}}{\frac{21}{5}}$ $\dfrac{4}{9}$

13. $6 \div \dfrac{2}{3}$ 9

14. $7 \div \dfrac{1}{4}$ 28

15. $\dfrac{15}{\frac{2}{3}}$ $22\dfrac{1}{2}$

16. $\dfrac{\frac{6}{5}}{\frac{5}{8}}$ $9\dfrac{3}{5}$

17. $\dfrac{\frac{4}{7}}{8}$ $\dfrac{1}{14}$

18. $\dfrac{\frac{11}{5}}{3}$ $\dfrac{11}{15}$

✍ Writing   ◉ Conceptual   ▲ Challenging   ≈ Estimation

Solve each of the following application problems by using division.

19. Ms. Shaffer has a piece of property with an area that is $\frac{8}{9}$ acre. She wishes to divide it into 4 equal parts for her children. How many acres of land will each child get?

$\frac{2}{9}$ acre

20. Joyce Chen wants to make vests to sell at a craft fair. Each vest requires $\frac{3}{4}$ yard of material. She has 36 yards of material. Find the number of vests she can make.

48 vests

21. Lisa Fuller wants to measure 4 cups of rice and only has a $\frac{1}{4}$-cup measuring cup. How many of these cups does she need to make 4 cups?

16 measuring cups

22. Robert Cockrill has 10 quarts of lubricating oil. If each lubricating reservoir holds $\frac{1}{3}$ quart of oil, how many reservoirs can be filled?

30 reservoirs

23. How many $\frac{1}{8}$-ounce eye drop dispensers can be filled with 11 ounces of eye drops?

88 dispensers

24. Each guest at a party will eat $\frac{5}{16}$ pound of peanuts. How many guests may be served with 10 pounds of peanuts?

32 guests

25. Pam Trizlia had a small pickup truck that would carry $\frac{2}{3}$-cord of firewood. Find the number of trips needed to deliver 40 cords of wood.

60 trips

26. Manuel Servin has a 200-yard roll of weather stripping material. Find the number of pieces of weather stripping $\frac{5}{8}$-yard in length that may be cut from the roll.

320 pieces

27. A batch of double chocolate chip cookies requires $\frac{3}{4}$ pound of chocolate chips. If you have 9 pounds of chocolate chips, how many batches of cookies can be made?

12 batches

28. Find the number of $\frac{4}{5}$-pound cartons that can be filled with 56 pounds of baking soda.

70 cartons

29. Your classmate is confused on how to divide by a fraction. Write a short note telling how this should be done.

You can divide two fractions by inverting the second fraction (divisor) and multiplying.

30. If you multiply common fractions the answer is smaller than the fractions multiplied. When you divide by a common fraction is the answer smaller than the numbers in the problem? Prove your answer with examples.

Sometimes the answer is smaller and sometimes it is larger.

$$\frac{1}{4} \div \frac{7}{8} = \frac{1}{\underset{1}{\cancel{4}}} \times \frac{\overset{2}{\cancel{8}}}{7} = \frac{2}{7} \text{ (smaller)}$$

$$\frac{1}{2} \div \frac{1}{4} = \frac{1}{\underset{1}{\cancel{2}}} \times \frac{\overset{2}{\cancel{4}}}{1} = \frac{2}{1} = 2 \text{ (larger)}$$

31. An airplane has flown $\frac{7}{8}$ of the distance to Phoenix. If it has flown 756 miles so far, find the number of miles remaining until it reaches its destination.

108 miles

32. Sheila has been working on a job for 63 hours. The job is $\frac{7}{9}$ finished. How many *more* hours must she work to finish the job?

18 hours

33. The Bridge Lighting Committee has raised $\frac{7}{8}$ of the funds necessary for their lighting project. If this amounts to $840,000, how much additional money must be raised?

$120,000

34. A mountain guide has used pack animals for $\frac{14}{15}$ of a trip and must finish the trip on foot. The distance covered with pack animals is 98 miles. Find the number of miles to be completed on foot.

7 miles

Review and Prepare

*Write each mixed number as an improper fraction. (For help, see **Section 2.2**.)*

35. $2\frac{3}{8}$ $\frac{19}{8}$

36. $4\frac{1}{4}$ $\frac{17}{4}$

37. $12\frac{1}{2}$ $\frac{25}{2}$

38. $26\frac{5}{8}$ $\frac{213}{8}$

39. $120\frac{4}{5}$ $\frac{604}{5}$

40. $182\frac{5}{6}$ $\frac{1097}{6}$

2.8 Multiplication and Division of Mixed Numbers

OBJECTIVE ▶ 1 When multiplying mixed numbers, it is a good idea to estimate the answer first. Then multiply the mixed numbers by using the following steps.

Multiplying Mixed Numbers
Step 1 **Change** each mixed number to an improper fraction.
Step 2 **Multiply** as fractions.
Step 3 Write the answer in lowest terms and change to a mixed number or whole number where possible.

To estimate the answer, round each mixed number to the nearest whole number. If the numerator is *half* of the denominator or *more*, round up the whole number part. If the numerator is *less* than half the denominator, leave the whole number as it is.

$$1\frac{5}{8} \quad \begin{matrix}\leftarrow \text{ 5 is more than 4} \\ \leftarrow \text{ Half of 8 is 4}\end{matrix} \qquad 1\frac{5}{8} \text{ rounds up to 2}$$

$$3\frac{2}{5} \quad \begin{matrix}\leftarrow \text{ 2 is less than } 2\frac{1}{2} \\ \leftarrow \text{ Half of 5 is } 2\frac{1}{2}\end{matrix} \qquad 3\frac{2}{5} \text{ rounds to 3}$$

> **WORK PROBLEM I AT THE SIDE.** ▶▶

E X A M P L E I Multiplying Mixed Numbers

First estimate the answer. Then multiply to get an exact answer. Write all answers in lowest terms.

(a) $2\frac{1}{2} \cdot 3\frac{1}{5}$

Estimate the answer by rounding the mixed numbers.

$$2\frac{1}{2} \text{ rounds to 3} \quad \text{and} \quad 3\frac{1}{5} \text{ rounds to 3}$$

$$3 \cdot 3 = 9 \quad \text{Estimated answer}$$

To find the exact answer change each mixed number to an improper fraction.

$$\text{Step 1} \qquad 2\frac{1}{2} = \frac{5}{2} \quad \text{and} \quad 3\frac{1}{5} = \frac{16}{5}$$

Next, multiply.

$$2\frac{1}{2} \cdot 3\frac{1}{5} = \frac{5}{2} \cdot \frac{16}{5} = \frac{\overset{1}{\cancel{5}}}{\underset{1}{\cancel{2}}} \cdot \frac{\overset{8}{\cancel{16}}}{\underset{1}{\cancel{5}}} = \frac{1 \cdot 8}{1 \cdot 1} = \frac{8}{1} = 8$$

The estimated answer is 9 and the exact answer is 8. The exact answer is reasonable.

CONTINUED ON NEXT PAGE

OBJECTIVES

▶ 1 Estimate the answer and multiply mixed numbers.

▶ 2 Estimate the answer and divide mixed numbers.

▶ 3 Solve application problems with mixed numbers.

FOR EXTRA HELP

Tutorial   Tape 4   SSM, Sec. 2.8

1. Round each mixed number to the nearest whole number.

(a) $2\frac{3}{4}$

(b) $6\frac{3}{8}$

(c) $4\frac{2}{3}$

(d) $1\frac{7}{10}$

(e) $3\frac{1}{2}$

(f) $5\frac{4}{9}$

ANSWERS

1. (a) 3 **(b)** 6 **(c)** 5 **(d)** 2 **(e)** 4 **(f)** 5

2. ≈ First estimate the answer. Then multiply to find the exact answer. Write answers in lowest terms.

(a) $2\frac{1}{4}$ • $7\frac{1}{3}$

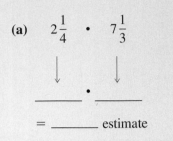

= _____ estimate

(b) $4\frac{1}{2}$ • $1\frac{2}{3}$

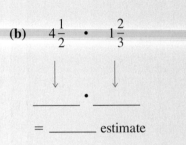

= _____ estimate

(c) $3\frac{3}{5}$ • $4\frac{4}{9}$

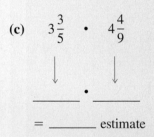

= _____ estimate

(d) $3\frac{1}{5}$ • $5\frac{3}{8}$

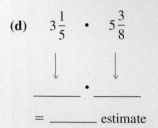

= _____ estimate

(b) $3\frac{5}{8} \cdot 4\frac{4}{5}$

$3\frac{5}{8}$ rounds to 4   and $4\frac{4}{5}$ rounds to 5

4 • 5 = 20   Estimated answer

Now find the exact answer.

Step 1   Step 2

$$3\frac{5}{8} \cdot 4\frac{4}{5} = \frac{29}{8} \cdot \frac{24}{5} = \frac{29}{\overset{}{\underset{1}{8}}} \cdot \frac{\overset{3}{24}}{5} = \frac{29 \cdot 3}{1 \cdot 5} = \frac{87}{5}$$

As a mixed number,

Step 3

$$\frac{87}{5} = 17\frac{2}{5}.$$

The estimate was 20, so the exact answer is reasonable.

(c) $1\frac{3}{5} \cdot 3\frac{1}{3}$

$1\frac{3}{5}$ rounds to 2   and $3\frac{1}{3}$ rounds to 3

2 • 3 = 6   Estimated answer

The exact answer is

$$1\frac{3}{5} \cdot 3\frac{1}{3} = \frac{8}{\underset{1}{5}} \cdot \frac{\overset{2}{10}}{3} = \frac{8 \cdot 2}{1 \cdot 3} = \frac{16}{3} = 5\frac{1}{3}.$$

The estimate was 6, so the exact answer is reasonable.

◀◀ **WORK PROBLEM 2 AT THE SIDE.**

ANSWERS

2. (a) estimate: 2 • 7 = 14; exact: $16\frac{1}{2}$
   (b) estimate: 5 • 2 = 10; exact: $7\frac{1}{2}$
   (c) estimate: 4 • 4 = 16; exact: 16
   (d) estimate: 3 • 5 = 15; exact: $17\frac{1}{5}$

OBJECTIVE 2 When dividing mixed numbers it is a good idea to estimate the answer first. Then divide the mixed numbers using the following steps.

> **Dividing Mixed Numbers**
>
> *Step 1* **Change** each mixed number to an improper fraction.
>
> *Step 2* **Invert** the second fraction (divisor).
>
> *Step 3* **Multiply.**
>
> *Step 4* Write the answer in lowest terms and change to a mixed number or whole number where possible.

E X A M P L E 2   Dividing Mixed Numbers

First estimate the answer. Then divide to find the exact answer. Write answers in lowest terms.

(a) $2\frac{2}{5} \div 1\frac{1}{2}$

First estimate the answer by rounding each mixed number to the nearest whole number.

$$2\frac{2}{5} \;\div\; 1\frac{1}{2}$$

$$\downarrow \text{ Rounded } \downarrow$$

$$2 \;\div\; 2 = 1 \quad \text{Estimate}$$

To find the exact answer, first change each mixed number to an improper fraction.

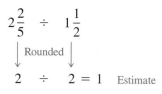

$$2\frac{2}{5} \div 1\frac{1}{2} = \frac{12}{5} \div \frac{3}{2}$$

Next, invert the second fraction and multiply.

$$\frac{12}{5} \div \frac{3}{2} = \frac{\overset{4}{\cancel{12}}}{5} \cdot \frac{2}{\underset{1}{\cancel{3}}} = \frac{4 \cdot 2}{5 \cdot 1} = \frac{8}{5} = 1\frac{3}{5}$$

Step 2 Step 3 Step 4

↑— Inverted

The estimate was 1 so the exact answer is reasonable.

(b) $8 \div 3\frac{3}{5}$

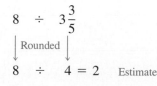

$$8 \;\div\; 3\frac{3}{5}$$

$$\downarrow \text{ Rounded } \downarrow$$

$$8 \;\div\; 4 = 2 \quad \text{Estimate}$$

Now find the exact answer.

— CONTINUED ON NEXT PAGE

3. ≈ First estimate the answer. Then divide to find the exact answer. Write answers in lowest terms.

(a) $6\frac{1}{4} \div 3\frac{1}{3}$

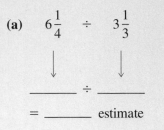

= _____ estimate

(b) $3\frac{3}{8} \div 2\frac{4}{7}$

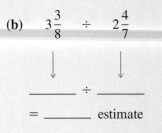

= _____ estimate

(c) $8 \div 5\frac{1}{3}$

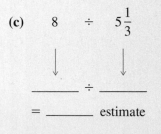

= _____ estimate

(d) $4\frac{1}{2} \div 6$

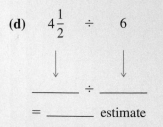

= _____ estimate

Inverted

$$8 \div 3\frac{3}{5} = \frac{8}{1} \div \frac{18}{5} = \frac{8}{1} \cdot \frac{5}{\cancel{18}_9} = \frac{20}{9} = 2\frac{2}{9}$$

Write 8 as $\frac{8}{1}$

The estimate was 2, so the exact answer is reasonable.

(c) $4\frac{3}{8} \div 5$

$$4\frac{3}{8} \div 5$$

Rounded

$$4 \div 5 = \frac{4}{1} \div \frac{5}{1} = \frac{4}{1} \times \frac{1}{5} = \frac{4}{5} \quad \text{Estimate}$$

The exact answer is

Inverted

$$4\frac{3}{8} \div 5 = \frac{35}{8} \div \frac{5}{1} = \frac{\cancel{35}^7}{8} \cdot \frac{1}{\cancel{5}_1} = \frac{7}{8}.$$

Write 5 as $\frac{5}{1}$

The estimate was $\frac{4}{5}$, so the exact answer is reasonable.

◀◀ **WORK PROBLEM 3 AT THE SIDE.**

OBJECTIVE ▶ The next two examples show how to solve application problems involving mixed numbers.

EXAMPLE 3 Applying Multiplication Skills

Suppose 11 building contractors each donate $3\frac{1}{4}$ days of labor to a community building project. How many days of labor will be donated in all?

Approach Because several contractors each donate the same amount of labor, multiply to get the total amount donated.

Solution Multiply the number of contractors and the amount of labor that each donates.

Estimate the answer.

$$11 \quad \bullet \quad 3\frac{1}{4}$$

Rounded

$$11 \quad \bullet \quad 3 = 33 \quad \text{Estimate}$$

CONTINUED ON NEXT PAGE ─

ANSWERS

3. **(a)** estimate: $6 \div 3 = 2$; exact: $1\frac{7}{8}$

  (b) estimate: $3 \div 3 = 1$; exact: $1\frac{5}{16}$

  (c) estimate: $8 \div 5 = 1\frac{3}{5}$; exact: $1\frac{1}{2}$

  (d) estimate: $5 \div 6 = \frac{5}{6}$; exact: $\frac{3}{4}$

Now find the exact answer.

$$11 \cdot 3\frac{1}{4} = 11 \cdot \frac{13}{4}$$

$$= \frac{11}{1} \cdot \frac{13}{4} = \frac{143}{4} = 35\frac{3}{4} \qquad \text{Close to estimate}$$

The community building project will receive $35\frac{3}{4}$ days of donated labor.

WORK PROBLEM 4 AT THE SIDE. ▶▶

E X A M P L E   4   Applying Division Skills

One tent requires $7\frac{1}{4}$ yards of nylon cloth. How many tents can be made from $65\frac{1}{4}$ yards of the cloth?

Approach   Division must be used to find the number of times one number is in another number.

Solution   Divide the number of yards of cloth by the number of yards needed for one tent.

Estimate the answer.

$$65\frac{1}{4} \quad \div \quad 7\frac{1}{4}$$

Rounded

$$65 \quad \div \quad 7 \approx 9 \qquad \text{Estimate}$$

The exact answer is

$$65\frac{1}{4} \div 7\frac{1}{4} = \frac{261}{4} \div \frac{29}{4}$$

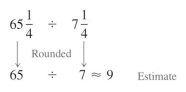

$$= \frac{\cancel{261}^{9}}{\cancel{4}_{1}} \cdot \frac{\cancel{4}^{1}}{\cancel{29}_{1}} = \frac{9}{1} = 9. \qquad \text{Matches estimate}$$

9 tents can be made from $65\frac{1}{4}$ yards of cloth.

WORK PROBLEM 5 AT THE SIDE. ▶▶

Note
When rounding mixed numbers to estimate the answer to a problem, the estimated answer will often vary from the exact answer. However, the importance of the estimated answer is that it will show you whether your answer is reasonable or not.

---

4. ≈ First estimate the answer. Then multiply to find the exact answer. Write answers in lowest terms.

(a) Suppose a dress requires $2\frac{3}{4}$ yards of material. How much material would be needed for 7 dresses?

(b) Clare earns $\$9\frac{1}{4}$ per hour. How much would she earn in $6\frac{1}{2}$ hours? Write the answer as a mixed number.

5. ≈ First estimate the answer. Then divide to find the exact answer. Write answers in lowest terms.

(a) An airplane needs $2\frac{3}{8}$ pounds of a special metal. How many airplanes could be built from $28\frac{1}{2}$ pounds of the metal?

(b) Student help is paid $\$6\frac{1}{4}$ per hour. Find the number of hours of student help that can be paid for with $150.

ANSWERS
4. **(a)** estimate: $3 \cdot 7 = 21$;
       exact: $19\frac{1}{4}$ yards
   (b) estimate: $9 \cdot 7 = 63$;
       exact: $\$60\frac{1}{8}$
5. **(a)** estimate: $29 \div 2 \approx 14$;
       exact: 12 airplanes
   (b) estimate: $150 \div 6 = 25$;
       exact: 24 hours

NUMBERS IN THE

Real World

collaborative investigations

Many food packages have useful recipes on them. The side of a corn starch box is shown below.

1. Suppose you work at a daycare center, or run one in your home. You decide to mix up 3 pounds of Fun-Time Dough for the children. The recipe on the box makes how many pounds? **2 pounds**

You will need to multiply each ingredient by some number that will give you 3 pounds of dough. Multiplying by 2 would give you 4 pounds of dough. What should you multiply by to get 3 pounds of dough? $1\frac{1}{2}$

$$\text{Multiply by } 1\tfrac{1}{2} \text{ because } 3 \div 2 = \frac{3}{1} \cdot \frac{1}{2} = \frac{3}{2} = 1\tfrac{1}{2}$$

Now figure out how much of each ingredient you should use to make 3 pounds of dough. Show your work and record your answers

corn starch	**flour**
$2\frac{1}{4}$ cups	$\frac{3}{4}$ cup
water	**cream of tartar**
3 cups	3 teaspoons
salt	**vegetable oil**
$1\frac{1}{2}$ cups	$1\frac{1}{2}$ Tablespoons

FARMER'S

CORN STARCH

FAVORITE RECIPES

Fun-Time Dough

1½ cups Farmer's Corn Starch
½ cup flour
2 cups water
2 tsp cream of tartar
1 cup salt
1 T. vegetable oil

Mix all ingredients together in saucepan. Cook over medium heat, stirring constantly, until mixture gathers on the stirring spoon and forms dough. This will take about 6 minutes. Dump onto waxed paper until cool enough to handle and knead to form a pliable mass. Store in covered container or plastic bag. Food coloring may be added to make different colors.
Makes about 2 lbs. of Fun-Time Dough.

Great Gravy

3 T. bacon fat or meat drippings
2 T. Farmer's Corn Starch
1½ cups water
½ tsp. salt
⅛ tsp. pepper

Blend fat and Farmer's Corn Starch over low heat until it is a rich brown color, stirring constantly. Gradually add water, salt and pepper. Heat to boiling over direct heat and then boil gently 2 minutes, stirring constantly. Makes 1½ cups.

— **Satisfaction Guaranteed** —

2. Suppose you decide to use the recipe on the box to make gravy for Thanksgiving dinner. How much gravy does the recipe make? $1\frac{1}{2}$ cups

If you need 4 cups of gravy, what will you do to the recipe amounts? **Multiply by $2\frac{2}{3}$**

Figure out the amount of each ingredient needed to make 4 cups of gravy. Show your work and record your answers below.

8 Tablespoons bacon fat **2 tsp. salt**

$5\frac{1}{3}$ Tablespoons corn starch **$\frac{1}{3}$ tsp. pepper**

4 cups water

154

2.8 Exercises

≈*First estimate the answer. Then multiply to find the exact answer. Write answers as mixed numbers or whole numbers.*

1. *exact*

$2\dfrac{1}{4} \cdot 3\dfrac{1}{2}$ $7\dfrac{7}{8}$

estimate

$\underline{\quad 2 \quad} \cdot \underline{\quad 4 \quad} = \underline{\quad 8 \quad}$

2. *exact*

$1\dfrac{1}{2} \cdot 3\dfrac{3}{4}$ $5\dfrac{5}{8}$

estimate

$\underline{\quad 2 \quad} \cdot \underline{\quad 4 \quad} = \underline{\quad 8 \quad}$

3. *exact*

$1\dfrac{2}{3} \cdot 2\dfrac{7}{10}$ $4\dfrac{1}{2}$

estimate

$\underline{\quad 2 \quad} \cdot \underline{\quad 3 \quad} = \underline{\quad 6 \quad}$

4. *exact*

$1\dfrac{1}{4} \cdot 2\dfrac{1}{2}$ $3\dfrac{1}{8}$

estimate

$\underline{\quad 1 \quad} \cdot \underline{\quad 3 \quad} = \underline{\quad 3 \quad}$

5. *exact*

$3\dfrac{1}{9} \cdot 1\dfrac{2}{7}$ 4

estimate

$\underline{\quad 3 \quad} \cdot \underline{\quad 1 \quad} = \underline{\quad 3 \quad}$

6. *exact*

$6\dfrac{1}{4} \cdot 3\dfrac{1}{5}$ 20

estimate

$\underline{\quad 6 \quad} \cdot \underline{\quad 3 \quad} = \underline{\quad 18 \quad}$

7. *exact*

$10 \cdot 7\dfrac{1}{4}$ $72\dfrac{1}{2}$

estimate

$\underline{\quad 10 \quad} \cdot \underline{\quad 7 \quad} = \underline{\quad 70 \quad}$

8. *exact*

$6 \cdot 2\dfrac{1}{3}$ 14

estimate

$\underline{\quad 6 \quad} \cdot \underline{\quad 2 \quad} = \underline{\quad 12 \quad}$

9. *exact*

$4\dfrac{1}{2} \cdot 2\dfrac{1}{5} \cdot 5$ $49\dfrac{1}{2}$

estimate

$\underline{\quad 5 \quad} \cdot \underline{\quad 2 \quad} \cdot \underline{\quad 5 \quad}$

$= \underline{\quad 50 \quad}$

10. *exact*

$2\dfrac{2}{3} \cdot 4\dfrac{1}{2} \cdot 3\dfrac{1}{4}$ 39

estimate

$\underline{\quad 3 \quad} \cdot \underline{\quad 5 \quad} \cdot \underline{\quad 3 \quad}$

$= \underline{\quad 45 \quad}$

11. *exact*

$3 \cdot 1\dfrac{1}{2} \cdot 2\dfrac{2}{3}$ 12

estimate

$\underline{\quad 3 \quad} \cdot \underline{\quad 2 \quad} \cdot \underline{\quad 3 \quad}$

$= \underline{\quad 18 \quad}$

12. *exact*

$\dfrac{2}{3} \cdot 3\dfrac{2}{3} \cdot \dfrac{6}{11}$ $1\dfrac{1}{3}$

estimate

$\underline{\quad 1 \quad} \cdot \underline{\quad 4 \quad} \cdot \underline{\quad 1 \quad}$

$= \underline{\quad 4 \quad}$

🖉 Writing   ◉ Conceptual   ▲ Challenging   ≈ Estimation

≈ *First estimate the answer. Then divide to find the exact answer. Write answers as mixed numbers or whole numbers.*

13. *exact*

$3\frac{1}{4} \div 2\frac{5}{8}$ $1\frac{5}{21}$

estimate

___3___ ÷ ___3___ = ___1___

14. *exact*

$2\frac{1}{4} \div 1\frac{1}{8}$ 2

estimate

___2___ ÷ ___1___ = ___2___

15. *exact*

$2\frac{1}{2} \div 3$ $\frac{5}{6}$

estimate

___3___ ÷ ___3___ = ___1___

16. *exact*

$5\frac{1}{2} \div 4$ $1\frac{3}{8}$

estimate

___6___ ÷ ___4___ = ___$1\frac{1}{2}$___

17. *exact*

$6 \div 1\frac{1}{4}$ $4\frac{4}{5}$

estimate

___6___ ÷ ___1___ = ___6___

18. *exact*

$5 \div 1\frac{7}{8}$ $2\frac{2}{3}$

estimate

___5___ ÷ ___2___ = ___$2\frac{1}{2}$___

19. *exact*

$\frac{1}{2} \div 2\frac{1}{4}$ $\frac{2}{9}$

estimate

___1___ ÷ ___2___ = ___$\frac{1}{2}$___

20. *exact*

$\frac{3}{4} \div 2\frac{1}{2}$ $\frac{3}{10}$

estimate

___1___ ÷ ___3___ = ___$\frac{1}{3}$___

21. *exact*

$1\frac{7}{8} \div 6\frac{1}{4}$ $\frac{3}{10}$

estimate

___2___ ÷ ___6___ = ___$\frac{1}{3}$___

22. *exact*

$7\frac{1}{2} \div \frac{2}{3}$ $11\frac{1}{4}$

estimate

___8___ ÷ ___1___ = ___8___

23. *exact*

$5\frac{2}{3} \div 6$ $\frac{17}{18}$

estimate

___6___ ÷ ___6___ = ___1___

24. *exact*

$5\frac{3}{4} \div 2$ $2\frac{7}{8}$

estimate

___6___ ÷ ___2___ = ___3___

$\approx$ *First estimate the answer. Then solve each of the following application problems by using multiplication or division to find the exact answer.*

25. Shirley Cicero wants to make 16 holiday wreaths to sell at the craft fair. Each wreath needs $2\frac{1}{4}$ yards of ribbon. How many yards does she need?

estimate: **16 • 2 = 32 yards**
exact: **36 yards**

26. Babbette Gray worked $36\frac{1}{2}$ hours at $9 per hour. How much money did she make?

estimate: **37 • \$9 = \$333**
exact: **\$328$\frac{1}{2}$**

27. Each home of a certain design needs $109\frac{1}{2}$ yards of prefinished baseboard. How many homes can be fitted with baseboard if there are 1314 yards of baseboard available?

estimate: **1314 ÷ 110 $\approx$ 12 homes**
exact: **12 homes**

28. For 1 acre of a crop, $7\frac{1}{2}$ gallons of fertilizer must be applied. How many acres can be fertilized with 1200 gallons of fertilizer?

estimate: **1200 ÷ 8 = 150 acres**
exact: **160 acres**

29. An insect spray manufactured by Dutch Chemicals Incorporated, is a mixture of $1\frac{3}{4}$ ounces of chemical per gallon of water. How many ounces of chemical are needed for $12\frac{1}{2}$ gallons of water?

estimate: **2 • 13 = 26 ounces**
exact: **21$\frac{7}{8}$ ounces**

30. Each home requires $37\frac{3}{4}$ pounds of roofing nails. How many pounds of roofing nails are needed for 36 homes?

estimate: **38 • 36 = 1368 pounds**
exact: **1359 pounds**

31. Write the rule for multiplying mixed numbers. Use your own words to write the 3 steps.

The answer should include
Step 1 **Change mixed numbers to improper fractions.**
Step 2 **Multiply the fractions.**
Step 3 **Write the answer in lowest terms changing to mixed or whole numbers where possible.**

32. Look at Exercise 31. In your own words, write the additional step that must be added to the rule for multiplying mixed numbers to make it the rule for dividing mixed numbers.

The additional step is to invert the second fraction (divisor).

33. A fishing boat anchor requires $10\frac{3}{8}$ pounds of steel. Find the number of anchors that can be manufactured with 25,730 pounds of steel.

estimate: **25,730 ÷ 10 = 2573 anchors**
exact: **2480 anchors**

34. Each apartment unit requires $62\frac{1}{2}$ square yards of carpet. Find the number of apartment units that can be carpeted with 6750 square yards of carpet.

estimate: **6750 ÷ 63 $\approx$ 107 units**
exact: **108 units**

35. A manufacturer of bird feeders needs spacers that are to be cut from a tube that is $9\frac{3}{4}$ inches long. How many spacers can be cut from the tube if each spacer has to be $\frac{3}{4}$-inch thick?

estimate: **10 ÷ 1 = 10 spacers**
exact: **13 spacers**

36. A building contractor must move 12 tons of sand. If his truck can carry $\frac{3}{4}$ ton of sand, how many trips must be made to move the sand?

estimate: **12 ÷ 1 = 12 trips**
exact: **16 trips**

▲ **37.** A photographer uses $12\frac{3}{4}$ rolls of film at a wedding and $7\frac{1}{8}$ rolls of film at a retirement party. Find the total number of rolls needed for 28 weddings and 16 retirement parties.

estimate: **13 · 28 =   364**
               7 · 16 = + 112
                             476 rolls
exact: **471 rolls**

▲ **38.** One necklace can be completed in $6\frac{1}{2}$ minutes, while a bracelet takes $3\frac{1}{8}$ minutes. Find the total time that it takes to complete 36 necklaces and 22 bracelets.

estimate: **7 · 36 =   252**
               3 · 22 = + 66
                            318 minutes
exact: **$302\frac{3}{4}$ minutes**

▲ **39.** Sandi Goldstein bought some stock in Telex Chile for $\$8\frac{3}{8}$ per share. If she paid $5025 for the stock, how many shares did she buy?

estimate: **5025 ÷ 8 ≈ 628 shares**
exact: **600 shares**

▲ **40.** Ms. Nishimoto bought 12 shares of stock at $\$18\frac{3}{4}$ per share, 24 shares at $\$36\frac{3}{8}$ per share, and 16 shares at $\$74\frac{1}{8}$ per share. Her broker charged her a commission of $12. Find the total amount that she paid.

estimate: **12 · 19 =   228**
             24 · 36 =   864
             16 · 74 = 1184
                         +   12
                           $2288
exact: **$2296**

Review and Prepare

*Write each fraction in lowest terms. (For help, see **Section 2.4**.)*

41. $\frac{6}{8}$ $\frac{3}{4}$

42. $\frac{5}{10}$ $\frac{1}{2}$

43. $\frac{35}{50}$ $\frac{7}{10}$

44. $\frac{27}{45}$ $\frac{3}{5}$

45. $\frac{56}{64}$ $\frac{7}{8}$

46. $\frac{36}{63}$ $\frac{4}{7}$

2.1	**numerator**	The number above the division bar in a fraction is called the numerator. It shows how many of the equivalent parts are being considered.
	denominator	The number below the division bar in a fraction is called the denominator. It shows the number of equal parts in a whole.
	proper fraction	In a proper fraction, the numerator is smaller than the denominator. The fraction is less than one.
	improper fraction	In an improper fraction, the numerator is greater than or equal to the denominator. The fraction is equal to or greater than one.
2.2	**mixed number**	A mixed number includes a fraction and a whole number written together.
2.3	**factors**	Numbers that are multiplied to give a product are factors.
	composite number	A composite number has at least one factor other than itself and 1.
	prime number	A prime number is a whole number other than 0 and 1 that has exactly two factors, itself and 1.
	factorizations	The numbers that can be multiplied to give a specific number (product) are factorizations of that number.
	prime factorization	In a prime factorization every factor is a prime number.
2.4	**common factor**	A common factor is a number that can be divided into two or more whole numbers.
	lowest terms	A fraction is written in lowest terms when its numerator and denominator have no common factor other than 1.
2.5	**cancellation**	When multiplying or dividing fractions, the process of dividing a numerator and denominator by a common factor is called cancellation.

QUICK REVIEW

Concepts	Examples
2.1 Types of Fractions	
Proper Numerator smaller than denominator.	**Proper** $\dfrac{2}{3}, \dfrac{3}{4}, \dfrac{15}{16}, \dfrac{1}{8}$
Improper Numerator equal to or greater than denominator.	**Improper** $\dfrac{17}{8}, \dfrac{19}{12}, \dfrac{11}{2}, \dfrac{5}{3}, \dfrac{7}{7}$
2.2 Converting Fractions	
Mixed to improper Multiply denominator by whole number, add numerator, and place over denominator.	**Mixed to improper** $7\dfrac{2}{3} = \dfrac{23}{3} \leftarrow 3 \times 7 + 2$ Same denominator
Improper to mixed Divide numerator by denominator and place remainder over denominator.	**Improper to mixed** $\dfrac{17}{5} = 3\dfrac{2}{5}$ Same denominator
2.3 Prime Numbers	
Determine whether a whole number is evenly divisible only by itself and 1. (By definition, 0 and 1 are not prime.)	The prime numbers less than 100 are 2, 3, 5, 7, 11, 13, 17, 19, 23, 29, 31, 37, 41, 43, 47, 53, 59, 61, 67, 71, 73, 79, 83, 89, and 97.

Concepts	Examples
2.3 Finding the Prime Factorization of a Number Divide each factor by a prime number by using a diagram that forms the shape of tree branches.	Find the prime factorization of 30. Use a factor tree. $$30$$ $$②\swarrow\qquad\searrow 15$$ $$③\swarrow\qquad\searrow⑤$$ Prime factors are circled. $30 = 2 \cdot 3 \cdot 5$
2.4 Writing Fractions in Lowest Terms Divide the numerator and denominator by the same number.	$$\frac{30}{42} = \frac{30 \div 6}{42 \div 6} = \frac{5}{7}$$
2.5 Multiplying Fractions **1.** Multiply numerators and denominators. **2.** Reduce answers to lowest terms if cancelling was not done.	$$\frac{6}{11} \cdot \frac{7}{8} = \frac{\overset{3}{\cancel{6}}}{11} \cdot \frac{7}{\underset{4}{\cancel{8}}} = \frac{3 \cdot 7}{11 \cdot 4} = \frac{21}{44}$$
2.7 Dividing Fractions Invert the divisor (second fraction) and multiply as fractions.	$$\frac{25}{36} \div \frac{15}{18} = \frac{\overset{5}{\cancel{25}}}{\underset{2}{\cancel{36}}} \cdot \frac{\overset{1}{\cancel{18}}}{\underset{3}{\cancel{15}}} = \frac{5 \cdot 1}{2 \cdot 3} = \frac{5}{6}$$
2.8 Multiplying Mixed Numbers First estimate the answers. Then follow these steps. **1. Change** each mixed number to an improper fraction. **2. Multiply.** **3.** Write the answer in lowest terms and change the answer to a mixed number if desired.	*estimate* *exact* $$1\frac{3}{5} \quad \cdot \quad 3\frac{1}{3} \qquad 1\frac{3}{5} \cdot 3\frac{1}{3} = \frac{8}{\underset{1}{\cancel{5}}} \cdot \frac{\overset{2}{\cancel{10}}}{3}$$ Rounded $$2 \quad \cdot \quad 3 = 6 \qquad\qquad = \frac{8 \cdot 2}{1 \cdot 3}$$ $$= \frac{16}{3} = 5\frac{1}{3}$$ Close to estimate
2.8 Dividing Mixed Numbers First estimate the answer. Then follow these steps. **1. Change** each mixed number to an improper fraction. **2. Invert** the divisor (second fraction). **3. Multiply.** **4.** Write the answer in lowest terms and change the answer to a mixed number when possible.	*estimate* *exact* $$3\frac{5}{9} \quad \div \quad 2\frac{2}{5} \qquad 3\frac{5}{9} \div 2\frac{2}{5} = \frac{32}{9} \div \frac{12}{5}$$ Rounded $$4 \quad \div \quad 2 = 2 \qquad\qquad = \frac{\overset{8}{\cancel{32}}}{9} \cdot \frac{5}{\underset{3}{\cancel{12}}} = \frac{40}{27}$$ $$= 1\frac{13}{27}$$ Close to estimate

[2.1] *Write the fraction that represents the shaded area.*

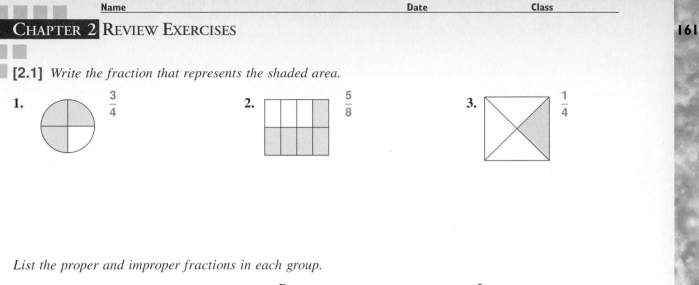

1. $\dfrac{3}{4}$ 2. $\dfrac{5}{8}$ 3. $\dfrac{1}{4}$

List the proper and improper fractions in each group.

	Proper	*Improper*
4. $\dfrac{1}{4}, \dfrac{3}{2}, \dfrac{5}{8}, \dfrac{4}{4}, \dfrac{2}{3}$	$\dfrac{1}{4}, \dfrac{5}{8}, \dfrac{2}{3}$	$\dfrac{3}{2}, \dfrac{4}{4}$
5. $\dfrac{6}{5}, \dfrac{15}{16}, \dfrac{16}{13}, \dfrac{1}{8}, \dfrac{5}{3}$	$\dfrac{15}{16}, \dfrac{1}{8}$	$\dfrac{6}{5}, \dfrac{16}{13}, \dfrac{5}{3}$

[2.2] *Write each mixed number as an improper fraction. Write each improper fraction as a mixed number.*

6. $4\dfrac{3}{8}$ $\dfrac{35}{8}$ 7. $10\dfrac{4}{5}$ $\dfrac{54}{5}$ 8. $\dfrac{21}{4}$ $5\dfrac{1}{4}$ 9. $\dfrac{63}{5}$ $12\dfrac{3}{5}$

[2.3] *Find all factors of each number.*

10. 8   11. 18   12. 55   13. 90
   1, 2, 4, 8   1, 2, 3, 6, 9, 18   1, 5, 11, 55   1, 2, 3, 5, 6, 9, 10,
                                                            15, 18, 30, 45, 90

Write the prime factorization of each number by using exponents.

14. 16 2^4 15. 150 $2 \cdot 3 \cdot 5^2$ 16. 225 $3^2 \cdot 5^2$

✍ Writing   ◉ Conceptual   ▲ Challenging   ≈ Estimation

Solve each of the following.

17. 5^2 **25**

18. $3^2 \cdot 2^3$ **72**

19. $8^2 \cdot 3^3$ **1728**

20. $4^3 \cdot 2^5$ **2048**

[2.4] *Write each fraction in lowest terms.*

21. $\dfrac{12}{16}$ $\dfrac{3}{4}$

22. $\dfrac{35}{40}$ $\dfrac{7}{8}$

23. $\dfrac{75}{80}$ $\dfrac{15}{16}$

Write the numerator and denominator of each fraction as a product of prime factors. Next, write the fraction in lowest terms.

24. $\dfrac{25}{60}$ $\dfrac{\overset{1}{\cancel{5}} \cdot 5}{2 \cdot 2 \cdot 3 \cdot \underset{1}{\cancel{5}}} ; \dfrac{5}{12}$

25. $\dfrac{384}{96}$ $\dfrac{\overset{1}{\cancel{2}} \cdot \overset{1}{\cancel{2}} \cdot \overset{1}{\cancel{2}} \cdot \overset{1}{\cancel{2}} \cdot \overset{1}{\cancel{2}} \cdot 2 \cdot 2 \cdot 3}{\underset{1}{\cancel{2}} \cdot \underset{1}{\cancel{2}} \cdot \underset{1}{\cancel{2}} \cdot \underset{1}{\cancel{2}} \cdot \underset{1}{\cancel{2}} \cdot 3} ; 4$

Decide whether the following pairs of fractions are equivalent or not equivalent, using cross multiplication.

26. $\dfrac{4}{5}$ and $\dfrac{72}{90}$ **equivalent**

27. $\dfrac{3}{4}$ and $\dfrac{42}{58}$ **not equivalent**

[2.5–2.8] *Multiply. Write all answers in lowest terms, and as mixed numbers or whole numbers where possible.*

28. $\dfrac{2}{3} \cdot \dfrac{3}{4}$ $\dfrac{1}{2}$

29. $\dfrac{4}{5} \cdot \dfrac{5}{12}$ $\dfrac{1}{3}$

30. $\dfrac{70}{175} \cdot \dfrac{5}{14}$ $\dfrac{1}{7}$

31. $\dfrac{44}{63} \cdot \dfrac{3}{11}$ $\dfrac{4}{21}$

32. $\dfrac{5}{16} \cdot 48$ **15**

33. $\dfrac{5}{8} \cdot 1000$ **625**

Divide. Write answers in lowest terms, and as mixed numbers or whole numbers where possible.

34. $\dfrac{1}{4} \div \dfrac{1}{2}$ $\dfrac{1}{2}$

35. $\dfrac{5}{6} \div \dfrac{1}{2}$ $\dfrac{5}{3} = 1\dfrac{2}{3}$

36. $\dfrac{\frac{15}{18}}{\frac{10}{30}}$ $\dfrac{5}{2} = 2\dfrac{1}{2}$

37. $\dfrac{\frac{3}{10}}{\frac{6}{40}}$ **2**

38. $5 \div \dfrac{5}{8}$ **8**

39. $18 \div \dfrac{3}{4}$ **24**

40. $\dfrac{7}{8} \div 2$ $\dfrac{7}{16}$

41. $\dfrac{2}{3} \div 5$ $\dfrac{2}{15}$

42. $\dfrac{\frac{12}{13}}{3}$ $\dfrac{4}{13}$

Find the area of each rectangle.

43. $\frac{1}{2}$ yard $\dfrac{15}{32}$ **square yard**

$\frac{15}{16}$ yard

44. $\frac{2}{3}$ inch $\dfrac{7}{12}$ **square inch**

$\frac{7}{8}$ inch

45. Find the area of a rectangle having a length of 15 feet and a width of $\frac{2}{3}$ foot.

10 square feet

46. Find the area of a rectangle having a length of 48 yards and a width of $\frac{3}{4}$ yard.

36 square yards

$\approx$ *First estimate the answer. Then multiply to find the exact answer. Write answers in lowest terms, and as mixed numbers where possible.*

47. *exact*

$2\dfrac{3}{8} \cdot 1\dfrac{1}{2}$ $3\dfrac{9}{16}$

48. *exact*

$2\dfrac{1}{4} \cdot 7\dfrac{1}{8} \cdot 1\dfrac{1}{3}$ $21\dfrac{3}{8}$

estimate

$\underline{\quad 2 \quad} \cdot \underline{\quad 2 \quad} = \underline{\quad 4 \quad}$

estimate

$\underline{\quad 2 \quad} \cdot \underline{\quad 7 \quad} \cdot \underline{\quad 1 \quad} = \underline{\quad 14 \quad}$

≈ *First estimate the answer. Then divide to find the exact answer. Write answers in lowest terms, and as mixed numbers where possible.*

49. *exact*

$15\frac{1}{2} \div 3$ $5\frac{1}{6}$

estimate

$\underline{\quad 16 \quad} \div \underline{\quad 3 \quad} = \underline{\quad 5\frac{1}{3} \quad}$

50. *exact*

$3\frac{1}{8} \div 5\frac{5}{7}$ $\frac{35}{64}$

estimate

$\underline{\quad 3 \quad} \div \underline{\quad 6 \quad} = \underline{\quad \frac{1}{2} \quad}$

Solve each of the following application problems by using multiplication or division.

51. Find the number of $\frac{3}{4}$-pound bags of sunflower seeds that can be filled with 225 pounds of sunflower seeds.

300 bags

52. An estate is divided so that each of 5 children receives equal shares of $\frac{2}{3}$ of the estate. What fraction of the total estate will each receive?

$\frac{2}{15}$ **of the estate**

≈**53.** Find the number of window blind pull cords that can be made from $157\frac{1}{2}$ yards of cord if $4\frac{3}{8}$ yards of cord are needed for each blind. First estimate and then find the exact answer.

estimate: **158 ÷ 4 ≈ 40 pull cords**
exact: **36 pull cords**

≈ **54.** Working as a bookkeeper, Neta Fitzgerald is paid $8\frac{1}{2}$ per hour. Find her earnings for a week in which she works 38 hours. First estimate and then find the exact answer.

estimate: **9 · 38 = $342**
exact: **$323**

55. Ebony Wilson purchased 100 pounds of detergent at the food co-op. After selling $\frac{1}{2}$ of this to her neighbor, she gives $\frac{2}{5}$ of the remaining detergent to her parents. How many pounds of detergent does she have left?

30 pounds

56. Mary received a check for $1200. After paying $\frac{3}{8}$ of this amount for room and board, she paid $\frac{1}{2}$ of the remaining amount for school fees. How much money does she have left after paying the fees?

$375

57. The Springvale Parish will divide $\frac{5}{8}$ of its library budget evenly among the 4 largest parish libraries. What fraction of the total library budget will each of these libraries receive?

$\frac{5}{32}$ **of the budget**

58. Play It Now Sports Center has decided to divide $\frac{2}{3}$ of the company's profit sharing funds evenly among the top 8 store managers. What fraction of the total profit sharing amount will each receive?

$\frac{1}{12}$ **of the total**

MIXED REVIEW EXERCISES

Multiply or divide as indicated. Write answers in lowest terms, and as mixed numbers or whole numbers where possible.

59. $\dfrac{2}{3} \cdot \dfrac{1}{2}$ $\dfrac{1}{3}$

60. $\dfrac{1}{4} \cdot \dfrac{2}{3}$ $\dfrac{1}{6}$

61. $10\dfrac{1}{4} \cdot 2\dfrac{1}{2}$ $25\dfrac{5}{8}$

62. $12\dfrac{1}{2} \cdot 2\dfrac{1}{4}$ $28\dfrac{1}{8}$

63. $\dfrac{\frac{7}{8}}{6}$ $\dfrac{7}{48}$

64. $\dfrac{\frac{5}{8}}{4}$ $\dfrac{5}{32}$

65. $\dfrac{15}{31} \cdot 62$ 30

66. $3\dfrac{1}{4} \div 1\dfrac{1}{2}$ $2\dfrac{1}{6}$

Write each mixed number as an improper fraction. Write each improper fraction as a mixed number.

67. $\dfrac{8}{5}$ $1\dfrac{3}{5}$

68. $\dfrac{137}{3}$ $45\dfrac{2}{3}$

69. $5\dfrac{2}{3}$ $\dfrac{17}{3}$

70. $38\dfrac{3}{8}$ $\dfrac{307}{8}$

Write the numerator and denominator of each fraction as a product of prime factors, then write the fraction in lowest terms.

71. $\dfrac{8}{12}$ $\dfrac{\overset{1}{\cancel{2}} \cdot \overset{1}{\cancel{2}} \cdot 2}{\underset{1}{\cancel{2}} \cdot \underset{1}{\cancel{2}} \cdot 3} = \dfrac{2}{3}$

72. $\dfrac{108}{210}$ $\dfrac{\overset{1}{\cancel{2}} \cdot 2 \cdot 3 \cdot 3 \cdot 3}{\underset{1}{\cancel{2}} \cdot 3 \cdot 5 \cdot 7} = \dfrac{18}{35}$

Write each fraction in lowest terms.

73. $\dfrac{36}{48}$ $\dfrac{3}{4}$

74. $\dfrac{28}{84}$ $\dfrac{1}{3}$

75. $\dfrac{44}{110}$ $\dfrac{2}{5}$

76. $\dfrac{87}{261}$ $\dfrac{1}{3}$

Solve each of the following application problems.

≈ **77.** The directions on a can of fabric glue say to apply $3\frac{1}{2}$ ounces of glue to each square yard. How many ounces are needed for $43\frac{5}{9}$ square yards? First estimate and then find the exact answer.

   estimate: 4 • 44 = 176 ounces

   exact: $152\frac{4}{9}$ ounces

≈ **78.** Jack Anderson Trucking purchased some diesel fuel additive. The instructions say to use $7\frac{1}{4}$ quarts of additive with each in-ground tank of fuel. How many quarts are needed for $25\frac{1}{2}$ in-ground tanks? First estimate and then find the exact answer.

   estimate: 7 • 26 = 182 quarts

   exact: $184\frac{7}{8}$ quarts

79. A postage stamp is $\frac{2}{3}$ inch by $\frac{3}{4}$ inch. Find its area.

   $\frac{1}{2}$ square inch

80. A hospital tray is $\frac{1}{2}$ meter by $\frac{7}{8}$ meter. Find its area.

   $\frac{7}{16}$ square meter

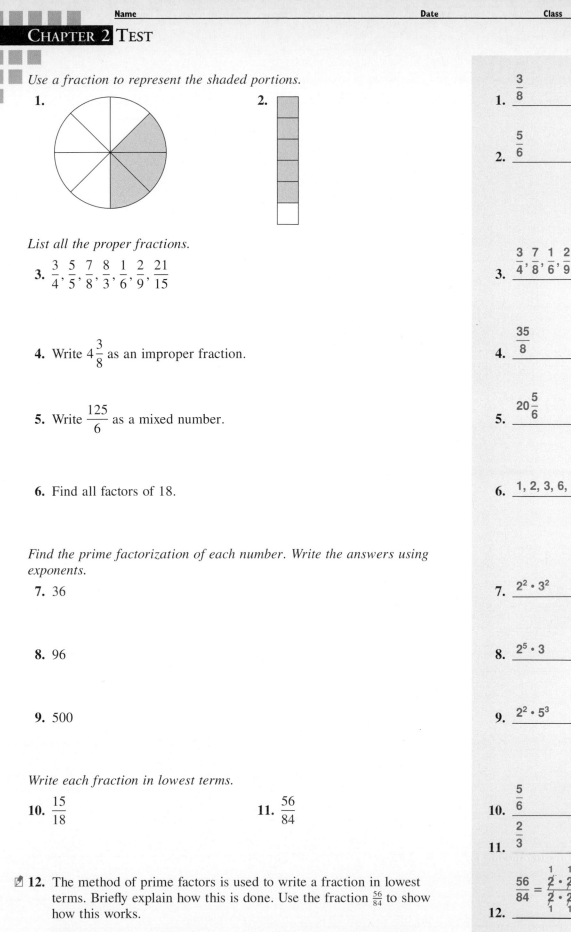

Use a fraction to represent the shaded portions.

1.

2.

List all the proper fractions.

3. $\dfrac{3}{4}, \dfrac{5}{5}, \dfrac{7}{8}, \dfrac{8}{3}, \dfrac{1}{6}, \dfrac{2}{9}, \dfrac{21}{15}$

4. Write $4\dfrac{3}{8}$ as an improper fraction.

5. Write $\dfrac{125}{6}$ as a mixed number.

6. Find all factors of 18.

Find the prime factorization of each number. Write the answers using exponents.

7. 36

8. 96

9. 500

Write each fraction in lowest terms.

10. $\dfrac{15}{18}$ 11. $\dfrac{56}{84}$

✎ 12. The method of prime factors is used to write a fraction in lowest terms. Briefly explain how this is done. Use the fraction $\frac{56}{84}$ to show how this works.

Write the prime factorization of both numerator and denominator. Use cancellation to divide numerator and denominator by any common factors. Multiply the remaining factors in numerator and denominator.

1. $\dfrac{3}{8}$ _____

2. $\dfrac{5}{6}$ _____

3. $\dfrac{3}{4}, \dfrac{7}{8}, \dfrac{1}{6}, \dfrac{2}{9}$ _____

4. $\dfrac{35}{8}$ _____

5. $20\dfrac{5}{6}$ _____

6. 1, 2, 3, 6, 9, 18 _____

7. $2^2 \cdot 3^2$ _____

8. $2^5 \cdot 3$ _____

9. $2^2 \cdot 5^3$ _____

10. $\dfrac{5}{6}$ _____

11. $\dfrac{2}{3}$ _____

12. $\dfrac{56}{84} = \dfrac{\overset{1}{\cancel{2}} \cdot \overset{1}{\cancel{2}} \cdot 2 \cdot \overset{1}{\cancel{7}}}{\underset{1}{\cancel{2}} \cdot \underset{1}{\cancel{2}} \cdot 3 \cdot \underset{1}{\cancel{7}}} = \dfrac{2}{3}$

13. _____

13. Explain how to multiply fractions. What additional step must be taken when dividing fractions?

Multiply fractions by multiplying the numerators and multiplying the denominators. Divide two fractions by inverting the second fraction (divisor) and multiplying.

Multiply or divide. Write answers in lowest terms, and as mixed numbers or whole numbers where possible.

14. $\dfrac{1}{2}$ _____

15. 18 _____

14. $\dfrac{5}{8} \cdot \dfrac{4}{5}$

15. $24 \cdot \dfrac{3}{4}$

16. $\dfrac{3}{8}$ square meter _____

16. Find the area of a kitchen grill measuring $\frac{3}{4}$ meter by $\frac{1}{2}$ meter.

17. 7392 students _____

17. There are 8448 students at the Metro Community College campus. If $\frac{7}{8}$ of the students work either full time or part time, find the total number of students who work.

18. $\dfrac{5}{6}$ _____

19. $15\dfrac{3}{4}$ _____

18. $\dfrac{5}{8} \div \dfrac{3}{4}$

19. $\dfrac{\frac{7}{4}}{9}$

20. 100 vehicles _____

20. There are 60 tanks of hydraulic fluid in the airport supply depot. If each maintenance vehicle has a container that holds $\frac{3}{5}$ of a tank of fluid, how many maintenance vehicles can be filled?

21. *estimate:* $5 \cdot 3 = 15$
exact: $17\dfrac{23}{32}$
_____

22. *estimate:* $2 \cdot 4 = 8$
exact: $7\dfrac{17}{18}$
_____

23. *estimate:* $5 \div 1 = 5$
exact: $4\dfrac{4}{15}$
_____

24. *estimate:* $9 \div 2 = 4\dfrac{1}{2}$
exact: $5\dfrac{1}{10}$
_____

25. *estimate:* $3 \cdot 12 = 36$
exact: $30\dfrac{5}{8}$ grams
_____

$\approx$ *First estimate the answer and then either multiply or divide to find the exact answer in each of the following problems. Write the answers as mixed numbers.*

21. $5\dfrac{1}{4} \cdot 3\dfrac{3}{8}$

22. $1\dfrac{5}{6} \cdot 4\dfrac{1}{3}$

23. $4\dfrac{4}{5} \div 1\dfrac{1}{8}$

24. $\dfrac{8\frac{1}{2}}{1\frac{2}{3}}$

25. A new vaccine is synthesized at the rate of $2\frac{1}{2}$ grams per day. How many grams can be synthesized in $12\frac{1}{4}$ days?

CUMULATIVE REVIEW EXERCISES CHAPTERS 1–2

Name the digit that has the given place value in each of the following problems.

1. 718

    hundreds **7**
    tens **1**

2. 6,748,215

    millions **6**
    ten thousands **4**

Add, subtract, multiply, or divide as indicated.

3.
$$\begin{array}{r} 27 \\ 43 \\ 85 \\ +\ 11 \\ \hline \mathbf{166} \end{array}$$

4.
$$\begin{array}{r} 82{,}121 \\ 5\ 468 \\ 316 \\ +\ 61{,}294 \\ \hline \mathbf{149{,}199} \end{array}$$

5.
$$\begin{array}{r} 2628 \\ -\ 1056 \\ \hline \mathbf{1572} \end{array}$$

6.
$$\begin{array}{r} 4{,}819{,}604 \\ -\ 1{,}597{,}783 \\ \hline \mathbf{3{,}221{,}821} \end{array}$$

7.
$$\begin{array}{r} 96 \\ \times\ \ 8 \\ \hline \mathbf{768} \end{array}$$

8. $6 \cdot 3 \cdot 5$ **90**

9.
$$\begin{array}{r} 3784 \\ \times\ \ \ 573 \\ \hline \mathbf{2{,}168{,}232} \end{array}$$

10.
$$\begin{array}{r} 629 \\ \times\ 700 \\ \hline \mathbf{440{,}300} \end{array}$$

11. $\dfrac{54}{6}$

    9

12. $18\overline{)136{,}458}$ **7 581**

13. $16{,}942 \div 4$

    4235 R2

14. $492\overline{)10{,}850}$ **22 R26**

Round each of the following to the nearest ten, nearest hundred, and nearest thousand.

	Ten	Hundred	Thousand
15. 8626	8630	8600	9000
16. 85,462	85,460	85,500	85,000

Simplify each problem by using the order of operations.

17. $5^2 - 9 \cdot 2$ **7**

18. $\sqrt{36} - 2 \cdot 3 + 5$ **5**

Solve each application problem.

19. The Quick-Stop Mini-Mart purchased 6 cases of medium-size drink cups for $35 per case and 9 cases of large-size drink cups for $45 per case. Find the total cost of the 15 cases of drink cups.

    $615

20. Home blood-pressure monitors sell for as little as $20 and as much as $150. If Scott bought the cheapest model and Jenn the most expensive model, how much more did Jenn pay than Scott?

    $130

21. A typical adult loses 100 hairs a day out of approximately 120,000 hairs. If the lost hairs were not replaced, find the number of hairs remaining after 2 years. (1 year = 365 days).

47,000 hairs

22. The cost of renting a group camp for one week is $3150. If the cost is to be split evenly among 18 families, find the cost for each family for the week.

$175

23. A lamp base is made of marble and measures $\frac{3}{4}$ foot by $\frac{7}{12}$ foot. Find its area.

$\frac{7}{16}$ **square foot**

24. The Municipal Utility District says that the cost of operating a hair dryer is $\frac{1}{5}$¢ per minute. Find the cost of operating the hair dryer for $\frac{1}{2}$ hour. (*Hint:* There are 30 minutes in a half hour.)

6¢

Write proper *or* improper *for each fraction.*

25. $\frac{3}{4}$ proper

26. $\frac{9}{9}$ improper

27. $\frac{7}{16}$ proper

Write each mixed number as an improper fraction. Write each improper fraction as a mixed number.

28. $2\frac{1}{2}$ **$\frac{5}{2}$**

29. $7\frac{1}{3}$ **$\frac{22}{3}$**

30. $\frac{12}{7}$ **$1\frac{5}{7}$**

31. $\frac{103}{8}$ **$12\frac{7}{8}$**

Find the prime factorization of each number. Write answers by using exponents.

32. 50 **$2 \cdot 5^2$**

33. 80 **$2^4 \cdot 5$**

34. 350 **$2 \cdot 5^2 \cdot 7$**

Simplify each of the following.

35. $2^2 \cdot 3^2$ **36**

36. $3^3 \cdot 5^2$ **675**

37. $2^3 \cdot 4^2 \cdot 5$ **640**

Write each fraction in lowest terms.

38. $\frac{35}{40}$ **$\frac{7}{8}$**

39. $\frac{16}{24}$ **$\frac{2}{3}$**

40. $\frac{30}{54}$ **$\frac{5}{9}$**

Multiply or divide as indicated. Write all answers in lowest terms, and as mixed numbers or whole numbers where possible.

41. $\frac{3}{4} \cdot \frac{1}{3}$ **$\frac{1}{4}$**

42. $30 \cdot \frac{2}{3} \cdot \frac{3}{5}$ **12**

43. $7\frac{1}{2} \cdot 3\frac{1}{3}$ **25**

44. $\frac{3}{8} \div \frac{2}{3}$ **$\frac{9}{16}$**

45. $\frac{3}{8} \div 1\frac{1}{4}$ **$\frac{3}{10}$**

46. $3 \div 1\frac{1}{4}$ **$2\frac{2}{5}$**

Adding and Subtracting Fractions

3

3.1 Adding and Subtracting Like Fractions

In Chapter 2 we looked at the basics of fractions and then practiced with multiplication and division of common fractions and mixed numbers. In this chapter we will work with addition and subtraction of common fractions and mixed numbers.

OBJECTIVE ▶1 Fractions with the same denominators are **like fractions.** Fractions with different denominators are **unlike fractions.**

E X A M P L E I   **Identifying Like and Unlike Fractions**

(a) $\frac{3}{4}, \frac{1}{4}, \frac{5}{4}, \frac{6}{4}$, and $\frac{4}{4}$ are **like** fractions.
↑ ↑ ↑ ↑   ↑ ⎽⎽⎽ All denominators are the same.

(b) $\frac{7}{12}$ and $\frac{12}{7}$ are **unlike** fractions.
↑ ⎽⎽⎽⎽↑ Different denominators

WORK PROBLEM I AT THE SIDE. ▶▶

Note
Like fractions have the same denominator.

OBJECTIVE ▶2 The following figures show you how to add the fractions $\frac{2}{7}$ and $\frac{4}{7}$.

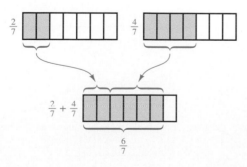

As the figures show,

$$\frac{2}{7} + \frac{4}{7} = \frac{6}{7}.$$

www.mathnotes.com

OBJECTIVES
▶1 Define like and unlike fractions.
▶2 Add like fractions.
▶3 Subtract like fractions.

FOR EXTRA HELP

Tutorial   Tape 5   SSM, Sec. 3.1

1. Write *like* or *unlike* for each pair of fractions.

(a) $\frac{3}{4}$ $\frac{1}{4}$

(b) $\frac{2}{3}$ $\frac{2}{5}$

(c) $\frac{11}{12}$ $\frac{9}{12}$

(d) $\frac{7}{3}$ $\frac{7}{4}$

ANSWERS
1. (a) like (b) unlike (c) like (d) unlike

171

2. Add. Write answers in lowest terms.

(a) $\dfrac{4}{8} + \dfrac{1}{8}$

(b) $\begin{array}{r} \dfrac{5}{9} \\[4pt] +\ \dfrac{2}{9} \\ \hline \end{array}$

(c) $\dfrac{1}{8} + \dfrac{3}{8}$

(d) $\dfrac{3}{10} + \dfrac{1}{10} + \dfrac{4}{10}$

Add like fractions as follows.

Adding Like Fractions

Step 1 Find the numerator of the *sum* (the answer) by adding the numerators of the fractions.

Step 2 Write the denominator of the like fractions as the denominator of the sum.

Step 3 Write the answer in lowest terms.

EXAMPLE 2   Adding Like Fractions

Add. Write answers in lowest terms.

(a) $\dfrac{1}{5} + \dfrac{2}{5}$

$$\dfrac{1}{5} + \dfrac{2}{5} = \dfrac{1+2}{5} = \dfrac{3}{5} \quad \begin{array}{l}\leftarrow \text{Add numerators.} \\ \leftarrow \text{Same denominator}\end{array}$$

(b) $\dfrac{1}{12} + \dfrac{7}{12} + \dfrac{1}{12}$

Step 1 $\dfrac{1+7+1}{12} \quad \leftarrow \text{Add numerators.}$

Step 2 $= \dfrac{9}{12} \quad \begin{array}{l}\leftarrow \text{Sum} \\ \leftarrow \text{Same denominator}\end{array}$

Step 3 $= \dfrac{9 \div 3}{12 \div 3} = \dfrac{3}{4} \quad \text{In lowest terms}$

Note
Fractions may be added *only* if they have like denominators.

◀◀ **WORK PROBLEM 2 AT THE SIDE.**

OBJECTIVE 3▶ The figures show $\frac{7}{8}$ broken into $\frac{4}{8}$ and $\frac{3}{8}$.

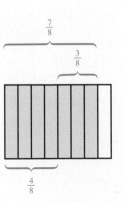

ANSWERS

2. (a) $\dfrac{5}{8}$ (b) $\dfrac{7}{9}$ (c) $\dfrac{1}{2}$ (d) $\dfrac{4}{5}$

Subtracting $\frac{3}{8}$ from $\frac{7}{8}$ gives the answer $\frac{4}{8}$, or

$$\frac{7}{8} - \frac{3}{8} = \frac{4}{8}.$$

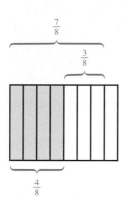

Write $\frac{4}{8}$ in lowest terms.

$$\frac{7}{8} - \frac{3}{8} = \frac{4 \div 4}{8 \div 4} = \frac{1}{2}$$

The steps for subtracting like fractions are very similar to those for adding like fractions.

Subtracting Like Fractions

Step 1 Find the numerator of the *difference* (the answer) by subtracting the numerators.

Step 2 Write the denominator of the like fractions as the denominator of the difference.

Step 3 Write the answer in lowest terms.

EXAMPLE 3 Subtracting Like Fractions

Subtract. Write answers in lowest terms or as a mixed number when possible.

(a) $\dfrac{11}{12} - \dfrac{7}{12}$

Step 1
$$\frac{11}{12} - \frac{7}{12} = \frac{11 - 7}{12} \quad \leftarrow \text{Subtract numerators.}$$

Step 2
$$= \frac{4}{12} \quad \begin{array}{l} \leftarrow \text{Difference} \\ \leftarrow \text{Same denominator} \end{array}$$

Write in lowest terms.

Step 3
$$= \frac{4 \div 4}{12 \div 4} = \frac{1}{3} \quad \text{In lowest terms}$$

CONTINUED ON NEXT PAGE

3. Subtract. Write answers in lowest terms and as mixed numbers where possible.

(a) $\dfrac{11}{15} - \dfrac{4}{15}$

(b) $\quad\dfrac{8}{9}$

$\quad -\dfrac{5}{9}$

(c) $\dfrac{27}{8} - \dfrac{14}{8}$

(d) $\quad\dfrac{103}{108}$

$\quad -\dfrac{48}{108}$

(b) $\dfrac{15}{3} - \dfrac{5}{3}$

$$\dfrac{15}{3} - \dfrac{5}{3} = \dfrac{15 - 5}{3} \quad \leftarrow \text{Subtract numerators.}$$
$$\phantom{\dfrac{15}{3} - \dfrac{5}{3}} \quad\quad\quad \leftarrow \text{Same denominator}$$
$$= \dfrac{10}{3}$$

Write as a mixed number.

$$\dfrac{10}{3} = 3\dfrac{1}{3}$$

Note
Fractions may be subtracted *only* if they have like denominators.

◀◀ **WORK PROBLEM 3 AT THE SIDE.**

ANSWERS

3. (a) $\dfrac{7}{15}$ (b) $\dfrac{1}{3}$ (c) $1\dfrac{5}{8}$ (d) $\dfrac{55}{108}$

3.1 Exercises

Add. Write answers in lowest terms and as mixed numbers where possible.

Example: $\dfrac{2}{9} + \dfrac{4}{9}$ **Solution:** Add numerators. $\dfrac{2}{9} + \dfrac{4}{9} = \dfrac{2+4}{9} = \dfrac{6}{9}$

Write in lowest terms. $\dfrac{6 \div 3}{9 \div 3} = \dfrac{2}{3}$

1. $\dfrac{3}{8} + \dfrac{2}{8}$ $\dfrac{5}{8}$

2. $\dfrac{1}{6} + \dfrac{4}{6}$ $\dfrac{5}{6}$

3. $\dfrac{7}{10} + \dfrac{2}{10}$ $\dfrac{9}{10}$

4. $\dfrac{9}{11} + \dfrac{1}{11}$ $\dfrac{10}{11}$

5. $\dfrac{1}{4} + \dfrac{1}{4}$ $\dfrac{1}{2}$

6. $\dfrac{1}{8} + \dfrac{1}{8}$ $\dfrac{1}{4}$

7. $\begin{array}{r} \dfrac{14}{12} \\[4pt] + \dfrac{1}{12} \\[2pt] \hline 1\dfrac{1}{4} \end{array}$

8. $\begin{array}{r} \dfrac{14}{10} \\[4pt] + \dfrac{1}{10} \\[2pt] \hline 1\dfrac{1}{2} \end{array}$

9. $\begin{array}{r} \dfrac{2}{9} \\[4pt] + \dfrac{1}{9} \\[2pt] \hline \dfrac{1}{3} \end{array}$

10. $\dfrac{8}{15} + \dfrac{4}{15}$ $\dfrac{4}{5}$

11. $\dfrac{6}{20} + \dfrac{4}{20} + \dfrac{3}{20}$ $\dfrac{13}{20}$

12. $\dfrac{1}{7} + \dfrac{2}{7} + \dfrac{3}{7}$ $\dfrac{6}{7}$

13. $\dfrac{3}{17} + \dfrac{2}{17} + \dfrac{5}{17}$ $\dfrac{10}{17}$

14. $\dfrac{5}{11} + \dfrac{1}{11} + \dfrac{4}{11}$ $\dfrac{10}{11}$

15. $\dfrac{3}{8} + \dfrac{7}{8} + \dfrac{2}{8}$ $1\dfrac{1}{2}$

16. $\dfrac{4}{9} + \dfrac{1}{9} + \dfrac{7}{9}$ $1\dfrac{1}{3}$

17. $\dfrac{2}{54} + \dfrac{8}{54} + \dfrac{12}{54}$ $\dfrac{11}{27}$

18. $\dfrac{7}{120} + \dfrac{9}{120} + \dfrac{18}{120}$ $\dfrac{17}{60}$

Subtract. Write answers in lowest terms and as mixed numbers where possible.

Example: $\dfrac{9}{12} - \dfrac{5}{12}$ **Solution:** Subtract numerators.

$$\dfrac{9}{12} - \dfrac{5}{12} = \dfrac{9-5}{12} = \dfrac{4}{12}$$

Write in lowest terms. $\dfrac{4 \div 4}{12 \div 4} = \dfrac{1}{3}$

19. $\dfrac{4}{5} - \dfrac{1}{5}$ $\dfrac{3}{5}$

20. $\dfrac{2}{3} - \dfrac{1}{3}$ $\dfrac{1}{3}$

21. $\dfrac{11}{15} - \dfrac{4}{15}$ $\dfrac{7}{15}$

22. $\dfrac{28}{32} - \dfrac{19}{32}$ $\dfrac{9}{32}$

23. $\dfrac{9}{10} - \dfrac{3}{10}$ $\dfrac{3}{5}$

24. $\dfrac{7}{8} - \dfrac{5}{8}$ $\dfrac{1}{4}$

25. $\begin{array}{r} \dfrac{31}{21} \\[4pt] - \dfrac{7}{21} \\[2pt] \hline 1\dfrac{1}{7} \end{array}$

26. $\begin{array}{r} \dfrac{43}{24} \\[4pt] - \dfrac{13}{24} \\[2pt] \hline 1\dfrac{1}{4} \end{array}$

27. $\begin{array}{r} \dfrac{27}{40} \\[4pt] - \dfrac{19}{40} \\[2pt] \hline \dfrac{1}{5} \end{array}$

28. $\dfrac{38}{55} - \dfrac{16}{55}$ $\dfrac{2}{5}$

   ✍ Writing   ◉ Conceptual   ▲ Challenging   ≈ Estimation

29. $\dfrac{103}{72} - \dfrac{7}{72}$ $1\dfrac{1}{3}$ **30.** $\dfrac{181}{100} - \dfrac{31}{100}$ $1\dfrac{1}{2}$ **31.** $\dfrac{87}{144} - \dfrac{71}{144}$ $\dfrac{1}{9}$ **32.** $\dfrac{356}{220} - \dfrac{235}{220}$ $\dfrac{11}{20}$ **33.** $\dfrac{746}{400} - \dfrac{506}{400}$ $\dfrac{3}{5}$

34. Describe in your own words the difference between like fractions and unlike fractions. Give three examples of each type.

> Like fractions have the same denominator. Unlike fractions have denominators that are different. Some examples are:
>
> like: $\dfrac{3}{8}, \dfrac{1}{8}, \dfrac{7}{8},$ unlike: $\dfrac{1}{2}, \dfrac{3}{4}, \dfrac{5}{8}$

35. In your own words, write an explanation of either how to add or subtract like fractions. Consider using three steps in your explanation.

> Three steps to add like fractions are:
> 1. Add the numerators of the fractions to find the numerator of the sum (the answer).
> 2. Use the denominator of the fractions as the denominator of the sum.
> 3. Write the answer in lowest terms.

Solve each application problem. Write answers in lowest terms.

36. Frank and Helen Ortiz have saved $\dfrac{3}{10}$ of the amount needed for an ocean cruise. If they save another $\dfrac{3}{10}$ of the amount needed, find the total fraction of the amount needed that they have saved. $\dfrac{3}{5}$

37. Captain Sisko has been ordered to investigate the wreckage of an enemy space ship. Sisko leads a search team $\dfrac{5}{12}$ mile down a ravine and then $\dfrac{1}{12}$ mile along a creek bed to the crash site. How far does the search team travel? $\dfrac{1}{2}$ mile

38. The Gerards owe $\dfrac{5}{9}$ of a loan for last year's vacation. If they pay $\dfrac{2}{9}$ of it this month, what fraction of the loan will they still owe? $\dfrac{1}{3}$

39. Sam must inspect $\dfrac{11}{16}$ of a mile of high voltage line. He has already inspected $\dfrac{5}{16}$ of a mile. How much additional line must he inspect? $\dfrac{3}{8}$ mile

▲ 40. A flower grower purchased $\dfrac{9}{10}$ acre of land one year and $\dfrac{3}{10}$ acre the next year. She then sold $\dfrac{7}{10}$ acre of land. How much land does she now have?

$\dfrac{1}{2}$ acre

▲ 41. A forester planted $\dfrac{5}{12}$ acre in seedlings in the morning and $\dfrac{11}{12}$ acre in the afternoon. If $\dfrac{7}{12}$ acre of seedlings were destroyed by frost, how many acres remained?

$\dfrac{3}{4}$ acre

Review and Prepare

Find the prime factorization. Do not use exponents when writing the answer. (For help, see Section 2.3.)

42. 6 $2 \cdot 3$ **43.** 10 $2 \cdot 5$ **44.** 30 $2 \cdot 3 \cdot 5$

45. 100 $2 \cdot 2 \cdot 5 \cdot 5$ **46.** 45 $3 \cdot 3 \cdot 5$ **47.** 75 $3 \cdot 5 \cdot 5$

48. 81 $3 \cdot 3 \cdot 3 \cdot 3$ **49.** 125 $5 \cdot 5 \cdot 5$ **50.** 200 $2 \cdot 2 \cdot 2 \cdot 5 \cdot 5$

3.2 Least Common Multiples

Only *like* fractions can be added or subtracted. Because of this, *unlike* fractions must be rewritten as *like* fractions before adding or subtracting.

OBJECTIVE 1 Unlike fractions can be written as like fractions by finding the *least common multiple* of the denominators.

Least Common Multiple

The **least common multiple (LCM)** of two whole numbers is the smallest whole number divisible by both those numbers.

EXAMPLE 1 Finding the Multiples of a Number

The list shows you multiples of 6.

$$6, 12, 18, 24, 30, \ldots$$

(The three dots show the list continues in the same pattern without stopping.) The next list shows multiples of 9.

$$9, 18, 27, 36, 45, \ldots$$

The smallest number found in both lists is 18, so 18 is the **least common multiple** of 6 and 9; the number 18 is the smallest whole number divisible by both 6 and 9.

Multiples of 6   6, 12, **18**, 24, 30, . . .

Multiples of 9   9, **18**, 27, 36, 45, . . .

18 is the smallest number found in both lists.

WORK PROBLEM 1 AT THE SIDE. ▶▶

OBJECTIVE 2 The least common multiple of small numbers can sometimes be determined by inspection. Can you think of a number that can be divided evenly by both 3 and 4? What about 6 or 8; perhaps 10 or 12? The number 12 will work; it is the least common multiple of the numbers 3 and 4. A method that works well to find the least common multiple is to write multiples of the larger number. In this case, write the multiples of 4:

$$4, 8, 12, 16, 20, \ldots$$

Now check each multiple of 4 to see if it is divisible by 3.

4 is *not* divisible by 3.

8 is *not* divisible by 3.

12 *is* divisible by 3.

The first one that works is 12, so 12 is the least common multiple.

EXAMPLE 2 Finding the Least Common Multiple

Use multiples of the larger number to find the least common multiple of 6 and 9.

We start by writing the first few multiples of 9.

$$9, 18, 27, 36, 45, 54, \ldots \leftarrow \text{Multiples of 9}$$

CONTINUED ON NEXT PAGE

OBJECTIVES

1 ▶ Find the least common multiple.

2 ▶ Find the least common multiple by using multiples of the largest number.

3 ▶ Find the least common multiple by using prime factorization and a table.

4 ▶ Find the least common multiple by using an alternative method.

5 ▶ Write a fraction with an indicated denominator.

FOR EXTRA HELP

Tutorial   Tape 5   SSM, Sec. 3.2

1. (a) List the multiples of 8.

8, _____, _____, _____,

_____, _____, . . .

(b) List the multiples of 10.

10, _____, _____, _____,

_____, . . .

(c) Find the least common multiple of 8 and 10.

ANSWERS

1. (a) 16, 24, 32, 40, 48, . . .

(b) 20, 30, 40, 50, . . .

(c) 40

177

2. Use multiples of the larger number to find the least common multiple in each set of numbers.

(a) 4 and 6

(b) 3 and 7

(c) 6 and 8

(d) 5 and 9

3. Find the least common multiple of 36 and 54.

(a) Find the prime factorization of each number.

(b) Complete this table.

Prime	2	3
36 =		
54 =		

(c) Identify the largest product in each column.

Prime	2	3
36 =	2 • 2 •	3 • 3
54 =	2 •	3 • 3 • 3

(d) Find the least common multiple.
$2 \cdot 2 \cdot \underline{\quad} \cdot \underline{\quad} \cdot \underline{\quad} =$

ANSWERS

2. (a) 12 **(b)** 21 **(c)** 24 **(d)** 45

3. (a) 36 = 2 • 2 • 3 • 3
     54 = 2 • 3 • 3 • 3
(b)

	2	3
36 =	2 • 2 •	3 • 3
54 =	2 •	3 • 3 • 3

(c) 2 • 2; 3 • 3 • 3 **(d)** 3, 3, 3; 108

Now, check each multiple of 9 to see if it is divisible by 6. The first multiple of 9 that is divisible by 6 is 18.

9, 18, 27, 36, 45, 54, . . . ← Multiples of 9

First multiple divisible by 6
(18 ÷ 6 = 3)

The least common multiple of the numbers 6 and 9 is 18.

◄◄ WORK PROBLEM 2 AT THE SIDE.

OBJECTIVE 3 Example 2 shows how to find the least common multiple by making a list of the multiples of the larger number. Although this method works for smaller numbers, it is usually easier to find the least common multiple for larger numbers by using prime factorization, as shown in the next example.

E X A M P L E 3 **Applying Prime Factorization Knowledge**

Use prime factorization to find the least common multiple of 18 and 60.

We start by finding the prime factorization of each number.

$$18 = 2 \cdot 3 \cdot 3 \qquad 60 = 2 \cdot 2 \cdot 3 \cdot 5$$

Next, place the factorizations in a table, as shown below.

Prime	2	3	5
18 =	2 •	3 • 3	
60 =	2 • 2 •	3 •	5

Then circle the largest product in each column and write this product in the bottom row of the table.

Prime	2	3	5
18 =	2 •	⟨3 • 3⟩	
60 =	⟨2 • 2⟩ •	3 •	⟨5⟩
LCM =	⟨2 • 2⟩	⟨3 • 3⟩	⟨5⟩

Now multiply the circled products to find the least common multiple.

least common multiple (LCM) = **2 • 2 • 3 • 3 • 5 = 180**

The smallest whole number divisible by both 18 and 60 is 180.

◄◄ WORK PROBLEM 3 AT THE SIDE.

E X A M P L E 4 **Using Prime Factorization**

Find the least common multiple of 12, 18, and 40.

Write each prime factorization.

$$12 = 2 \cdot 2 \cdot 3 \qquad 18 = 2 \cdot 3 \cdot 3 \qquad 40 = 2 \cdot 2 \cdot 2 \cdot 5$$

CONTINUED ON NEXT PAGE

Prepare the following table.

Prime	2	3	5
12 =	2 • 2 •	3	
18 =	2 •	3 • 3	
40 =	2 • 2 • 2 •		5

Circle the largest product in each column.

Prime	2	3	5
12 =	2 • 2 •	3	
18 =	2 •	⟨3 • 3⟩	
40 =	⟨2 • 2 • 2⟩ •		⑤
LCM =	⟨2 • 2 • 2⟩	⟨3 • 3⟩	⑤

Now multiply the circled products.

    least common multiple (LCM) = 2 • 2 • 2 • 3 • 3 • 5 = 360

The smallest whole number divisible by 12, 18, and 40 is 360.

Note
If two of the products in a column are equal, circle either one but not both. Only one product in each column will be used.

WORK PROBLEM 4 AT THE SIDE. ▶▶

E X A M P L E   5   Find the Least Common Multiple

Find the least common multiple for each set of numbers.

(a) 5 and 35

    Write each prime factorization.

               5 = 5   35 = 5 • 7

Prepare the table.

Prime	5	7
5 =	5	
35 =	5 •	7

Circle the largest product in each column, in this case, the only prime number in the column.

Prime	5	7
5 =	5	
35 =	⑤ •	⑦
LCM =	⑤	⑦

      least common multiple (LCM) = 5 • 7 = 35

(b) 20, 24, 42

    Write each prime factorization

        20 = 2 • 2 • 5   24 = 2 • 2 • 2 • 3   42 = 2 • 3 • 7

CONTINUED ON NEXT PAGE

---

4. Find the least common multiple of the denominators in these fractions.

(a) $\dfrac{2}{3}$ and $\dfrac{1}{10}$

(b) $\dfrac{3}{8}$ and $\dfrac{6}{5}$

(c) $\dfrac{5}{6}$ and $\dfrac{1}{14}$

(d) $\dfrac{5}{18}$ and $\dfrac{7}{24}$

ANSWERS

4. (a) 30 **(b)** 40 **(c)** 42 **(d)** 72

5. Find the least common multiple for each set of numbers.

(a) 12, 15

(b) 8, 9, 12

(c) 18, 20, 30

(d) 15, 20, 30, 40

Prepare the table.

Prime	2	3	5	7
20 =	2 • 2 •	5		
24 =	2 • 2 • 2 •	3		
42 =	2 •	3 •		7

Circle the largest product in each column.

Prime	2	3	5	7
20 =	2 • 2 •		⑤	
24 =	(2 • 2 • 2) •	3		
42 =	2 •	③ •		⑦
LCM =	(2 • 2 • 2)	③	⑤	⑦

least common multiple (LCM) = 2 • 2 • 2 • 3 • 5 • 7 = **840**

◀◀ **WORK PROBLEM 5 AT THE SIDE.**

OBJECTIVE 4 Some people like the following *alternative method* for finding the least common multiple. Try both methods, and *use the one you prefer.* As a review, a list of the first few primes follows.

$$2, 3, 5, 7, 11, 13, 17$$

E X A M P L E   6   Alternative Method for Finding the Least Common Multiple

Find the least common multiple of each set of numbers.

(a) 14 and 21

Start by trying to divide 14 and 21 by the prime numbers listed above. Use the following shortcut.

Divide by 2, the first prime.

$$\begin{array}{r|rr} 2 & 14 & 2\!\!\!/1 \\ \hline & 7 & 21 \end{array}$$

Because 21 cannot be divided evenly by 2, cross 21 out and bring it down. Divide by 3, the second prime.

$$\begin{array}{r|rr} 2 & 14 & 2\!\!\!/1 \\ 3 & 7 & 21 \\ \hline & 7 & 7 \end{array}$$

Since 7 cannot be divided evenly by 5, the third prime, skip 5, and divide by the next prime, 7.

Divide by 7, the fourth prime.

$$\begin{array}{r|rr} 2 & 14 & 2\!\!\!/1 \\ 3 & 7 & 21 \\ 7 & 7 & 7 \\ \hline & 1 & 1 \end{array}$$ All quotients are 1.

CONTINUED ON NEXT PAGE

ANSWERS

5. (a) 60 **(b)** 72 **(c)** 180 **(d)** 120

When all quotients are 1, multiply the prime numbers on the left side.

$$\text{least common multiple} = 2 \cdot 3 \cdot 7 = 42$$

The least common multiple of 14 and 21 is 42.

(b) 6, 15, 18

Divide by 2.

$$
\begin{array}{c|ccc}
2 & 6 & \cancel{15} & 18 \\
  & 3 & 15 & 9
\end{array}
$$

Cross out 15 and bring it down.

Divide by 3.

$$
\begin{array}{c|ccc}
2 & 6 & \cancel{15} & 18 \\
3 & 3 & 15 & 9 \\
  & 1 & 5 & 3
\end{array}
$$

Divide by 3 again.

$$
\begin{array}{c|ccc}
2 & 6 & \cancel{15} & 18 \\
3 & 3 & 15 & 9 \\
3 & \cancel{1} & \cancel{5} & 3 \\
  & 1 & 5 & 1
\end{array}
$$

Finally, divide by 5.

$$
\begin{array}{c|ccc}
2 & 6 & \cancel{15} & 18 \\
3 & 3 & 15 & 9 \\
3 & \cancel{1} & \cancel{5} & 3 \\
5 & \cancel{1} & 5 & \cancel{1} \\
  & 1 & 1 & 1
\end{array}
$$

All quotients are 1.

Multiply the prime numbers on the side.

$$2 \cdot 3 \cdot 3 \cdot 5 = 90 \leftarrow \text{Least common multiple}$$

WORK PROBLEMS 6 AND 7 AT THE SIDE. ▶▶

OBJECTIVE 5 ▶ When adding and subtracting unlike fractions, the least common multiple is used as the denominator of the fractions.

┌ **E X A M P L E 7**   **Writing a Fraction with an Indicated Denominator**

Write the fraction $\frac{2}{3}$ by using a denominator of 15.

Find a numerator, so that

$$\frac{2}{3} = \frac{}{15}.$$

To find the new numerator, first divide **15** by **3**.

$$\frac{2}{3} = \frac{}{15} \qquad 15 \div 3 = 5$$

Multiply both numerator and denominator of the fraction $\frac{2}{3}$ by 5.

$$\frac{2}{3} = \frac{2 \cdot 5}{3 \cdot 5} = \frac{10}{15}$$

6. In the problems below, the divisions have already been worked out. Multiply the prime numbers on the left to find the least common multiple.

(a)
$$
\begin{array}{c|cc}
2 & 6 & \cancel{15} \\
3 & 3 & 15 \\
5 & \cancel{1} & 5 \\
  & 1 & 1
\end{array}
$$

(b)
$$
\begin{array}{c|cc}
2 & 20 & 36 \\
2 & 10 & 18 \\
3 & \cancel{5} & 9 \\
3 & \cancel{5} & 3 \\
5 & 5 & \cancel{1} \\
  & 1 & 1
\end{array}
$$

7. Find the least common multiple of each set of numbers.

(a) 9 and 24

(b) 25 and 30

(c) 4, 8, and 12

(d) 25, 20, 35

ANSWERS

6. (a) 30 **(b)** 180

7. (a) 72 **(b)** 150 **(c)** 24 **(d)** 700

8. Write each fraction by using the indicated denominator.

(a) $\dfrac{1}{3} = \dfrac{}{12}$

(b) $\dfrac{7}{9} = \dfrac{}{27}$

(c) $\dfrac{4}{5} = \dfrac{}{50}$

(d) $\dfrac{6}{11} = \dfrac{}{55}$

This process is just the opposite of writing a fraction in lowest terms. Check the answer by writing $\frac{10}{15}$ in lowest terms; you should get $\frac{2}{3}$.

E X A M P L E   8   Changing to a New Denominator

Write each fraction with the indicated denominator.

(a) $\dfrac{3}{8} = \dfrac{}{48}$

Divide 48 by 8, getting 6. Now multiply both numerator and denominator of $\frac{3}{8}$ by 6.

$$\frac{3}{8} = \frac{3 \cdot 6}{8 \cdot 6} = \frac{18}{48} \qquad \text{Multiply numerator and denominator by 6.}$$

That is, $\frac{3}{8} = \frac{18}{48}$. As a check, write $\frac{18}{48}$ in lowest terms.

(b) $\dfrac{5}{6} = \dfrac{}{42}$

Divide 42 by 6, getting 7. Next, multiply both numerator and denominator of $\frac{5}{6}$ by 7.

$$\frac{5}{6} = \frac{5 \cdot 7}{6 \cdot 7} = \frac{35}{42} \qquad \text{Multiply numerator and denominator by 7.}$$

This shows that $\frac{5}{6} = \frac{35}{42}$. As a check, write $\frac{35}{42}$ in lowest terms.

Note
In Example 7 the fraction $\frac{2}{3}$ was multiplied by $\frac{5}{5}$. In Example 8 the fraction $\frac{3}{8}$ was multiplied by $\frac{6}{6}$ and the fraction $\frac{5}{6}$ was multiplied by $\frac{7}{7}$. The fractions $\frac{5}{5}$, $\frac{6}{6}$, and $\frac{7}{7}$ are all equal to 1.

$$\frac{5}{5} = 1 \qquad \frac{6}{6} = 1 \qquad \frac{7}{7} = 1$$

Recall that any number multiplied by 1 is the number itself.

◄◄ **WORK PROBLEM 8 AT THE SIDE.**

ANSWERS
8. (a) $\dfrac{4}{12}$ (b) $\dfrac{21}{27}$ (c) $\dfrac{40}{50}$ (d) $\dfrac{30}{55}$

3.2 Exercises

Use multiples of the larger number to find the least common multiple in each set of numbers.

Example: 8 and 10   **Solution:** Write the first few multiples of 10.

10, 20, 30, 40, 50, . . . ← Multiples of 10

⌐ First multiple divisible by 8

$(40 \div 8 = 5)$

The least common multiple of the numbers 8 and 10 is 40.

1. 5 and 10   10

2. 6 and 12   12

3. 4 and 6   12

4. 3 and 7   21

5. 4 and 9   36

6. 6 and 8   24

7. 2 and 7   14

8. 5 and 8   40

9. 4 and 10   20

10. 6 and 10   30

11. 20 and 50   100

12. 25 and 75   75

Find the least common multiple of each set of numbers.

Example: 18, 30   **Solution:** Complete a table and circle the largest product in each column.

Prime	2	3	5
18 =	2 ·	③ · ③	
30 =	② ·	3 ·	⑤
LCM =	②	③ · ③	⑤

The least common multiple (LCM) = $2 \cdot 3 \cdot 3 \cdot 5 = 90$.

✍ Writing   ◎ Conceptual   ▲ Challenging   ≈ Estimation

Alternative Method: Divide by 2   2 |18   30

Divide by 3   3 | 9   15

Divide by 3   3 | 3   5̸

Divide by 5   5 | 1̸   5

1   1

Multiply the prime numbers on the side.

2 • 3 • 3 • 5 = **90** = least common multiple

13. 6, 9   18

14. 15, 20   60

15. 18, 24   72

16. 6, 14   42

17. 18, 20, 24   360

18. 20, 24, 30   120

19. 6, 8, 10, 12   120

20. 8, 9, 12, 18   72

21. 12, 15, 18, 20   180

22. 6, 9, 27, 36   108

23. 15, 20, 30, 40   120

24. 5, 18, 25, 30   450

Rewrite each of the following fractions, so that it has a denominator of 24.

Example: $\dfrac{7}{12} = \dfrac{}{24}$ **Solution:** Divide 24 by 12, getting 2. Next, multiply both numerator and denominator by 2.

$$\frac{7}{12} = \frac{7 \cdot 2}{12 \cdot 2} = \frac{14}{24}$$

This shows that $\frac{7}{12} = \frac{14}{24}$. As a check, write $\frac{14}{24}$ in lowest terms.

25. $\dfrac{1}{2} = \dfrac{12}{24}$

26. $\dfrac{1}{4} = \dfrac{6}{24}$

27. $\dfrac{3}{4} = \dfrac{18}{24}$

28. $\dfrac{5}{12} = \dfrac{10}{24}$

29. $\dfrac{3}{8} = \dfrac{9}{24}$

30. $\dfrac{5}{8} = \dfrac{15}{24}$

Rewrite each of the following fractions with the indicated denominators.

31. $\dfrac{1}{2} = \dfrac{2}{4}$

32. $\dfrac{3}{8} = \dfrac{6}{16}$

33. $\dfrac{9}{10} = \dfrac{36}{40}$

34. $\dfrac{7}{11} = \dfrac{28}{44}$

35. $\dfrac{7}{8} = \dfrac{28}{32}$

36. $\dfrac{5}{12} = \dfrac{20}{48}$

37. $\dfrac{5}{6} = \dfrac{55}{66}$

38. $\dfrac{7}{8} = \dfrac{84}{96}$

39. $\dfrac{9}{8} = \dfrac{45}{40}$

40. $\dfrac{6}{5} = \dfrac{48}{40}$

41. $\dfrac{9}{7} = \dfrac{72}{56}$

42. $\dfrac{3}{2} = \dfrac{96}{64}$

43. $\dfrac{8}{3} = \dfrac{136}{51}$

44. $\dfrac{9}{5} = \dfrac{216}{120}$

45. $\dfrac{8}{11} = \dfrac{96}{132}$

46. $\dfrac{7}{15} = \dfrac{98}{210}$

47. $\dfrac{3}{16} = \dfrac{27}{144}$

48. $\dfrac{9}{32} = \dfrac{63}{224}$

49. There are several methods for finding the least common multiple (LCM). Will you use the method using multiples of the largest number or the method using prime factors? Why? Would you ever use the other method?

It probably depends on how large the numbers are. If the numbers are small, the method using multiples of the largest number seems best. If numbers are larger, or there are more of them, then the factorization method will be better.

50. Explain in your own words how to write a fraction with an indicated denominator. As part of your explanation, show how to change $\frac{3}{4}$ to a fraction having 12 as a denominator.

Divide the desired denominator by the given denominator. Next, multiply both the given numerator and denominator by this number.

Divide 12 by 4 to get 3. Then multiply.

$$\frac{3}{4} = \frac{3 \cdot 3}{4 \cdot 3} = \frac{9}{12}$$

▲ *Find the least common multiple of the denominators of the following fractions.*

51. $\dfrac{17}{800}, \dfrac{23}{3600}$ 7200

52. $\dfrac{53}{288}, \dfrac{115}{1568}$ 14,112

53. $\dfrac{109}{1512}, \dfrac{23}{392}$ 10,584

54. $\dfrac{61}{810} \cdot \dfrac{37}{1170}$ 10,530

Review and Prepare

Write each improper fraction as a mixed number. (For help, see Section 2.2.)

55. $\dfrac{8}{5}$ $1\dfrac{3}{5}$

56. $\dfrac{9}{4}$ $2\dfrac{1}{4}$

57. $\dfrac{15}{8}$ $1\dfrac{7}{8}$

58. $\dfrac{14}{9}$ $1\dfrac{5}{9}$

59. $\dfrac{27}{7}$ $3\dfrac{6}{7}$

60. $\dfrac{28}{9}$ $3\dfrac{1}{9}$

3.3 *Adding and Subtracting Unlike Fractions*

OBJECTIVES

1 ▶ Add unlike fractions.

2 ▶ Add fractions vertically.

3 ▶ Subtract unlike fractions.

FOR EXTRA HELP

Tutorial   Tape 5   SSM, Sec. 3.3

OBJECTIVE 1 ▶ To add unlike fractions we must first change them to like fractions (fractions with the same denominator). For example, the diagrams show $\frac{3}{8}$ and $\frac{1}{4}$.

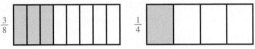

These fractions can be added by changing them to like fractions. Make like fractions by changing $\frac{1}{4}$ to the equivalent fraction $\frac{2}{8}$.

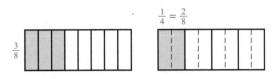

Next, add.

$$\frac{3}{8} + \frac{1}{4} = \frac{3}{8} + \frac{2}{8} = \frac{5}{8}$$

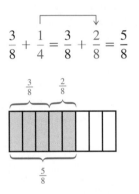

Unlike fractions may be added or subtracted using the following steps.

Adding or Subtracting Unlike Fractions

Step 1 Rewrite the **unlike fractions** as **like fractions** having the least common multiple as a denominator. This new denominator is called the **least common denominator (LCD).**

Step 2 Add or subtract as you did like fractions.

Step 3 Write the answer in lowest terms.

┌ **E X A M P L E I**   **Adding Unlike Fractions**

Add $\frac{2}{3}$ and $\frac{1}{9}$.

The least common multiple of 3 and 9 is 9, so write the fractions as like fractions with a denominator of 9. This denominator is called the **least common denominator** of 3 and 9. First,

Step 1 $\frac{2}{3} = \frac{}{9}$.

            Divide 9 by 3, getting 3. Next, multiply numerator and denominator by 3.

$$\frac{2}{3} = \frac{2 \cdot 3}{3 \cdot 3} = \frac{6}{9}$$

└ **CONTINUED ON NEXT PAGE**

1. Add.

(a) $\dfrac{1}{2} + \dfrac{1}{4}$

(b) $\dfrac{1}{8} + \dfrac{3}{4}$

(c) $\dfrac{3}{10} + \dfrac{2}{5}$

(d) $\dfrac{1}{12} + \dfrac{5}{6}$

2. Add. Write answers in lowest terms.

(a) $\dfrac{1}{5} + \dfrac{3}{10}$

(b) $\dfrac{1}{3} + \dfrac{1}{12}$

(c) $\dfrac{1}{10} + \dfrac{1}{3} + \dfrac{1}{6}$

ANSWERS

1. (a) $\dfrac{3}{4}$ (b) $\dfrac{7}{8}$ (c) $\dfrac{7}{10}$ (d) $\dfrac{11}{12}$

2. (a) $\dfrac{1}{2}$ (b) $\dfrac{5}{12}$ (c) $\dfrac{3}{5}$

Now, add the like fractions $\frac{6}{9}$ and $\frac{1}{9}$.

Step 2 $\dfrac{2}{3} + \dfrac{1}{9} = \dfrac{6}{9} + \dfrac{1}{9} = \dfrac{6+1}{9} = \dfrac{7}{9}$

Step 3 Step 3 is not needed because $\frac{7}{9}$ is already in lowest terms.

◀◀ **WORK PROBLEM 1 AT THE SIDE.**

E X A M P L E   2   Adding Fractions

Add the following fractions using the three steps. Write all answers in lowest terms.

(a) $\dfrac{1}{3} + \dfrac{1}{6}$

The least common multiple of 3 and 6 is 6. Write both fractions as fractions with a least common denominator of 6.

Rewrite as like fractions.

Step 1 $\dfrac{1}{3} + \dfrac{1}{6} = \dfrac{2}{6} + \dfrac{1}{6}$

Add numerators.

Step 2 $\dfrac{2}{6} + \dfrac{1}{6} = \dfrac{3}{6}$

Step 3 $\dfrac{3}{6} = \dfrac{1}{2}$ ⟵ Lowest terms

(b) $\dfrac{6}{15} + \dfrac{3}{10}$

The least common multiple of 15 and 10 is 30, so write both fractions with a least common denominator of 30.

Rewrite as like fractions.

Step 1 $\dfrac{6}{15} + \dfrac{3}{10} = \dfrac{12}{30} + \dfrac{9}{30}$

Add numerators.

Step 2 $\dfrac{12}{30} + \dfrac{9}{30} = \dfrac{21}{30}$

Step 3 $\dfrac{21}{30} = \dfrac{7}{10}$ ⟵ Lowest terms

◀◀ **WORK PROBLEM 2 AT THE SIDE.**

OBJECTIVE 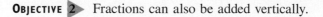 **2** Fractions can also be added vertically.

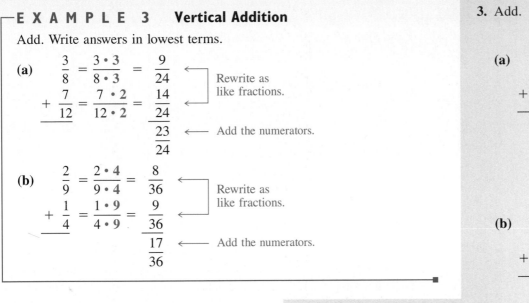

E X A M P L E 3   Vertical Addition

Add. Write answers in lowest terms.

(a)

$$\frac{3}{8} = \frac{3 \cdot 3}{8 \cdot 3} = \frac{9}{24}$$ ← Rewrite as like fractions.

$$+\frac{7}{12} = \frac{7 \cdot 2}{12 \cdot 2} = \frac{14}{24}$$ ←

$$\frac{23}{24}$$ ← Add the numerators.

(b)

$$\frac{2}{9} = \frac{2 \cdot 4}{9 \cdot 4} = \frac{8}{36}$$ ← Rewrite as like fractions.

$$+\frac{1}{4} = \frac{1 \cdot 9}{4 \cdot 9} = \frac{9}{36}$$ ←

$$\frac{17}{36}$$ ← Add the numerators.

> **WORK PROBLEM 3 AT THE SIDE.** ▶▶

OBJECTIVE ▸ **3** ▸ The next example shows subtraction of unlike fractions.

E X A M P L E 4   Subtracting Unlike Fractions

Subtract the following fractions. Write answers in lowest terms. As with addition, rewrite unlike fractions with a least common denominator.

(a) $\frac{3}{4} - \frac{3}{8}$

Rewrite as like fractions.

Step 1 $\frac{3}{4} - \frac{3}{8} = \frac{6}{8} - \frac{3}{8}$

Subtract numerators.

Step 2 $\frac{6}{8} - \frac{3}{8} = \frac{6 - 3}{8} = \frac{3}{8}$

Step 3   Not needed. ($\frac{3}{8}$ is in lowest terms.)

(b) $\frac{3}{4} - \frac{5}{9}$

Rewrite as like fractions.

Step 1 $\frac{3}{4} - \frac{5}{9} = \frac{27}{36} - \frac{20}{36}$

Subtract numerators.

Step 2 $\frac{27}{36} - \frac{20}{36} = \frac{27 - 20}{36} = \frac{7}{36}$

Step 3   Not needed. ($\frac{7}{36}$ is in lowest terms.)

Note

Step 3 was not needed in Example 4 because the answers $\frac{3}{8}$ and $\frac{7}{36}$ are both in lowest terms.

> **WORK PROBLEM 4 AT THE SIDE.** ▶▶

---

3. Add.

(a)

$$\frac{1}{8}$$

$$+\frac{3}{4}$$

(b)

$$\frac{2}{3}$$

$$+\frac{2}{9}$$

4. Subtract. Write answers in lowest terms.

(a) $\frac{1}{2} - \frac{3}{8}$

(b) $\frac{7}{8} - \frac{5}{6}$

(c)

$$\frac{17}{18}$$

$$-\frac{20}{27}$$

ANSWERS

3. (a) $\frac{7}{8}$ **(b)** $\frac{8}{9}$

4. (a) $\frac{1}{8}$ **(b)** $\frac{1}{24}$ **(c)** $\frac{11}{54}$

Numbers in the

Real World *collaborative investigations*

The time signature at the beginning of a piece of music looks like a fraction. Commonly used time signatures are $\frac{2}{4}$, $\frac{3}{4}$, $\frac{4}{4}$, and $\frac{6}{8}$. Musicians use the time signature to tell how long to hold each note. The values of different notes can be written as fractions:

𝅝 = 1   𝅗𝅥 = $\frac{1}{2}$ ♩ = $\frac{1}{4}$ ♪ = $\frac{1}{8}$ 𝅘𝅥𝅯 = $\frac{1}{16}$

Music is divided into measures. In $\frac{4}{4}$ time, each measure contains notes that add up to $\frac{4}{4}$ (or 1). In $\frac{2}{4}$ time the notes in each measure add up to $\frac{2}{4}$ (or $\frac{1}{2}$), and so on for $\frac{3}{4}$ time and $\frac{6}{8}$ time.

Write one or more notes in each measure to make it add up to its time signature. Use as many different kinds of notes as possible. **There are many correct answers. Some possibilities are:**

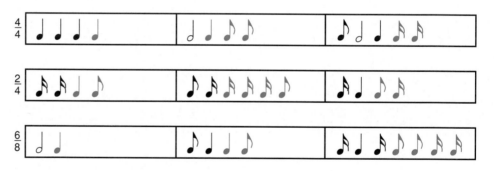

Below are excerpts from "Jingle Bells" and "The Star-Spangled Banner." Divide each line of music into measures.

3.3 Exercises

Add the following fractions. Write answers in lowest terms.

Example: $\frac{2}{3} + \frac{1}{6}$

Solution:

Step 1
Rewrite as
like fractions.

Step 2
Add
numerators.

Step 3
Lowest
terms

$$\frac{2}{3} + \frac{1}{6} = \frac{4}{6} + \frac{1}{6} = \frac{4+1}{6} = \frac{5}{6}$$

Least common
denominator is 6.

1. $\frac{3}{4} + \frac{1}{8}$ $\frac{7}{8}$

2. $\frac{1}{3} + \frac{1}{2}$ $\frac{5}{6}$

3. $\frac{1}{14} + \frac{3}{7}$ $\frac{1}{2}$

4. $\frac{2}{9} + \frac{2}{3}$ $\frac{8}{9}$

5. $\frac{9}{20} + \frac{3}{10}$ $\frac{3}{4}$

6. $\frac{5}{8} + \frac{1}{4}$ $\frac{7}{8}$

7. $\frac{3}{5} + \frac{3}{8}$ $\frac{39}{40}$

8. $\frac{5}{7} + \frac{3}{14}$ $\frac{13}{14}$

9. $\frac{2}{9} + \frac{5}{12}$ $\frac{23}{36}$

10. $\frac{5}{8} + \frac{1}{12}$ $\frac{17}{24}$

11. $\frac{1}{3} + \frac{3}{5}$ $\frac{14}{15}$

12. $\frac{2}{5} + \frac{3}{7}$ $\frac{29}{35}$

13. $\frac{1}{4} + \frac{2}{9} + \frac{1}{3}$ $\frac{29}{36}$

14. $\frac{3}{7} + \frac{2}{5} + \frac{1}{10}$ $\frac{13}{14}$

15. $\frac{3}{10} + \frac{2}{5} + \frac{3}{20}$ $\frac{17}{20}$

16. $\frac{1}{3} + \frac{3}{8} + \frac{1}{4}$ $\frac{23}{24}$

17. $\frac{4}{15} + \frac{1}{6} + \frac{1}{3}$ $\frac{23}{30}$

18. $\frac{5}{12} + \frac{2}{9} + \frac{1}{6}$ $\frac{29}{36}$

19. $\begin{array}{r} \frac{1}{3} \\ + \frac{1}{4} \\ \hline \frac{7}{12} \end{array}$

20. $\begin{array}{r} \frac{1}{12} \\ + \frac{5}{8} \\ \hline \frac{17}{24} \end{array}$

21. $\begin{array}{r} \frac{5}{12} \\ + \frac{1}{16} \\ \hline \frac{23}{48} \end{array}$

22. $\begin{array}{r} \frac{3}{7} \\ + \frac{1}{3} \\ \hline \frac{16}{21} \end{array}$

⊙ Conceptual   ▲ Challenging   ≈ Estimation

Subtract the following fractions. Write answers in lowest terms.

Example: $\dfrac{3}{5} - \dfrac{1}{2}$ **Solution:**

	Step 1	Step 2	Step 3
	Rewrite as like fractions.	Subtract numerators.	Lowest terms

$$\frac{3}{5} - \frac{1}{2} = \frac{6}{10} - \frac{5}{10} = \frac{6-5}{10} = \frac{1}{10}$$

Least common denominator is 10.

23. $\dfrac{3}{4} - \dfrac{1}{8}$ $\dfrac{5}{8}$

24. $\dfrac{7}{8} - \dfrac{1}{4}$ $\dfrac{5}{8}$

25. $\dfrac{2}{3} - \dfrac{1}{6}$ $\dfrac{1}{2}$

26. $\dfrac{7}{8} - \dfrac{1}{2}$ $\dfrac{3}{8}$

27. $\dfrac{5}{12} - \dfrac{1}{4}$ $\dfrac{1}{6}$

28. $\dfrac{5}{6} - \dfrac{7}{9}$ $\dfrac{1}{18}$

29. $\dfrac{11}{12} - \dfrac{3}{4}$ $\dfrac{1}{6}$

30. $\dfrac{5}{7} - \dfrac{1}{3}$ $\dfrac{8}{21}$

31. $\dfrac{8}{9} - \dfrac{7}{15}$ $\dfrac{19}{45}$

32.
$$\begin{array}{r} \dfrac{7}{8} \\ -\ \dfrac{2}{3} \\ \hline \dfrac{5}{24} \end{array}$$

33.
$$\begin{array}{r} \dfrac{4}{5} \\ -\ \dfrac{2}{3} \\ \hline \dfrac{2}{15} \end{array}$$

34.
$$\begin{array}{r} \dfrac{5}{8} \\ -\ \dfrac{1}{3} \\ \hline \dfrac{7}{24} \end{array}$$

35.
$$\begin{array}{r} \dfrac{5}{12} \\ -\ \dfrac{1}{16} \\ \hline \dfrac{17}{48} \end{array}$$

36.
$$\begin{array}{r} \dfrac{3}{4} \\ -\ \dfrac{5}{9} \\ \hline \dfrac{7}{36} \end{array}$$

Solve each of the following application problems.

Example: Two prospectors are loading packs on an uncooperative mule. They secure $\frac{1}{6}$ of the load in a half hour and $\frac{1}{4}$ of the load in another 20 minutes. What fraction of their supplies have they secured in 50 minutes?

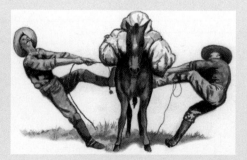

Approach: To find the fraction loaded in 50 minutes, add the fraction loaded in a half hour to the fraction loaded in another 20 minutes.

Solution: To find the total amount loaded, add:

$$\frac{1}{6} \text{ in a half hour } + \frac{1}{4} \text{ in another 20 minutes.}$$

$$\frac{1}{6} + \frac{1}{4} = \frac{2}{12} + \frac{3}{12} = \frac{5}{12}$$

By the end of 50 minutes, $\frac{5}{12}$ of the supplies had been secured.

37. The owner of Racy's Feed Store ordered $\frac{1}{3}$ cubic yard of corn, $\frac{3}{8}$ cubic yard of oats, and $\frac{1}{4}$ cubic yard of washed medium mesh gravel. Find the total cubic yards of products ordered.

$\frac{23}{24}$ cubic yard

38. Aaron Lee paid $\frac{1}{4}$ of his savings for tuition, $\frac{1}{10}$ of his savings for books, $\frac{1}{6}$ of his savings for rent, and $\frac{1}{12}$ of his savings for food. What fraction of his total savings has been spent?

$\frac{3}{5}$ spent

39. The Weiner Works has $\frac{3}{4}$ acre of land. If $\frac{1}{6}$ acre must remain as a green belt and the remainder is buildable, find the amount of land that is buildable.

$\frac{7}{12}$ acre

40. Della Daniel wants to open a day care center and has saved $\frac{2}{5}$ of the amount needed for start-up costs. If she saves another $\frac{1}{8}$ of the amount needed and then $\frac{1}{6}$ more, find the total portion of the start-up costs she has saved.

$\frac{83}{120}$ of the start-up costs

41. The warrior princess, Xena, urges local peasants to fortify their village against a warlord's attack. Find the total distance around (perimeter of) this parcel of land.

$\frac{41}{48}$ mile

42. When installing cabinets, Pete Phelps must be certain that the proper type and size of mounting hardware is used. Find the total length of the bolt below.

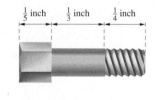

$\frac{47}{60}$ inch

43. A hydraulic jack contains $\frac{7}{8}$ gallon of hydraulic fluid. A cracked seal resulted in a loss of $\frac{1}{6}$ gallon of fluid in the morning and another $\frac{1}{3}$ gallon in the afternoon. Find the amount of fluid remaining.

$\frac{3}{8}$ gallon

44. Adrian Ortega drives a tanker for the British Petroleum Company. He leaves the refinery with his tanker filled to $\frac{7}{8}$ of capacity. If he delivers $\frac{1}{4}$ of the tank capacity at the first stop and $\frac{1}{3}$ of the tank capacity at the second stop, find the fraction of the tanker's capacity remaining.

$\frac{7}{24}$ of the tank capacity

45. Step 1 in adding or subtracting unlike fractions is to rewrite the fractions so they have the least common multiple as a denominator. Explain in your own words why this is necessary.

You cannot add or subtract until all the fractional pieces are the same size. For example, halves are larger than fourths, so you cannot add $\frac{1}{2} + \frac{1}{4}$ until you rewrite $\frac{1}{2}$ as $\frac{2}{4}$.

46. Briefly list the three steps used for addition and subtraction of unlike fractions.

Step 1 **Rewrite the unlike fractions as like fractions.**

Step 2 **Add or subtract like fractions.**

Step 3 **Write the answer in lowest terms.**

Refer to the circle graph to answer exercises 47–50.

THE DAY OF THE STUDENT

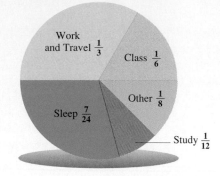

Work and Travel $\frac{1}{3}$

Class $\frac{1}{6}$

Other $\frac{1}{8}$

Sleep $\frac{7}{24}$

Study $\frac{1}{12}$

47. What fraction of the day was spent in class and study?

$\frac{1}{4}$

48. What fraction of the day was spent in work and travel and other?

$\frac{11}{24}$

▲ **49.** In which activity was the greatest amount of time spent? How many hours did this activity take?

work and travel; 8 hours

▲ **50.** In which activity was the least amount of time spent? How many hours did this activity take?

study; 2 hours

▲ **51.** A hazardous waste dump site will require $\frac{7}{8}$ mile of security fencing. The site has four sides with three of the sides measuring $\frac{1}{4}$ mile, $\frac{1}{6}$ mile, and $\frac{3}{8}$ mile. Find the length of the fourth side.

$\frac{1}{12}$ mile

▲ **52.** Chakotay is fitting a turquoise stone into a bear claw pendant. Find the diameter of the hole in the pendant. (The diameter is the distance across the center of the hole.)

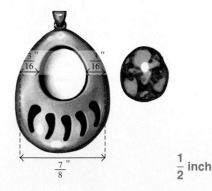

$\frac{3}{16}$" $\qquad$ $\frac{3}{16}$"

$\frac{7}{8}$"

$\frac{1}{2}$ inch

Review and Prepare

*Multiply or divide as indicated. Write answers as whole or mixed numbers. (For help, see **Section 2.8**.)*

53. $1\frac{3}{4} \cdot 2\frac{1}{2}$ $4\frac{3}{8}$

54. $5\frac{1}{4} \cdot 2\frac{1}{3}$ $12\frac{1}{4}$

55. $5 \cdot 2\frac{7}{10}$ $13\frac{1}{2}$

56. $5\frac{1}{2} \div 2\frac{3}{4}$ 2

57. $1\frac{1}{2} \div 3\frac{3}{4}$ $\frac{2}{5}$

58. $5\frac{3}{4} \div 2$ $2\frac{7}{8}$

3.4 *Adding and Subtracting Mixed Numbers*

Recall that a mixed number is the sum of a whole number and a fraction. For example,

$$3\frac{2}{5} \quad \text{means} \quad 3 + \frac{2}{5}.$$

OBJECTIVE 1 Add or subtract mixed numbers by adding or subtracting the fraction parts and then the whole number parts. It is a good idea to estimate the answer first, as we did when multiplying and dividing mixed numbers in **Section 2.8.**

WORK PROBLEM I AT THE SIDE. ▶▶

┌─
E X A M P L E I   **Adding and Subtracting Mixed Numbers**

≈ First estimate the answer. Then add or subtract to find the exact answer.

(a) $16\frac{1}{8} + 5\frac{5}{8}$

	estimate		*exact*
	16	$\xleftarrow{\text{Rounds to}}$	$16\frac{1}{8}$
	$+\quad 6$	$\longleftarrow$	$+\quad 5\frac{5}{8}$
	22		$21\frac{6}{8}$ ← Sum of fractions
			↑ Sum of whole numbers.

In lowest terms $\frac{6}{8}$ is $\frac{3}{4}$, so the final answer is $21\frac{3}{4}$. This is reasonable because it is close to the estimate of 22.

(b) $8\frac{5}{8} - 3\frac{1}{12}$

	estimate		*exact*
	9	$\xleftarrow{\text{Rounds to}}$	$8\frac{5}{8} = 8\frac{15}{24}$ ←
	$-\ 3$	$\longleftarrow$	$-\ 3\frac{1}{12} = 3\frac{2}{24}$ ← Least common denominator
	6		$5\frac{13}{24}$ ← Subtract fractions.
			↑ Subtract whole numbers.

$5\frac{13}{24}$ is reasonable because it is close to the estimated answer of 6. Just as before, check by adding $5\frac{13}{24}$ and $3\frac{1}{12}$; the sum should be $8\frac{5}{8}$.
─┘

> **Note**
> When estimating, if the numerator is *half* of the denominator or *more*, round up the whole number part. If the numerator is *less* than *half* the denominator, leave the whole number part as it is.

OBJECTIVES

1 ▶ Estimate an answer, then add or subtract mixed numbers.

2 ▶ Estimate an answer, then subtract mixed numbers by using borrowing.

3 ▶ Add or subtract mixed numbers using an alternate method.

FOR EXTRA HELP

Tutorial   Tape 6   SSM, Sec. 3.4

1. As a review of mixed numbers, convert mixed numbers to improper fractions and improper fractions to mixed numbers.

   (a) $\frac{5}{2}$

   (b) $\frac{14}{3}$

   (c) $6\frac{3}{4}$

   (d) $4\frac{5}{8}$

ANSWERS

1. **(a)** $2\frac{1}{2}$ **(b)** $4\frac{2}{3}$ **(c)** $\frac{27}{4}$ **(d)** $\frac{37}{8}$

195

2. First estimate, and then add or subtract to find the exact answer.

◀◀ **WORK PROBLEM 2 AT THE SIDE.**

(a) *estimate*   *exact*

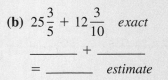

$$5 \xleftarrow{\text{Rounds to}} 4\frac{5}{8}$$

$$+\ 3 \longleftarrow +\ 3\frac{1}{4}$$

(b) $25\frac{3}{5} + 12\frac{3}{10}$ *exact*

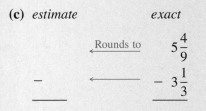

_____ + _____

= _____ *estimate*

(c) *estimate*   *exact*

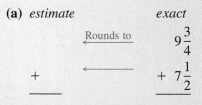

$$\xleftarrow{\text{Rounds to}} 5\frac{4}{9}$$

$$-\ \longleftarrow -\ 3\frac{1}{3}$$

3. First estimate, and then add to find the exact answer.

(a) *estimate*   *exact*

$$\xleftarrow{\text{Rounds to}} 9\frac{3}{4}$$

$$+ \longleftarrow +\ 7\frac{1}{2}$$

(b) *estimate*   *exact*

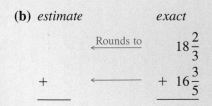

$$\xleftarrow{\text{Rounds to}} 18\frac{2}{3}$$

$$+ \longleftarrow +\ 16\frac{3}{5}$$

ANSWERS

2. (a) $5 + 3 = 8; 7\frac{7}{8}$

   (b) $26 + 12 = 38; 37\frac{9}{10}$

   (c) $5 - 3 = 2; 2\frac{1}{9}$

3. (a) $10 + 8 = 18; 17\frac{1}{4}$

   (b) $19 + 17 = 36; 35\frac{4}{15}$

When you add the fraction parts of mixed numbers, the answer may be greater than 1. If this happens, **carrying** from the fraction column to the whole number is the best procedure. (You wouldn't want a whole number along with an improper fraction.)

EXAMPLE 2   **Carrying When Adding Mixed Numbers**

≈ First estimate and then add $9\frac{5}{8} + 13\frac{7}{8}$.

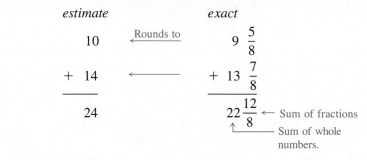

estimate   *exact*

$$10 \xleftarrow{\text{Rounds to}} 9\ \frac{5}{8}$$

$$+\ 14 \longleftarrow +\ 13\ \frac{7}{8}$$

$$\underline{\hphantom{+\ 14}} \qquad \underline{\hphantom{+\ 13\ \frac{7}{8}}}$$

$$24 \qquad\qquad 22\frac{12}{8} \leftarrow \text{Sum of fractions}$$

↑ Sum of whole numbers.

The improper fraction $\frac{12}{8}$ can be written in lowest terms as $\frac{3}{2}$. Because $\frac{3}{2} = 1\frac{1}{2}$, the sum is

$$22\frac{12}{8} = 22 + \frac{12}{8} = 22 + 1\frac{1}{2} = 23\frac{1}{2}.$$

The estimate was 24, so the exact answer is reasonable.

| **Note** |
| When adding mixed numbers, first add the fraction parts, then add the whole number parts. Then combine the two answers. |

◀◀ **WORK PROBLEM 3 AT THE SIDE.**

OBJECTIVE 2▶ Borrowing is sometimes necessary when subtracting mixed numbers.

EXAMPLE 3   **Borrowing When Subtracting Mixed Numbers**

≈ First estimate and then subtract.

(a) $7 - 2\frac{5}{6}$

estimate   *exact*

$$7 \xleftarrow{\text{Rounds to}} 7$$

$$-\ 3 \longleftarrow -\ 2\frac{5}{6}$$

There is no fraction from which to subtract $\frac{5}{6}$.

$$\underline{\hphantom{-\ 3}} \qquad \underline{\hphantom{-\ 2\frac{5}{6}}}$$

$$4$$

CONTINUED ON NEXT PAGE —

It is not possible to subtract $\frac{5}{6}$ without borrowing from the whole number 7 first.

$$\overset{\text{Borrow 1.}}{7 = 6 + \mathbf{1}}$$

$$1 = \frac{6}{6}$$

$$= 6 + \frac{\mathbf{6}}{\mathbf{6}}$$

$$= 6\frac{6}{6}$$

Next, subtract.

$$\begin{array}{r} 7 \phantom{\frac{5}{6}} = 6\frac{6}{6} \\ - 2\frac{5}{6} = 2\frac{5}{6} \\ \hline 4\frac{1}{6} \end{array}$$

The estimate was 4, so the exact answer is reasonable.

(b) $8\frac{1}{3} - 4\frac{3}{5}$

$$\begin{array}{ccc} \textit{estimate} & & \textit{exact} \\ 8 & \xleftarrow{\text{Rounds to}} & 8\frac{1}{3} = 8\frac{5}{15} \leftarrow \\ - 5 & \longleftarrow & - 4\frac{3}{5} = 4\frac{9}{15} \leftarrow \\ \hline 3 & & \end{array}$$

Least common denominator

It is not possible to subtract $\frac{9}{15}$ from $\frac{5}{15}$, so borrow from the whole number **8**.

$$\overset{\text{Borrow 1.}}{8\frac{5}{15} = 8 + \frac{5}{15} = 7 + 1 + \frac{5}{15}}$$

$$1 \text{ is } \frac{15}{15}.$$

$$= 7 + \frac{\mathbf{15}}{\mathbf{15}} + \frac{5}{15}$$

$$= 7 + \frac{\mathbf{20}}{\mathbf{15}} \longleftarrow \frac{15}{15} + \frac{5}{15}$$

$$= 7\frac{20}{15}$$

Next, subtract.

$$\begin{array}{r} 8\frac{1}{3} = 8\frac{5}{15} = 7\frac{20}{15} \\ - 4\frac{3}{5} = 4\frac{9}{15} = 4\frac{9}{15} \\ \hline 3\frac{11}{15} \end{array}$$

The exact answer is $3\frac{11}{15}$ (lowest terms), which is reasonable because it is close to the estimate of 3.

WORK PROBLEM 4 AT THE SIDE. ▶▶

4. First estimate and then subtract to find the exact answer.

(a)

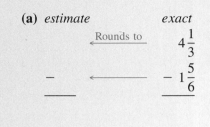

$$\begin{array}{ccc} \textit{estimate} & & \textit{exact} \\ & \xleftarrow{\text{Rounds to}} & 4\frac{1}{3} \\ - & \longleftarrow & - 1\frac{5}{6} \\ \hline & & \hline \end{array}$$

(b)

$$\begin{array}{ccc} \textit{estimate} & & \textit{exact} \\ & \xleftarrow{\text{Rounds to}} & 2\frac{5}{8} \\ - & \longleftarrow & - 1\frac{15}{16} \\ \hline & & \hline \end{array}$$

(c)

$$\begin{array}{ccc} \textit{estimate} & & \textit{exact} \\ & \xleftarrow{\text{Rounds to}} & 25\frac{1}{6} \\ - & \longleftarrow & - 18\frac{11}{15} \\ \hline & & \hline \end{array}$$

ANSWERS

4. (a) $4 - 2 = 2$; $2\frac{1}{2}$

(b) $3 - 2 = 1$; $\frac{11}{16}$

(c) $25 - 19 = 6$; $6\frac{13}{30}$

5. Add or subtract by changing mixed numbers to improper fractions. Write answers as mixed numbers in lowest terms.

(a) $1\dfrac{3}{4}$
     $+\ 3\dfrac{1}{8}$
     ──────

OBJECTIVE 3 ▶ An alternate method for adding or subtracting mixed numbers is to first change the mixed numbers to improper fractions. Then rewrite the unlike fractions as like fractions. Finally, add or subtract the numerators and write the answer in lowest terms.

E X A M P L E   4   Adding or Subtracting Mixed Numbers

Add or subtract.

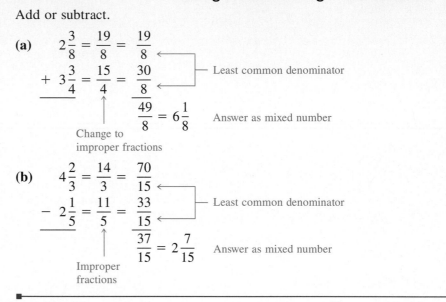

(a)
$$2\dfrac{3}{8} = \dfrac{19}{8} = \dfrac{19}{8}$$ ← Least common denominator
$$+\ 3\dfrac{3}{4} = \dfrac{15}{4} = \dfrac{30}{8}$$ ←
─────────
$$\dfrac{49}{8} = 6\dfrac{1}{8}$$ Answer as mixed number

Change to improper fractions

(b) $6\dfrac{1}{2}$
     $+\ 2\dfrac{3}{5}$
     ──────

(b)
$$4\dfrac{2}{3} = \dfrac{14}{3} = \dfrac{70}{15}$$ ← Least common denominator
$$-\ 2\dfrac{1}{5} = \dfrac{11}{5} = \dfrac{33}{15}$$ ←
─────────
$$\dfrac{37}{15} = 2\dfrac{7}{15}$$ Answer as mixed number

Improper fractions

◀◀ **WORK PROBLEM 5 AT THE SIDE.**

> **Note**
> The advantage of this alternate method of adding or subtracting mixed numbers is that it eliminates the need to carry when adding or to borrow when subtracting. However, if the mixed numbers are large, then the numerators of the improper fractions become so large that they are difficult to work with. In such cases you may want to keep the numbers as mixed numbers.

(c) $7\dfrac{2}{3}$
     $-\ 5\dfrac{1}{4}$
     ──────

(d) $8\dfrac{1}{2}$
     $-\ 3\dfrac{1}{6}$
     ──────

ANSWERS

5. (a) $\dfrac{39}{8} = 4\dfrac{7}{8}$ **(b)** $\dfrac{91}{10} = 9\dfrac{1}{10}$

 (c) $\dfrac{29}{12} = 2\dfrac{5}{12}$ **(d)** $\dfrac{32}{6} = 5\dfrac{2}{6} = 5\dfrac{1}{3}$

3.4 Exercises

≈ *First estimate the answer. Then add to find the exact answer. Write answers as mixed numbers.*

Example:

$$2\frac{3}{5}$$
$$+ \ 9\frac{2}{3}$$

Solution:

estimate   *exact*

3 ←Rounds to— $2\frac{3}{5} = 2\frac{9}{15}$

$+ \ 10$ ←———— $+ 9\frac{2}{3} = 9\frac{10}{15}$

13 $11\frac{19}{15}$ $\frac{19}{15} = 1\frac{4}{15}$, so

$$11\frac{19}{15} = 11 + 1\frac{4}{15}$$

The exact answer is close to the estimate. $= 12\frac{4}{15}$

---

1. *estimate*   *exact*

6 ←Rounds to— $5\frac{1}{2}$

$+ \ 3$ ←———— $+ 3\frac{1}{3}$

9 $8\frac{5}{6}$

2. *estimate*   *exact*

7 $7\frac{1}{10}$

$+ \ 2$ $+ 2\frac{3}{10}$

9 $9\frac{2}{5}$

3. *estimate*   *exact*

10 $10\frac{1}{6}$

$+ \ 5$ $+ 5\frac{1}{3}$

15 $15\frac{1}{2}$

4. *estimate*   *exact*

6 $6\frac{3}{8}$

$+ \ 15$ $+ 15\frac{1}{4}$

21 $21\frac{5}{8}$

5. *estimate*   *exact*

27 $26\frac{5}{8}$

$+ \ 9$ $+ 9\frac{1}{12}$

36 $35\frac{17}{24}$

6. *estimate*   *exact*

83 $82\frac{3}{5}$

$+ \ 16$ $+ 15\frac{4}{5}$

99 $98\frac{2}{5}$

7. *estimate*   *exact*

25 $24\frac{5}{6}$

$+ \ 19$ $+ 18\frac{5}{6}$

44 $43\frac{2}{3}$

8. *estimate*   *exact*

15 $14\frac{6}{7}$

$+ \ 16$ $+ 15\frac{1}{2}$

31 $30\frac{5}{14}$

9. *estimate*   *exact*

34 $33\frac{3}{5}$

$+ \ 19$ $+ 18\frac{1}{2}$

53 $52\frac{1}{10}$

10. *estimate*   *exact*

69 $68\frac{3}{5}$

$+ \ 25$ $+ 25\frac{3}{8}$

94 $93\frac{39}{40}$

11. *estimate*   *exact*

23 $22\frac{3}{4}$

$+ \ 15$ $+ 15\frac{3}{7}$

38 $38\frac{5}{28}$

12. *estimate*   *exact*

7 $7\frac{1}{4}$

$+ \ 26$ $+ 25\frac{7}{8}$

33 $33\frac{1}{8}$

✎ Writing   ◉ Conceptual   ▲ Challenging   ≈ Estimation

13.

estimate	exact
19	$18\dfrac{3}{5}$
48	$47\dfrac{7}{10}$
+ 26	$+\ 25\dfrac{8}{15}$
93	$91\dfrac{5}{6}$

14.

estimate	exact
28	$28\dfrac{1}{4}$
24	$23\dfrac{3}{5}$
+ 20	$+\ 19\dfrac{9}{10}$
72	$71\dfrac{3}{4}$

15.

estimate	exact
33	$32\dfrac{3}{4}$
6	$6\dfrac{1}{3}$
+ 15	$+\ 14\dfrac{5}{8}$
54	$53\dfrac{17}{24}$

≈*First estimate the answer. Then subtract to find the exact answer. Write answers as mixed numbers.*

Example:

$$6$$
$$-\ 4\dfrac{7}{8}$$

Solution:

estimate		exact
6	$\xleftarrow{\text{Rounds to}}$	6
− 5	⟵	$-\ 4\dfrac{7}{8}$
1		

Borrow. $6 = 5 + 1$
$= 5 + \dfrac{8}{8}$
$= 5\dfrac{8}{8}$

Subtract. $6 = 5\dfrac{8}{8}$
$-\ 4\dfrac{7}{8} = 4\dfrac{7}{8}$
$\mathbf{1\dfrac{1}{8}}$

The exact answer is close to the estimate.

16.

estimate		exact
17	$\xleftarrow{\text{Rounds to}}$	$16\dfrac{3}{4}$
− 12	⟵	$-\ 12\dfrac{3}{8}$
5		$4\dfrac{3}{8}$

17.

estimate	exact
19	$19\dfrac{3}{8}$
− 16	$-\ 16\dfrac{1}{4}$
3	$3\dfrac{1}{8}$

18.

estimate	exact
13	$12\dfrac{2}{3}$
− 1	$-\ 1\dfrac{1}{5}$
12	$11\dfrac{7}{15}$

19.

estimate	exact
11	$11\dfrac{9}{20}$
− 5	$-\ 4\dfrac{3}{5}$
6	$6\dfrac{17}{20}$

20.

estimate	exact
28	$28\dfrac{3}{10}$
− 6	$-\ 6\dfrac{1}{15}$
22	$22\dfrac{7}{30}$

21.

estimate	exact
15	$15\dfrac{7}{20}$
− 6	$-\ 6\dfrac{1}{8}$
9	$9\dfrac{9}{40}$

22.

estimate	exact
19	19
− 9	$-\ 8\dfrac{7}{8}$
10	$10\dfrac{1}{8}$

23.

estimate	exact
35	35
− 17	$-\ 17\dfrac{3}{8}$
18	$17\dfrac{5}{8}$

24.

estimate	exact
68	$68\dfrac{3}{8}$
− 7	$-\ 6\dfrac{4}{5}$
61	$61\dfrac{23}{40}$

25.
estimate	exact
47	$47\frac{3}{8}$
$-\ 7$	$-\ 6\frac{7}{12}$
40	$40\frac{19}{24}$

26.
estimate	exact
26	$25\frac{13}{24}$
$-\ 19$	$-\ 18\frac{15}{16}$
7	$6\frac{29}{48}$

27.
estimate	exact
26	$26\frac{5}{18}$
$-\ 12$	$-\ 12\frac{11}{24}$
14	$13\frac{59}{72}$

28.
estimate	exact
157	157
$-\ 87$	$-\ 86\frac{14}{15}$
70	$70\frac{1}{15}$

29.
estimate	exact
374	374
$-\ 212$	$-\ 211\frac{5}{6}$
162	$162\frac{1}{6}$

30.
estimate	exact
430	$429\frac{15}{16}$
$-\ 57$	$-\ 57$
373	$372\frac{15}{16}$

Add or subtract by changing mixed numbers to improper fractions. Write answers as mixed numbers.

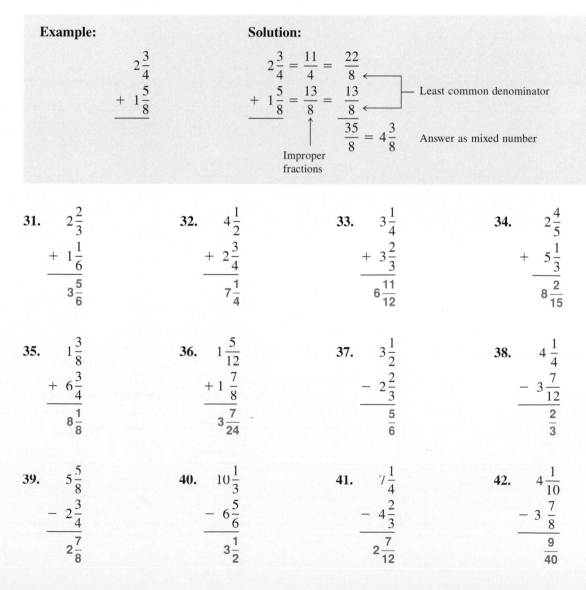

Example:

$$2\frac{3}{4}$$
$$+\ 1\frac{5}{8}$$

Solution:

$$2\frac{3}{4} = \frac{11}{4} = \frac{22}{8}$$
$$+\ 1\frac{5}{8} = \frac{13}{8} = \frac{13}{8}$$

⟵ ⟵ Least common denominator

↑ Improper fractions

$$\frac{35}{8} = 4\frac{3}{8}$$ Answer as mixed number

31.
$$2\frac{2}{3}$$
$$+\ 1\frac{1}{6}$$
$$3\frac{5}{6}$$

32.
$$4\frac{1}{2}$$
$$+\ 2\frac{3}{4}$$
$$7\frac{1}{4}$$

33.
$$3\frac{1}{4}$$
$$+\ 3\frac{2}{3}$$
$$6\frac{11}{12}$$

34.
$$2\frac{4}{5}$$
$$+\ 5\frac{1}{3}$$
$$8\frac{2}{15}$$

35.
$$1\frac{3}{8}$$
$$+\ 6\frac{3}{4}$$
$$8\frac{1}{8}$$

36.
$$1\frac{5}{12}$$
$$+\ 1\frac{7}{8}$$
$$3\frac{7}{24}$$

37.
$$3\frac{1}{2}$$
$$-\ 2\frac{2}{3}$$
$$\frac{5}{6}$$

38.
$$4\frac{1}{4}$$
$$-\ 3\frac{7}{12}$$
$$\frac{2}{3}$$

39.
$$5\frac{5}{8}$$
$$-\ 2\frac{3}{4}$$
$$2\frac{7}{8}$$

40.
$$10\frac{1}{3}$$
$$-\ 6\frac{5}{6}$$
$$3\frac{1}{2}$$

41.
$$7\frac{1}{4}$$
$$-\ 4\frac{2}{3}$$
$$2\frac{7}{12}$$

42.
$$4\frac{1}{10}$$
$$-\ 3\frac{7}{8}$$
$$\frac{9}{40}$$

43. In your own words, explain the steps you would take to add two large mixed numbers.

Find the least common denominator. Change the fraction parts so that they have the same denominator. Add the fraction parts. Add the whole number parts. Write the answer as a mixed number.

44. When subtracting mixed numbers, explain when you need to borrow. Explain how to borrow using your own example.

You need to borrow when the fraction in the minuend (top number) is smaller than the fraction in the subtrahend (bottom number). One example is

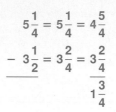

$$5\frac{1}{4} = 5\frac{1}{4} = 4\frac{5}{4}$$
$$-3\frac{1}{2} = 3\frac{2}{4} = 3\frac{2}{4}$$
$$\overline{\qquad\qquad 1\frac{3}{4}}$$

≈ *First estimate the answer. Then solve each application problem.*

45. Hot on the trail of a giant blood-sucking fluke, Mulder and Scully worked a combined total of $15\frac{1}{8}$ hours over the weekend. If they worked $6\frac{1}{2}$ hours in the forensics lab on Saturday and the rest of the time slogging through the sewers of New York City on Sunday, find the number of hours spent in the sewers.

estimate: 15 − 7 = 8 hours

exact: $8\frac{5}{8}$ hours

46. The gas tank on a Jeep Cherokee has a capacity of $21\frac{3}{8}$ gallons. Scott started with a full tank, then used $8\frac{1}{2}$ gallons of gasoline. Find the number of gallons that remain.

estimate: 21 − 9 = 12 gallons

exact: $12\frac{7}{8}$ gallons

47. A carpenter has two pieces of oak trim. One piece of trim is $12\frac{1}{2}$ feet long and the other is $8\frac{2}{3}$ feet in length. How many feet of oak trim does he have in all?

estimate: 13 + 9 = 22 ft

exact: $21\frac{1}{6}$ ft

48. On Monday, $5\frac{3}{4}$ tons of cans were recycled, and $9\frac{3}{5}$ tons were recycled on Tuesday. How many tons were recycled in total on these two days?

estimate: 6 + 10 = 16 tons

exact: $15\frac{7}{20}$ tons

49. Mike Putnam works at Round Table Pizza. He worked $4\frac{1}{2}$ hours on Monday, $6\frac{3}{8}$ hours on Tuesday, $5\frac{1}{4}$ hours on Wednesday, $3\frac{3}{4}$ hours on Thursday, and 7 hours on Friday. How many hours did he work altogether?

estimate: 5 + 6 + 5 + 4 + 7 = 27 hours

exact: $26\frac{7}{8}$ hours

50. On a recent vacation to Canada, Erin Gavin drove for $7\frac{3}{4}$ hours on the first day of her vacation, $5\frac{1}{4}$ hours on the second day, $6\frac{1}{2}$ hours on the third day, and 9 hours on the fourth day. How many hours did she drive altogether?

estimate: 8 + 5 + 7 + 9 = 29 hours

exact: $28\frac{1}{2}$ hours

51. A craftsperson must attach a lead strip around all four sides of a stained glass window before it is installed. Find the length of lead stripping needed.

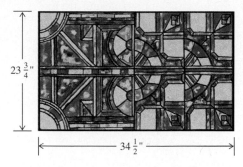

$23\frac{3}{4}''$

$34\frac{1}{2}''$

estimate: 24 + 35 + 24 + 35 = 118 inches

exact: $116\frac{1}{2}$ inches

52. To complete a custom order, Zak Morten of Home Depot must find the number of inches of brass trim needed to go around the four sides of the lamp base plate shown below. Find the length of brass trim needed.

$5\frac{1}{8}''$

$9\frac{7}{8}''$

estimate: 10 + 5 + 10 + 5 = 30 inches
exact: 30 inches

53. A landscaper has $9\frac{5}{8}$ cubic yards of peat moss in a truck. If he unloads $1\frac{1}{2}$ cubic yards at the first stop, $2\frac{3}{4}$ cubic yards at the second stop, and 3 cubic yards at the third stop, how much peat moss remains in the truck?

estimate: 10 − 2 − 3 − 3 = 2 cubic yards

exact: $2\frac{3}{8}$ cubic yards

54. Marv Levenson bought 15 yards of Italian silk fabric. He made two tops with $3\frac{3}{4}$ yards of the material, a suit for his wife with $4\frac{1}{8}$ yards, and a jacket with $3\frac{7}{8}$ yards. Find the number of yards of material remaining.

estimate: 15 − 4 − 4 − 4 = 3 yards

exact: $3\frac{1}{4}$ yards

55. The exercise yard at the correction center has four sides and is enclosed with $527\frac{1}{24}$ feet of security fencing around it. If three sides of the yard measure $107\frac{2}{3}$ feet, $150\frac{3}{4}$ feet, and $138\frac{5}{8}$ feet, find the length of the fourth side.

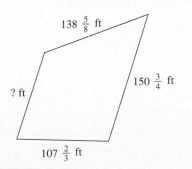

$138\frac{5}{8}$ ft

$150\frac{3}{4}$ ft

? ft

$107\frac{2}{3}$ ft

estimate: 527 − 108 − 151 − 139 = 129 ft
exact: 130 ft

56. Three sides of a parking lot are $108\frac{1}{4}$ feet, $162\frac{3}{8}$ feet, and $143\frac{1}{2}$ feet. If the distance around the lot is $518\frac{3}{4}$ feet, find the length of the fourth side.

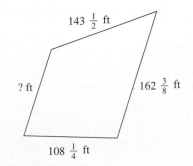

$143\frac{1}{2}$ ft

? ft

$162\frac{3}{8}$ ft

$108\frac{1}{4}$ ft

estimate: 519 − 108 − 162 − 144 = 105 ft
exact: $104\frac{5}{8}$ ft

57. The head keeper at Jurassic Park had to transfer several dinosaurs to new enclosures. A $3\frac{1}{4}$ ton stegosaurus was moved on Monday, a $7\frac{1}{2}$ ton adult tyrannosaurus and a $2\frac{3}{8}$ ton juvenile on Wednesday, and a $1\frac{5}{6}$ ton triceratops on Friday. Find the total number of tons of dinosaurs moved during the week.

estimate: **3 + 8 + 2 + 2 = 15 tons**

exact: **$14\frac{23}{24}$ tons**

58. Comet Auto Supply sold $16\frac{1}{2}$ cases of generic brand oil last week, $12\frac{1}{8}$ cases of Havoline Oil, $8\frac{3}{4}$ cases of Valvoline Oil, and $12\frac{5}{8}$ cases of Castrol Oil. Find the total number of cases of oil that Comet Auto Supply sold during the week.

estimate: **17 + 12 + 9 + 13 = 51 cases**
exact: **50 cases**

▲ *Find the length of the section represented by* x *in the following figures.*

59.

$4\frac{11}{16}$ in.

x

$2\frac{3}{8}$ in. $2\frac{3}{8}$ in.

$9\frac{7}{16}$ in.

60.

$14\frac{9}{20}$ in.

x

$6\frac{9}{10}$ in. $6\frac{9}{10}$ in.

$28\frac{1}{4}$ in.

61.

$6\frac{1}{4}$ in. x $1\frac{7}{8}$ in.

$21\frac{3}{8}$ in.

$29\frac{1}{2}$ in.

62.

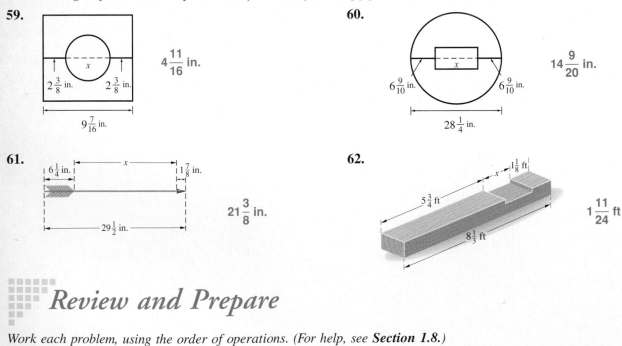

$1\frac{1}{8}$ ft

x

$5\frac{3}{4}$ ft

$1\frac{11}{24}$ ft

$8\frac{1}{3}$ ft

Review and Prepare

*Work each problem, using the order of operations. (For help, see **Section 1.8**.)*

63. $6^2 + 3 - 9$ **30**

64. $3 \cdot 8 - 5$ **19**

65. $4 \cdot 1 + 8 \cdot 7 + 3$ **63**

66. $8 + 9 \div 3 + 6 \cdot 2$ **23**

67. $3^2 \cdot (5 - 2)$ **27**

68. $8 \cdot 4 - (15 - 8)$ **25**

69. $(16 - 6) \cdot 2^3 - (3 \cdot 9)$ **53**

70. $(15 - 9) \cdot 3^2 - (8 \cdot 5)$ **41**

3.5 Order Relations and the Order of Operations

Fractions, like whole numbers, can be located on a number line. For fractions, divide the space between whole numbers into equal parts.

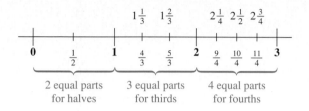

OBJECTIVE 1 To compare the size of two numbers, place the two numbers on a number line.

Comparing the Size of Two Numbers

The number farther to the left on the number line is always smaller and the number farther to the right on the number line is always larger.

For example, on the number line above, $\frac{1}{2}$ is to the left of $\frac{4}{3}$ $\left(1\frac{1}{3}\right)$, so $\frac{1}{2}$ is smaller than $\frac{4}{3}$ $\left(1\frac{1}{3}\right)$.

WORK PROBLEM I AT THE SIDE. ▶▶

Write *order relations* by using the following symbols.

Symbols for Less Than and Greater Than

$<$ is less than

$>$ is greater than

E X A M P L E   I   **Using Less Than and Greater Than Symbols**

Write the following using $<$ and $>$ symbols.

(a) $\frac{1}{2}$ is less than $\frac{4}{3}$

$\frac{1}{2}$ is less than $\frac{4}{3}$ is written as $\frac{1}{2} < \frac{4}{3}$

(b) $\frac{9}{4}$ is greater than 1

$\frac{9}{4}$ is greater than 1 is written as $\frac{9}{4} > 1$

(c) $\frac{5}{3}$ is less than $\frac{11}{4}$

$\frac{5}{3}$ is less than $\frac{11}{4}$ is written as $\frac{5}{3} < \frac{11}{4}$

OBJECTIVES

1 ▶ Identify the greater of two fractions.

2 ▶ Use exponents with fractions.

3 ▶ Use the order of operations.

FOR EXTRA HELP

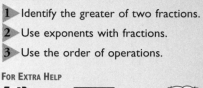

Tutorial   Tape 6   SSM, Sec. 3.5

1. Locate each fraction on the number line.

(a) $\frac{1}{4}$

(b) $\frac{2}{3}$

(c) $1\frac{1}{2}$

(d) $2\frac{3}{4}$

ANSWERS

1.

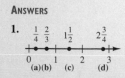

205

2. Use the number line in the text to help place < or > in each blank to make a true statement.

(a) $2 \underline{\hspace{1cm}} \dfrac{9}{4}$

(b) $\dfrac{11}{4} \underline{\hspace{1cm}} \dfrac{4}{3}$

(c) $0 \underline{\hspace{1cm}} 2$

(d) $\dfrac{11}{4} \underline{\hspace{1cm}} \dfrac{5}{3}$

3. Place < or > in each blank to make a true statement.

(a) $\dfrac{3}{8} \underline{\hspace{1cm}} \dfrac{7}{12}$

(b) $\dfrac{11}{18} \underline{\hspace{1cm}} \dfrac{5}{9}$

(c) $\dfrac{17}{24} \underline{\hspace{1cm}} \dfrac{5}{6}$

(d) $\dfrac{13}{15} \underline{\hspace{1cm}} \dfrac{8}{9}$

> **Note**
> Using the number line helps with Example 1. A number line is a very useful tool when working with order relations.

◀◀ **WORK PROBLEM 2 AT THE SIDE.**

The fraction $\frac{7}{8}$ represents 7 of 8 equivalent parts, while $\frac{3}{8}$ means 3 of 8 equivalent parts. Because $\frac{7}{8}$ represents more of the equivalent parts, $\frac{7}{8}$ is greater than $\frac{3}{8}$, or

$$\frac{7}{8} > \frac{3}{8}.$$

To identify the greater fraction use the following steps.

Identifying the Greater Fraction

Step 1 Write the fractions as like fractions.

Step 2 Compare the numerators. The fraction with the greater numerator is the greater fraction.

E X A M P L E   2   Identifying the Greater Fraction

Decide which fraction in each pair is greater.

(a) $\dfrac{7}{8}, \dfrac{9}{10}$

First, write the fractions as like fractions. The least common multiple for 8 and 10 is 40, so

$$\frac{7}{8} = \frac{7 \cdot 5}{8 \cdot 5} = \frac{35}{40} \quad \text{and} \quad \frac{9}{10} = \frac{9 \cdot 4}{10 \cdot 4} = \frac{36}{40}.$$

Look at the numerators. Because 36 is greater than 35, $\frac{36}{40}$ is greater than $\frac{35}{40}$. Because $\frac{36}{40}$ is equivalent to $\frac{9}{10}$,

$$\frac{9}{10} > \frac{7}{8} \quad \text{or} \quad \frac{7}{8} < \frac{9}{10}.$$

The greater fraction is $\frac{9}{10}$.

(b) $\dfrac{8}{5}, \dfrac{23}{15}$

The least common multiple of 5 and 15 is 15.

$$\frac{8}{5} = \frac{8 \cdot 3}{5 \cdot 3} = \frac{24}{15} \quad \text{and} \quad \frac{23}{15} = \frac{23}{15}$$

This shows that $\frac{8}{5}$ is greater than $\frac{23}{15}$, or

$$\frac{8}{5} > \frac{23}{15}.$$

◀◀ **WORK PROBLEM 3 AT THE SIDE.**

ANSWERS

2. (a) < (b) > (c) < (d) >

3. (a) < (b) > (c) < (d) <

OBJECTIVE 2 Exponents were used in Chapter 1 to write repeated products. For example,

$$3^2 = \underbrace{3 \cdot 3}_{\text{Two factors of 3}} = 9 \quad \text{and} \quad 5^3 = \underbrace{5 \cdot 5 \cdot 5}_{\text{Three factors of 5}} = 125.$$

The next example shows exponents used with fractions.

E X A M P L E 3 Using Exponents with Fractions

Simplify each of the following.

(a) $\left(\dfrac{1}{2}\right)^3$

$$\left(\dfrac{1}{2}\right)^3 = \overbrace{\dfrac{1}{2} \cdot \dfrac{1}{2} \cdot \dfrac{1}{2}}^{\text{Three factors of } \frac{1}{2}} = \dfrac{1}{8}$$

(b) $\left(\dfrac{5}{8}\right)^2$

$$\left(\dfrac{5}{8}\right)^2 = \overbrace{\dfrac{5}{8} \cdot \dfrac{5}{8}}^{\text{Two factors of } \frac{5}{8}} = \dfrac{25}{64}$$

(c) $\left(\dfrac{3}{4}\right)^2 \cdot \left(\dfrac{2}{3}\right)^3$

$$\left(\dfrac{3}{4}\right)^2 \cdot \left(\dfrac{2}{3}\right)^3 = \left(\dfrac{3}{4} \cdot \dfrac{3}{4}\right) \cdot \left(\dfrac{2}{3} \cdot \dfrac{2}{3} \cdot \dfrac{2}{3}\right)$$

$$= \dfrac{\overset{1}{\cancel{3}} \cdot \overset{1}{\cancel{3}} \cdot \overset{1}{\cancel{2}} \cdot \overset{1}{\cancel{2}} \cdot \overset{1}{\cancel{2}}}{\underset{2}{\cancel{4}} \cdot \underset{2}{\cancel{4}} \cdot \underset{1}{\cancel{3}} \cdot \underset{1}{\cancel{3}} \cdot 3} \quad \text{Use cancellation.}$$

$$= \dfrac{1}{6}$$

WORK PROBLEM 4 AT THE SIDE.

OBJECTIVE 3 Recall the *order of operations* from Chapter 1.

Order of Operations

1. Do all operations inside **parentheses.**
2. Simplify any expressions with **exponents** and find any **square roots.**
3. **Multiply** or **divide** from left to right.
4. **Add** or **subtract** from left to right.

---

4. Simplify each of the following.

(a) $\left(\dfrac{1}{2}\right)^2$

(b) $\left(\dfrac{7}{8}\right)^2$

(c) $\left(\dfrac{2}{3}\right)^2 \cdot \left(\dfrac{1}{2}\right)^3$

(d) $\left(\dfrac{1}{4}\right)^2 \cdot \left(\dfrac{8}{3}\right)^2$

ANSWERS

4. (a) $\dfrac{1}{4}$ (b) $\dfrac{49}{64}$ (c) $\dfrac{1}{18}$ (d) $\dfrac{4}{9}$

5. Simplify by using the order of operations.

(a) $\dfrac{2}{3} - \dfrac{5}{9} \cdot \dfrac{3}{4}$

(b) $\dfrac{3}{5} \cdot \left(\dfrac{3}{4} - \dfrac{1}{3}\right)$

(c) $\dfrac{3}{4} \cdot \dfrac{2}{3} - \left(\dfrac{1}{2}\right)^2$

(d) $\dfrac{\left(\dfrac{5}{6}\right)^2}{\dfrac{4}{3}}$

The next example shows you how to apply the order of operations with fractions.

EXAMPLE 4   Using Order of Operations with Fractions

Simplify by using the order of operations.

(a) $\dfrac{1}{3} + \dfrac{1}{2} \cdot \dfrac{4}{5}$

Multiply first.

$$\dfrac{1}{3} + \dfrac{1}{\underset{1}{2}} \cdot \dfrac{\overset{2}{4}}{5} = \dfrac{1}{3} + \dfrac{2}{5}$$

Next, add. The least common denominator of 3 and 5 is 15.

$$\dfrac{1}{3} + \dfrac{2}{5} = \dfrac{5}{15} + \dfrac{6}{15} = \dfrac{11}{15}$$

(b) $\dfrac{3}{8} \cdot \left(\dfrac{1}{2} + \dfrac{1}{3}\right)$

$$\dfrac{3}{8} \cdot \left(\dfrac{1}{2} + \dfrac{1}{3}\right) = \dfrac{3}{8} \cdot \underbrace{\left(\dfrac{3}{6} + \dfrac{2}{6}\right)}_{}$$

Work in parentheses first.

$$= \dfrac{3}{8} \cdot \dfrac{5}{6}$$

$$= \dfrac{\overset{1}{3}}{8} \cdot \dfrac{5}{\underset{2}{6}} \qquad \text{Use cancellation.}$$

$$= \dfrac{5}{16} \qquad \text{Multiply.}$$

(c) $\left(\dfrac{2}{3}\right)^2 - \dfrac{4}{5} \cdot \dfrac{1}{2}$

$$\left(\dfrac{2}{3}\right)^2 - \dfrac{4}{5} \cdot \dfrac{1}{2} = \dfrac{4}{9} - \dfrac{4}{5} \cdot \dfrac{1}{2} \qquad \begin{array}{l}\text{Evaluate exponential} \\ \text{expression first.} \\ \dfrac{2}{3} \cdot \dfrac{2}{3} \text{ is } \dfrac{4}{9}\end{array}$$

$$= \dfrac{4}{9} - \dfrac{\overset{2}{4}}{5} \cdot \dfrac{1}{\underset{1}{2}} \qquad \text{Next, multiply.}$$

$$= \dfrac{4}{9} - \dfrac{2}{5}$$

$$= \dfrac{20}{45} - \dfrac{18}{45} \qquad \begin{array}{l}\text{Subtract. (Least common} \\ \text{denominator is 45.)}\end{array}$$

$$= \dfrac{2}{45}$$

5. (a) $\dfrac{1}{4}$ (b) $\dfrac{1}{4}$ (c) $\dfrac{1}{4}$ (d) $\dfrac{25}{48}$

◀◀ WORK PROBLEM 5 AT THE SIDE.

3.5 Exercises

Locate each fraction on the number line.

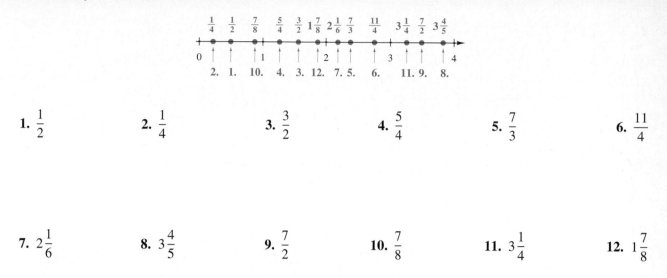

1. $\dfrac{1}{2}$

2. $\dfrac{1}{4}$

3. $\dfrac{3}{2}$

4. $\dfrac{5}{4}$

5. $\dfrac{7}{3}$

6. $\dfrac{11}{4}$

7. $2\dfrac{1}{6}$

8. $3\dfrac{4}{5}$

9. $\dfrac{7}{2}$

10. $\dfrac{7}{8}$

11. $3\dfrac{1}{4}$

12. $1\dfrac{7}{8}$

Write < or > to make a true statement.

Example: $\dfrac{7}{4}$ _____ $\dfrac{13}{6}$

Solution: The least common multiple of 4 and 6 is 12.

$$\frac{7}{4} = \frac{21}{12} \qquad \frac{13}{6} = \frac{26}{12}$$

Because 21 is smaller than 26, $\frac{21}{12}$ or $\frac{7}{4}$ is smaller, so write $\frac{7}{4} \lneq \frac{13}{16}$.

13. $\dfrac{1}{4}$ __<__ $\dfrac{1}{3}$

14. $\dfrac{3}{4}$ __>__ $\dfrac{1}{2}$

15. $\dfrac{5}{6}$ __<__ $\dfrac{11}{12}$

16. $\dfrac{5}{6}$ __>__ $\dfrac{2}{3}$

17. $\dfrac{3}{8}$ __<__ $\dfrac{5}{12}$

18. $\dfrac{7}{15}$ __>__ $\dfrac{9}{20}$

19. $\dfrac{7}{12}$ __<__ $\dfrac{11}{18}$

20. $\dfrac{19}{24}$ __>__ $\dfrac{17}{36}$

21. $\dfrac{19}{27}$ __<__ $\dfrac{13}{18}$

22. $\dfrac{21}{40}$ __<__ $\dfrac{17}{30}$

23. $\dfrac{37}{50}$ __>__ $\dfrac{13}{20}$

24. $\dfrac{7}{12}$ __>__ $\dfrac{11}{20}$

📝 Writing   ◉ Conceptual   ▲ Challenging   ≈ Estimation

Evaluate each of the following.

Example: $\left(\dfrac{2}{3}\right)^4$ **Solution:** $\left(\dfrac{2}{3}\right)^4 = \dfrac{2}{3} \cdot \dfrac{2}{3} \cdot \dfrac{2}{3} \cdot \dfrac{2}{3} = \dfrac{16}{81}$

25. $\left(\dfrac{3}{8}\right)^2$ $\dfrac{9}{64}$

26. $\left(\dfrac{3}{4}\right)^2$ $\dfrac{9}{16}$

27. $\left(\dfrac{5}{7}\right)^2$ $\dfrac{25}{49}$

28. $\left(\dfrac{3}{8}\right)^2$ $\dfrac{9}{64}$

29. $\left(\dfrac{2}{3}\right)^3$ $\dfrac{8}{27}$

30. $\left(\dfrac{3}{5}\right)^3$ $\dfrac{27}{125}$

31. $\left(\dfrac{5}{6}\right)^3$ $\dfrac{125}{216}$

32. $\left(\dfrac{4}{7}\right)^3$ $\dfrac{64}{343}$

33. $\left(\dfrac{3}{2}\right)^4$ $\dfrac{81}{16} = 5\dfrac{1}{16}$

34. $\left(\dfrac{4}{3}\right)^4$ $\dfrac{256}{81} = 3\dfrac{13}{81}$

35. $\left(\dfrac{1}{2}\right)^5$ $\dfrac{1}{32}$

36. $\left(\dfrac{2}{3}\right)^5$ $\dfrac{32}{243}$

37. Describe in your own words what a number line is, and draw a picture of one. Be sure to include how it works and how it can be used.

A number line is a horizontal line with a range of numbers placed on it. The lowest number is on the left and the highest number is on the right. It can be used to compare the size or value of numbers.

$$\overset{}{\underset{0 \quad \frac{1}{2} \quad 1 \quad 1\frac{1}{2} \quad 2 \quad 2\frac{1}{2} \quad 3}{\longmapsto\!\!\!\!\!\!\!\!\!\!\longrightarrow}}$$

38. You have used the order of operations in two chapters now. List from memory the steps in the order of operations.

1. Do all operations inside parentheses.
2. Simplify any expressions with exponents or square roots.
3. Multiply or divide from left to right.
4. Add or subtract from left to right.

Use the order of operations to simplify each of the following.

Example: $\left(\dfrac{2}{3}\right)^2 \cdot \left(\dfrac{1}{2} + \dfrac{1}{4}\right)$

Solution: $= \left(\dfrac{2}{3}\right)^2 \cdot \left(\dfrac{3}{4}\right)$ Work in parentheses first.

$= \dfrac{\overset{1}{\cancel{4}}}{\underset{3}{\cancel{9}}} \cdot \dfrac{\overset{1}{\cancel{3}}}{\underset{1}{\cancel{4}}}$ Evaluate exponential expression.

$= \dfrac{1}{3}$ Multiply.

39. $4 + 2 - 2^2$

2

40. $3^2 + 3 \cdot 2$

15

41. $5 \cdot 2^2 - \dfrac{12}{3}$

16

42. $4 \cdot 3^2 - \dfrac{8}{2}$

32

43. $\left(\dfrac{1}{2}\right)^2 \cdot 4$

1

44. $3 \cdot \left(\dfrac{1}{3}\right)^2$

$\dfrac{1}{3}$

45. $\left(\dfrac{3}{4}\right)^2 \cdot \left(\dfrac{1}{3}\right)$

$\dfrac{3}{16}$

46. $\left(\dfrac{2}{3}\right)^3 \cdot \left(\dfrac{1}{2}\right)$

$\dfrac{4}{27}$

47. $\left(\dfrac{3}{4}\right)^2 \cdot \left(\dfrac{2}{3}\right)^2$

$\dfrac{1}{4}$

48. $\left(\dfrac{5}{8}\right)^2 \cdot \left(\dfrac{4}{25}\right)^2$

$\dfrac{1}{100}$

49. $6 \cdot \left(\dfrac{2}{3}\right)^2 \cdot \left(\dfrac{1}{2}\right)^3$

$\dfrac{1}{3}$

50. $9 \cdot \left(\dfrac{1}{3}\right)^3 \cdot \left(\dfrac{4}{3}\right)^2$

$\dfrac{16}{27}$

51. $\dfrac{4}{3} \cdot \dfrac{3}{8} + \dfrac{3}{4} \cdot \dfrac{1}{4}$

$\dfrac{11}{16}$

52. $\dfrac{3}{4} \cdot \dfrac{2}{5} + \dfrac{1}{3} \cdot \dfrac{3}{5}$

$\dfrac{1}{2}$

53. $\dfrac{1}{2} + \left(\dfrac{1}{2}\right)^2 - \dfrac{3}{8}$

$\dfrac{3}{8}$

54. $\dfrac{2}{3} + \left(\dfrac{1}{3}\right)^2 - \dfrac{5}{9}$

$\dfrac{2}{9}$

55. $\left(\dfrac{1}{3} + \dfrac{1}{6}\right) \cdot \dfrac{1}{2}$

$\dfrac{1}{4}$

56. $\left(\dfrac{3}{5} - \dfrac{3}{20}\right) \cdot \dfrac{4}{3}$

$\dfrac{3}{5}$

57. $\dfrac{9}{8} \div \left(\dfrac{2}{3} + \dfrac{1}{12}\right)$

$1\dfrac{1}{2}$

58. $\dfrac{6}{5} \div \left(\dfrac{3}{5} - \dfrac{3}{10}\right)$

4

59. $\left(\dfrac{3}{5} - \dfrac{1}{10}\right) \div \dfrac{5}{2}$

$\dfrac{1}{5}$

60. $\left(\dfrac{8}{5} - \dfrac{7}{10}\right) \div \dfrac{3}{5}$

$\dfrac{3}{2} = 1\dfrac{1}{2}$

61. $\dfrac{3}{8} \cdot \left(\dfrac{1}{4} + \dfrac{1}{2}\right) \cdot \dfrac{32}{3}$

3

62. $\dfrac{1}{3} \cdot \left(\dfrac{4}{5} - \dfrac{3}{10}\right) \cdot \dfrac{4}{2}$

$\dfrac{1}{3}$

63. $\left(\dfrac{3}{4}\right)^2 - \left(\dfrac{3}{4} - \dfrac{1}{8}\right) \div \dfrac{7}{4}$

$\dfrac{23}{112}$

64. $\left(\dfrac{2}{3}\right)^2 - \left(\dfrac{4}{5} - \dfrac{3}{10}\right) \div \dfrac{5}{4}$

$\dfrac{2}{45}$

65. $\left(\dfrac{7}{8} - \dfrac{1}{4}\right) - \left(\dfrac{3}{4}\right)^2 \cdot \dfrac{2}{3}$

$\dfrac{1}{4}$

66. $\left(\dfrac{5}{6} - \dfrac{7}{12}\right) - \left(\dfrac{1}{3}\right)^2 \cdot \dfrac{3}{4}$

$\dfrac{1}{6}$

▲ 67. $\left(\dfrac{3}{4}\right)^2 \cdot \left(\dfrac{2}{3} - \dfrac{5}{9}\right) - \dfrac{1}{4} \cdot \dfrac{1}{8}$

$\dfrac{1}{32}$

▲ 68. $\left(\dfrac{2}{3}\right)^2 \cdot \left(\dfrac{1}{2} - \dfrac{1}{8}\right) - \dfrac{2}{3} \cdot \dfrac{1}{8}$

$\dfrac{1}{12}$

Review and Prepare

*Rewrite the following numbers in words. (For help, see **Section 1.1**.)*

69. 5728   five thousand, seven hundred twenty-eight

70. 625,115   six hundred twenty-five thousand, one hundred fifteen

71. 4,071,280   four million, seventy-one thousand, two hundred eighty

72. 220,518,315   two hundred twenty million, five hundred eighteen thousand, three hundred fifteen

---

KEY TERMS

3.1	**like fractions**	Fractions with the same denominator are called like fractions.
	unlike fractions	Fractions with different denominators are called unlike fractions.
3.2	**least common multiple**	Given two or more whole numbers, the least common multiple is the smallest whole number that is divisible by all the numbers.
3.3	**least common denominator**	When unlike fractions are rewritten as like fractions having the least common multiple as the denominator, the new denominator is the least common denominator.
3.4	**carrying**	The method used when the sum of the fractions of mixed numbers is greater than 1 is called carrying. Carry from the fraction to the whole number.
	borrowing	The method used in subtracting mixed numbers when the fraction part of the minuend is too small.

QUICK REVIEW

Concepts

Examples

3.1 Adding Like Fractions

Add numerators and write in lowest terms.

$$\frac{3}{4} + \frac{1}{4} + \frac{5}{4} = \frac{3 + 1 + 5}{4} = \frac{9}{4} = 2\frac{1}{4}$$

3.1 Subtracting Like Fractions

Subtract numerators and write in lowest terms.

$$\frac{7}{8} - \frac{5}{8} = \frac{7 - 5}{8} = \frac{2 \div 2}{8 \div 2} = \frac{1}{4}$$

3.2 Finding the Least Common Multiple

Method of using multiples of the largest number: List the first few multiples of the larger number. Check each one until you find the multiple that is divisible by the smaller number.

$$\frac{1}{3} + \frac{1}{4}$$

4, 8, 12, 16, . . . ← Multiples of 4

↑ First multiple divisible by 3 ($12 \div 3 = 4$)

The least common multiple of the numbers 3 and 4 is 12.

3.2 Finding the Least Common Multiple

Method of prime numbers: Use prime numbers to find the least common multiple.

$$\frac{1}{3} + \frac{1}{4} + \frac{1}{10}$$

Prime	2	3	5
3 =		③	
4 =	②•②		
10 =	2 •		⑤
LCM =	②•②	③	⑤

least common multiple (LCM) = $2 \cdot 2 \cdot 3 \cdot 5 = 60$

Concepts	Examples
3.3 Adding Unlike Fractions **1.** Find the least common multiple (LCM). **2.** Rewrite fractions with the least common multiple as the denominator. **3.** Add numerators, placing the answer over the least common denominator.	$\dfrac{1}{3}$ + $\dfrac{1}{4}$ + $\dfrac{1}{10}$ LCM = 60 $\dfrac{1}{3} = \dfrac{20}{60}$, $\dfrac{1}{4} = \dfrac{15}{60}$, $\dfrac{1}{10} = \dfrac{6}{60}$ $\dfrac{20}{60} + \dfrac{15}{60} + \dfrac{6}{60} = \dfrac{41}{60}$
3.3 Subtracting Unlike Fractions **1.** Find the least common multiple (LCM). **2.** Rewrite fractions with the least common multiple as the denominator. **3.** Subtract numerators, placing the difference over the common denominator.	$\dfrac{5}{8}$ − $\dfrac{1}{3}$ LCM = 24 $\dfrac{5}{8} = \dfrac{15}{24}$, $\dfrac{1}{3} = \dfrac{8}{24}$ $\dfrac{15}{24} - \dfrac{8}{24} = \dfrac{7}{24}$
3.4 Adding Mixed Numbers Round the numbers and estimate the answer. Then find the exact answer. **1.** Add fractions. **2.** Add whole numbers. **3.** Combine the sums of whole numbers and fractions. Write the answer in lowest terms. Compare the exact answer to the estimate to see if it is reasonable.	*estimate* *exact* 10 Rounds to $9\dfrac{2}{3} = 9\dfrac{8}{12}$ $+\,7$ ⟵ $+\,6\dfrac{3}{4} = 6\dfrac{9}{12}$ ──── ──────── 17 $15\dfrac{17}{12} = 16\dfrac{5}{12}$ Exact answer is reasonable because it is close to the estimate of 17.
3.4 Subtracting Mixed Numbers Round the numbers and estimate the answer. Then find the exact answer. **1.** Subtract fractions by using borrowing if necessary. **2.** Subtract whole numbers. **3.** Combine the differences of whole numbers and fractions. Write the answer in lowest terms. Compare the exact answer to the estimate to see if it is reasonable.	*estimate* *exact* 9 Rounds to $8\dfrac{5}{8} = 8\dfrac{15}{24} = 7\dfrac{39}{24}$ $-\,4$ ⟵ $-\,3\dfrac{11}{12} = 3\dfrac{22}{24} = 3\dfrac{22}{24}$ ──── ──────── 5 $4\dfrac{17}{24}$ Exact answer is reasonable because it is close to the estimate of 5.

Concepts	Examples
3.4 Adding or Subtracting Mixed Numbers Using an Alternate Method	Add.
1. Change the mixed numbers to improper fractions.	$$2\frac{2}{3} = \frac{8}{3} = \frac{64}{24}$$ $\longleftarrow$
2. Rewrite the unlike fractions as like fractions.	$$+\ 1\frac{3}{8} = \frac{11}{8} = \frac{33}{24}$$ $\longleftarrow$ — Least common denominator
3. Add or subtract the numerators. Write the answer as a mixed number in lowest terms.	$$\frac{97}{24} = 4\frac{1}{24} \quad \text{Answer as mixed number}$$
	Improper fractions
	Subtract.
	$$8\frac{2}{3} = \frac{26}{3} = \frac{104}{12}$$ $\longleftarrow$
	$$-\ 5\frac{3}{4} = \frac{23}{4} = \frac{69}{12}$$ $\longleftarrow$ — Least common denominator
	$$\frac{35}{12} = 2\frac{11}{12} \quad \text{Answer as mixed number}$$
	Improper fractions
3.5 Identifying the Larger of Two Fractions	Identify the larger fraction.
With unlike fractions, change to like fractions first. The fraction with the greater numerator is the greater fraction.	$$\frac{7}{8}, \frac{9}{10}$$
$<$ is less than	$$\frac{7}{8} = \frac{7 \cdot 5}{8 \cdot 5} = \frac{35}{40}$$
$>$ is greater than	$$\frac{9}{10} = \frac{9 \cdot 4}{10 \cdot 4} = \frac{36}{40}$$
	$\frac{35}{40}$ is smaller than $\frac{36}{40}$, so $\frac{7}{8} < \frac{9}{10}$ or $\frac{9}{10} > \frac{7}{8}$. $\frac{9}{10}$ is greater.
3.5 Using the Order of Operations with Fractions	Simplify by using the order of operations.
Follow the order of operations.	$\dfrac{1}{2} \cdot \dfrac{2}{3} - \left(\dfrac{1}{4}\right)^2$ Simplify exponents.
1. Do all operations inside parentheses.	
2. Simplify any expressions with exponents and find any square roots.	$= \dfrac{1}{\underset{1}{\cancel{2}}} \cdot \dfrac{\overset{1}{\cancel{2}}}{3} - \dfrac{1}{16}$ Next, multiply.
3. Multiply or divide from left to right.	
4. Add or subtract from left to right.	$= \dfrac{1}{3} - \dfrac{1}{16}$
	$= \dfrac{16}{48} - \dfrac{3}{48}$ Change to common denominator and subtract.
	$= \dfrac{13}{48}$

Numbers in the

Real World *collaborative investigations*

One of the ways people invest their money is to buy stock in a company. The price of each share of stock changes frequently, and many newspapers list the prices. You could buy stock in hundreds of different companies. The listing here shows the price for just five companies.

Stock prices use fractions to show parts of a dollar. For example:

$\frac{1}{4}$ = $0.25 $\frac{1}{2}$ = $0.50 $\frac{3}{4}$ = $0.75

From this information, figure out the value of 1/8, 3/8, 5/8, and 7/8.
$0.125, $0.375, $0.625, $0.875
Each one is halfway between, for example, 1/8 is halfway between $0.00 and $0.25.

	52 Week		**Close**	**Net Chg**
	High	**Low**		
AT&T	43 3/4	30 3/4	31 1/4	+1/2
AppleC	28 7/8	15 1/8	17 1/2	−3/8
MusicLd	5 1/8	11/16	1 1/4	−1/8
RenoAir	14	6 3/8	8 1/8	+3/4
Spiegel	13 1/4	5 7/8	6	−1/8

Market Review

The first two numbers after each company name show the highest and lowest price for the stock over the last 52 weeks (one year). Then comes the current price, ("Close") and finally, how the current price has changed from yesterday ("Net Chg"). Use the fractions or mixed numbers in the list to answer these questions.

What is the difference between the highest and lowest price for RenoAir during the last year? **7 5/8**

How does the current price of RenoAir compare to its lowest price during the past year? **1 3/4 higher**

Use your estimating skills to find the company that had the greatest difference between its highest and lowest price during the past year. Then find the exact difference for that company. **AppleC 13 3/4. Estimates of differences, in order, are 13, 14, 4, 8, and 7.**

How much would it cost to buy 10 shares of MusicLd stock at the current price? **12 1/2**

How much would it have cost to buy 10 shares of MusicLd at the lowest price during the past year? **6 7/8**

If AT&T stock is 1/2 dollar higher than it was yesterday (+1/2), what was its price yesterday? Find the price of each of the other stocks yesterday.
**AT&T 31 1/4 − 1/2 = 30 3/4   AppleC 17 1/2 + 3/8 = 17 7/8   MusicLd 1 1/4 + 1/8 = 1 3/8
RenoAir 8 1/8 − 3/4 = 7 3/8   Spiegel 6 + 1/8 = 6 1/8**

If you have $100 to invest in one of these companies, how many shares could you buy of each company's stock? You can only buy whole shares, not parts of a share. How much would be left over? (Ignore broker fees.)
**AT&T 3 shares; $6 1/4 left over   RenoAir 12 shares; $2 1/2 left over
AppleC 5 shares, $12 1/2 left over   Spiegel 16 shares; $4 left over
MusicLd 80 shares; no money left over**

[3.1] *Add or subtract. Write answers in lowest terms.*

1. $\frac{2}{8} + \frac{5}{8}$

   $\frac{7}{8}$

2. $\frac{1}{5} + \frac{2}{5}$

   $\frac{3}{5}$

3. $\frac{1}{8} + \frac{3}{8} + \frac{2}{8}$

   $\frac{3}{4}$

4. $\frac{7}{12} - \frac{2}{12}$

   $\frac{5}{12}$

5. $\frac{3}{10} - \frac{1}{10}$

   $\frac{1}{5}$

6. $\frac{5}{16} - \frac{1}{16}$

   $\frac{1}{4}$

7. $\frac{36}{62} - \frac{10}{62}$

   $\frac{13}{31}$

8. $\frac{79}{108} - \frac{47}{108}$

   $\frac{8}{27}$

Solve each application problem. Write answers in lowest terms.

9. Tyrone milled $\frac{3}{16}$ of the lumber on the first day and $\frac{5}{16}$ of the lumber on the second day. What fraction of the lumber did he mill on these two days?

   $\frac{1}{2}$ of the lumber

10. Dominique Moceanu completed $\frac{7}{10}$ of her workout in the morning and $\frac{1}{10}$ of her workout in the afternoon. How much less of a workout did she have in the afternoon than in the morning?

   $\frac{3}{5}$ less

[3.2] *Find the least common multiple of each set of numbers.*

11. 4, 3

   12

12. 8, 5

   40

13. 10, 12, 20

   60

14. 9, 20, 15

   180

15. 6, 8, 5, 15

   120

16. 24, 5, 16

   240

☑ Writing   ◉ Conceptual   ▲ Challenging   ≈ Estimation

Rewrite each of the following fractions by using the indicated denominators.

17. $\dfrac{3}{4} = \dfrac{12}{16}$

18. $\dfrac{2}{3} = \dfrac{10}{15}$

19. $\dfrac{2}{5} = \dfrac{10}{25}$

20. $\dfrac{5}{9} = \dfrac{45}{81}$

21. $\dfrac{7}{16} = \dfrac{63}{144}$

22. $\dfrac{3}{22} = \dfrac{12}{88}$

[3.1–3.3] *Add or subtract. Write answers in lowest terms.*

23. $\dfrac{1}{4} + \dfrac{1}{3}$ $\dfrac{7}{12}$

24. $\dfrac{1}{5} + \dfrac{3}{10} + \dfrac{3}{8}$ $\dfrac{7}{8}$

25.
$$\begin{array}{r} \dfrac{9}{16} \\ + \dfrac{1}{12} \\ \hline \dfrac{31}{48} \end{array}$$

26. $\dfrac{4}{5} - \dfrac{1}{4}$ $\dfrac{11}{20}$

27.
$$\begin{array}{r} \dfrac{3}{4} \\ - \dfrac{1}{3} \\ \hline \dfrac{5}{12} \end{array}$$

28.
$$\begin{array}{r} \dfrac{11}{12} \\ - \dfrac{4}{9} \\ \hline \dfrac{17}{36} \end{array}$$

Solve each of the following application problems.

29. A dump truck contains $\frac{1}{4}$ cubic yard of fine gravel, $\frac{1}{3}$ cubic yard of pea gravel, and $\frac{3}{8}$ cubic yard of coarse gravel. How many cubic yards of gravel are on the truck?

$\dfrac{23}{24}$ cubic yard

30. Rachel is saving money for a birthday bash for Ross. Monica has raised $\frac{2}{5}$ of the amount needed through a bake sale; Joey has earned $\frac{1}{3}$ of the amount needed from an acting gig, and Phoebe has raised another $\frac{1}{4}$ singing at the local coffee shop. Find the portion of the total that has been raised.

$\dfrac{59}{60}$ of the amount needed

$\approx$ *First estimate the answer. Then add or subtract to find the exact answer. Write answers as mixed numbers.*

31. *estimate*   *exact*

$$26 \xleftarrow{\text{Rounds to}} 25\frac{3}{4}$$

$$\underline{+\ 16} \xleftarrow{\hspace{1.5cm}} \underline{+\ 16\frac{3}{8}}$$

$$42 \qquad\qquad 42\frac{1}{8}$$

32. *estimate*   *exact*

$$78 \qquad 78\frac{3}{7}$$

$$\underline{+\ 18} \qquad \underline{+\ 17\frac{6}{7}}$$

$$96 \qquad 96\frac{2}{7}$$

33. *estimate*   *exact*

$$13 \qquad 12\frac{3}{5}$$

$$9 \qquad 8\frac{5}{8}$$

$$\underline{+\ 10} \qquad \underline{+\ 10\frac{5}{16}}$$

$$32 \qquad 31\frac{43}{80}$$

34. *estimate*   *exact*

$$18 \qquad 18\frac{1}{3}$$

$$\underline{-\ 13} \qquad \underline{-\ 12\frac{3}{4}}$$

$$5 \qquad\ \ 5\frac{7}{12}$$

35. *estimate*   *exact*

$$74 \qquad 73\frac{1}{2}$$

$$\underline{-\ 56} \qquad \underline{-\ 55\frac{2}{3}}$$

$$18 \qquad 17\frac{5}{6}$$

36. *estimate*   *exact*

$$215 \qquad 215\frac{7}{16}$$

$$\underline{-\ 136} \qquad \underline{-\ 136}$$

$$79 \qquad 79\frac{7}{16}$$

Add or subtract by changing mixed numbers to improper fractions. Write answers as mixed numbers in lowest terms.

37. $3\frac{1}{4}$

$$\underline{+\ 2\frac{1}{2}}$$

$$5\frac{3}{4}$$

38. $2\frac{1}{3}$

$$\underline{+\ 3\frac{3}{4}}$$

$$6\frac{1}{12}$$

39. $3\frac{3}{5}$

$$\underline{+\ 2\frac{2}{3}}$$

$$6\frac{4}{15}$$

40. $4\frac{1}{4}$

$$\underline{-\ 1\frac{5}{12}}$$

$$2\frac{5}{6}$$

41. $8\frac{1}{3}$

$$\underline{-\ 2\frac{5}{6}}$$

$$5\frac{1}{2}$$

42. $5\frac{5}{12}$

$$\underline{-\ 2\frac{5}{8}}$$

$$2\frac{19}{24}$$

≈ *First estimate the answer. Then solve each application problem.*

43. The lab had $14\frac{2}{3}$ gallons of distilled water. If $5\frac{1}{2}$ gallons were used in the morning and $6\frac{3}{4}$ gallons were used in the afternoon, find the number of gallons remaining.

estimate: 15 − 6 − 7 = 2 gallons

exact: $2\frac{5}{12}$ gallons

44. The ecology club collected $14\frac{3}{4}$ tons of cardboard on Saturday and $18\frac{2}{3}$ tons on Sunday. Find the total amount of cardboard collected.

estimate: 15 + 19 = 34 tons

exact: $33\frac{5}{12}$ tons

45. At birth, the Bolton triplets weigh $5\frac{3}{4}$ pounds, $4\frac{7}{8}$ pounds, and $5\frac{1}{3}$ pounds. Find their total weight.

estimate: 6 + 5 + 5 = 16 pounds

exact: $15\frac{23}{24}$ pounds

46. A developer wants to build a shopping center. She bought two parcels of land, one, $1\frac{11}{16}$ acres, and the other, $2\frac{3}{4}$ acres. If she needs a total of $8\frac{1}{2}$ acres for the center, how much additional land does she need to buy?

estimate: 9 − 2 − 3 = 4 acres

exact: $4\frac{1}{16}$ acres

[3.5] *Locate each fraction on the number line.*

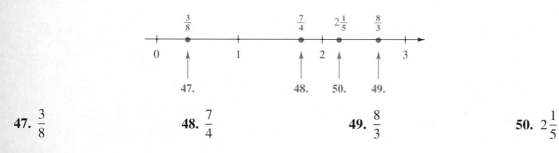

47. $\dfrac{3}{8}$ **48.** $\dfrac{7}{4}$ **49.** $\dfrac{8}{3}$ **50.** $2\dfrac{1}{5}$

Write < or > to make a true statement.

51. $\dfrac{3}{4}$ __<__ $\dfrac{7}{8}$ **52.** $\dfrac{5}{8}$ __<__ $\dfrac{2}{3}$ **53.** $\dfrac{2}{3}$ __>__ $\dfrac{8}{15}$ **54.** $\dfrac{7}{10}$ __>__ $\dfrac{8}{15}$

55. $\dfrac{5}{12}$ __<__ $\dfrac{8}{18}$ **56.** $\dfrac{7}{20}$ __>__ $\dfrac{8}{25}$ **57.** $\dfrac{19}{36}$ __<__ $\dfrac{29}{54}$ **58.** $\dfrac{19}{132}$ __>__ $\dfrac{7}{55}$

Simplify each of the following.

59. $\left(\dfrac{1}{3}\right)^2$ $\dfrac{1}{9}$ **60.** $\left(\dfrac{3}{4}\right)^2$ $\dfrac{9}{16}$ **61.** $\left(\dfrac{3}{5}\right)^3$ $\dfrac{27}{125}$ **62.** $\left(\dfrac{3}{8}\right)^4$ $\dfrac{81}{4096}$

Simplify by using the order of operations.

63. $5 \cdot \left(\dfrac{1}{4}\right)^2$ $\dfrac{5}{16}$

64. $\left(\dfrac{3}{4}\right)^2 \cdot 20$ $11\dfrac{1}{4}$

65. $\left(\dfrac{3}{4}\right)^2 \cdot \left(\dfrac{8}{9}\right)^2$ $\dfrac{4}{9}$

66. $\dfrac{3}{5} \div \left(\dfrac{1}{10} + \dfrac{1}{5}\right)$ 2

67. $\left(\dfrac{1}{2}\right)^2 \cdot \left(\dfrac{1}{4} + \dfrac{1}{2}\right)$ $\dfrac{3}{16}$

68. $\left(\dfrac{1}{4}\right)^3 + \left(\dfrac{5}{8} + \dfrac{3}{4}\right)$ $1\dfrac{25}{64}$

MIXED REVIEW EXERCISES

Solve by using the order of operations as necessary. Write answers in lowest terms or as mixed numbers.

69. $\dfrac{7}{8} - \dfrac{3}{8}$ $\dfrac{1}{2}$

70. $\dfrac{2}{3} - \dfrac{1}{4}$ $\dfrac{5}{12}$

71. $\dfrac{75}{86} - \dfrac{4}{43}$ $\dfrac{67}{86}$

72. $\dfrac{1}{4} + \dfrac{1}{8} + \dfrac{5}{16}$ $\dfrac{11}{16}$

73. $\begin{array}{r} 5\dfrac{2}{3} \\ -\ 2\dfrac{1}{2} \\ \hline 3\dfrac{1}{6} \end{array}$

74. $\begin{array}{r} 9\dfrac{1}{2} \\ +\ 16\dfrac{3}{4} \\ \hline 26\dfrac{1}{4} \end{array}$

75. $\begin{array}{r} 7 \\ -\ 1\dfrac{5}{8} \\ \hline 5\dfrac{3}{8} \end{array}$

76. $\begin{array}{r} 2\dfrac{3}{5} \\ 8\dfrac{5}{8} \\ +\ \dfrac{5}{16} \\ \hline 11\dfrac{43}{80} \end{array}$

77. $\begin{array}{r} 92\dfrac{5}{16} \\ -\ 27 \\ \hline 65\dfrac{5}{16} \end{array}$

78. $\dfrac{7}{22} + \dfrac{3}{22} + \dfrac{3}{11}$ $\dfrac{8}{11}$

79. $\left(\dfrac{1}{4}\right)^2 \cdot \left(\dfrac{2}{5}\right)^3$ $\dfrac{1}{250}$

80. $\dfrac{1}{4} \div \left(\dfrac{1}{3} + \dfrac{1}{6}\right)$ $\dfrac{1}{2}$

81. $\left(\dfrac{2}{3}\right)^2 \cdot \left(\dfrac{1}{3} + \dfrac{1}{6}\right)$ $\dfrac{2}{9}$

82. $\left(\dfrac{2}{3}\right)^3 + \left(\dfrac{2}{3} - \dfrac{5}{9}\right)$ $\dfrac{11}{27}$

Write < or > to make a true statement.

83. $\dfrac{7}{8}$ ——$>$—— $\dfrac{13}{16}$ **84.** $\dfrac{7}{10}$ ——$>$—— $\dfrac{13}{20}$ **85.** $\dfrac{19}{40}$ ——$<$—— $\dfrac{29}{60}$ **86.** $\dfrac{5}{8}$ ——$>$—— $\dfrac{17}{30}$

Find the least common multiple of each set of numbers.

87. 18, 24   72   **88.** 10, 15, 20, 25   300   **89.** 8, 9, 12, 18   72

Rewrite each of the following fractions by using the indicated denominators.

90. $\dfrac{3}{8} = \dfrac{18}{48}$ **91.** $\dfrac{9}{12} = \dfrac{108}{144}$ **92.** $\dfrac{3}{7} = \dfrac{180}{420}$

$\approx$ *First estimate the answer. Then solve each application problem.*

93. A carpet layer needs $13\frac{1}{2}$ feet of carpet for a bedroom and $22\frac{3}{8}$ feet of carpet for a living room. If the roll from which the carpet layer is cutting is $92\frac{3}{4}$ feet long, find the number of feet remaining after the two rooms have been carpeted.

estimate: 93 − 14 − 22 = 57 feet

exact: $56\dfrac{7}{8}$ feet

94. Quark's Place sold $2\frac{3}{8}$ liters of Klingon blood-wine and $4\frac{1}{2}$ liters of Regalian Ale. If Quark had stocked a total of 10 liters, find the number of liters that remain.

estimate: 10 − 2 − 5 = 3 liters

exact: $3\dfrac{1}{8}$ liters

Add or subtract. Write answers in lowest terms.

1. $\dfrac{3}{8} + \dfrac{1}{8}$

2. $\dfrac{5}{10} + \dfrac{1}{10}$

3. $\dfrac{5}{8} - \dfrac{1}{8}$

4. $\dfrac{7}{12} - \dfrac{5}{12}$

Find the least common multiple of each set of numbers.

5. 3, 4, 6

6. 7, 15, 3, 5

7. 6, 9, 27, 36

Add or subtract. Write answers in lowest terms.

8. $\dfrac{2}{3} + \dfrac{1}{4}$

9. $\dfrac{2}{9} + \dfrac{5}{12}$

10. $\dfrac{7}{8} - \dfrac{2}{3}$

11. $\dfrac{3}{8} - \dfrac{1}{5}$

$\approx$ *First estimate the answer. Then add or subtract to find the exact answer. Write answers as mixed numbers.*

12. $5\dfrac{1}{6} + 6\dfrac{2}{3}$

13. $16\dfrac{2}{5} - 11\dfrac{2}{3}$

14. $18\dfrac{3}{4} + 9\dfrac{2}{5} + 12\dfrac{1}{3}$

15. $24 - 18\dfrac{3}{8}$

1. $\dfrac{1}{2}$ _____

2. $\dfrac{3}{5}$ _____

3. $\dfrac{1}{2}$ _____

4. $\dfrac{1}{6}$ _____

5. 12 _____

6. 105 _____

7. 108 _____

8. $\dfrac{11}{12}$ _____

9. $\dfrac{23}{36}$ _____

10. $\dfrac{5}{24}$ _____

11. $\dfrac{7}{40}$ _____

estimate: 5 + 7 = 12
exact: $11\dfrac{5}{6}$
12. _____

estimate: 16 − 12 = 4
exact: $4\dfrac{11}{15}$
13. _____

estimate: 19 + 9 + 12 = 40
exact: $40\dfrac{29}{60}$
14. _____

estimate: 24 − 18 = 6
exact: $5\dfrac{5}{8}$
15. _____

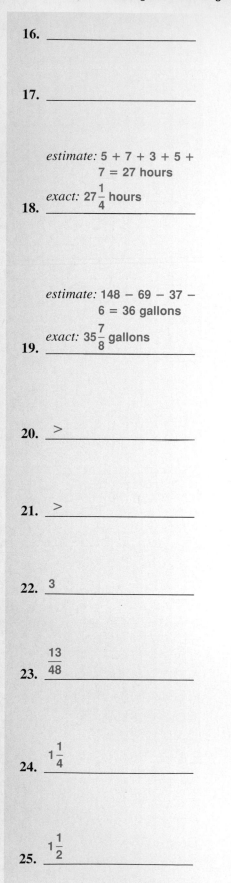

16. _____

17. _____

estimate: 5 + 7 + 3 + 5 +
7 = 27 hours

exact: $27\frac{1}{4}$ hours

18. _____

estimate: 148 − 69 − 37 −
6 = 36 gallons

exact: $35\frac{7}{8}$ gallons

19. _____

20. _> _____

21. _> _____

22. __3_____

23. __$\frac{13}{48}$_____

24. __$1\frac{1}{4}$_____

25. __$1\frac{1}{2}$_____

☑ **16.** Most of my students say that "addition and subtraction of fractions
⊙   is more difficult than multiplication and division of fractions." Why
do you think they say this? Do you agree with these students?

**Probably addition and subtraction of fractions is more difficult
because you have to find the least common denominator and then
change the fractions to the same denominator.**

☑ **17.** Devise and explain a method of estimating an answer to addition and
⊙   subtraction problems involving mixed numbers. Might your estimated
answer vary from the exact answer? If it did, what would the estimation accomplish?

**Round mixed numbers to the nearest whole number. Then add,
subtract, multiply, or divide to estimate the answer. The estimate
may vary from the exact answer but it lets you know if your answer
is reasonable.**

≈First estimate the answer. Then solve the following application problems.

18. Ann-Marie Sargent is training for an upcoming bicycle race. She
rides $4\frac{5}{6}$ hours on Monday, $6\frac{2}{3}$ hours on Tuesday, $3\frac{1}{4}$ hours on
Wednesday, $5\frac{1}{3}$ hours on Thursday, and $7\frac{1}{6}$ hours on Friday. Find the
total number of hours that she trained.

19. A commercial painting contractor arrived at a 6-unit apartment
complex with $147\frac{1}{2}$ gallons of exterior paint. If his crew sprayed
$68\frac{1}{2}$ gallons on the wood siding, rolled $37\frac{3}{8}$ gallons on the masonry
exterior and brushed $5\frac{3}{4}$ gallons on the trim, find the number of
gallons of paint remaining.

Write < or > to make a true statement.

20. $\frac{2}{3}$ _____ $\frac{13}{20}$ **21.** $\frac{19}{24}$ _____ $\frac{17}{36}$

Simplify each of the following. Use the order of operations as needed.

22. $\left(\frac{1}{2}\right)^3 \cdot 24$ **23.** $\left(\frac{3}{4}\right)^2 - \left(\frac{7}{8} \cdot \frac{1}{3}\right)$

24. $\left(\frac{5}{6} - \frac{5}{12}\right) \cdot 3$ **25.** $\frac{2}{3} + \frac{5}{8} \cdot \frac{4}{3}$

Name the digit that has the given place value in each of the following problems.

1. 583
  hundreds   5
  ones   3

2. 2,785,476
  millions   2
  thousands   5

Round each of the following to the nearest ten, nearest hundred, and nearest thousand.

	Ten	**Hundred**	**Thousand**
3. 1746	1750	1700	2000
4. 59,803	59,800	59,800	60,000

≈ _Round the numbers in each problem so there is only one non-zero digit. Then add, subtract, multiply, or divide as indicated to estimate the answer. Finally, solve for the exact answer._

5. _estimate_   _exact_
  10,000 ←Rounds to— 9 834
    300 ←———— 279
  50,000 ←———— 51,506
  + 50,000 ←———— + 51,702
  110,300   113,321

6. _estimate_   _exact_
  20,000   24,276
  − 10,000   − 9 887
  10,000   14,389

7. _estimate_   _exact_
  2000   2375
  × 400   × 370
  800,000   878,750

8. _estimate_   _exact_
    2 500   3 211
  40)100,000   35)112,385

Add, subtract, multiply, or divide as indicated.

9.   3
    9
    4
  + 8
   24

10. 375,899
    521,742
  + 357,968
   1,255,609

11. 1479
   − 1187
    292

12. 3,896,502
  − 1,094,807
   2,801,695

13. 5 × 9 × 3   135

14. 7 × 2 × 8   112

15. 9 × 4 × 6   216

16. 68
  × 7
  476

17.   962
  × 384
  369,408

18.   450
  × 60
  27,000

19. 158
  9)1422

20. 13,467 ÷ 5
   2693 R2

21. 32 R166
  506)16,358

≈ *First round so there is only one non-zero digit and estimate the answer. Then solve each application problem.*

22. The Americans With Disabilities Act provides the single parking space design below. Find the perimeter of (distance around) this parking space including the accessible aisle.

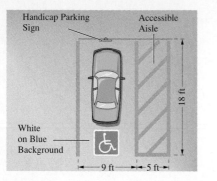

Handicap Parking Sign

Accessible Aisle

18 ft

White on Blue Background

9 ft — 5 ft

estimate: 20 + 9 + 5 + 20 + 9 + 5 = 68 feet
exact: 64 feet

23. The single parking space design in Exercise 22 measures 18 feet by 14 feet. Find its area.

estimate: 20 • 10 = 200 square feet
exact: 252 square feet

24. How many 32-ounce cans of oil can be filled from a tank holding 20,160 ounces of the oil?

estimate: 20,000 ÷ 30 ≈ 667 cans
exact: 630 cans

25. A fan blade makes 1800 revolutions in one minute. How many revolutions would it make in 50 minutes?

estimate: 2000 × 50 = 100,000 revolutions
exact: 90,000 revolutions

Round the mixed number in each problem to the nearest whole number and estimate the answer. Then find the exact answer.

26. A rectangular game table is $1\frac{3}{4}$ yards by $2\frac{2}{3}$ yards. Find its area.

estimate: 2 • 3 = 6 square yards

exact: $4\frac{2}{3}$ square yards

27. A wildlife refuge is $3\frac{1}{2}$ miles wide by $5\frac{3}{8}$ miles long. How many square miles are in the wildlife refuge?

estimate: 4 • 5 = 20 square miles

exact: $18\frac{13}{16}$ square miles

28. Larry Foxworthy cuts, splits, and delivers firewood. If his truck, when fully loaded, holds $5\frac{1}{4}$ cords of firewood, find the number of cords he could deliver in $3\frac{1}{2}$ loads.

estimate: 5•4 = 20 cords

exact: $18\frac{3}{8}$ cords

29. The Sears Tower in Chicago is 110 stories tall. If the total height of the building is $1536\frac{7}{8}$ feet including a flagpole at the top of the building which is $82\frac{1}{2}$ feet tall, find the height of the building itself.

estimate: 1537 − 83 = 1454 feet

exact: $1454\frac{3}{8}$ feet

Find the prime factorization of each number. Write answers by using exponents.

30. 20 $2^2 \cdot 5$

31. 144 $2^4 \cdot 3^2$

32. 250 $2 \cdot 5^3$

Solve each of the following.

33. $4^2 \cdot 2^3$ 128

34. $2^4 \cdot 3^2$ 144

35. $4^2 \cdot 3^3$ 432

Find each square root.

36. $\sqrt{25}$ 5

37. $\sqrt{49}$ 7

38. $\sqrt{144}$ 12

Simplify each of the following by using the order of operations.

39. $6^2 - 3 \cdot 7$ 15

40. $\sqrt{25} + 5 \cdot 9 - 6$ 44

41. $\left(\dfrac{3}{8} - \dfrac{1}{3}\right) \cdot \dfrac{1}{2}$ $\dfrac{1}{48}$

42. $\dfrac{3}{4} \div \left(\dfrac{1}{3} + \dfrac{1}{2}\right)$ $\dfrac{9}{10}$

43. $\dfrac{2}{3} + \left(\dfrac{7}{8}\right)^2 - \dfrac{1}{4}$ $1\dfrac{35}{192}$

Write proper *or* improper *for each fraction.*

44. $\dfrac{2}{3}$ proper

45. $\dfrac{5}{5}$ improper

46. $\dfrac{8}{7}$ improper

Write each fraction in lowest terms.

47. $\dfrac{35}{50}$ $\dfrac{7}{10}$

48. $\dfrac{38}{50}$ $\dfrac{19}{25}$

49. $\dfrac{105}{300}$ $\dfrac{7}{20}$

Add, subtract, multiply, or divide as indicated. Write answers in lowest terms and as mixed numbers where possible.

50. $\dfrac{2}{3} \times \dfrac{3}{4}$ $\dfrac{1}{2}$

51. $\dfrac{9}{11} \cdot \dfrac{5}{18}$ $\dfrac{5}{22}$

52. $42 \times \dfrac{7}{8}$ $36\dfrac{3}{4}$

53. $\dfrac{4}{5} \div \dfrac{2}{3}$ $1\dfrac{1}{5}$

54. $\dfrac{25}{40} \div \dfrac{10}{35}$ $2\dfrac{3}{16}$

55. $9 \div \dfrac{2}{3}$ $13\dfrac{1}{2}$

56. $\dfrac{5}{8} + \dfrac{1}{3}$ $\dfrac{23}{24}$

57. $\dfrac{5}{16} + \dfrac{1}{4} + \dfrac{3}{8}$ $\dfrac{15}{16}$

58. $\dfrac{11}{18} - \dfrac{5}{12}$ $\dfrac{7}{36}$

≈ *First estimate the answer. Then add or subtract to find the exact answer. Write answers as mixed numbers.*

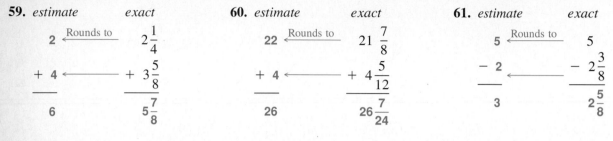

59. *estimate*   *exact*

$$\begin{array}{r} 2 \\ +\,4 \\ \hline 6 \end{array} \xleftarrow{\text{Rounds to}} \begin{array}{r} 2\frac{1}{4} \\ +\,3\frac{5}{8} \\ \hline 5\frac{7}{8} \end{array}$$

60. *estimate*   *exact*

$$\begin{array}{r} 22 \\ +\,4 \\ \hline 26 \end{array} \xleftarrow{\text{Rounds to}} \begin{array}{r} 21\frac{7}{8} \\ +\,4\frac{5}{12} \\ \hline 26\frac{7}{24} \end{array}$$

61. *estimate*   *exact*

$$\begin{array}{r} 5 \\ -\,2 \\ \hline 3 \end{array} \xleftarrow{\text{Rounds to}} \begin{array}{r} 5 \\ -\,2\frac{3}{8} \\ \hline 2\frac{5}{8} \end{array}$$

Find the least common multiple of each set of numbers.

62. 12, 18   36

63. 15, 20, 50   300

64. 12, 16, 18   144

Write each fraction by using the indicated denominator.

65. $\dfrac{7}{8} = \dfrac{35}{40}$

66. $\dfrac{7}{12} = \dfrac{77}{132}$

67. $\dfrac{9}{15} = \dfrac{81}{135}$

68. $\dfrac{5}{7} = \dfrac{60}{84}$

Locate each fraction on the number line.

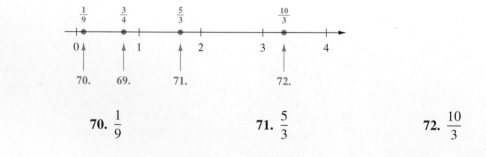

69. $\dfrac{3}{4}$

70. $\dfrac{1}{9}$

71. $\dfrac{5}{3}$

72. $\dfrac{10}{3}$

Write < or > to make a true statement.

73. $\dfrac{7}{10} \underline{\ <\ } \dfrac{37}{50}$

74. $\dfrac{19}{25} \underline{\ <\ } \dfrac{23}{30}$

75. $\dfrac{7}{12} \underline{\ <\ } \dfrac{11}{18}$

Decimals

Fractions are used to represent parts of a whole. In this chapter, **decimals** (DES-i-muls) are used as another way to show parts of a whole. For example, our money system is based on decimals. One dollar is divided into 100 equivalent parts. One cent ($0.01) is one of the parts, and a dime ($0.10) is 10 of the parts.

4.1 Reading and Writing Decimals

OBJECTIVE 1 ▶ Decimals are used when a whole is divided into 10 equivalent parts or into 100 or 1000 or 10,000 equivalent parts. In other words, decimals are fractions with denominators that are a power of 10. For example, the square below is cut into 10 equivalent parts. Written as a fraction, each part is $\frac{1}{10}$ of the whole. Written as a decimal, each part is 0.1. Both are read as "*one tenth.*"

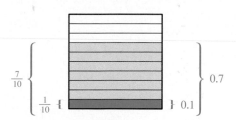

The dot in 0.1 is called the **decimal point.**

Decimal point

The square above has 7 of its 10 parts shaded.

Written as a fraction, $\frac{7}{10}$ of the square is shaded.

Written as a decimal, **0.7** of the square is shaded.

Both are read as "*seven tenths.*"

INTERNET
www.mathnotes.com

OBJECTIVES

1 ▶ Write parts of a whole as decimals.
2 ▶ Find the place value of a digit.
3 ▶ Read decimals.
4 ▶ Write decimals as fractions.

FOR EXTRA HELP

Tutorial   Tape 6   SSM, Sec. 4.1

1. There are 10 dimes in one dollar. Each dime is $\frac{1}{10}$ of a dollar. Write a fraction, a decimal, and the words that name the yellow shaded portion of each dollar.

(a)

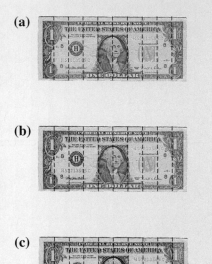

(b)

(c)

◀◀ **WORK PROBLEM I AT THE SIDE.**

The square below is cut into 100 equivalent parts. Written as a fraction, each part is $\frac{1}{100}$ of the whole.

$\frac{1}{100}$ ← → 0.01

Written as a decimal, each part is

0.01 of the whole.

↑
Read "one hundredth."

The square has 87 parts shaded.

Written as a fraction, $\frac{87}{100}$ of the total area is shaded.

Written as a decimal, **0.87** of the total area is shaded.

Both are read as *"eighty-seven hundredths."*

◀◀ **WORK PROBLEM 2 AT THE SIDE.**

2. Write the portion of each square that is shaded as a fraction, as a decimal, and in words.

(a)

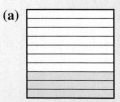

(b)

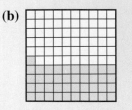

The example below shows several numbers written as both fractions and decimals.

E X A M P L E I   Using the Decimal Forms of Fractions

	Fraction	*Decimal*	*Read As*
(a)	$\frac{3}{10}$	0.3	three tenths
(b)	$\frac{9}{100}$	0.09	nine hundredths
(c)	$\frac{71}{100}$	0.71	seventy-one hundredths
(d)	$\frac{8}{1000}$	0.008	eight thousandths
(e)	$\frac{45}{1000}$	0.045	forty-five thousandths
(f)	$\frac{832}{1000}$	0.832	eight hundred thirty-two thousandths

ANSWERS

1. (a) $\frac{1}{10}$; 0.1; one tenth **(b)** $\frac{3}{10}$; 0.3; three tenths

(c) $\frac{9}{10}$; 0.9; nine tenths

2. (a) $\frac{3}{10}$; 0.3; three tenths **(b)** $\frac{41}{100}$; 0.41; forty-one hundredths

WORK PROBLEM 3 AT THE SIDE. ▶▶

OBJECTIVE 2 The decimal point separates the *whole number part* from the *fractional part* in a decimal number. In the chart below, you see that the **place value names** for fractional parts are similar to those on the whole number side but end in "*ths.*"

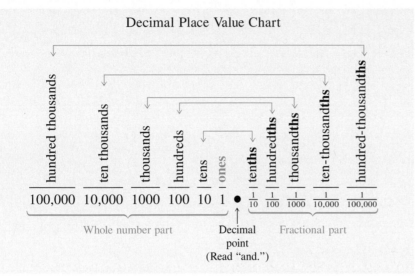

Decimal Place Value Chart

Notice that the ones place is at the center. (There is no "oneths" place.) Also notice that each place is 10 times the value of the place to its right.

> **Note**
> In this chapter, if a number does *not* have a decimal point, it is a *whole number*. A whole number has no fractional part. If you want to show the decimal point in a whole number, it is just to the *right* of the digit in the ones place. For example:
>
> $$8 = 8. \qquad 306 = 306.$$
>
> Decimal point   Decimal point

E X A M P L E 2   Identifying the Place Value of a Digit

Give the place values of the digits in each decimal.

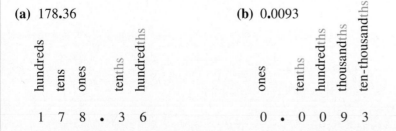

(a) 178.36   **(b)** 0.0093

Notice in Example 2(b) that we do *not* use commas on the right side of the decimal point.

WORK PROBLEM 4 AT THE SIDE. ▶▶

3. Write each decimal as a fraction.

(a) 0.7

(b) 0.9

(c) 0.03

(d) 0.69

(e) 0.047

(f) 0.351

4. Identify the place value of each digit in these decimals.

(a) 971.54

(b) 0.4

(c) 5.60

(d) 0.0835

ANSWERS

3. (a) $\dfrac{7}{10}$ (b) $\dfrac{9}{10}$ (c) $\dfrac{3}{100}$

(d) $\dfrac{69}{100}$ (e) $\dfrac{47}{1000}$ (f) $\dfrac{351}{1000}$

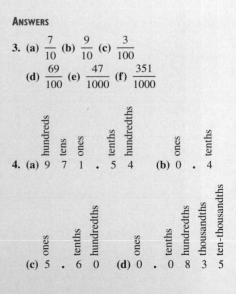

4. (a) 9 7 1 . 5 4   (b) 0 . 4

(c) 5 . 6 0   (d) 0 . 0 8 3 5

5. Write each decimal in words.

(a) 0.3

(b) 0.46

(c) 0.09

(d) 0.409

(e) 0.0003

(f) 0.0703

(g) 0.088

ANSWERS

5. (a) three tenths
  (b) forty-six hundredths
  (c) nine hundredths
  (d) four hundred nine thousandths
  (e) three ten-thousandths
  (f) seven hundred three ten-thousandths
  (g) eighty-eight thousandths

OBJECTIVE 3 A decimal is read according to its form as a fraction. We read 0.9 as "nine tenths" because 0.9 is the same as $\frac{9}{10}$. Notice that 0.9 ends in the tenths place.

ones
tenths

0. 9

We read 0.02 as "two hundredths" because 0.02 is the same as $\frac{2}{100}$. Notice that 0.02 ends in the hundredths place.

ones
tenths
hundredths

0. 0 2

E X A M P L E 3   **Reading a Decimal Number**

Write each decimal in words.

(a) 0.3

    Because $0.3 = \frac{3}{10}$, write the decimal as three ten**ths**.

(b) 0.49   Write it as forty-nine hundred**ths**.

(c) 0.08   Write it as eight hundred**ths**.

(d) 0.918   Write it as nine hundred eighteen thousand**ths**.

(e) 0.0106   Write it as one hundred six ten-thousand**ths**.

◀◀ **WORK PROBLEM 5 AT THE SIDE.**

Reading a Decimal Number

Step 1 Read any whole number part to the *left* of the decimal point as you normally would.

Step 2 Read the decimal point as "*and*."

Step 3 Read the part of the number to the *right* of the decimal point as if it was an ordinary whole number.

Step 4 Finish with the place value name of the right-most digit; these names all end in "*ths*."

Note
If there is *no whole number part,* you will use only Steps 3 and 4.

E X A M P L E 4   **Reading a Decimal**

Read each decimal.

(a)
          → 9 is in tenths place
    16.9

sixteen **and** nine **tenths** ←

16.9 is read "sixteen and nine tenths."

(b)
          → 5 is in hundredths place
    482.35

four hundred eighty-two **and** thirty-five **hundredths** ←

482.35 is read "four hundred eighty-two and thirty-five hundredths."

CONTINUED ON NEXT PAGE ─

→3 is in thousandths place
(c) 0.063 is sixty-three **thousandths**. (No whole number part.)

(d) 11.1085 is eleven **and** one thousand eighty-five **ten-thousandths**

> **Note**
> Use "and" *only* when reading a decimal point. A common mistake is to read the whole number 405 as "four hundred *and* five." But there is no decimal point shown in 405, so it is read "four hundred five."

WORK PROBLEM 6 AT THE SIDE. ▶▶

OBJECTIVE 4 ▶ Knowing how to read decimals will help you when writing decimals as fractions.

Writing Decimals as Fractions or Mixed Numbers

Step 1 The digits to the right of the decimal point are the numerator of the fraction.

Step 2 The denominator is 10 for tenths, 100 for hundredths, 1000 for thousandths, 10,000 for ten-thousandths, and so on.

Step 3 If the decimal has a whole number part, the fraction will be a mixed number with the same whole number part.

E X A M P L E 5   Writing a Decimal as a Fraction or Mixed Number

Write each decimal as a fraction or mixed number.

(a) 0.19

The digits to the right of the decimal point, 19, are the numerator of the fraction. The denominator is 100 for hundredths because the right-most digit is in the hundredths place.

$$0.19 = \frac{19}{100} \leftarrow 100 \text{ for hundredths.}$$
↑
Hundredths place

(b) 0.863

$$0.863 = \frac{863}{1000} \leftarrow 1000 \text{ for thousandths.}$$
↑
Thousandths place

(c) 4.0099

The whole number part stays the same.
$$4.0099 = 4\frac{99}{10,000} \leftarrow 10,000 \text{ for ten-thousandths.}$$
↑
Ten-thousandths place

WORK PROBLEM 7 AT THE SIDE. ▶▶

6. Write each decimal in words.

(a) 3.8

(b) 15.1

(c) 0.72

(d) 64.309

7. Write each decimal as a fraction or mixed number.

(a) 0.7

(b) 9.89

(c) 0.101

(d) 0.007

(e) 1.3717

ANSWERS
6. (a) three and eight tenths
(b) fifteen and one tenth
(c) seventy-two hundredths
(d) sixty-four and three hundred
nine thousandths

7. (a) $\frac{7}{10}$ **(b)** $9\frac{89}{100}$ **(c)** $\frac{101}{1000}$
(d) $\frac{7}{1000}$ **(e)** $1\frac{3717}{10,000}$

8. Write each decimal as a fraction or mixed number in lowest terms.

(a) 0.2

(b) 12.6

(c) 0.85

(d) 3.05

(e) 0.225

(f) 420.0802

Note

After you write a decimal as a fraction or a mixed number, check to see if the fraction is in lowest terms.

E X A M P L E   6   Writing a Decimal as a Fraction or Mixed Number

Write each decimal as a fraction or mixed number in lowest terms.

(a) $0.4 = \dfrac{4}{10}$ ← 10 for tenths.

Write $\dfrac{4}{10}$ in lowest terms. $\dfrac{4}{10} = \dfrac{4 \div 2}{10 \div 2} = \dfrac{2}{5}$

(b) $0.75 = \dfrac{75}{100} = \dfrac{75 \div 25}{100 \div 25} = \dfrac{3}{4}$ Lowest terms

(c) $18.105 = 18\dfrac{105}{1000} = 18\dfrac{105 \div 5}{1000 \div 5} = 18\dfrac{21}{200}$ Lowest terms

(d) $42.8085 = 42\dfrac{8085}{10{,}000} = 42\dfrac{8085 \div 5}{10{,}000 \div 5} = 42\dfrac{1617}{2000}$ Lowest terms

◀◀ **WORK PROBLEM 8 AT THE SIDE.**

▦ *Calculator Tip:*   In this book you'll notice that we use a zero in the ones place for decimal fractions. We write 0.45 instead of just .45, to emphasize that there is no whole number. Your calculator shows these zeros also. Enter ⬚ 4 5 . Notice that the display automatically shows 0.45 even though you did not press 0. For comparison, enter the whole number 45 by pressing 4 5 + and notice where the decimal point is shown in the display. (It automatically appears to the *right* of the 5.)

ANSWERS

8. (a) $\dfrac{1}{5}$ (b) $12\dfrac{3}{5}$ (c) $\dfrac{17}{20}$ (d) $3\dfrac{1}{20}$

(e) $\dfrac{9}{40}$ (f) $420\dfrac{401}{5000}$

4.1 Exercises

Name the digit that has the given place value.

Example: 3406.251
Solution: hundreds **4**
         hundredths **5**
         thousandths **1**

Example: 324.078
Solution: ones **4**
         tenths **0**
         tens **2**

1. 37.602
    ones   7
    tenths   6
    tens   3

2. 135.296
    ones   5
    tenths   2
    tens   3

3. 0.2518
    hundredths   5
    thousandths   1
    ten-thousandths   8

4. 0.9347
    hundredths   3
    thousandths   4
    ten-thousandths   7

5. 93.01472
    thousandths   4
    ten-thousandths   7
    tenths   0

6. 0.51968
    tenths   5
    ten-thousandths   6
    hundredths   1

7. 314.658
    tens   1
    tenths   6
    hundreds   3

8. 51.325
    tens   5
    tenths   3
    hundredths   2

9. 149.0832
    hundreds   1
    hundredths   8
    ones   9

10. 3458.712
    hundreds   4
    hundredths   1
    tenths   7

11. 6285.7125
    thousands   6
    thousandths   2
    hundredths   1

12. 5417.6832
    thousands   5
    thousandths   3
    ones   7

Write the decimal number that has the specified place values.

Example: 3 tenths, 5 ones, 9 thousandths, 0 hundredths
Solution: **5.309**

13. 0 ones, 5 hundredths, 1 ten, 4 hundreds, 2 tenths

410.25

14. 7 tens, 9 tenths, 3 ones, 6 hundredths, 8 hundreds

873.96

15. 3 thousandths, 4 hundredths, 6 ones, 2 ten-thousandths, 5 tenths

6.5432

16. 8 ten-thousandths, 4 hundredths, 0 ones, 2 tenths, 6 thousandths

0.2468

✐ Writing   ◉ Conceptual   ▲ Challenging   ≈ Estimation

17. 4 hundredths, 4 hundreds, 0 tens, 0 tenths, 5 thousandths, 5 thousands, 6 ones

5406.045

18. 7 tens, 7 tenths, 6 thousands, 6 thousandths, 3 hundreds, 3 hundredths, 2 ones

6372.736

Write each decimal as a fraction or mixed number in lowest terms.

Example: 0.68

Solution: $0.68 = \dfrac{68}{100} = \dfrac{17}{25}$ (Lowest terms)

Example: 4.005

Solution: $4.005 = 4\dfrac{5}{1000}$

$= 4\dfrac{1}{200}$ (Lowest terms)

19. 0.7 $\dfrac{7}{10}$ **20.** 0.1 $\dfrac{1}{10}$ **21.** 13.4 $13\dfrac{2}{5}$ **22.** 9.8 $9\dfrac{4}{5}$ **23.** 0.35 $\dfrac{7}{20}$

24. 0.85 $\dfrac{17}{20}$ **25.** 0.66 $\dfrac{33}{50}$ **26.** 0.33 $\dfrac{33}{100}$ **27.** 10.17 $10\dfrac{17}{100}$ **28.** 31.99 $31\dfrac{99}{100}$

29. 0.06 $\dfrac{3}{50}$ **30.** 0.08 $\dfrac{2}{25}$ **31.** 0.205 $\dfrac{41}{200}$ **32.** 0.805 $\dfrac{161}{200}$

33. 5.002 $5\dfrac{1}{500}$ **34.** 4.008 $4\dfrac{1}{125}$ **35.** 0.686 $\dfrac{343}{500}$ **36.** 0.492 $\dfrac{123}{250}$

Write each decimal in words.

Example: **Solution:**

16.028

8 is in thousandths place.

16 . 028

sixteen **and** twenty-eight **thousandths**

37. 0.5

five tenths

38. 0.9

nine tenths

39. 0.78

seventy-eight hundredths

40. 0.55

fifty-five hundredths

41. 0.105

one hundred five thousandths

42. 0.609

six hundred nine thousandths

43. 12.04

twelve and four hundredths

44. 86.09

eighty-six and nine hundredths

45. 1.075

one and seventy-five thousandths

46. 4.025

four and twenty-five thousandths

Write each decimal in numbers.

47. six and seven tenths

6.7

48. eight and twelve hundredths

8.12

49. thirty-two hundredths

0.32

50. one hundred eleven thousandths

0.111

51. four hundred twenty and eight thousandths

420.008

52. two hundred and twenty-four thousandths

200.024

53. seven hundred three ten-thousandths

0.0703

54. eight hundred and six hundredths

800.06

55. seventy-five and thirty thousandths

75.030

56. sixty and fifty hundredths

60.50

57. Anne read the number 4302 as "four thousand three hundred and two." Explain what is wrong with the way Anne read the number.

Anne should not say "and" because that denotes a decimal point.

58. Jerry read the number 9.0106 as "nine and one hundred and six ten-thousandths." Explain the error he made.

Jerry used "and" twice; only the first "and" is correct.

Suppose your job is to take phone orders for precision parts. Use the table below. In Exercises 59–62, write the correct part number that matches what you hear the customer say over the phone. In Exercises 63–64, write the words you would say to the customer.

Part Number	Size in Centimeters
3-A	0.06
3-B	0.26
3-C	0.6
3-D	0.86
4-A	1.006
4-B	1.026
4-C	1.06
4-D	1.6
4-E	1.602

59. "Please send the six-tenths centimeter bolt."

Part number ___3-C___ .

60. "The part missing from our order was the one and six hundredths size."

Part number ___4-C___ .

61. "The size we need is one and six thousandths centimeters."

Part number ___4-A___ .

62. "Do you still stock the twenty-six hundredths centimeter bolt?"

Part number ___3-B___ .

63. "What size is part number 4-E?" Write your answer in words.

One and six hundred two thousandths centimeters

64. "What size is part number 4-B?" Write your answer in words.

One and twenty-six thousandths centimeters

65. Look back at the Decimal Place Value Chart in this section. What do you think would be the names of the next four places to the *right* of hundred-thousandths? What information did you use to come up with these names?

millionths, ten-millionths, hundred-millionths, billionths; these match the words on the left side of the chart with "ths" added.

66. A common mistake is to think that the first place to the right of the decimal point is "oneths" and the second place is "tenths." Why might someone make that mistake? How would you explain why there is no "oneths" place?

First place to left of decimal point is ones, so first place to right could be one*ths*, like tens and ten*ths*. But anything that is 1 or more is to the left of the decimal point.

67. Write 0.72436955 in words.

Seventy-two million four hundred thirty-six thousand nine hundred fifty-five hundred-millionths

68. Write 0.000678554 in words.

Six hundred seventy-eight thousand five hundred fifty-four billionths

69. Write 8006.500001 in words.

Eight thousand six and five hundred thousand one millionths

70. Write 20,060.000505 in words.

Twenty thousand sixty and five hundred five millionths

Review and Prepare

Round each of the following to the nearest ten, nearest hundred, and nearest thousand. (For help, see **Section 1.7.***)*

	Ten	Hundred	Thousand
71. 8235	8240	8200	8000
72. 3565	3570	3600	4000
73. 19,705	19,710	19,700	20,000
74. 89,604	89,600	89,600	90,000

4.2 *Rounding Decimals*

Section 1.7 showed how to round whole numbers. For example, 89 rounded to the nearest ten is 90, and 8512 rounded to the nearest hundred is 8500.

OBJECTIVE 1 It is also important to be able to **round** decimals. For example, a store is selling 2 candy mints for $0.75 but you want only one mint. The price of each mint is $0.75 ÷ 2, which is $0.375, but you cannot pay part of a cent. Is $0.375 closer to $0.37 or to $0.38? Actually, it's exactly halfway between. When this happens in everyday situations, the rule is to round *up*. The store will charge you $0.38 for the mint.

OBJECTIVES

1 ▶ Learn the rules for rounding decimals.

2 ▶ Round decimals to any given place.

3 ▶ Round money amounts to the nearest cent or nearest dollar.

FOR EXTRA HELP

Tutorial   Tape 6   SSM, Sec. 4.2

Rounding Decimals

Step 1   Find the place to which the rounding is being done. Draw a "cut-off" line **after** that place to show that you are cutting off and dropping the rest of the digits.

Step 2   Look **only** at the **first** digit you are cutting off.

Step 3A If this digit is **less than 5,** the part of the number you are keeping **stays the same.**

Step 3B If this digit is **5 or more,** you must **round up** the part of the number you are keeping.

Step 4   Use the "≈" sign to indicate that the rounded number is now an approximation (close, but not exact). "≈" means "is approximately equal to."

Note
Do **not** move the decimal point when rounding.

OBJECTIVE 2 These examples show you how to round decimals.

EXAMPLE 1   Rounding a Decimal Number

Round 14.39652 to the nearest thousandth. Is it closer to 14.396 or to 14.397?

Step 1 Draw a "cut-off" line after the thousandths place.

$$1\ 4\ .\ 3\ 9\ 6\ |\ 5\ 2$$

Thousandths ⟶   You are cutting off the 5 and 2. They will be dropped.

Step 2 Look *only* at the *first* digit you are cutting off. Ignore the other digits you are cutting off.

$$1\ 4\ .\ 3\ 9\ 6\ |\ 5\ 2$$

Look only at the 5. Ignore the 2.

Step 3 If the first digit you are cutting off is 5 or more, round up the part of the number you are keeping.

First digit cut is 5 or more, so round up by adding 1 thousandth to the part you are keeping.

$$\begin{array}{r} 1\ 4\ .\ 3\ 9\ 6\ |\ 5\ 2 \\ +\ \ \ 0\ .\ 0\ 0\ 1 \\ \hline 1\ 4\ .\ 3\ 9\ 7 \end{array}$$

So, 14.39652 rounded to the nearest thousandth is 14.397. Write it ≈14.397.

Note
When rounding whole numbers in **Section 1.7** you kept all the digits, but changed some to zeros. With decimals, you cut off and *drop the extra digits.* In the example above, 14.39652 rounds to 14.397 **not** 14.39700.

1. Round to the nearest thousandth. Write your answers using the "≈" sign.

(a) 0.33492

(b) 8.00851

(c) 265.42068

(d) 10.70180

◀◀ **WORK PROBLEM I AT THE SIDE.**

In Example 1, the rounded number 14.397 had *three* **decimal places.** Decimal places are the number of digits to the *right* of the decimal point. The first decimal place is tenths, the second is hundredths, the third is thousandths, and so on.

E X A M P L E 2   **Rounding Decimals to Different Places**

Round to the place indicated.

(a) 5.3496 to the nearest tenth (Is it closer to 5.3 or to 5.4?)

Step 1 Draw a cut-off line after the tenths place.

$$5 . 3 \overset{\text{✂}}{\cancel{9}} 4 \ 9 \ 6$$
Tenths ⟶   You will be cutting off the 4, 9, and 6.

Step 2
Look only at the 4.
$$5 . 3 \cancel{9} 4 \ 9 \ 6$$
Ignore these digits.

Step 3
$$5 . 3 \cancel{9} 4 \ 9 \ 6$$ First digit cut is less than 5 so the part you are keeping stays the same.
$$5 . 3 \ \leftarrow \text{Stays the same}$$

5.3496 rounded to the nearest tenth is 5.3 (one decimal place for tenths). Write it ≈5.3. Notice that it does **not** round to 5.3000 which would be ten-thousandths.

(b) 0.69738 to the nearest hundredth (Is it closer to 0.69 or to 0.70?)

Step 1
$$0 . 6 \ 9 | 7 \ 3 \ 8$$
Hundredths   Draw a cut-off line after the hundredths place.

Step 2
Look only at the 7.
$$0 . 6 \ 9 | 7 \ 3 \ 8$$

Step 3
$$0 . 6 \ 9 | 7 \ 3 \ 8$$ First digit cut is 5 or more, so round up by adding 1 hundredth to the part you are keeping.

$$
\begin{array}{r}
0 . 6 \ 9 \\
+ \ 0 . 0 \ 1 \\
\hline
0 . 7 \ 0
\end{array}
$$
⟵ Keep this part.
⟵ To round up, add 1 hundredth.
⟵ 9 + 1 is 10; write 0 and carry 1 to the 6 in the tenths place.

0.69738 rounded to the nearest hundredth is 0.70. Hundredths is *two* decimal places so you *must* write the 0 in the hundredths place. Write it ≈0.70.

(c) 0.01806 to the nearest thousandth (Is it closer to 0.018 or to 0.019?)

$$0 . 0 \ 1 \ 8 | 0 \ 6$$ First digit cut is less than 5 so the part you are keeping stays the same.
$$0 . 0 \ 1 \ 8$$

0.01806 rounded to the nearest thousandth is 0.018 (three decimal places for thousandths). Write it ≈0.018.

CONTINUED ON NEXT PAGE

ANSWERS

1. (a) ≈0.335 (b) ≈8.009 (c) ≈265.421
   (d) ≈10.702

(d) 57.976 to the nearest tenth (Is it closer to 57.9 or to 58.0?)

$$57.9 \mid \overset{\downarrow}{7}6 \quad \begin{array}{l} \text{First digit cut is 5 or more so round up} \\ \text{by adding 1 tenth to the part you are keeping.} \end{array}$$

$$\begin{array}{r} 57.9 \\ + \quad 0.1 \\ \hline 58.0 \end{array} \quad \leftarrow \begin{array}{l} 9 + 1 \text{ is } 10; \text{ write the 0 and carry 1} \\ \text{to the 7 in the ones place.} \end{array}$$

57.976 rounded to the nearest tenth is 58.0. Write it ≈58.0.
You *must* write the zero in the tenths place to show that the number was rounded to the nearest tenth.

Note
Check that your rounded answer shows **exactly** the number of decimal places called for, even if a zero is in that place. Be sure your answer shows one decimal place if you rounded to tenths, two decimal places for hundredths, or three decimal places for thousandths.

WORK PROBLEM 2 AT THE SIDE. ▶▶

OBJECTIVE 3 When you are shopping in a store, money amounts are usually rounded to the nearest cent. There are 100 cents in a dollar.

$$\text{Each cent is } \frac{1}{100} \text{ of a dollar.}$$

Another way to write $\frac{1}{100}$ is 0.01. So rounding to the *nearest cent* is the same as rounding to the *nearest hundredth of a dollar*.

E X A M P L E 3 **Rounding to the Nearest Cent**

Round each of these money amounts to the nearest cent.

(a) $2.4238 (Is it closer to $2.42 or to $2.43?)

$$\$2.42 \mid \overset{\downarrow}{3}8 \quad \begin{array}{l} \text{Less than 5 so the part you are} \\ \text{keeping stays the same.} \end{array}$$

$$\$2.42 \leftarrow \text{You pay}$$

(b) $0.695 (Is it closer to $0.69 or to $0.70?)

$$\$0.69 \mid \overset{\downarrow}{5} \quad \text{5 or more; round up}$$

$$\begin{array}{r} \$0.69 \\ + \quad \$0.01 \\ \hline \$0.70 \end{array} \quad \begin{array}{l} \leftarrow \text{To round up, add 1 hundredth (1 cent)} \\ \leftarrow \text{You pay} \end{array}$$

WORK PROBLEM 3 AT THE SIDE. ▶▶

It is also common to round money amounts to the nearest dollar. You can do that on your federal and state income tax, for example, to make the calculations easier.

---

2. Round to the place indicated.

(a) 0.8988 to the nearest hundredth

(b) 5.8903 to the nearest hundredth

(c) 11.0299 to the nearest thousandth

(d) 0.545 to the nearest tenth

3. Round each of the following money amounts to the nearest cent.

(a) $14.595

(b) $578.0663

(c) $0.849

(d) $0.0548

ANSWERS

2. (a) ≈0.90 **(b)** ≈5.89 **(c)** ≈11.030
    (d) ≈0.5
3. (a) ≈$14.60 **(b)** ≈$578.07
    (c) ≈$0.85 **(d)** ≈$0.05

4. Round to the nearest dollar.

(a) $29.10

(b) $136.49

(c) $990.91

(d) $5949.88

(e) $49.60

(f) $0.55

(g) $1.08

E X A M P L E 4 — Rounding to the Nearest Dollar

Round to the nearest dollar.

(a) $48.69 (Is it closer to $48 or to $49?)

$$\$48.\underbrace{|69}$$

First digit cut is 5 or more so round up by adding $1.

$$\begin{array}{r} \$48 \\ +\quad 1 \\ \hline \$49 \end{array}$$

Note
$48.69 rounded to the nearest dollar is $49. Write the answer as $49 to show that the rounding is to the *nearest dollar*. Writing $49.00 would show rounding to the nearest *cent*.

(b) $594.36 (Is it closer to $594 or to $595?)

$$\underbrace{\$594.}|36$$

Less than 5 so the part you keep stays the same.

$$\$594$$

$594.36 rounded to the nearest dollar is $594.

(c) $349.88 (Is it closer to $349 or to $350?)

$$\underbrace{\$349.}|88$$

5 or more, so round up by adding $1.

$$\begin{array}{r} \$349 \\ +\quad 1 \\ \hline \$350 \end{array}$$

$349.88 rounded to the nearest dollar is $350.

(d) $2689.50 rounded to the nearest dollar is $2690.

(e) $0.61 rounded to the nearest dollar is $1.

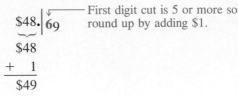 *Calculator Tip:* Accountants and other people who work with money amounts often set their calculators to automatically round to 2 decimal places (nearest cent) or to round to 0 decimal places (nearest dollar). Your calculator may have this feature.

◀◀ **WORK PROBLEM 4 AT THE SIDE.**

ANSWERS
4. (a) ≈$29 **(b)** ≈$136 **(c)** ≈$991
(d) ≈$5950 **(e)** ≈$50 **(f)** ≈$1
(g) ≈$1

4.2 Exercises

Round each number to the place indicated. Write your answers using the "≈" sign.

Example:	**Solution:**

5.7061 to the nearest hundredth

Draw cut-off line after hundredths place.
First digit cut is 5 or more so round up the part you are keeping.

5.70│6₁
↑
Hundredths

5.70 ← Keep this part.
+ 0.01 ← Add 1 hundredth.
≈5.71 ← Check that answer has exactly 2 decimal places for hundredths.

1. 16.8974 to the nearest tenth

≈16.9

2. 193.845 to the nearest hundredth

≈193.85

3. 0.95647 to the nearest thousandth

≈0.956

4. 96.81584 to the nearest ten-thousandth

≈96.8158

5. 0.799 to the nearest hundredth

≈0.80

6. 0.952 to the nearest tenth

≈1.0

7. 3.66062 to the nearest thousandth

≈3.661

8. 1.5074 to the nearest hundredth

≈1.51

9. 793.988 to the nearest tenth

≈794.0

10. 476.1196 to the nearest thousandth

≈476.120

11. 0.09804 to the nearest ten-thousandth

≈0.0980

12. 176.004 to the nearest tenth

≈176.0

13. 48.512 to the nearest one

≈49

14. 3.385 to the nearest one

≈3

15. 9.0906 to the nearest hundredth

≈9.09

16. 30.1290 to the nearest thousandth

≈30.129

17. 82.000151 to the nearest ten-thousandth

≈82.0002

18. 0.400594 to the nearest ten-thousandth

≈0.4006

Nardos is grocery shopping. The store will round the amount she pays for each item to the nearest cent. Write the rounded amounts.

19. Soup is 3 cans for $2.45, so one can is $0.81666. Nardos pays __$0.82__

20. Orange juice is 2 cartons for $2.69, so one carton is $1.345. Nardos pays __$1.35__

21. Facial tissue is 4 boxes for $4.89, so one box is $1.2225. Nardos pays __$1.22__

22. Muffin mix is 3 packages for $1.75, so one package is $0.58333. Nardos pays __$0.58__

23. Candy bars are 6 for $2.99, so one bar is $0.4983. Nardos pays __$0.50__

24. Boxes of spaghetti are 4 for $3.59, so one box is $0.8975. Nardos pays __$0.90__

📝 Writing　　◉ Conceptual　　▲ Challenging　　≈ Estimation

As she gets ready to do her income tax return, Ms. Chen rounds each amount to the nearest dollar. Write the rounded amounts.

25. Income from job, $17,249.70

≈$17,250

26. Income from interest on bank account, $69.58

≈$70

27. Union dues, $310.08

≈$310

28. Federal withholding, $2150.49

≈$2150

29. Donations to charity, $378.82

≈$379

30. Medical expenses, $609.38

≈$609

31. Explain what happens when you round $0.499 to the nearest dollar. Why does this happen?

Rounds to $0 (zero dollars) because $0.499 is closer to $0 than to $1.

32. Explain what happens when you round $0.0015 to the nearest cent. Why does this happen?

Rounds to $0.00 (zero cents) because $0.0015 is closer to $0.00 than to $0.01.

33. Look again at Exercise 31 above. How else could you round $0.499 that would be more helpful? What kind of guideline does this suggest about rounding to the nearest dollar?

Round amounts less than $1.00 to nearest cent instead of nearest dollar.

34. Suppose you want to know which of these amounts is less, so you round them both to the nearest cent.

$0.5968 $0.6014

Explain what happens. Describe what you could do instead of rounding to the nearest cent.

Both round to $0.60. Rounding to nearest thousandth (tenth of a cent) would allow you to identify $0.597 as less than $0.601.

▲ *Round each of these money amounts.*

35. $499.98 to the nearest dollar.

≈$500

36. $9899.59 to the nearest dollar.

≈$9900

37. $0.996 to the nearest cent.

≈$1.00

38. $0.09929 to the nearest cent.

≈$0.10

39. $999.73 to the nearest dollar.

≈$1000

40. $9999.80 to the nearest dollar.

≈$10,000

Review and Prepare

≈ *Round each number so there is only one non-zero digit and* **estimate** *the total. Then add to get the* **exact** *answer. (For help, see* **Sections 1.2 and 1.7.***)*

41. *estimate*   *exact*

8000	Rounds to ←	7929
6000	←	6076
+ 8000	←	+ 8218
22,000		22,223

42. *estimate*   *exact*

2000	←	2078
200	←	183
200	←	231
+ 7000	←	+ 7209
9400		9701

43. $\underline{80,000}$ + $\underline{100}$ + $\underline{800}$ = $\underline{80,900}$ *estimate*

$\uparrow$ $\uparrow$ $\uparrow$

81,976 + 98 + 785 = $\underline{82,859}$ *exact*

44. $\underline{2000}$ + $\underline{20,000}$ + $\underline{900}$ = $\underline{22,900}$ *estimate*

$\uparrow$ $\uparrow$ $\uparrow$

1750 + 18,763 + 918 = $\underline{21,431}$ *exact*

4.3 Adding Decimals

OBJECTIVES

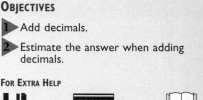

1 ▶ Add decimals.

2 ▶ Estimate the answer when adding decimals.

FOR EXTRA HELP

Tutorial   Tape 7   SSM, Sec. 4.3

OBJECTIVE 1 ▶ When adding *whole* numbers (**Section 1.2**), you lined up the numbers in columns so that you were adding ones to ones, tens to tens, and so on. A similar idea applies to adding *decimal* numbers. With decimals you line up the decimal points to make sure you are adding tenths to tenths, hundredths to hundredths, and so on.

Adding Decimals

Step 1 Write the numbers in columns with the decimal points lined up.

Step 2 Add the numbers as if they were whole numbers.

Step 3 Line up the decimal point in the answer directly below the decimal points in the problem.

EXAMPLE 1   Adding Decimal Numbers

Add.

(a) 16.92 and 48.34

Write the numbers in columns with the decimal points lined up.

$$\begin{array}{r} \text{tens ones . tenths hundredths} \\ 1\,6\,.\,9\,2 \\ +\ 4\,8\,.\,3\,4 \\ \hline \end{array}$$

— Decimal points are lined up.

Add as if these were whole numbers. Then line up the decimal point in the answer under the decimal points in the problem.

$$\begin{array}{r} \overset{11}{} \\ 16.92 \\ +\ 48.34 \\ \hline 65.26 \end{array}$$

Decimal point in answer is lined up under decimal points in problem.

(b) 5.897 + 4.632 + 12.174

Write the numbers vertically with decimal points lined up. Next, add.

$$\begin{array}{r} \overset{11\ \ 21}{} \\ 5.897 \\ 4.632 \\ +\ 12.174 \\ \hline 22.703 \end{array}$$

— Decimal points are lined up.

▶▶ **WORK PROBLEM 1 AT THE SIDE.** ▶▶

In Example 1(a), both numbers had *two* decimal places (two digits to the right of the decimal point). In Example 1(b), all the numbers had *three decimal places* (three digits to the right of the decimal point). That made it easy to add tenths to tenths, hundredths to hundredths, and so on.

If the number of decimal places does *not* match, you can write in zeros as placeholders to make them match. This is shown in Example 2.

1. Find each sum.

(a) 2.86 + 7.09

(b) 13.761 + 8.325

(c) 0.319 + 56.007 + 8.252

(d) 39.4 + 0.4 + 177.2

ANSWERS

1. (a) 9.95 **(b)** 22.086 **(c)** 64.578
(d) 217.0

245

2. Find each of the following sums.

(a) $6.54 + 9.8$

(b) $0.831 + 222.2 + 10$

(c) $8.64 + 39.115 + 3.0076$

(d) $5 + 429.823 + 0.76$

3. $\approx$ First, round each number so there is only one non-zero digit and estimate the answer. Then add to find the exact answer.

(a) $2.83 + 5.009 + 76.1$

▦ (b) $398.81 + 47.658 + 4158.7$

▦ (c) $3217.6 + 5.4 + 37.288$

ANSWERS

2. (a) 16.34 **(b)** 233.031 **(c)** 50.7626
   (d) 435.583
3. (a) $3 + 5 + 80 = 88$; 83.939
   (b) $400 + 50 + 4000 = 4450$;
     4605.168
   (c) $3000 + 5 + 40 = 3045$; 3260.288

E X A M P L E 2   **Writing Zeros as Placeholders Before Adding**

Add.

(a) $7.3 + 0.85$

There are two decimal places in 0.85 (tenths and hundredths), so write a zero in the hundredths place in 7.3 so that it has two decimal places also.

$$\begin{array}{r} 7.30 \\ +\ 0.85 \\ \hline 8.15 \end{array}$$ ← One 0 is written in.

7.30 is equivalent to 7.3 because

$7\dfrac{30}{100}$ in lowest terms is $7\dfrac{3}{10}$

(b) $6.42 + 9 + 2.576$

Make all the addends have three decimal places. Notice how the whole number 9 is written with the decimal point at the *far right* side. (If you put the decimal point on the *left* side of the 9, you would turn it into the decimal fraction 0.9.)

$$\begin{array}{r} 6.4\,2\,0 \\ 9.0\,0\,0 \\ +\ \ 2.5\,7\,6 \\ \hline 17.9\,9\,6 \end{array}$$

6.4 2 0 ← One 0 is written in.
9.0 0 0 ← 9 is a whole number; decimal point and three 0's are written in.
+ 2.5 7 6 ← No 0's are needed.

Note
Writing zeros to the right of a *decimal* number does *not* change the value of the number.

◄◄ **WORK PROBLEM 2 AT THE SIDE.**

OBJECTIVE ▶2 A common error in working decimal problems by hand is to misplace the decimal point in the answer. Or, when using a calculator, you may accidentally press the wrong key. **Estimating** (ES-tih-may-ting) the answer will help you avoid these mistakes. Start by rounding each number so there is only one non-zero digit (as you did in **Section 1.7**). Here are several examples. Notice that in the rounded numbers only the left-most digit is something other than zero.

3.25 rounds to 3   6.812 rounds to 7

532.6 rounds to 500   26.397 rounds to 30

E X A M P L E 3   **Estimating a Decimal Answer**

Round each number so there is only one non-zero digit. Then add the rounded numbers to get an estimated answer. Finally, find the exact answer. Add 194.2 and 6.825.

$$\begin{array}{cc} \textit{estimate} & \textit{exact} \\ 200 \xleftarrow{\text{Rounds to}} & 194.200 \\ +\ \ \ 7 \xleftarrow{\text{Rounds to}} & +\ \ \ 6.825 \\ \hline 207 & 201.025 \end{array}$$

The estimate goes out to the hundreds place (three places to the *left* of the decimal point), and so does the exact answer. Therefore, the decimal point is probably in the right place in the exact answer.

◄◄ **WORK PROBLEM 3 AT THE SIDE.**

4.3 Exercises

Find each sum.

Example: $0.28 + 5 + 38.152$ **Solution:**

Line up decimal points.

$$\begin{array}{r} 0.280 \\ 5.000 \\ + \; 38.152 \\ \hline \mathbf{43.432} \end{array}$$

Use zeros as placeholders.

1.
$$\begin{array}{r} 5.69 \\ 0.24 \\ + \; 11.79 \\ \hline 17.72 \end{array}$$

2.
$$\begin{array}{r} 372.1 \\ 33.7 \\ + \; 42.3 \\ \hline 448.1 \end{array}$$

3.
$$\begin{array}{r} 8224.008 \\ 0.995 \\ + \; 96.409 \\ \hline 8321.412 \end{array}$$

4.
$$\begin{array}{r} 0.7759 \\ 306.2602 \\ + \; 9.8883 \\ \hline 316.9244 \end{array}$$

5.
$$\begin{array}{r} 8.763 \\ 0.5 \\ + \; 339.25 \\ \hline 348.513 \end{array}$$

6.
$$\begin{array}{r} 76.5 \\ 89.39 \\ + \; 0.506 \\ \hline 166.396 \end{array}$$

7.
$$\begin{array}{r} 0.38 \\ 7 \\ + \; 4.6 \\ \hline 11.98 \end{array}$$

8.
$$\begin{array}{r} 3.7 \\ 0.812 \\ + \; 55 \\ \hline 59.512 \end{array}$$

9. $14.23 + 8 + 74.63 + 18.715 + 0.286$ **115.861**

10. $197.4 + 0.72 + 17.43 + 25 + 1.4$ **241.95**

11. $27.65 + 18.714 + 9.749 + 3.21$ **59.323**

12. $58.546 + 19.2 + 8.735 + 14.58$ **101.061**

13. $39.76005 + 182 + 4.799 + 98.31 + 5.9999$ **330.86895**

14. $489.76 + 0.9993 + 38 + 8.55087 + 80.697$ **618.00717**

15. Explain and correct the error that a student made when he added $0.72 + 6 + 39.5$ this way:

$$\begin{array}{r} 0.72 \\ 6 \\ + \; 39.50 \\ \hline 40.28 \end{array}$$

6 should be written 6.00; sum is 46.22.

16. Explain and correct the error that a student made when she added $7.21 + 65 + 13.15$ this way:

$$\begin{array}{r} 7.21 \\ .65 \\ + \; 13.15 \\ \hline 21.01 \end{array}$$

65 should be written 65.00; sum is 85.36.

✍ Writing   ◉ Conceptual   ▲ Challenging   ≈ Estimation

≈ *Round each number so there is only one non-zero digit and estimate the sum. Then add to find the exact answer.*

Example: 56.9 + 0.82 + 12.06   **Solution:** *estimate*   *exact*

estimate	Rounds to	*exact*
60	←	56.90
1	←	0.82
+ 10	←	+ 12.06
71		69.78

17. *estimate*   *exact*

40	←	37.25
20	←	18.9
+ 8	←	+ 7.5
68		63.65

18. *estimate*   *exact*

20	24.83
20	19.7
+ 50	+ 46.19
90	90.72

19. *estimate*   *exact*

400	392.7
1	0.865
+ 20	+ 21.08
421	414.645

20. *estimate*   *exact*

40	38.55
8	7.716
+ 1	+ 0.6
49	46.866

21. *estimate*   *exact*

60	62.8173
500	539.99
+ 6	+ 5.629
566	608.4363

22. *estimate*   *exact*

300	332.607
10	12.5
+ 800	+ 823.3949
1110	1168.5019

23. *estimate*   *exact*

400	382.504
600	591.089
+ 600	+ 612.715
1600	1586.308

24. *estimate*   *exact*

8000	8159.76
9000	9382.54
+ 7000	+ 7179.18
24,000	24,721.48

≈ *Round each number so there is only one non-zero digit and estimate the answer. Then find the exact answer for each application problem.*

25. Mrs. Little Owl put two checks in the deposit envelope at the automated teller machine. There was a $310.14 paycheck and a $0.95 refund check. How much did she deposit in her account?

estimate: $300 + $1 = $301
exact: $311.09

26. Rodney Green's paycheck stub showed wages of $274.19 at the regular rate of pay and $72.94 at the overtime rate. What were his total wages?

estimate: $300 + $70 = $370
exact: $347.13

27. Chris Howard worked at Blockblaster Video 4.5 days one week, 6.25 days another week, and 3.74 days a third week. How many days did he work altogether?

estimate: 5 + 6 + 4 = 15 days
exact: 14.49 days

28. The tallest known land mammal is a prehistoric ancestor of the rhino measuring 6.4 m. Find the combined heights of these NBA basketball stars: Charles Barkley at 1.98 m, Karl Malone at 2.06 m, and David Robinson at 2.16 m. Is their combined height greater or less than the prehistoric rhino?

6.4 m

estimate: 2 + 2 + 2 = 6 m
exact: 6.2 m which is less than 6.4 m

29. At a bakery, Sue Chee bought $7.42 worth of muffins and $10.09 worth of croissants for a staff party and a $0.69 cookie for herself. How much money did she spend altogether?

estimate: **$7 + $10 + $1 = $18**
exact: **$18.20**

30. Jeff McGee wrote checks for $172.15, $0.75, $9.06, and $122.24. Find the total of the checks.

estimate: **$200 + $1 + $9 + $100 = $310**
exact: **$304.20**

31. At the beginning of Lilia's trip to Dallas, her car odometer read 7942.1 miles. The distance to Dallas is 154.8 miles. What should the odometer read after driving to Dallas *and back?*

estimate: **8000 + 200 + 200 = 8400 miles**
exact: **8251.7 miles**

32. Gonzalo runs his own package delivery service. On one trip he started from Atlanta and drove 226.6 miles to Charlotte, then 153.8 miles to Roanoke, and finally, 341.3 miles back to Atlanta. Find the total length of his trip.

estimate: **200 + 200 + 300 = 700 miles**
exact: **721.7 miles**

Yiangos works part-time at a factory. His time card for last week is shown below. Use the time card to solve Exercises 33–35.

Day	Date	Hours
Mon	6/1	4.5
Tue	6/2	0
Wed	6/3	0
Thr	6/4	6.2
Fri	6/5	5
Sat	6/6	9.5
Sun	6/7	4.8

33. Yiangos is paid a higher hourly wage for working on weekends. How many weekend hours did he work?

estimate: **10 + 5 = 15 hours**
exact: **14.3 hours**

34. How many hours did Yiangos work on weekdays?

estimate: **5 + 6 + 5 = 16 hours**
exact: **15.7 hours**

35. How many hours did Yiangos work in all?

estimate: **5 + 6 + 5 + 10 + 5 = 31 hours**
exact: **30 hours**

The accountant at Top Notch Lumber had the list of expenses shown below for March. Use the list to solve Exercises 36–38.

Expenses for March	
payroll	$4919.20
utilities	$ 732.44
radio ads	$1864.02
newspaper ads	$1015.16
TV ads	$2890
mill payments	$31,941.84

36. Find the amount spent on advertising during the month.

estimate: **$2000 + $1000 + $3000 = $6000**
exact: **$5769.18**

37. Find the amount spent on payroll and utilities in March.

estimate: **$5000 + $700 = $5700**
exact: **$5651.64**

38. Find the total expenses for the month.

estimate: **$5000 + $700 + $2000 + $1000 + $3000 + 30,000 = $41,700**
exact: **$43,362.66**

39. Show why 0.3 is equivalent to 0.3000.

$$0.3000 = \frac{3000 \div 1000}{10,000 \div 1000} = \frac{3}{10} = 0.3$$

40. Explain why 7 could be written as 7.0 but not as 0.7.

0.7 is $\frac{7}{10}$; 7.0 is $7\frac{0}{10}$.

▲ *Solve each application problem. There may be extra information in the problem, or you may need to do several steps.*

41. Tameka keeps track of her business mileage so her company will pay for her travel. She is not paid for trips to lunch or for travel to and from home. Today she drove 12.6 miles to work, 35.4 miles to visit a client, 14.9 miles to visit another client, 8 miles to lunch, 40 miles to attend a business meeting, and 12.6 miles home. How many miles will her company pay for?

90.3 miles

42. Tony wrote a lot of checks today. His tuition at the community college was $476.44 and textbooks were $80.06. He also paid $17.99 for an oil change on his car, $20.75 at the grocery store, and $31.62 for brushes and paint for an art class he is taking at the college. What were his total school expenses?

$588.12

43. In the 1996 Olympics, gymnast Dominique Dawes' score in the vault finals was 9.649. Her score in the floor exercise was 0.188 more. What was her total score for the two events?

19.486

44. James jogged 3.25 kilometers this morning. His friend Anthony jogged with him and kept going another 1.4 kilometers after James stopped. What was the total distance run by the two men?

7.9 kilometers

Find the perimeter of (distance around) these figures by adding the lengths of the sides.

45.

19.75 inches

6.3 inches   6.3 inches

19.75 inches

estimate: **20 + 6 + 20 + 6 = 52 inches**
exact: **52.1 inches**

46.

2 meters   1 meter

0.9 meter

1.7 meters

1.18 meters

0.86 meter

2.095 meters

estimate: **2 + 1 + 2 + 1 + 2 + 1 + 1 = 10 meters**
exact: **9.735 meters**

Review and Prepare

≈ *Round the numbers so there is only one non-zero digit and estimate each answer. Then find the exact answer. (For help, see Sections 1.3 and 1.7.)*

47. *estimate*   *exact*

$$
\begin{array}{r} 300 \\ -\ 100 \\ \hline 200 \end{array}
\quad \xleftarrow{\text{Rounds to}} \quad
\begin{array}{r} 301 \\ -\ 104 \\ \hline 197 \end{array}
$$

48. *estimate*   *exact*

$$
\begin{array}{r} 600 \\ -\ 400 \\ \hline 200 \end{array}
\qquad
\begin{array}{r} 553 \\ -\ 386 \\ \hline 167 \end{array}
$$

49.   **7000** − **100** = **6900**   *estimate*

    6708 − **139** = **6569**   *exact*

50.   **70,000** − **900** = **69,100**   *estimate*

    71,000 − **856** = **70,144**   *exact*

4.4 Subtracting Decimals

OBJECTIVE 1 Subtraction of decimals is done in much the same way as addition of decimals. Use the following steps.

Subtracting Decimals

Step 1 Write the numbers in columns with the decimal points lined up.

Step 2 If necessary, write in zeros so both numbers have the same number of decimal places. Then subtract as if they were whole numbers.

Step 3 Line up the decimal point in the answer directly below the decimal points in the problem.

EXAMPLE 1 Subtracting Decimal Numbers

Subtract each of the following. Check your answer using addition.

(a) 15.82 from 28.93

Step 1
$$28.93$$
$$-\ 15.82$$
Line up decimal points. Then you will be subtracting hundredths from hundredths and tenths from tenths.

Step 2
$$28.93$$
$$-\ 15.82$$
$$13\ 11$$
Both numbers have two decimal places; no need to write in zeros.
Subtract as if they were whole numbers.

Step 3
$$28.93$$
$$-\ 15.82$$
$$13.11$$
Decimal point in answer lined up.

Check the answer by adding 13.11 and 15.82. If the subtraction is done correctly, the sum will be 28.93.

(b) 146.35 minus 58.98
Borrowing is needed here.

Line up decimal points.

$$
\begin{array}{r}
\overset{0\ \ 13\,15\ \ \downarrow\ 12\,15}{1\,4\,6\,.\,3\,5} \\
-\ \ \ 5\,8\,.\,9\,8 \\
\hline
8\,7\,.\,3\,7
\end{array}
$$

Check the answer by adding 87.37 and 58.98. If you did the subtraction correctly, the sum will be 146.35. (If it *isn't*, you need to rework the problem.)

WORK PROBLEM 1 AT THE SIDE. ▶▶

EXAMPLE 2 Writing Zeros as Placeholders Before Subtracting

Subtract each of the following.

(a) 16.5 from 28.362

Use the same steps as above, remembering to write in zeros so both numbers have three decimal places.

$$
\begin{array}{r}
28.362 \\
-\ 16.500 \\
\hline
11.862
\end{array}
$$

Line up decimal points.
← Write two 0's.
← Subtract as usual.

— **CONTINUED ON NEXT PAGE**

OBJECTIVES

1 ▶ Subtract decimals.

2 ▶ Estimate the answer when subtracting decimals.

FOR EXTRA HELP

Tutorial   Tape 7   SSM, Sec. 4.4

1. Subtract. Check your answers by addition.

(a) 22.7 from 72.9

(b) 6.425 from 11.813

(c) 20.15 − 19.67

ANSWERS

1. (a) 50.2; 50.2 + 22.7 = 72.9
   (b) 5.388; 5.388 + 6.425 = 11.813
   (c) 0.48; 0.48 + 19.67 = 20.15

2. Subtract. Check your answers by addition.

(a) 18.651 from 25.3

(b) 5.816 − 4.98

(c) 40 less 3.66

(d) 1 − 0.325

3. ≈ Round the numbers so there is only one non-zero digit and estimate the answer. Then subtract to find an exact answer.

(a) 11.365 from 38

(b) 214.603 − 53.4

(c) $19.28 less $1.53

(d) Find the difference between 12.837 meters and 46.091 meters

(b) 59.7 − 38.914

$$\begin{array}{r} 59.700 \leftarrow \text{Write two 0's} \\ -\ 38.914 \\ \hline 20.786 \leftarrow \text{Subtract as usual.} \end{array}$$

(c) 12 less 5.83

$$\begin{array}{r} 12.00 \leftarrow \text{Write a decimal point and two 0's.} \\ -\ 5.83 \\ \hline 6.17 \leftarrow \text{Subtract as usual.} \end{array}$$

◀◀ **WORK PROBLEM 2 AT THE SIDE.**

OBJECTIVE 2 *Estimating* the answer will help you check that the problem is set up properly and the decimal point is correctly placed in the answer.

E X A M P L E   3   Estimating a Decimal Answer

Estimate the answer. Then subtract to find the exact answer.

(a) $69.42 − $13.78

Estimate the answer by rounding each number so there is only one non-zero digit (as you did in **Section 4.3**).

estimate		*exact*	
$70	⟵ Rounds to	$69.42	
− 10	⟵ Rounds to	− 13.78	Answer is close to estimate, so the
$60		$55.64	problem is probably set up correctly.

(b) Find the difference between 0.92 feet and 8 feet.

Use subtraction to find the difference between two numbers. The larger number, 8, is written on top.

estimate		*exact*	
8	⟵ Rounds to	8.00	← Write a decimal point and two 0's.
−1	⟵ Rounds to	− 0.92	
7		7.08 feet	Answer is close to estimate.

(c) 1.8614 from 7.3

estimate		*exact*	
7	⟵ Rounds to	7.3000	← Write three 0's.
−2	⟵ Rounds to	− 1.8614	Answer is close to
5		5.4386	estimate.

◀◀ **WORK PROBLEM 3 AT THE SIDE.**

▦ *Calculator Tip:* If you are *adding* numbers, you can enter them in any order on your calculator. Try these; jot down the answers.

9.82 ⊞ 1.86 ⊟ _____   1.86 ⊞ 9.82 ⊟ _____

The answers are the same because addition is *commutative* (See **Section 1.2**). But subtraction is *not* commutative. It *does* matter which number you enter first. Try these:

9.82 ⊟ 1.86 ⊟ _____   1.86 ⊟ 9.82 ⊟ _____

The second answer has a negative sign (−) next to it. A negative number is less than zero. If it's in your checkbook, you'd be "in the hole" by $7.96. (See **Section 9.1** for more about negative numbers.)

ANSWERS

2. (a) 6.649; 6.649 + 18.651 = 25.3
   (b) 0.836; 0.836 + 4.98 = 5.816
   (c) 36.34; 36.34 + 3.66 = 40
   (d) 0.675; 0.675 + 0.325 = 1
3. (a) 40 − 10 = 30; 26.635
   (b) 200 − 50 = 150; 161.203
   (c) $20 − $2 = $18; $17.75
   (d) 50 − 10 = 40; 33.254 meters

4.4 Exercises

Subtract. Check your answer by addition.

Example:
71 − 0.352

Solution:
71.000 ← Write decimal point and three 0's.
− 0.352 }→ Add these numbers. If the sum
70.648 } is 71.000, the problem is done correctly.

1. 73.5
− 19.2
54.3

2. 47.8
− 36.5
11.3

3. 58.413
− 25.847
32.566

4. 27.905
− 18.176
9.729

5. 58.254
− 19.7
38.554

6. 47.658
− 20.9
26.758

7. 21
− 0.896
20.104

8. 9
− 1.183
7.817

9. 15.7
− 2.852
12.848

10. 36.9
− 14.582
22.318

11. 90.5 − 0.8

89.7

12. 303.72 − 0.68

303.04

13. 0.4 − 0.291

0.109

14. 0.35 − 0.088

0.262

15. 6 − 5.09

0.91

16. 80 − 16.3

63.7

17. 15 − 8.339

6.661

18. 44 − 0.08

43.92

🖉 **19.** Explain and correct
the error that a student
made when he subtracted
7.45 from 15.32 this way:

7.45
− 15.32
12.13

15.32 should be on top; correct answer is 7.87.

🖉 **20.** Explain the difference between saying
"subtract 2.9 from 8" and saying
"2.9 minus 8."

The two problems are done in a different order:
8 − 2.9 is not the same as 2.9 − 8 because
subtraction is not commutative.

≈ *Round the numbers so there is only one non-zero digit and estimate the answer. Then subtract to find the exact answer.*

Example: Subtract 4.962 from 7.3.

Solution: *estimate*   *exact*

7   Rounds to   7.300
− 5   Rounds to   − 4.962
2   **2.338**

21. *estimate*   *exact*
$20 $19.74
− 7   − 6.58
$13   **$13.16**

22. *estimate*   *exact*
$30 $27.96
− 8   − 8.39
$22   **$19.57**

23. What is 8.6 less 3.751?

estimate: *exact:*
9 − 4 = 5   **4.849**

24. What is 31.7 less 4.271?

estimate: *exact:*
30 − 4 = 26   **27.429**

🖉 Writing   ◉ Conceptual   ▲ Challenging   ≈ Estimation

25. Find the difference between 1.981 inches and 2 inches.

estimate:
2 − 2 = 0

exact:
0.019 inch

26. Find the difference between 13.582 meters and 28 meters.

estimate:
30 − 10 = 20

exact:
14.418 meters

27. What is 9.006 liters from 384.2 liters?

estimate:
400 − 9 = 391

exact:
375.194 liters

28. What is 23.607 kilograms from 786.1 kilograms?

estimate:
800 − 20 = 780

exact:
762.493 kilograms

*Use your estimation skills to pick the most reasonable answer for each example. Do **not** solve the problems. Circle your choice.*

29. 12 − 11.725

2.75   (0.275)   27.5

30. 20 − 1.37

0.1863   1.863   (18.63)

31. 6.5 − 0.007

(6.493)   0.6493   64.93

32. 9.67 − 0.09

0.958   (9.58)   0.00958

33. 456.71 − 454.9

18.1   181   (1.81)

34. 803.25 − 0.6

(802.65)   0.80265   8.0265

35. 6004.003 − 52.7172

59.512858   595.12858   (5951.2858)

36. 128.35 − 97.0093

313.407   (31.3407)   0.313407

≈ *Round the numbers so there is only one non-zero digit and estimate the answer. Then find the exact answer to each application problem.*

37. Tom has agreed to work 42.5 hours a week as a car wash attendant. So far this week he has worked 16.35 hours. How many more hours must he work?

estimate: 40 − 20 = 20 hours
exact: 26.15 hours

38. The U.S. population was 262.82 million in 1995. The Census Bureau estimates that it will be 393.93 million in the year 2050. The increase in population during that 55-year period is how many millions of people?

estimate: 400 − 300 = 100 million people
exact: 131.11 million people

39. Steven One Feather gave the cashier a $20 bill to pay for $9.12 worth of groceries. How much change did he get?

estimate: $20 − $9 = $11
exact: $10.88

40. The cost of Julie's tennis racket, with tax, is $41.09. She gave the clerk two $20 bills and a $10 bill. What amount of change did Julie receive?

estimate: $50 − $40 = $10
exact: $8.91

41. Namiko is comparing two boxes of chicken nuggets. One box weighs 9.85 ounces and the other weighs 10.5 ounces. What is the difference in the weight of the two boxes?

estimate: **11 − 10 = 1 ounce**
exact: **0.65 ounce**

42. Sammy works in a veterinarian's office. He weighed two newborn kittens. One was 3.9 ounces and the other was 4.05 ounces. What was the difference in the weight of the two kittens?

estimate: **4 − 4 = 0 ounce**
exact: **0.15 ounce**

Maria DeRisi kept track of her expenses for one month. Use her list to solve Exercises 43–48.

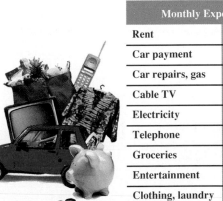

Monthly Expenses	
Rent	$515
Car payment	$190.78
Car repairs, gas	$105
Cable TV	$19.95
Electricity	$42.10
Telephone	$27.36
Groceries	$95.81
Entertainment	$57.75
Clothing, laundry	$52

43. What were Maria's total expenses for the month?

$1105.75

44. How much did Maria pay for electricity, telephone, and cable TV?

$89.41

45. What was the difference in the amounts spent for groceries and for the car payment?

$94.97

46. Compare the amount Maria spent on entertainment to the amount spent on car repairs and gas. What is the difference?

$47.25

47. How much more did Maria spend on rent than on all her car expenses?

$219.22

48. How much less did Maria spend on clothing and laundry than on all her car expenses?

$243.78

49. Mitch Albers had a checking account balance of $129.86 on September 1. During the month, he deposited an additional $1749.82 to the account, and wrote checks totaling $1802.15. The bank charged him a $2 service charge. Find the amount in the account at the end of the month.

$75.53

50. On February 1, Lynn Fiorentino had $1009.24 in her checking account. During the month she deposited a tax refund check of $704.42 and her paycheck of $1258.94. She wrote checks totaling $1389.54 and had $200 transferred to her savings account. Find her checking account balance at the end of the month.

$1383.06

▲ ≈ *Solve each application problem. First round the numbers so there is only one non-zero digit and estimate the answer. Then find the exact answer.*

51. The manual for Jason's car says the gas tank holds 16.6 gallons. Jason knows that the tank actually holds an extra 1.4 gallons. The gas station pump showed that Jason bought 8.628 gallons of gas to fill the tank. How much gas was in the tank before he filled it?

estimate: **20 + 1 = 21; 21 − 9 = 12 gallons**
exact: **9.372 gallons**

52. Tamara's rectangular garden plot is 4.75 meters on each long side and 2.9 meters on each short side. She has 20 meters of fencing. How much fencing will be left after Tamara puts fencing around all four sides of the garden?

estimate: **5 + 5 + 3 + 3 = 16; 20 − 16 = 4 meters**
exact: **4.7 meters**

Find the length of the dashed line in each rectangle or circle.

53.

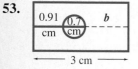

$b = 1.39$ **cm**

54.

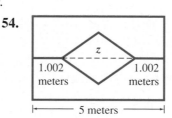

$z = 2.996$ **meters**

55.

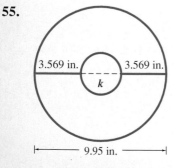

$k = 2.812$ **inches**

56.

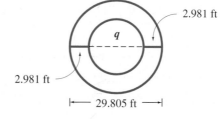

$q = 23.843$ **feet**

▧ *Review and Prepare*

≈ *First round the numbers so there is only one non-zero digit and estimate each answer. Then find the exact answer. (For help, see **Sections 1.4 and 1.7**.)*

57. *estimate*   *exact*

$$\begin{array}{r} 80 \\ \times\ 30 \\ \hline 2400 \end{array} \xleftarrow{\text{Rounds to}} \begin{array}{r} 83 \\ \times\ 28 \\ \hline 2324 \end{array}$$

58. *estimate*   *exact*

$$\begin{array}{r} 70 \\ \times\ 70 \\ \hline 4900 \end{array} \qquad \begin{array}{r} 67 \\ \times\ 72 \\ \hline 4824 \end{array}$$

59. $\underline{\quad 4000 \quad} \times \underline{\quad 200 \quad} = \underline{\quad 800{,}000 \quad}$ *estimate*

$\qquad\quad 3789 \quad \times \quad 205 \quad = \underline{\quad 776{,}745 \quad}$ *exact*

60. $\underline{\quad 6000 \quad} \times \underline{\quad 700 \quad} = \underline{\quad 4{,}200{,}000 \quad}$ *estimate*

$\qquad\quad 6381 \quad \times \quad 709 \quad = \underline{\quad 4{,}524{,}129 \quad}$ *exact*

4.5 Multiplying Decimals

OBJECTIVE 1 The decimals 0.3 and 0.07 can be multiplied by writing them as fractions.

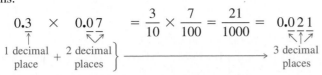

$$0.3 \times 0.07 = \frac{3}{10} \times \frac{7}{100} = \frac{21}{1000} = 0.021$$

1 decimal place + 2 decimal places → 3 decimal places

Can you see a way to multiply decimals without writing them as fractions? Try these steps. Remember that each number in a multiplication problem is called a *factor*.

Multiplying Decimals

Step 1 Multiply the numbers (the factors) as if they were whole numbers.

Step 2 Find the *total* number of decimal places in *both* factors.

Step 3 Write the decimal point in the answer (the product) so it has the same number of decimal places as the total from Step 2. You may need to write in extra zeros on the left side of the product in order to get the correct number of decimal places.

Note
When multiplying decimals, you do **not** need to line up decimal points. (You **do** need to line up decimal points when adding or subtracting decimals.)

E X A M P L E 1   Multiplying Decimal Numbers

Multiply 8.34 times 4.2.

Step 1 Multiply the numbers as if they were whole numbers.

```
      8.3 4
  ×    4.2
    1 6 6 8
  3 3 3 6
  3 5 0 2 8
```

Step 2 Count the total number of decimal places in both factors.

```
      8.3 4  ← 2 decimal places
  ×    4.2  ← 1 decimal place
    1 6 6 8     3 total decimal places
  3 3 3 6
  3 5 0 2 8
```

Step 3 Count over 3 places and write the decimal point in the answer. Count from *right to left*.

```
      8.3 4  ← 2 decimal places
  ×    4.2  ← 1 decimal place
    1 6 6 8     3 total decimal places
  3 3 3 6
  3 5.0 2 8  ← 3 decimal places in answer
```
Count over 3 places from right to left to position the decimal point.

WORK PROBLEM 1 AT THE SIDE. ▶▶

OBJECTIVES

1 ▶ Multiply decimals.

2 ▶ Estimate the answer when multiplying decimals.

FOR EXTRA HELP

Tutorial   Tape 7   SSM, Sec. 4.5

1. Multiply.

   (a)   2.6
        × 0.4

   (b)   45.2
        × 0.25

   (c)   0.104 ← 3 decimal places
        ×   7 ← 0 decimal places
                 ← 3 decimal places in the answer

   (d)   3.18
        × 2.23

   (e)   611
        × 3.7

ANSWERS

1. (a) 1.04 (b) 11.300 (c) 0.728
   (d) 7.0914 (e) 2260.7

2. Multiply.

  (a) 0.04×0.09

  (b) $0.2 \cdot 0.008$

  (c) $(0.063)(0.04)$

  (d) $0.0081 \cdot 0.003$

  (e) $(0.11)(0.0005)$

3. $\approx$ First round the numbers and estimate the answer. Then multiply to find the exact answer.

  (a) $(11.62)(4.01)$

  (b) $(5.986)(33)$

  (c) $8.31 \cdot 4.2$

  (d) $58.6 \cdot 17.4$

E X A M P L E 2   Writing Zeros as Placeholders in the Answer

Multiply 0.042 by 0.03.

Start by multiplying and counting decimal places.

$$
\begin{array}{r}
0.042 \leftarrow \text{3 decimal places} \\
\times \quad 0.03 \leftarrow \text{2 decimal places} \\
\hline
126 \leftarrow \text{5 decimal places needed in answer}
\end{array}
$$

The answer has only three decimal places, but five are needed. So write two zeros on the *left* side of the answer.

$$
\begin{array}{r}
0.042 \\
\times \quad 0.03 \\
\hline
00126
\end{array}
\qquad
\begin{array}{r}
0.042 \leftarrow \text{3 decimal places} \\
\times \quad 0.03 \leftarrow \text{2 decimal places} \\
\hline
.00126 \leftarrow \text{5 decimal places}
\end{array}
$$

Write two 0's on *left* side of answer.   Now count over 5 places and write in the decimal point.

The final answer is 0.00126, which has five decimal places.

◄◄ **WORK PROBLEM 2 AT THE SIDE.**

OBJECTIVE 2 If you are doing multiplication problems by hand, estimating the answer helps you check that the decimal point is in the right place. When you are using a calculator, estimating helps you catch an error like pressing the ÷ key instead of the × key.

E X A M P L E 3   Estimating before Multiplying

First estimate $76.34 \cdot 12.5$. Round each number so there is only one non-zero digit. Then multiply to find the exact answer.

estimate

$$
\begin{array}{r}
80 \\
\times \ 10 \\
\hline
800
\end{array}
$$

Rounds to

exact

$$
\begin{array}{r}
7\,6.3\,4 \leftarrow \text{2 decimal places} \\
\times \quad 1\,2.5 \leftarrow \text{1 decimal place} \\
\hline
3\,8\,1\,7\,0 \\
1\,5\,2\,6\,8 \\
7\,6\,3\,4 \\
\hline
9\,5\,4.2\,5\,0
\end{array}
$$

3 decimal places are in answer.

Both the estimate and the exact answer go out to the hundreds, so the decimal point in 954.250 is probably in the correct place.

◄◄ **WORK PROBLEM 3 AT THE SIDE.**

Calculator Tip: When working with money amounts, you need to write a zero in your answer. For example, try multiplying $\$3.54 \times 5$ on your calculator. Write down the result.

3.54 × 5 = _____

Notice the result is 17.7 which is *not* the way to write a money amount. You have to add the zero in the hundredths place: $\$17.70$ is correct. The calculator does not show the "extra" zero because:

$$17.70 \text{ or } 17\frac{70}{100} \text{ reduces to } 17\frac{7}{10} \text{ or } 17.7.$$

So keep an eye on your calculator—it doesn't know when you're working with money amounts.

ANSWERS

2. (a) 0.0036 **(b)** 0.0016 **(c)** 0.00252
   (d) 0.0000243 **(e)** 0.000055
3. (a) $10 \cdot 4 = 40$; 46.5962
   (b) $6 \cdot 30 = 180$; 197.538
   (c) $8 \cdot 4 = 32$; 34.902
   (d) $60 \cdot 20 = 1200$; 1019.64

4.5 Exercises

Multiply.

Example:	0.093	**Solution:**	$0.093 \leftarrow$ 3 decimal places
	$\times \quad 0.6$		$\times \quad 0.6 \leftarrow$ 1 decimal place
			$\mathbf{0.0558} \leftarrow$ 4 decimal places in answer
			└──Write a zero in order to get 4 decimal places.

1.   0.042
   $\times \quad 3.2$
   0.1344

2.   0.571
   $\times \quad 2.9$
   1.6559

3.   21.5
   $\times \quad 7.4$
   159.10

4.   85.4
   $\times \quad 3.5$
   298.90

5.   23.4
   $\times \ 0.666$
   15.5844

6.   0.896
   $\times \ 0.799$
   0.715904

7. $51.88
   $\times \quad 665$
   $34,500.20

8. $736.75
   $\times \quad 118$
   $86,936.50

◉ *Use the fact that* $72 \times 6 = 432$ *to help you solve Exercises 9–16 by simply counting decimal places.*

9. 72×0.6

   43.2

10. 7.2×6

   43.2

11. $(7.2)(0.06)$

   0.432

12. $(0.72)(0.6)$

   0.432

13. 0.72×0.06

   0.0432

14. 72×0.0006

   0.0432

15. 0.0072×0.6

   0.00432

16. 0.072×0.006

   0.000432

Multiply.

17. $(0.006)(0.0052)$

   0.0000312

18. $(0.0052)(0.009)$

   0.0000468

19. $0.003 \cdot 0.002$

   0.000006

20. $0.0079 \cdot 0.006$

   0.0000474

21. Do these multiplications:

   $(5.96)(10)$ $(3.2)(10)$
   $(0.476)(10)$ $(80.35)(10)$
   $(722.6)(10)$ $(0.9)(10)$

   What pattern do you see? Write a "rule" for multiplying by 10. What do you think the rule is for multiplying by 100? By 1000? Write the rules and try them out on the numbers above.

   Multiplying by 10, decimal point moves one place to the right; by 100, two places to the right; by 1000, three places to the right.

22. Do these multiplications:

   $(59.6)(0.1)$ $(3.2)(0.1)$
   $(0.476)(0.1)$ $(80.35)(0.1)$
   $(65)(0.1)$ $(523)(0.1)$

   What pattern do you see? Write a "rule" for multiplying by 0.1. What do you think the rule is for multiplying by 0.01? By 0.001? Write the rules and try them out on the numbers above.

   Multiplying by 0.1, decimal point moves one place to the left; by 0.01, two places to the left; by 0.001, three places to the left.

◷ Writing   ◉ Conceptual   ▲ Challenging   ≈ Estimation

≈ *First round the numbers so there is only one non-zero digit and estimate the answer. Then multiply to find the exact answer.*

23. estimate   exact

```
        Rounds to
  40  ←──────────  39.6
× 5  ←──────────  × 4.8
─────            ───────
 200              190.08
```

24. estimate   exact

```
  20              18.7
× 2             × 2.3
─────           ───────
 40              43.01
```

25. estimate   exact

```
  40              37.1
× 40            × 42
─────           ───────
1600             1558.2
```

26. estimate   exact

```
   5              5.08
× 70            × 71
─────           ───────
 350             360.68
```

27. estimate   exact

```
   7              6.53
× 5             × 4.6
─────           ───────
  35             30.038
```

28. estimate   exact

```
   8              7.51
× 8             × 8.2
─────           ───────
  64             61.582
```

29. estimate   exact

```
   3              2.809
× 7             × 6.85
─────           ───────
  21             19.24165
```

30. estimate   exact

```
  70              73.52
× 20            × 22.34
─────           ─────────
1400             1642.4368
```

◉ *Even with most of the problem missing, you can tell whether or not these answers are reasonable. Circle "reasonable" or "unreasonable." If the answer is unreasonable, move the decimal point or insert a decimal point to make the answer reasonable.*

31. How much was his car payment? $18.90
reasonable
(unreasonable,) should be __$189.00__

32. How many hours did she work today? 25 hours
reasonable
(unreasonable,) should be __2.5 hours__

33. How tall is her son? 60.5 inches
(reasonable)
unreasonable, should be _____

34. How much does he pay for rent now? $4.92
reasonable
(unreasonable,) should be __$492__

35. What is the price of one gallon of milk? $319
reasonable
(unreasonable,) should be __$3.19__

36. How long is the living room? 16.8 feet
(reasonable)
unreasonable, should be _____

37. How much did the baby weigh? 0.095 pounds
reasonable
(unreasonable,) should be __9.5 pounds__

38. What was the sale price of the jacket? $1.49
reasonable
(unreasonable,) should be __$14.90 or $149__

Solve. If the problem involves money, round to the nearest cent, if necessary.

39. LaTasha worked 50.5 hours over the last two weeks. She earns $11.73 per hour. How much did she make?

≈$592.37

40. Michael's time card shows 42.2 hours at $10.03 per hour. What are his gross earnings?

≈$423.27

41. Sid needs 0.6 meter of canvas material to make a carry-all bag that fits on his wheelchair. If canvas is $4.09 per meter, how much will Sid spend? (*Note:* $4.09 *per* meter means $4.09 for *one* meter.)

≈$2.45

42. How much will Mrs. Nguyen pay for 3.5 yards of lace trim that costs $0.87 per yard?

≈$3.05

43. Michelle pumped 18.65 gallons of gas into her pickup truck. The price was $1.45 per gallon. How much did she pay for gas?

≈$27.04

44. Spicy chicken wings were on sale for $0.98 per pound. Juma bought 1.7 pounds of wings. How much did the chicken wings cost?

≈$1.67

45. Ms. Rolack is a real estate broker who helps people sell their homes. Her fee is 0.07 times the price of the home. What was her fee for selling a $125,300 home?

$8771.00

46. Alex Rodriguez, shortstop for the Seattle Mariners, has a 1996 batting average of 0.358. If he went to bat 601 times, how many hits did he make? (*Hint:* Multiply his batting average by the number of times at bat.) Round to the nearest whole number.

≈215 hits

47. Judy Lewis pays $28.96 per month for cable TV. How much will she pay for cable over one year?

$347.52

48. Chuck's car payment is $220.27 per month for three years. How much will he pay altogether?

$7929.72

49. Paper for the copy machine at the library costs $0.015 per sheet. How much will the library pay for 5100 sheets?

$76.50

50. A student group collected 2200 pounds of plastic as a fund raiser. How much will they make if the recycling center pays $0.142 per pound?

$312.40

⊞ *Use the list of prices below from the Look Smart mail order catalog to solve Exercises 51 and 52.*

Knit Shirt Ordering Information		
43–2A	short sleeve, solid colors	$14.75 each
43–2B	short sleeve, stripes	$16.75 each
43–3A	long sleeve, solid colors	$18.95 each
43–3B	long sleeve, stripes	$21.95 each
Extra-large size, add $2 per shirt.		

51. Find the total cost of four long-sleeve, solid-color shirts and two short-sleeve, striped shirts, all in the extra-large size.

$121.30

52. What is the total cost of eight long-sleeve shirts, five in solid colors and three striped?

$160.60

▲ **53.** Jack Burgess used 3.5 gallons of fertilizer on each of his 158.2 acres of corn. After he finished, how much fertilizer was left in a storage tank that originally contained 600 gallons?

46.3 gallons

▲ **54.** Stan Johnson bought 7.8 yards of a Hawaiian print fabric at $5.62 per yard. He paid for it with three $20 bills. Find the amount of his change. (Ignore sales tax.)

≈$16.16

▲ **55.** Ms. Sanchez paid $29.95 a day to rent a car, plus $0.29 per mile. Find the cost of her rental for a four-day trip of 926 miles.

$388.34

▲ **56.** The Bell family rented a motor home for $375 per week plus $0.35 per mile. What was the rental cost for their three-week vacation trip of 2650 miles?

$2052.50

▲ **57.** Barry bought 16.5 meters of rope at $0.47 per meter and three meters of wire at $1.05 per meter. How much change did he get from three $5 bills?

≈$4.09

▲ **58.** Susan bought a VCR that cost $229.88. She paid $45 down and $37.98 per month for six months. How much could she have saved by paying cash?

$43

▦ *Review and Prepare*

≈ *First round the numbers and estimate each answer. Round so there is only one non-zero digit. Then find the exact answer using long division and an **R** to express a remainder. (For help, see Sections 1.5, 1.6, and 1.7.)*

59. estimate   exact

$$\frac{200}{5)1000} \qquad \frac{190\ R4}{5)954}$$

60. estimate   exact

$$\frac{500}{4)2000} \qquad \frac{555\ R3}{4)2223}$$

61. estimate   exact

$$\frac{1000}{20)20,000} \qquad \frac{905\ R15}{21)19,020}$$

62. estimate   exact

$$\frac{3000}{30)90,000} \qquad \frac{3343\ R17}{28)93,621}$$

4.6 Dividing Decimals

There are two kinds of decimal division problems; those in which a decimal is divided by a whole number, and those in which a decimal is divided by a decimal. First recall the parts of a division problem from **Section 1.5.**

$$
\begin{array}{r}
8 \leftarrow \text{Quotient} \\
\text{Divisor} \rightarrow 4\overline{)33} \leftarrow \text{Dividend} \\
\underline{32} \\
1 \leftarrow \text{Remainder}
\end{array}
$$

OBJECTIVE 1 When the divisor is a whole number, use these steps.

Dividing Decimals by Whole Numbers

Step 1 Write the decimal point in the quotient (answer) directly above the decimal point in the dividend.

Step 2 Divide as if both numbers were whole numbers.

EXAMPLE 1 Dividing a Decimal by a Whole Number

Divide.

(a) 21.93 by 3

Dividend ⌣ ⌐ Divisor

Rewrite the division problem. $3\overline{)21.93}$

Write the decimal point in the answer directly above the decimal point in the dividend.

$3\overline{)21.93}$ — Decimal points lined up

Divide as if the numbers were whole numbers.

$\begin{array}{r} 7.31 \\ 3\overline{)21.93} \end{array}$

Check by multiplying the quotient times the divisor.

$\begin{array}{r} 7.31 \\ \times\quad 3 \\ \hline 21.93 \end{array}$ Matches, so 7.31 is correct.

The quotient (answer) is 7.31.

(b) $9\overline{)470.7}$

Divisor ⌐ Dividend

Write the decimal point in the answer above the decimal point in the dividend. Then divide as if the numbers were whole numbers.

Decimal points lined up

$\begin{array}{r} 52.3 \\ 9\overline{)470.7} \\ \underline{45} \\ 20 \\ \underline{18} \\ 27 \\ \underline{27} \\ 0 \end{array}$

Check

$\begin{array}{r} 52.3 \\ \times\quad 9 \\ \hline 470.7 \end{array}$

Matches

WORK PROBLEM 1 AT THE SIDE. ▶▶

OBJECTIVES

1 ▶ Divide a decimal by a whole number.

2 ▶ Divide a decimal by a decimal.

3 ▶ Estimate the answer when dividing decimals.

4 ▶ Use the order of operations with decimals.

FOR EXTRA HELP

Tutorial   Tape 7   SSM, Sec. 4.6

1. Divide. Check your answers by multiplying.

   (a) $4\overline{)93.6}$

   (b) $6\overline{)6.804}$

   (c) $11\overline{)278.3}$

   (d) $0.51835 \div 5$

   (e) $213.45 \div 15$

ANSWERS

1. (a) 23.4; 23.4 • 4 = 93.6
   (b) 1.134; 1.134 • 6 = 6.804
   (c) 25.3; 25.3 • 11 = 278.3
   (d) 0.10367; 0.10367 • 5 = 0.51835
   (e) 14.23; 14.23 • 15 = 213.45

263

2. Divide. Check your answers by multiplying.

(a) $5\overline{)6.4}$

(b) $30.87 \div 14$

(c) $\dfrac{259.5}{30}$

(d) $0.3 \div 8$

E X A M P L E   2   Writing Extra Zeros to Complete a Division

Divide 1.5 by 8.

Keep dividing until the remainder is zero, or until the digits in the answer begin to repeat in a pattern. In Example 1(b), you ended up with a remainder of 0. But sometimes you run out of digits in the dividend before that happens. If so, write extra zeros on the right side of the dividend so you can continue dividing.

$$
\begin{array}{r}
0.1 \\
8\overline{)1.5} \\
\underline{8} \\
7
\end{array}
$$
← All digits have been used.
← Remainder is not yet 0.

Write a zero after the 5 so you can continue dividing. Keep writing more zeros in the dividend if needed. Recall that writing zeros to the right of a decimal number does **not** change its value.

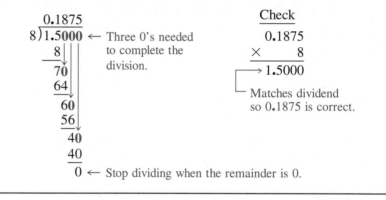

Three 0's needed to complete the division.

Check
$$
\begin{array}{r}
0.1875 \\
\times \quad 8 \\
\hline
1.5000
\end{array}
$$
Matches dividend so 0.1875 is correct.

Stop dividing when the remainder is 0.

 Calculator Tip: When *multiplying* numbers, you can enter them in any order because multiplication is commutative (see **Section 1.4**). But division is *not* commutative. It *does* matter which number you enter first. Try Example 2 both ways; jot down your answers.

$1.5 \boxed{\div} 8 \boxed{=}$ _____ $8 \boxed{\div} 1.5 \boxed{=}$ _____

Notice that the first answer, 0.1875, matches the result from Example 2 above. But the second answer is much different: 5.333333333. Be careful to enter the dividend first.

Also notice that in decimals the dividend may *not* be the larger number, as it was in whole numbers. In Example 2 the dividend is 1.5, which is *smaller* than 8.

◀◀ **WORK PROBLEM 2 AT THE SIDE.**

The next example shows a quotient (answer) that must be rounded because you will never get a remainder of zero.

ANSWERS

2. (a) 1.28; 1.28 • 5 = 6.40 or 6.4
    (b) 2.205; 2.205 • 14 = 30.870 or 30.87
    (c) 8.65; 8.65 • 30 = 259.50 or 259.5
    (d) 0.0375; 0.0375 • 8 = 0.3000 or 0.3

EXAMPLE 3 **Rounding a Decimal Quotient**

Divide 4.7 by 3. Round to the nearest thousandth.

Write extra zeros in the dividend so you can continue dividing.

$$
\begin{array}{r}
1.5\,6\,6\,6 \\
3\overline{)4.7\,0\,0\,0} \quad \leftarrow \text{Three 0's added so far} \\
\underline{3} \\
1\,7 \\
\underline{1\,5} \\
2\,0 \\
\underline{1\,8} \\
2\,0 \\
\underline{1\,8} \\
2\,0 \\
\underline{1\,8} \\
2 \quad \leftarrow \text{Remainder is still not 0.}
\end{array}
$$

Notice that the digit 6 in the answer is repeating. It will continue to do so. The remainder will never be zero. There are two ways to show that the answer is a **repeating decimal** that goes on forever. Write three dots after the answer, or, write a bar above the digits that repeat (in this case, the 6).

$$
\underbrace{1.5666 \ldots}_{\text{Three dots}} \quad \text{or} \quad 1.5\overline{6} \quad \overset{\leftarrow \text{ Bar above}}{\text{repeating digit}}
$$

When repeating decimals occur, round the answer according to the directions in the problem. In this example, to round to thousandths, divide out one *more* place, to ten-thousandths.

$$4.7 \div 3 = 1.5666 \ldots \text{ rounds to } 1.567$$

Check the answer by multiplying 1.567 by 3. Because 1.567 is a rounded answer, the check will not give exactly 4.7, but it should be very close.

$$1.567 \cdot 3 = 4.701 \quad \overset{\leftarrow \text{ Does not equal exactly 4.7}}{\text{because 1.567 was rounded.}}$$

Note

When checking answers that you've rounded, the check will not match the dividend exactly, but it should be very close.

WORK PROBLEM 3 AT THE SIDE. ▶▶

OBJECTIVE 2 To divide by a *decimal* divisor, first change the divisor to a whole number. Then divide as before. To see how this is done, write the problem in fraction form. For example:

$$1.2\overline{)6.36} \quad \text{can be written} \quad \frac{6.36}{1.2}.$$

In **Section 3.2** you learned that multiplying the numerator and denominator by the same number gives an equivalent fraction. We want the divisor (1.2) to be a whole number. Multiplying by 10 will accomplish that.

$$\frac{6.36}{1.2} = \frac{6.36 \cdot 10}{1.2 \cdot 10} = \frac{63.6}{12}$$

3. Divide. Round answers to the nearest thousandth. If it is a repeating decimal, also write the answer using a bar. Check your answers by multiplying.

(a) $13\overline{)267.01}$

(b) $6\overline{)20.5}$

(c) $\dfrac{10.22}{9}$

(d) $16.15 \div 3$

(e) $116.3 \div 7$

ANSWERS

3. **(a)** ≈ 20.539; no repeating digits visible on calculator;
$20.539 \cdot 13 = 267.007$
(b) ≈ 3.417; $3.41\overline{6}$
$3.417 \cdot 6 = 20.502$
(c) ≈ 1.136; $1.13\overline{5}$
$1.136 \cdot 9 = 10.224$
(d) ≈ 5.383; $5.38\overline{3}$
$5.383 \cdot 3 = 16.149$
(e) ≈ 16.614; starts repeating in eighth decimal place as $16.6\overline{142857}$;
$16.614 \cdot 7 = 116.298$

4. Divide. If the quotient does not come out even, round to the nearest hundredth.

(a) $0.2\overline{)1.04}$

(b) $0.06\overline{)1.8072}$

(c) $0.005\overline{)32}$

(d) $8.1 \div 0.025$

(e) $\dfrac{7}{1.3}$

(f) $5.3091 \div 6.2$

The short way to multiply by 10 is to move the decimal point one place to the right in both the divisor and the dividend.

$$1.2\overline{)6.36} \quad \text{is equivalent to} \quad 12\overline{)63.6}$$

Note

Moving the decimal points the **same** number of places in **both** the divisor and dividend will **not** change the answer.

Dividing by Decimals

Step 1 Count the number of decimal places in the divisor and move the decimal point that many places to the *right*. (This changes the divisor to a whole number.)

Step 2 Move the decimal point in the dividend the *same* number of places to the *right*. (Write in extra zeros if needed.)

Step 3 Write the decimal point in the answer directly above the decimal point in the dividend. Then divide as usual.

EXAMPLE 4   Dividing by a Decimal

(a) $0.003\overline{)27.69}$

Move the decimal point in the divisor *three* places to the *right* so 0.003 becomes the whole number 3. In order to move the decimal point in the dividend the same number of places, write in an extra zero.

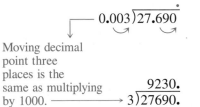

Moving decimal point three places is the same as multiplying by 1000. $\longrightarrow 3\overline{)27690.}$

Move decimal points in divisor and dividend. Then line up decimal point in answer.

$9230.$

Divide as usual.

(b) Divide 5 by 4.2. Round to the nearest hundredth.

Move the decimal point in the divisor one place to the right so 4.2 becomes the whole number 42. The decimal point in the dividend starts on the right side of 5 and is also moved one place to the right.

$$
\begin{array}{r}
1.1\,9\,0 \\
4.2\overline{)5.0\,0\,0\,0} \\
4\,2 \\
\hline
8\,0 \\
4\,2 \\
\hline
3\,8\,0 \\
3\,7\,8 \\
\hline
2\,0
\end{array}
$$

← In order to round to hundredths, divide out one *more* place, to thousandths.

Round the quotient. It is ≈ 1.19 (nearest hundredth).

◀◀ **WORK PROBLEM 4 AT THE SIDE.**

OBJECTIVE ③ Estimating the answer to a division problem helps you catch errors. Compare the estimate to your exact answer. If they are very different, do the division again.

ANSWERS

4. (a) 5.2 (b) 30.12 (c) 6400 (d) 324
(e) ≈5.38 (f) ≈0.86

EXAMPLE 5 Estimating before Dividing

First round the numbers so there is only one non-zero digit and estimate the answer. Then divide to find the exact answer.

$$580.44 \div 2.8$$

Here is how one student solved this problem. She rounded 580.44 to 600 and 2.8 to 3 to estimate the answer.

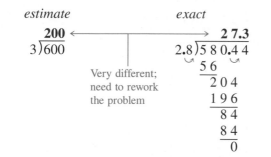

estimate

$$\begin{array}{r} \mathbf{200} \\ 3\overline{)600} \end{array}$$

Very different; need to rework the problem

exact

$$\begin{array}{r} \mathbf{2\,7.3} \\ 2.8\overline{)5\,8\,0.4\,4} \\ \underline{5\,6} \\ 2\,0\,4 \\ \underline{1\,9\,6} \\ 8\,4 \\ \underline{8\,4} \\ 0 \end{array}$$

Notice that the estimate, which is in the hundreds, is very different from the exact answer, which is only in the tens. This tells the student that she needs to rework the problem. Can you find the error? (The exact answer should be 207.3, which fits with the estimate of 200.)

WORK PROBLEM 5 AT THE SIDE. ▶▶

OBJECTIVE 4 Use the order of operations when a decimal problem involves more than one operation.

Order of Operations

1. Do all operations inside parentheses.
2. Simplify any expressions with exponents and find any square roots.
3. Multiply or divide from left to right.
4. Add or subtract from left to right.

EXAMPLE 6 Using the Order of Operations

Simplify by using the order of operations.

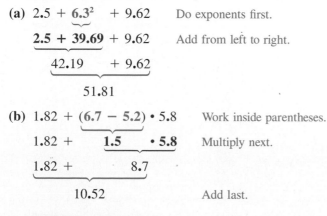

(a) $2.5 + \mathbf{6.3^2} + 9.62$ Do exponents first.

$\mathbf{2.5 + 39.69} + 9.62$ Add from left to right.

$42.19 \quad + 9.62$

51.81

(b) $1.82 + \mathbf{(6.7 - 5.2)} \cdot 5.8$ Work inside parentheses.

$1.82 + \quad \mathbf{1.5} \quad \mathbf{\cdot 5.8}$ Multiply next.

$1.82 + \quad \quad 8.7$

10.52 Add last.

CONTINUED ON NEXT PAGE

---

5. ≈ Decide if each answer is reasonable by rounding the numbers and estimating the answer. If the exact answer is *not* reasonable, find and correct the error.

(a) $42.75 \div 3.8 = 1.125$
estimate:

(b) $807.1 \div 1.76 = 458.580$
to nearest thousandth
estimate:

(c) $48.63 \div 52 = 93.519$
to nearest thousandth
estimate:

(d) $9.0584 \div 2.68 = 0.338$
estimate:

ANSWERS

5. (a) Estimate is $40 \div 4 = 10$; exact answer not reasonable, should be 11.25
(b) Estimate is $800 \div 2 = 400$; exact answer is reasonable.
(c) Estimate is $50 \div 50 = 1$; exact answer is not reasonable, should be 0.935.
(d) Estimate is $9 \div 3 = 3$; exact answer is not reasonable, should be 3.38.

6. Simplify by using the order of operations.

(a) $4.6 - 0.79 + 1.5^2$

(b) $3.64 \div 1.3 \cdot 3.6$

(c) $0.08 + 0.6 \cdot (3 - 2.99)$

(d) $10.85 - 2.3 \cdot 5.2 \div 3.2$

(c) $\underline{3.7^2}\ -\ 1.8 \times 5.1 \div 1.5$ Do exponents first.

$13.69 -\ \underline{1.8 \times 5.1} \div 1.5$ Multiply and divide from left to right.

$13.69 -\ \ \underline{9.18\ \ \div\ 1.5}$

$\underline{13.69 -\ \ \ \ \ \ \ \ \ \ \ 6.12}$

$\ \ \ \ \ \ \ \ \ \ \ 7.57$ Subtract last.

◀◀ **WORK PROBLEM 6 AT THE SIDE.**

Calculator Tip: Most scientific calculators that have parentheses keys ☐ ☐ can handle calculations like those in Example 6 just by entering the numbers in the order given. For example, the keystrokes for Example 6(b) are:

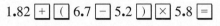

$1.82\ \boxed{+}\ \boxed{(}\ 6.7\ \boxed{-}\ 5.2\ \boxed{)}\ \boxed{\times}\ 5.8\ \boxed{=}$

Standard, four-function calculators generally do *not* give the correct answer if you enter the numbers in the order given. Check the instruction manual that came with your calculator for information on "order of calculations" to see if your machine has the rules for order of operations built into it. For a quick check, try entering this problem:

$2\ \boxed{+}\ 2\ \boxed{\times}\ 2\ \boxed{=}$

If the result is 6, the calculator follows the order of operations. If the result is 8, it does not have the rules built into it.

ANSWERS

6. (a) 6.06 **(b)** 10.08 **(c)** 0.086
(d) 7.1125

4.6 *Exercises*

Divide.

Example: $3 \div 0.08$ **Solution:**

$$
\begin{array}{r}
37.5 \\
0.08\overline{)3.00\ 0} \\
\underline{2\ 4} \\
60 \\
\underline{56} \\
4\ 0 \\
\underline{4\ 0} \\
0
\end{array}
$$

1. $7\overline{)27.3}$ **3.9**

2. $8\overline{)50.4}$ **6.3**

3. $\dfrac{4.23}{9}$ **0.47**

4. $\dfrac{1.62}{6}$ **0.27**

5. $0.05\overline{)20.01}$ **400.2**

6. $0.08\overline{)16.04}$ **200.5**

7. $1.5\overline{)54}$ **36**

8. $2.4\overline{)132}$ **55**

⊙ *Use the fact that* $108 \div 18 = 6$ *to solve Exercises 9–12 simply by moving decimal points.*

9. $0.108 \div 1.8$ **0.06**

10. $10.8 \div 18$ **0.6**

11. $0.018\overline{)108}$ **6000**

12. $0.18\overline{)1.08}$ **6**

In Exercises 13–16, round your answers to the nearest hundredth, if necessary.

13. $4.6\overline{)116.38}$ **25.3**

14. $2.6\overline{)4.992}$ **1.92**

15. $\dfrac{3.1}{0.006}$ **≈516.67**

16. $\dfrac{1.7}{0.09}$ **≈18.89**

🖩 *In Exercises 17–20, round your answers to the nearest thousandth.*

17. $240 \div 9.88$ **≈24.291**

18. $7643 \div 5.36$ **≈1425.933**

19. $0.034\overline{)342.81}$ **≈10,082.647**

20. $0.043\overline{)1748.4}$ **≈40,660.465**

🖉 **21.** Do these division problems:
⊙
$3.77 \div 10$	$9.1 \div 10$
$0.886 \div 10$	$30.19 \div 10$
$406.5 \div 10$	$6625.7 \div 10$

What pattern do you see? Write a "rule" for dividing by 10. What do you think the rule is for dividing by 100? By 1000? Write the rules and try them out on the numbers above.

Dividing by 10, decimal point moves one place to the left; by 100, two places to the left; by 1000, three places to the left.

🖉 **22.** Do these division problems:
⊙
$40.2 \div 0.1$	$7.1 \div 0.1$
$0.339 \div 0.1$	$15.77 \div 0.1$
$46 \div 0.1$	$873 \div 0.1$

What pattern do you see? Write a "rule" for dividing by 0.1. What do you think the rule is for dividing by 0.01? By 0.001? Write the rules and try them out on the numbers above.

Dividing by 0.1, decimal point moves one place to the right; by 0.01, two places to the right; by 0.001, three places to the right.

🖉 Writing   ⊙ Conceptual   ▲ Challenging   ≈ Estimation

≈ *Decide if each answer is reasonable or unreasonable by rounding the numbers and estimating the answer. If the exact answer is not reasonable, find the correct answer.*

23. 37.8 ÷ 8 = 47.25   **unreasonable**

estimate:

40 ÷ 8 = 5   8)37.8   4.725

24. 345.6 ÷ 3 = 11.52   **unreasonable**

estimate:

300 ÷ 3 = 100   3)345.6   115.2

25. 54.6 ÷ 48.1 = 1.135   **reasonable**

estimate:

50 ÷ 50 = 1

26. 2428.8 ÷ 4.8 = 50.6   **unreasonable**

estimate:

2000 ÷ 5 = 400   4.8)2428.8   506

27. 307.02 ÷ 5.1 = 6.2   **unreasonable**

estimate:

300 ÷ 5 = 60   5.1)307.02   60.2

28. 395.415 ÷ 5.05 = 78.3   **reasonable**

estimate:

400 ÷ 5 = 80

29. 9.3 ÷ 1.25 = 0.744   **unreasonable**

estimate: 7.44
9 ÷ 1 = 9   1.25)9.3

30. 78 ÷ 14.2 = 0.182   **unreasonable**

estimate: 5.493
80 ÷ 10 = 8   14.2)78

Solve each application problem. Round money answers to the nearest cent, if necessary.

31. Alfred has discovered that Batman's favorite brand of superhero tights are on sale at six pairs for $23.98, but he's been told to buy only one pair for Robin. How much will he pay for one pair?

≈$4.00

32. The bookstore is selling four notepads for $1.69. How much did Randall pay for one notepad?

≈$0.42

33. It will take 21 months for Aimee to pay off her charge account balance of $408.66. How much is she paying each month?

$19.46

34. Marcella Anderson bought 2.6 meters of suede fabric for $18.19. How much did she pay per meter?

≈$7.00 per meter

35. Adrian Webb bought 619 bricks to build a barbecue pit, paying $185.70. Find the cost per brick. (*Hint:* Cost *per* brick means the cost for *one* brick.)

$0.30

36. Lupe Wilson is a newspaper distributor. Last week she paid the newspaper $130.51 for 842 copies. Find the cost per copy.

≈$0.16

37. Darren Jackson earned $235.60 for 40 hours of work. Find his earnings per hour.

$5.89 per hour

38. At a record manufacturing company, 400 records cost $289. Find the cost per record.

≈$0.72

39. It took 16.35 gallons of gas to fill Kim's car gas tank. She had driven 346.2 miles since her last fill-up. How many miles per gallon did she get? Round to the nearest tenth.

≈21.2 miles per gallon

40. Mr. Rodriquez pays $53.19 each month to Household Finance. How many months will it take him to pay off $1436.13?

27 months

Use the table of 1996 Olympic long jumps to solve Exercises 41–46. To find an average, add up the values you are interested in and then divide the sum by the number of values. Round your answer to the nearest hundredth. Some of the other exercises may require subtraction or multiplication.

Country	Length
U.S.	8.50 meters
Jamaica	8.29 meters
U.S.	8.24 meters
France	8.19 meters
U.S.	8.17 meters
Slovenia	8.11 meters
Belarus	8.07 meters

41. Find the average length of the long jumps made by U.S. athletes.

≈8.30 meters

42. Find the average length of all the long jumps listed in the table.

≈8.22 meters

43. How much longer was the second place jump than the third place jump?

0.05 meter

44. If the first place athlete made six jumps of the same length, what was the total distance jumped?

51.00 or 51 meters

45. What was the total length jumped by the top three athletes?

25.03 meters

46. How much less was the last place jump than the next to last jump?

0.04 meter

Simplify by using the order of operations.

Example: $\underbrace{5.2^2}$ + 7.9 • 6.3   Do exponents first.

27.04 + $\underbrace{7.9 • 6.3}$ Multiply next.

$\underbrace{27.04 + 49.77}$ Add last.

76.81

47. $7.2 - 5.2 + 3.5^2$

14.25

48. $6.2 + 4.3^2 - 9.72$

14.97

49. $38.6 + 11.6 • (13.4 - 10.4)$

73.4

50. $2.25 - 1.06 • (4.85 - 3.95)$

1.296

51. $8.68 - 4.6 • 10.4 \div 6.4$

1.205

52. $25.1 + 11.4 \div 7.5 • 3.75$

30.8

53. $33 - 3.2 • (0.68 + 9) - 1.3^2$

0.334

54. $0.6 + (1.89 + 0.11) \div 0.004 • 0.5$

250.6

▲ *Solve each application problem.*

55. Soup is on sale at six cans for $3.25, or you can purchase individual cans for $0.57. How much will you save per can if you buy six cans? Round to the nearest cent.

≈$0.03

56. Nadia's diet says she can eat 3.5 ounces of chicken nuggets. The package weighs 10.5 ounces and contains 15 nuggets. How many nuggets can Nadia eat?

5 nuggets

57. The annual premium for Jenny's auto insurance policy is $938. She can pay it in four quarterly installments, if she adds a $2.75 service fee to each payment. Find the amount of each quarterly payment.

$237.25

58. Lock and Store charges rent of $936 per year for 200 square feet of storage space. To pay the rent monthly, $1.25 must be added to each payment. Find the amount of each monthly payment.

$79.25

Review and Prepare

*Write < or > in each blank to make a true statement. (For help, see **Section 3.5**.)*

59. $\dfrac{7}{12}$ __<__ $\dfrac{3}{4}$

60. $\dfrac{5}{8}$ __<__ $\dfrac{11}{16}$

61. $\dfrac{5}{6}$ __>__ $\dfrac{7}{9}$

62. $\dfrac{7}{8}$ __<__ $\dfrac{11}{12}$

63. $\dfrac{13}{24}$ __<__ $\dfrac{23}{36}$

64. $\dfrac{9}{20}$ __>__ $\dfrac{11}{30}$

4.7 *Writing Fractions as Decimals*

Writing fractions as equivalent decimals can help you do calculations more easily or compare the size of two numbers.

OBJECTIVE ▶ Recall that a fraction is one way to show division (see **Section 1.5**). For example, $\frac{3}{4}$ means $3 \div 4$. If you are doing the division by hand, write it as $4\overline{)3}$. When you do the division, you will get the decimal equivalent of $\frac{3}{4}$.

> **Writing Fractions as Decimals**
>
> *Step 1* Divide the numerator of the fraction by the denominator.
>
> *Step 2* If necessary, round the answer to the place indicated.

WORK PROBLEM I AT THE SIDE. ▶▶

E X A M P L E I   Writing a Fraction or Mixed Number as a Decimal

(a) Write $\frac{1}{8}$ as a decimal.

$\frac{1}{8}$ means $1 \div 8$. Write it as $8\overline{)1}$. The decimal point in the dividend is on the right side of the 1. Write extra zeros in the dividend so you can continue dividing until the remainder is 0.

```
               ↙——————— Decimal points lined up.
        0.125
     8)1.000   ← Three extra 0's needed.
        8
        ——
        20
        16
        ——
         40
         40
         ——
          0   ← Remainder is 0.
```

Therefore, $\frac{1}{8} = 0.125$. To check this, write 0.125 as a fraction, then change it to lowest terms.

$$0.125 = \frac{125}{1000} \quad \text{To write in lowest terms} \quad \frac{125 \div 125}{1000 \div 125} = \frac{1}{8}$$

▦ *Calculator Tip:* When changing fractions to decimals on your calculator, enter the numbers from the top down. Remember that the order in which you enter the numbers *does* matter in division. Example 1(a) works like this:

$$\frac{1}{8} \downarrow \text{Top down} \qquad \text{Enter } 1 \boxed{\div} 8 \boxed{=}$$

What happens if you enter $8 \boxed{\div} 1 \boxed{=}$? Do you see why that cannot possibly be the answer?

(b) Write $2\frac{3}{4}$ as a decimal.

One method is to divide 3 by 4 to get 0.75 for the fraction part. Then add the whole number part to 0.75.

```
     0.75
   4)3.00        Whole number part →   2.00
     2 8         Fraction part →     + 0.75
     ——                              ——————
      20                               2.75
      20
      ——
       0
```

CONTINUED ON NEXT PAGE

OBJECTIVES

1 ▶ Change a fraction to a decimal.

2 ▶ Compare the size of fractions and decimals.

FOR EXTRA HELP

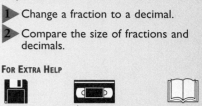

Tutorial   Tape 8   SSM, Sec. 4.7

1. Rewrite each fraction so you could do the division by hand. Do **not** complete the division.

(a) $\frac{1}{9}$ is written $9\overline{)}$

(b) $\frac{2}{3}$ is written $\overline{)}$

(c) $\frac{5}{4}$ is written $\overline{)}$

(d) $\frac{3}{10}$ is written $\overline{)}$

(e) $\frac{21}{16}$ is written $\overline{)}$

(f) $\frac{1}{50}$ is written $\overline{)}$

ANSWERS

1. **(a)** $9\overline{)1}$ **(b)** $3\overline{)2}$ **(c)** $4\overline{)5}$
   (d) $10\overline{)3}$ **(e)** $16\overline{)21}$ **(f)** $50\overline{)1}$

2. Write each fraction or mixed number as a decimal.

(a) $\dfrac{1}{4}$

(b) $2\dfrac{1}{2}$

(c) $\dfrac{5}{8}$

(d) $4\dfrac{3}{5}$

(e) $\dfrac{7}{8}$

So, $2\frac{3}{4} = 2.75$ Check: $2.75 = 2\frac{75}{100} = 2\frac{3}{4}$ Lowest terms

Whole number parts match.

A second method is to write $2\frac{3}{4}$ as an improper fraction.

$$2\frac{3}{4} = \frac{11}{4} \quad \leftarrow \frac{11}{4} \text{ means } 11 \div 4 \text{ or } 4\overline{)11}$$

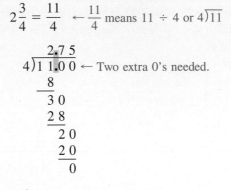

```
        2.7 5
  4)1 1.0 0   ← Two extra 0's needed.
    8
    ---
    3 0
    2 8
    ---
      2 0
      2 0
      ---
        0
```

Whole number parts match.

So, $2\dfrac{3}{4} = \mathbf{2.75}$

$\frac{3}{4}$ is equivalent to $\frac{75}{100}$ or 0.75.

◀◀ **WORK PROBLEM 2 AT THE SIDE.**

E X A M P L E 2   **Changing to a Decimal and Rounding**

Write $\frac{2}{3}$ as a decimal and round to the nearest thousandth.

$\frac{2}{3}$ means $2 \div 3$. To round to thousandths, divide out one *more* place, to ten-thousandths.

```
        0.6666
  3)2.0000   ← Four 0's needed for ten-thousandths.
    1 8
    ---
    20
    18
    --
     20
     18
     --
      20
      18
      --
       2
```

Written as a repeating decimal, $\frac{2}{3} = 0.\overline{6}$. Rounded to the nearest thousandth, $\frac{2}{3} \approx 0.667$.

🖩 *Calculator Tip:*   Try Example 2 on your calculator. Enter 2 ÷ 3. Which answer do you get?

| 0.666666667 | or | 0.6666666 |

Many scientific calculators will show a 7 as the last digit. Because the 6's keep on repeating forever, the calculator automatically rounds in the last decimal place it has room to show. If you have a 10-digit display space, the calculator is rounding like this:

0.6666666666 (11 digits)   rounds to   0.666666667.

ANSWERS

2. (a) 0.25 (b) 2.5 (c) 0.625
(d) 4.6 (e) 0.875

Other calculators, especially standard, four-function ones, may *not* round. They just cut off, or truncate, the extra digits. Such a calculator would show 0.6666666 in the display.

Would this difference in calculators show up when changing $\frac{1}{3}$ to a decimal? Why not?

WORK PROBLEM 3 AT THE SIDE. ▶▶

OBJECTIVE 2 You can use a number line to compare fractions and decimals. For example, the number line below shows the space between 0 and 1. The locations of some commonly used fractions are marked, along with their decimal equivalents.

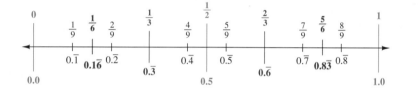

The next number line shows the locations of some commonly used fractions between 0 and 1 that are equivalent to repeating decimals. The decimal equivalents use a bar above repeating digits.

E X A M P L E 3 Using a Number Line to Compare Numbers

Use the number lines above to decide whether to write >, <, or = in the blank between each pair of numbers.

(a) 0.6875 _____ 0.625

You learned in **Section 3.5** that the number farther to the right on the number line is the greater number. On the first number line, 0.6875 is to the *right* of 0.625, so use the > symbol.

$$0.6875 \underbrace{\text{ is greater than }}_{} 0.625 \qquad 0.6875 > 0.625$$

(b) $\frac{3}{4}$ _____ 0.75

On the first number line, $\frac{3}{4}$ and 0.75 are at the same point on the number line. They are equivalent.

$$\frac{3}{4} = 0.75$$

CONTINUED ON NEXT PAGE

3. Write as decimals. Round to the nearest thousandth.

(a) $\frac{1}{3}$

(b) $2\frac{7}{9}$

(c) $\frac{10}{11}$

(d) $\frac{3}{7}$

(e) $3\frac{5}{6}$

ANSWERS

3. (a) ≈0.333 **(b)** ≈2.778
   (c) ≈0.909 **(d)** ≈0.429
   (e) ≈3.833

4. Use the number lines in the text to help you decide whether to write $<$, $>$, or $=$ in each blank.

(a) 0.4375 _____ 0.5

(b) 0.75 _____ 0.6875

(c) 0.625 _____ 0.0625

(d) $\dfrac{2}{8}$ _____ 0.375

(e) $0.8\overline{3}$ _____ $\dfrac{5}{6}$

(f) $\dfrac{1}{2}$ _____ $0.\overline{5}$

(g) $0.\overline{1}$ _____ $0.1\overline{6}$

(h) $\dfrac{8}{9}$ _____ $0.\overline{8}$

(i) $0.\overline{7}$ _____ $\dfrac{4}{6}$

(j) $\dfrac{1}{4}$ _____ 0.25

5. Arrange in order from smallest to largest.

(a) 0.7, 0.703, 0.7029

(b) 6.39, 6.309, 6.4, 6.401

(c) 1.085, $1\dfrac{3}{4}$, 0.9

(d) $\dfrac{1}{4}$, $\dfrac{2}{5}$, $\dfrac{3}{7}$, 0.428

ANSWERS

4. (a) $<$ (b) $>$ (c) $>$ (d) $<$ (e) $=$
(f) $<$ (g) $<$ (h) $=$ (i) $>$ (j) $=$
5. (a) 0.7, 0.7029, 0.703
(b) 6.309, 6.39, 6.4, 6.401
(c) 0.9, 1.085, $1\dfrac{3}{4}$
(d) $\dfrac{1}{4}$, $\dfrac{2}{5}$, 0.428, $\dfrac{3}{7}$

(c) 0.5 _____ $0.\overline{5}$

On the second number line, 0.5 is to the *left* of $0.\overline{5}$ (which is actually 0.555 . . .) so use the $<$ symbol.

$$0.5 \underbrace{\text{ is less than }} 0.\overline{5} \qquad 0.5 < 0.\overline{5}$$

(d) $\dfrac{2}{6}$ _____ $0.\overline{3}$

Write $\dfrac{2}{6}$ in lowest terms as $\dfrac{1}{3}$.
On the second number line you can see that $\dfrac{1}{3} = 0.\overline{3}$.

■

◀◀ **WORK PROBLEM 4 AT THE SIDE.**

Fractions can also be compared by first writing each one as a decimal. The decimals can then be compared by writing each one with the same number of decimal places.

E X A M P L E 4   Arranging Numbers in Order

Write the following numbers in order, from smallest to largest.

(a) 0.49   0.487   0.4903.

It is easier to compare decimals if they are all tenths, or all hundredths, and so on. Because 0.4903 has four decimal places (ten-thousandths), write zeros to the right of 0.49 and 0.487 so they also have four decimal places. Writing zeros to the right of a decimal number does *not* change its value (see **Section 4.3**). Now find the smallest and largest number of ten-thousandths.

$0.49 = 0.4900 =$ **4900** ten-thousandths ← 4900 is in the middle.
$0.487 = 0.4870 =$ **4870** ten-thousandths ← 4870 is smallest.
$0.4903 =$ **4903** ten-thousandths ← 4903 is largest.

From smallest to largest the correct order is:

0.487   0.49   0.4903.

(b) $2\dfrac{5}{8}$ 2.63   2.6

Write $2\dfrac{5}{8}$ as $\dfrac{21}{8}$ and divide $8\overline{)21}$ to get the decimal form, 2.625. Then, because 2.625 has three decimal places, write zeros so all the numbers have three decimal places.

$2\dfrac{5}{8} = 2.625 = 2$ and **625** thousandths ← 625 is in the middle.

$2.63 = 2.630 = 2$ and **630** thousandths ← 630 is largest.

$2.6 = 2.600 = 2$ and **600** thousandths ← 600 is smallest.

From smallest to largest, the correct order is:

2.6 $2\dfrac{5}{8}$ 2.63.

■

◀◀ **WORK PROBLEM 5 AT THE SIDE.**

4.7 Exercises

Write each fraction or mixed number as a decimal. Round to the nearest thousandth, if necessary.

Example: $\frac{9}{16}$ **Solution:** $\begin{array}{r} 0.5625 \\ 16\overline{)9.0000} \\ \underline{8\ 0} \\ 1\ 00 \\ \underline{96} \\ 40 \\ \underline{32} \\ 80 \\ \underline{80} \\ 0 \end{array}$ Rounds to $\approx\mathbf{0.563}$
← Four 0's needed to complete the division.

1. $\frac{1}{2}$ **0.5**

2. $\frac{1}{4}$ **0.25**

3. $\frac{3}{4}$ **0.75**

4. $\frac{1}{10}$ **0.1**

5. $\frac{3}{10}$ **0.3**

6. $\frac{7}{10}$ **0.7**

7. $\frac{9}{10}$ **0.9**

8. $\frac{4}{5}$ **0.8**

9. $\frac{3}{5}$ **0.6**

10. $\frac{2}{5}$ **0.4**

11. $\frac{7}{8}$ **0.875**

12. $\frac{3}{8}$ **0.375**

13. $2\frac{1}{4}$ **2.25**

14. $1\frac{1}{2}$ **1.5**

15. $14\frac{7}{10}$ **14.7**

16. $23\frac{3}{5}$ **23.6**

17. $3\frac{5}{8}$ **3.625**

18. $2\frac{7}{8}$ **2.875**

🖩 19. $\frac{1}{3}$ **≈0.333**

20. $\frac{2}{3}$ **≈0.667**

21. $\frac{5}{6}$ **≈0.833**

22. $\frac{1}{6}$ **≈0.167**

23. $1\frac{8}{9}$ **≈1.889**

24. $5\frac{4}{7}$ **≈5.571**

25. Explain and correct the error that Keith made when changing a fraction to an equivalent decimal.

$$\frac{5}{9} = 5\begin{array}{r} 1.8 \\ \overline{)9.0} \\ \underline{5} \\ 4\ 0 \\ \underline{4\ 0} \\ 0 \end{array} \quad \text{so} \quad \frac{5}{9} = 1.8$$

$\frac{5}{9}$ means $5 \div 9$ or $9\overline{)5}$ so correct answer is ≈0.556.

26. Explain and correct the error Sandra made when writing $2\frac{7}{20}$ as a decimal.

$$2\frac{7}{20} = 20\begin{array}{r} 0.35 \\ \overline{)7.00} \\ \underline{6\ 0} \\ 1\ 00 \\ \underline{1\ 00} \\ 0 \end{array} \quad \text{so} \quad 2\frac{7}{20} = 2.035$$

Adding the whole number part gives 2 + 0.35 which is 2.35 not 2.035.

📝 Writing ⊙ Conceptual ▲ Challenging ≈ Estimation

27. Ving knows that $\frac{3}{8} = 0.375$. How can he write $1\frac{3}{8}$ as a decimal *without* having to do a division? How can he write $3\frac{3}{8}$ as a decimal? $295\frac{3}{8}$? Explain your answer.

Just add the whole number part to 0.375. So
$1\frac{3}{8} = 1.375;\ 3\frac{3}{8} = 3.375;\ 295\frac{3}{8} = 295.375.$

28. Iris has found a shortcut for writing mixed numbers as decimals:

$$2\frac{7}{10} = 2.7 \qquad 1\frac{13}{100} = 1.13$$

Does her shortcut work for all mixed numbers? Explain.

It works only when the fraction part has a denominator of 10, 100, 1000, and so on.

Find the decimal or fraction equivalent for each of the following. Write fractions in lowest terms.

Example:		
Fraction	*Decimal*	
_____	0.375	

Solution: $0.375 = \dfrac{375}{1000} = \dfrac{375 \div 125}{1000 \div 125} = \dfrac{3}{8}$ (In lowest terms)

Thousandths place so write 1000 in the denominator

	Fraction	*Decimal*		*Fraction*	*Decimal*
29.	$\frac{2}{5}$	0.4	**30.**	$\frac{3}{4}$	0.75
31.	$\frac{5}{8}$	0.625	**32.**	$\frac{111}{1000}$	0.111
33.	$\frac{7}{20}$	0.35	**34.**	$\frac{9}{10}$	0.9
35.	$\frac{7}{20}$	0.35	**36.**	$\frac{1}{40}$	0.025
37.	$\frac{1}{25}$	0.04	**38.**	$\frac{13}{25}$	0.52
39.	$\frac{3}{20}$	0.15	**40.**	$\frac{17}{20}$	0.85
41.	$\frac{1}{5}$	0.2	**42.**	$\frac{1}{8}$	0.125
43.	$\frac{9}{100}$	0.09	**44.**	$\frac{1}{50}$	0.02

Solve each application problem.

45. The label on the bottle of vitamins says that each capsule contains 0.5 gram of calcium. When checked, each capsule had 0.505 gram of calcium. Was there too much or too little calcium? What was the difference?

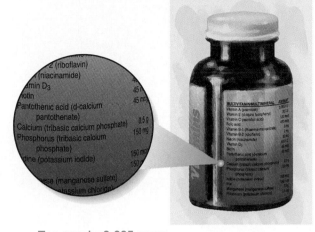

Too much; 0.005 gram

46. The patient in room 830 is supposed to get 8.3 milligrams of medicine. She was actually given 8.03 milligrams. Did she get too much or too little medicine? What was the difference?

Too little; 0.27 milligram

47. The average length of a newborn baby is 20.8 inches. Charlene's baby is 20.08 inches long. Is her baby longer or shorter than the average? By how much?

Shorter; 0.72 inch

48. The glass mirror of the Hubble telescope had to be repaired in space in 1993 because it would not focus properly. The problem was that the mirror's outer edge had a thickness of 0.6248 cm when it was supposed to be 0.625 cm. Was the edge too thick or too thin? By how much?

Too thin; 0.0002 cm

49. Precision Medical Parts makes an artificial heart valve that must measure between 0.998 centimeter and 1.002 centimeters. Circle the lengths that are acceptable:

1.01 cm, (0.9991 cm,) (1.0007 cm,) 0.99 cm.

50. The mice in a medical experiment must start out weighing between 2.95 ounces and 3.05 ounces. Circle the weights that can be used: (3.0 ounces,) (2.995 ounces,) 3.055 ounces, (3.005 ounces.)

51. Ginny Brown hoped her crops would get $3\frac{3}{4}$ inches of rain this month. The newspaper said the area received 3.8 inches of rain. Was that more or less than Ginny hoped for? By how much?

More; 0.05 inch

52. The mice in the experiment gained $\frac{3}{8}$ ounce. They were expected to gain 0.3 ounce. Was their actual gain more or less than expected? By how much?

More; 0.075 ounce

Arrange in order from smallest to largest.

Example: 0.8075, 0.875, 0.88, 0.808

Solution:
Write zeros so all the numbers have four decimal places.

$$0.8075 = 8075 \text{ ten-thousandths} \leftarrow 8075 \text{ is smallest}$$
$$0.875 = 0.8750 = 8750 \text{ ten-thousandths}$$
$$0.88 = 0.8800 = 8800 \text{ ten-thousandths} \leftarrow 8800 \text{ is largest}$$
$$0.808 = 0.8080 = 8080 \text{ ten-thousandths}$$

(smallest) **0.8075   0.808   0.875   0.88** (largest)

53. 0.54, 0.5455, 0.5399

0.5399, 0.54, 0.5455

54. 0.76, 0.7, 0.7006

0.7, 0.7006, 0.76

55. 5.8, 5.79, 5.0079, 5.804

5.0079, 5.79, 5.8, 5.804

56. 12.99, 12.5, 13.0001, 12.77

12.5, 12.77, 12.99, 13.0001

57. 0.628, 0.62812, 0.609, 0.6009

0.6009, 0.609, 0.628, 0.62812

58. 0.27, 0.281, 0.296, 0.3

0.27, 0.281, 0.296, 0.3

59. 5.8751, 4.876, 2.8902, 3.88

2.8902, 3.88, 4.876, 5.8751

60. 0.98, 0.89, 0.904, 0.9

0.89, 0.9, 0.904, 0.98

61. 0.043, 0.051, 0.006, $\frac{1}{20}$

0.006, 0.043, $\frac{1}{20}$, 0.051

62. 0.629, $\frac{5}{8}$, 0.65, $\frac{7}{10}$

$\frac{5}{8}$, 0.629, 0.65, $\frac{7}{10}$

63. $\frac{3}{8}$, $\frac{2}{5}$, 0.37, 0.4001

0.37, $\frac{3}{8}$, $\frac{2}{5}$, 0.4001

64. 0.1501, 0.25, $\frac{1}{10}$, $\frac{1}{5}$

$\frac{1}{10}$, 0.1501, $\frac{1}{5}$, 0.25

Some rulers show each inch divided into tenths. Use this scale drawing for Exercises 65–70. Change the measurements on the drawing to decimals and round them to the nearest tenth of an inch.

65. length (a) is _____ **≈1.4 in.**

66. length (b) is _____ **≈1.1 in.**

67. length (c) is _____ **≈0.3 in.**

68. length (d) is _____ **0.5 in.**

69. length (e) is _____ **≈0.4 in.**

70. length (f) is _____ **≈0.7 in.**

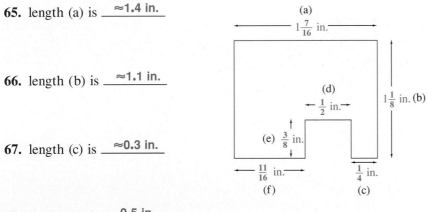

Arrange in order from smallest to largest.

71. $\frac{6}{11}, \frac{5}{9}, \frac{4}{7}, 0.571$

$\frac{6}{11}, \frac{5}{9}, 0.571, \frac{4}{7}$

72. $\frac{8}{13}, \frac{10}{17}, 0.615, \frac{11}{19}$

$\frac{11}{19}, \frac{10}{17}, 0.615, \frac{8}{13}$

73. $\frac{3}{11}, \frac{4}{15}, 0.25, \frac{1}{3}$

$0.25, \frac{4}{15}, \frac{3}{11}, \frac{1}{3}$

74. $0.223, \frac{2}{11}, \frac{2}{9}, \frac{1}{4}$

$\frac{2}{11}, \frac{2}{9}, 0.223, \frac{1}{4}$

75. $\frac{3}{16}, \frac{1}{6}, \frac{1}{5}, 0.188$

$\frac{1}{6}, \frac{3}{16}, 0.188, \frac{1}{5}$

76. $\frac{7}{20}, \frac{1}{3}, \frac{3}{8}, 0.375$

$\frac{1}{3}, \frac{7}{20}, 0.375, \frac{3}{8}$

or $\frac{1}{3}, \frac{7}{20}, \frac{3}{8}, 0.375$

Review and Prepare

*Write each fraction in lowest terms. (For help, see **Section 2.4**.)*

77. $\frac{9}{12}$ $\frac{3}{4}$

78. $\frac{30}{60}$ $\frac{1}{2}$

79. $\frac{60}{80}$ $\frac{3}{4}$

80. $\frac{40}{75}$ $\frac{8}{15}$

81. $\frac{96}{132}$ $\frac{8}{11}$

82. $\frac{26}{98}$ $\frac{13}{49}$

4.1	**decimals**	Decimals, like fractions, are used to show parts of a whole.
	decimal point	The dot that is used to separate the whole number part from the fractional part of a decimal number.
	place value	The value assigned to each place to the right or left of the decimal point. Whole numbers, such as ones and tens, are to the *left* of the decimal point. Fractional parts, such as tenths and hundredths, are to the *right* of the decimal point.
4.2	**rounding**	"Cutting off" a number after a certain place, such as rounding to the nearest hundredth. The rounded number is less accurate than the original number, so write the symbol "≈" in front of it to mean "approximately equal to."
	decimal places	The number of digits to the *right* of the decimal point; for example, 6.37 has two decimal places, 4.706 has three decimal places.
4.3	**estimating**	The process of rounding the numbers in a problem and getting an approximate answer. This helps you check that the decimal point is in the correct place in the exact answer.
4.6	**repeating decimal**	A decimal with one or more digits that repeat forever, such as the 6 in 0.1666 . . . . Use three dots to indicate that it is a repeating decimal; it never terminates (ends). You can also write it with a bar above the repeating digits, as in $0.1\overline{6}$.

Concepts	*Examples*
4.1 Reading and Writing Decimals	Write each decimal in words.
thousands hundreds tens ones • tenths hundredths thousandths ten-thousandths 5 8 4 6 • 0 7 3 2 Whole number part — Fractional part	→ 8 is in hundredths place → 15.38 ↓ fifteen **and** thirty-eight **hundredths** ← → 3 is in ten-thousandths place → 0.0103 one hundred three **ten-thousandths** ←
4.1 Writing Decimals as Fractions The digits to the right of the decimal point are the numerator. The place value of the right-most digit determines the denominator. Reduce to lowest terms.	Write 0.45 as a fraction. The numerator is 45. The right-most digit, 5, is in the hundredths place, so the denominator is 100. Then reduce. $$\frac{45}{100} = \frac{45 \div 5}{100 \div 5} = \frac{9}{20} \quad \text{Lowest terms}$$

Concepts	*Examples*
4.2 Rounding Decimals Find the place to which you are rounding. Draw a cut-off line to the right of that place; the rest of the digits will be dropped. Look *only* at the first digit being cut. If it is *less than 5*, the part you are keeping stays the same. If it is *5 or more*, the part you are keeping rounds up. Do not move the decimal point when rounding. Write "≈" in front of the rounded number to mean "approximately equal to."	Round 0.17952 to the nearest thousandth. Is it closer to 0.179 or to 0.180? ┌─ First digit cut is 5 or more so round up by adding 1 thousandth to the part you are keeping. $0.179 \mid 52$ $\underline{0.179}$ $+\ 0.001$ ◄── To round up, add 1 thousandth. $\overline{0.180}$ 0.17952 rounds to 0.180. Write it as ≈0.180.
4.3 and 4.4 Adding and Subtracting Decimals Estimate the answer by rounding each number so there is only one non-zero digit. To find the exact answer, line up the decimal points. If needed, write in zeros as placeholders. Add or subtract as if they were whole numbers. Line up the decimal point in the answer.	Add 5.68 + 785.3 + 12 + 2.007. *estimate* *exact* 6 ◄── 5.680 Use zeros as place- 800 ◄── 785.300 holders so all 10 ◄── 12.000 numbers have + 2 ◄── + 2.007 three decimal places. $\overline{818}$ $\overline{804.987}$ Line up decimal points. The estimate and exact answer are both in hundreds, so the decimal point is probably in the correct place.
4.5 Multiplying Decimals 1. Multiply as you would for whole numbers. 2. Count the total number of decimal places in both factors. 3. Write the decimal point in the answer so it has the same number of decimal places as the total from Step 2. You may need to write extra zeros on the left side of the product in order to get enough decimal places in the answer.	Multiply 0.169 × 0.21. 0.169 ← 3 decimal places × 0.21 ← 2 decimal places $\overline{169}$ 5 total decimal places 338 $\overline{.03549}$ ← 5 decimal places in answer Write in a zero so you can count over 5 decimal places. Final answer is 0.03549.
4.6 Dividing by a Decimal 1. Change the divisor to a whole number by moving the decimal point to the right. 2. Move the decimal point in the dividend the same number of places to the right. 3. Write the decimal point in the answer directly above the decimal point in the dividend. 4. Divide as with whole numbers.	Divide 52.8 by 0.75. 70.4 $0.75\,)\overline{52.800}$ Move decimal point two $\underline{525}$ places to the right in divisor and dividend. 300 Write zeros in the dividend so $\underline{300}$ you can move the decimal 0 point and continue dividing until the remainder is zero. To check your answer, multiply 70.4 times 0.75. If the result matches the dividend (52.8), you solved the problem correctly.

Concepts	Examples
4.7 Writing Fractions as Decimals Divide the numerator by the denominator. If necessary, round to the place indicated.	Write $\frac{1}{8}$ as a decimal. $\frac{1}{8}$ means $1 \div 8$. Write it as $8\overline{)1}$. The decimal point is on the right side of 1. $$\begin{array}{r} 0.125 \\ 8\overline{)1.000} \\ \underline{8} \\ 20 \\ \underline{16} \\ 40 \\ \underline{40} \\ 0 \end{array}$$ ← Decimal point and three zeros written in so you can continue dividing. Therefore, $\frac{1}{8}$ is equivalent to 0.125.
4.7 Comparing the Size of Fractions and Decimals 1. Write any fractions as decimals. 2. Write zeros so that all the numbers being compared have the same number of decimal places. 3. Use < to mean "is less than," > to mean "is greater than," or list the numbers from smallest to largest.	Arrange in order from smallest to largest. $\qquad 0.505 \qquad \frac{1}{2} \qquad 0.55$ $0.505 = 505$ thousandths $\frac{1}{2} = 0.5 = 0.500 = 500$ thousandths ← 500 is smallest. $0.55 = 0.550 = 550$ thousandths ← 550 is largest. (smallest) $\frac{1}{2}$ 0.505 0.55 (largest)

Numbers in the Real World

Real World collaborative investigations

Gotta Be Green

A lot's being said about personal responsibility these days, and the idea seems to be ending up on the front lawn—literally! Each spring, homeowners across the country gear up to green up their lawns, and the increased use of fertilizer has a lot of environmentalists concerned about the potential effects of chemical runoff into nearby rivers and streams.

Each year, according to a study conducted by the University of Minnesota's Department of Agriculture, each household in the Minneapolis/St. Paul metro area uses an average of 36 pounds of lawn fertilizer. That adds up to 25,529,295 pounds, or 12,765 tons. Add to that another 193,000 pounds of weed killer and you're looking at the total picture for keeping it green in the Twin Cities.

Source: Minneapolis Star Tribune

1. According to the article, how many pounds of lawn fertilizer are used each year in the *entire metro area*? **25,529,295 pounds**

Use division to find the number of *households* in the metro area.
25,529,295 ÷ 36 = 709,147.0833

Would it make sense to round your answer? If so, how would you round it?
Answers vary. Two rounding possibilities are nearest whole number (709,147) or nearest thousand (709,000).

2. There are 2000 pounds in one ton. Find the number of tons equivalent to 25,529,295 pounds. **25,529,295 ÷ 2000 = 12,764.6475**

Does your answer match the figure given in the article? **no**
If not, what did the writer of the article do to get 12,765 tons? **Rounded to nearest whole number.**

Is the author's figure accurate? Why or why not? **Answers vary, but it is probably accurate enough for a newspaper article.**

3. The article states that "each household in the Minneapolis/St. Paul metro area uses an average of 36 pounds of lawn fertilizer" each year. What mathematical operations do you do to find an average? **Add all values, then divide by the number of values.**

When the calculations were done to find the average, the answer probably was not *exactly* 36 pounds. List six different answers that are *less than* 36 that would round to 36. List two answers that have one decimal place, two answers with two decimal places, and two answers with three decimal places.
Some possibilities are: 35.5 and 35.8; 35.55 and 35.62; 35.905 and 35.544

List six answers that are *greater than* 36 that would round to 36. Use one, two, and three decimal places. **Some possibilities are 36.1 and 36.4; 36.01 and 36.49; 36.001 and 36.485**

What is the *smallest* number that rounds to 36? **35.5** The *largest* number? **36.499 (if limited to 3 decimal places) or 36.49̄**

Use the memory keys on your calculator to find the average number of pounds of *weed killer* used by each household each year. Start by dividing to find the number of households, as you did in exercise 1 above. Pressing the [STO] and [1] keys stores the answer in your calculator's first memory.

25,529,925 [÷] 36 [=] [STO] [1] (The calculator display will show 709147.0833 and a small M1 in the corner to show that the number is stored in the first memory.)

Now enter the number of pounds of weed killer. To divide by the number of households, use the [RCL] key to recall the number of households without having to enter the number again.

193,000 [÷] [RCL] [1] [=] What is your answer? **0.272157927**

How will you round your answer? Why? **Answers vary.**

[4.1] *Name the digit that has the given place value.*

1. 243.059
tenths **0**
hundredths **5**

2. 0.6817
ones **0**
tenths **6**

3. $5824.39
hundreds **8**
hundredths **9**

4. 896.503
tenths **5**
tens **9**

5. 20.73861
tenths **7**
ten-thousandths **6**

Write each decimal as a fraction or mixed number in lowest terms.

6. 0.5 $\dfrac{1}{2}$

7. 0.75 $\dfrac{3}{4}$

8. 4.05 $4\dfrac{1}{20}$

9. 0.875 $\dfrac{7}{8}$

10. 0.027 $\dfrac{27}{1000}$

11. 27.8 $27\dfrac{4}{5}$

Write each decimal in words.

12. 0.8   **eight tenths**

13. 400.29   **four hundred and twenty-nine hundredths**

14. 12.007   **twelve and seven thousandths**

15. 0.0306   **three hundred six ten-thousandths**

Write each decimal in numbers.

16. eight and three tenths   **8.3**

17. two hundred five thousandths   **0.205**

18. seventy and sixty-six ten-thousandths   **70.0066**

19. thirty hundredths   **0.30**

[4.2] *Round to the place indicated. Write your answers using the "≈" sign.*

20. 275.635 to the nearest tenth   **≈275.6**

21. 72.789 to the nearest hundredth   **≈72.79**

22. 0.1604 to the nearest thousandth   **≈0.160**

23. 0.0905 to the nearest thousandth   **≈0.091**

24. 0.98 to the nearest tenth   **≈1.0**

Round to the nearest cent.

25. $15.8333   **≈$15.83**

26. $0.698   **≈$0.70**

27. $17,625.7906   **≈$17,625.79**

Round each income or expense item to the nearest dollar.

28. Income from pancake breakfast was $350.48.   **≈$350**

29. Members paid $129.50 in dues.   **≈$130**

30. Refreshments cost $99.61.   **≈$100**

31. Bank charges were $29.37.   **≈$29**

📝 Writing   ◎ Conceptual   ▲ Challenging   ≈ Estimation

[4.3] ≈*First round the numbers so there is only one non-zero digit and estimate the answer. Then add to find the exact answer.*

32.
estimate	exact
6	5.81
400	423.96
+ 20	+ 15.09
426	444.86

33.
estimate	exact
80	75.6
1	1.29
100	122.045
1	0.88
+ 30	+ 33.7
212	233.515

[4.4] ≈*First round the numbers so there is only one non-zero digit and estimate the answer. Then subtract to find the exact answer.*

34.
estimate	exact
300	308.5
− 20	− 17.8
280	290.7

35.
estimate	exact
9	9.2
− 8	− 7.9316
1	1.2684

[4.3–4.4] ≈*First round the numbers so there is only one non-zero digit and estimate the answer. Then find the exact answer.*

36. Tim agreed to donate 12.5 hours of work at his children's school. He has already worked 9.75 hours. How many more hours will he work?

estimate: 13 − 10 = 3 hours
exact: 2.75 hours

37. Today Jasmin wrote a check to the daycare center for $215.53 and a check for $44.47 at the grocery store. What was the total of the two checks?

estimate: $200 + $40 = $240
exact: $260.00

38. Joey spent $1.59 for toothpaste, $5.33 for a gift, and $18.94 for a toaster. He gave the clerk three $10 bills. How much change did he get?

estimate: $2 + $5 + $20 = $27;
$30 − $27 = $3
exact: $4.14

39. Roseanne is training for a wheelchair race. She raced 2.3 kilometers on Monday, 4 kilometers on Wednesday, and 5.25 kilometers on Friday. How far did she race altogether?

estimate: 2 + 4 + 5 = 11 kilometers
exact: 11.55 kilometers

[4.5] ≈*First round the numbers so there is only one non-zero digit and estimate an answer. Then multiply to find the exact answer.*

40.
estimate	exact
6	6.138
× 4	× 3.7
24	22.7106

41.
estimate	exact
40	42.9
× 3	× 3.3
120	141.57

Multiply.

42. $(5.6)(0.002)$ **0.0112**

43. $(0.071)(0.005)$ **0.000355**

[4.6] $\approx$*Decide if each answer is reasonable by rounding the numbers and estimating the answer. If the exact answer is not reasonable, find the correct answer.*

44. $706.2 \div 12 = 58.85$ **reasonable**
estimate:

700 ÷ 10 = 70

45. $26.6 \div 2.8 = 0.95$ **unreasonable**
estimate:

$$\begin{array}{r} 9.5 \\ 30 \div 3 = 10 \quad 2.8\overline{)26.6} \end{array}$$

Divide. Round to the nearest thousandth, if necessary.

46. $3\overline{)43.4}$ $\approx$**14.467**

47. $\dfrac{72}{0.06}$

1200

48. $0.00048 \div 0.0012$

0.4

[4.5–4.6] *Solve these application problems.*

49. Adrienne worked 36.5 hours this week. Her hourly wage is $9.59. Find her total earnings to the nearest dollar.

$\approx$**$350**

50. A book of 12 tickets costs $23.89 at the amusement park. What is the cost per ticket, to the nearest cent?

$\approx$**$1.99**

51. Stock in MathTronic sells for $3.75 per share. Kenneth is thinking of investing $500. How many whole shares could he buy?

$\approx$**133 shares**

52. Hamburger meat is on sale at $0.89 per pound. How much will Ms. Lee pay for 3.5 pounds of hamburger, to the nearest cent?

$\approx$**$3.12**

Simplify by using the order of operations.

53. $3.5^2 + 8.7 \cdot 1.95$

29.215

54. $11 - 3.06 \div (3.95 - 0.35)$

10.15

[4.7] *Write each fraction as a decimal. Round to the nearest thousandth, if necessary.*

55. $3\dfrac{4}{5}$

3.8

56. $\dfrac{16}{25}$

0.64

57. $1\dfrac{7}{8}$

1.875

58. $\dfrac{1}{9}$

$\approx$**0.111**

Arrange in order from smallest to largest.

59. $3.68,\ 3.806,\ 3.6008$

3.6008, 3.68, 3.806

60. $0.215,\ 0.22,\ 0.209,\ 0.2102$

0.209, 0.2102, 0.215, 0.22

61. $0.17,\ \dfrac{3}{20},\ \dfrac{1}{8},\ 0.159$

$\dfrac{1}{8},\ \dfrac{3}{20},$ **0.159, 0.17**

MIXED REVIEW EXERCISES

Solve each problem.

62. 89.19 + 0.075 + 310.6 + 5

    404.865

63. 72.8 × 3.5

    254.8

64. 1648.3 ÷ 0.46   Round to thousandths.

    ≈3583.261

65. 30 − 0.9102

    29.0898

66. (4.38)(0.007)

    0.03066

67. $0.005\overline{)0.047}$ with quotient 9.4

68. 72.105 + 8.2 + 95.37

    175.675

69. 81.36 ÷ 9

    9.04

70. (5.6 − 1.22) + 4.8 • 3.15

    19.50

71. 0.455 × 18

    8.19

72. 1.6 • 0.58

    0.928

73. $0.218\overline{)7.63}$ with quotient 35

74. 21.059 − 20.8

    0.259

75. 18.3 − 3² ÷ 0.5

    0.3

Use the information in the ad to solve Exercises 76–80. Round money answers to the nearest cent. (Disregard any sales tax.)

Grand Opening Sale!
Save on Clothing for the Entire Family!!

Jeans for Teens
only $19.95 each
women's sizes $24.99

Athletic Shoes
regularly priced
$89.99 to $149.50
NOW just $71 to $119.60!

Men's socks NOW 3 pairs for $8.99
Children's socks 6 pairs for $5

Hurry in — *TWO DAYS ONLY!!*

76. How much would one pair of men's socks cost?

    ≈$3.00

77. How much more would one pair of men's socks cost than one pair of children's socks?

    ≈$2.17

78. How much would Fernando pay for a dozen pair of men's socks?

    $35.96

79. How much would Akiko pay for five pairs of teen jeans and four pairs of women's jeans?

    $199.71

80. What is the difference between the cheapest sale price for athletic shoes and the highest regular price?

    $78.50

Write each decimal as a fraction or mixed number in lowest terms.

1. 18.4   **2.** 0.075

Write each decimal in words.

3. 60.007   **4.** 0.0208

≈*Round to the place indicated.*

5. 725.6089 to the nearest tenth

6. 0.62951 to the nearest thousandth

7. $1.4945 to the nearest cent

8. $7859.51 to the nearest dollar

≈*Round the numbers so there is only one non-zero digit and estimate each answer. Then find the exact answer.*

9. 7.6 + 82.0128 + 39.59

10. 79.1 − 3.602

11. 5.79 • 1.2

12. 20.04 ÷ 4.8

Solve.

13. 53.1 + 4.631 + 782 + 0.031

14. 670 − 0.996

15. (0.0069)(0.007)

16. 0.15)‾72

1. $18\frac{2}{5}$

2. $\frac{3}{40}$

3. sixty and seven thousandths

4. two hundred eight ten-thousandths

5. ≈725.6

6. ≈0.630

7. ≈$1.49

8. ≈$7860

9. estimate: 8 + 80 + 40 = 128
exact: 129.2028

10. estimate: 80 − 4 = 76
exact: 75.498

11. estimate: 6 • 1 = 6
exact: 6.948

12. estimate: 20 ÷ 5 = 4
exact: 4.175

13. 839.762

14. 669.004

15. 0.0000483

16. 480

17. 2.625 _____

17. Write $2\frac{5}{8}$ as a decimal. Round to the nearest thousandth, if necessary.

Arrange in order from smallest to largest.

18. $0.44, \dfrac{9}{20}, 0.4506, 0.451$ _____

18. $0.44, 0.451, \dfrac{9}{20}, 0.4506$

Use the order of operations to simplify.

19. 35.49 _____

19. $6.3^2 - 5.9 + 3.4 \cdot 0.5$

Solve these application problems.

20. $446.87 _____

20. Jennifer had $71.15 in her checking account. Yesterday her account earned $0.95 interest for the month, and she deposited a paycheck for $390.77. The bank charged her $16 for new checks. What is the new balance in her account?

21. Davida, by 0.441 minute _____

21. Davida ran a race in 3.059 minutes. Angela ran the race in 3.5 minutes. Who won? By how much?

22. ≈$5.35 _____

22. Mr. Yamamoto bought 1.85 pounds of cheese at $2.89 per pound. How much did he pay for the cheese, to the nearest cent?

23. 2.8 degrees _____

23. Loren's baby had a temperature of 102.7 degrees. Later in the day it was 99.9 degrees. How much had the temperature dropped?

24. $4.55 per meter _____

24. Pat bought 3.4 meters of fabric. She paid $15.47. What was the cost per meter?

25. Answer varies. _____

25. Write your own application problem using decimals. Make it different from problems 20–24. Then show how to solve your problem.

Name the digit that has the given place value.

1. 19,076,542
   hundreds **5**
   millions **9**
   ones **2**

2. 83.0754
   tenths **0**
   thousandths **5**
   tens **8**

Round each number as indicated.

3. 499,501 to the nearest thousand.

   ≈500,000

4. 602.4937 to the nearest hundredth.

   ≈602.49

5. $709.60 to the nearest dollar.

   ≈$710

6. $0.0528 to the nearest cent.

   ≈$0.05

≈Round the numbers in each problem so there is only one non-zero digit. Then add, subtract, multiply, or divide the rounded numbers, as indicated, to estimate the answer. Finally, solve for the exact answer.

7. *estimate*   *exact*
   4000   3672
   600   589
   + 9000   **+ 9078**
   13,600   **13,339**

8. *estimate*   *exact*
   4   4.06
   20   15.7
   + 1   **+ 0.923**
   25   **20.683**

9. *estimate*   *exact*
   5000   5018
   − 2000   **− 1809**
   3000   **3209**

10. *estimate*   *exact*
    50   51.6
    − 7   **− 7.094**
    43   **44.506**

11. *estimate*   *exact*
    3000   3317
    × 200   **× 166**
    600,000   **550,622**

12. *estimate*   *exact*
    7   6.82
    × 7   **× 7.3**
    49   **49.786**

13. *estimate*   *exact*
    2000   **2690**
    50)¯100,000   46)¯123,740

14. *estimate*   *exact*
    5   **4.5**
    8)¯40   8.4)¯37.8

15. *estimate*

    __**2**__ • __**4**__ = __**8**__

    exact

    $1\frac{9}{10} \cdot 3\frac{3}{4}$ $7\frac{1}{8}$

16. *estimate*

    __**2**__ ÷ __**1**__ = __**2**__

    exact

    $2\frac{1}{3} \div \frac{5}{6}$ $2\frac{4}{5}$

17. *estimate*

    __**2**__ + __**2**__ = __**4**__

    exact

    $1\frac{4}{5} + 1\frac{2}{3}$ $3\frac{7}{15}$

18. *estimate*

    __**5**__ − __**2**__ = __**3**__

    exact

    $4\frac{1}{2} - 1\frac{7}{8}$ $2\frac{5}{8}$

Add, subtract, multiply, or divide as indicated.

19. $10 - 0.329$

9.671

20. $2\dfrac{3}{5} \cdot \dfrac{5}{9}$

$1\dfrac{4}{9}$

21. $9 + 72{,}417 + 799$

73,225

22. $11\dfrac{1}{5} \div 8$

$1\dfrac{2}{5}$

23. $5006 - 92$

4914

24. $0.7 + 85 + 7.903$

93.603

25. Write your answer using R for the remainder.

$$7\overline{)2831} \quad \text{404 R3}$$

26. $\dfrac{5}{6} + \dfrac{7}{8}$

$1\dfrac{17}{24}$

27. 332×704

233,728

28. $(0.006)(5.44)$

0.03264

29. 3.2×2.5

8

30. $25.2 \div 0.56$

45

31. $\dfrac{2}{3} \div 5\dfrac{1}{6}$

$\dfrac{4}{31}$

32. $5\dfrac{1}{4} - 4\dfrac{7}{12}$

$\dfrac{2}{3}$

33. $4.7 \div 9.3$

Round to nearest hundredth.

≈ 0.51

Simplify by using the order of operations.

34. $10 - 4 \div 2 \cdot 3$

4

35. $\sqrt{36} + 3 \cdot 8 - 4^2$

14

36. $\dfrac{2}{3} \cdot \left(\dfrac{7}{8} - \dfrac{1}{2} \right)$

$\dfrac{1}{4}$

37. $0.9^2 + 10.6 \div 0.53$

20.81

38. Solve $4^3 \cdot 3^2$.   576

39. Find $\sqrt{196}$.   14

40. Find the prime factorization of 200. Write your answer using exponents.

$2^3 \cdot 5^2$

41. Write 40.035 in words.

     forty and thirty-five thousandths

42. Write three hundred six ten-thousandths in numbers.

     0.0306

Write each decimal as a fraction or mixed number in lowest terms.

43. 0.125 $\dfrac{1}{8}$

44. 3.08 $3\dfrac{2}{25}$

Write each fraction or mixed number as a decimal. Round to the nearest thousandth, if necessary.

45. $2\dfrac{3}{5}$ 2.6

46. $\dfrac{7}{11}$ ≈0.636

47. Write $<$ or $>$ in the blank to make a true statement: $\dfrac{5}{8}$ ___$>$___ $\dfrac{4}{9}$.

Arrange in order from smallest to largest.

48. 7.005, 7.5005, 7.5, 7.505

     7.005, 7.5, 7.5005, 7.505

49. $\dfrac{7}{8}$, 0.8, $\dfrac{21}{25}$, 0.8015

     0.8, 0.8015, $\dfrac{21}{25}$, $\dfrac{7}{8}$

50. In the 1996 Olympics, Finland's Heli Rantanen threw the javelin 67.95 meters. If a local newspaper reported the distance as 67.905 meters, would that figure be too long or too short? By how much?

     Too short, by 0.045 meter

51. About $\dfrac{7}{8}$ of all children are right-handed. How many of the 96 children in the daycare center would be expected to be right-handed?

     84 children

≈First round the numbers and estimate the answer to each application problem. Then find the exact answer.

52. Lameck had two $10 bills. He spent $7.96 on gasoline and $0.87 for a candy bar at the convenience store. How much money does he have left?

estimate: $20 − $8 − $1 = $11
exact: $11.17

53. Manuela's daughter is 50 inches tall. Last year she was $46\frac{5}{8}$ inches tall. How much has she grown?

estimate: 50 − 47 = 3 inches
exact: $3\frac{3}{8}$ inches

54. Sharon records textbooks on tape for students who are blind. Her hourly wage is $8.73. How much did she earn working 16.5 hours last week, to the nearest cent?

estimate: $9 × 17 = $153
exact: ≈$144.05

55. The Farnsworth Elementary School has eight classrooms with 22 students in each one and 12 classrooms with 26 students in each one. How many students attend the school?

estimate: (8 × 20) + (10 × 30) = 160 + 300
= 460 students
exact: 488 students

56. Toshihiro bought $2\frac{1}{3}$ yards of cotton fabric and $3\frac{7}{8}$ yards of wool fabric. How many yards did he buy in all?

estimate: 2 + 4 = 6 yards
exact: $6\frac{5}{24}$ yards

57. Kimberly had $29.44 in her checking account. She wrote a check for $40 and deposited a $220.06 paycheck into her account, but not in time to prevent an $18 overdraft charge. What is the new balance in her account?

estimate: $30 + $200 − $40 − $20 = $170
exact: $191.50

58. Paulette bought 2.7 pounds of grapes for $2.56. What was the cost per pound, to the nearest cent?

estimate: $3 ÷ 3 = $1 per pound
exact: ≈$0.95 per pound

59. Carter Community College received a $78,000 grant from a local computer company to help students pay tuition for computer classes. How much money could be given to each of 107 students? Round to the nearest dollar.

estimate: $80,000 ÷ 100 = $800
exact: ≈$729

Ratio and Proportion 5

A **ratio** (RAY-show) compares two quantities. You can compare two numbers, such as 8 and 4, or two measurements, such as 3 days and 12 days.

5.1 Ratios

OBJECTIVE ▶ A ratio can be written in three ways.

Writing a Ratio

The ratio of $7 to $3 can be written:

$$7 \text{ to } 3 \quad \text{or} \quad 7{:}3 \quad \text{or} \quad \frac{7}{3} \leftarrow \text{Fraction bar indicates } \textbf{to}$$

"**:**" indicates **to**

Writing a ratio as a fraction is the most common method, and the one we will use here. All three ways are read, "the ratio of 7 to 3." The word **to** separates the quantities being compared.

Writing a Ratio as a Fraction

Order is important when writing a ratio. The quantity mentioned **first** is the **numerator**. The quantity mentioned **second** is the **denominator**. For example:

The ratio of **5** to **12** is written $\dfrac{5}{12}$.

www.mathnotes.com

OBJECTIVES

1 ▶ Write ratios as fractions.

2 ▶ Solve ratio problems involving decimals or mixed numbers.

3 ▶ Solve ratio problems after converting units.

FOR EXTRA HELP

Tutorial   Tape 8   SSM, Sec. 5.1

1. Shane spent $14 on meat, $5 on milk, and $7 on fresh fruit. Write these ratios as fractions.

(a) The ratio of amount spent on fruit to amount spent on milk.

(b) The ratio of amount spent on milk to amount spent on meat.

(c) The ratio of amount spent on meat to amount spent on milk.

EXAMPLE 1   Writing a Ratio

The Anasazi, ancestors of the Pueblo Indians, built multi-story apartment towns in New Mexico about 1100 years ago. A room might measure 14 feet long, 11 feet wide, and 15 feet high.

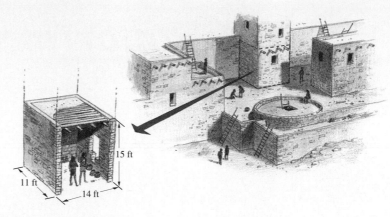

Write these ratios using the room measurements:

(a) ratio of length to width

$$\text{The ratio of } \textbf{length to width} \text{ is } \frac{14 \text{ feet}}{11 \text{ feet}} = \frac{14}{11}.$$

Numerator (mentioned first)   Denominator (mentioned second)

You can cancel common *units* just like you canceled common *factors* when writing fractions in lowest terms. (See **Section 2.4**.)

(b) ratio of width to height

$$\text{The ratio of width to height is } \frac{11 \text{ feet}}{15 \text{ feet}} = \frac{11}{15}.$$

Note
Remember, the *order* of the numbers is important in a ratio. Look for the words "ratio of $\underline{a}$ to $\underline{b}$." Write the ratio as $\frac{a}{b}$, **not** $\frac{b}{a}$. The quantity mentioned first is the numerator.

◀◀ **WORK PROBLEM 1 AT THE SIDE.**

Any ratio can be written as a fraction. Therefore, you can write a ratio in lowest terms, just as you do with any fraction.

EXAMPLE 2   Writing a Ratio in Lowest Terms

Write each ratio in lowest terms.

(a) 60 days to 20 days.

The ratio is $\frac{60}{20}$. Write this ratio in lowest terms by dividing numerator and denominator by 20.

$$\frac{60}{20} = \frac{60 \div 20}{20 \div 20} = \frac{3}{1} \quad \begin{cases} \text{Ratio in} \\ \text{lowest terms} \end{cases}$$

Note
In the fractions chapters you would have rewritten $\frac{3}{1}$ as 3. But a *ratio* compares *two* quantities, so you need to keep both parts of the ratio and write it as $\frac{3}{1}$.

ANSWERS

1. (a) $\frac{7}{5}$ (b) $\frac{5}{14}$ (c) $\frac{14}{5}$

CONTINUED ON NEXT PAGE ──

(b) 50 ounces of medicine to 120 ounces of medicine.

The ratio is $\frac{50}{120}$. Divide numerator and denominator by 10.

$$\frac{50}{120} = \frac{50 \div 10}{120 \div 10} = \frac{5}{12} \qquad \left\{ \begin{array}{l} \text{Ratio in} \\ \text{lowest terms} \end{array} \right.$$

(c) 18 people in a large van to 8 people in a small van.

$$\text{The ratio is } \frac{18}{8} = \frac{18 \div 2}{8 \div 2} = \frac{9}{4} \qquad \left\{ \begin{array}{l} \text{Ratio in} \\ \text{lowest terms} \end{array} \right.$$

Note

Although $\frac{9}{4} = 2\frac{1}{4}$, ratios are *not* written as mixed numbers. Nevertheless, in Example 2(c), the ratio $\frac{9}{4}$ does mean the large van holds $2\frac{1}{4}$ times as many people as the small van.

> **WORK PROBLEM 2 AT THE SIDE.** ▶▶

OBJECTIVE ▶ Sometimes a ratio compares two decimal numbers or two fractions. It is easier to understand if we rewrite the ratio as a ratio of two whole numbers.

E X A M P L E   3   Using Decimal Numbers in a Ratio

The price of a Sunday newspaper increased from \$1.20 to \$1.50. Find the ratio of the <u>increase in price</u> **to** the <u>original price</u>.

Approach The words <u>increase in price</u> are mentioned first, so the increase will be the numerator. How much did the price go up? Use subtraction.

$$\begin{array}{ccccc} \text{new price} & - & \text{original price} & = & \text{increase} \\ \$1.50 & - & \$1.20 & = & \$0.30 \end{array}$$

The words <u>original price</u> are mentioned second, so the original price is the denominator.

Solution The ratio of <u>increase in price</u> **to** <u>original price</u> is

$$\frac{0.30}{1.20} \begin{array}{l} \leftarrow \text{increase} \\ \leftarrow \text{original price} \end{array}$$

Now we rewrite the ratio as a ratio of whole numbers. Recall that if you multiply both the numerator and denominator of a fraction by the same number, you get an equivalent fraction. The decimals in this example are hundredths, so multiply by 100 to get whole numbers. (If the decimals are tenths, multiply by 10. If thousandths, multiply by 1000.) Then write the ratio in lowest terms.

$$\frac{0.30}{1.20} = \frac{0.30 \cdot 100}{1.20 \cdot 100} = \frac{30}{\underbrace{120}} = \frac{30 \div 30}{120 \div 30} = \frac{1}{4} \qquad \left\{ \begin{array}{l} \text{Ratio in} \\ \text{lowest terms} \end{array} \right.$$

$$\text{Ratio as two} \\ \text{whole numbers}$$

> **WORK PROBLEM 3 AT THE SIDE.** ▶▶

---

2. Write each ratio as a fraction in lowest terms.

(a) 9 hours to 12 hours

(b) 100 meters to 50 meters

(c) Write the ratio of width to length for this rectangle.

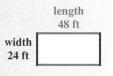

3. Write each ratio as a ratio of whole numbers in lowest terms.

(a) The price of Tamar's favorite brand of lipstick increased from \$3.75 to \$4.25. Find the ratio of the increase in price to the original price.

(b) Last week Lance worked 4.5 hours each day. This week he cut back to 3 hours each day. Find the ratio of the decrease in hours to the original number of hours.

ANSWERS

2. (a) $\frac{3}{4}$ (b) $\frac{2}{1}$ (c) $\frac{1}{2}$

3. (a) $\dfrac{0.50 \times 100}{3.75 \times 100} = \dfrac{50 \div 25}{375 \div 25} = \dfrac{2}{15}$

   (b) $\dfrac{1.5 \times 10}{4.5 \times 10} = \dfrac{15 \div 15}{45 \div 15} = \dfrac{1}{3}$

4. Write each ratio as a ratio of whole numbers in lowest terms.

(a) $3\frac{1}{2}$ to 4

(b) $5\frac{5}{8}$ pounds to $3\frac{3}{4}$ pounds

(c) $3\frac{1}{2}$ inches to $\frac{7}{8}$ inch

E X A M P L E   4   Using Mixed Numbers in a Ratio

Write each ratio as a comparison of whole numbers in lowest terms.

(a) 2 days to $2\frac{1}{4}$ days

Write the ratio as follows. Cancel the common units.

$$\frac{2 \text{ days}}{2\frac{1}{4} \text{ days}} = \frac{2}{2\frac{1}{4}}$$

Next, write 2 as $\frac{2}{1}$ and $2\frac{1}{4}$ as the improper fraction $\frac{9}{4}$.

$$\frac{2}{2\frac{1}{4}} = \frac{\frac{2}{1}}{\frac{9}{4}}$$

Now rewrite the problem using the "÷" symbol for division. Finally, invert and multiply, as you did in **Section 2.7.**

$$\frac{\frac{2}{1}}{\frac{9}{4}} = \frac{2}{1} \div \frac{9}{4} = \frac{2}{1} \cdot \frac{4}{9} = \frac{8}{9}$$

The ratio, in lowest terms, is $\frac{8}{9}$.

(b) $3\frac{1}{4}$ to $1\frac{1}{2}$

Write the ratio as $\frac{3\frac{1}{4}}{1\frac{1}{2}}$. Then write $3\frac{1}{4}$ and $1\frac{1}{2}$ as improper fractions.

$$3\frac{1}{4} = \frac{13}{4} \quad \text{and} \quad 1\frac{1}{2} = \frac{3}{2}$$

The ratio is

$$\frac{3\frac{1}{4}}{1\frac{1}{2}} = \frac{\frac{13}{4}}{\frac{3}{2}}.$$

Write as a division problem using the "÷" symbol. Invert and multiply.

$$\frac{13}{4} \div \frac{3}{2} = \frac{13}{\overset{2}{\cancel{4}}} \cdot \frac{\overset{1}{\cancel{2}}}{3} = \frac{13}{6} \quad \begin{cases} \text{Ratio in} \\ \text{lowest terms} \end{cases}$$

◄◄ WORK PROBLEM 4 AT THE SIDE.

ANSWERS

4. (a) $\frac{7}{8}$ (b) $\frac{3}{2}$ (c) $\frac{4}{1}$

OBJECTIVE ▶ When a ratio compares measurements, both measurements must be in the *same* units. For example, *feet* must be compared to *feet*, *hours* to *hours*, *pints* to *pints*, and *inches* to *inches*.

E X A M P L E 5   Ratio Applications Using Measurement

(a) Write the ratio of the length of the board on the left to the board on the right. Compare in inches.

First, express 2 feet in inches. Because 1 foot has 12 inches, 2 feet is

$$2 \cdot \textbf{12 inches} = \textbf{24 inches}.$$

The length of the board on the left is 24 inches, so the ratio of the lengths is

$$\frac{24 \text{ inches}}{30 \text{ inches}} = \frac{24}{30}.$$

Write the ratio in lowest terms.

$$\frac{24}{30} = \frac{24 \div 6}{30 \div 6} = \frac{4}{5} \quad \left\{ \begin{array}{l} \text{Ratio in} \\ \text{lowest terms} \end{array} \right.$$

The shorter board is $\frac{4}{5}$ the length of the longer board.

> **Note**
> Notice that we wrote the ratio using the smaller unit (inches are smaller than feet). Using the smaller unit will help you avoid working with fractions. If we wrote the ratio using feet, then:
>
> $$30 \text{ inches} = 2\frac{1}{2} \text{ feet}.$$
>
> So the ratio is:
>
> $$\frac{2 \text{ feet}}{2\frac{1}{2} \text{ feet}} = \frac{2}{1} \div \frac{5}{2} = \frac{2}{1} \cdot \frac{2}{5} = \frac{4}{5}.$$
>
> The ratio is the same, but it takes more steps to get the answer. Using the smaller unit is usually easier.

(b) Write the ratio of 28 days to 3 weeks. Compare in days.

First express 3 weeks in days. Because 1 week has 7 days, 3 weeks is

$$3 \cdot \textbf{7 days} = 21 \text{ days}.$$

So the ratio in days is

$$\frac{28 \text{ days}}{21 \text{ days}} = \frac{28}{21} = \frac{28 \div 7}{21 \div 7} = \frac{4}{3}. \quad \leftarrow \text{Lowest terms}$$

5. Write each ratio as a fraction in lowest terms.

(a) 9 inches to 6 feet
Compare in inches.

(b) 2 days to 8 hours
Compare in hours.

(c) 7 yards to 14 feet
Compare in feet.

(d) 3 quarts to 3 gallons
Compare in quarts.

(e) 25 minutes to 2 hours
Compare in minutes.

(f) 4 pounds to 12 ounces
Compare in ounces.

The following table will help you set up ratios that compare measurements. You will work with these measurements again in Chapter 7.

Length	Capacity (Volume)
1 foot = 12 inches	1 pint = 2 cups
1 yard = 3 feet	1 quart = 2 pints
1 mile = 5280 feet	1 gallon = 4 quarts

Weight	Time
1 pound = 16 ounces	1 week = 7 days
1 ton = 2000 pounds	1 day = 24 hours
	1 hour = 60 minutes
	1 minute = 60 seconds

◀◀ **WORK PROBLEM 5 AT THE SIDE.**

ANSWERS

5. (a) $\frac{1}{8}$ (b) $\frac{6}{1}$ (c) $\frac{3}{2}$ (d) $\frac{1}{4}$ (e) $\frac{5}{24}$ (f) $\frac{16}{3}$

5.1 Exercises

Write each ratio as a fraction in lowest terms.

1. 8 to 9 $\dfrac{8}{9}$ 　　　　**2.** 11 to 15 $\dfrac{11}{15}$ 　　　　**3.** $100 to $50 $\dfrac{2}{1}$ 　　　　**4.** 35¢ to 7¢ $\dfrac{5}{1}$

5. 30 minutes to 90 minutes $\dfrac{1}{3}$ 　　　　**6.** 9 pounds to 36 pounds $\dfrac{1}{4}$

7. 80 miles to 50 miles $\dfrac{8}{5}$ 　　　　**8.** 300 people to 450 people $\dfrac{2}{3}$

9. 6 hours to 16 hours $\dfrac{3}{8}$ 　　　　**10.** 45 books to 35 books $\dfrac{9}{7}$

Write each ratio as a ratio of whole numbers in lowest terms.

11. $4.50 to $3.50 $\dfrac{9}{7}$ 　　　　**12.** $0.08 to $0.06 $\dfrac{4}{3}$ 　　　　**13.** 15 to $2\dfrac{1}{2}$ $\dfrac{6}{1}$

14. 5 to $1\dfrac{1}{4}$ $\dfrac{4}{1}$ 　　　　**15.** $1\dfrac{1}{4}$ to $1\dfrac{1}{2}$ $\dfrac{5}{6}$ 　　　　**16.** $2\dfrac{1}{3}$ to $2\dfrac{2}{3}$ $\dfrac{7}{8}$

Write each ratio as a fraction in lowest terms. For help, use the table of measurement relationships in Example 5.

Example:	**Solution:**
40 ounces to 2 pounds Compare in ounces.	1 pound has 16 ounces; 2 • 16 ounces = 32 ounces $\dfrac{40 \text{ ounces}}{32 \text{ ounces}} = \dfrac{40 \div 8}{32 \div 8} = \dfrac{5}{4} \leftarrow$ Lowest terms

17. 4 feet to 30 inches $\dfrac{8}{5}$
Compare in inches. 　　　　**18.** 8 feet to 4 yards $\dfrac{2}{3}$
Compare in feet.

19. 5 minutes to 1 hour $\dfrac{1}{12}$
Compare in minutes. 　　　　**20.** 8 quarts to 5 pints $\dfrac{16}{5}$
Compare in pints.

21. 15 hours to 2 days $\dfrac{5}{16}$
Compare in hours. 　　　　**22.** 3 pounds to 6 ounces $\dfrac{8}{1}$
Compare in ounces.

23. 5 gallons to 5 quarts $\dfrac{4}{1}$
Compare in quarts. 　　　　**24.** 3 cups to 3 pints $\dfrac{1}{2}$
Compare in cups.

✍ Writing 　　◉ Conceptual 　　▲ Challenging 　　≈ Estimation

Solve each application problem. Write each ratio as a fraction in lowest terms.

25. A 440-pound tiger may consume 60 pounds of food in one meal. Find the ratio of the tiger's weight to the weight of its meal.

$\dfrac{22}{3}$

26. When the Concorde jet lifts off the runway it is flying at 250 miles per hour. Fifty minutes later it is cruising at a supersonic speed of 1350 miles per hour. Write the ratio of the jet's takeoff speed to its cruising speed.

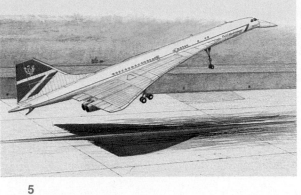

$\dfrac{5}{27}$

27. Our math class has 16 women and 20 men. What is the ratio of men to women?

$\dfrac{5}{4}$

28. Cherise sells souvenirs at baseball games. She sold 30 red hats and 40 blue hats. What is the ratio of blue hats to red hats?

$\dfrac{4}{3}$

29. The Sanchez Company made 400 washing machines. Four of them had defects. What is the ratio of defective washers to the total number of washers?

$\dfrac{1}{100}$

30. Andrew spends $500 per month on rent and $120 per month on utilities. Find the ratio of the amount spent on utilities to the amount spent on rent.

$\dfrac{6}{25}$

The table below shows the number of Americans who play various instruments. Use the information in the table to complete Exercises 31–34.

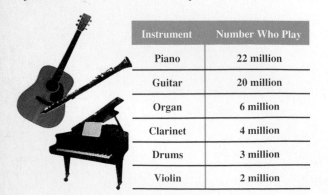

Instrument	Number Who Play
Piano	22 million
Guitar	20 million
Organ	6 million
Clarinet	4 million
Drums	3 million
Violin	2 million

31. Find the ratio of piano players to violin players.

$\dfrac{11}{1}$

32. Find the ratio of drum players to organ players.

$\dfrac{1}{2}$

33. Find the ratio of organ players to guitar players.

$\dfrac{3}{10}$

34. Find the ratio of guitar players to clarinet players.

$\dfrac{5}{1}$

35. Would you prefer that the ratio of your income to your friend's income be 1 to 3 or 3 to 1? Explain your answer.

A ratio of 3 to 1 means your income is 3 times your friend's income.

36. Amelia said that the ratio of her age to her mother's age is 5 to 3. Is this possible? Explain your answer.

It is not possible. Amelia would have to be older than her mother to have a ratio of 5 to 3.

Use the circle graph below of one family's monthly budget to complete Exercises 37–40.
Write each ratio as a fraction in lowest terms.

37. Find the ratio of taxes to transportation. $\dfrac{2}{1}$

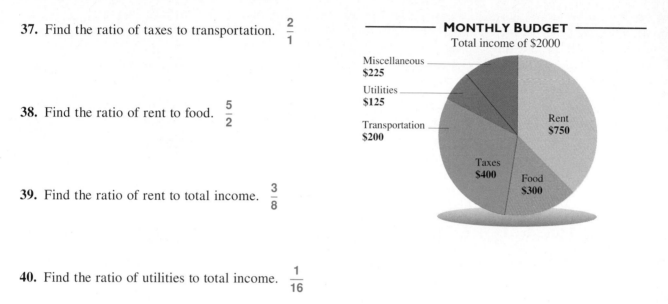

MONTHLY BUDGET
Total income of $2000

Miscellaneous **$225**
Utilities **$125**
Transportation **$200**
Rent **$750**
Taxes **$400**
Food **$300**

38. Find the ratio of rent to food. $\dfrac{5}{2}$

39. Find the ratio of rent to total income. $\dfrac{3}{8}$

40. Find the ratio of utilities to total income. $\dfrac{1}{16}$

For each figure, find the ratio of the length of the longest side to the length of the shortest
side. Write each ratio as a fraction in lowest terms.

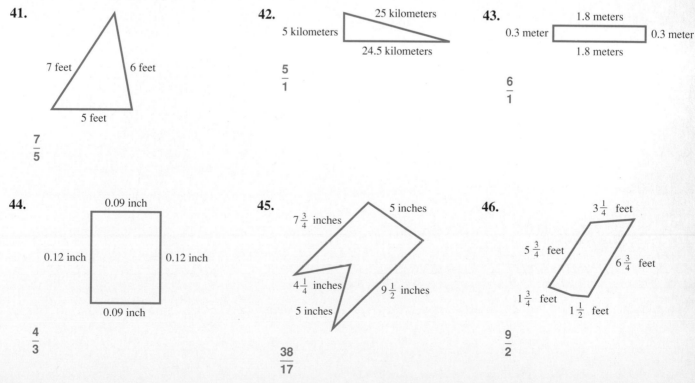

41.

7 feet   6 feet
5 feet

$\dfrac{7}{5}$

42.
25 kilometers
5 kilometers
24.5 kilometers

$\dfrac{5}{1}$

43.
1.8 meters
0.3 meter ☐ 0.3 meter
1.8 meters

$\dfrac{6}{1}$

44.
0.09 inch
0.12 inch   0.12 inch
0.09 inch

$\dfrac{4}{3}$

45.
5 inches
$7\frac{3}{4}$ inches
$4\frac{1}{4}$ inches $9\frac{1}{2}$ inches
5 inches

$\dfrac{38}{17}$

46.
$3\frac{1}{4}$ feet
$5\frac{3}{4}$ feet
$6\frac{3}{4}$ feet
$1\frac{3}{4}$ feet
$1\frac{1}{2}$ feet

$\dfrac{9}{2}$

▲ **47.** The price of oil recently went from $6.60 to $9.90 per case of 12 quarts. Find the ratio of the increase in price to the original price.

$\dfrac{1}{2}$

▲ **48.** The price of an antibiotic decreased from $8.80 to $5.60 for a bottle of 100 tablets. Find the ratio of the decrease in price to the original price.

$\dfrac{4}{11}$

▲ **49.** The first time a movie was made in Minnesota, the cast and crew spent $59\frac{1}{2}$ days filming winter scenes. The next year, another movie was filmed in $8\frac{3}{4}$ weeks. Find the ratio of the first movie's filming time to the second movie's time. Compare in weeks.

$\dfrac{34}{35}$

▲ **50.** The percheron, a large draft horse, measures about $5\frac{3}{4}$ feet at the shoulder. The prehistoric ancestor of the horse measured only $15\frac{3}{4}$ inches at the shoulder. Find the ratio of the percheron's height to its prehistoric ancestor's height. Compare in inches.

$\dfrac{92}{21}$

◉ **51.** The ratio of John's age to his sister's age is 4 to 5. One possibility is that John is 4 years old and his sister is 5 years old. Find six other possibilities that fit the 4 to 5 ratio.

Answer varies. Some possibilities are:
$$\dfrac{4}{5} = \dfrac{8}{10} = \dfrac{12}{15} = \dfrac{16}{20} = \dfrac{20}{25} = \dfrac{24}{30} = \dfrac{28}{35}.$$

◉ **52.** In this painting, what is the ratio of the length of the longest side to the length of the shortest side? What other measurements could the painting have and still maintain the same ratio?

$\dfrac{1}{1}$; **as long as the sides all have the same length, any measurement you choose will maintain the ratio.**

Review and Prepare

*Divide. Round to the nearest thousandth, if necessary. (For help, see **Section 4.6**.)*

53. $7\overline{)0.65}$ ≈0.093   **54.** $3\overline{)7.33}$ ≈2.443   **55.** $4\overline{)4.1}$ 1.025   **56.** $0.95\overline{)41.8}$ 44   🖩 **57.** $0.71\overline{)6.72}$ ≈9.465   **58.** $4.6\overline{)116.38}$ 25.3

5.2 Rates

A ratio compares two measurements with the same type of units, such as 9 feet to 12 feet (both length measurements). But many of the comparisons we make use measurements with different types of units, such as:

40 dollars **for** 8 hours (money to time)

450 miles **on** 18 gallons (distance to capacity)

This type of comparison is called a **rate.**

OBJECTIVE ▶ For example, suppose you hiked 18 miles **in** 4 hours. The **rate** at which you hiked can be written as a fraction in lowest terms.

$$\frac{18 \text{ miles}}{4 \text{ hours}} = \frac{18 \text{ miles} \div 2}{4 \text{ hours} \div 2} = \frac{9 \text{ miles}}{2 \text{ hours}} \Bigg\} \text{Lowest terms}$$

In a rate, you often find these words separating the quantities you are comparing:

in   for   on   per   from

Note
When writing a rate, always include the units. Because the units in a rate are different, they do *not* cancel.

E X A M P L E I   Write a Rate in Lowest Terms

Write each rate as a fraction in lowest terms.

(a) 5 gallons of chemical **for** $60.

$$\frac{5 \text{ gallons} \div 5}{60 \text{ dollars} \div 5} = \frac{1 \text{ gallon}}{12 \text{ dollars}}$$

(b) $1500 wages **in** 10 weeks

$$\frac{1500 \text{ dollars} \div 10}{10 \text{ weeks} \div 10} = \frac{150 \text{ dollars}}{1 \text{ week}}$$

(c) 2225 miles **on** 75 gallons of gas

$$\frac{2225 \text{ miles} \div 25}{75 \text{ gallons} \div 25} = \frac{89 \text{ miles}}{3 \text{ gallons}}$$

WORK PROBLEM I AT THE SIDE. ▶▶

OBJECTIVE ▶ When the *denominator* of a rate is 1, it is called a **unit rate.** We use unit rates frequently. For example, you earn $8.75 for *1 hour* of work. This unit rate is written:

$8.75 **per** hour   or $8.75/hour.

Use **per** or a / mark when writing unit rates. You drive 28 miles on *1 gallon* of gas. This unit rate is written 28 miles **per** gallon, or 28 miles/gallon.

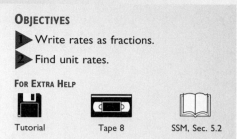
1. Write each rate as a fraction in lowest terms.

(a) $6 for 30 packages

(b) 500 miles in 10 hours

(c) 4 teachers for 90 students

(d) 1270 bushels on 30 acres

ANSWERS

1. (a) $\dfrac{\$1}{5 \text{ packages}}$ (b) $\dfrac{50 \text{ miles}}{1 \text{ hour}}$

   (c) $\dfrac{2 \text{ teachers}}{45 \text{ students}}$ (d) $\dfrac{127 \text{ bushels}}{3 \text{ acres}}$

307

2. Find each unit rate.

    (a) $4.35 for 3 pounds of cheese

    (b) 304 miles on 9.5 gallons of gas

    (c) $850 in 5 days

    (d) 24-pound turkey for 15 people

EXAMPLE 2   Finding a Unit Rate

Find each unit rate.

(a) 337.5 miles on 13.5 gallons of gas

    Write the rate as a fraction.

$$\frac{337.5 \text{ miles}}{13.5 \text{ gallons}} \leftarrow \text{Fraction bar indicates division}$$

Divide 337.5 by 13.5 to find the unit rate.

$$13.5\overline{)337.5} \quad \begin{array}{r} 2\,5. \end{array}$$

$$\frac{337.5 \text{ miles} \div 13.5}{13.5 \text{ gallons} \div 13.5} = \frac{25 \text{ miles}}{1 \text{ gallon}}$$

The unit rate is 25 miles/gallon.

(b) 549 miles in 18 hours

$$\frac{549 \text{ miles}}{18 \text{ hours}} \qquad \text{Divide } 18\overline{)549.0} \quad \begin{array}{r} 30.5 \end{array}$$

The unit rate is 30.5 miles/hour.

(c) $810 in 6 days

$$\frac{810 \text{ dollars}}{6 \text{ days}} \qquad \text{Divide } 6\overline{)810} \quad \begin{array}{r} 135 \end{array}$$

The unit rate is $135/day.

◀◀ **WORK PROBLEM 2 AT THE SIDE.**

Cost per Unit

Cost per unit is a rate that tells how much you pay for *one* item or *one* unit. Examples are $1.25 per gallon, $47 per shirt, and $2.98 per pound. When shopping, you can save money by finding the lowest cost per unit.

EXAMPLE 3   Determining the Best Buy

The local store charges the following prices for pancake syrup. Find the best buy.

Size	Price
12 ounces	$1.28
24 ounces	$1.81
36 ounces	$2.73

CONTINUED ON NEXT PAGE

ANSWERS

2. **(a)** $1.45/pound **(b)** 32 miles/gallon
   (c) $170/day **(d)** 1.6 pounds/person

Approach The best buy is the container with the *lowest* cost per unit. All the containers are measured in *ounces*, so you first need to find the *cost per ounce* for each one. Divide the price of the container by the number of ounces in it. Round to the nearest thousandth, if necessary.

Solution

Size	Cost per Unit (Rounded)
12 ounces	$\dfrac{\$1.28}{12 \text{ ounces}} \approx \$0.107$ per ounce (highest)
24 ounces	$\dfrac{\$1.81}{24 \text{ ounces}} \approx \$0.075$ per ounce (lowest)
36 ounces	$\dfrac{\$2.73}{36 \text{ ounces}} \approx \$0.076$ per ounce

The lowest cost per ounce is $0.075, so the 24-ounce container is the best buy.

Note
Earlier we rounded money amounts to the nearest hundredth (nearest cent). But when comparing unit costs, rounding to the nearest thousandth will help you see the difference between very similar unit costs. Notice that the 24-ounce and 36-ounce containers would both have rounded to $0.08 if we had rounded to hundredths.

WORK PROBLEM 3 AT THE SIDE. ▶▶

Calculator Tip: When using a calculator to find unit prices, remember that division is *not* commutative. In Example 3 you wanted to find cost per ounce. Let the *order* of the *words* help you enter the numbers in the correct order.

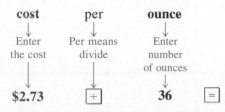

(If you entered 36 ÷ 2.73 = you'd get the number of *ounces* per *dollar*. How could you use that information to find the best buy?)

Finding the "best buy" is sometimes a complicated process. Things that affect the cost per unit can include "cents off" coupons and differences in how much use you'll get out of each unit.

3. Find the best buy (lowest cost per unit) for each purchase.

(a) 2 quarts for $3.25
3 quarts for $4.95
4 quarts for $6.48

(b) 6 cans of cola for $1.99
12 cans of cola for $3.49
24 cans of cola for $7

ANSWERS
3. (a) 4 quarts, at $1.62 per quart
(b) 12 cans, at ≈$0.291 per can

4. (a) Some batteries claim to last longer than others. If you believe these claims, which brand is the "best buy"?

Four-pack of AA size batteries for $2.79.

One AA size battery for $1.19. Lasts twice as long.

(b) Which tube of toothpaste is the better buy? You have a coupon for 85¢ off Brand C and a coupon for 20¢ off Brand D.

Brand C is $3.89 for 6 ounces

Brand D is $1.59 for 2.5 ounces

EXAMPLE 4   Solving Best Buy Applications

(a) There are many brands of liquid laundry detergent. If you feel they all do a good job of cleaning your clothes, you can base your purchase on cost per unit. But some brands are now "concentrated" so you use less detergent for each load of clothes. Which choice is the "best buy"?

To find Sudzy's unit cost, divide $3.99 by 64 ounces, not 50 ounces. You're getting as many clothes washed as if you bought 64 ounces. Similarly, to find White-O's unit cost, divide $9.89 by 256 ounces (twice 128 ounces, or 2 · 128 ounces = 256 ounces).

$$\text{Sudzy} \qquad \frac{\$3.99}{64 \text{ ounces}} \approx \$0.062 \text{ per ounce}$$

$$\text{White-O} \qquad \frac{\$9.89}{256 \text{ ounces}} \approx \$0.039 \text{ per ounce}$$

White-O has the lower cost per ounce and is the better buy. (However, if you try it and it really doesn't get out all the stains, Sudzy may be worth the extra cost.)

(b) "Cents-off" coupons also affect the best buy. Suppose you are looking at these choices for "extra strength" aspirin.

Brand X is $2.29 for 50 tablets

Brand Y is $10.75 for 200 tablets

You have a 40¢ coupon for Brand X and a 75¢ coupon for Brand Y. To find the better buy, first subtract the coupon amounts, then divide to find the lower cost per ounce.

$$\text{Brand X costs } \$2.29 - \$0.40 = \$1.89$$

$$\frac{\$1.89}{50 \text{ tablets}} \approx \$0.038 \text{ per tablet}$$

$$\text{Brand Y costs } \$10.75 - \$0.75 = \$10.00$$

$$\frac{\$10.00}{200 \text{ tablets}} = \$0.05 \text{ per tablet}$$

Brand X has the lower cost per tablet and is the better buy.

ANSWERS

4. (a) One battery that lasts twice as long (like getting two) is the better buy. The cost per unit is $0.595 per battery. The four-pack is ≈$0.698 per battery.
(b) Brand C with the 85¢ coupon is the better buy at ≈$0.507 per ounce. Brand D with the 20¢ coupon is $0.556 per ounce.

◀◀ **WORK PROBLEM 4 AT THE SIDE.**

5.2 Exercises

Write each rate as a fraction in lowest terms.

1. 10 cups for 6 people $\dfrac{5 \text{ cups}}{3 \text{ people}}$

2. $12 for 30 pens $\dfrac{\$2}{5 \text{ pens}}$

3. 15 feet in 35 seconds $\dfrac{3 \text{ feet}}{7 \text{ seconds}}$

4. 100 miles in 30 hours $\dfrac{10 \text{ miles}}{3 \text{ hours}}$

5. 14 people for 28 dresses $\dfrac{1 \text{ person}}{2 \text{ dresses}}$

6. 12 wagons for 48 horses $\dfrac{1 \text{ wagon}}{4 \text{ horses}}$

7. 25 letters in 5 minutes $\dfrac{5 \text{ letters}}{1 \text{ minute}}$

8. 68 pills for 17 people $\dfrac{4 \text{ pills}}{1 \text{ person}}$

9. $63 for 6 visits $\dfrac{\$21}{2 \text{ visits}}$

10. 25 doctors for 310 patients $\dfrac{5 \text{ doctors}}{62 \text{ patients}}$

11. 72 miles on 4 gallons $\dfrac{18 \text{ miles}}{1 \text{ gallon}}$

12. 132 miles on 8 gallons $\dfrac{33 \text{ miles}}{2 \text{ gallons}}$

Find each unit rate.

Example: $49.35 for 15 boxes

Solution: Divide $15\overline{)49.35}$ with quotient 3.29

The unit rate is **$3.29 per box**

13. $60 in 5 hours

$12 per hour or $12/hour

14. $2500 in 20 days

$125 per day or $125/day

15. 50 eggs from 10 chickens

5 eggs per chicken or 5 eggs/chicken

16. 36 children from 12 families

3 children per family or 3 children/family

17. 7.5 pounds for 6 people

1.25 pounds/person

18. 44 bushels from 8 trees

5.5 bushels/tree

19. $413.20 for 4 days

$103.30/day

20. $74.25 for 9 hours

$8.25/hour

✐ Writing   ◉ Conceptual   ▲ Challenging   ≈ Estimation

Earl kept the record shown below of the gas he bought for his car. For each entry, find the number of miles he traveled and the unit rate. Round your answers to the nearest tenth.

	Date	Odometer at Start	Odometer at End	Miles Traveled	Gallons Purchased	Miles per Gallon
21.	2/4	27,432.3	27,758.2	**325.9**	15.5	**≈21.0**
22.	2/9	27,758.2	28,058.1	**299.9**	13.4	**≈22.4**
23.	2/16	28,058.1	28,396.7	**338.6**	16.2	**≈20.9**
24.	2/20	28,396.7	28,704.5	**307.8**	13.3	**≈23.1**

Find the best buy (based on the cost per unit) for each of the following.

25. black pepper
4 ounces for $0.89
8 ounces for $2.13

4 oz for $0.89

26. shampoo
8 ounces for $0.99
12 ounces for $1.47

12 ounces for $1.47

27. cereal
15 ounces for $2.60
17 ounces for $2.89
21 ounces for $3.79

17 ounces for $2.89

28. soup
2 cans for $0.75
3 cans for $1.17
5 cans for $1.79

5 cans for $1.79

29. chunky peanut butter
12 ounces for $1.09
18 ounces for $1.41
28 ounces for $2.29
40 ounces for $3.19

18 ounces for $1.41

30. pork and beans
8 ounces for $0.37
16 ounces for $0.77
21 ounces for $0.99
31 ounces for $1.50

8 ounces for $0.37

31. Suppose you are choosing between two brands of chicken noodle soup. Brand A is $0.38 per can and Brand B is $0.48 per can. But Brand B has more chunks of chicken in it. Which soup is the better buy? Explain your choice.

You might choose Brand B because you like more chicken, so the cost per chicken chunk may actually be the same or less than Brand A.

32. A small bag of potatoes costs $0.19 per pound. A large bag costs $0.15 per pound. But there are only two people in your family, so half the large bag would probably rot before you use it up. Which bag is the better buy? Explain.

If you use only half of the larger bag, you really pay $0.30 per pound, so the smaller bag is the better buy.

Solve each application problem.

33. Makesha lost 10.5 pounds in six weeks. What was her rate of loss in pounds per week?

1.75 pounds/week

34. Enrique's taco recipe uses four pounds of meat to feed 10 people. Give the rate in pounds per person.

0.4 pounds/person

35. Russ works 7 hours to earn $85.82. What is his rate per hour?

$12.26/hour

36. Find the cost of 1 gallon of gas if 18 gallons cost $20.88.

$1.16/gallon

37. Ms. Johnson bought 150 shares of stock for $1725. Find the cost of one share.

$11.50/share

38. A company pays $6450 in dividends for the 2500 shares of its stock. Find the dividend per share.

$2.58/share

39. In 1996, Michael Johnson ran the 200-meter sprint in a record time of approximately 20 seconds (actually 19.32 seconds). Give his rate in seconds per meter and in meters per second. Use 20 seconds as the time.

0.1 second/meter or $\frac{1}{10}$ second/meter

10 meters/second

40. Sofia can clean and adjust five hearing aids in four hours. Give her rate in hearing aids per hour and in hours per hearing aid.

1.25 hearing aids/hour or $1\frac{1}{4}$ aids/hour

0.8 hour/hearing aid or $\frac{4}{5}$ hour/aid

41. The 4.6 yards of fabric needed for a dress coat cost $51.75. Find the cost of 1 yard of fabric.

$11.25/yard

42. The cost to lay 42.4 square yards of carpet is $691.12. Find the cost of 1 square yard of carpet.

$16.30/square yard

43. If you believe the claims that some batteries last longer, which is the better buy: one AAA battery for $1.79 that lasts three times as long, or an eight-pack of AAA batteries for $4.99?

    One battery for $1.79; like getting 3 batteries so $1.79 ÷ 3 ≈ $0.597 per battery

44. Which is the better buy, assuming these laundry detergents both clean equally well: 64 fluid ounces for $5.99, concentrated so you can wash twice as many loads as usual; or 150 fluid ounces for $7.29 (not concentrated).

    64 fluid ounces; 2 · 64 = 128 ounces $5.99 ÷ 128 ounces ≈ $0.047

▲ 45. Three brands of cornflakes are available. Brand G is priced at $2.39 for 10 ounces. Brand K is $3.99 for 20.3 ounces and Brand P is $3.39 for 16.5 ounces. You have a coupon for 50¢ off Brand P and a coupon for 60¢ off Brand G. Which cereal is the best buy based on cost per unit?

    Brand P with the 50¢ coupon is the best buy. $3.39 − $0.50 = $2.89 ÷ 16.5 ounces ≈ $0.175/ounce

▲ 46. Two brands of facial tissue are available. Brand K is on special at three boxes of 175 tissues each for $5. Brand S is priced at $1.29 per box of 125 tissues. You have a coupon for 20¢ off one box of Brand S and a coupon for 45¢ off one box of Brand K. How can you get the best buy on one box of tissue?

    1 box Brand K with 45¢ coupon $5 ÷ 3 ≈ $1.67 per box − 0.45 = $1.22 ÷ 175 tissues ≈ $0.007/tissue

Review and Prepare

*Multiply. Write your answers as whole or mixed numbers. (For help, see **Section 2.8**.)*

47. $4 \cdot 2\frac{3}{4}$ 11

48. $12 \cdot 5\frac{2}{3}$ 68

49. $5\frac{2}{5} \cdot 20$ 108

50. $3\frac{5}{8} \cdot 6$ $21\frac{3}{4}$

51. $1\frac{1}{6} \cdot 3$ $3\frac{1}{2}$

52. $12 \cdot 2\frac{5}{8}$ $31\frac{1}{2}$

5.3 Proportions

OBJECTIVE 1 A **proportion** (proh-POR-shun) states that two ratios (or rates) are equivalent. For example,

$$\frac{\$20}{4 \text{ hours}} = \frac{\$40}{8 \text{ hours}}$$

is a proportion that says the rate $\frac{\$20}{4 \text{ hours}}$ is equivalent to the rate $\frac{\$40}{8 \text{ hours}}$.
As the amount of money doubles, the number of hours also doubles. This proportion is read:

20 dollars **is to** 4 hours **as** 40 dollars **is to** 8 hours.

EXAMPLE 1 Writing a Proportion
Write each of the following proportions.

(a) 6 feet is to 11 feet as 18 feet is to 33 feet

$$\frac{6 \text{ feet}}{11 \text{ feet}} = \frac{18 \text{ feet}}{33 \text{ feet}} \quad \text{so} \quad \frac{6}{11} = \frac{18}{33}$$
The common units (feet) cancel and are not written.

(b) $9 is to 6 liters as $3 is to 2 liters

$$\frac{\$9}{6 \text{ liters}} = \frac{\$3}{2 \text{ liters}}$$
Units must be written.

> **WORK PROBLEM 1 AT THE SIDE.** ▶▶

OBJECTIVE 2 There are two ways to see whether a proportion is true. One way is to *write both of the ratios in lowest terms.*

EXAMPLE 2 Writing Both Ratios in Lowest Terms
Are the following proportions true?

(a) $\frac{5}{9} = \frac{18}{27}$ Write each ratio in lowest terms.

$$\frac{5}{9} \leftarrow \begin{array}{l}\text{Already in} \\ \text{lowest terms}\end{array} \qquad \frac{18 \div 9}{27 \div 9} = \frac{2}{3} \leftarrow \begin{array}{l}\text{Lowest} \\ \text{terms}\end{array}$$

Because $\frac{5}{9}$ is *not* equivalent to $\frac{2}{3}$, the proportion is *false*.

(b) $\frac{16}{12} = \frac{28}{21}$ Write each ratio in lowest terms.

$$\frac{16 \div 4}{12 \div 4} = \frac{4}{3} \quad \text{and} \quad \frac{28 \div 7}{21 \div 7} = \frac{4}{3}$$

Both ratios are equivalent to $\frac{4}{3}$, so the proportion is *true*.

> **WORK PROBLEM 2 AT THE SIDE.** ▶▶

OBJECTIVE 3 The ratios in a proportion are written as fractions, so another way to test whether they are equivalent is to use cross multiplication, as you did in **Section 2.4.**

OBJECTIVES

1 ▶ Write proportions.

2 ▶ Decide whether proportions are true or false.

3 ▶ Find cross products.

FOR EXTRA HELP

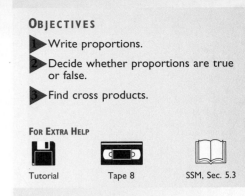

Tutorial   Tape 8   SSM, Sec. 5.3

1. Write each proportion.

   (a) $7 is to 3 cans as $28 is to 12 cans

   (b) 9 meters is to 16 meters as 18 meters is to 32 meters

   (c) 5 is to 7 as 35 is to 49

   (d) 10 is to 30 as 60 is to 180

2. Are these proportions true or false?

   (a) $\frac{6}{12} = \frac{15}{30}$

   (b) $\frac{20}{24} = \frac{3}{4}$

   (c) $\frac{25}{40} = \frac{30}{48}$

   (d) $\frac{35}{45} = \frac{12}{18}$

   (e) $\frac{21}{45} = \frac{56}{120}$

ANSWERS

1. **(a)** $\frac{\$7}{3 \text{ cans}} = \frac{\$28}{12 \text{ cans}}$ **(b)** $\frac{9}{16} = \frac{18}{32}$
   (c) $\frac{5}{7} = \frac{35}{49}$ **(d)** $\frac{10}{30} = \frac{60}{180}$

2. **(a)** true **(b)** false **(c)** true **(d)** false
   (e) true

3. Cross multiply to see whether the following proportions are true or false.

(a) $\dfrac{5}{9} = \dfrac{10}{18}$

(b) $\dfrac{32}{15} = \dfrac{16}{8}$

(c) $\dfrac{10}{17} = \dfrac{20}{34}$

(d) $\dfrac{2.4}{6} = \dfrac{5}{12}$ $\quad$ $6 \cdot 5 =$ $\quad$ $2.4 \cdot 12 =$

(e) $\dfrac{3}{4.25} = \dfrac{24}{34}$

(f) $\dfrac{1\frac{1}{6}}{2\frac{1}{3}} = \dfrac{4}{8}$

Deciding Whether a Proportion Is True or False

To see whether a proportion is true, first multiply along one diagonal, then multiply along the other diagonal, as shown here.

$$5 \cdot 4 = 20$$
$$\dfrac{2}{5} = \dfrac{4}{10}$$
$$2 \cdot 10 = 20$$

Cross products are equal

In this case the **cross products** are both 20. When cross products are *equal*, the proportion is *true*. If the cross products are *unequal*, the proportion is *false*.

The cross products test is based on rewriting both fractions with the common denominator of $5 \cdot 10$ or 50.

$$\dfrac{2 \cdot 10}{5 \cdot 10} = \dfrac{20}{50} \qquad \dfrac{4 \cdot 5}{10 \cdot 5} = \dfrac{20}{50}$$

We take a shortcut by comparing only the two numerators ($20 = 20$).

E X A M P L E   3   Using Cross Products

Use cross multiplication to see whether the following proportions are true or false.

(a) $\dfrac{3}{5} = \dfrac{12}{20}$ $\quad$ Cross multiply one way and then the other way.

$$5 \cdot 12 = 60$$
$$\dfrac{3}{5} = \dfrac{12}{20}$$
$$3 \cdot 20 = 60$$

Equal

The cross products are equal, so the proportion is *true*.

(b) $\dfrac{2\frac{1}{3}}{3\frac{1}{3}} = \dfrac{9}{16}$ $\quad$ Cross multiply.

Changed to improper fractions

$$3\frac{1}{3} \cdot 9 = \dfrac{10}{3} \cdot \dfrac{\overset{3}{\cancel{9}}}{1} = \dfrac{30}{1} = 30$$

$$\dfrac{2\frac{1}{3}}{3\frac{1}{3}} = \dfrac{9}{16}$$

$$2\frac{1}{3} \cdot 16 = \dfrac{7}{3} \cdot \dfrac{16}{1} = \dfrac{112}{3} = 37\frac{1}{3}$$

Unequal

The cross products are unequal, so the proportion is *false*.

Note
The numbers in a proportion do *not* have to be whole numbers.

◀◀ **WORK PROBLEM 3 AT THE SIDE.**

ANSWERS

3. (a) true (b) false (c) true (d) false
(e) true (f) true

5.3 Exercises

Write each proportion.

1. $9 is to 12 cans as $18 is to 24 cans

$$\frac{\$9}{12 \text{ cans}} = \frac{\$18}{24 \text{ cans}}$$

2. 28 people is to 7 cars as 16 people is to 4 cars

$$\frac{28 \text{ people}}{7 \text{ cars}} = \frac{16 \text{ people}}{4 \text{ cars}}$$

3. 200 adults is to 450 children as 4 adults is to 9 children

$$\frac{200 \text{ adults}}{450 \text{ children}} = \frac{4 \text{ adults}}{9 \text{ children}}$$

4. 150 trees is to 1 acre as 1500 trees is to 10 acres

$$\frac{150 \text{ trees}}{1 \text{ acre}} = \frac{1500 \text{ trees}}{10 \text{ acres}}$$

5. 120 feet is to 150 feet as 8 feet is to 10 feet

$$\frac{120}{150} = \frac{8}{10}$$

6. $6 is to $9 as $10 is to $15

$$\frac{6}{9} = \frac{10}{15}$$

7. 2.2 hours is to 3.3 hours as 3.2 hours is to 4.8 hours

$$\frac{2.2}{3.3} = \frac{3.2}{4.8}$$

8. 4 meters is to 4.75 meters as 6 meters is to 7.125 meters

$$\frac{4}{4.75} = \frac{6}{7.125}$$

9. $1\frac{1}{2}$ is to $4\frac{1}{2}$ as 6 is to 18
$$\frac{1\frac{1}{2}}{4\frac{1}{2}} = \frac{6}{18}$$

10. 8 is to $2\frac{2}{3}$ as 32 is to $10\frac{2}{3}$
$$\frac{8}{2\frac{2}{3}} = \frac{32}{10\frac{2}{3}}$$

Write each ratio in lowest terms in order to decide whether the following proportions are true or false.

Example: $\frac{35}{15} - \frac{21}{9}$

Solution: Write both ratios in lowest terms.

$$\frac{35 \div 5}{15 \div 5} = \frac{7}{3} \quad \text{and} \quad \frac{21 \div 3}{9 \div 3} = \frac{7}{3}$$

Both ratios are equivalent to $\frac{7}{3}$ so the proportion is **true.**

11. $\frac{6}{10} = \frac{3}{5}$ true

12. $\frac{1}{4} = \frac{9}{36}$ true

13. $\frac{5}{8} = \frac{25}{40}$ true

14. $\frac{2}{3} = \frac{20}{27}$ false

15. $\frac{150}{200} = \frac{200}{300}$ false

16. $\frac{100}{120} = \frac{75}{100}$ false

🖉 Writing ◉ Conceptual ▲ Challenging ≈ Estimation

17. $\dfrac{42}{15} = \dfrac{28}{10}$ true

18. $\dfrac{18}{16} = \dfrac{36}{32}$ true

19. $\dfrac{32}{18} = \dfrac{48}{27}$ true

20. $\dfrac{15}{48} = \dfrac{10}{24}$ false

21. $\dfrac{7}{6} = \dfrac{54}{48}$ false

22. $\dfrac{28}{21} = \dfrac{44}{33}$ true

Use cross multiplication to decide whether the following proportions are true or false. Circle the correct answer.

Example: **Solution:**

$\dfrac{10.2}{15.3} = \dfrac{4}{6}$

$15.3 \cdot 4 = \mathbf{61.2}$ ⎤
                              ⎥ Equal
$10.2 \cdot 6 = \mathbf{61.2}$ ⎦

(True) False   Cross products are equal, so proportion is true.

23. $\dfrac{2}{9} = \dfrac{6}{27}$

(True)   False

24. $\dfrac{20}{25} = \dfrac{4}{5}$

(True)   False

25. $\dfrac{20}{28} = \dfrac{12}{16}$

True   (False)

26. $\dfrac{16}{40} = \dfrac{22}{55}$

(True)   False

27. $\dfrac{110}{18} = \dfrac{160}{27}$

True   (False)

28. $\dfrac{600}{420} = \dfrac{20}{14}$

(True)   False

29. $\dfrac{3.5}{4} = \dfrac{7}{8}$

(True)   False

30. $\dfrac{36}{23} = \dfrac{9}{5.75}$

(True)   False

31. $\dfrac{18}{16} = \dfrac{2.8}{2.5}$

True   (False)

32. $\dfrac{0.26}{0.39} = \dfrac{1.3}{1.9}$

True   (False)

33. $\dfrac{6}{3\frac{2}{3}} = \dfrac{18}{11}$

(True)   False

34. $\dfrac{16}{13} = \dfrac{2}{1\frac{5}{8}}$

(True)   False

35. $\dfrac{2\frac{5}{8}}{3\frac{1}{4}} = \dfrac{21}{26}$

(True)   False

36. $\dfrac{28}{17} = \dfrac{9\frac{1}{3}}{5\frac{2}{3}}$

(True)   False

37. Suppose Jerome Walton of the Atlanta Braves had 16 hits in 50 times at bat, and Mariano Duncan of the New York Yankees was at bat 400 times and got 128 hits. Paul is trying to convince Jamie that the two men hit equally well. Show how you could use a proportion and cross products to see if Paul is correct.

$\dfrac{16 \text{ hits}}{50 \text{ at bats}} = \dfrac{128 \text{ hits}}{400 \text{ at bats}}$ $50 \cdot 128 = 6400$
$16 \cdot 400 = 6400$

Cross products are equal so the proportion is true; they hit equally well.

38. Jay worked 3.5 hours and packed 91 cartons. Craig packed 126 cartons in 5.25 hours. To see if the men worked equally fast, Barry set up this proportion:

$$\frac{3.5}{91} = \frac{126}{5.25}.$$

Explain what is wrong with Barry's proportion and write a correct one. Is the correct proportion true or false?

Left-hand ratio compares hours to cartons but right-hand ratio compares cartons to hours. Correct proportion is:

$$\frac{3.5 \text{ hours}}{91 \text{ cartons}} = \frac{5.25 \text{ hours}}{126 \text{ cartons}}.$$

Cross products are not equal so the proportion is false.

▲ *Decide whether each proportion is true or false. Circle the correct answer.*

39. $\dfrac{\frac{2}{3}}{2} = \dfrac{2.7}{8}$

     True   (False)

40. $\dfrac{3.75}{1\frac{1}{4}} = \dfrac{7.5}{2\frac{1}{2}}$

     (True)   False

Review and Prepare

Write each fraction in lowest terms. Write your answers as mixed numbers when possible.
(For help, see Sections 2.2 and 2.4.)

41. $\dfrac{32}{40}$ $\dfrac{4}{5}$

42. $\dfrac{35}{42}$ $\dfrac{5}{6}$

43. $\dfrac{60}{48}$ $1\dfrac{1}{4}$

44. $\dfrac{20}{15}$ $1\dfrac{1}{3}$

45. $\dfrac{36}{63}$ $\dfrac{4}{7}$

46. $\dfrac{30}{48}$ $\dfrac{5}{8}$

47. $\dfrac{65}{10}$ $6\dfrac{1}{2}$

48. $\dfrac{38}{8}$ $4\dfrac{3}{4}$

5.4 Solving Proportions

OBJECTIVES

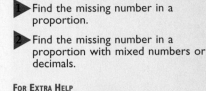

1 Find the missing number in a proportion.

2 Find the missing number in a proportion with mixed numbers or decimals.

FOR EXTRA HELP

Tutorial   Tape 9   SSM, Sec. 5.4

OBJECTIVE 1 Four numbers are used in a proportion. If any three of these numbers are known, the fourth can be found. For example, find the missing number that will make this proportion true.

$$\frac{3}{5} = \frac{x}{40}$$

The x represents the unknown number. Start by finding the cross products.

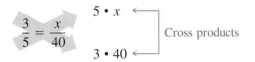

$$\begin{array}{l} 5 \cdot x \\ 3 \cdot 40 \end{array} \Bigg\} \text{Cross products}$$

To make the proportion true, the cross products must be equal.

$$5 \cdot x = 3 \cdot 40$$
$$5 \cdot x = 120$$

The equal sign says that $5 \cdot x$ and 120 are equivalent. If both $5 \cdot x$ and 120 are divided by 5, the results will still be equivalent.

$$\frac{5 \cdot x}{5} = \frac{120}{5} \quad \leftarrow \text{Divide both sides by 5.}$$

Cancel 5 in numerator and denominator. $\qquad \dfrac{\overset{1}{\cancel{5}} \cdot x}{\underset{1}{\cancel{5}}} = 24 \qquad$ Divide 120 by 5.

Multiplying by 1 does *not* change a number, so $1 \cdot x$ is the same as x.

$$\frac{x}{1} = 24$$

Dividing by 1 does *not* change a number, so $\frac{x}{1}$ is the same as x.

$$x = 24$$

The missing number in the proportion is 24. The complete proportion is shown below.

$$\frac{3}{5} = \frac{24}{40} \quad \leftarrow x \text{ is 24.}$$

Check by finding the cross products. If they are equal, you solved the problem correctly. If they are unequal, rework the problem.

$$\frac{3}{5} \bowtie \frac{24}{40} \qquad \begin{array}{l} 5 \cdot 24 = \mathbf{120} \\ 3 \cdot 40 = \mathbf{120} \end{array} \Bigg\} \text{Equal; proportion is true}$$

The cross products are equal, so the solution, $x = 24$, is correct.

> **Note**
> The solution is 24, which is the missing number in the proportion. 120 is **not** the solution; it is the cross product you get when checking the solution.

Solve a proportion for a missing number with the following steps.

Finding a Missing Number in a Proportion

Step 1 Find the cross products.

Step 2 Show that the cross products are equivalent.

Step 3 Divide both products by the number multiplied by x (the number next to x).

E X A M P L E I   Solving for a Missing Number

Find the missing number in each proportion. Round to hundredths, if necessary.

(a) $\dfrac{16}{x} = \dfrac{32}{20}$

Recall that ratios can be rewritten in lowest terms. If desired, you can do that before finding the cross products. In this example, write $\frac{32}{20}$ in lowest terms ($\frac{8}{5}$) to get $\frac{16}{x} = \frac{8}{5}$.

Step 1 $\dfrac{16}{x} \bowtie \dfrac{8}{5}$ $\begin{array}{l} x \cdot 8 \leftarrow \\[4pt] 16 \cdot 5 \leftarrow \end{array}$ Find cross products.

Step 2 $x \cdot 8 = 16 \cdot 5$ ← Show that cross products are equivalent.
$$x \cdot 8 = \underbrace{}\ \ 80$$

Step 3 $\dfrac{x \cdot \overset{1}{\cancel{8}}}{\underset{1}{\cancel{8}}} = \dfrac{80}{8}$ ← Divide both sides by 8.

$x = 10$ ← Find x. (No rounding necessary.)

Write the complete proportion and check by finding the cross products.

$\dfrac{16}{10} \bowtie \dfrac{8}{5}$ $\begin{array}{l} 10 \cdot 8 = \mathbf{80} \leftarrow \\[4pt] 16 \cdot 5 = \mathbf{80} \leftarrow \end{array}$ Equal; proportion is true

The cross products are equal, so 10 is the solution.

Note

It is not necessary to write the ratios in lowest terms before solving. However, if you do, you will have smaller numbers to work with.

(b) $\dfrac{7}{12} = \dfrac{15}{x}$

$\dfrac{7}{12} \bowtie \dfrac{15}{x}$ $\begin{array}{l} 12 \cdot 15 = 180 \leftarrow \\[4pt] 7 \cdot x \ \ \leftarrow \end{array}$ Find cross products.

Show that cross products are equivalent.

$$7 \cdot x = 180$$

CONTINUED ON NEXT PAGE

Divide both sides by 7.

$$\frac{\overset{1}{7} \cdot x}{\underset{1}{7}} = \frac{180}{7}$$

$x \approx 25.71$ (rounded to nearest hundredth)

When the division does not come out even, check for directions on how to round your answer. Divide out one more place, then round.

$$\begin{array}{r} 25.714 \\ 7\overline{)180.000} \end{array} \quad \begin{array}{l} \leftarrow \text{Divide out to thousandths.} \\ \text{Round to hundredths.} \end{array}$$

Write the complete proportion and check by finding cross products.

$$\frac{7}{12} = \frac{15}{25.71}$$

$12 \cdot 15 = \mathbf{180} \leftarrow$

$\qquad$ Very close but not equal

$7 \cdot 25.71 = \mathbf{179.97} \leftarrow$

The cross products are slightly different because you rounded the value of x. However, they are close enough to see that the problem was done correctly and 25.71 is the solution.

WORK PROBLEM I AT THE SIDE. ▶▶

OBJECTIVE ▶ The following examples show how the numbers in a proportion can be mixed numbers or decimals.

E X A M P L E 2   Using Mixed Numbers and Decimals

Find the missing number in each proportion.

(a) $\dfrac{2\frac{1}{5}}{6} = \dfrac{x}{10}$ Cross multiply.

$$\frac{2\frac{1}{5}}{6} = \frac{x}{10}$$

$6 \cdot x$

$2\frac{1}{5} \cdot 10$

Find $2\frac{1}{5} \cdot 10$.

$$2\frac{1}{5} \cdot 10 = \frac{11}{5} \cdot \frac{10}{1} = \frac{11}{\underset{1}{\cancel{5}}} \cdot \frac{\overset{2}{\cancel{10}}}{1} = \frac{22}{1} = 22$$

Changed to improper fraction

CONTINUED ON NEXT PAGE

---

1. Find the missing numbers. Round to hundredths, if necessary. Check your answers by finding cross products.

(a) $\dfrac{1}{2} = \dfrac{x}{12}$

(b) $\dfrac{6}{10} = \dfrac{15}{x}$

(c) $\dfrac{28}{x} = \dfrac{21}{9}$

(d) $\dfrac{x}{8} = \dfrac{3}{5}$

(e) $\dfrac{14}{11} = \dfrac{x}{3}$

ANSWERS

1. **(a)** $x = 6$ **(b)** $x = 25$ **(c)** $x = 12$
   (d) $x = 4.8$
   (e) $x \approx 3.82$ (rounded to nearest hundredth)

2. Find the missing numbers. Round to hundredths on the decimal problems, if necessary. Check your answers by finding cross products.

(a) $\dfrac{3\frac{1}{4}}{2} = \dfrac{x}{8}$

(b) $\dfrac{x}{3} = \dfrac{1\frac{2}{3}}{5}$

(c) $\dfrac{0.06}{x} = \dfrac{0.3}{0.4}$

(d) $\dfrac{2.2}{5} = \dfrac{13}{x}$

(e) $\dfrac{x}{6} = \dfrac{0.5}{1.2}$

(f) $\dfrac{0}{2} = \dfrac{x}{7.092}$

Show that the cross products are equivalent.

$$6 \cdot x - 22$$

Divide both sides by 6.

$$\dfrac{\overset{1}{\cancel{6}} \cdot x}{\underset{1}{\cancel{6}}} = \dfrac{22}{6}$$

Write answer as a mixed number in lowest terms.

$$x = \dfrac{22 \div 2}{6 \div 2} = \dfrac{11}{3} = 3\dfrac{2}{3}$$

Write the complete proportion and check by finding cross products.

$$\dfrac{2\frac{1}{5}}{6} = \dfrac{3\frac{2}{3}}{10}$$

$$6 \cdot 3\dfrac{2}{3} = \dfrac{\overset{2}{\cancel{6}}}{1} \cdot \dfrac{11}{\underset{1}{\cancel{3}}} = \dfrac{22}{1} = \mathbf{22}$$

$$2\dfrac{1}{5} \cdot 10 = \dfrac{11}{\underset{1}{\cancel{5}}} \cdot \dfrac{\overset{2}{\cancel{10}}}{1} = \dfrac{22}{1} = \mathbf{22}$$

Equal

The cross products are equal, so $3\frac{2}{3}$ is the correct solution.

(b) $\dfrac{1.5}{0.6} = \dfrac{2}{x}$

Show that cross products are equivalent.

$$1.5 \cdot x = 0.6 \cdot 2$$
$$1.5 \cdot x = 1.2$$

Divide both sides by 1.5.

$$\dfrac{\overset{1}{\cancel{1.5}} \cdot x}{\underset{1}{\cancel{1.5}}} = \dfrac{1.2}{1.5}$$

$$x = \dfrac{1.2}{1.5}$$

Complete the division.

$$1.5\overline{)1.20}^{\ .8}$$

So the missing number is 0.8. Check by finding cross products.

$$\dfrac{1.5}{0.6} = \dfrac{2}{0.8}$$

$$0.6 \cdot 2 = \mathbf{1.2}$$
$$1.5 \cdot 0.8 = \mathbf{1.2}$$

Equal

The cross products are equal, so 0.8 is the correct solution.

◀◀ **WORK PROBLEM 2 AT THE SIDE.**

ANSWERS

2. (a) $x = 13$ (b) $x = 1$ (c) $x = 0.08$
(d) $x \approx 29.55$ (rounded to nearest hundredth)
(e) $x = 2.5$ (f) $x = 0$

5.4 Exercises

Find the missing number in each proportion. Round your answers to hundredths, if necessary.
Check your answers by finding cross products.

Example:

$$\frac{x}{5} = \frac{6}{15}$$

Solution:

$x \cdot 15 = \underline{5 \cdot 6}$ Show that cross products
are equivalent.

$x \cdot 15 = 30$

$$\frac{x \cdot \overset{1}{\cancel{15}}}{\cancel{15}_{1}} = \frac{30}{15}$$ Divide both sides by 15.

$$x = 2$$

Check:

x is 2 → $\dfrac{2}{5} \bowtie \dfrac{6}{15}$

$5 \cdot 6 = 30$
$2 \cdot 15 = 30$ } Equal

Cross products are equal, so 2 is the correct solution.

1. $\dfrac{1}{3} = \dfrac{x}{12}$ **4**

2. $\dfrac{x}{6} = \dfrac{15}{18}$ **5**

3. $\dfrac{15}{10} = \dfrac{3}{x}$ **2**

4. $\dfrac{5}{x} = \dfrac{20}{8}$ **2**

5. $\dfrac{x}{11} = \dfrac{32}{4}$ **88**

6. $\dfrac{12}{9} = \dfrac{8}{x}$ **6**

7. $\dfrac{42}{x} = \dfrac{18}{39}$ **91**

8. $\dfrac{49}{x} = \dfrac{14}{18}$ **63**

9. $\dfrac{x}{25} = \dfrac{4}{20}$ **5**

10. $\dfrac{6}{x} = \dfrac{4}{8}$ **12**

11. $\dfrac{8}{x} = \dfrac{24}{30}$ **10**

12. $\dfrac{32}{5} = \dfrac{x}{10}$ **64**

13. $\dfrac{99}{55} = \dfrac{44}{x}$ **≈24.44**

14. $\dfrac{x}{12} = \dfrac{101}{147}$ **≈8.24**

15. $\dfrac{0.7}{9.8} = \dfrac{3.6}{x}$ **50.4**

16. $\dfrac{x}{3.6} = \dfrac{4.5}{6}$ **2.7**

17. $\dfrac{250}{24.8} = \dfrac{x}{1.75}$ **≈17.64**

18. $\dfrac{4.75}{17} = \dfrac{43}{x}$ **≈153.89**

⌨ Writing   ◉ Conceptual   ▲ Challenging   ≈ Estimation

⊙ *These proportions are* not *true. Change any* one *of the numbers in each proportion to make them true.*

19. $\dfrac{10}{4} = \dfrac{5}{3}$

$\dfrac{6.67}{4} = \dfrac{5}{3}$ or $\dfrac{10}{6} = \dfrac{5}{3}$ or $\dfrac{10}{4} = \dfrac{7.5}{3}$ or $\dfrac{10}{4} = \dfrac{5}{2}$

20. $\dfrac{6}{8} = \dfrac{24}{30}$

$\dfrac{6.4}{8} = \dfrac{24}{30}$ or $\dfrac{6}{7.5} = \dfrac{24}{30}$ or $\dfrac{6}{8} = \dfrac{22.5}{30}$ or $\dfrac{6}{8} = \dfrac{24}{32}$

Find the missing number in each proportion. Write your answers as whole or mixed numbers when possible.

21. $\dfrac{15}{1\frac{2}{3}} = \dfrac{9}{x}$ 1

22. $\dfrac{x}{\frac{3}{10}} = \dfrac{2\frac{2}{9}}{1}$ $\dfrac{2}{3}$

23. $\dfrac{2\frac{1}{3}}{1\frac{1}{2}} = \dfrac{x}{2\frac{1}{4}}$ $3\frac{1}{2}$

24. $\dfrac{1\frac{5}{6}}{x} = \dfrac{\frac{3}{14}}{\frac{6}{7}}$ $7\frac{1}{3}$

▲ *Solve these proportions two different ways. First change all the numbers to decimal form and solve. Then change all the numbers to fraction form and solve; write your answers in lowest terms.*

25. $\dfrac{\frac{1}{2}}{x} = \dfrac{2}{0.8}$ 0.2 or $\dfrac{1}{5}$

26. $\dfrac{\frac{3}{20}}{0.1} = \dfrac{0.03}{x}$ 0.02 or $\dfrac{1}{50}$

27. $\dfrac{x}{\frac{3}{50}} = \dfrac{0.15}{1\frac{4}{5}}$ 0.005 or $\dfrac{1}{200}$

28. $\dfrac{8\frac{4}{5}}{1\frac{1}{10}} = \dfrac{x}{0.4}$ 3.2 or $3\frac{1}{5}$

Review and Prepare

*Write each set of rates as a proportion and use cross multiplication to decide whether it is true or false. Circle the correct answer. (For help, see **Sections 5.2 and 5.3**.)*

29. 25 feet in 18 seconds
15 feet in 10 seconds

True   (False)

$\dfrac{25 \text{ feet}}{18 \text{ sec}} = \dfrac{15 \text{ feet}}{10 \text{ sec}}$

30. 50 children to 70 adults
15 children to 21 adults

(True)   False

$\dfrac{50 \text{ children}}{70 \text{ adults}} = \dfrac{15 \text{ children}}{21 \text{ adults}}$

31. 170 miles on 6.8 gallons
330 miles on 13.2 gallons

(True)   False

$\dfrac{170 \text{ miles}}{6.8 \text{ gallons}} = \dfrac{330 \text{ miles}}{13.2 \text{ gallons}}$

32. \$14.75 for 2 hours
\$33.25 for 4.5 hours

True   (False)

$\dfrac{\$14.75}{2 \text{ hours}} = \dfrac{\$33.25}{4.5 \text{ hours}}$

5.5 *Applications of Proportions*

OBJECTIVE ▶ Use proportions to solve application problems.

FOR EXTRA HELP

Tutorial Tape 9 SSM, Sec. 5.5

OBJECTIVE ▶ Proportions can be used to solve a wide variety of problems. Watch for problems in which you are given a ratio or rate and then asked to find part of a corresponding ratio or rate. Remember that a ratio or rate compares two quantities and often includes one of these indicator words:

in for on per from to

When setting up the proportion, use a letter to represent the unknown number. We have used the letter x, but you may use any letter you like.

E X A M P L E I **Using a Proportion**

Mike's car can go 163 **miles on** 6.4 **gallons** of gas. How far can it go on a full tank of 14 **gallons** of gas? Round to the nearest whole mile.

Approach Decide what is being compared. This example compares **miles** to **gallons**. Write the two rates described in the example. Be sure that *both* rates compare miles to gallons in the same order. In other words, miles is in both numerators and gallons is in both denominators. Use a letter to represent the missing number.

compares **miles** $\left\{ \dfrac{163 \text{ miles}}{6.4 \text{ gallons}} = \dfrac{x \text{ miles}}{14 \text{ gallons}} \right\}$ compares **miles**
to **gallons**   to **gallons**

Solution Both rates compare **miles** to **gallons**, so you can set them up as a proportion.

> **Note**
> Do **not** mix up the units in the rates.
>
>
>
> compares **miles** $\left\{ \dfrac{163 \text{ miles}}{6.4 \text{ gallons}} \quad \dfrac{14 \text{ gallons}}{x \text{ miles}} \right\}$ compares **gallons**
> to **gallons**   to **miles**
>
>
>
> These rates do **not** compare things in the same order and **cannot** be set up as a proportion.

With the proportion set up correctly, solve for the missing number.

— Matching units

$\dfrac{163 \text{ miles}}{6.4 \text{ gallons}} = \dfrac{x \text{ miles}}{14 \text{ gallons}}$

— Matching units

Ignore the units while finding the cross products and dividing both sides by 6.4.

$$6.4 \cdot x = 163 \cdot 14 \qquad \text{Show that cross products are equivalent.}$$

$$6.4 \cdot x = 2282$$

$$\frac{6.4 \cdot x}{6.4} = \frac{2282}{6.4} \qquad \text{Divide both sides by 6.4.}$$

$$x = 356.5625$$

Rounded to the nearest mile, the car can go ≈ 357 **miles** on a full tank of gas. Be sure to *include the units* in your answer.

---

1. Set up and solve a proportion for each problem.

(a) If 2 pounds of fertilizer will cover 50 square feet of garden, how many pounds are needed for 225 square feet?

(b) A U.S. map has a scale of 1 inch to 75 miles. Lake Superior is 4.75 inches long on the map. What is the lake's actual length in miles?

(c) Cough syrup is to be given at the rate of 30 milliliters for each 100 pounds of body weight. How much should be given to a 34-pound child? Round to the nearest whole milliliter.

ANSWERS

1. (a) $\dfrac{2 \text{ pounds}}{50 \text{ sq feet}} = \dfrac{x \text{ pounds}}{225 \text{ sq feet}}$
$x = 9$ pounds

(b) $\dfrac{1 \text{ inch}}{75 \text{ miles}} = \dfrac{4.75 \text{ inches}}{x \text{ miles}}$
$x = 356.25$ miles or $x \approx 356$ miles

(c) $\dfrac{30 \text{ milliliters}}{100 \text{ pounds}} = \dfrac{x \text{ milliliters}}{34 \text{ pounds}}$
$x \approx 10$ milliliters

WORK PROBLEM I AT THE SIDE. ▶▶

2. Solve each problem to find a reasonable answer. Then flip one side of your proportion to see what answer you get with an incorrect setup. Explain why the second answer is unreasonable.

(a) A survey showed that 2 out of 3 people would like to lose weight. At this rate, how many people in a group of 150 want to lose weight?

(b) In one state, 3 out of 5 college students receive financial aid. At this rate, how many of the 4500 students at Central Community College receive financial aid?

(c) An advertisement says that 9 out of 10 dentists recommend sugarless gum. If the ad is true, how many of the 60 dentists in our city would recommend sugarless gum?

EXAMPLE 2 More Proportion Applications

A newspaper report says that 7 out of 10 people surveyed watch the news on TV. At that rate, how many of the 3200 people in town would you expect to watch the news?

Approach   You are comparing people who watch the news to people surveyed. Write the two rates described in the example. Be sure that both rates make the same comparison. "People who watch the news" is mentioned first, so it should be in the numerator of *both* ratios.

Solution   Set up the two rates as a proportion and solve for the missing number.

$$\text{People who watch news} \rightarrow \frac{7}{10} = \frac{x}{3200} \leftarrow \text{People who watch news}$$
$$\text{Total group} \rightarrow \quad\quad\quad \leftarrow \text{Total group}$$
$$\text{(people surveyed)} \quad\quad\quad\quad \text{(people in town)}$$

$$10 \cdot x = 7 \cdot 3200 \quad\quad \text{Show that cross products are equivalent.}$$

$$10 \cdot x = 22{,}400$$

$$\frac{\overset{1}{\cancel{10}} \cdot x}{\underset{1}{\cancel{10}}} = \frac{22{,}400}{10} \quad\quad \text{Divide both sides by 10.}$$

$$x = 2240$$

You would expect 2240 people in town to watch the news on TV.

Note

To check the answer to an application problem, do *two* things:
1. Check that the answer is reasonable.
2. Put the answer back into the proportion and make sure the cross products are equal.

For example, if you had set up the last proportion like this:

$$\frac{7}{10} = \frac{3200}{x} \quad\quad \leftarrow \text{Incorrect setup}$$

$$7 \cdot x = 10 \cdot 3200$$

$$\frac{\overset{1}{\cancel{7}} \cdot x}{\underset{1}{\cancel{7}}} = \frac{32{,}000}{7}$$

$$x \approx 4571 \text{ people.} \leftarrow \text{Unreasonable answer}$$

This answer is unreasonable because there are only 3200 people in the town; it is not possible for 4571 people to watch the news.

Note

Always check that your answer is reasonable. If it isn't, look at the way your proportion is set up. Be sure you have matching units in the numerators and matching units in the denominators.

◀◀ **WORK PROBLEM 2 AT THE SIDE.**

ANSWERS
2. (a) 100 people (reasonable); incorrect setup gives 225 people (only 150 people in the group).
   (b) 2700 students (reasonable); incorrect setup gives 7500 students (only 4500 students at the college).
   (c) 54 dentists (reasonable); incorrect setup gives ≈67 dentists (only 60 dentists in the city).

5.5 Exercises

Set up and solve a proportion for each problem.

Example:

8 pounds of vegetables cost $5. Find the cost of 20 pounds.

Solution:

You are comparing pounds to dollars. Set up a proportion.

Matching units

$$\frac{8 \text{ pounds}}{5 \text{ dollars}} = \frac{20 \text{ pounds}}{x \text{ dollars}} \qquad \frac{8}{5} = \frac{20}{x}$$

Matching units

Ignore the units while solving for x.

$$8 \cdot x = 5 \cdot 20 \qquad \text{Show that cross products are equivalent.}$$

$$8 \cdot x = 100$$

$$\frac{\overset{1}{8} \cdot x}{\underset{1}{8}} = \frac{100}{8} \qquad \text{Divide both sides by 8.}$$

$$x = 12.5$$

The cost is $12.50. Write $ and 0 for money.

1. Caroline can sketch 4 cartoon strips in five hours. How long will it take her to sketch 18 strips?

 22.5 hours

2. The Cosmic Toads recorded 8 songs on their first album in 26 hours. How long will it take them to record 14 songs for their second album?

 45.5 hours

3. 60 newspapers cost $27. Find the cost of 16 newspapers.

 $7.20

4. 22 guitar lessons cost $330. Find the cost of 12 lessons.

 $180

5. Five pounds of grass seed cover 3500 square feet of ground. How many pounds are needed for 4900 square feet?

 7 pounds

6. Anna earns $1242.08 in 14 days. How much does she earn in 260 days?

 $23,067.20

7. Tom makes $255.75 in 5 days. How much does he make in 3 days?

 $153.45

8. If 5 ounces of a medicine must be mixed with 11 ounces of water, how many ounces of medicine would be mixed with 99 ounces of water?

 45 ounces

✍ Writing   ◉ Conceptual   ▲ Challenging   ≈ Estimation

Use the floor plan shown below to complete Exercises 9–12. On the plan, one inch represents four feet.

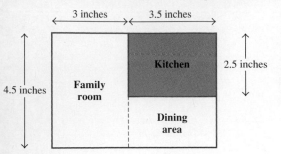

9. What is the actual length and width of the kitchen?

14 feet, 10 feet

10. What is the actual length and width of the family room?

18 feet, 12 feet

11. What is the actual length and width of the dining area?

14 feet, 8 feet

12. What is the actual length and width of the entire floor plan?

26 feet, 18 feet

13. The Cardinals' pitcher gave up 78 runs in 234 innings. At that rate, how many runs will he give up in a 9-inning game?

3 runs

14. A quarterback completed 198 out of 318 passes last season. If he tries 30 passes in today's game, how many would you expect him to complete? Round to the nearest whole number of passes.

≈19 passes

Set up a proportion to solve each problem. Check to see if your answer is reasonable. Then flip one side of your proportion to see what answer you get with an incorrect setup. Explain why the second answer is unreasonable.

15. About 7 out of 10 people entering our community college need to take a refresher math course. If we have 950 entering students, how many will probably need refresher math?

**665 students (reasonable);
≈1357 students with incorrect setup
(only 950 students in the group).**

16. In a survey, only 3 out of 100 people like their eggs poached. At that rate, how many of the 60 customers at Soon-Won's restaurant ordered poached eggs this morning? Round to the nearest whole person.

**≈2 people (reasonable);
2000 people with incorrect set up (only 60 people at the restaurant).**

17. Nearly 4 out of 5 people choose vanilla as their favorite ice cream flavor. If 238 people attend an ice cream social, how many would you expect to choose vanilla? Round to the nearest whole person.

**≈190 people (reasonable);
≈298 people with incorrect setup
(only 238 people attended).**

18. In a test of 200 sewing machines, only one had a defect. At that rate, how many of the 5600 machines shipped from the factory have defects?

**28 sewing machines (reasonable);
1,120,000 machines with incorrect setup
(only 5600 machines shipped).**

Solve each application problem.

19. The tax on a $20 item is $1. Find the tax on a $110 item.

$5.50

20. A carpenter charges $195.50 to install a deck railing 10 feet long. How much would he charge to install a deck railing 18 feet long?

$351.90

21. The stock market report says that 5 stocks went up for every 6 stocks that went down. If 750 stocks went down yesterday, how many went up?

625 stocks

22. Raoul paid $15 for 14 cans of oil. How much would 8 cans cost? Round to the nearest cent.

≈$8.57

23. Terry's boat traveled 65 miles in 3 hours. At that rate, how long will it take her to travel 100 miles? Round to the nearest tenth.

≈4.6 hours

24. The human body contains 90 pounds of water for every 100 pounds of body weight. How many pounds of water are in a child who weighs 80 pounds?

72 pounds

25. The ratio of the length of an airplane wing to its width is 8 to 1. If the length of a wing is 32.5 meters, how wide must it be? Round to the nearest hundredth.

≈4.06 meters

26. The Rosebud School District wants a student-to-teacher ratio of 19 to 1. How many teachers are needed for 1850 students? Round to the nearest whole number.

≈97 teachers

27. At 3 P.M., Coretta's shadow is 1.05 meters long. Her height is 1.68 meters. At the same time, a tree's shadow is 6.58 meters long. How tall is the tree? Round to the nearest hundredth.

≈10.53 meters

28. Refer to Exercise 27. Later in the day, the same woman had a shadow that was 2.95 meters long. How long a shadow did the tree have at that time? Round to the nearest hundredth.

≈18.49 meters

29. Can you set up a proportion to solve this problem? Explain why or why not. Jim is 25 years old and weighs 180 pounds. How much will he weigh when he is 50 years old?

You cannot solve this problem using a proportion because the ratio of age to weight is not constant. As Jim's age increases, his weight may decrease, stay the same, or increase.

30. Write your own application problem that can be solved by setting up a proportion. Also show the proportion and the steps needed to solve your problem.

Answers will vary; Exercises 1–28 are all examples of application problems.

▲ *A box of instant mashed potatoes has the list of ingredients shown below. Use this information to find the amount of each ingredient you would need to make 15 servings.*

Ingredient	For 12 Servings
Water	$3\frac{1}{2}$ cups
Margarine	6 tablespoons
Milk	$1\frac{1}{2}$ cups
Potato flakes	4 cups

31. Amount of water for 15 servings. $4\frac{3}{8}$ **cups**

32. Amount of milk for 15 servings. $1\frac{7}{8}$ **cups**

33. Amount of margarine for 15 servings.

$7\frac{1}{2}$ **tbsp**

34. Amount of potato flakes for 15 servings.

5 cups

▲ **35.** A survey of college students shows that 4 out of 5 drink coffee. Of the students who drink coffee, 1 out of 8 adds cream to it. How many of the 38,000 students at the University of Minnesota would be expected to use cream in their coffee?

3800 students

▲ **36.** Nearly 9 out of 10 adults think it's a good idea to exercise regularly. But of the ones who think it is a good idea, only 1 in 6 actually exercise at least three times a week. At this rate, how many of the 300 employees in our company exercise regularly?

45 employees

Review and Prepare

*Multiply or divide as indicated. (For help, see **Sections 4.5 and 4.6.**)*

37. 0.06×100 **6**

38. 6.1×100 **610**

39. 2.87×1000 **2870**

40. $25.8 \div 100$ **0.258**

41. $1.93 \div 100$ **0.0193**

42. $5 \div 1000$ **0.005**

KEY TERMS

5.1	**ratio**	A ratio compares two quantities. For example, the ratio of 6 apples to 11 apples is written in fraction form as $\frac{6}{11}$.
5.2	**rate**	A rate compares two measurements with different types of units. Examples are 96 dollars for 8 hours or 450 miles on 18 gallons.
	unit rate	A unit rate has 1 in the denominator.
	cost per unit	Cost per unit is a rate that tells how much you pay for one item or one unit. The lowest cost per unit is the best buy.
5.3	**proportion**	A proportion states that two ratios or rates are equivalent.
	cross products	Cross multiply to get the cross products of a proportion. If the cross products are equal, the proportion is true.

QUICK REVIEW

Concepts	Examples
5.1 Writing a Ratio A ratio compares two quantities. A ratio is usually written as a fraction with the number that is mentioned first in the numerator. The common units cancel. Check that the fraction is in lowest terms.	Write this ratio as a fraction in lowest terms. 60 ounces of medicine **to** 160 ounces of medicine $\dfrac{60 \text{ ounces}}{160 \text{ ounces}} = \dfrac{60 \div 20}{160 \div 20} = \dfrac{3}{8} \leftarrow$ Lowest terms ↑ Common units cancel
5.1 Using Mixed Numbers in a Ratio If a ratio has mixed numbers, change the mixed numbers to improper fractions. Rewrite the problem using the "÷" symbol for division. Finally, invert the divisor and multiply.	Write as a ratio of whole numbers in lowest terms. $2\frac{1}{2}$ to $3\frac{3}{4}$ $\dfrac{2\frac{1}{2}}{3\frac{3}{4}}$ Ratio in mixed numbers $= \dfrac{\frac{5}{2}}{\frac{15}{4}}$ Ratio in improper fractions $= \dfrac{5}{2} \div \dfrac{15}{4} = \dfrac{5}{2} \cdot \dfrac{4}{15}$ Invert and multiply. $= \dfrac{\overset{1}{5}}{\underset{1}{2}} \cdot \dfrac{\overset{2}{4}}{\underset{3}{15}} = \dfrac{2}{3} \leftarrow$ Ratio in lowest terms

Concepts	Examples
5.1 Using Measurements in Ratios When a ratio compares measurements, both measurements must be in the *same* units. It is usually easier to compare the measurements using the smaller unit, for example, inches instead of feet.	Write as a ratio in lowest terms. Compare in inches. 8 inches to 6 feet Because 1 foot has 12 inches, 6 feet is $6 \cdot 12$ inches $= 72$ inches. The ratio is $$\frac{8 \text{ inches}}{72 \text{ inches}} = \frac{8 \div 8}{72 \div 8} = \frac{1}{9}.$$ Common units cancel
5.2 Writing Rates A rate compares two measurements with different types of units. The units do *not* cancel, so you must write them as part of the rate.	Write the rate as a fraction in lowest terms. 475 miles in 10 hours $$\frac{475 \text{ miles} \div 5}{10 \text{ hours} \div 5} = \frac{95 \text{ miles}}{2 \text{ hours}}$$ Must write units
5.2 Finding a Unit Rate A unit rate has 1 in the denominator. To find the unit rate, divide the numerator by the denominator. Write unit rates using the word "per" or a / mark.	Write as a unit rate: $1278 in 9 days. $$\frac{\$1278}{9 \text{ days}}$$ ← Fraction bar indicates division $9)\overline{1278} = 142$ so $\dfrac{\$1278 \div 9}{9 \text{ days} \div 9} = \dfrac{\$142}{1 \text{ day}}$ Write answer as $142 per day or $142/day.
5.2 Finding the Best Buy The best buy is the item with the lowest cost per unit. Divide the price by the number of units. Round to thousandths, if necessary. Then compare to find the lowest cost per unit.	Find the best buy on cheese. 2 pounds for $2.25 3 pounds for $3.40 Find cost per unit (cost per pound). $$\frac{\$2.25}{2} = \$1.125 \text{ per pound}$$ $$\frac{\$3.40}{3} \approx \$1.133 \text{ per pound}$$ The lower cost per pound is $1.125, so 2 pounds for $2.25 is the better buy.

Concepts	Examples

5.3 Writing Proportions

A proportion states that two ratios or rates are equivalent. This proportion,

$$\frac{5}{6} = \frac{25}{30},$$

is read as "5 is to 6 as 25 is to 30."

To see whether a proportion is true or false, cross multiply one way, then cross multiply the other way. If the two products are equal, the proportion is true. If the two products are unequal, the proportion is false.

Write as a proportion: 8 is to 40 as 32 is to 160.

$$\frac{8}{40} = \frac{32}{160}$$

Cross multiply to see whether the following proportion is true or false.

$$\frac{6}{8\frac{1}{2}} = \frac{24}{34}$$

Cross multiply.

$$8\frac{1}{2} \cdot 24 = \frac{17}{2} \cdot \frac{\overset{12}{24}}{1} = \mathbf{204}$$

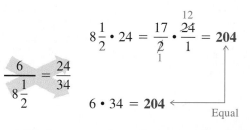

$$\frac{6}{8\frac{1}{2}} = \frac{24}{34}$$

$$6 \cdot 34 = \mathbf{204} \quad \text{Equal}$$

Cross products are equal, so the proportion is true.

5.4 Solving Proportions

Solve for a missing number in a proportion by using these steps.

Step 1 Find the cross products. (If desired, you can rewrite the ratios in lowest terms before finding the cross products.)

Step 2 Show that the cross products are equivalent.

Step 3 Divide both products by the number multiplied by x (the number next to x).

Check your answer by writing the complete proportion and finding the cross products.

Find the missing number.

$$\frac{12}{x} = \frac{6}{8}$$

$$\frac{12}{x} = \frac{3}{4} \quad \text{Lowest terms}$$

Step 1

$$\frac{12}{x} = \frac{3}{4}$$

$$x \cdot 3 \quad \\ 12 \cdot 4 \quad$$ Find cross products.

Step 2 $x \cdot 3 = \underbrace{12 \cdot 4}$ Show that cross products are equivalent.

$x \cdot 3 = \quad 48$

Step 3 $\dfrac{x \cdot \overset{1}{\cancel{3}}}{\underset{1}{\cancel{3}}} = \dfrac{48}{3}$ Divide both sides by 3.

$$x = 16$$

Check

x is 16 →

$$\frac{12}{16} = \frac{6}{8}$$

$$16 \cdot 6 = \mathbf{96} \\ 12 \cdot 8 = \mathbf{96}$$ Equal

Cross products are equal, so 16 is the correct solution.

Concepts	Examples
5.5 Applications of Proportions Decide what is being compared, for example, pounds to square feet. Write the two rates described in the problem. Be sure that *both* rates compare things in the *same order*. Use a letter, like *x*, to represent the missing number. Set up a proportion. Check that the numerators have matching units and the denominators have matching units. Solve for the missing number.	If 3 pounds of grass seed cover 450 square feet of lawn, how much seed is needed for 1500 square feet of lawn? Matching units $$\frac{3 \text{ pounds}}{450 \text{ square feet}} = \frac{x \text{ pounds}}{1500 \text{ square feet}}$$ Matching units Both sides compare pounds to square feet. Ignore the units while finding cross products. $450 \cdot x = 3 \cdot 1500$ Show that cross products are equivalent. $450 \cdot x = 4500$ $$\frac{\overset{1}{\cancel{450}} \cdot x}{\underset{1}{\cancel{450}}} = \frac{4500}{450}$$ Divide both sides by 450. $x = 10$ 10 pounds of seed are needed.

CHAPTER 5 REVIEW EXERCISES

[5.1] *Write each ratio as a fraction in lowest terms. Change to the same units when necessary, using the list of relationships in* **Section 5.1.**

1. 3 oranges to 11 oranges $\dfrac{3}{11}$

2. 19 miles to 7 miles $\dfrac{19}{7}$

3. 9 doughnuts to 6 doughnuts $\dfrac{3}{2}$

4. 90 feet to 50 feet $\dfrac{9}{5}$

5. \$2.50 to \$1.25 $\dfrac{2}{1}$

6. \$0.30 to \$0.45 $\dfrac{2}{3}$

7. $1\dfrac{2}{3}$ cups to $\dfrac{2}{3}$ cup $\dfrac{5}{2}$

8. $2\dfrac{3}{4}$ miles to $16\dfrac{1}{2}$ miles $\dfrac{1}{6}$

9. 5 hours to 100 minutes
Compare in minutes. $\dfrac{3}{1}$

10. 9 inches to 2 feet
Compare in inches. $\dfrac{3}{8}$

11. 1 ton to 1500 pounds
Compare in pounds. $\dfrac{4}{3}$

12. 8 hours to 3 days
Compare in hours. $\dfrac{1}{9}$

13. Jake sold \$350 worth of kachina figures. Ramona sold \$500 worth of pottery. What is the ratio of her sales to his?

$\dfrac{10}{7}$

14. Ms. Wei's new car gets 35 miles per gallon. Her old car got 25 miles per gallon. Find the ratio of the new car's mileage to the old car's mileage.

$\dfrac{7}{5}$

15. This fall, 60 students are taking math and 72 students are taking English. Find the ratio of math students to English students.

$\dfrac{5}{6}$

16. There are 9 players on a baseball team and 5 players on a basketball team. What is the ratio of basketball players to baseball players?

$\dfrac{5}{9}$

[5.2] *Write each rate as a fraction in lowest terms.*

17. \$88 for 8 dozen $\dfrac{\$11}{1\ dozen}$

18. 96 children in 40 families $\dfrac{12\ children}{5\ families}$

KVCC Arcadia Commons Campus Library

✍ Writing ◉ Conceptual ▲ Challenging ≈ Estimation

✏ **19.** Explain the similarities and differences be-
◉ tween a ratio and a rate. Give an example of
each.

Both compare two things. In a ratio the
common units cancel, but in a rate the units are
different and must be written. Examples are:

$$(\text{ratio}) \; \frac{5 \; \cancel{\text{feet}}}{10 \; \cancel{\text{feet}}} = \frac{1}{2} \qquad \frac{55 \; \text{miles}}{1 \; \text{hour}} \; (\text{rate})$$

✏ **20.** In your own words, explain the term "unit
◉ rate." Give three examples of unit rates.

A unit rate has 1 in the denominator. Examples are
55 miles in 1 hour, $440 in 1 week, or 30 miles on
1 gallon of gas. We usually write them using "per"
or a slash mark: 55 miles per hour, etc.

21. In his keyboarding class, Patrick can type
4 pages in 20 minutes. Give his rate in pages
per minute and minutes per page.

0.2 page/minute or $\frac{1}{5}$ page/minute

5 minutes/page

22. Elena made $24 in 3 hours. Give her earnings
in dollars per hour and hours per dollar.

$8/hour

0.125 hour/dollar or $\frac{1}{8}$ hour/dollar

Find the best buy.

23. minced onion
13 ounces for $2.29
8 ounces for $1.45
3 ounces for $0.95

13 ounces for $2.29

24. dog food; you have a coupon for $1 off on
25 pounds or more.
50 pounds for $19.95
25 pounds for $10.40
8 pounds for $3.40

25 pounds for $10.40 − $1 coupon

[5.3] *Write each proportion.*

25. 5 is to 10 as 20 is to 40.

$$\frac{5}{10} = \frac{20}{40}$$

26. 7 is to 2 as 35 is to 10.

$$\frac{7}{2} = \frac{35}{10}$$

27. $1\frac{1}{2}$ is to 6 as $2\frac{1}{4}$ is to 9.

$$\frac{1\frac{1}{2}}{6} = \frac{2\frac{1}{4}}{9}$$

*Use the method of writing in lowest terms or cross multiplication to decide whether the
following proportions are true or false.*

28. $\frac{6}{10} = \frac{9}{15}$ true

29. $\frac{16}{48} = \frac{9}{36}$ false

30. $\frac{47}{10} = \frac{98}{20}$ false

31. $\frac{64}{36} = \frac{96}{54}$ true

32. $\frac{1.5}{2.4} = \frac{2}{3.2}$ true

33. $\frac{3\frac{1}{2}}{2\frac{1}{3}} = \frac{6}{4}$ true

[5.4] *Find the missing number in each proportion. Round to hundredths, if necessary.*

34. $\frac{4}{42} = \frac{150}{x}$ 1575

35. $\frac{16}{x} = \frac{12}{15}$ 20

36. $\frac{100}{14} = \frac{x}{56}$ 400

37. $\frac{5}{8} = \frac{x}{20}$ 12.5

38. $\frac{x}{24} = \frac{11}{18}$ ≈14.67

39. $\frac{7}{x} = \frac{18}{21}$ ≈8.17

40. $\frac{x}{3.6} = \frac{9.8}{0.7}$ 50.4

🖩 **41.** $\frac{13.5}{1.7} = \frac{4.5}{x}$ ≈0.57

42. $\frac{0.82}{1.89} = \frac{x}{5.7}$ ≈2.47

[5.5] *Set up and solve a proportion for each application problem.*

43. The ratio of cats to dogs at the animal shelter is 3 to 5. If there are 45 dogs, how many cats are there?

27 cats

44. Danielle had 8 hits in 28 times at bat during last week's games. If she continues to hit at the same rate, how many hits will she get in 161 times at bat?

46 hits

45. If 3.5 pounds of steak cost $13.79, what will 5.6 pounds cost? Round to the nearest cent.

≈$22.06

46. About 4 out of 10 students are expected to vote in campus elections. There are 8247 students. How many are expected to vote? Round to the nearest whole number.

≈3299 students

47. The scale on Brian's model railroad is 1 inch to 16 feet. One of the scale model boxcars is 4.25 inches long. What is the length of a real boxcar in feet?

68 feet

48. In the hospital pharmacy, Michiko sees that a certain medicine is to be given at the rate of 3.5 milligrams for every 50 pounds of body weight. How much medicine should be given to a patient who weighs 210 pounds?

14.7 milligrams

49. Damien earns $91 for 14 hours of part-time work at the convenience store. How long must he work to earn $520 to pay his tuition?

80 hours

50. Marvette makes necklaces to sell at a local gift shop. She made 2 dozen necklaces in $16\frac{1}{2}$ hours. How long will it take her to make 40 necklaces?

$27\frac{1}{2}$ hours or 27.5 hours

MIXED REVIEW EXERCISES

Find the missing number in each proportion. Round to hundredths, if necessary.

51. $\dfrac{x}{45} = \dfrac{70}{30}$ **105**

52. $\dfrac{x}{52} = \dfrac{0}{20}$ **0**

53. $\dfrac{64}{10} = \dfrac{x}{20}$ **128**

54. $\dfrac{15}{x} = \dfrac{65}{100}$ **≈23.08**

55. $\dfrac{7.8}{3.9} = \dfrac{13}{x}$ **6.5**

56. $\dfrac{34.1}{x} = \dfrac{0.77}{2.65}$ **≈117.36**

Use cross multiplication to decide whether the following proportions are true or false. Circle the correct answer.

57. $\dfrac{55}{18} = \dfrac{80}{27}$

True   (False)

58. $\dfrac{5.6}{0.6} = \dfrac{18}{1.94}$

True   (False)

59. $\dfrac{\frac{1}{5}}{2} = \dfrac{1\frac{1}{6}}{11\frac{2}{3}}$

(True)   False

Write each ratio as a fraction in lowest terms. Change to the same units when necessary.

60. 4 dollars to 10 quarters
Compare in quarters.

$\dfrac{8}{5}$

61. $4\frac{1}{8}$ inches to 10 inches

$\dfrac{33}{80}$

62. 10 yards to 8 feet
Compare in feet.

$\dfrac{15}{4}$

63. $3.60 to $0.90 $\dfrac{4}{1}$

64. 12 eggs to 15 eggs $\dfrac{4}{5}$

65. 37 meters to 7 meters $\dfrac{37}{7}$

66. 3 pints to 4 quarts $\dfrac{3}{8}$
Compare in pints.

67. 15 minutes to 3 hours $\dfrac{1}{12}$
Compare in minutes.

68. $4\dfrac{1}{2}$ miles to $1\dfrac{3}{10}$ miles $\dfrac{45}{13}$

69. Nearly 7 out of 8 fans buy something to drink at the ballpark. How many of the 28,500 fans at today's game would be expected to buy a beverage? Round to the nearest hundred fans.

≈24,900 fans

70. Emily spent $150 on car repairs and $400 on car insurance. What is the ratio of amount spent on insurance to amount spent on repairs?

$\dfrac{8}{3}$

71. Antonio is choosing among three packages of plastic wrap. Is the best buy 25 feet for $0.78; 75 feet for $1.99; or 100 feet for $2.59? He has a coupon for 50¢ off that is good for either of the larger two packages.

75 feet for $1.99 − $0.50 coupon

72. On this scale drawing of a backyard patio, 1 inch represents 6 feet. If the patio measures 2.75 inches long on the drawing, what will the actual length of the patio be when it is built?

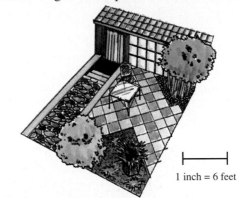

1 inch = 6 feet

16.5 feet

73. An antibiotic is to be given at the rate of $1\frac{1}{2}$ teaspoons for every 24 pounds of body weight. How much should be given to an infant who weighs 8 pounds?

$\dfrac{1}{2}$ teaspoon or 0.5 teaspoon

74. Charles made 251 points during 169 minutes of playing time last year. If he plays 14 minutes in tonight's game, how many points would you expect him to make? Round to the nearest whole number.

≈21 points

75. Refer to Exercise 73. Explain each step you took in solving the problem. Be sure to tell how you decided which way to set up the proportion and how you checked your answer.

Set up the proportion to compare teaspoons to pounds on both sides.

$$\dfrac{1.5 \text{ tsp}}{24 \text{ pounds}} = \dfrac{x \text{ tsp}}{8 \text{ pounds}}$$

Show that cross products are equal.

$$24 \cdot x = 1.5 \cdot 8$$

Divide both sides by 24.

$$\dfrac{\overset{1}{\cancel{24}} \cdot x}{\underset{1}{\cancel{24}}} = \dfrac{12}{24} \quad \text{so } x = \dfrac{1}{2} \text{ tsp or } 0.5 \text{ tsp}$$

76. A lawn mower uses 0.8 gallon of gas every 3 hours. The gas tank holds 2 gallons. How long can the mower run on a full tank?

7.5 hours or $7\dfrac{1}{2}$ hours

CHAPTER 5 TEST

Write each rate or ratio as a fraction in lowest terms. Change to the same units when necessary.

1. 16 fish to 20 fish

1. $\dfrac{4}{5}$ _____

2. 300 miles on 15 gallons

2. $\dfrac{20 \text{ miles}}{1 \text{ gallon}}$ _____

3. $15 for 75 minutes

3. $\dfrac{\$1}{5 \text{ minutes}}$ _____

4. The little theater has 320 seats. The auditorium has 1200 seats. Find the ratio of auditorium seats to theater seats.

4. $\dfrac{15}{4}$ _____

5. 3 quarts to 60 gallons
Compare in quarts.

5. $\dfrac{1}{80}$ _____

6. 3 hours to 40 minutes
Compare in minutes.

6. $\dfrac{9}{2}$ _____

7. Find the best buy on spaghetti sauce. You have a coupon for 75¢ off Brand X and a coupon for 25¢ off Brand Y.
28 ounces of Brand X for $3.89
18 ounces of Brand Y for $1.89
13 ounces of Brand Z for $1.29

7. 18 ounces for
$1.89 – $0.25 coupon _____

8. Suppose the ratio of your income last year to your income this year is 3 to 2. Explain what this means. Give an example of the dollars earned last year and this year that fits the 3 to 2 ratio.

8. You earned less this year.

An example is:

Last year → $15,000 $\quad 3$
This year → $10,000 $\quad \overline{2}$ _____

Decide whether the following proportions are true or false.

9. $\dfrac{6}{14} = \dfrac{18}{45}$

10. $\dfrac{8.4}{2.8} = \dfrac{2.1}{0.7}$

9. False _____

10. True _____

11. 25

12. ≈2.67

13. 325

14. $10\frac{1}{2}$

15. 576 words

16. 6.4 hours

17. ≈87 students

18. No, 4875 cannot be correct because there are only 650 students in the whole school.

19. ≈23.8 grams

20. 60 feet

Find the missing number in each proportion. Round to hundredths, if necessary.

11. $\dfrac{5}{9} = \dfrac{x}{45}$

12. $\dfrac{3}{1} = \dfrac{8}{x}$

13. $\dfrac{x}{20} = \dfrac{6.5}{0.4}$

14. $\dfrac{2\frac{1}{3}}{x} = \dfrac{\frac{8}{9}}{4}$

Set up and solve a proportion for each application problem.

15. Pedro types 240 words in 5 minutes. How many words can he type in 12 minutes?

16. A boat travels 75 miles in 4 hours. At that rate, how long will it take to travel 120 miles?

17. About 2 out of every 15 people are left-handed. How many of the 650 students in our school would you expect to be left-handed? Round to the nearest whole number.

18. A student set up the proportion for Exercise 17 this way and arrived at an answer of 4875.

$$\frac{2}{15} = \frac{650}{x} \qquad \text{Check:} \quad \frac{2}{15} \bowtie \frac{650}{4875} \qquad \begin{array}{l} 15 \cdot 650 = 9750 \\[6pt] 2 \cdot 4875 = 9750 \end{array}$$

Because the cross products are equal, the student said the answer is correct. Is the student right? Explain why or why not.

19. A medication is given at the rate of 8.2 grams for every 50 pounds of body weight. How much should be given to a 145-pound person? Round to the nearest tenth.

20. On a scale model, 1 inch represents 8 feet. If a building in the model is 7.5 inches tall, what is the actual height of the building in feet?

Name the digit that has the given place value.

1. 216,475,038
thousands **5**
tens **3**
millions **6**

2. 340.6915
hundredths **9**
ones **0**
ten-thousandths **5**

Round each number as indicated.

3. 9903 to the nearest hundred.

≈ **9900**

4. 617.0519 to the nearest tenth.

≈ **617.1**

5. $99.81 to the nearest dollar.

≈ **$100**

6. $3.0555 to the nearest cent.

≈ **$3.06**

≈*Round the numbers in each problem so there is only one non-zero digit. Then add, subtract, multiply, or divide the rounded numbers, as indicated, to estimate the answer. Finally, solve for the exact answer.*

7. estimate   exact

$$\begin{array}{r} 30 \\ 5000 \\ +\ \ 400 \\ \hline 5430 \end{array} \quad \begin{array}{r} 28 \\ 5206 \\ +\ \ 351 \\ \hline 5585 \end{array}$$

8. estimate   exact

$$\begin{array}{r} 60 \\ -\ \ 6 \\ \hline 54 \end{array} \quad \begin{array}{r} 63.1 \\ -\ \ 5.692 \\ \hline 57.408 \end{array}$$

9. estimate   exact

$$\begin{array}{r} 5000 \\ \times\ \ 800 \\ \hline 4{,}000{,}000 \end{array} \quad \begin{array}{r} 4716 \\ \times\ \ 804 \\ \hline 3{,}791{,}664 \end{array}$$

10. estimate exact

$$\begin{array}{r} 1 \\ \times\ 18 \\ \hline 18 \end{array} \quad \begin{array}{r} 0.982 \\ \times\ 17.8 \\ \hline 17.4796 \end{array}$$

11. estimate   exact

$$50\overline{)50{,}000} \quad \begin{array}{c} 907 \\ 53\overline{)48{,}071} \end{array}$$

$$\begin{array}{c} 1000 \end{array}$$

12. estimate exact

$$5\overline{)2000} \quad 4.5\overline{)1638}$$

$$\begin{array}{c} 400 \end{array} \quad \begin{array}{c} 364 \end{array}$$

13. estimate   exact

$$\frac{\ \ }{2} \cdot \frac{\ \ }{4} = \frac{\ \ }{8} \quad 1\frac{5}{6} \cdot 3\frac{3}{5} \quad 6\frac{3}{5}$$

14. estimate   exact

$$\frac{\ \ }{5} \div \frac{\ \ }{1} = \frac{\ \ }{5} \quad 5\frac{1}{4} \div \frac{7}{8} \quad 6$$

15. estimate   exact

$$\frac{\ \ }{3} - \frac{\ \ }{2} = \frac{\ \ }{1} \quad 2\frac{4}{5} - 1\frac{5}{6} \quad \frac{29}{30}$$

16. estimate   exact

$$\frac{\ \ }{3} + \frac{\ \ }{11} = \frac{\ \ }{14} \quad 2\frac{9}{10} + 10\frac{1}{2} \quad 13\frac{2}{5}$$

Add, subtract, multiply, or divide as indicated.

17. 988 + 373,422 + 6

374,416

18. 30 − 0.66

29.34

19. Write your answer using R for the remainder.

$$\begin{array}{c} 610\ \textbf{R27} \\ 33\overline{)20{,}157} \end{array}$$

20. $(1.9)(0.004)$

0.0076

21. $3020 - 708$

2312

22. $0.401 + 62.98 + 5$

68.381

23. $1.39 \div 0.025$

55.6

24. $(6392)(5609)$

35,852,728

Simplify by using the order of operations.

25. $36 + 18 \div 6$

39

26. $8 \div 4 + (10 - 3^2) \cdot 4^2$

18

27. $88 \div \sqrt{121} \cdot 2^3$

64

28. $(16.2 - 5.85) - 2.35 \cdot 4$

0.95

29. Write 0.0105 in words.

one hundred five ten-thousandths

30. Write sixty and seventy-one thousandths in numbers.

60.071

Write each fraction or mixed number as a decimal. Round to the nearest thousandth, if necessary.

31. $\dfrac{5}{16}$ 0.313

32. $4\dfrac{7}{9}$ 4.778

Arrange in order from smallest to largest.

33. 0.0711, 0.7, 0.707, 0.07

0.07, 0.0711, 0.7, 0.707

34. $\dfrac{3}{8}, \dfrac{7}{20}, 0.305, \dfrac{1}{3}$

$0.305, \dfrac{1}{3}, \dfrac{7}{20}, \dfrac{3}{8}$

Write each rate or ratio as a fraction in lowest terms. Change to the same units when necessary.

35. 20 cars to 5 cars $\dfrac{4}{1}$

36. $39 for 6 hours $\dfrac{\$13}{2 \text{ hours}}$

37. 20 minutes to 4 hours
Compare in minutes. $\dfrac{1}{12}$

38. 8 inches to 2 feet
Compare in inches. $\dfrac{1}{3}$

39. Ray is 25 years old. His father is 55 years old. Find the ratio of the father's age to Ray's age.

$\dfrac{11}{5}$

40. Find the best buy on instant mashed potatoes. You have a coupon for 50¢ off on either the 36-serving or 48-serving box.
a box that makes 20 servings for $1.59
a box that makes 36 servings for $3.24
a box that makes 48 servings for $4.99

36 servings for $3.24 − $0.50 coupon

Find the missing number in each proportion. Round your answers to the nearest hundredth, if necessary.

41. $\dfrac{9}{12} = \dfrac{x}{28}$ **21**

42. $\dfrac{7}{12} = \dfrac{10}{x}$ **≈17.14**

43. $\dfrac{x}{\frac{3}{4}} = \dfrac{2\frac{1}{2}}{\frac{1}{6}}$ **$11\frac{1}{4}$**

44. $\dfrac{6.7}{x} = \dfrac{62.8}{9.15}$ **≈0.98**

Solve each application problem.

45. The honor society has a goal of collecting 1500 pounds of food to fill Thanksgiving baskets. So far they've collected $\frac{5}{6}$ of their goal. How many more pounds do they need?

250 pounds

46. Tara has a photo that is 10 centimeters wide by 15 centimeters long. If the photo is enlarged to a length of 40 centimeters, find the new width, to the nearest tenth.

≈26.7 centimeters

Use the circle graph below of one college's enrollment to complete Exercises 47–50. Where indicated, first round the numbers so there is only one non-zero digit and estimate the answer. Then find the exact answer.

─────── **COLLEGE ENROLLMENT** ───────

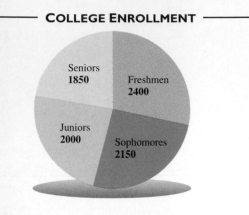

Seniors 1850
Freshmen 2400
Juniors 2000
Sophomores 2150

47. Find the total enrollment at the college.

estimate: 2000 + 2000 + 2000 + 2000
    = 8000 students
exact: 8400 students

48. The college has budgeted $186,400 for freshman orientation. How much is spent on each freshman, to the nearest dollar?

estimate: $200,000 ÷ 2000 = $100
exact: ≈$78

49. Write the ratio of freshmen to the total enrollment as a fraction in lowest terms.

$\frac{2}{7}$

50. The college collects a $3.75 technology fee from each student to support the computer lab. What total amount is collected?

estimate: $4 · 8000 = $32,000
exact: $31,500

51. The distance around Dunning Pond is $1\frac{1}{10}$ miles. Norma ran around the pond 4 times in the morning and $2\frac{1}{2}$ times in the afternoon. How far did she run in all?

estimate: 4 + 3 = 7; 7 · 1 = 7 miles
exact: $7\frac{3}{20}$ miles

52. Rodney bought 49.8 gallons of gas for his truck while driving 896.5 miles on a vacation. How many miles per gallon did he get, rounded to the nearest tenth?

estimate: 900 miles ÷ 50 gallons
    = 18 miles per gallon
exact: ≈18.0 miles per gallon

53. In a survey, 5 out of 6 apartment residents said they are sometimes bothered by noise from their neighbors. How many of the 240 residents at Harris Towers would you expect to be bothered by noise?

200 residents

54. The directions on a can of plant food call for $\frac{1}{2}$ teaspoon in two quarts of water. How much plant food is needed for five quarts?

$1\frac{1}{4}$ teaspoons

Percent 6

6.1 Basics of Percent

Notice that the figure below has one hundred squares of equal size. Eleven of the squares are shaded. The shaded portion is $\frac{11}{100}$, or 0.11, of the total figure.

www.mathnotes.com

OBJECTIVES

1. Learn the meaning of percent.
2. Write percents as decimals.
3. Write decimals as percents.
4. Understand 100% and 50%.

FOR EXTRA HELP

Tutorial   Tape 9   SSM, Sec. 6.1

The shaded portion is also 11% of the total, or "eleven parts out of 100 parts." Read **11%** as "eleven percent."

OBJECTIVE 1 As shown above, a percent is a ratio with a denominator of 100.

> **The Meaning of Percent**
>
> **Percent** (per-SENT) means *per one hundred*. The "%" sign is used to show the number of parts out of one hundred parts.

E X A M P L E 1   Understanding Percent

(a) If 43 out of 100 students are men, then 43 per (out of) 100 or $\frac{43}{100}$ or **43%** of the students are men.

(b) If a person pays a tax of $7 on every $100 of purchases, then the tax rate is $7 per $100. The ratio is $\frac{7}{100}$ and the percent of tax is 7%.

WORK PROBLEM 1 AT THE SIDE. ▶▶

OBJECTIVE 2 If 8% means 8 parts out of 100 parts or $\frac{8}{100}$, then $p\%$ means p parts out of 100 parts or $\frac{p}{100}$. Because $\frac{p}{100}$ is another way to write the division $p \div 100$, we have

$$p\% = \frac{p}{100} = p \div 100.$$

1. Write as percents.

(a) In a group of 100 people, 63 are unmarried. What percent are unmarried?

(b) The tax is $14 per $100. What percent is this?

(c) Out of 100 students, 36 are attending school full time. What percent attend full time?

ANSWERS

1. (a) 63% **(b)** 14% **(c)** 36%

2. Write as a decimal.

(a) 53%

(b) 27%

(c) 38.6%

(d) 150%

Write a Percent as a Decimal

$$p\% = \frac{p}{100} \quad or \quad p\% = p \div 100$$

As a fraction   As a decimal

E X A M P L E   2   Writing a Percent as a Decimal

Write each percent as a decimal.

(a) 47%

Because $p\% = p \div 100$,

$$47\% = 47 \div 100 = 0.47 \quad \text{Decimal form}$$

(b) 76% $76\% = 76 \div 100 = 0.76$ Decimal form

(c) 28.2% $28.2\% = 28.2 \div 100 = 0.282$ Decimal form

(d) 100% $100\% = 100 \div 100 = 1.00$ Decimal form

Note

In Example 2(d) notice 100% is 1.00, or 1, which is a whole number. Whenever you have a percent that is 100% or higher, the equivalent decimal number will be a number of 1 or higher.

◄◄ **WORK PROBLEM 2 AT THE SIDE.**

The resulting answers in Example 2 suggest the following rule for writing a percent as a decimal.

Writing a Percent as a Decimal

Step 1 Drop the percent sign.

Step 2 Divide by 100.

Note

A quick way to divide a number by 100 is to move the decimal point two places to the left.

E X A M P L E   3   Changing to a Decimal by Moving the Decimal Point

Write each percent as a decimal by moving the decimal point two places to the left.

(a) 17%

$$17\% = 17.\% \quad \text{Decimal point starts at far right side.}$$
$$0.17 \leftarrow \text{Percent sign is dropped. (Step 1)}$$
Decimal point is moved two places to the left. (Step 2)
$$17\% = 0.17$$

(b) 160%

$$160\% = 160.\% = 1.60 \text{ or } 1.6 \quad \text{Decimal starts at far right side.}$$
(1.60 is equivalent to 1.6.)

CONTINUED ON NEXT PAGE

ANSWERS

2. (a) 0.53 (b) 0.27 (c) 0.386 (d) 1.5

(c) 4.9%

0.049% 0 is attached so the decimal point can be moved two places to the left.

$4.9\% = 0.049$

(d) 0.6%

$0.6\% = 0.006$ 0 is attached so the decimal point can be moved.

> **Note**
> Look at Example 3(d) where 0.6% is less than 1%. Because 1% is equivalent to 0.01 or $\frac{1}{100}$, any fraction of a percent smaller than 1% is less than 0.01.

WORK PROBLEM 3 AT THE SIDE. ▶▶

OBJECTIVE 3 You can write a decimal as a percent. For example, the decimal 0.78 is the same as the fraction

$$\frac{78}{100}.$$

This fraction means 78 of 100 parts, or 78%. The following steps give the same result.

Writing a Decimal as a Percent

Step 1 Multiply by 100.

Step 2 Attach a percent sign.

> **Note**
> A quick way to multiply a number by 100 is to move the decimal point two places to the right.
>
> Decimal $\xleftarrow{\text{Divide by 100} \atop \text{Move 2 places left}}$ Percent
>
> Decimal $\xrightarrow{\text{Multiply by 100} \atop \text{Move 2 places right}}$ Percent

E X A M P L E 4 **Changing to Percent by Moving the Decimal Point**

Write each decimal as a percent.

(a) 0.21

0.21% ← Percent sign is attached. (Step 1)

 Decimal point is moved two places to the right. (Step 2)

$0.21 = 21\%$

 Decimal point is not written with whole number percents.

(b) $0.529 = 52.9\%$

(c) $1.92 = 192\%$

CONTINUED ON NEXT PAGE

---

3. Write each percent as a decimal.

(a) 88%

(b) 4%

(c) 21.6%

(d) 0.8%

ANSWERS

3. (a) 0.88 **(b)** 0.04 **(c)** 0.216 **(d)** 0.008

4. Write as a percent.

(a) 0.95   (b) 0.18

(c) 0.09   (d) 0.617

(e) 0.834   (f) 5.34

(g) 2.8   (h) 4

5. Fill in the blanks.

(a) 100% of $3.95 is _____.

(b) 100% of 3000 students is _____.

(c) 100% of 7 pages is _____.

(d) 100% of 305 miles is _____.

(e) 100% of $10\frac{1}{2}$ hours is _____.

6. Fill in the blanks.

(a) 50% of $10 is _____.

(b) 50% of 36 cookies is _____.

(c) 50% of 6000 women is _____.

(d) 50% of 8 hours is _____.

(e) 50% of $2.50 is _____.

(d) 2.5

     2.50%   0 is attached so the decimal point can be moved two places to the right.

     2.5 = 250%

(e) 3

     3. = 3.00% so 3 = 300%

Note

Look at Examples 4(c), 4(d), and 4(e) where 1.92, 2.5, and 3 are greater than 1. Because the number 1 is equivalent to 100%, all numbers greater than 1 will be 100% or larger.

◀◀ **WORK PROBLEM 4 AT THE SIDE.**

OBJECTIVE 4 When working with percents, it is helpful to have several reference points. 100% and 50% are two helpful reference points.

100% means 100 parts out of 100 parts. That's *all* of the parts. If you pay 100% of a $45 dentist bill, you pay $45 (*all* of it).

E X A M P L E 5   **Finding 100% of a Number**

Fill in the blanks.

(a) 100% of $34 is _____.
100% is *all* of the money.
So, 100% of $34 is ___$34___.

(b) 100% of 4 cats is _____.
100% is *all* of the cats.
So, 100% of 4 cats is ___4 cats___.

◀◀ **WORK PROBLEM 5 AT THE SIDE.**

50% means 50 parts out of 100 parts, which is *half* of the parts $(\frac{50}{100} = \frac{1}{2})$. 50% of $12 is $6 (*half* of the money).

E X A M P L E 6   **Finding 50% of a Number**

Fill in the blanks.

(a) 50% of $20 is _____.
50% is *half* of the money.
So, 50% of $20 is ___$10___.

(b) 50% of 280 miles is _____.
50% is *half* of the miles.
So, 50% of 280 miles is ___140 miles___.

◀◀ **WORK PROBLEM 6 AT THE SIDE.**

ANSWERS

4. (a) 95% **(b)** 18% **(c)** 9% **(d)** 61.7%
   (e) 83.4% **(f)** 534% **(g)** 280%
   (h) 400%

5. (a) $3.95 **(b)** 3000 students
   (c) 7 pages **(d)** 305 miles
   (e) $10\frac{1}{2}$ hours

6. (a) $5 **(b)** 18 cookies **(c)** 3000 women
   (d) 4 hours **(e)** $1.25

6.1 Exercises

Write each percent as a decimal.

> **Examples:** 42%   1.4%
>
> **Solutions:** 0 42. ← Percent sign is dropped.   0 01.4
>   ‿   ‿
>   └── Decimal point is moved   └─ Two places left
>   two places to the left.
>
> 42% = **0.42**   1.4% = **0.014**

1. 25%   **0.25**

2. 35%   **0.35**

3. 30%   **0.30 or 0.3**

4. 20%   **0.20 or 0.2**

5. 55%   **0.55**

6. 77%   **0.77**

7. 140%   **1.40 or 1.4**

8. 250%   **2.50 or 2.5**

9. 7.8%   **0.078**

10. 6.7%   **0.067**

11. 100%   **1.00 or 1**

12. 600%   **6.00 or 6**

13. 0.5%   **0.005**

14. 0.2%   **0.002**

15. 0.35%   **0.0035**

16. 0.076%   **0.00076**

Write each decimal as a percent.

> **Examples:** 0.23   8.6
>
> **Solutions:** 0.23 % ← Percent sign is attached.   8.60%
>   ‿   ‿
>   └── Decimal point is moved
>   two places to the right.
>
> 0.23 = **23%**   8.6 = **860%**

17. 0.5   **50%**

18. 0.6   **60%**

19. 0.62   **62%**

20. 0.18   **18%**

21. 0.03   **3%**

22. 0.09   **9%**

23. 0.125   **12.5%**

24. 0.875   **87.5%**

25. 0.629   **62.9%**

26. 0.494   **49.4%**

27. 2   **200%**

28. 5   **500%**

29. 2.6   **260%**

30. 1.8   **180%**

31. 0.0312   **3.12%**

32. 0.0625   **6.25%**

33. 4.162   **416.2%**

34. 8.715   **871.5%**

35. 0.0017   **0.17%**

36. 0.0032   **0.32%**

37. Fractions, decimals, and percents are all used to describe a part of something. The use of percent is much more common than fractions and decimals. Why do you suppose this is true?

Possible answers:
No common denominators are needed with percents. The denominator is always 100 with percent which makes comparisons easier to understand.

38. List five uses of percent that are or will be part of your life. Consider the activities of working, shopping, saving, and planning for the future.

Some answers might be:
When using discounts on purchases, calculating sales tax, figuring interest on loans, examining investments, finding tips in restaurants, calculating interest on savings, and doing math problems in this book.

In each of the following, write percents as decimals and decimals as percents.

39. In Folsum, the sales tax rate is 8%.

0.08

40. At College of DuPage, 82% of the students work part time.

0.82

41. In one company, 65% of the salespeople are women.

0.65

42. Only 38.6% of those registered actually voted.

0.386

43. The property tax rate in Alpine County is 0.035.

3.5%

44. A church building fund has 0.49 of the money needed.

49%

45. The success rate in CPR training this session is 2 times that of the last session.

200%

46. Attendance at the picnic this year is 3 times last year's attendance.

300%

47. Only 0.005 of the total population has this genetic defect.

0.5%

48. Defects in cellular phone production has fallen to 0.0061 of total production.

0.61%

49. The patient's blood pressure was 153.6% of normal.

1.536

50. Success with the diet was 248.7% greater than anticipated.

2.487

Fill in the blanks. Remember that 100% is all *of something and 50% is* half *of it.*

51. 100% of $78 is ___$78___ .

52. 100% of $14\frac{1}{2}$ hours is ___$14\frac{1}{2}$ hours___ .

53. There are 20 children in the preschool class. 100% of the children are served breakfast and lunch. How many children are served both meals? ___20 children___ .

54. The company owns 345 vans. 100% of the vans are painted white with blue lettering. How many vans are painted white with blue lettering? ___345 vans___ .

55. 50% of 180 miles is ___90 miles___ .

56. 50% of $900 is ___$450___ .

57. John owes $285 for tuition. Financial aid will pay 50% of the cost.

Financial aid will pay _$142.50_ .

58. The Animal Humane Society took in 20,000 animals last year. About 50% of them were dogs. The number of dogs taken in was _10,000_ .

59. 50% of 8200 college students is _4100 students_ .

60. 100% of 8200 college students is _8200 students_ .

61. Describe a shortcut way to find 100% of a number.

Since 100% means 100 parts out of 100 parts, 100% is all of the number.

62. Describe a shortcut way to find 50% of a number.

50% means 50 parts out of 100 parts. That's half of the number. A shortcut for finding 50% of a number is to divide the number by 2.

▲ *Write a percent for both the shaded and unshaded part of each figure.*

63.

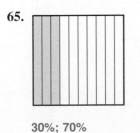

95%; 5%

64.

20%; 80%

65.

30%; 70%

66.

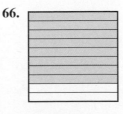

80%; 20%

67.

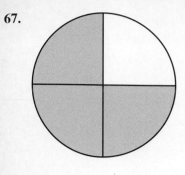

75%; 25%

68.

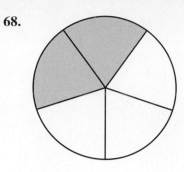

40%; 60%

69.

55%; 45%

70.

37%; 63%

Review and Prepare

*Change each of the following fractions to decimals. (For help, see **Section 4.7.**)*

71. $\frac{2}{5}$ 0.4

72. $\frac{3}{10}$ 0.3

73. $\frac{3}{4}$ 0.75

74. $\frac{1}{2}$ 0.5

75. $\frac{7}{8}$ 0.875

76. $\frac{5}{8}$ 0.625

77. $\frac{4}{5}$ 0.8

78. $\frac{7}{10}$ 0.7

6.2 *Percents and Fractions*

OBJECTIVE 1 Percents can be written as fractions by using what we learned in the previous section.

OBJECTIVES
1. Write percents as fractions.
2. Write fractions as percents.
3. Use the table of percent equivalents.

FOR EXTRA HELP

Tutorial   Tape 9   SSM, Sec. 6.2

Write a Percent as a Fraction

$$p\% = \frac{p}{100}, \quad \text{as a fraction.}$$

E X A M P L E   I   Writing a Percent as a Fraction

Write each percent as a fraction or mixed number in lowest terms.

(a) 45%

As we saw in the last section, 45% can be written as a decimal.

$$45\% = 45 \div 100 = 0.45 \quad \text{(Percent sign dropped)}$$

Because 0.45 means 45 hundredths,

$$0.45 = \frac{45 \div 5}{100 \div 5} = \frac{9}{20}. \quad \text{(Lowest terms)}$$

It is not necessary, however, to write 45% as a decimal first. Just write

$$45\% = \frac{45}{100} \quad \text{(45 per 100)}$$

$$= \frac{9}{20}. \quad \text{(Lowest terms)}$$

(b) 76%

The percent becomes the numerator.

Write 76% as $\frac{76}{100}$.

The *denominator* is always 100 because percent means parts per 100.

Write $\frac{76}{100}$ in lowest terms.

$$\frac{76 \div 4}{100 \div 4} = \frac{19}{25} \quad \text{(Lowest terms)}$$

(c) 150%

$$150\% = \frac{150}{100} = \frac{3}{2} = 1\frac{1}{2} \quad \text{(Mixed number)}$$

Note
Remember that percent means **per 100.**

WORK PROBLEM I AT THE SIDE. ▶▶

1. Write each percent as a fraction or mixed number in lowest terms.

(a) 50%

(b) 22%

(c) 68%

(d) 61%

(e) 120%

(f) 210%

ANSWERS

1. (a) $\frac{1}{2}$ (b) $\frac{11}{50}$ (c) $\frac{17}{25}$ (d) $\frac{61}{100}$ (e) $1\frac{1}{5}$

(f) $2\frac{1}{10}$

2. Write as fractions in lowest terms.

(a) 18.5%

(b) 87.5%

(c) 6.5%

(d) $66\frac{2}{3}\%$

(e) $12\frac{1}{3}\%$

(f) $62\frac{1}{2}\%$

The next example shows how to write decimal and fraction percents as fractions.

E X A M P L E   2   Writing a Decimal or Fraction Percent as a Fraction

Write each percent as a fraction in lowest terms.

(a) 15.5%

Place 15.5 over 100.

$$15.5\% = \frac{15.5}{100}$$

To get a whole number in the numerator, we must multiply the numerator and denominator by 10. (Multiplying by $\frac{10}{10}$ is the same as multiplying by 1.)

$$\frac{15.5}{100} = \frac{15.5 \cdot 10}{100 \cdot 10} = \frac{155}{1000}$$

Write in lowest terms.

$$\frac{155 \div 5}{1000 \div 5} = \frac{31}{200}$$

(b) $33\frac{1}{3}\%$

Place $33\frac{1}{3}$ over 100.

$$33\frac{1}{3}\% = \frac{33\frac{1}{3}}{100}$$

When we have a mixed number in the numerator we must write the mixed number as an improper fraction.

Here, we write $33\frac{1}{3}$ as the improper fraction $\frac{100}{3}$. Now,

$$\frac{33\frac{1}{3}}{100} = \frac{\frac{100}{3}}{100}.$$

Rewrite the division problem in a horizontal form. Then invert the divisor and multiply.

$$\frac{\frac{100}{3}}{100} = \frac{100}{3} \div 100 = \frac{100}{3} \div \frac{100}{1} = \frac{\cancel{100}}{3} \cdot \frac{1}{\cancel{100}} = \frac{1}{3}$$

Invert and multiply.

Note

In Example 2(a) we could have changed 15.5% to $15\frac{1}{2}\%$ and then written it as the improper fraction $\frac{31}{2}$ over 100 as was done in part (b). But it is usually best to leave decimal percents as they are.

◀◀ **WORK PROBLEM 2 AT THE SIDE.**

ANSWERS

2. (a) $\frac{37}{200}$ **(b)** $\frac{7}{8}$ **(c)** $\frac{13}{200}$ **(d)** $\frac{2}{3}$ **(e)** $\frac{37}{300}$

(f) $\frac{5}{8}$

OBJECTIVE 2 ▶ We may use the formula given at the beginning of this section to write fractions as percents.

$$p\% = \frac{p}{100}$$

E X A M P L E 3   Writing a Fraction as a Percent

Write each fraction as a percent. Round to the nearest tenth if necessary.

(a) $\frac{3}{5}$

Write this fraction as a percent by solving for p in the proportion

$$\frac{3}{5} = \frac{p}{100}.$$

Find cross products and show that they are equivalent.

$$5 \cdot p = 3 \cdot 100$$
$$5 \cdot p = 300$$

Divide both sides by 5.

$$\frac{\overset{1}{\cancel{5}} \cdot p}{\underset{1}{\cancel{5}}} = \frac{300}{5}$$

$$p = 60$$

This result means that $\frac{3}{5} = \frac{60}{100}$ or 60%.

Note
Solving proportions can be reviewed in **Section 5.4.**

(b) $\frac{7}{8}$

Write a proportion.

$$\frac{7}{8} = \frac{p}{100}$$

$$8 \cdot p = 7 \cdot 100 \qquad \text{Show that cross products}$$
$$8 \cdot p = 700 \qquad \text{are equivalent.}$$

$$\frac{\overset{1}{\cancel{8}} \cdot p}{\underset{1}{\cancel{8}}} = \frac{700}{8} \qquad \text{Divide both sides by 8.}$$

$$p = 87.5$$

Finally, $\frac{7}{8} = 87.5\%$.

Note
If you think of $\frac{700}{8}$ as an improper fraction, changing it to a mixed number gives an answer of $87\frac{1}{2}$. So $\frac{7}{8} = 87.5\%$ or $87\frac{1}{2}\%$.

CONTINUED ON NEXT PAGE

3. Write as percents. Round to the nearest tenth if necessary.

(a) $\dfrac{3}{4}$

(b) $\dfrac{7}{25}$

(c) $\dfrac{9}{10}$

(d) $\dfrac{3}{8}$

(e) $\dfrac{1}{6}$

(f) $\dfrac{2}{9}$

(c) $\dfrac{5}{6}$

Start with a proportion.

$$\frac{5}{6} = \frac{p}{100}$$

$$6 \cdot p = 5 \cdot 100 \qquad \text{Show that cross products}$$
$$6 \cdot p = 500 \qquad\quad \text{are equivalent.}$$

$$\frac{\overset{1}{\cancel{6}} \cdot p}{\underset{1}{\cancel{6}}} = \frac{500}{6} \qquad \text{Divide both sides by 6.}$$

$$p \approx 83.3 \qquad\qquad \text{Round to the nearest tenth.}$$

Solving this proportion shows

$$\frac{5}{6} \approx 83.3\% \quad \text{(rounded).}$$

Or, treat $\frac{500}{6}$ as an improper fraction to get an exact answer of $83\frac{1}{3}\%$.

■◀◀ **WORK PROBLEM 3 AT THE SIDE.**

OBJECTIVE 3 The more you work with common fractions and mixed numbers and their decimal and percent equivalents, the more familiar you will become with them.

The table on the next two pages shows common and not so common fractions and mixed numbers and their decimal and percent equivalents.

E X A M P L E 4   Using a Conversion Table

Read the following from the table.

(a) $\frac{1}{12}$ as a percent

Find $\frac{1}{12}$ in the "fraction" column. The percent is $\approx 8.33\%$ or exactly $8\frac{1}{3}\%$.

(b) 0.375 as a fraction

Look in the "decimal" column for 0.375. The fraction is $\frac{3}{8}$.

(c) $\frac{13}{16}$ as a percent

Find $\frac{13}{16}$ in the "fraction" column. The percent is 81.25% or $81\frac{1}{4}\%$.

⊞ *Calculator Tip:* Example 4(c) could be solved on the calculator as

$$13 \;\boxed{\div}\; 16 \;\boxed{=}\; 0.8125 \;\boxed{\times}\; 100 \;\boxed{=}\; 81.25.$$

Multiplying by 100 changes the decimal to percent.

> **Note**
> When a fraction like $\frac{1}{12}$ is changed to a decimal, it is a repeating decimal that goes on forever, 0.083333 . . . . In the chart these decimals are rounded to the nearest thousandth. When the decimal is then changed to a percent, it will be to the nearest tenth. Decimals that do not go on forever are not rounded.

ANSWERS

3. (a) 75% **(b)** 28% **(c)** 90% **(d)** 37.5%

(e) $\approx 16.7\%$ or exactly $16\frac{2}{3}\%$

(f) $\approx 22.2\%$ or exactly $22\frac{2}{9}\%$

WORK PROBLEM 4 AT THE SIDE. ▶▶

Percent, Decimal, and Fraction Equivalents

Percent (rounded to tenths when necessary)	Decimal	Fraction
1%	0.01	$\frac{1}{100}$
5%	0.05	$\frac{1}{20}$
6.25% or $6\frac{1}{4}\%$	0.0625	$\frac{1}{16}$
≈8.3% or exactly $8\frac{1}{3}\%$	≈0.083	$\frac{1}{12}$
10%	0.1	$\frac{1}{10}$
12.5% or $12\frac{1}{2}\%$	0.125	$\frac{1}{8}$
≈16.7% or exactly $16\frac{2}{3}\%$	≈0.167	$\frac{1}{6}$
18.75% or $18\frac{3}{4}\%$	0.1875	$\frac{3}{16}$
20%	0.2	$\frac{1}{5}$
25%	0.25	$\frac{1}{4}$
30%	0.3	$\frac{3}{10}$
31.25% or $31\frac{1}{4}\%$	0.3125	$\frac{5}{16}$
≈33.3% or exactly $33\frac{1}{3}\%$	≈0.333	$\frac{1}{3}$
37.5% or $37\frac{1}{2}\%$	0.375	$\frac{3}{8}$
40%	0.4	$\frac{2}{5}$
43.75% or $43\frac{3}{4}\%$	0.4375	$\frac{7}{16}$
50%	0.5	$\frac{1}{2}$
56.25% or $56\frac{1}{4}\%$	0.5625	$\frac{9}{16}$
60%	0.6	$\frac{3}{5}$

4. Read the following common fractions, decimals, and percents from the table. If you already know the answer or can solve the answer quickly, don't use the table.

(a) $\frac{1}{2}$ as a percent

(b) $12\frac{1}{2}\%$ as a fraction

(c) ≈0.667 as a fraction

(d) 20% as a fraction

(e) $\frac{7}{8}$ as a percent

(f) $\frac{1}{4}$ as a percent

(g) $33\frac{1}{3}\%$ as a fraction

(h) $\frac{4}{5}$ as a percent

ANSWERS

4. (a) 50% **(b)** $\frac{1}{8}$ **(c)** $\frac{2}{3}$ **(d)** $\frac{1}{5}$ **(e)** 87.5%

(f) 25% **(g)** $\frac{1}{3}$ **(h)** 80%

Percent, Decimal, and Fraction Equivalents (continued)

Percent (rounded to tenths when necessary)	Decimal	Fraction
62.5% or $62\frac{1}{2}\%$	0.625	$\frac{5}{8}$
$\approx 66.7\%$ or exactly $66\frac{2}{3}\%$	≈ 0.667	$\frac{2}{3}$
68.75% or $68\frac{3}{4}\%$	0.6875	$\frac{11}{16}$
70%	0.7	$\frac{7}{10}$
75%	0.75	$\frac{3}{4}$
80%	0.8	$\frac{4}{5}$
81.25% or $81\frac{1}{4}\%$	0.8125	$\frac{13}{16}$
$\approx 83.3\%$ or exactly $83\frac{1}{3}\%$	≈ 0.833	$\frac{5}{6}$
87.5% or $87\frac{1}{2}\%$	0.875	$\frac{7}{8}$
90%	0.9	$\frac{9}{10}$
93.75% or $93\frac{3}{4}\%$	0.9375	$\frac{15}{16}$
100%	1.0	1
110%	1.1	$1\frac{1}{10}$
125%	1.25	$1\frac{1}{4}$
$\approx 133.3\%$ or exactly $133\frac{1}{3}\%$	≈ 1.333	$1\frac{1}{3}$
150%	1.5	$1\frac{1}{2}$
$\approx 166.7\%$ or exactly $166\frac{2}{3}\%$	≈ 1.667	$1\frac{2}{3}$
175%	1.75	$1\frac{3}{4}$
200%	2.0	2

6.2 Exercises

Write each percent as a fraction or mixed number in lowest terms.

Example: 20%   **Solution:** $\overbrace{20\%} = \dfrac{20}{\underset{\uparrow}{100}} = \dfrac{1}{5}$ ← In lowest terms

Denominator is always 100.

1. 20% $\dfrac{1}{5}$ **2.** 40% $\dfrac{2}{5}$ **3.** 50% $\dfrac{1}{2}$ **4.** 75% $\dfrac{3}{4}$

5. 55% $\dfrac{11}{20}$ **6.** 35% $\dfrac{7}{20}$ **7.** 37.5% $\dfrac{3}{8}$ **8.** 87.5% $\dfrac{7}{8}$

9. 6.25% $\dfrac{1}{16}$ **10.** 43.75% $\dfrac{7}{16}$ **11.** $16\dfrac{2}{3}\%$ $\dfrac{1}{6}$ **12.** $83\dfrac{1}{3}\%$ $\dfrac{5}{6}$

13. $6\dfrac{2}{3}\%$ $\dfrac{1}{15}$ **14.** $46\dfrac{2}{3}\%$ $\dfrac{7}{15}$ **15.** 0.5% $\dfrac{1}{200}$ **16.** 0.8% $\dfrac{1}{125}$

17. 130% $1\dfrac{3}{10}$ **18.** 175% $1\dfrac{3}{4}$ **19.** 250% $2\dfrac{1}{2}$ **20.** 325% $3\dfrac{1}{4}$

📝 Writing   ⊙ Conceptual   ▲ Challenging   ≈ Estimation

Write each fraction as a percent. Round percents to the nearest tenth if necessary.

Examples: $\dfrac{1}{4}$ $\dfrac{3}{7}$

Solutions: $\dfrac{1}{4} = \dfrac{p}{100}$ $\dfrac{3}{7} = \dfrac{p}{100}$

$4 \cdot p = 1 \cdot 100$ $7 \cdot p = 3 \cdot 100$

$4 \cdot p = 100$ $7 \cdot p = 300$

$\dfrac{\overset{1}{\cancel{4}} \cdot p}{\underset{1}{\cancel{4}}} = \dfrac{100}{4}$ $\dfrac{\overset{1}{\cancel{7}} \cdot p}{\underset{1}{\cancel{7}}} = \dfrac{300}{7}$

$p = 25$ $p = 42.9$ Round to nearest tenth.

$\dfrac{1}{4} = \mathbf{25\%}$ $\dfrac{3}{7} \approx \mathbf{42.9\%}$ (Rounded)

21. $\dfrac{1}{4}$ **25%**

22. $\dfrac{1}{5}$ **20%**

23. $\dfrac{3}{10}$ **30%**

24. $\dfrac{9}{10}$ **90%**

25. $\dfrac{3}{5}$ **60%**

26. $\dfrac{3}{4}$ **75%**

27. $\dfrac{37}{100}$ **37%**

28. $\dfrac{63}{100}$ **63%**

29. $\dfrac{5}{8}$ **62.5%**

30. $\dfrac{1}{8}$ **12.5%**

31. $\dfrac{7}{8}$ **87.5%**

32. $\dfrac{3}{8}$ **37.5%**

33. $\dfrac{11}{25}$ **44%**

34. $\dfrac{9}{25}$ **36%**

35. $\dfrac{23}{50}$ **46%**

36. $\dfrac{17}{20}$ **85%**

37. $\dfrac{1}{20}$ **5%**

38. $\dfrac{1}{50}$ **2%**

39. $\dfrac{5}{6}$ **≈83.3%**

40. $\dfrac{1}{6}$ **≈16.7%**

41. $\dfrac{5}{9}$ **≈55.6%**

42. $\dfrac{7}{9}$ **≈77.8%**

43. $\dfrac{1}{7}$ **≈14.3%**

44. $\dfrac{5}{7}$ **≈71.4%**

Complete this chart. Round decimals to the nearest thousandth and percents to the nearest tenth if necessary.

Example:	*Fraction*	*Decimal*	*Percent*
	$\dfrac{3}{50}$	**0.06**	6%

	Fraction	Decimal	Percent
45.	$\dfrac{3}{4}$	0.75	75%
46.	$\dfrac{1}{50}$	0.02	2%
47.	$\dfrac{7}{8}$	0.875	87.5%
48.	$\dfrac{3}{8}$	0.375	37.5%
49.	$\dfrac{4}{5}$	0.8	80%
50.	$\dfrac{3}{5}$	0.6	60%
51.	$\dfrac{1}{6}$	≈0.167	≈16.7%
52.	$\dfrac{1}{3}$	≈0.333	≈33.3%
53.	$\dfrac{1}{4}$	0.25	25%
54.	$\dfrac{1}{2}$	0.5	50%
55.	$\dfrac{1}{8}$	0.125	12.5%
56.	$\dfrac{5}{8}$	0.625	62.5%
57.	$\dfrac{2}{3}$	≈0.667	≈66.7%
58.	$\dfrac{5}{6}$	≈0.833	≈83.3%

	Fraction	*Decimal*	*Percent*
59.	$\frac{1}{10}$	0.1	10%
60.	$\frac{1}{5}$	0.2	20%
61.	$\frac{1}{100}$	0.01	1%
62.	1	1.00 or 1	100%
63.	$\frac{1}{200}$	0.005	0.5%
64.	$\frac{7}{500}$	0.014	1.4%
65.	$2\frac{1}{2}$	2.5	250%
66.	$1\frac{7}{10}$	1.7	170%
67.	$3\frac{1}{4}$	3.25	325%
68.	$2\frac{4}{5}$	2.8	280%

69. Select a decimal percent and write it as a fraction. Select a fraction and write it as a percent. Write an explanation of each step of your work.

There are many possible answers. Examples 2 and 3 show the steps that students should include in their answers.

70. Prepare a table showing fraction, decimal, and percent equivalents for five common fractions and mixed numbers of your choice.

There are many correct answers. The table in Example 4 shows some of the possibilities.

In the following application problems, write the answer as a fraction in lowest terms, as a decimal, and as percent.

71. The license applicant answered 76 of 100 questions correctly. What portion was correct?

$\frac{19}{25}$; 0.76; 76%

72. Of 80 videos available, 20 are for children. What portion are for children?

$\frac{1}{4}$; 0.25; 25%

73. The regular price of a laser jet printer was $750. It was reduced $150. By what portion was the price reduced?

$\frac{1}{5}$; 0.2; 20%

74. Dick McIntosh must complete 200 hours of community service. So far, he has completed 75 hours of community service. What portion of the time has he completed?

$\frac{3}{8}$; 0.375; 37.5%

75. Of the fifteen people at the office, nine are single parents. What portion are single parents?

$\frac{3}{5}$; 0.6; 60%

76. A pizza shop has 25 employees. Of these, 14 are students. What portion are students?

$\frac{14}{25}$; 0.56; 56%

▲ **77.** An insurance office has 80 employees. If 64 of the employees have cellular phones, what portion of the employees do not have cellular phones?

$\frac{1}{5}$; 0.20; 20%

▲ **78.** A zoo has 125 animals, including 25 that are members of endangered species. What portion is not endangered?

$\frac{4}{5}$; 0.8; 80%

▲ **79.** An antibiotic is used to treat 380 people. If 342 people do not have a side reaction to the antibiotic, find the portion that do have a side reaction.

$\frac{1}{10}$; 0.1; 10%

▲ **80.** An apple grower's cooperative has 250 members. If 100 of the growers use a certain insecticide, find the portion that do not use the insecticide.

$\frac{3}{5}$; 0.6; 60%

The circle graph below shows the type of transportation used by 4200 students at Metro-Community College. It also shows the number of people using each type of transportation. Use this graph to answer Exercises 81–84, giving the answer as a fraction, as a decimal, and as a percent.

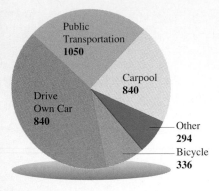

81. What portion of the students use public transportation?

$\frac{1}{4}$; 0.25; 25%

82. What portion of the students use a bicycle?

$\frac{2}{25}$; 0.08; 8%

83. Find the portion that drive their own cars.

$\frac{2}{5}$; 0.4; 40%

84. Find the portion that carpool.

$\frac{1}{5}$; 0.2; 20%

Review and Prepare

*Find the missing number in each proportion. (For help, see **Section 5.4**.)*

85. $\frac{8}{4} = \frac{x}{15}$ 30

86. $\frac{n}{50} = \frac{8}{20}$ 20

87. $\frac{4}{y} = \frac{12}{15}$ 5

88. $\frac{5}{x} = \frac{12}{24}$ 10

89. $\frac{42}{30} = \frac{14}{b}$ 10

90. $\frac{6}{22} = \frac{a}{220}$ 60

6.3 The Percent Proportion

There are two ways to solve percent problems. One method uses ratios and is discussed in this section, while the percent equation method is explained in **Section 6.6.**

OBJECTIVE 1 We have seen that a statement of two equivalent ratios is called a proportion.

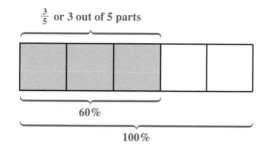

$\frac{3}{5}$ or 3 out of 5 parts

60%

100%

For example, the fraction $\frac{3}{5}$ is the same as the ratio 3 to 5, and 60% is the ratio 60 to 100. As the figure above shows, these two ratios are equivalent and make a proportion.

WORK PROBLEM I AT THE SIDE. ▶▶

The **percent proportion** (per-SENT proh-POR-shun) can be used to solve percent problems.

The Percent Proportion

Amount is to *base* as percent is to 100.

$$\frac{\text{amount}}{\text{base}} = \frac{\text{percent}}{100} \quad \leftarrow \text{Always 100 because percent means per 100.}$$

Use the letter a for *amount* and b for *base*.

$$\frac{a}{b} = \frac{p}{100} \quad \textit{Percent proportion}$$

The final statement in the box is the **percent proportion.** In the figure at the top of the page, the **base** is 5 (the entire quantity), the **amount** is 3 (the part of the whole), and the **percent** is 60. Write the percent proportion as follows.

$$\frac{a}{b} \to \frac{3}{5} = \frac{60}{100} \leftarrow \frac{p}{100}$$

OBJECTIVE 2 As shown in **Section 5.4,** if any two of the three values in the percent proportion are known, the third can be found by solving the proportion.

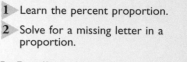
1. As a review of proportions, use the method of cross products to decide whether these proportions are *true* or *false*.

(a) $\dfrac{3}{4} = \dfrac{75}{100}$

(b) $\dfrac{3}{8} = \dfrac{75}{200}$

(c) $\dfrac{4}{5} = \dfrac{108}{140}$

(d) $\dfrac{29}{83} = \dfrac{145}{415}$

(e) $\dfrac{104}{37} = \dfrac{515}{185}$

ANSWERS

1. (a) true (b) true (c) false
   (d) true (e) false

2. Use the percent proportion
($\frac{a}{b} = \frac{p}{100}$) and solve for the value
of the missing letter.

(a) $a = 15, p = 20$

(b) $a = 50, b = 200$

(c) $b = 175, p = 12$

(d) $b = 5000, p = 27$

(e) $a = 74, b = 185$

E X A M P L E I   Using the Percent Proportion

Use the percent proportion and solve for the missing number.

(a) $a = 12, p = 25$, find b

Replace a with 12 and p with 25.

$$\frac{a}{b} = \frac{p}{100} \qquad \text{Percent proportion}$$

$$\frac{12}{b} = \frac{25}{100} \quad \text{or} \quad \frac{12}{b} = \frac{1}{4} \qquad \text{(Lowest terms)}$$

Find the cross products to solve this proportion.

$$\frac{12}{b} = \frac{1}{4} \qquad \begin{array}{l} b \cdot 1 \\ \\ 12 \cdot 4 \end{array}$$

Show that the cross products are equivalent.

$$b \cdot 1 = 12 \cdot 4$$
$$b = 48$$

The base is 48.

(b) $a = 30, b = 50$, find p

Use the percent proportion.

$$\frac{30}{50} = \frac{p}{100} \qquad \text{Percent proportion}$$

$$\frac{3}{5} = \frac{p}{100} \qquad \text{(Lowest terms)}$$

$$5 \cdot p = 3 \cdot 100 \qquad \text{Cross products}$$

$$5 \cdot p = 300$$

$$\frac{\overset{1}{5} \cdot p}{\underset{1}{5}} = \frac{300}{5} \qquad \text{Divide both sides by 5.}$$

$$p = 60$$

The percent is 60, written as 60%.

(c) $b = 150, p = 18$, find a

$$\frac{a}{150} = \frac{18}{100} \quad \text{or} \quad \frac{a}{150} = \frac{9}{50} \qquad \text{(Lowest terms)}$$

$$a \cdot 50 = 150 \cdot 9 \qquad \text{Cross products}$$

$$a \cdot 50 = 1350$$

$$\frac{a \cdot \overset{1}{50}}{\underset{1}{50}} = \frac{1350}{50} \qquad \text{Divide both sides by 50.}$$

$$a = 27$$

The amount is 27.

◄◄ WORK PROBLEM 2 AT THE SIDE.

ANSWERS

2. **(a)** $b = 75$
(b) $p = 25$ (so, the percent is 25%)
(c) $a = 21$ **(d)** $a = 1350$
(e) $p = 40$ (so, the percent is 40%)

6.3 Exercises

Find the value of the missing letter in the percent proportion $\frac{a}{b} = \frac{p}{100}$. Round to the nearest tenth if necessary. If the answer is percent (p), be sure to include a percent sign (%).

Examples:

$a = 10$, $p = 50$, find b $a = 80$, $b = 120$, find p $b = 90$, $p = 75$, find a

Solutions:

$$\frac{10}{b} = \frac{50}{100}$$

$$\frac{80}{120} = \frac{p}{100}$$

$$\frac{a}{90} = \frac{75}{100}$$

$$\frac{10}{b} = \frac{1}{2} \leftarrow \text{Lowest terms}$$

$$\text{Lowest terms} \rightarrow \frac{2}{3} = \frac{p}{100}$$

$$\frac{a}{90} = \frac{3}{4} \leftarrow \text{Lowest terms}$$

$$b \cdot 1 = 10 \cdot 2$$

$$3 \cdot p = 2 \cdot 100$$

$$a \cdot 4 = 90 \cdot 3$$

$$b = \mathbf{20}$$

$$3 \cdot p = 200$$

$$a \cdot 4 = 270$$

$$\frac{\overset{1}{\cancel{3}} \cdot p}{\underset{1}{\cancel{3}}} = \frac{200}{3}$$

$$\frac{a \cdot \overset{1}{\cancel{4}}}{\underset{1}{\cancel{4}}} = \frac{270}{4}$$

$$p \approx \mathbf{66.7} \text{ (rounded)}$$

$$a = \mathbf{67.5}$$

The percent is **66.7%**.

1. $a = 10$, $p = 20$

50

2. $a = 35$, $p = 25$

140

3. $a = 60$, $p = 50$

120

4. $a = 40$, $p = 25$

160

5. $a = 24$, $p = 30$

80

6. $a = 11$, $p = 5$

220

7. $a = 25$, $p = 6$

≈416.7

8. $a = 61$, $p = 12$

≈508.3

9. $a = 55$, $b = 110$

50%

10. $a = 15$, $b = 60$

25%

11. $a = 105$, $b = 35$

300%

12. $a = 36$, $b = 24$

150%

13. $a = 1.5$, $b = 4.5$

≈33.3%

14. $a = 9.25$, $b = 27.75$

≈33.3%

15. $b = 52$, $p = 50$

26

16. $b = 160$, $p = 35$

56

17. $b = 72$, $p = 30$

21.6

18. $b = 112$, $p = 38$

≈42.6

19. $b = 47.2$, $p = 28$

≈13.2

20. $b = 79.6$, $p = 13$

≈10.3

✎ Writing   ◎ Conceptual   ▲ Challenging   ≈ Estimation

21. Give two examples of your own choosing— one showing a true proportion and why it is true and the other showing a false proportion and why it is false.

One answer is:

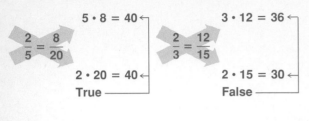

$5 \cdot 8 = 40$

$\dfrac{2}{5} = \dfrac{8}{20}$

$2 \cdot 20 = 40$

True

$3 \cdot 12 = 36$

$\dfrac{2}{3} = \dfrac{12}{15}$

$2 \cdot 15 = 30$

False

22. Make up a problem that uses the percent proportion and that has an answer of $p = 15$. (*Hint:* Many different values may be used for a and b to get the result $p = 15$.)

This is one of many answers.

$\dfrac{a \to 12}{b \to 80} = \dfrac{p}{100} \leftarrow \dfrac{p}{100}$

$\dfrac{3}{20} = \dfrac{p}{100}$ Lowest terms

$20 \cdot p = 3 \cdot 100$ Cross products

$20p = 300$

$\dfrac{\overset{1}{\cancel{20}}p}{\underset{1}{\cancel{20}}} = \dfrac{300}{20}$ Divide both sides by 20.

$p = 15 = 15\%$

Solve each of the following problems. If the answer is percent (p), be sure to include a percent sign (%).

23. Find b if a is 89 and p is 25.

356

24. p is 45 and b is 160. Find a.

72

25. b is 5000 and a is 20. Find p.

0.4%

26. Suppose a is 15 and b is 2500. Find p.

0.6%

▲ 27. Find p if b is 2150 and a is 53.75.

2.5%

▲ 28. What is a, if p is $25\frac{1}{2}$ and b is 2800?

714

29. b is 6480 and a is 19.44. Find p.

0.3%

▲ 30. Suppose a is 281.25 and p is $1\frac{1}{4}$. Find b.

22,500

Review and Prepare

*Write each fraction as a percent. Round the percent to the nearest tenth if necessary. (For help, see **Section 6.2**.)*

31. $\dfrac{1}{2}$ 50%

32. $\dfrac{1}{10}$ 10%

33. $\dfrac{57}{100}$ 57%

34. $\dfrac{4}{5}$ 80%

35. $\dfrac{7}{8}$ 87.5%

36. $\dfrac{1}{6}$ ≈16.7%

37. $\dfrac{2}{3}$ ≈66.7%

38. $\dfrac{17}{25}$ 68%

6.4 Identifying the Parts in a Percent Problem

In this section you will learn how to solve percent problems. As a help in solving these problems it is good to remember what is involved in percent problems.

Percent Problems

All percent problems involve a comparison between a part of something and the whole.

Solving these problems requires identifying the three parts of a percent proportion: amount (*a*), base (*b*), and percent (*p*).

OBJECTIVE 1 Look for *p*, percent, first. It is the easiest to identify.

Percent

The **percent** is the ratio of a part to a whole, with 100 as the denominator. In a problem, the percent *p* appears with the word "percent" or with the symbol "%" after it.

E X A M P L E I   Finding Percent in a Percent Problem

Find *p* in each of the following.

(a) 32% of the 900 men were retired.

    ↓
    p

p is 32. The number 32 appears with the symbol %.

(b) $150 is 25 percent of what number?

    ↓
    p

p is 25 because 25 appears with the word "percent."

(c) What percent of the 350 women will go?

    ↓
    p   (An unknown)

The word "percent" has no number with it, so the percent is the unknown part of the problem.

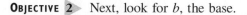 **WORK PROBLEM I AT THE SIDE.** ▶▶

OBJECTIVE 2 Next, look for *b*, the base.

Base

The **base** is the entire quantity, or the whole. In a problem, the base often appears after the word **of.**

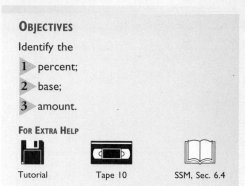

OBJECTIVES

Identify the

1 ▶ percent;

2 ▶ base;

3 ▶ amount.

FOR EXTRA HELP

Tutorial   Tape 10   SSM, Sec. 6.4

1. Identify *p*.

(a) Of the $2000, 15% will be spent on a washing machine.

(b) Of the 750 employees, 8% will retire.

(c) Find the sales tax by multiplying $590 and $6\frac{1}{2}$ percent.

(d) 105 is 3% of what number?

(e) What percent of the 110 rental cars will be rented today?

ANSWERS

1. (a) 15 **(b)** 8 **(c)** $6\frac{1}{2}$ **(d)** 3

    (e) *p* is unknown.

371

2. Identify b.

(a) Of the $2000, 15% will be spent on a washing machine.

(b) Of the 750 employees, 8% will retire.

(c) Find the amount of sales tax by multiplying sales of $590 and $6\frac{1}{2}$ percent.

(d) $105 is 3% of what number?

(e) What percent of the 110 rental cars will be rented today?

3. Identify a.

(a) Of the $2000, 15% will be spent on a $300 washing machine.

(b) Of the 750 employees, 8%, or 60 employees, will retire.

(c) Find the sales tax of $38.35 by multiplying $590 and $6\frac{1}{2}$ percent.

(d) $105 is 3% of what number?

(e) 90% of the 110 rental cars will be rented today.

ANSWERS
2. (a) $2000 **(b)** 750 **(c)** $590
   (d) what number (an unknown)
   (e) 110
3. (a) $300 **(b)** 60 **(c)** $38.35 **(d)** $105
   (e) unknown

E X A M P L E 2   Finding Base in a Percent Problem

Identify b in each of the following.

(a) 32% **of** the 900 men were too large for the imported car.
           $\downarrow$
           b

b is 900. The number 900 appears after the word *of*.

(b) $150 is 25 percent **of** what number?
                              $\downarrow$
                              b The base is the unknown part of the problem.

(c) 85% **of** 7000 is what number?
              $\downarrow$
              b

◄◄ WORK PROBLEM 2 AT THE SIDE.

OBJECTIVE 3 Finally, look for a, the amount.

Amount

The **amount** is the part being compared with the whole.

Note
If you have trouble identifying amount, find base and percent first. The remaining number is amount.

E X A M P L E 3   Finding Amount in a Percent Problem

Identify a in each of the following.

(a) 54% **of** 700 students is 378 students.
      Find p and b.

54% of 700 students is 378 students.
 $\downarrow$ $\downarrow$
 p b
With % sign   Whole; follows "of"

54% of 700 students is 378 students.
 $\downarrow$ $\downarrow$ $\downarrow$
 p b a Part

The amount, a, is 378.

(b) $150 is 25% **of** what number?
       $\downarrow$ $\downarrow$
       p b (The unknown part of the proportion)

150 is the remaining number, so $a = 150$.

(c) 85% **of** $7000 is what number?
       $\downarrow$ $\downarrow$ $\downarrow$
       p b a (The unknown part of the proportion)

◄◄ WORK PROBLEM 3 AT THE SIDE.

Name _____ Date _____ Class _____

373

6.4 Exercises

Identify p, b, and a in each of the following. Do not try to solve for any unknowns.

> **Example:** $\underbrace{60}$ is $\underbrace{75\%}$ of $\underbrace{\text{what number}}$?
>
> **Solution:** $\uparrow$ $\uparrow$ $\uparrow$
> a p b is unknown
>
> **Example:** Of the $\underbrace{592}$ tomato plants, $\underbrace{75\%,}$ or $\underbrace{444,}$ are ready to be sold.
>
> **Solution:** $\uparrow$ $\uparrow$ $\uparrow$
> b p a

	p	b	a
1. 20% of how many salespeople is 80 salespeople?	20	unknown	80
2. 58% of how many preschoolers is 203 preschoolers?	58	unknown	203
3. 80% of \$950 is \$760.	80	950	760
4. 93% of \$1500 is \$1395.	93	1500	1395
5. What is 15% of \$75?	15	75	unknown
6. What is 61% of 830 homes?	61	830	unknown
7. 9 is 36% of what number of guests?	36	unknown	9
8. 92 is 26% of what number of servings?	26	unknown	92
9. 34 trophies is 50% of 68 trophies.	50	68	34
10. 208 sacks is $33\frac{1}{3}\%$ of 624 sacks.	$33\frac{1}{3}$	624	208
11. What percent of \$296 is \$177.60?	unknown	296	177.6
12. What percent of \$47.50 is \$21.75?	unknown	47.5	21.75

	p	*b*	*a*
13. 27.17 is 6.5% of what number?	6.5	unknown	27.17
14. 16.74 is 11.9% of what number?	11.9	unknown	16.74
15. 0.68% of 487 is what number?	0.68	487	unknown
16. What number is 12.42% of 1408.7?	12.42	1408.7	unknown

17. Identify the three parts in a percent problem. In your own words, write one sentence telling how you will identify each of these three parts.

p(percent)—the ratio of the part to the whole. It appears with the word *percent* or "%" after it.
b(base)—the entire quantity or whole. Often appears after the word *of*.
a(amount)—the part being compared with the whole.

18. Write one short sentence or statement using numbers and words. The statement should include a percent, a base, and an amount. Identify each of these three parts.

A possible sentence is:
Of the 580 cars entering the parking lot, 464 cars, or 80% had parking stickers on their windshield.

$p = 80; b = 580; a = 464$

Find p, b, and a in the following application problems. Do not try to solve for any unknowns.

19. In a tree-planting project, 640 of the 810 trees planted were still living one year later. What percent of the trees planted were still living?

p is unknown; 810; 640

20. A popular bar of soap is $99\frac{44}{100}$% pure. If the bar of soap weighs 9 ounces, how many ounces are pure?

$99\frac{44}{100}$; 9; *a* is unknown

21. Sales tax of $0.90 is charged on a compact disc costing $15. What percent sales tax is charged?

p is unknown; 15; 0.90

22. On her first check from the Pizza Hut Restaurant, 15% is withheld from total earnings of $225. What amount is withheld?

15; 225; *a* is unknown

23. Of the lunch and dinner customers, 23% prefer a fat-free salad dressing. If the total number of customers is 610, find the number of customers who prefer fat-free dressing.

23; 610; *a* is unknown

24. There are 590 quarts of grape juice in a vat holding a total of 1700 quarts of fruit juice. What percent is grape juice?

p is unknown; 1700; 590

25. Of the total candy bars contained in a vending machine, 240 bars have been sold. If 25% of the bars have been sold, find the total number of candy bars that were in the machine.

25; *b* is unknown; 240

26. There have been 36 cups of coffee served from a banquet-sized coffee pot. If this is 30% of the capacity of the pot, find the capacity of the pot.

30; *b* is unknown; 36

27. A survey of 650 students found that 78% of the students considered a detachable face CD player a high priority in their car. Find the number of students who consider this a high priority.

78; 650; *a* is unknown

28. A student needs 64 credits to graduate. If 48 of the credits needed have already been completed, what percent of the credits have already been completed?

p is unknown; 64; 48

29. At a recent health fair 32% of the people tested were found to have high blood cholesterol levels. If 272 people were tested, find the number having high blood cholesterol.

32; 272; *a* is unknown

30. The sales tax on a new car is $820. If the sales tax rate is 5%, find the price of the car before the sales tax is added.

5; *b* is unknown; 820

31. A medical clinic found that 16.8% of the patients were late for their appointments. The number of patients who were late was 504. Find the total number of patients.

16.8; *b* is unknown; 504

32. 1848 automobiles are tested for exhaust emissions and 231 do not pass the test. Find the percent that do not pass.

p is unknown; 1848; 231

Review and Prepare

Write a proportion for each of the following and then find the missing number. (For help, see Section 5.4.)

	Proportion	*Missing Number*
33. 8 is to x as 15 is to 30.	$\dfrac{8}{x} = \dfrac{15}{30}$	16
34. 35 is to 20 as y is to 100.	$\dfrac{35}{20} = \dfrac{y}{100}$	175
35. r is to 36 as $\frac{4}{3}$ is to 12.	$\dfrac{r}{36} = \dfrac{\frac{4}{3}}{12}$	4
36. 1.2 is to 10 as 3.6 is to s.	$\dfrac{1.2}{10} = \dfrac{3.6}{s}$	30

6.5 Using Proportions to Solve Percent Problems

In the percent proportion,

$$\frac{\text{amount}}{\text{base}} = \frac{\text{percent}}{100}$$

or

$$\frac{a}{b} = \frac{p}{100}$$

three letters, a, b, and p are used. As discussed, if the values of any two of these letters are known, the third can be found.

> **Note**
> Remember that in the percent proportion, b (base) is the entire quantity or whole, a (amount) is part of the whole, and p is the percent.

OBJECTIVE 1 The first example shows the percent proportion used to find a, the amount. (Remember: the amount is a part of the whole.)

E X A M P L E 1 **Finding Amount with the Percent Proportion**

Find 15% of $160.

Here p (percent) is 15 and b (base) is 160. (Recall that the base often comes after the word *of*.) Now find a (amount).

$$\frac{a}{b} = \frac{p}{100} \quad \text{so} \quad \frac{a}{160} = \frac{15}{100} \quad \text{or} \quad \frac{a}{160} = \frac{3}{20} \quad \text{(Lowest terms)}$$

Find the cross products in the proportion.

$$a \cdot 20 = 160 \cdot 3 \qquad \text{Cross products}$$

$$a \cdot 20 = 480$$

$$\frac{a \cdot \overset{1}{\cancel{20}}}{\underset{1}{\cancel{20}}} = \frac{480}{20} \qquad \text{Divide both sides by 20.}$$

$$a = 24 \qquad \text{Amount}$$

15% of $160 is $24.

WORK PROBLEM 1 AT THE SIDE. ▶▶

Just as with the application problems given earlier, the word *of* is an indicator word meaning *multiply*. For example:

$$15\% \text{ of } 160$$
$$\downarrow$$
$$15\% \cdot 160.$$

Because of this, there is another way to find the amount, a.

OBJECTIVES

1. Use the percent proportion to solve for a (amount).
2. Solve for b (base) using the percent proportion.
3. Find p (percent) using the percent proportion.

FOR EXTRA HELP

Tutorial   Tape 10   SSM, Sec. 6.5

1. Use the percent proportion to find the following.

 (a) Find 20% of 1800 calories.

 (b) Find 25% of $2032.

 (c) Find 9% of 3250 miles.

 (d) Find 78% of 610 meters.

ANSWERS

1. (a) 360 calories (b) $508
   (c) 292.5 miles (d) 475.8 meters

377

2. Use multiplication to find a (amount).

(a) Find 45% of 6000 hogs.

(b) Find 18% of 80 feet.

(c) Find 125% of 78 acres.

(d) Find 0.6% of $120.

Finding Amount

To find amount (a):

Step 1 Find p. Write the percent as a decimal.

Step 2 Multiply this decimal and b.

E X A M P L E   2   Finding Amount by Using Multiplication

Use multiplication to find a.

(a) Find 42% of 830 yards.

   Step 1 Here p is 42. Write 42% as the decimal 0.42.

   Step 2 Multiply 0.42 and b, which is 830.

$$a = 0.42 \cdot 830$$
$$a = 348.6 \text{ yards}$$

$\approx$It is a good idea to estimate the answer, to make sure no mistakes were made with decimal points. Estimate 42% as 40% or 0.4, and estimate 830 as 800. Next, 40% of 800 is

$$0.4 \cdot 800 = 320,$$

so that 348.6 is a reasonable answer.

(b) Find 25% of 1680 cars.

   Identify p as 25. Write 25% in decimal form as 0.25. Now, multiply 0.25 and 1680.

$$a = 0.25 \cdot 1680 = 420 \text{ cars} \quad \text{Multiply.}$$

$\approx$We can use a shortcut to estimate the answer. Since 25% means 25 parts out of 100 parts, this is the same as $\frac{1}{4}$ of the parts ($\frac{25}{100} = \frac{1}{4}$). Do you see a shortcut here? You can find $\frac{1}{4}$ of a number by dividing the number by 4. So, this shortcut gives us the exact answer, $1680 \div 4 = 420$.

(c) Find 140% of 60 miles.

   In this problem, p is 140. Write 140% as the decimal 1.40. Next, multiply 1.40 and 60.

$$a = 1.40 \cdot 60 = 84 \text{ miles} \quad \text{Multiply.}$$

$\approx$We can estimate because 140% is close to 150% (which is $1\frac{1}{2}$) and $1\frac{1}{2}$ times 60 is 90. So, 84 miles is a reasonable answer.

(d) Find 0.4% of 50 kilometers.

$$a = 0.004 \cdot 50 = 0.2 \text{ kilometers} \quad \text{Multiply.}$$
$$\uparrow \text{ Write 0.4% as a decimal.}$$

An estimate would not be very useful here.

◀◀ **WORK PROBLEM 2 AT THE SIDE.**

ANSWERS

2. (a) 2700 hogs **(b)** 14.4 feet
   (c) 97.5 acres **(d)** $0.72

E X A M P L E   3   **Solving for Amount by Using Multiplication**

Video Production has 850 employees. Of these employees, 28% are students. How many of the employees are students?

Approach   Look for the word *of* as an indicator word for multiplication.

Solution

28% **of** the employees are students

↑ Indicator word

The total number of employees is 850, so $b = 850$. The percent is 28 ($p = 28$). Find a to find the number of students.

$$a = 0.28 \cdot 850 = 238 \qquad \text{Multiply.}$$
↑— Write 28% as a decimal.

Video Productions has 238 student employees.

$\approx$ The answer in Example 3 can be estimated using 25% since 28% is very close to 25%. Remember 25% is 25 parts out of 100 which is equivalent to $\frac{1}{4}$.

$$850 \div 4 \approx 213 \qquad \text{Divide by 4 to find } \tfrac{1}{4}.$$

So, 238 is a reasonable answer.

WORK PROBLEM 3 AT THE SIDE. ▶▶

🖩 *Calculator Tip:*   If you are using a calculator you could solve the problem like this.

.28 ☒ 850 ▭ 238

Or, you can use this alternate approach on calculators with a % key.

850 ☒ 28 ▣% ▭ 238

OBJECTIVE ▶2▶   The next example shows how to use the percent proportion to find b, the base.

> **Note**
> Remember, the base is the entire quantity, or the whole.

E X A M P L E   4   **Finding *b* (Base) with the Percent Proportion**

(a) 8 tables is 4% of what number of tables?

   Here $p = 4$, b is unknown, and $a = 8$. Use the percent proportion to find b.

$$\frac{a}{b} = \frac{p}{100} \quad \text{so} \quad \frac{8}{b} = \frac{4}{100} \quad \text{or} \quad \frac{8}{b} = \frac{1}{25} \qquad \text{(Lowest terms)}$$

Find cross products.

$$b \cdot 1 = 8 \cdot 25$$
$$b = 200$$

8 tables is 4% of 200 tables.

— **CONTINUED ON NEXT PAGE**

3. Use multiplication to find a.

   (a) In a club of 1460 people, 35% play tennis. How many people play tennis?

   (b) There are 9250 students enrolled at the campus. If 28% of the students use tobacco products, find the number of tobacco users.

ANSWERS

3. (a) 511 people **(b)** 2590 students

4. Use the percent proportion to find the missing base.

(a) 75 bidders is 20% of what number of bidders?

(b) 30 lines is 15% of what number of lines?

(c) 774 employees is 72% of what number of employees?

(d) 97.5 miles is 12.5% of what number of miles?

5. (a) A freeze resulted in a loss of 52% of an avocado crop. If the loss was 182 tons, find the total number of tons in the crop.

(b) A metal alloy contains 8% zinc. The alloy contains 450 pounds of zinc. Find the total weight of the alloy.

(b) 135 tourists is 15% of what number of tourists?
$p = 15$ and $a = 135$, so

$$\frac{135}{b} = \frac{15}{100}$$

$$\frac{135}{b} = \frac{3}{20} \qquad \text{(Lowest terms)}$$

$$b \cdot 3 = 135 \cdot 20 \qquad \text{Cross products}$$

$$b \cdot 3 = 2700$$

$$\frac{b \cdot \overset{1}{\cancel{3}}}{\underset{1}{\cancel{3}}} = \frac{2700}{3} \qquad \text{Divide both sides by 3.}$$

$$b = 900.$$

135 tourists is 15% of 900 tourists.

◀◀ **WORK PROBLEM 4 AT THE SIDE.**

E X A M P L E 5 Applying the Percent Proportion

At Newark Salt Works, 78 employees are absent because of illness. If this is 5% of the total number of employees, how many employees does the company have?

Approach From the information in the problem, the percent is 5 ($p = 5$) and the amount, or *part* of the total number of employees is 78 ($a = 78$). The total number of employees or entire quantity, which is the base (b), must be found.

Solution We can use the percent proportion to find b (the total number of employees).

$$\frac{78}{b} = \frac{5}{100}$$

$$\frac{78}{b} = \frac{1}{20} \qquad \text{(Lowest terms)}$$

Find cross products.

$$b \cdot 1 = 78 \cdot 20$$

$$b = 1560$$

$\approx$ Estimate the answer. 78 is approximately 80, and 5% is equivalent to the fraction $\frac{1}{20}$. Because 80 is $\frac{1}{20}$ of 1600, or

$$80 \cdot 20 = 1600,$$

1560 is a reasonable answer. The company has 1560 employees.

> **Note**
> To estimate the answer to Example 5 the 5% was changed to its fraction equivalent $\frac{1}{20}$. Because 80 (rounded) is $\frac{1}{20}$ of the total employees, 80 was multiplied by 20 to get 1600, the estimated answer.

◀◀ **WORK PROBLEM 5 AT THE SIDE.**

ANSWERS
4. (a) 375 bidders **(b)** 200 lines
(c) 1075 employees **(d)** 780 miles
5. (a) 350 tons **(b)** 5625 pounds

OBJECTIVE **3** Finally, if the amount and base are known, the percent proportion can be used to find p, the percent.

EXAMPLE 6 **Using the Percent Proportion to Find p**

(a) 13 roofs is what percent of 52 roofs?
The base is $b = 52$ (52 follows *of*) and $a = 13$. Next, find p.

$$\frac{a}{b} = \frac{p}{100}$$

$$\frac{13}{52} = \frac{p}{100}$$

$$\frac{1}{4} = \frac{p}{100} \quad \text{(Lowest terms)}$$

Find cross products.

$$4 \cdot p = 1 \cdot 100$$

$$\frac{\overset{1}{\cancel{4}} \cdot p}{\underset{1}{\cancel{4}}} = \frac{100}{4} \quad \text{Divide both sides by 4.}$$

$$p = 25$$

13 roofs is 25% of 52 roofs.

(b) What percent of $500 is $100?
$b = 500$ (follows *of*), $a = 100$, so

$$\frac{100}{500} = \frac{p}{100}$$

$$\frac{1}{5} = \frac{p}{100} \quad \text{(Lowest terms)}$$

$$5 \cdot p = 1 \cdot 100 \quad \text{Cross products}$$

$$5 \cdot p = 100$$

$$\frac{\overset{1}{\cancel{5}} \cdot p}{\underset{1}{\cancel{5}}} = \frac{100}{5} \quad \text{Divide both sides by 5.}$$

$$p = 20$$

20% of $500 is $100.

Note
When finding p (percent), be sure to label your answer with the percent symbol (%).

WORK PROBLEM 6 AT THE SIDE. ▶▶

---

6. (a) $7 is what percent of $28?

(b) What percent of 150 athletes is 30 athletes?

(c) What percent of 2280 court trials is 1026 trials?

(d) 72 sales is what percent of 18 sales?

ANSWERS

6. (a) 25% **(b)** 20% **(c)** 45% **(d)** 400%

7. (a) The Cruisers Club has raised $578 of the $850 needed for an annual picnic. What percent of the total has been raised?

E X A M P L E 7   Applying the Percent Proportion

A roof is expected to last 20 years before needing replacement. If the roof is now 15 years old, what percent of the roof's life has been used?

Approach   The expected life of the roof is the entire quantity or base ($b = 20$). The part of the roof that is already used is the amount ($a = 15$). You need to find the percent (p) of the roof's life that is already used.

Solution   Use the percent proportion to find p, the percent of roof life used.

$$\frac{15}{20} = \frac{p}{100} \quad \text{or} \quad \frac{3}{4} = \frac{p}{100} \qquad \text{(Lowest terms)}$$

Find cross products.

$$4 \cdot p = 3 \cdot 100$$
$$4 \cdot p = 300$$
$$\frac{\overset{1}{\cancel{4}} \cdot p}{\underset{1}{\cancel{4}}} = \frac{300}{4} \qquad \text{Divide both sides by 4.}$$
$$p = 75$$

75% of the roof's life has been used.

(b) A late-model domestic car gets 38 miles per gallon on the highway and 32.3 miles per gallon around town. What percent of the highway mileage does the car get around town?

◀◀ **WORK PROBLEM 7 AT THE SIDE.**

---

8. (a) The number of students who usually take this class is 300. If 450 students are taking this class now, find the percent of the usual number who are now taking the class.

E X A M P L E 8   Applying the Percent Proportion

Rainfall this year was 33 inches while normal rainfall is only 30 inches. What percent of normal rainfall is this year's rainfall?

Approach   The normal rainfall is the base ($b = 30$). This year's rainfall is all of last year's rainfall and more, or 33 ($a = 33$). You need to find the percent (p) that this year's rainfall is of last year's rainfall.

Solution

$$\frac{33}{30} = \frac{p}{100} \quad \text{or} \quad \frac{11}{10} = \frac{p}{100} \qquad \text{(Lowest terms)}$$

Find cross products.

$$10 \cdot p = 11 \cdot 100$$
$$10 \cdot p = 1100$$
$$\frac{\overset{1}{\cancel{10}} \cdot p}{\underset{1}{\cancel{10}}} = \frac{1100}{10}$$
$$p = 110$$

This year's rainfall is 110% of last year's rainfall.

(b) The workers group set a production goal of 480 units. If their production amounted to 600 units, find the percent of their production goal that they achieved.

◀◀ **WORK PROBLEM 8 AT THE SIDE.**

ANSWERS

7. (a) 68% **(b)** 85%

8. (a) 150% **(b)** 125%

6.5 Exercises

Find the amount using the multiplication shortcut.

Example: 47% of 5000 bicycles

Solution: $0.47 \cdot 5000 = 2350$

$a = \textbf{2350}$ bicycles

1. 5% of 1040 bowlers

52 bowlers

2. 10% of 3000 runners

300 runners

3. 35% of 2340 volunteers

819 volunteers

4. 6% of 1750 words

105 words

5. 4% of 120 feet

4.8 feet

6. 9% of $150

$13.50

7. 125% of 108 folders

135 folders

8. 130% of 60 trees

78 trees

9. 22.5% of 1100 boxes

247.5 boxes

10. 38.2% of 4250 loads

1623.5 loads

11. 2% of $164

$3.28

12. 6% of $434

$26.04

13. 250% of 740 sales

1850 sales

14. 125% of 920 students

1150 students

15. 15.5% of 275 pounds

42.625 pounds

16. 46.1% of 843 kilograms

388.623 kilograms

17. 0.9% of $2400

$21.60

18. 0.3% of $1400

$4.20

✎ Writing   ◉ Conceptual   ▲ Challenging   ≈ Estimation

Find the base using the percent proportion.

> **Example:** 64% of what number is 1600?
>
> **Solution:** $p = 64$ and $a = 1600$, so
>
> $$\frac{1600}{b} = \frac{64}{100}$$
>
> $$\frac{1600}{b} = \frac{16}{25} \quad \text{(Lowest terms)}$$
>
> Solve, to get $b = \mathbf{2500}.$

19. 40 printers is 25% of what number of printers?

160 printers

20. 8 musicians is 4% of what number of musicians?

200 musicians

21. 20% of what number is 34?

170

22. 65% of what number is 182?

280

23. 495 successful students is 90% of what number of students?

550 students

24. 84 letters is 28% of what number of letters?

300 letters

25. 748 books is 110% of what number of books?

680 books

26. 154 bicycles is 140% of what number of bicycles?

110 bicycles

27. $12\frac{1}{2}\%$ of what number is 350?
(*Hint:* $12\frac{1}{2}\% = 12.5\%$)

2800

28. $5\frac{1}{2}\%$ of what number is 176?

3200

Find the percent using the percent proportion. Round your answers to the nearest tenth if necessary.

Example: What percent of 80 is 15?

Solution: $b = 80$ and $a = 15$, so

$$\frac{15}{80} = \frac{p}{100}$$

$$80 \cdot p = 1500$$

$$\frac{\overset{1}{\cancel{80}} \cdot p}{\underset{1}{\cancel{80}}} = \frac{1500}{80}$$

$$p = 18.75 \qquad \text{Round to the nearest tenth.}$$

$$p \approx 18.8$$

$$\approx \textbf{18.8\%} \text{ of 80 is 15.}$$

29. 16 pepperoni pizzas is what percent of 32 pizzas?

50%

30. 35 hours is what percent of 140 hours?

25%

31. 13 tables is what percent of 25 tables?

52%

32. 325 bottles is what percent of 500 bottles?

65%

33. 16 servings is what percent of 200 servings?

8%

34. 7 bridges is what percent of 350 bridges?

2%

35. 9 rolls is what percent of 600 rolls?

1.5%

36. 60 cartons is what percent of 2400 cartons?

2.5%

37. What percent of $172 is $32?

≈18.6%

38. What percent of $398 is $14?

≈3.5%

39. What percent of 500 wheels is 46 wheels?

9.2%

40. What percent of 105 employees is 54 employees?

≈51.4%

41. A student turned in the following answers on a test. You can tell that two of the answers are incorrect without even working the problems. Find the incorrect answers and explain how you identified them (without actually solving the problems).

50% of $84 is __$42__
150% of $30 is __$20__
25% of $16 is __$32__
100% of $217 is __$217__

150% of $30 cannot be less than $30 because 150% is greater than 1 (100%). The answer must be greater than $30.
25% of $16 cannot be greater than $16 because 25% is less than 1 (100%). The answer must be less than $16.

42. Write a percent problem on any topic you choose. Be sure to include only two of the three parts so that you can solve for the third part. Identify each part of the problem and then solve it.

One answer is:
There are 600 vehicles in the parking lot and 45 of them are pickup trucks. What percent are pickup trucks?
Total vehicles, 600, is *b,* and the number of pickup trucks, 45, is *a.*

$$\frac{45}{600} = \frac{p}{100}$$

$$600 \cdot p = 4500$$

$$\frac{\overset{1}{\cancel{600}} \cdot p}{\underset{1}{\cancel{600}}} = \frac{4500}{600}$$

$$p = 7.5 \text{ or } 7.5\%$$

Solve each application problem. Round percent answers to the nearest tenth if necessary.

43. Robert Garrett, who works part-time, earns $110 per week and has 18% of this amount withheld for taxes, social security, and medicare. Find the amount withheld.

$19.80

44. Sharon McDonald needs 64 credits to graduate. If McDonald has already completed 75% of the necessary credits, find the number of credits completed.

48 credits

45. A survey at an intersection found that of 2200 drivers, 38% were wearing seat belts. How many drivers in the survey were wearing seat belts?

836 drivers

46. A home valued at $95,000 will gain 6% in value this year. Find the gain in value this year.

$5700

47. This year, there are 550 scholarship applicants. If 40% of the applicants will receive a scholarship, find the number of students who will receive a scholarship.

220 students

48. A U.S. Food and Drug Administration (FDA) biologist found that canned tuna is "relatively clean." Extraneous matter was found in 5% of the 1600 cans of tuna tested. How many cans of tuna contained extraneous matter?

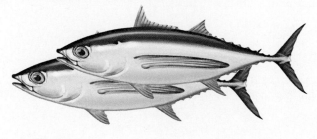

80 cans

▲ **49.** The average industry-wide profit on all new cars and vans is 6.7% of the dealer's cost. If a new Plymouth Voyager costs the dealer $21,200, find the selling price of the van after the dealer has added the profit to the cost.

$22,620.40

▲ **50.** The Saturn automobile dealers use a one price, "no haggle" selling policy. Saturn dealers average 13% profit on new car sales. If a dealer pays $15,600 for a Saturn SC, find the selling price after adding the profit to the dealer's cost.

$17,628

51. Meadow Vista Bottled Water estimates that $117,000 will be spent this year on delivery costs alone. If total sales are estimated at $755,000, what percent of total sales will be spent on delivery?

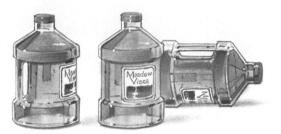

≈15.5%

52. There are 55,000-plus words in Webster's Dictionary, but most educated people can identify only 20,000 of these words. What percent of the words in the dictionary can these people identify?

≈36.4%

53. The size of Cuba's labor force is 3.8 million people. If 61% of the labor force is male, find **(a)** the percent who are female and **(b)** the number of workers who are male.

(a) 39% female
(b) ≈2.3 million male workers

54. In the United States there are 132 million people in the labor force. If 54% of the labor force is male, find **(a)** the percent of the labor force who are female and **(b)** the number of workers who are male.

(a) 46% female
(b) ≈71.3 million male workers

55. This month's sales goal for Easy Writer Pen Company is 2,380,000 ball-point pens. If sales of 2,618,000 pens have been made, what percent of the goal has been reached?

110%

56. The number of apartment unit vacancies was predicted to be 2112 while the actual number of vacancies was 2640. The actual number was what percent of the predicted number of vacancies?

125%

57. Americans who are 65 years of age or older make up 12.7% of the total population. If there are 31.5 million Americans in this group, find the total U.S. population. (Round to the nearest tenth of a million.)

248.0 million

58. About 61% of the 43,000,000 people who receive social security benefits are paid with a direct deposit to their bank. How many of the people receiving benefits are paid with a direct deposit?

26,230,000 people

The graph (pictograph) below shows the percent of chicken noodle soup sold during the cold and flu season. Use this information to do Exercises 59–62.

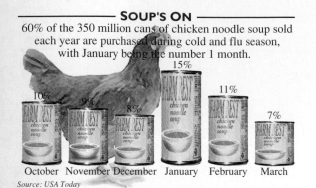

——————— **SOUP'S ON** ———————
60% of the 350 million cans of chicken noodle soup sold each year are purchased during cold and flu season, with January being the number 1 month.

15%

10% 9% 8% 11% 7%

October November December January February March

Source: USA Today

59. Which of the flu season months had the lowest sales of chicken noodle soup?

March

60. What percent of the chicken noodle soup sales take place in the *non-flu* season months?

40%

61. Find the number of cans of soup sold in the highest sales month.

52.5 million cans

62. How many more cans of soup were sold in October than in November?

3.5 million cans

▲ **63.** A collection agency, specializing in collecting past-due child support, charges $25 as an application fee plus 20% of the amount collected. What is the total charge for collecting $3100 past-due child support?

$645

▲ **64.** Raw steel production by the nation's steel mills decreased by 2.5% from last week. The decrease amounted to 50,475 tons. Find the steel production last week.

2,019,000 tons

▲ **65.** Marketing Intelligence Service says that there were 15,401 new products introduced last year. If 86% of the products introduced last year failed to reach their business objectives, find the number of products that did reach their objectives. (Round to the nearest whole number.)

2156 products

▲ **66.** A family of four with a monthly income of $2900 spends 90% of its earnings and saves the balance. Find **(a)** the monthly savings and **(b)** the annual savings of this family.

(a) $290
(b) $3480

Review and Prepare

*Multiply or divide as indicated. (For help, see **Sections 4.5 and 4.6**.)*

67.   38.04
   × 0.52
    19.7808

68.   0.728
   × 0.015
    0.01092

69.   218.7
   × 0.042
    9.1854

70.   51.81
   × 0.021
    1.08801

71. $306 \div 0.085$ **3600**   **72.** $24 \div 0.01$ **2400**   **73.** $172 \div 688$ **0.25**   **74.** $120 \div 24$ **5**

6.6 *The Percent Equation*

In the last section you were shown how to use a proportion to solve percent problems. This section shows another way to solve these problems by using the *percent equation.* The percent equation is just a rearrangement of the percent proportion.

> ### Percent Equation
>
> $$\text{amount} = \text{percent} \cdot \text{base}$$
>
> Be sure to write the percent as a decimal before using the equation.

In the percent proportion we did not have to write the percent as a decimal because of the 100 in the denominator. However, because there is no 100 in the percent *equation,* it is necessary for us first to write the percent as a decimal by dividing by 100.

Some of the examples solved earlier will be reworked by using the percent equation. If you want to, you can look back at **Section 6.5** to see how some of these same problems were solved using proportions. This will give you a comparison of the two methods.

OBJECTIVE 1 The first example shows how to find the amount, *a*.

E X A M P L E 1   Solving for the Amount (*a*)

(a) Find 15% of $160.

Write 15% as the decimal 0.15. The base (the whole, which often comes after the word *of*) is 160. Next, use the percent equation.

$$\text{amount} = \text{percent} \cdot \text{base}$$
$$\text{amount} = 0.15 \cdot 160$$

Multiply 0.15 and 160 to get

$$\text{amount} = 24.$$

15% of $160 is $24.

(b) Find 110% of 80 cases.

Write 110% as the decimal 1.10. The base is 80.

$$\text{amount} = \text{percent} \cdot \text{base}$$
$$a = 1.10 \cdot 80$$
$$a = 88$$

110% of 80 cases is 88 cases.

(c) Find 0.4% of 250 patients.

Write 0.4% as the decimal 0.004. The base is 250.

$$\text{amount} = \text{percent} \cdot \text{base}$$
$$a = 0.004 \cdot 250$$
$$a = 1$$

0.4% of 250 patients is 1 patient.

> **Note**
>
> When using the percent equation, the percent must always be changed to a decimal before multiplying.

WORK PROBLEM 1 AT THE SIDE. ▶▶

OBJECTIVES

1. ▶ Use the percent equation to solve for amount (*a*).

2. ▶ Solve for base (*b*) using the percent equation.

3. ▶ Find percent (*p*) using the percent equation.

FOR EXTRA HELP

| Tutorial | Tape 10 | SSM, Sec. 6.6 |

1. Find each of the following.

(a) 28% of 1050 garments

(b) 19% of 360 pages

(c) 125% of $84

(d) 145% of $580

(e) 0.5% of 600 samples

(f) 0.25% of 160 pounds

ANSWERS

1. (a) 294 garments **(b)** 68.4 pages
(c) $105 **(d)** $841 **(e)** 3 samples
(f) 0.4 pound

OBJECTIVE ▶2▶   The next example shows how to use the percent equation to find the base, *b*.

> **Note**
> Remember that the word *of* is an indicator word for *multiply*.

E X A M P L E   2   Solving for the Base (*b*)

(a) 8 tables is 4% of what number of tables?

    The amount is 8 and the percent is 4% or the decimal 0.04. The base is unknown.

$$8 \quad \text{is} \quad 4\% \text{ of what number?}$$

         ↑   Indicator word

Next, use the percent equation.

$$\textbf{amount} = \textbf{percent} \cdot \textbf{base}$$

$$8 = 0.04 \cdot b \qquad b \text{ is the base.}$$

$$\frac{8}{0.04} = \frac{\overset{1}{\cancel{0.04}} \cdot b}{\underset{1}{\cancel{0.04}}} \qquad \text{Divide both sides by 0.04.}$$

$$200 = b \qquad \text{Base}$$

8 tables is 4% of 200 tables.

(b) 135 tourists is 15% of what number of tourists?

    Write 15% as 0.15. The amount is 135. Next, use the percent equation to find the base.

$$\textbf{amount} = \textbf{percent} \cdot \textbf{base}$$

$$135 = 0.15 \cdot b$$

$$\frac{135}{0.15} = \frac{\overset{1}{\cancel{0.15}} \cdot b}{\underset{1}{\cancel{0.15}}} \qquad \text{Divide both sides by 0.15.}$$

$$900 = b \qquad \text{Base}$$

135 tourists is 15% of 900 tourists.

(c) $8\frac{1}{2}\%$ of what number is 102?

    Write $8\frac{1}{2}\%$ as 8.5%, or the decimal 0.085. The amount is 102. Use the percent equation.

$$\textbf{amount} = \textbf{percent} \cdot \textbf{base}$$

$$102 = 0.085 \cdot b$$

$$\frac{102}{0.085} = \frac{\overset{1}{\cancel{0.085}} \cdot b}{\underset{1}{\cancel{0.085}}} \qquad \text{Divide both sides by 0.085.}$$

$$1200 = b \qquad \text{Base}$$

102 is $8\frac{1}{2}\%$ of 1200.

Note
In Example 2(c) the $8\frac{1}{2}\%$ was changed to 8.5%, which is the decimal form of $8\frac{1}{2}\%$ ($8\frac{1}{2} = 8.5$). The percent sign still remained in 8.5%. Then 8.5% was changed to the decimal 0.085 before dividing.

WORK PROBLEM 2 AT THE SIDE. ▶▶

OBJECTIVE 3 The final example shows how to use the percent equation to find the percent, p.

E X A M P L E 3   Solving For Percent (p)

(a) 13 roofs is what percent of 52 roofs?
    Because 52 follows *of*, the base is 52. The amount is 13, and the percent is unknown. Use the percent equation,

$$\text{amount} = \text{percent} \cdot \text{base}$$
$$13 = p \cdot 52$$

$$\frac{13}{52} = \frac{p \cdot \overset{1}{\cancel{52}}}{\underset{1}{\cancel{52}}} \qquad \text{Divide both sides by 52.}$$

$$0.25 = p$$
$$0.25 \text{ is } 25\% \qquad \text{Write the decimal as percent.}$$

13 roofs is 25% of 52 roofs.
    The equation can also be set up by using *of* as an indicator word for multiplication, and *is* as an indicator word for "is equal to."

$$13 \text{ is what percent of } 52?$$
$$\downarrow \downarrow \qquad\quad \downarrow \quad \downarrow \; \downarrow$$
$$13 = \qquad\quad p \quad \cdot 52$$
$$13 = p \cdot 52$$

(b) What percent of $500 is $100?
    The base is 500 and the amount is 100. Let p be the unknown percent.

$$\text{amount} = \text{percent} \cdot \text{base}$$
$$100 = p \cdot 500$$

$$\frac{100}{500} = \frac{p \cdot \overset{1}{\cancel{500}}}{\underset{1}{\cancel{500}}} \qquad \text{Divide both sides by 500.}$$

$$0.20 = p$$
$$0.20 = 20\% \qquad \text{Write decimal as percent.}$$

20% of $500 is $100.

CONTINUED ON NEXT PAGE

2. Find the base.

(a) 24 patients is 15% of what number of patients?

(b) 22.5 boxes is 18% of what number of boxes?

(c) 270 lab tests is 45% of what number of lab tests?

(d) $5\frac{1}{2}\%$ of what number of policies is 66 policies?

ANSWERS

2. (a) 160 patients **(b)** 125 boxes
    (c) 600 lab tests **(d)** 1200 policies

3. (a) What percent of 90 pallets is 18 pallets?

(c) What percent of $300 is $390?

The base is 300 and the amount is 390. Let p be the unknown percent.

$$\textbf{amount} = \textbf{percent} \cdot \textbf{base}$$

$$390 = p \cdot 300$$

$$\frac{390}{300} = \frac{p \cdot \overset{1}{\cancel{300}}}{\underset{1}{\cancel{300}}} \qquad \text{Divide both sides by 300.}$$

$$1.3 = p$$

$$1.3 = 130\% \qquad \text{Write decimal as percent.}$$

130% of $300 is $390.

(b) 68 cartons is what percent of 170 cartons?

(d) 6 ladders is what percent of 1200 ladders?

Since 1200 follows *of*, the base is 1200. The amount is 6. Let p be the unknown percent.

$$\textbf{amount} = \textbf{percent} \cdot \textbf{base}$$

$$6 = p \cdot 1200$$

$$\frac{6}{1200} = \frac{p \cdot \overset{1}{\cancel{1200}}}{\underset{1}{\cancel{1200}}} \qquad \text{Divide both sides by 1200.}$$

$$0.005 = p$$

$$0.005 = 0.5\% \qquad \text{Write decimal as percent.}$$

6 ladders is 0.5% of 1200 ladders.

(c) What percent of 460 orders is 644 orders?

Note

When you use the percent equation to solve for an unknown percent, the answer will always be in decimal form. Notice that in Example 3(a), (b), (c), and (d) the decimal answer had to be changed to a percent by multiplying by 100. The answers became: (a) 0.25 = 25%; (b) 0.20 = 20%; (c) 1.3 = 130%; and (d) 0.005 = 0.5%.

◀◀ **WORK PROBLEM 3 AT THE SIDE.**

(d) 3 sacks is what percent of 375 sacks?

ANSWERS

3. (a) 20% (b) 40% (c) 140% (d) 0.8%

6.6 Exercises

Find the amount using the percent equation.

Example: 14% of $750

Solution: amount = percent • base

amount = 0.14 • 750   Write 14% as the decimal 0.14.

amount = 105

14% of $750 is **$105.**

1. 35% of 660 programs

231 programs

2. 55% of 740 cannisters

407 cannisters

3. 65% of 1300 species

845 species

4. 75% of 360 dosages

270 dosages

5. 32% of 260 quarts

83.2 quarts

6. 44% of 430 liters

189.2 liters

7. 140% of 500 tablets

700 tablets

8. 175% of 540 patients

945 patients

9. 12.4% of 8300 meters

1029.2 meters

10. 13.2% of 9400 acres

1240.8 acres

11. 0.8% of $520

$4.16

12. 0.3% of $480

$1.44

Find the base using the percent equation.

Example: 48% of what number of shirts is 2496 shirts?

Solution: amount = percent • base

$2496 = 0.48 • b$ Write 48% as the decimal 0.48.

$$\frac{2496}{0.48} = \frac{\overset{1}{\cancel{0.48}} • b}{\underset{1}{\cancel{0.48}}}$$ Divide both sides by 0.48.

$5200 = b$ base

48% of **5200 shirts** is 2496 shirts.

13. 142 employees is 50% of what number of employees?

284 employees

14. 16 books is 20% of what number of books?

80 books

15. 40% of what number of salads is 130 salads?

325 salads

16. 75% of what number of wrenches is 675 wrenches?

900 wrenches

17. 476 circuits is 70% of what number of circuits?

680 circuits

18. 621 tons is 45% of what number of tons?

1380 tons

✎ Writing   ◉ Conceptual   ▲ Challenging   ≈ Estimation

19. $12\frac{1}{2}\%$ of what number of people is 135 people?

1080 people

20. $6\frac{1}{2}\%$ of what number of bottles is 130 bottles?

2000 bottles

21. $1\frac{1}{4}\%$ of what number of gallons is 3.75 gallons?

300 gallons

22. $2\frac{1}{4}\%$ of what number of files is 9 files?

400 files

Find the percent using the percent equation.

Example: What percent of 80 inches is 20 inches?

Solution: amount = percent • base

$20 = p \cdot 80$ p is the unknown percent.

$$\frac{20}{80} = \frac{p \cdot \overset{1}{\cancel{80}}}{\underset{1}{\cancel{80}}}$$ Divide both sides by 80.

$0.25 = p$

$0.25 = 25\%$ Write the decimal as percent.

25% of 80 inches is 20 inches.

23. 70 truckloads is what percent of 140 truckloads?

50%

24. 30 crew members is what percent of 75 crew members?

40%

25. 38 styles is what percent of 50 styles?

76%

26. 75 offices is what percent of 125 offices?

60%

27. What percent of $264 is $330?

125%

28. What percent of $480 is $696?

145%

29. What percent of 160 liters is 2.4 liters?

1.5%

30. What percent of 600 is 7.5?

1.25%

31. 999 is what percent of 740?

135%

32. 1224 is what percent of 850?

144%

33. When using the percent equation the percent must always be changed to a decimal before doing any calculations. Show and explain how to change a fraction percent to a decimal. Use $2\frac{1}{2}\%$ in your explanation.

You must first change the fraction in the percent to a decimal, then divide the percent by 100 to change it to a decimal.

$$2\frac{1}{2}\% = 2.5\% = \underline{0.025}$$

Change $\frac{1}{2}$ to decimal $\qquad$ $2\frac{1}{2}\%$ as a decimal

Solve each application problem.

35. The product known as WD-40 is used in 79% of U.S. homes. If there are 104.2 million homes in the U.S., find the number of homes that have WD-40 in them. Round to the nearest tenth of a million.

≈82.3 million homes

37. In a recent survey of 1100 employers, it was found that 84% offer only one health plan to employees. How many of these employers offer only one health plan?

924 employers

39. In a test by *Consumer Reports,* six of the 123 cans of tuna that it analyzed contained more than the 30 microgram intake limit of mercury. What percent of the cans contained this level of mercury? Round to the nearest tenth of a percent.

≈4.9%

34. Suppose a problem on your homework assignment was, "Find $\frac{1}{2}\%$ of $1300." Your classmates got answers of $0.65, $6.50, $65, and $650. Which answer is correct? How and why are they getting all of these answers? Explain.

The correct answer is $6.50. The error is in changing $\frac{1}{2}\%$ to a decimal.

$$\frac{1}{2}\% = 0.5\% = 0.005; \quad 0.005 \cdot \$1300 = \$6.50 \qquad \text{Correct}$$

The incorrect answers and how your classmates got them are these.

$$\frac{1}{2}\% = 0.0005; \quad 0.0005 \cdot \$1300 = \$0.65$$

$$\frac{1}{2}\% = 0.05; \qquad 0.05 \quad \cdot \$1300 = \$65 \qquad \text{Incorrect}$$

$$\frac{1}{2}\% = 0.5; \qquad 0.5 \quad \cdot \$1300 = \$650$$

36. Most shampoos contain 75% to 90% water. If a 16-ounce bottle of shampoo contains 78% water, find the number of ounces of water in the 16-ounce bottle. Round to the nearest tenth of an ounce.

≈12.5 ounces

38. For a tour of the eastern United States, a travel agent promised a trip of 3300 miles. Exactly 35% of the trip was by air. How many miles would be traveled by air?

1155 miles

40. According to industry figures there are 44,500 hotels and motels in America. Economy hotels and motels account for 16,910 of this total. What percent of the total are economy hotels and motels?

38%

41. Chemical Banking Corporation made $338 million worth of mortgage loans to minorities last year. If this represented 18.6% of all their mortgages, find the total value of all mortgages that they made last year. (Round to the nearest tenth of a million.)

≈$1817.2 million

42. Julie Ward has 8.5% of her earnings deposited into the credit union. If this amounts to $131.75 per month, find her annual earnings.

$18,600 (12 · $1550)

43. The Chevy Camaro was introduced in 1967. Sales that year were 220,917 which was 46.2% of the number of Ford Mustangs sold in the same year. Find the number of Mustangs sold in the same year. Round to the nearest whole number.

≈478,175 Mustangs

44. Unemployment reported by the Bureau of Labor Statistics states that the number of unemployed people is 8.9 million or 7.1% of all workers. What is the size of the total workforce? Round to the nearest tenth of a million.

≈125.4 million

▲ **45.** J & K Mustang has increased the sale of auto parts by $32\frac{1}{2}\%$ over last year. If the sale of parts last year amounted to $385,200, find the volume of sales this year.

$510,390

▲ **46.** An ad for steel-belted radial tires promises 15% better mileage. If mileage has been 25.6 miles per gallon in the past, what mileage could be expected after new tires are installed? (Round to the nearest tenth of a mile.)

≈29.4 miles per gallon

▲ **47.** A Polaris Vac-Sweep 380 is priced at $559 with an allowed trade-in of $125 on an old unit. If sales tax of $7\frac{3}{4}\%$ is charged on the price of the new Polaris unit before the trade-in, find the total cost to the customer after receiving the trade-in. (Round to the nearest cent.)

≈$477.32

▲ **48.** A Hewlett Packard Desk Jet 820 CSE color printer priced at $499 is marked down 17%. Find the price of the printer after the markdown.

$414.17

Review and Prepare

Identify p, b, and a in each of the following. Do not try to solve for any unknowns. (For help, see Section 6.4.)

	p	*b*	*a*
49. 23% of 500 hinges is 115 hinges.	23	500	115
50. What is 8% of 425 smokers?	8	425	unknown
51. 72 jockeys is 18% of what number of jockeys?	18	unknown	72
52. 3% of $2448 is $73.44.	3	2448	73.44
53. What percent of $830 is $128.65?	unknown	830	128.65
54. 182 cages is what percent of 546 cages?	unknown	546	182

6.7 *Applications of Percent*

Percent has many applications in our daily lives. This section discusses percent as it applies to sales tax, commissions, discounts, and the percent of change (increase and decrease).

OBJECTIVE 1 States, counties, and cities often collect taxes on sales to customers. The **sales tax** is a percent of the total sale. You may use the following formula to find sales tax.

> **Sales Tax Formula**
>
> amount of sales tax = rate of tax • cost of item

E X A M P L E 1 Solving for Sales Tax

Fit and Fine Cyclery sold a mountain bicycle for $374. If the sales tax rate is 5%, how much tax was paid? What was the total cost of the bicycle?

Approach Use the formula above with the cost of the bicycle ($374) as cost of the item and the tax rate (5%) to find the amount of sales tax.

Solution

$$\text{amount of sales tax} = \text{rate of tax} \cdot \text{cost of item}$$
$$= 5\% \cdot \$374$$
$$= 0.05 \cdot \$374$$
$$= \$18.70 \quad \text{Sales tax}$$

The tax paid on the bicycle is $18.70. The customer buying the bicycle would pay a total cost of $374 + $18.70 = $392.70.

WORK PROBLEM 1 AT THE SIDE. ▶▶

E X A M P L E 2 Finding the Sales Tax Rate

The sales tax on a $14,800 pickup truck is $962. Find the rate of the sales tax.

Approach Use the sales tax formula:

$$\text{sales tax} = \text{rate of tax} \cdot \text{cost of item}$$

The rate of tax is the percent.

Solution Solve for the rate of tax. The cost of the pickup truck is $14,800, and the amount of sales tax is $962. Use r for the rate of tax (the percent).

$$\text{sales tax} = \text{rate of tax} \cdot \text{cost of item}$$
$$\$962 = r \cdot \$14,800$$

$$\frac{962}{14,800} = \frac{\overset{1}{\cancel{14,800}} \cdot r}{\underset{1}{\cancel{14,800}}} \quad \text{Divide both sides by 14,800.}$$

$$0.065 = r$$
$$0.065 = 6.5\% \quad \text{Write the decimal as percent.}$$

The sales tax rate is 6.5% or $6\frac{1}{2}\%$.

OBJECTIVES

1 ▶ Find sales tax.
2 ▶ Find commissions.
3 ▶ Find the discount and sale price.
4 ▶ Find the percent of change.

FOR EXTRA HELP

Tutorial   Tape 11   SSM, Sec. 6.7

1. Suppose the sales tax rate in your state is 6%. Find the amount of the tax and the total you would pay for each item.

   (a) $59 AM-FM cassette radio

   (b) $495 camcorder

   (c) $1287 sofa and chair

   (d) $21,400 automobile

ANSWERS

1. **(a)** $3.54; $62.54 **(b)** $29.70; $524.70
   (c) $77.22; $1364.22
   (d) $1284; $22,684

2. Find the rate of sales tax.

(a) The tax on a $380 desk is $19.

(b) The tax on a $12 pair of running shorts is $0.78.

(c) The tax on a $12,320 refrigeration case is $862.40.

3. Find the amount of commission.

(a) Freida Mobley sells office products at a commission rate of 7% and has sales for the month of $48,350.

(b) Last month the appliance sales for Angie Gragg were $62,500 with a commission rate of 3%.

4. Find the rate of commission.

(a) A commission of $450 is earned on one sale of computer products worth $22,500.

(b) Jamal Story earns $2898 for selling office furniture worth $32,200.

ANSWERS

2. (a) 5% (b) 6.5% or $6\frac{1}{2}$% (c) 7%
3. (a) $3384.50 (b) $1875
4. (a) 2% (b) 9%

Note
You can use the sales tax formula to find the amount of sales tax, the cost of an item, or the rate of sales tax (the percent).

◀◀ **WORK PROBLEM 2 AT THE SIDE.**

OBJECTIVE 2 Many salespeople are paid by **commission** rather than an hourly wage. In this method you are paid a certain percent of the total sales dollars. Use the following formula to find the commission.

> **Commission Formula**
>
> amount of commission
> = rate or percent of commission • amount of sales

EXAMPLE 3   Determining the Amount of Commission

Jill Beauteo sold dental tools worth $19,500. If her commission rate is 11%, find the amount of her commission.

Approach   Use the commission formula: the rate of commission (11%) is multiplied by the amount of sales ($19,500).

Solution

amount of commission = rate of commission • amount of sales

$$= 11\% \cdot \$19,500$$
$$= 0.11 \cdot \$19,500$$
$$= \$2145 \quad \text{Amount of commission}$$

She earned a commission of $2145 for selling the dental tools.

◀◀ **WORK PROBLEM 3 AT THE SIDE.**

EXAMPLE 4   Finding the Rate of Commission

A salesperson earned $510 for selling $17,000 worth of paper products. Find the rate of commission.

Approach   You could use the commission formula. Or another approach is to use the percent proportion with $b = 17,000$, $a = 510$, and p unknown. (The rate of commission is the percent.)

Solution

$$\frac{a}{b} = \frac{p}{100}$$
$$\frac{510}{17,000} = \frac{p}{100}$$
$$17,000 \cdot p = 510 \cdot 100 \quad \text{Cross multiply}$$
$$\frac{\overset{1}{\cancel{17,000}} \cdot p}{\underset{1}{\cancel{17,000}}} = \frac{51,000}{17,000} \quad \text{Divide both sides by 17,000.}$$
$$p = 3$$

The rate of commission is 3%.

◀◀ **WORK PROBLEM 4 AT THE SIDE.**

OBJECTIVE 3 Most of us prefer buying things when they are on sale. A store will reduce prices, or **discount,** to attract additional customers. Use the following formula to find the discount and the sale price.

> **Discount Formula and Sale Price Formula**
>
> amount of discount = rate of discount • original price
>
> sale price = original price − amount of discount

E X A M P L E 5   Application of a Sales Discount

The Oak Mill Furniture Store has an oak entertainment center with an original price of $840 on sale at 15% off. Find the sale price of the entertainment center.

Approach   This problem is solved in two steps. First, find the amount of the discount, that is, the amount that will be "taken off" (subtracted) by multiplying the original price ($840) by the rate of the discount (15%). The second step is to subtract the amount of discount from the original price. This gives you the sale price, what you will actually pay for the entertainment center.

Solution   First find the amount of the discount.

amount of discount = rate of discount • original price

$$= 0.15 \cdot \$840 \quad \text{Write 15\% as a decimal.}$$

$$= \$126 \quad \text{Amount of discount}$$

Find the sale price of the entertainment center by subtracting the amount of the discount ($126) from the original price.

sale price = original price − amount of discount

$$= \$840 - \$126$$

$$= \$714 \quad \text{Sale price}$$

During the sale, you can buy the entertainment center for $714.

> **WORK PROBLEM 5 AT THE SIDE.** ▶▶

Calculator Tip: In Example 5, you can use a calculator to find the amount of discount and subtract the discount from the original price.

$$840 \boxed{-} .15 \boxed{\times} 840 \boxed{=} 714$$

Original price   Amount of discount   Sale price

Your scientific calculator observes the order of operations, so it will automatically do the multiplication before the subtraction.

OBJECTIVE 4 We are often interested in looking at increases or decreases in sales, production, population, and many other areas. This type of problem involves finding the percent of change. Use the following steps to find the **percent of increase.**

> **Finding Percent of Increase**
>
> *Step 1* Use subtraction to find the amount of increase.
>
> *Step 2* Use the percent proportion to find the percent of increase.
>
> $$\frac{\text{amount of increase}}{\text{original value}} = \frac{p}{100}$$

---

5. Find the amount of the discount and the sale price.

(a) An Easy-Boy leather recliner originally priced at $950 is offered at a 35% discount.

(b) Eastside Department Store has women's swimsuits on sale at 40% off. One swimsuit was originally priced at $34.

ANSWERS

5. (a) $332.50; $617.50
   (b) $13.60; $20.40

6. Find the percent of increase.

(a) Production of aluminum boats increased from 9400 units last year to 12,690 this year.

(b) The number of flu cases rose from 496 cases last week to 620 this week.

E X A M P L E 6   Finding the Percent of Increase

Attendance at county parks climbed from 18,300 last month to 56,730 this month. Find the percent of increase.

Approach   Subtract the attendance last month (18,300) from the attendance this month (56,730) to find the amount of increase in attendance. Next, use the percent proportion, with $b = 18,300$ (last month's original attendance), $a = 38,430$ (amount of increase in attendance), and p as the unknown percent.

Solution

Step 1 $56,730 - 18,300 = 38,430$ Amount of increase in attendance

Step 2 $$\frac{38,430}{18,300} = \frac{p}{100}$$ Percent proportion

Solve this proportion to find that $p = 210$, so the percent of increase is 210%.

■

◀◀ WORK PROBLEM 6 AT THE SIDE.

Use the following steps to find the **percent of decrease.**

Finding Percent of Decrease

Step 1 Use subtraction to find the amount of decrease.

Step 2 Use the percent proportion to find the percent of decrease.

$$\frac{\textbf{amount of decrease}}{\textbf{original value}} = \frac{p}{100}$$

---

7. Find the percent of decrease.

(a) The number of employees absent this week fell to 285 from 380 last week.

(b) The number of workers applying for unemployment fell from 4850 last month to 3977 this month.

E X A M P L E 7   Finding the Percent of Decrease

The number of production employees this week fell to 1406 people from 1480 people last week. Find the percent of decrease.

Approach   Subtract the number of employees this week (1406) from the number of employees last week (1480) to find the amount of decrease. Next, use the percent proportion, with $b = 1480$ (last week's original number of employees), $a = 74$ (decrease in employees), and p as the unknown percent.

Solution

Step 1 $1480 - 1407 = 74$ Decrease in number of employees

Step 2 $$\frac{74}{1480} = \frac{p}{100}$$ Percent proportion

Solve this proportion to find that $p = 5$, so the percent of decrease is 5%.

■

Note
When solving for percent of increase or decrease, the base is always the original value or value before the change occurred. The amount is the change in values, that is, how much something went up or went down.

◀◀ WORK PROBLEM 7 AT THE SIDE.

Answers

6. (a) 35% **(b)** 25%

7. (a) 25% **(b)** 18%

6.7 Exercises

Find the amount of the sales tax or tax rate and the total cost (amount of sale + amount of tax = total cost). Round money answers to the nearest cent.

Example: The cost of a table is $235 and the sales tax rate is 5%. Find the amount of sales tax and the total cost.

Solution:

$$\text{sales tax} = \text{rate of tax} \cdot \text{cost of item}$$
$$= 5\% \cdot \$235$$
$$= 0.05 \cdot \$235$$
$$= \$11.75 \quad \text{Sales tax}$$

Sales tax is **$11.75** and the total cost is **$246.75** ($235 + $11.75).

	Amount of Sale	Tax Rate	Amount of Tax	Total Cost
1.	$100	6%	$6	$106
2.	$200	4%	$8	$208
3.	$68	3%	$2.04	$70.04
4.	$185	5%	$9.25	$194.25
5.	$365	6%	$21.90	$386.90
6.	$28	7%	$1.96	$29.96
7.	$220	$5\frac{1}{2}\%$	$12.10	$232.10
8.	$780	$7\frac{1}{2}\%$	$58.50	$838.50

✍ Writing   ◉ Conceptual   ▲ Challenging   ≈ Estimation

Find the commission earned, or the rate of commission. Round money answers to the nearest cent if necessary.

Example: The sales are $9850 and the commission rate is 3%. Find the amount of commission.

Solution: commission = rate of commission · sales
= 3% · $9850
= 0.03 · $9850
= $295.50   Amount of commission

The amount of commission is **$295.50**.

	Sales	Rate of Commission	Commission
9.	$300	7%	$21
10.	$650	8%	$52
11.	$4800	25%	$1200
12.	$3250	12%	$390
13.	$5783	3%	$173.49
14.	$2275	7%	$159.25
15.	$45,250	10%	$4525
16.	$65,300	5%	$3265

Find the amount or rate of discount and the amount paid after the discount. Round money answers to the nearest cent if necessary.

Example: A Casio QV-30 Digital Camera is normally priced at $898. If it is discounted 20%, find the amount of discount and the sale price.	**Solution:** discount = rate of discount • original price
	= 20% • $898
	= 0.2 • $898
	− $179.60 Amount of discount
	The amount of discount is **$179.60** and the sale price is **$718.40** ($898 − $179.60).

	Original Price	Rate of Discount	Amount of Discount	Sale Price
17.	$100	15%	$15	$85
18.	$200	20%	$40	$160
19.	$180	30%	$54	$126
20.	$38	25%	$9.50	$28.50
21.	$17.50	25%	$4.38	$13.12
22.	$76	60%	$45.60	$30.40
23.	$37.50	10%	$3.75	$33.75
24.	$49.90	40%	$19.96	$29.94

25. You are trying to decide between Company A paying a 10% commission and Company B paying an 8% commission. For which company would you prefer to work? Are there considerations other than commission rate that would be important to you? What would they be?

On the basis of commission alone I would choose Company A. Other considerations might be: reputation of the company; expense allowances; other fringe benefits; travel; promotion and training, to name a few.

26. Give four examples of where you might use the percent of increase or the percent of decrease in your own personal activities. Think in terms of work, school, home, hobbies, and sports. Write an increase or a decrease problem about one of these four examples, then show how to solve it.

Some answers might be:
Calculating percent pay increases or decreases; changes in the cost of utilities, groceries, gasoline, and insurance; changes in the value of investments; the economy (inflation or deflation); to name a few.
   The price of a loaf of bread increased from $1.50 to $1.65. Find the percent of increase.

$1.65 − $1.50 = $0.15   Increase

$$\frac{0.15}{1.5} = \frac{p}{100}$$ **Percent proportion**

$$10 = p$$

The percent of increase is 10%.

Solve the following application problems. Round money answers to the nearest cent and rates to the nearest tenth of a percent if necessary.

27. The sales tax rate is 5% and the sales at Fort Bragg Gifts are $1050. Find the amount of sales tax.

$52.50

28. An Exer-Cycle Machine sells for $590 plus 7% sales tax. Find the amount of sales tax.

$41.30

29. The Diamond Center sells diamond jewelry at 40% off the regular price. Find the sale price of a diamond ring normally priced at $3850.

$2310

30. Stephen Louis can purchase a new car at 8% below sticker price. Find his cost on a car with a window sticker price of $17,650.

$16,238

31. An Anderson wood frame French door is priced at $1980 with a sales tax of $99. Find the rate of sales tax.

5%

32. Textbooks for three classes cost $135 plus sales tax of $8.10. Find the sales tax rate.

6%

33. Students were charged $1449 as tuition this quarter. If the tuition was $1228 last quarter, find the percent of increase.

≈18.0%

34. Americans are eating more fish. This year the average American will eat $15\frac{1}{2}$ pounds compared to only $12\frac{1}{2}$ pounds per year a decade ago. Find the percent of increase. (*Hint:* $15\frac{1}{2} = 15.5$; $12\frac{1}{2} = 12.5$.)

24%

35. The number of industrial accidents this month fell to 989 accidents from 1276 accidents last month. Find the percent of decrease.

≈22.5%

36. The average number of hours worked in manufacturing jobs last week fell from 41.1 to 40.9. Find the percent of decrease.

≈0.5%

37. A "super 45% off sale" begins today. What is the sale price of a ski parka normally priced at $135?

$74.25

38. What is the sale price of a $549 Maytag dishwasher with a discount of 35%?

$356.85

39. Tara Chatard is a sales representative for a cosmetic company. If she was paid a commission of $459 on sales of $7650, find her rate of commission.

6%

40. Easthills Ski Center has just been sold for $1,692,804. The real estate agent selling the Center earned a commission of $47,400. Find the rate of commission.

≈2.8%

41. An 8-millimeter camcorder normally priced at $590 is on sale for 18% off. Find the discount and the sales price.

$106.20; $483.80

42. This week minivans are offered at 15% off manufacturers' suggested price. Find the discount and the sale price of a minivan originally priced at $23,500.

$3525; $19,975

43. The price per share of Toys R Us stock fell from $35.50 to close at $33.50. Find the percent of decrease in price.

≈5.6%

44. In the past five years, the cost of generating electricity from the sun has been brought down from 24 cents per kilowatt hour to 8 cents (less than the newest nuclear power plants). Find the percent of decrease.

≈66.7%

▲ **45.** College students are offered a 6% discount on a dictionary that sells for $18.50. If the sales tax is 6%, find the cost of the dictionary including the sales tax.

≈$18.43

▲ **46.** A FAX machine priced at $398 is marked down 7% to promote the new model. If the sales tax is also 7%, find the cost of the FAX machine including sales tax.

≈$396.05

47. A real estate agent sells a house for $129,605. A sales commission of 6% is charged. The agent gets 55% of this commission. How much money does the agent get?

≈$4276.97

▲ 48. The local real estate agents' association collects a fee of 2% on all money received by its members. The members charge 6% of the selling price of a property as their fee. How much does the association get, if its members sell property worth a total of $8,680,000?

$10,416

▲ 49. What is the total price of a boat with an original price of $13,905, if it is sold at an 18% discount? Sales tax is $4\frac{3}{4}\%$.

≈$11,943.70

▲ 50. A commercial security alarm system originally priced at $10,800 is discounted 22%. Find the total price of the system if the sales tax rate is $7\frac{1}{4}\%$.

$9034.74

Review and Prepare

Use the percent equation (amount = percent • base) to find amount, base, or percent.
*(For help, see **Section 6.6**.)*

51. 10% of 780 brackets

78 brackets

52. 6.2% of $2075

$128.65

53. 0.2% of $1920

$3.84

54. 0.5% of 700 barrels

3.5 barrels

55. $6\frac{1}{4}\%$ of what number is 50?

800

56. $5\frac{1}{2}\%$ of what number is 66?

1200

57. 147.2 meters is what percent of 460 meters?   32%

58. 125.8 yards is what percent of 740 yards?   17%

6.8 Simple Interest

When we open a savings account we are actually lending money to the financial institution. It will in turn lend this money to individuals and businesses. These people then become borrowers. The financial institution pays a fee to the savings account holders and charges a higher fee to its borrowers. This fee is called interest.

Interest is a fee paid or a charge made for lending or borrowing money. The amount of money borrowed is called the **principal.** The charge for interest is often given as a percent, called the interest rate or **rate of interest.** The rate of interest is assumed to be *per year,* unless stated otherwise.

OBJECTIVE 1 In most cases interest is computed on the original principal and is called **simple interest.** We use the following **interest formula** to find simple interest.

> **Formula for Simple Interest**
>
> interest = principal • rate • time
>
> The formula is usually written in letters.
>
> $$I = p \cdot r \cdot t$$

Note
Simple interest is used for most short-term business loans, most real estate loans, and many automobile and consumer loans.

EXAMPLE 1   Finding Interest for a Year

Find the interest on $2000 at 6% for 1 year.

The amount borrowed, or principal (p), is $2000. The interest rate (r) is 6%, which is 0.06 as a decimal, and the time of the loan (t) is 1 year. Use this formula.

$$I = p \quad \cdot \quad r \quad \cdot t$$
$$I = 2000 \cdot (0.06) \cdot 1$$
$$I = 120$$

The interest is $120.

WORK PROBLEM 1 AT THE SIDE. ▶▶

EXAMPLE 2   Finding Interest for More Than a Year

Find the interest on $4200 at 8% for three and a half years.

The principal (p) is $4200. The rate ($r$) is 8% or 0.08 as a decimal, and the time (t) is $3\frac{1}{2}$ or 3.5 years. Use the formula.

$$I = p \quad \cdot \quad r \quad \cdot \quad t$$
$$I = 4200 \cdot (0.08) \cdot (3.5)$$
$$I = 1176$$

The interest charge is $1176.

WORK PROBLEM 2 AT THE SIDE. ▶▶

Note
It is best to change fractions in percents or fractions of years to their decimal form. In Example 2, the $3\frac{1}{2}$ years becomes 3.5 years.

OBJECTIVES

1 ▶ Find the simple interest on a loan.
2 ▶ Find the total amount due on a loan.

FOR EXTRA HELP

Tutorial   Tape 11   SSM, Sec. 6.8

1. Find the interest.

   (a) $500 at 4% for 1 year

   (b) $1850 at 6% for 1 year

2. Find the interest.

   (a) $340 at 5% for $3\frac{1}{2}$ years

   (b) $2450 at 8% for $3\frac{1}{4}$ years

   (c) $14,200 at 6% for $2\frac{3}{4}$ years

ANSWERS
1. **(a)** $20 **(b)** $111
2. **(a)** $59.50 **(b)** $637 **(c)** $2343

3. Find the interest.

(a) $1500 at 7% for 4 months

(b) $25,000 at $10\frac{1}{2}$% for 3 months

Interest rates are given *per year*. For loan periods of less than one year, be careful to express time as a fraction of a year.

If time is given in months, for example, use a denominator of 12, because there are 12 months in a year. A loan of 9 months would be for $\frac{9}{12}$ of a year.

E X A M P L E 3   Finding Interest for Less Than 1 Year

Find the interest on $840 at $8\frac{1}{2}$% for 9 months.

The principal is $840. The rate is $8\frac{1}{2}$% or 0.085, and the time is $\frac{9}{12}$ of a year. Use the formula $I = p \cdot r \cdot t$.

$$I = \underbrace{840 \cdot (0.085)} \cdot \frac{9}{12} \qquad \text{9 months} = \frac{9}{12} \text{ of a year.}$$

$$= \quad 71.4 \quad \cdot \frac{3}{4} \qquad \frac{9}{12} \text{ in lowest terms is } \frac{3}{4}.$$

$$= \frac{(71.4) \cdot 3}{4}$$

$$= \frac{214.2}{4} = 53.55$$

The interest is $53.55.

4. Find the total amount due on a loan of

(a) $2500 at $7\frac{1}{2}$% for 6 months

Calculator Tip: The calculator solution to Example 3 uses chain calculations.

$$840 \boxed{\times} .085 \boxed{\times} 9 \boxed{\div} 12 \boxed{=} 53.55$$

◀◀ **WORK PROBLEM 3 AT THE SIDE.**

OBJECTIVE 2▶ When a loan is repaid, the interest is added to the original principal to find the total amount due.

Formula for Amount Due

amount due = principal + interest

(b) $10,800 at 6% for 4 years

E X A M P L E 4   Calculating the Total Amount Due

A loan of $1080 has been made at 8% for three months. Find the total amount due.

First find the interest, then add the principal and the interest to find the total amount due.

(c) $4300 at 10% for $2\frac{1}{2}$ years

$$I = 1080 \cdot (0.08) \cdot \frac{3}{12} \qquad \text{3 months} = \frac{3}{12} \text{ of a year.}$$

$$I = 21.60$$

The interest is $21.60.

$$\textbf{amount due} = \text{principal} + \text{interest}$$
$$= \$1080 + \$21.60 = \$1101.60$$

The total amount due is $1101.60.

ANSWERS

3. (a) $35 **(b)** $656.25
4. (a) $2593.75 **(b)** $13,392 **(c)** $5375

◀◀ **WORK PROBLEM 4 AT THE SIDE.**

6.8 Exercises

Find the interest.

> **Example:** Find the interest on $750 at 5% for 2 years.
>
> **Solution:** $I = p \cdot r \cdot t$
> $= 750 \cdot (0.05) \cdot 2$
> $= 75$
>
> The interest is **$75.**

	Principal	Rate	Time in Years	Interest
1.	$200	4%	1	$8
2.	$100	5%	1	$5
3.	$600	6%	4	$144
4.	$800	7%	2	$112
5.	$240	4%	3	$28.80
6.	$190	3%	2	$11.40
7.	$2300	$8\frac{1}{2}$%	$2\frac{1}{2}$	$488.75
8.	$4700	$5\frac{1}{2}$%	$1\frac{1}{2}$	$387.75
9.	$9400	$6\frac{1}{2}$%	$1\frac{1}{4}$	$763.75
10.	$10,000	$7\frac{1}{2}$%	$3\frac{1}{4}$	$2437.50

✎ Writing   ◉ Conceptual   ▲ Challenging   ≈ Estimation

Find the interest. Round to the nearest cent if necessary.

Example: Find the interest on $980 at 7% for 6 months.

Solution: $I = p \cdot r \cdot t$

$$= \underbrace{980 \cdot (0.07)} \cdot \frac{6}{12} \qquad \text{6 months} = \tfrac{6}{12} \text{ of a year.}$$

$$= \quad 68.6 \quad \cdot \frac{1}{2} \qquad \tfrac{6}{12} \text{ in lowest terms is } \tfrac{1}{2}.$$

$$= \frac{68.6}{2} = 34.3$$

The interest is **$34.30.**

	Principal	Rate	Time in Months	Interest
11.	$200	6%	6	$6
12.	$500	7%	9	$26.25
13.	$750	5%	12	$37.50
14.	$920	6%	18	$82.80
15.	$820	8%	24	$131.20
16.	$92	4%	5	≈$1.53
17.	$1250	$5\frac{1}{2}$%	3	≈$17.19
18.	$2440	$6\frac{1}{4}$%	3	≈$38.13

	Principal	Rate	Time in Months	Interest
19.	$15,000	$7\frac{1}{4}\%$	7	≈$634.38
20.	$11,700	$4\frac{1}{2}\%$	5	≈$219.38

Find the total amount due on the following loans.

Example: A loan of $550 was made at 6% for 8 months. Find the total amount due.

Solution: $I = p \cdot r \cdot t$

$$= \underbrace{550 \cdot 0.06} \cdot \frac{8}{12}$$

$$= 33 \cdot \frac{2}{3} \qquad \frac{8}{12} \text{ in lowest terms is } \frac{2}{3}.$$

$$= \frac{33 \cdot 2}{3} = \frac{66}{3} = 22$$

The interest is $22.

amount due = principal + interest

= $550 + $22 = $572

The total amount due is **$572.**

	Principal	Rate	Time	Total Amount Due
21.	$300	4%	1 year	$312
22.	$600	5%	6 months	$615
23.	$740	6%	9 months	$773.30
24.	$1180	3%	2 years	$1250.80
25.	$1500	10%	18 months	$1725

	Principal	Rate	Time	Total Amount Due
26.	$3000	5%	5 months	$3062.50
27.	$2450	7%	6 months	$2535.75
28.	$5400	4%	1 year	$5616
29.	$17,800	$7\frac{1}{2}\%$	9 months	$18,801.25
30.	$20,500	$5\frac{1}{2}\%$	6 months	$21,063.75

31. The amount of interest paid on savings accounts and charged on loans can vary from one institution to another. However, when the amount of interest is calculated, three factors are used in the calculation. Name these three factors and describe them in your own words.

The answer should include:
Amount of principal—This is the amount of money borrowed or loaned.
Interest rate—This is the percent used to calculate the interest.
Time of loan—The length of time that money is loaned or borrowed is an important factor in determining interest.

32. Interest rates are usually given as a rate per year (annual rate). Explain what must be done when time is given in months. Write your own problem where time is given in months and then show how to solve it.

When time is given in months, the number of months are placed over 12. This becomes a fraction of a year. For example,
6 months = $\frac{6}{12}$ year = $\frac{1}{2}$ year = 0.5 year.

Solve the following application problems. Round to the nearest cent if necessary.

33. Gil Eckern deposits $4850 at 6% for 1 year. How much interest will he earn?

$291

34. The Jidobu family invests $18,000 at 9% for 6 months. What amount of interest will the family earn?

$810

35. A bank in New York City loans $50,000 to a business at 9% for 18 months. How much interest will the bank earn?

$6750

36. Pat Martin, a retiree, deposits $80,000 at 7% for 3 years. How much interest will be earned?

$16,800

37. A student borrows $1000 at 10% for 3 months to pay tuition. Find the total amount due.

$1025

38. A loan of $1350 will be paid back with 12% interest at the end of 9 months. Find the total amount due.

$1471.50

39. Norell Williams deposits $7840 in a credit union account for 9 months. If the credit union pays $5\frac{1}{2}$% interest, find the amount of interest she will earn.

$323.40

40. Silvo Di Loreto, owner of Sunset Realtors, borrows $27,000 to update his office computer system. If the loan is for 24 months at $7\frac{1}{4}$%, find the amount of interest he will owe.

$3915

▲ **41.** An investment fund pays $7\frac{1}{4}\%$ interest. If Beverly Habecker deposits $8800 in her account for $\frac{1}{4}$ year, find the amount of interest she will earn.

$159.50

▲ **42.** Ms. Henderson owes $1900 in taxes. She is charged a penalty of $12\frac{1}{4}\%$ annual interest and pays the taxes and penalty after 6 months. Find the total amount she must pay.

≈$2016.38

▲ **43.** A gift shop owner has additional profits of $11,500 that are invested at $8\frac{3}{4}\%$ interest for $\frac{3}{4}$ of a year. Find the total amount in the account at the end of this time.

≈$12,254.69

▲ **44.** A pawn shop owner lends $35,400 to another business for $\frac{1}{2}$ of a year at an interest rate of 14.9%. How much interest will be earned on the loan?

$2637.30

Review and Prepare

Rewrite each of the following fractions with the indicated denominator. (For help, see Section 3.2.)

45. $\dfrac{2}{3} = \dfrac{8}{12}$

46. $\dfrac{7}{8} = \dfrac{21}{24}$

47. $\dfrac{4}{5} = \dfrac{48}{60}$

48. $\dfrac{3}{8} = \dfrac{27}{72}$

49. $\dfrac{5}{12} = \dfrac{25}{60}$

50. $\dfrac{4}{5} = \dfrac{80}{100}$

51. $\dfrac{15}{19} = \dfrac{60}{76}$

52. $\dfrac{7}{15} = \dfrac{98}{210}$

6.9 *Compound Interest*

The interest we studied in **Section 6.8** was *simple interest* (interest only on the original principal). A common type of interest used with savings accounts and most investments is **compound interest,** or interest paid on past interest as well as on the principal.

OBJECTIVE 1 Suppose that you make a single deposit of $1000 in a savings account that earns 5% per year. What will happen to your savings over three years? At the end of the first year, one year's interest on the original deposit is found. Use the simple interest formula.

$$\text{Interest} = \text{principal} \cdot \text{rate} \cdot \text{time}$$

Year 1 $1000 \cdot 0.05 \cdot 1 = 50
Add the interest to the $1000 to find the amount in your account at the end of the first year. $1000 + $50 = 1050
The interest for the second year is found on $1050; that is, the interest is **compounded.**

Year 2 $1050 \cdot 0.05 \cdot 1 = 52.50
Add this interest to the $1050 to find the amount in your account at the end of the second year. **$1050 + $52.50 = 1102.50**
The interest for the third year is found on $1102.50.

Year 3 **$1102.50 \cdot 0.05 \cdot 1 \approx 55.13**
Add this interest to the **$1102.50.** ($1102.50 + $55.13 = 1157.63)

At the end of three years, you will have **$1157.63** in your savings account. The $1157.63 that you have in your account is called the **compound amount.**

If you had earned only *simple* interest, then, for 3 years,

$$I = 1000 \cdot 0.05 \cdot 3$$
$$= 150$$

and you would have $1000 + $150 = $1150 in your account. Compounding the interest increased your earnings by $7.63. ($1157.63 − $1150)

With *compound* interest, the interest earned during the second year is greater than that earned during the first year, and the interest earned during the third year is greater than that earned during the second year.

This happens because the interest earned each year is *added* to the principal, and the new total is used to find the amount of interest in the next year.

Compound Interest

Interest paid on principal plus past interest is **compound interest.**

OBJECTIVES

1 ▸ Understand compound interest.

2 ▸ Understand compound amount.

3 ▸ Solve for the compound amount.

4 ▸ Use a compound interest table.

5 ▸ Find the compound amount and the amount of compound interest.

FOR EXTRA HELP

Tutorial   Tape 11   SSM, Sec. 6.9

1. Find the compound amount given the following deposits.

   (a) $500 at 4% for 2 years

   (b) $1200 at 3% for 3 years

2. Find the compound amount by multiplying the original deposited by 100% plus the compound interest rate in each of the following.

   (a) $1500 at 5% for 3 years

   $1500 \cdot 1.05 \cdot 1.05 \cdot 1.05$

   = _____

   (b) $900 at 3% for 2 years

   (c) $2900 at 6% for 4 years

ANSWERS
1. (a) $540.80 (b) $1311.27
2. (a) ≈$1736.44 (b) $954.81
   (c) ≈$3661.18

OBJECTIVE 2 Find the compound amount as follows.

E X A M P L E I   Finding the Compound Amount

Nancy Wegener deposits $3400 in an account that pays 6% interest compounded annually for 4 years. Find the compound amount. Round to the nearest cent when necessary.

Year	Interest	Compound Amount
1	$3400 \cdot 0.06 \cdot 1 = \$204$	
	$3400 + \$204 = \$3604$	
2	$3604 \cdot 0.06 \cdot 1 = \$216.24$	
	$3604 + \$216.24 = \$3820.24$	
3	$3820.24 \cdot 0.06 \cdot 1 \approx \$229.21$	
	$3820.24 + \$229.21 = \$4049.45$	
4	$4049.45 \cdot 0.06 \cdot 1 \approx \$242.97$	
	$4049.45 + \$242.97 = \$4292.42$	

The compound amount is $4292.42.

◄◄ WORK PROBLEM I AT THE SIDE.

OBJECTIVE 3 A more efficient way of finding the compound amount is to add the interest rate to 100% and then multiply by the original deposit. Notice that in Example 1, at the end of the first year, you will have $3400 (100% of the original deposit) plus 6% (of the original deposit) or 106% (100% + 6% = 106%).

E X A M P L E 2   Finding the Compound Amount

Find the compound amount in Example 1 using multiplication.

Year 1 Year 2 Year 3 Year 4

$$\$3400 \cdot \underbrace{1.06 \cdot 1.06 \cdot 1.06 \cdot 1.06}_{} \approx \$4292.42$$

Original deposit   100% + 6% = 106% = 1.06   Compound amount

Our answer, $4292.42 is the same as Example 1.

◄◄ WORK PROBLEM 2 AT THE SIDE.

Note

By adding the compound interest rate to 100% we can then multiply by the original deposit. This will give us the compound amount at the end of each compound interest period.

▦ *Calculator Tip:* If you use a calculator for Example 2 you can use the $\boxed{y^x}$ key (exponent key).

$$3400 \boxed{\times} 1.06 \boxed{y^x} 4 \boxed{=} 4292.42 \text{ (rounded)}$$

The 4 following the $\boxed{y^x}$ key represents the number of compound interest periods.

OBJECTIVE 4 The calculation of compound interest can be quite tedious. For this reason compound interest tables have been developed.

Suppose you deposit $1 in a savings account today that earns 4% compounded annually and you allow the deposit to remain for 3 years. The diagram below shows the compound amount at the end of each of the 3 years.

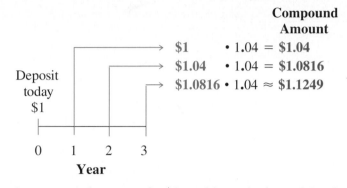

Using the compound amounts for $1, a table can be formed. Look at the table below and find the column headed 4%. The first three numbers for years 1, 2, and 3 are the same as those we have calculated for $1 at 4% for 3 years. This table giving the compound amounts on a $1 deposit for given lengths of time and interest rates can be used for finding the compound amount on any amount of deposit.

Compound Interest

Time Periods	3.00%	3.50%	4.00%	4.50%	5.00%	5.50%	6.00%	8.00%
1	1.0300	1.0350	1.0400	1.0450	1.0500	1.0550	1.0600	1.0800
2	1.0609	1.0712	1.0816	1.0920	1.1025	1.1130	1.1236	1.1664
3	1.0927	1.1087	**1.1249**	1.1412	1.1576	1.1742	1.1910	1.2597
4	1.1255	1.1475	1.1699	1.1925	1.2155	1.2388	1.2625	1.3605
5	1.1593	1.1877	1.2167	1.2462	1.2763	1.3070	1.3382	1.4693
6	1.1941	1.2293	1.2653	1.3023	1.3401	1.3788	1.4185	1.5869
7	1.2299	1.2723	1.3159	1.3609	1.4071	1.4547	1.5036	1.7138
8	1.2668	1.3168	1.3686	1.4221	1.4775	1.5347	1.5938	1.8509
9	1.3048	1.3629	1.4233	1.4861	1.5513	1.6191	1.6895	1.9990
10	1.3439	1.4106	1.4802	1.5530	**1.6289**	1.7081	1.7908	2.1589
11	1.3842	1.4600	1.5395	1.6229	1.7103	1.8021	1.8983	2.3316
12	1.4258	1.5111	1.6010	1.6959	1.7959	1.9012	2.0122	2.5182

EXAMPLE 3   Using a Compound Interest Table

Find the compound amount.

(a) $1 is deposited at a 5% interest rate for 10 years.

Look down the column headed 5%, and across to row 10 (10 years = 10 time periods). At the intersection of the column and row, read the compound amount, **1.6289**, which can be rounded to $1.63.

(b) $1 is deposited at $5\frac{1}{2}$% for 6 years.

The intersection of the $5\frac{1}{2}$% (5.50%) column and row 6 shows 1.3788 as the compound amount. Round this to $1.38.

> **WORK PROBLEM 3 AT THE SIDE.** ▶▶

---

3. Find the compound amount using the compound interest table. Round to the nearest cent.

(a) $1 at 4% for 12 years

(b) $1 at 3% for 5 years

(c) $1 at $5\frac{1}{2}$% for 8 years

ANSWERS

3. (a) ≈$1.60 **(b)** ≈$1.16 **(c)** ≈$1.53

▦ **4.** Find the compound amount and the interest.

  (a) $5000 at 6% for 12 years

  (b) $14,100 at $3\frac{1}{2}$% for 10 years

  (c) $25,600 at 8% for 11 years

OBJECTIVE 5 Find the compound amount and interest as follows.

> **Finding the Compound Amount and the Interest**
>
> **Compound Amount**
> Find the compound amount for any amount of principal by multiplying the principal by the compound amount for $1.
>
> **Interest**
> Find the amount of interest earned on a deposit by subtracting the amount originally deposited from the compound amount.

EXAMPLE 4   **Finding Compound Interest**

Find the compound amount and the interest.

(a) $1000 at $5\frac{1}{2}$% interest for 12 years

Look in the table for $5\frac{1}{2}$% (5.50%) and 12 periods; find the number 1.9012 but do not round it. Multiply this number and the principal of $1000.

$$\$1000 \cdot 1.9012 = \$1901.20$$

The account will contain $1901.20 after 12 years.

Find the amount of interest by subtracting the original deposit from the compound amount.

Compound amount ——————⌐   Original amount   ⌐—— Amount of interest

$$\$1901.20 - \$1000 = \$901.20$$

(b) $6400 at 8% for 7 years

Look in the table for 8% and 7 periods, finding 1.7138. Multiply.

$$\$6400 \cdot 1.7138 = \$10,968.32 \quad \text{Compound amount}$$

Subtract the original deposit from the compound amount.

$$\$10,968.32 - \$6400 = \$4568.32 \quad \text{Interest}$$

A total of $4568.32 in interest was earned.

◀◀ **WORK PROBLEM 4 AT THE SIDE.**

ANSWERS

4. (a) $10,061; $5061
  (b) $19,889.46; $5789.46
  (c) $59,688.96; $34,088.96

6.9 Exercises

Find the compound amount given the following deposits. Calculate the interest each year, then add it to the previous year's amount.

Example: $800 at 4% for 3 years

Solution:

Year	Interest	Compound Amount
1	$800 • 0.04 • 1 = $32	
	$800 + $32 =	$832
2	$832 • 0.04 • 1 = $33.28	
	$832 + $33.28 =	$865.28
3	$865.28 • 0.04 • 1 ≈ $34.61	
	$865.28 + $34.61 =	$899.89

The compound amount is **$899.89**

1. $1000 at 5% for 2 years

$1102.50

2. $700 at 3% for 2 years

$742.63

3. $500 at 4% for 3 years

$562.43

4. $2000 at 8% for 3 years

$2519.42

5. $3500 at 7% for 4 years

$4587.79

6. $5500 at 6% for 4 years

$6943.63

Find the compound amount by multiplying the original amount deposited by 100% plus the compound rate in each of the following.

Example: $1000 at 7% for 4 years

Solution:

Year 1 Year 2 Year 3 Year 4

$1000 • $\underbrace{\text{1.07} • \text{1.07} • \text{1.07} • \text{1.07}}_{100\% + 7\% = 107\% = 1.07}$ ≈ $1310.80

↑ Compound amount

The compound amount is **$1310.80.**

7. $800 at 4% for 3 years

≈$899.89

8. $600 at 6% for 3 years

≈$714.61

9. $1200 at 3% for 5 years

≈$1391.13

10. $1500 at 5% for 4 years

≈$1823.26

11. $1180 at 7% for 8 years

≈$2027.46

12. $12,800 at 6% for 7 years

≈$19,246.47

13. $9850 at 5% for 10 years

≈$16,044.61

14. $14,120 at 4% for 10 years

≈$20,901.05

Use the table on page 417 to find the compound amount and the interest. Interest is compounded annually. Round to the nearest cent if necessary.

> **Example:** $500 at 4% for 8 years   **Solution:** 4% column, row 8 of the table gives 1.3686.
> Multiply. $500 • 1.3686 = **$684.30**   **Compound amount**
> $684.30 − $500 = **$184.30**   **Interest**

15. $1000 at 6% for 4 years

$1262.50; $262.50

16. $10,000 at 3% for 5 years

$11,593; $1593

17. $4000 at 5% for 9 years

$6205.20; $2205.20

18. $5700 at 4% for 10 years

$8437.14; $2737.14

19. $8428.17 at $4\frac{1}{2}$% for 6 years

≈$10,976.01; $2547.84

20. $10,472.88 at $5\frac{1}{2}$% for 12 years

≈$19,911.04; $9438.16

21. Write a definition for compound interest. Describe in your own words what compound interest means to you.

Interest paid on past interest as well as on the principal.
Many people describe compound interest as "interest on interest."

22. What is the difference between the compound amount and compound interest?

Compound amount is the total amount, original deposit plus interest on deposit, at the end of the compound interest period.
Compound interest is found by subtracting the original deposit from the compound amount.

Use the table on page 417 to solve each application problem. Round to the nearest cent if necessary.

23. Glenda Wong deposits $5280 in an account that pays 8% interest, compounded annually. Find the amount she will have (compound amount) at the end of 5 years.

≈$7757.90

24. John Hendrick borrows $10,500 from his uncle to open Campus Bicycles. He will repay the loan at the end of 6 years with interest at 6% compounded annually. Find the amount he will repay.

$14,894.25

25. Al Granard lends $7500 to the owner of Rick's Limousine Service. He will be repaid at the end of 6 years at 8% interest compounded annually. Find **(a)** the total amount that he should be repaid and **(b)** the amount of interest earned.

(a) $11,901.75   (b) $4401.75

26. Sadie Simms has $28,500 in an Individual Retirement Account (IRA) that pays 5% interest compounded annually. Find **(a)** the total amount she will have at the end of 5 years and **(b)** the amount of interest earned.

(a) $36,374.55   (b) $7874.55

▲ **27.** Jennifer Boalt deposits $10,000 at 6% compounded annually. Two years after she makes the first deposit, she adds another $20,000, also at 6% compounded annually.
(a) What total amount will she have five years after her first deposit?
(b) What amount of interest will she have earned?

(a) ≈$37,202.08   (b) ≈$7202.08

▲ **28.** Scott Striver invests $15,000 at 8% compounded annually. Three years after he makes the first deposit, he adds another $15,000, also at 8% compounded annually.
(a) What total amount will he have five years after his first deposit?
(b) What amount of interest will he have earned?

(a) ≈$39,535.71   (b) ≈$9535.71

Review and Prepare

*Simplify by using the order of operations. (For help, see **Section 4.6**.)*

29. $12.6 \div 8.4 \cdot 6.2$

9.3

30. $18.304 \div 8.32 \cdot 3$

6.6

31. $19.3 + (6.7 - 5.2) \cdot 58$

106.3

32. $1.06 + (4.85 - 3.95) \cdot 2.25$

3.085

33. $5.34 - 2.6 \cdot 5.2 \div 2.6$

0.14

34. $61.5 - 22.8 \cdot 15 \div 5.7$

1.5

6.1	**percent**	Percent means per one hundred. A percent is a ratio with a denominator of 100.
6.3	**percent proportion**	The proportion $\dfrac{\text{amount}}{\text{base}} = \dfrac{\text{percent}}{100}$ or $\dfrac{a}{b} = \dfrac{p}{100}$ is used to solve percent problems.
6.4	**base**	The base in a percent problem is the entire quantity, the total, or the whole.
	amount	The amount in a percent problem is the part being compared with the whole.
6.6	**percent equation**	The percent equation is amount = percent • base. It is another way to solve percent problems.
6.7	**sales tax**	Sales tax is a percent of the total sales charged as a tax.
	commission	Commission is a percent of the dollar value of total sales paid to a salesperson.
	discount	Discount is often expressed as a percent of the original price; it is then deducted from the original price, resulting in the sale price.
	percent of increase or decrease	Percent of increase or decrease is the amount of increase or decrease expressed as a percent of the original amount.
6.8	**interest**	Interest is a fee paid or a charge for lending or borrowing money.
	interest formula	The interest formula is used to calculate interest. It is interest = principal • rate • time or $I = p \cdot r \cdot t$.
	simple interest	Interest that is computed on the original principal.
	principal	Principal is the amount of money on which interest is earned.
	rate of interest	Often referred to as "rate," it is the charge for interest and is given as a percent.
6.9	**compound interest**	Compound interest is interest paid on past interest as well as on principal.
	compound amount	The total amount in an account including compound interest and the original principal.
	compounding	Interest that is **compounded** once each year is compounded **annually.**

Concepts	*Examples*
6.1 Basics of Percent	
Writing a Percent as a Decimal To write a percent as a decimal, move the decimal point two places to the left and drop the % sign.	50% (.50%) = 0.50 or just 0.5 3% (.03%) = 0.03
Writing a Decimal as a Percent To write a decimal as a percent, move the decimal point two places to the right and attach a % sign.	0.75 (0.75) = 75% 3.6 (3.60) = 360%
6.2 Writing a Fraction as a Percent Use a proportion and solve for p to change a fraction to percent.	$\dfrac{2}{5} = \dfrac{p}{100}$ Proportion $5 \cdot p = 2 \cdot 100$ Cross products $5 \cdot p = 200$ $\dfrac{\overset{1}{\cancel{5}} \cdot p}{\underset{1}{\cancel{5}}} = \dfrac{200}{5}$ Divide both sides by 5. $p = 40$ $\dfrac{2}{5} = 40\%$ Attach % sign.

Concepts	Examples
6.3 Learning the Percent Proportion **Amount (a)** is to **base (b)** as percent (**p**) is to 100, or $$\frac{a}{b} = \frac{p}{100}.$$	Use the percent proportion to solve for the missing number. $a = 30$, $b = 50$, find p. $\dfrac{30}{50} = \dfrac{p}{100}$ Percent proportion $\dfrac{3}{5} = \dfrac{p}{100}$ Lowest terms $5 \cdot p = 3 \cdot 100$ Percent proportion $5 \cdot p = 300$ $\dfrac{\overset{1}{\cancel{5}} \cdot p}{\underset{1}{\cancel{5}}} = \dfrac{300}{5}$ Divide both sides by 5. $p = 60$ The percent is 60, which is 60%.

6.4 Identifying Percent (p), Base (b), and Amount (a) in a Percent Problem
 	
The percent (*p*) appears with the word **percent** or with the symbol %.
 	
The base (*b*) often appears after the word **of.** Base is the entire quantity or total.
 	
The amount (*a*) is the part of the total. If **p** and **b** are found first, the remaining number is **a.**
	Find *p*, *b*, and *a* in each of the following.
	
	10% **of** the 500 pies is how many pies?
	  p *b* *a* (unknown)
	
	20 cats is 5% **of** what number of cats?
	  a *p* *b* (unknown)
	
	What percent **of** \$220 is \$33?
	  p (unknown) *b* *a*

6.5 Applying the Percent Proportion Read the problem and identify *p*, *b*, and *a*. Use the percent proportion to solve for the unknown quantity *b*.	A tank contains 35% distilled water. 28 gallons of distilled water are in the tank when it is full. Find the volume of the tank. $$p = 35 \quad \text{and} \quad a = 28$$ Use the percent proportion to find *b*. $\dfrac{a}{b} = \dfrac{p}{100}$ $\dfrac{28}{b} = \dfrac{35}{100}$ $\dfrac{28}{b} = \dfrac{7}{20}$ Lowest terms $b \cdot 7 = 560$ Cross products $\dfrac{b \cdot \overset{1}{\cancel{7}}}{\underset{1}{\cancel{7}}} = \dfrac{560}{7}$ Divide both sides by 7. $b = 80$ The volume of the tank is 80 gallons.

Concepts	Examples
6.6 Using the Percent Equation The percent equation is amount = percent • base. Identify p, b, and a and solve for the unknown quantity. Always write the percent as a decimal before using the equation.	Solve each of the following. **(a)** Find 15% of 160 drivers. $$\text{amount} = \text{percent} \cdot \text{base}$$ $$a = 0.15 \cdot 160$$ $$a = 24$$ 15% of 160 drivers is 24 drivers. **(b)** 8 balls is 4% of what number of balls? $$\text{amount} = \text{percent} \cdot \textbf{base}$$ $$8 = 0.04 \cdot b$$ $$\frac{8}{0.04} = \frac{\overset{1}{\cancel{0.04}} \cdot b}{\underset{1}{\cancel{0.04}}}$$ $$b = 200$$ 8 balls is 4% of 200 balls. **(c)** \$13 is what percent of 52? $$\text{amount} = \textbf{percent} \cdot \text{base}$$ $$13 = p \cdot 52$$ $$\frac{13}{52} = \frac{p \cdot \overset{1}{\cancel{52}}}{\underset{1}{\cancel{52}}}$$ $$p = 0.25 = 25\%$$ \$13 is 25% of \$52.
6.7 Applications of Percent To solve for **sales tax,** use the formula amount of sales tax = rate of tax • cost of item. To find **commissions,** use the formula amount of commission = rate or percent of commission • amount of sales. To find the **discount** and the **sale price,** use these formulas amount of discount = rate of discount • original price **sale price** = original price − amount of discount.	The cost of an item is \$450, and the sales tax is 6%. Find the sales tax. amount of sales tax = 6% • \$450 = 0.06 • \$450 = \$27 The sales are \$92,000 with a commission rate of 3%. Find the commission. amount of commission = 3% • \$92,000 = 0.03 • \$92,000 = \$2760 A gas oven originally priced at \$480 is offered at a 25% discount. Find the amount of the discount and the sale price. discount = 0.25 • \$480 = **\$120** sale price = \$480 − \$120 = **\$360**

Concepts	*Examples*
6.7 Applications of Percent (continued) To find the **percent of change,** subtract to find the amount of change (increase or decrease), which is the amount (*a*). Base (*b*) is the original value or value before the change.	Enrollment rose from 3820 students to 5157 students. Find the percent of increase. $$5157 - 3820 = 1337 \quad \text{Increase}$$ $$\frac{1337}{3820} = \frac{p}{100}$$ Solve the proportion to find that $p = 35$ so the percent of increase is 35%.
6.8 Finding Simple Interest Use the formula $I = p \cdot r \cdot t$. **interest = principal • rate • time** Time (*t*) is in years. When the time is given in months, use a fraction with 12 in the denominator because there are 12 months in a year.	$2800 is deposited at 8% for 3 months. Find the amount of interest. $$I = p \cdot r \cdot t$$ $$= 2800 \cdot 0.08 \cdot \frac{3}{12}$$ $$= 224 \cdot \frac{1}{4} = \frac{224 \cdot 1}{4} = \$56$$

6.9 Finding Compound Amount and Compound Interest
 	
There are three methods for finding the compound amount.
 	
1. Calculate the interest for each compound interest period, then add it back to the principal.
 	
2. Multiply the original deposit by 100% plus the compound interest rate.
 	
3. Use the table to find the interest on $1. Then, multiply the table value by the principal.
 	
The compound interest is found with the formula
 	
$$\begin{array}{ccc} \textbf{compound} \\ \textbf{interest} \end{array} = \begin{array}{c} \textbf{compound} \\ \textbf{amount} \end{array} - \begin{array}{c} \textbf{original} \\ \textbf{deposit} \end{array}$$
	Find the compound amount and interest if $1500 is deposited at 5% interest for 3 years.
	
	                             Compound
	                      Interest **Amount**
	
	1. year 1 $1500 • 0.05 • 1 = **$75**
	               $1500 + $75 = **$1575**
	    year 2 $1575 • 0.05 • 1 = **$78.75**
	               $1575 + $78.75 = **$1653.75**
	    year 3 $1653.75 • 0.05 ≈ **$82.69**
	               $1653.75 + $82.69 = **$1736.44**
	
	2. $1500 • 1.05 • 1.05 • 1.05 ≈ $1736.44
	   ↑ ↑
	  Original Compound
	  deposit amount
	       100% + 5% = 105% = 1.05
	
	3. Locate 5% across the top of the table and 3 periods at the left. The table value is 1.1576.
	
	    compound amount
	       = $1500 • 1.1576 = $1736.40*
	
	    interest = $1736.44 − $1500 = $236.44
	
	--- 	
	*The difference in the compound amount results from rounding in the table.

[6.1] *Write each of the following percents as decimals and decimals as percents.*

1. 25% 0.25

2. 180% 1.8

3. 12.5% 0.125

4. 0.085% 0.00085

5. 2.65 265%

6. 0.02 2%

7. 0.875 87.5%

8. 0.002 0.2%

[6.2] *Write each percent as a fraction or mixed number in lowest terms and each fraction as percent.*

9. 12% $\frac{3}{25}$

10. 37.5% $\frac{3}{8}$

11. 250% $2\frac{1}{2}$

12. 0.25% $\frac{1}{400}$

13. $\frac{3}{4}$ 75%

14. $\frac{5}{8}$ 62.5% or $62\frac{1}{2}$%

15. $3\frac{1}{4}$ 325%

16. $\frac{1}{200}$ 0.5%

Complete this chart.

Fraction	Decimal	Percent
$\frac{1}{8}$	**17.** 0.125	**18.** 12.5%
19. $\frac{3}{20}$	0.15	**20.** 15%
21. $1\frac{4}{5}$	**22.** 1.8	180%

[6.3] *Find the value of the missing letter in the percent proportion $\frac{a}{b} = \frac{p}{100}$.*

23. $a = 50$, $p = 5$

1000

24. $b = 960$, $p = 10$

96

[6.4] *Identify percent, base, and amount in each of the following. Do not try to solve.*

25. 40% of 150 bulbs is 60 bulbs.

40; 150; 60

26. 73 brooms is what percent of 90 brooms?

unknown; 90; 73

27. Find 28% of 320 cabinets.

28; 320; unknown

28. 209 ratchets is 32% of what number of ratchets?

32; unknown; 209

29. A golfer lost 3 of his 8 golf balls. What percent were lost?

unknown; 8; 3

30. Only 88% of the door keys cut will operate properly. If there are 1280 keys cut, find the number of keys that will operate properly.

88; 1280; unknown

✍ Writing   ◉ Conceptual   ▲ Challenging   ≈ Estimation

[6.5] *Find the amount using the percent proportion or the multiplication shortcut.*

31. 12% of 450 telephones

54 telephones

32. 60% of 1450 reference books

870 reference books

33. 0.9% of 4800 miles

43.2 miles

34. 0.2% of 1400 kilograms

2.8 kilograms

Find the base using the percent proportion.

35. 35 athletes is 7% of what number of athletes?

500 athletes

36. 174 capsules is 15% of what number of capsules?

1160 capsules

37. 338.8 meters is 140% of what number of meters?

242 meters

38. 2.5% of what number of cases is 425 cases?

17,000 cases

Find the percent using the percent proportion. Round percent answers to the nearest tenth if necessary.

39. 345 lamps is what percent of 690 lamps?

50%

40. What percent of 1850 reams is 75 reams?

≈4.1%

41. What percent of 380 pairs is 36 pairs?

≈9.5%

42. What percent of 650 cans is 200 cans?

≈30.8%

[6.1–6.5] *Solve each application problem. Round to the nearest tenth if necessary.*

43. Last year 20 million Americans visited chiropractors. If 3% of these chiropractic patients were referred by a medical doctor, find the number of patients who were referred to chiropractors by medical doctors.

0.6 million or 600,000 patients

44. Scientists tell us that there are 9600 species of birds and that 1000 of these species are in danger of extinction. What percent of the bird species are in danger of extinction?

≈10.4%

[6.6] *Use the percent equation to find each of the following.*

45. 11% of $236

$25.96

46. 125% of 64 dumpsters

80 dumpsters

47. 0.128 ounces is what percent of 32 ounces?

0.4%

48. 304.5 meters is what percent of 174 meters?

175%

49. 33.6 miles is 28% of what number of miles?

120 miles

50. $46 is 8% of what number?

$575

[6.7] *Find the amount of sales tax or the tax rate and the total cost (amount of sale + amount of tax = total cost). Round to the nearest cent.*

	Amount of Sale	Tax Rate	Amount of Tax	Total Cost
51.	$210	4%	$8.40	$218.40
52.	$780	$7\frac{1}{2}$%	$58.50	$838.50

Find the commission earned or the rate of commission.

	Sales	Rate of Commission	Commission
53.	$2800	11%	$308
54.	$65,300	5%	$3265

Find the amount or rate of discount and the amount paid after the discount. Round to the nearest cent.

	Original Price	Rate of Discount	Amount of Discount	Sale Price
55.	$37.50	10%	$3.75	$33.75
56.	$252	25%	$63	$189

[6.8] *Find the simple interest due on each loan.*

	Principal	Rate	Time	Interest
57.	$100	5%	1 year	$5
58.	$960	6%	$1\frac{1}{4}$ years	$72

Find the simple interest paid on each investment.

	Principal	Rate	Time in Months	Interest
59.	$300	8%	6 months	$12
60.	$1280	$7\frac{1}{2}$%	18 months	$144

Find the total amount due on the following simple interest loans.

	Principal	Rate	Time	Total Amount Due
61.	$350	$4\frac{1}{2}$%	3 years	$397.25
62.	$1530	6%	9 months	$1598.85

[6.9] *Find the missing numbers in the following compound interest problems. Interest is compounded annually. You may use the table on page 417.*

	Principal	Rate	Time in Years	Compound Amount	Compound Interest
63.	$2000	5%	10	$3257.80	$1257.80
64.	$1530	4%	5	≈$1861.55	$331.55
65.	$3600	8%	3	$4534.92	$934.92
66.	$11,400	6%	6	$16,170.90	$4770.90

Find the value of the missing letter in the percent proportion $\frac{a}{b} = \frac{p}{100}$.

67. $b = 40, p = 30$

    12

68. $a = 574, p = 35$

    1640

Use the percent proportion or equation to find each of the following.

69. 24% of 97 meters

    23.28 meters

70. 327 cars is what percent of 218 cars?

    150%

71. 0.6% of $85

    $0.51

72. 198 students is 40% of what number of students?

    495 students

73. 76 chickens is what percent of 190 chickens?

    40%

74. 107.242 liters is 43% of what number of liters?

    249.4 liters

Write the percents as decimals and the decimals as percents.

75. 75%   0.75

76. 200%   2

77. 4   400%

78. 4.71   471%

79. 6.2%   0.062

80. 0.621   62.1%

81. 0.375%   0.00375

82. 0.0006   0.06%

Write each percent as a fraction in lowest terms and each fraction as a percent.

83. $\frac{1}{4}$ 25%

84. 38% $\frac{19}{50}$

85. 62.5% $\frac{5}{8}$

86. $\frac{3}{8}$ 37.5% or $37\frac{1}{2}$%

87. $32\frac{1}{2}$% $\frac{13}{40}$

88. $\frac{1}{5}$ 20%

89. 0.5% $\frac{1}{200}$

90. $2\frac{1}{4}$ 225%

Solve each of the following application problems. Round percent answers to the nearest tenth and money amounts to the nearest cent if necessary.

91. The owner of Fair Oaks Hardware deposits $8520 at $5\frac{1}{2}\%$ for 9 months. Find the amount of interest earned.

$351.45

92. Clarence Hanks borrows $1620 at 14% for 18 months to buy a toy train collection. Find the total amount due.

$1960.20

▲ 93. A Hotpoint refrigerator has a capacity of 11.5 cubic feet in the refrigerator and 5.5 cubic feet in the freezer. What percent of the total capacity is the capacity of the freezer?

≈32.4%

94. Tommy Downs invests the money he inherited from his aunt at 6% compounded annually for 4 years. If the amount of money invested is $12,500, find
   (a) the compound amount at the end of 4 years and
   (b) the amount of interest that he earned. Do not use the table.

(a) ≈$15,780.96   (b) $3280.96

95. Linda Freitas, a real estate agent, sold two properties, one for $105,000 and the other for $145,000. After all of her expenses she receives a commission of $1\frac{1}{2}\%$ of total sales. Find the commission that she earned.

$3750

96. Our mail carrier, Norm, saw his route expand from 481 residential stops to 520 residential stops. Find the percent of increase.

≈8.1%

97. The mileage on a car dropped from 32.8 miles per gallon to 28.5 miles per gallon. Find the percent of decrease.

≈13.1%

98. Ford Motor Company hopes to have annual sales of 200,000 cars in Japan by the year 2000. If this will amount to only 4% of the annual automobile sales in Japan, find the total number of cars sold annually in Japan.

5,000,000 cars

▲ 99. Suzanne and Walter Roig established a budget allowing 25% for rent, 30% for food, 8% for clothing, 20% for travel and recreation, and the remainder for savings. Walter takes home $1950 per month, and Suzanne takes home $28,500 per year. How much money will the couple save in a year?

$8823

100. A fax machine priced at $398 is marked down 7% to promote the new model. If the sales tax is also 7%, find the cost of the fax machine including sales tax.

≈$396.05

Write each percent as a decimal and each decimal as a percent.

1. 75%   **2.** 0.6

3. 1.8   **4.** 0.875

5. 300%   **6.** 0.05%

Write as fractions in lowest terms.

7. 62.5%   **8.** 0.25%

Write each fraction or mixed number as a percent.

9. $\dfrac{1}{4}$ **10.** $\dfrac{5}{8}$

11. $1\dfrac{3}{4}$

Solve each of the following.

12. 16 files is 5% of what number of files?

13. $250 is what percent of $1250?

14. Erica Green has saved 75% of the amount needed for a down payment on a condominium. If she has saved $14,625, find the total down payment needed.

15. The price of a copy machine is $2680 plus sales tax of $6\frac{1}{2}$%. Find the total cost of the copy machine including sales tax.

16. A vacuum cleaner company pays its salespeople a commission of 24% on all sales. Find the commission earned for selling a vacuum system for $1040.

Answers:

1. 0.75
2. 60%
3. 180%
4. 87.5% or $87\frac{1}{2}$%
5. 3.00 or 3
6. 0.0005
7. $\dfrac{5}{8}$
8. $\dfrac{1}{400}$
9. 25%
10. 62.5% or $62\frac{1}{2}$%
11. 175%
12. 320 files
13. 20%
14. $19,500
15. $2854.20
16. $249.60

17. _35%_

18.
$$\frac{\text{amount of increase}}{\text{last year's salary}} = \frac{p}{100}$$

$$I = p \cdot r \cdot t$$

9 months $1000 \cdot 0.05 \cdot \dfrac{9}{12} = \$37.50$

$2\frac{1}{2}$ years $1000 \cdot 0.05 \cdot 2.5 = \$125$

19. _____

20. _$3.84; $44.16_____

21. _$68.25; $113.25_____

22. _$262.50_____

23. _$52_____

24. _$4876_____

25. _(a) $10,668 (b) $1668_____

---

17. Enrollment in mathematics courses increased from 1440 students last semester to 1944 students this semester. Find the percent of increase.

18. A problem includes last year's salary, this year's salary, and asks for the percent of increase. Explain how you would identify the amount, the base, and the percent in the problem. Show the percent proportion that you would use.

   A possible answer is: Amount is the increase in salary.
   this year − last year = increase
   Base is last year's salary. Percent of increase is unknown.

19. Write the formula used to find interest. Explain the difference in what to do if the time is expressed in months or in years. Write a problem that involves finding interest for 9 months and another problem that involves finding interest for $2\frac{1}{2}$ years. Use your own numbers for the principal and the rate. Show how to solve your problems.

   The interest formula is $I = p \cdot r \cdot t$.
   If time is in months it is expressed as a fraction with 12 as the denominator.
   If time is expressed in years it is placed over 1 or shown as a decimal number.

Find the amount of discount and the sale price. Round answers to the nearest cent.

	Original Price	Rate of Discount
20.	$48	8%
21.	$182	37.5%

Find the interest on each of the following.

	Principal	Rate	Time
22.	$3500	5%	$1\frac{1}{2}$ years
23.	$5200	4%	3 months

24. A parent borrows $4600 to help her child finish college. The loan is for 6 months at 12% interest. Find the total amount due on the loan.

25. The River City School PTA Emergency Fund deposited $4000 at 6% compounded annually. Two years after the first deposit, they add another $5000, also at 6% compounded annually. Use the compound interest table.
   (a) What total amount will they have four years after their first deposit?
   (b) What amount of interest will they have earned?

≈ *Round the numbers in each problem so there is only one non-zero digit. Then add, subtract, multiply, or divide the rounded numbers, as indicated, to estimate the answer. Finally, solve for the exact answer.*

1. *estimate*   *exact*
$$
\begin{array}{r}
9000 \\
80 \\
+\ 500 \\
\hline
9580
\end{array}
\qquad
\begin{array}{r}
8702 \\
83 \\
+\ 549 \\
\hline
9334
\end{array}
$$

2. *estimate*   *exact*
$$
\begin{array}{r}
1 \\
40 \\
+\ 5 \\
\hline
46
\end{array}
\qquad
\begin{array}{r}
0.68 \\
36.531 \\
+\ 5.3 \\
\hline
42.511
\end{array}
$$

3. *estimate*   *exact*
$$
\begin{array}{r}
60,000 \\
-\ 50,000 \\
\hline
10,000
\end{array}
\qquad
\begin{array}{r}
61,033 \\
-\ 51,040 \\
\hline
9993
\end{array}
$$

4. *estimate*   *exact*
$$
\begin{array}{r}
6 \\
-\ 3 \\
\hline
3
\end{array}
\qquad
\begin{array}{r}
6.2 \\
-\ 2.7055 \\
\hline
3.4945
\end{array}
$$

5. *estimate*   *exact*
$$
\begin{array}{r}
7000 \\
\times\quad 700 \\
\hline
4,900,000
\end{array}
\qquad
\begin{array}{r}
6538 \\
\times\quad 708 \\
\hline
4,628,904
\end{array}
$$

6. *estimate*   *exact*
$$
\begin{array}{r}
70 \\
\times\quad 5 \\
\hline
350
\end{array}
\qquad
\begin{array}{r}
71.6 \\
\times\quad 4.5 \\
\hline
322.2
\end{array}
$$

7. *estimate*   *exact*

1000 $43\overline{)40,000}$ 902 $43\overline{)38,786}$

8. *estimate*   *exact*

250 $8\overline{)2000}$ 320 $7.6\overline{)2432}$

9. *estimate*   *exact*

7 $1\overline{)7}$ 8.45 $0.8\overline{)6.76}$

Simplify each problem by using the order of operations.

10. $8^2 - 5 \cdot 4$ **44**

11. $\sqrt{36} + 4 \cdot 8 - 7$ **31**

12. $9 + 6 \div 3 + 7 \cdot 4$ **39**

Round each number to the place shown.

13. 2356 to the nearest ten   ≈**2360**

14. 5,678,159 to the nearest hundred thousand   ≈**5,700,000**

15. $718.499 to the nearest dollar   ≈**$718**

16. $451.825 to the nearest cent   ≈**$451.83**

Add, subtract, multiply, or divide as indicated. Write answers in lowest terms and as whole or mixed numbers when possible.

17. $\dfrac{3}{8} + \dfrac{3}{4}$ $\quad 1\dfrac{1}{8}$

18. $\dfrac{1}{2} + \dfrac{2}{3}$ $\quad 1\dfrac{1}{6}$

19. $\quad 4\dfrac{5}{8}$
$\quad + \; 8\dfrac{3}{4}$
$\quad \overline{\quad 13\dfrac{3}{8}}$

20. $\dfrac{3}{4} - \dfrac{5}{8}$ $\quad \dfrac{1}{8}$

21. $\quad 5\dfrac{1}{2}$
$\quad - \; 2\dfrac{2}{3}$
$\quad \overline{\quad 2\dfrac{5}{6}}$

22. $\quad 26\dfrac{1}{3}$
$\quad - \; 17\dfrac{4}{5}$
$\quad \overline{\quad 8\dfrac{8}{15}}$

23. $\dfrac{7}{8} \cdot \dfrac{2}{3}$ $\quad \dfrac{7}{12}$

24. $7\dfrac{3}{4} \cdot 3\dfrac{3}{8}$ $\quad 26\dfrac{5}{32}$

25. $36 \cdot \dfrac{4}{5}$ $\quad 28\dfrac{4}{5}$

26. $\dfrac{5}{9} \div \dfrac{5}{8}$ $\quad \dfrac{8}{9}$

27. $10 \div \dfrac{2}{5}$ $\quad 25$

28. $2\dfrac{3}{4} \div 7\dfrac{1}{2}$ $\quad \dfrac{11}{30}$

Write $<$ or $>$ to make a true statement.

29. $\dfrac{2}{3} \underline{\quad < \quad} \dfrac{3}{4}$

30. $\dfrac{5}{12} \underline{\quad < \quad} \dfrac{7}{15}$

31. $\dfrac{7}{15} \underline{\quad > \quad} \dfrac{9}{20}$

Simplify each of the following. Use the order of operations as needed.

32. $\left(\dfrac{3}{4} - \dfrac{5}{8}\right) \cdot \dfrac{2}{3}$ $\dfrac{1}{12}$

33. $\dfrac{3}{4} \div \left(\dfrac{2}{5} + \dfrac{1}{5}\right)$ $1\dfrac{1}{4}$

34. $\left(\dfrac{5}{6} - \dfrac{5}{12}\right) - \left(\dfrac{1}{2}\right)^2 \cdot \dfrac{2}{3}$ $\dfrac{1}{4}$

Write each fraction as a decimal. Round to the nearest thousandth if necessary.

35. $\dfrac{5}{8}$ 0.625

36. $\dfrac{4}{5}$ 0.8

37. $\dfrac{17}{20}$ 0.85

38. $\dfrac{12}{14}$ ≈0.857

Write each of the following ratios in lowest terms. Be sure to make all necessary conversions.

39. 2 hours to 40 minutes

Compare in minutes. $\dfrac{3}{1}$

40. There are 12 cars and 15 parking places. What is the ratio of parking places to cars? $\dfrac{5}{4}$

41. $1\dfrac{5}{8}$ to 13 $\dfrac{1}{8}$

Use cross multiplication to decide whether the following proportions are true or false. Circle the correct answer.

42. $\dfrac{4}{10} = \dfrac{36}{90}$

(True)   False

43. $\dfrac{64}{144} = \dfrac{48}{108}$

(True)   False

Find the missing number in each proportion.

44. $\dfrac{1}{4} = \dfrac{x}{20}$ 5

45. $\dfrac{315}{45} = \dfrac{21}{x}$ 3

46. $\dfrac{7}{x} = \dfrac{81}{162}$ 14

47. $\dfrac{x}{120} = \dfrac{7.5}{30}$ 30

Write each of the following percents as decimals. Write each of the following decimals as percents.

48. 25%   0.25

49. 7%   0.07

50. 300%   3.00 or 3

51. 0.5%   0.005

52. 0.56   56%

53. 2.7   270%

54. 0.023   2.3%

Write each percent as a fraction or mixed number in lowest terms. Write each fraction as a percent.

55. 6% $\dfrac{3}{50}$

56. 62.5% $\dfrac{5}{8}$

57. 175% $1\dfrac{3}{4}$

58. $\dfrac{7}{8}$ 87.5% or $87\dfrac{1}{2}$%

59. $\dfrac{1}{20}$ 5%

60. $3\dfrac{1}{2}$ 350%

Solve these percent problems.

61. 45% of 1200 officers

540 officers

62. $8\dfrac{1}{2}$% of \$850 is how much?

\$72.25

63. 36 cans is 20% of what number of cans?

180 cans

64. $4\dfrac{1}{2}$% of what number of miles is 76.5 miles?

1700 miles

65. What percent of 328 trees is 164 trees?

50%

66. 72 hours is what percent of 180 hours?

40%

Find the amount of sales tax or the tax rate and the total cost (amount of sale + amount of tax = total cost). Round to the nearest cent, if necessary.

	Amount of Sale	Tax Rate	Amount of Tax	Total Cost
67.	$53.99	4%	≈$2.16	$56.15
68.	$460	6.5% or $6\frac{1}{2}$%	$29.90	$489.90

Find the commission earned or the rate of commission.

	Sales	Rate of Commission	Commission
69.	$14,622	5%	$731.10
70.	$225,300	2.5% or $2\frac{1}{2}$%	$5632.50

Find the amount or rate of discount and the amount paid after the discount. Round to the nearest cent, if necessary.

	Original Price	Rate of Discount	Amount of Discount	Sale Price
71.	$152	35%	$53.20	$98.80
72.	$1085	15%	$162.75	$922.25

Find the total amount due on the following loans. Round to the nearest cent, if necessary.

	Principal	Rate	Time	Total Amount To Be Repaid
73.	$714	9%	2 years	$842.52
74.	$18,350	11%	9 months	≈$19,863.88

Set up and solve a proportion for each problem.

75. 7 watches can be cleaned in 3 hours. Find the number of watches that can be cleaned in 12 hours.

28 watches

76. If 3.5 ounces of weed killer is needed to make 6 gallons of spray, how much weed killer is needed for 102 gallons of spray?

59.5 ounces

▥ *Solve the following application problems. Use the table below to answer Exercises 77–80. Round answers to the nearest tenth of a percent, if necessary.*

EXISTING HOME SALES

Region	Last Year	This Year
Northeast	32,000	36,000
Midwest	65,000	66,300
South	82,000	77,500
West	54,000	49,600

77. Find the percent of increase in sales in the northeastern region.

12.5%

78. Find the percent of increase in sales in the midwestern region.

2%

79. What is the percent of decrease in sales in the southern region?

≈5.5%

80. What is the percent of decrease in sales in the western region?

≈8.1%

81. After receiving a legal settlement, Joan Ong invests $13,440 in the ownership of a construction company. If this is 28% of the legal settlement, find the total amount that she received.

$48,000

82. Stephanie Hirata deposits $6000 in a savings account that pays 5% interest compounded annually. Find
(a) the amount that she will have in the account at the end of 4 years, and
(b) the amount of interest earned.
Do not use the table.

(a) ≈$7293.04 (b) $1293.04

Measurement

7

We measure things all the time: the distance traveled on vacation, the floor area we want to cover with carpet, the amount of milk in a recipe, the weight of the bananas we buy at the store, the number of hours we work, and many more.

In the United States we still use the **English system** of measurement for many everyday activities. Examples of English units are inches, feet, quarts, ounces, and pounds. However, the fields of science, medicine, sports, and manufacturing increasingly use the **metric system** (meters, liters, and grams). And, because the rest of the world uses only the metric system, U.S. businesses are beginning to change to the metric system in order to compete internationally.

7.1 The English System

OBJECTIVE 1 ▶ Until the switch to the metric system is complete, we still need to know how to use the English system of measurement. The table below lists the relationships you should memorize. The time relationships are used in both the English and metric systems.

LENGTH	WEIGHT
1 foot = 12 inches (in.)	1 pound (lb) = 16 ounces (oz)
1 yard (yd) = 3 feet (ft)	1 ton (T) = 2000 pounds (lb)
1 mile (mi) = 5280 feet (ft)	

CAPACITY	TIME
1 cup (c) = 8 fluid ounces	1 week (wk) = 7 days
1 pint (pt) = 2 cups	1 day = 24 hours (hr)
1 quart (qt) = 2 pints (pt)	1 hour (hr) = 60 minutes (min)
1 gallon (gal) = 4 quarts (qt)	1 minute (min) = 60 seconds (sec)

As you can see, there is no simple or "natural" way to convert among these various measures. The units evolved over hundreds of years and were based on a variety of "standards." For example, one yard was the distance from the tip of a king's nose to his thumb when his arm was outstretched. An inch was three dried barleycorns laid end to end.

www.mathnotes.com

OBJECTIVES

1 ▶ Know the basic units in the English system.

2 ▶ Convert among units.

3 ▶ Use unit fractions to convert among units.

FOR EXTRA HELP

Tutorial   Tape 11   SSM, Sec. 7.1

1. After memorizing the measurement conversions, answer these questions.

(a) 1 cup = _____ fluid ounces

(b) 1 gallon = _____ quarts

(c) 1 week = _____ days

(d) 1 yard = _____ feet

(e) 1 foot = _____ inches

(f) 1 pound = _____ ounces

(g) 1 ton = _____ pounds

(h) 1 hour = _____ minutes

(i) 1 pint = _____ cups

(j) 1 day = _____ hours

(k) 1 minute = _____ seconds

(l) 1 quart = _____ pints

(m) 1 mile = _____ feet

ANSWERS

1. (a) 8 **(b)** 4 **(c)** 7 **(d)** 3 **(e)** 12 **(f)** 16 **(g)** 2000 **(h)** 60 **(i)** 2 **(j)** 24 **(k)** 60 **(l)** 2 **(m)** 5280

E X A M P L E   I   Knowing English Measurement Units

Memorize the English measurement conversions. Then answer these questions.

(a) 7 days = _____ week   Answer: 1 week
(b) 1 yard = _____ feet   Answer: 3 feet

◀◀ **WORK PROBLEM I AT THE SIDE.**

OBJECTIVE 2 You often need to convert from one unit of measure to another. Two methods of converting measurements are shown here. Study each way and use the method you prefer. Some conversions can be done by deciding whether to multiply or divide.

> **Converting among Measurement Units**
> **1.** *Multiply* when converting from a larger unit to a smaller unit.
> **2.** *Divide* when converting from a smaller unit to a larger unit.

E X A M P L E   2   Converting from One Unit of Measure to Another

Convert each measurement.

(a) 7 feet to inches
You are converting from a larger unit to a smaller unit (feet to inches), so multiply.
Because *1 foot = 12 inches,* multiply by 12.
$$7 \text{ feet} = 7 \cdot 12 = 84 \text{ inches}$$

(b) $3\frac{1}{2}$ pounds to ounces
You are converting from a larger unit to a smaller unit, pounds to ounces, so multiply.
Because *1 pound = 16 ounces,* multiply by 16.
$$3\frac{1}{2} \text{ pounds} = 3\frac{1}{2} \cdot 16 = \frac{7}{2} \cdot \frac{\overset{8}{\cancel{16}}}{1} = \frac{56}{1} = 56 \text{ ounces}$$

(c) 20 quarts to gallons
You are converting from a smaller unit to a larger unit (quarts to gallons), so divide.
Because *4 quarts = 1 gallon,* divide by 4.
$$20 \text{ quarts} = \frac{20}{4} = 5 \text{ gallons}$$

(d) 45 minutes to hours
You are converting from a smaller unit to a larger unit (minutes to hours), so divide.
Because *60 minutes = 1 hour,* divide by 60 and write the fraction in lowest terms.
$$45 \text{ minutes} = \frac{45}{60} = \frac{45 \div 15}{60 \div 15} = \frac{3}{4} \text{ hour}$$

WORK PROBLEM 2 AT THE SIDE. ▶▶

OBJECTIVE ▶**3**▶ If you have trouble deciding whether to multiply or divide when converting units, using **unit fractions** will solve the problem. You'll also find this method useful in science classes. A unit fraction is equivalent to 1. For example:

$$\frac{12 \text{ inches}}{12 \text{ inches}} = \frac{\overset{1}{\cancel{12 \text{ inches}}}}{\underset{1}{\cancel{12 \text{ inches}}}} = 1.$$

You know that 12 inches is the same as 1 foot. So you can substitute 1 foot for 12 inches in the numerator, or you can substitute 1 foot for 12 inches in the denominator. This makes two useful unit fractions.

$$\frac{1 \text{ foot}}{12 \text{ inches}} = 1 \quad \text{or} \quad \frac{12 \text{ inches}}{1 \text{ foot}} = 1$$

To convert from one measurement unit to another, just multiply by the appropriate unit fraction. Remember, a unit fraction is equivalent to 1. Multiplying something by 1 does *not* change its value.

Use these guidelines to choose the correct unit fraction.

Choosing a Unit Fraction

The *numerator* should use the measurement unit you want in the *answer*.

The *denominator* should use the measurement unit you want to *change*.

EXAMPLE 3   Using Unit Fractions with Length Measurement

(a) Convert 60 inches to feet.

Use a unit fraction with feet (the unit for your answer) in the numerator, and inches (the unit being changed) in the denominator. Because *1 foot = 12 inches,* the necessary unit fraction is

$$\frac{1 \text{ foot}}{12 \text{ inches}} \begin{array}{l} \leftarrow \text{Unit for your answer} \\ \leftarrow \text{Unit being changed} \end{array}$$

Next, multiply 60 inches times this unit fraction. Write 60 inches as the fraction $\dfrac{60 \text{ inches}}{1}$. Then cancel units and numbers wherever possible.

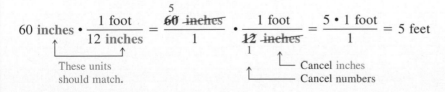

$$60 \text{ inches} \cdot \frac{1 \text{ foot}}{12 \text{ inches}} = \frac{\overset{5}{\cancel{60 \text{ inches}}}}{1} \cdot \frac{1 \text{ foot}}{\underset{1}{\cancel{12 \text{ inches}}}} = \frac{5 \cdot 1 \text{ foot}}{1} = 5 \text{ feet}$$

These units should match.   Cancel inches   Cancel numbers

─── **CONTINUED ON NEXT PAGE**

2. Convert each measurement.

(a) $5\frac{1}{2}$ feet to inches

(b) 64 ounces to pounds

(c) 6 yards to feet

(d) 2 tons to pounds

(e) 35 pints to quarts

(f) 20 minutes to hours

(g) 4 weeks to days

ANSWERS

2. (a) 66 inches **(b)** 4 pounds
(c) 18 feet **(d)** 4000 pounds
(e) $17\frac{1}{2}$ quarts **(f)** $\frac{1}{3}$ hour **(g)** 28 days

3. First write the unit fraction needed to make each conversion. Then complete the conversion.

(a) 36 inches to feet

unit fraction } $\dfrac{1\ \text{foot}}{12\ \text{inches}}$

(b) 14 feet to inches

unit fraction } $\dfrac{\text{inches}}{\text{foot}}$

(c) 60 inches to feet

unit fraction } _____

(d) 4 yards to feet

unit fraction } _____

(e) 39 feet to yards

unit fraction } _____

(f) 2 miles to feet

unit fraction } _____

ANSWERS

3. **(a)** 3 feet **(b)** $\dfrac{12\ \text{inches}}{1\ \text{foot}}$; 168 inches

(c) $\dfrac{1\ \text{foot}}{12\ \text{inches}}$; 5 feet

(d) $\dfrac{3\ \text{feet}}{1\ \text{yard}}$; 12 feet

(e) $\dfrac{1\ \text{yard}}{3\ \text{feet}}$; 13 yards

(f) $\dfrac{5280\ \text{feet}}{1\ \text{mile}}$; 10,560 feet

(b) Convert 9 feet to inches.

Select the correct unit fraction to change 9 feet to inches.

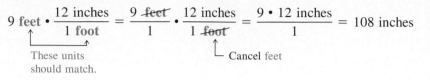

$$\dfrac{12\ \text{inches}}{1\ \text{foot}} \quad \leftarrow \text{Unit for your answer} \\ \quad\quad\quad\quad \leftarrow \text{Unit being changed}$$

Multiply 9 feet times the unit fraction.

$$9\ \textbf{feet} \cdot \dfrac{12\ \text{inches}}{1\ \textbf{foot}} = \dfrac{9\ \cancel{\text{feet}}}{1} \cdot \dfrac{12\ \text{inches}}{1\ \cancel{\text{foot}}} = \dfrac{9 \cdot 12\ \text{inches}}{1} = 108\ \text{inches}$$

These units should match. Cancel feet

> **Note**
> If no units will cancel, you made a mistake in choosing the unit fraction.

◀◀ **WORK PROBLEM 3 AT THE SIDE.**

E X A M P L E 4   Using Unit Fractions with Capacity and Weight Measurement

(a) Convert 9 pints to quarts.
Select the correct unit fraction.

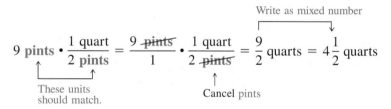

$$\dfrac{1\ \text{quart}}{2\ \text{pints}} \quad \leftarrow \text{Unit for your answer} \\ \quad\quad\quad\quad \leftarrow \text{Unit being changed}$$

Next multiply.

Write as mixed number

$$9\ \textbf{pints} \cdot \dfrac{1\ \text{quart}}{2\ \textbf{pints}} = \dfrac{9\ \cancel{\text{pints}}}{1} \cdot \dfrac{1\ \text{quart}}{2\ \cancel{\text{pints}}} = \dfrac{9}{2}\ \text{quarts} = 4\tfrac{1}{2}\ \text{quarts}$$

These units should match.   Cancel pints

(b) Convert $7\tfrac{1}{2}$ gallons to quarts.

Write as improper fraction

$$\dfrac{7\tfrac{1}{2}\ \cancel{\text{gallons}}}{1} \cdot \dfrac{4\ \text{quarts}}{1\ \cancel{\text{gallon}}} = \dfrac{15}{2} \cdot \dfrac{4}{1}\ \text{quarts}$$

$$= \dfrac{15}{\cancel{2}} \cdot \dfrac{\overset{2}{\cancel{4}}}{1}\ \text{quarts}$$

$$= 30\ \text{quarts}$$

(c) Convert 36 ounces to pounds.

$$\dfrac{\overset{9}{\cancel{36}}\ \cancel{\text{ounces}}}{1} \cdot \dfrac{1\ \text{pound}}{\underset{4}{\cancel{16}}\ \cancel{\text{ounces}}} = \dfrac{9}{4}\ \text{pounds} = 2\tfrac{1}{4}\ \text{pounds}$$

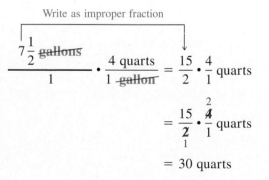

Note

In Example 4(c) you get $\frac{9}{4}$ pounds. Recall that $\frac{9}{4}$ means $9 \div 4$. If you do $9 \div 4$ on your calculator, you get 2.25 pounds. English measurements usually use fractions or mixed numbers, like $2\frac{1}{4}$ pounds, or in Example 4(a), $4\frac{1}{2}$ quarts. However, 2.25 pounds is also correct and is the way grocery stores often show weights of produce, meat, and cheese.

WORK PROBLEM 4 AT THE SIDE. ▶▶

E X A M P L E 5   Using Several Unit Fractions

Sometimes you may need to use two or three unit fractions in problems like these.

(a) Convert 63 inches to yards.

Use the unit fraction $\dfrac{1 \text{ foot}}{12 \text{ inches}}$ to change inches to feet and the unit fraction $\dfrac{1 \text{ yard}}{3 \text{ feet}}$ to change feet to yards. Notice how all the units cancel except yards, which is the unit you want in the answer.

$$\frac{63 \ \text{inches}}{1} \cdot \frac{1 \ \text{foot}}{12 \ \text{inches}} \cdot \frac{1 \ \text{yard}}{3 \ \text{feet}} = \frac{63}{36} \text{ yards} = 1\frac{3}{4} \text{ yards}$$

You can also cancel the numbers.

$$\frac{\overset{\overset{7}{\cancel{21}}}{\cancel{63}}}{1} \cdot \frac{1}{\underset{4}{\cancel{12}}} \cdot \frac{1}{\underset{1}{\cancel{3}}} = \frac{7}{4} = 1\frac{3}{4} \text{ yards}$$

Instead of changing $\frac{7}{4}$ to $1\frac{3}{4}$, you can divide $7 \div 4$ on your calculator to get 1.75 yards. Both answers are correct because 1.75 is equivalent to $1\frac{3}{4}$.

(b) Convert 2 days to seconds.

Use three unit fractions. The first one changes days to hours, the second one changes hours to minutes, and the third one changes minutes to seconds. All the units cancel except seconds, which is what you want in your answer.

$$\frac{2 \ \text{days}}{1} \cdot \frac{24 \ \text{hours}}{1 \ \text{day}} \cdot \frac{60 \ \text{minutes}}{1 \ \text{hour}} \cdot \frac{60 \ \text{seconds}}{1 \ \text{minute}} = 172{,}800 \text{ seconds}$$

WORK PROBLEM 5 AT THE SIDE. ▶▶

4. Convert using unit fractions.

  (a) 16 pints to quarts

  (b) 16 quarts to gallons

  (c) 3 cups to pints

  (d) $2\frac{1}{4}$ gallons to quarts

  (e) 48 ounces to pounds

  (f) $3\frac{1}{2}$ tons to pounds

  (g) $1\frac{3}{4}$ pounds to ounces

  (h) 4 ounces to pounds

5. Convert using two or three unit fractions.

  (a) 4 tons to ounces

  (b) 90 inches to yards

  (c) 3 miles to inches

  (d) 36 pints to gallons

  (e) 4 weeks to minutes

  (f) $1\frac{1}{2}$ gallons to cups

ANSWERS

4. (a) 8 quarts **(b)** 4 gallons
  (c) $1\frac{1}{2}$ pints **(d)** 9 quarts
  (e) 3 pounds **(f)** 7000 pounds
  (g) 28 ounces
  (h) $\frac{1}{4}$ pound or 0.25 pound

5. (a) 128,000 ounces
  (b) $2\frac{1}{2}$ yards or 2.5 yards
  (c) 190,080 inches
  (d) $4\frac{1}{2}$ gallons or 4.5 gallons
  (e) 40,320 minutes **(f)** 24 cups

SUNFLOWER

89¢
NET WT
5 g

MAMMOTH GREY STRIPE

Gigantic Flowers
Edible seed

MAMMOTH RUSSIAN SUNFLOWER *Helianthus annuus* Adults and children love this fast growing variety with enormous golden-yellow blossoms and edible seeds. Quickly grows to 8–10 feet with only minimal care. Great for a temporary hedge around the garden. Very easy to grow.

Type	Height	Planting Depth	Seed Spacing	Thinning Height	Thinning Spacing	Days to Germinate
Annual	8–10' 2–3 m	1" 3 cm	1' 30 cm	3" 8 cm	2' 61 cm	10–14

PLANTING: Select a location with full sun to light shade and average soil. May be started indoors 4 weeks before last frost date. Keep young plants watered and weeded until established. Mature plants are quite drought tolerant.
NOTE: Save seeds for winter eating by hanging heads upside down in a cool, dry area for a few weeks. The seeds will brush off the flower head when they are dry.

The front and back of a seed packet for sunflowers are shown. Look at the front of the packet first.

There were 49 seeds in the packet and 46 of the seeds sprouted. What was the cost per seed, to the nearest cent? **$0.02**
What percent of the seeds sprouted? **≈94%**
About how many seeds would weight 1 gram? **10 seeds**
How much does one seed weigh, rounded to the nearest tenth of a gram? **≈0.1 gram**

The back of the packet uses an apostrophe (') which is a symbol for feet, and " which is a symbol for inches.

How tall will the plants grow, in feet? **8 to 10 ft**
How tall will they grow in inches? **96 to 120 in.**
How tall will they grow in yards? **2 2/3 to 3 1/3 yd**

How many inches tall should the plants be when you thin them (remove less vigorous plants to give others room to grow)? **3 in.**
How tall is that in feet? **1/4 ft**

If you plant all 49 seeds in one long row, using the spacing given on the package, how long will your row be in feet? **48 ft (one foot less than the number of seeds)**

If you thin the sprouts according to the directions, how many sprouts will you have to remove? Assume that all seeds sprout, and that you leave as many sprouts as possible. **24 sprouts**

Using the information in the article, how many gum wrappers are needed to make 1 foot of chain **≈52 wrappers**
To make 1 inch of chain? **≈4 wrappers**

How many inches of chain are made from one wrapper? **≈0.23 inches**

Is the article correct in saying that 125 miles is 34 million gum wrappers? Find two different methods for calculating the comparison.

$$\frac{0.23 \text{ in.}}{1 \text{ wrapper}} \cdot \frac{34{,}000{,}000 \text{ wrappers}}{1} \cdot \frac{1 \text{ ft}}{12 \text{ in.}} \cdot \frac{1 \text{ mi}}{5280 \text{ ft}} \approx 123.4 \text{ miles}$$

$$\frac{6602 \text{ wrappers}}{128 \text{ ft}} \cdot \frac{5280 \text{ ft}}{1 \text{ mi}} \cdot \frac{125 \text{ mi}}{1} \approx 34{,}041{,}563 \text{ wrappers}$$

Chew on This!

Sometimes a person's life's work makes it into a museum. So it figures that Michael Knutson's 128-foot chain of gum wrappers now resides in the Yellow Medicine County Museum in Granite Falls, Minn.

According to the *Redwood Gazette*, the wrapper chain started in 1974 when Knutson, now a 37-year-old woodworker, became bored during study hall. "I've never been much of a gum chewer, but I found most of the wrappers on the streets of the city," he said. (FYI: It takes 6602 wrappers to make a 128-foot chain.)

Granite Falls is about 125 miles—or 34 million gum wrappers—west of the Twin Cities.

Source: Minneapolis Star Tribune

7.1 Exercises

Fill in the blanks with the measurement relationships you have memorized.

> **Example:** 1 hour = _____ minutes   **Solution:** 1 hour = __60__ minutes

1. 1 yard = __3__ feet

2. 1 foot = __12__ inches

3. __8__ fluid ounces = 1 cup

4. __4__ quarts = 1 gallon

5. 1 mile = __5280__ feet

6. 1 week = __7__ days

7. __2000__ pounds = 1 ton

8. __16__ ounces = 1 pound

9. 1 minute = __60__ seconds

10. 1 day = __24__ hours

Convert these measurements using unit fractions.

> **Example:** 3000 pounds = _____ tons
>
> **Solution:** $\dfrac{\overset{3}{\cancel{3000 \text{ pounds}}}}{1} \cdot \dfrac{1 \text{ ton}}{\underset{2}{\cancel{2000 \text{ pounds}}}} = \dfrac{3}{2}$ ton $= 1\dfrac{1}{2}$ **tons or 1.5 tons**
>
> └── Cancel pounds
> └── Cancel numbers

11. 120 seconds = __2__ minutes

12. 180 minutes = __3__ hours

13. 8 quarts = __2__ gallons

14. 24 quarts = __6__ gallons

15. An adult sperm whale may weigh 38 to 40 tons. How many pounds could it weigh?

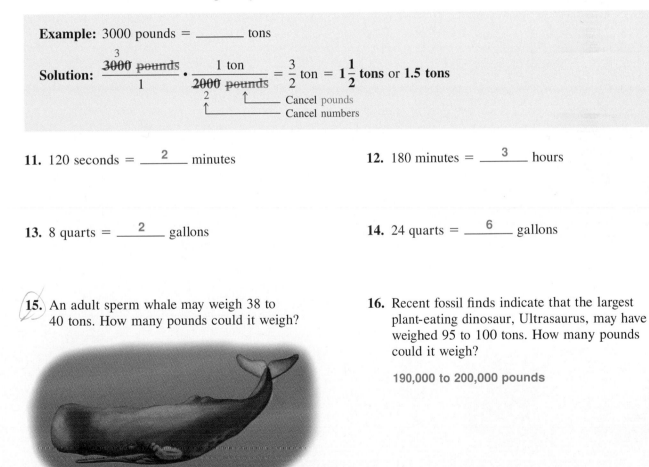

76,000 to 80,000 pounds

16. Recent fossil finds indicate that the largest plant-eating dinosaur, Ultrasaurus, may have weighed 95 to 100 tons. How many pounds could it weigh?

190,000 to 200,000 pounds

✎ Writing   ⊙ Conceptual   ▲ Challenging   ≈ Estimation

17. 12 feet = _____4_____ yards

18. 15 yards = _____45_____ feet

19. 7 pounds = _____112_____ ounces

20. 6 pounds = _____96_____ ounces

21. 5 quarts = _____10_____ pints

22. 13 quarts = _____26_____ pints

23. 90 minutes = ___$1\frac{1}{2}$ or 1.5___ hours

24. 45 seconds = ___$\frac{3}{4}$ or 0.75___ minutes

25. 3 inches = ___$\frac{1}{4}$ or 0.25___ feet

26. 30 inches = ___$2\frac{1}{2}$ or 2.5___ feet

27. 24 ounces = ___$1\frac{1}{2}$ or 1.5___ pounds

28. 36 ounces = ___$2\frac{1}{4}$ or 2.25___ pounds

29. 5 cups = ___$2\frac{1}{2}$ or 2.5___ pints

30. 15 quarts = ___$3\frac{3}{4}$ or 3.75___ gallons

31. Mr. Kashpaws worked for 12 hours doing traditional harvesting of wild rice. What part of the day did he work?

$\frac{1}{2}$ **day or 0.5 day**

32. Michelle prepares 4-ounce hamburgers at a fast-food restaurant. Each hamburger is what part of a pound?

$\frac{1}{4}$ **pound or 0.25 pound**

33. $2\frac{1}{2}$ tons = _____5000_____ pounds

34. $4\frac{1}{2}$ pints = _____9_____ cups

35. $4\frac{1}{4}$ gallons = _____17_____ quarts

36. $2\frac{1}{4}$ hours = _____135_____ minutes

37. Our premature baby weighed $2\frac{3}{4}$ pounds at birth. How many ounces did our baby weigh?

44 ounces

38. The NBA basketball star is $7\frac{1}{4}$ feet tall. What is his height in inches?

87 inches

Use two or three unit fractions to convert the following.

Example: 3 days = _____ minutes

Solution: $\dfrac{3 \text{ days}}{1} \cdot \dfrac{24 \text{ hours}}{1 \text{ day}} \cdot \dfrac{60 \text{ minutes}}{1 \text{ hour}} = $ **4320 minutes**

39. 6 yards = _____216_____ inches

40. 2 tons = _____64,000_____ ounces

41. 112 cups = _____28_____ quarts

42. 336 hours = _____2_____ weeks

43. 6 days = _____518,400_____ seconds

44. 5 gallons = _____80_____ cups

45. $1\frac{1}{2}$ tons = _____48,000_____ ounces

46. $3\frac{1}{3}$ yards = _____120_____ inches

⊙ **47.** The statement $8 = 2$ is *not* true. But with appropriate measurement units, it *is* true.

$$8 \text{ } quarts = 2 \text{ } gallons$$

Attach measurement units to these numbers to make the statements true.
(a) $1 = 16$ pound/ounces
(b) $10 = 20$ quarts/pints or pints/cups
(c) $120 = 2$ min/hr or sec/min
(d) $2 = 24$ feet/inches
(e) $6000 = 3$ pounds/tons
(f) $35 = 5$ days/weeks

⊡ **48.** Explain in your own words why you can add 2 feet + 12 inches to get 3 feet, but you cannot add 2 feet + 12 pounds.

Feet and inches both measure length, so you can add them once you've changed 2 feet into inches. But pounds measure weight and cannot be added to a length measurement.

▦ *Convert the following.*
▲

49. $2\frac{3}{4}$ miles = _____174,240_____ inches

50. $5\frac{3}{4}$ tons = _____184,000_____ ounces

51. $6\frac{1}{4}$ gallons = _____800_____ fluid ounces

52. $3\frac{1}{2}$ days = _____302,400_____ seconds

53. 24,000 ounces = ___0.75 or $\frac{3}{4}$___ ton

54. 57,024 inches = ___0.9 or $\frac{9}{10}$___ mile

55. 129,600 seconds = ___1.5 or $1\frac{1}{2}$___ days

56. 1952 fluid ounces = ___15.25 or $15\frac{1}{4}$___ gallons

▦▦
▦
Review and Prepare

*Place $<$ or $>$ in each blank to make a true statement. (For help, see **Section 3.5**.)*

57. 2 weeks __$<$__ 15 days

58. 72 hours __$<$__ 4 days

59. 4 hours __$>$__ 185 minutes

60. 2 years __$<$__ 28 months

61. 32 days __$>$__ 4 weeks

62. 14 minutes __$>$__ 780 seconds

7.2 The Metric System—Length

Around 1790, a group of French scientists developed the metric system of measurement. It is an organized system based on multiples of 10, like our number system and our money. After you are familiar with metric units, you will see that they are easier to use than the hodgepodge of English measurement relationships you used in **Section 7.1.**

OBJECTIVE 1 The basic unit of length in the metric system is the **meter** (also spelled *metre*). Use the symbol **m** for meter; do not put a period after it. If you put five of the pages from this textbook side by side, they would measure about 1 meter. Or, look at a yardstick—a meter is just a little longer. (A meter is about 39 inches long.)

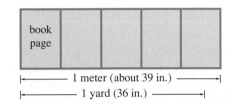

In the metric system you use meters for things like buying fabric for sewing projects, measuring the length of your living room, talking about heights of buildings, or describing track and field athletic events.

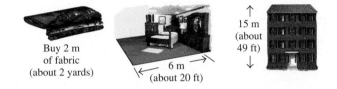

Buy 2 m of fabric (about 2 yards)

6 m (about 20 ft)

15 m (about 49 ft)

WORK PROBLEM I AT THE SIDE. ▶▶

To make longer or shorter length units in the metric system, **prefixes** are written in front of the word meter. For example, the prefix *kilo* means 1000, so a *kilo*meter is 1000 meters. The table below shows how to use the prefixes for length measurements. It is helpful to memorize the prefixes because they are also used with weight and capacity measurements. The colored boxes are the units you will use most often in daily life.

Prefix	kilo-meter	hecto-meter	deka-meter	meter	deci-meter	centi-meter	milli-meter
Meaning	1000 meters	100 meters	10 meters	1 meter	$\frac{1}{10}$ of a meter	$\frac{1}{100}$ of a meter	$\frac{1}{1000}$ of a meter
Symbol	km	hm	dam	m	dm	cm	mm

Here are some comparisons to help you get acquainted with the commonly used length units: km, m, cm, mm.

*Kilo*meters are used instead of miles. A kilometer is 1000 meters. It is about 0.6 mile (a little more than half a mile) or about 6 city blocks. If you participate in a 10 km run, you'll go about 6 miles.

OBJECTIVES

1 ▶ Know the basic metric units of length.

2 ▶ Use unit fractions to convert among units.

3 ▶ Move the decimal point to convert among units.

FOR EXTRA HELP

Tutorial   Tape 12   SSM, Sec. 7.2

1. Circle the items that measure about 1 meter.

   length of a pencil

   length of a baseball bat

   height of doorknob from the floor

   height of a house

   basketball player's arm length

   length of a paper clip

ANSWERS

1. baseball bat, height of doorknob, arm length

2. Write the most reasonable metric unit in each blank: km, m, cm, or mm.

(a) The woman's height is 168 _____ .

(b) The man's waist is 90 _____ around.

(c) Louise ran the 100 _____ dash in the track meet.

(d) A postage stamp is 22 _____ wide.

(e) Michael paddled his canoe 2 _____ down the river.

(f) The pencil lead is 1 _____ thick.

(g) A stick of gum is 7 _____ long.

(h) The highway speed limit is 90 _____ per hour.

(i) The classroom was 12 _____ long.

(j) A penny is about 18 _____ across.

A meter is divided into 100 smaller pieces called *centi*meters. Each centimeter is $\frac{1}{100}$ of a meter. Centimeters are used instead of inches. A centimeter is a little shorter than $\frac{1}{2}$ inch. The cover of this textbook is 21 cm wide. A nickel is about 2 cm across. Measure the width and length of your little finger on this centimeter ruler. The width of your little finger is probably about 1 centimeter.

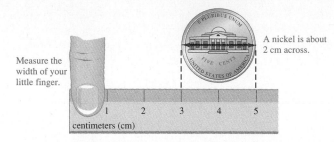

A meter is divided into 1000 smaller pieces called *milli*meters. Each millimeter is $\frac{1}{1000}$ of a meter. It takes 10 mm to equal 1 cm, so it is a very small length. The thickness of a dime is about 1 mm. Measure the width of your pen or pencil and the width of your little finger on this millimeter ruler.

Thickness of a dime is about 1 mm.

| 10 mm same as 1 cm | 50 mm same as 5 cm | 100 mm same as 10 cm |

EXAMPLE 1 Using Metric Length Units

Write the most reasonable metric unit in each blank. Choose from km, m, cm, and mm.

(a) The distance from home to work is 20 _____ .

   20 <u>km</u> because kilometers are used instead of miles. 20 km is about 12 miles.

(b) My wedding ring is 4 _____ wide.

   4 <u>mm</u> because the width of a ring is very small.

(c) The newborn baby is 50 _____ long.

   50 <u>cm</u>, which is half of a meter; a meter is about 39 inches so half a meter is around 20 inches.

◀◀ **WORK PROBLEM 2 AT THE SIDE.**

ANSWERS

2. **(a)** cm **(b)** cm **(c)** m **(d)** mm **(e)** km
   (f) mm **(g)** cm **(h)** km **(i)** m **(j)** mm

OBJECTIVE ▶ You can convert among metric length units using unit fractions. Keep these relationships in mind when setting up the unit fractions.

Metric Length Relationships

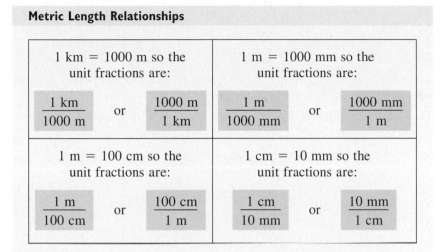

1 km = 1000 m so the unit fractions are:	1 m = 1000 mm so the unit fractions are:
$\dfrac{1\ km}{1000\ m}$ or $\dfrac{1000\ m}{1\ km}$	$\dfrac{1\ m}{1000\ mm}$ or $\dfrac{1000\ mm}{1\ m}$
1 m = 100 cm so the unit fractions are:	1 cm = 10 mm so the unit fractions are:
$\dfrac{1\ m}{100\ cm}$ or $\dfrac{100\ cm}{1\ m}$	$\dfrac{1\ cm}{10\ mm}$ or $\dfrac{10\ mm}{1\ cm}$

┌ **E X A M P L E   2   Using Unit Fractions with Length Measurement**

Convert the following.

(a) 5 km to m

Put the unit for the answer (meters) in the numerator of the unit fraction; the unit you want to change (km) in the denominator.

$$\text{Unit fraction equivalent to 1.} \begin{cases} \dfrac{1000\ m}{1\ km} \end{cases} \begin{array}{l} \leftarrow \text{Unit for answer} \\ \leftarrow \text{Unit being changed} \end{array}$$

Multiply. Cancel units where possible.

$$5\ km \cdot \frac{1000\ m}{1\ km} = \frac{5\ \cancel{km}}{1} \cdot \frac{1000\ m}{1\ \cancel{km}} = \frac{5 \cdot 1000\ m}{1} = 5000\ m$$

These units should match.

The answer makes sense because a kilometer is much longer than a meter, so 5 km will contain many meters.

(b) 18.6 cm to m

Multiply by a unit fraction that allows you to cancel centimeters.

$$\frac{18.6\ \cancel{cm}}{1} \cdot \overbrace{\frac{1\ m}{100\ \cancel{cm}}}^{\text{Unit fraction}} = \frac{18.6}{100}\ m = 0.186\ m$$

There are 100 cm in a meter, so 18.6 cm will be a small part of a meter. The answer makes sense.

WORK PROBLEM 3 AT THE SIDE. ▶▶

3. First write the unit fraction needed to make each conversion. Then complete the conversion.

(a) 3.67 m to cm

$$\text{unit fraction} \begin{cases} \dfrac{100\ cm}{1\ m} \end{cases}$$

(b) 92 cm to m

$$\text{unit fraction} \begin{cases} \dfrac{\quad m}{\quad cm} \end{cases}$$

(c) 432.7 cm to m

$$\text{unit fraction} \begin{cases} \underline{\qquad} \end{cases}$$

(d) 65 mm to cm

$$\text{unit fraction} \begin{cases} \underline{\qquad} \end{cases}$$

(e) 0.9 m to mm

$$\text{unit fraction} \begin{cases} \underline{\qquad} \end{cases}$$

(f) 2.5 cm to mm

$$\text{unit fraction} \begin{cases} \underline{\qquad} \end{cases}$$

ANSWERS

3. (a) 367 cm **(b)** $\dfrac{1\ m}{100\ cm}$; 0.92 m

(c) $\dfrac{1\ m}{100\ cm}$; 4.327 m

(d) $\dfrac{1\ cm}{10\ mm}$; 6.5 cm

(e) $\dfrac{1000\ mm}{1\ m}$; 900 mm

(f) $\dfrac{10\ mm}{1\ cm}$; 25 mm

4. Do each multiplication or division by hand or on a calculator. Compare your answer to the one obtained by moving the decimal point.

(a) 43.5 • 10 = _____

43.5 gives 435.

(b) 43.5 ÷ 10 = _____

43.5 gives _____

(c) 28 • 100 = _____

28.00 gives _____

(d) 28 ÷ 100 = _____

28. gives _____

(e) 0.7 • 1000 = _____

0.700 gives _____

(f) 0.7 ÷ 1000 = _____

000.7 gives _____

OBJECTIVE 3 By now you have probably noticed that conversions among metric units are made by multiplying or dividing by 10, by 100, or by 1000. A quick way to multiply by 10 is to move the decimal point one place to the *right*. Move it two places to multiply by 100, three places to multiply by 1000. Division is done by moving the decimal point to the *left*.

◀◀ **WORK PROBLEM 4 AT THE SIDE.**

An alternate conversion method to unit fractions is moving the decimal point using this **metric conversion line.**

Here are the steps for using the conversion line.

Using the Metric Conversion Line

1. Find the unit you are given on the metric conversion line.
2. Count the number of places to get from the unit you are given to the unit you want in the answer.
3. Move the decimal point the *same number of places*. Move in the *same direction* as you did on the conversion line.

E X A M P L E 3   Using a Metric Conversion

Use the metric conversion line to make the following conversions.

(a) 5.702 km to m

Find **km** on the metric conversion line. To get to **m**, you move *three places* to the *right*. So move the decimal point in 5.702 *three places* to the *right*.

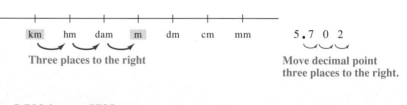

So 5.702 km = 5702 m.

(b) 69.5 cm to m

Find **cm** on the conversion line. To get to **m**, move *two places* to the *left*.

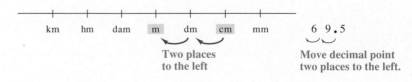

So 69.5 cm = 0.695 m.

CONTINUED ON NEXT PAGE ⎯

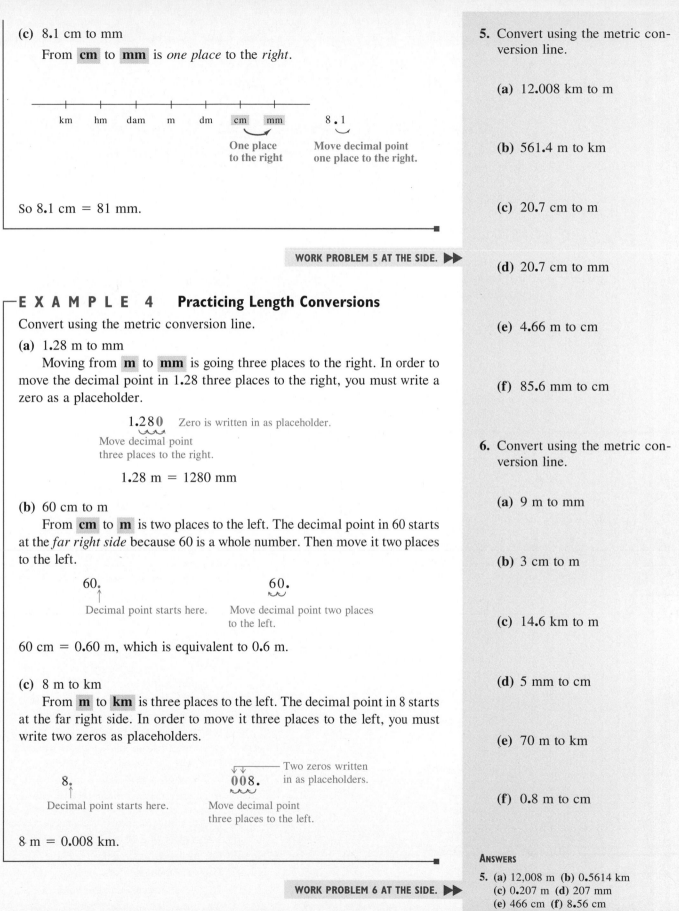

(c) 8.1 cm to mm

From **cm** to **mm** is *one place* to the *right*.

| km | hm | dam | m | dm | cm | mm | 8 . 1 |

One place
to the right

Move decimal point
one place to the right.

So 8.1 cm = 81 mm.

WORK PROBLEM 5 AT THE SIDE. ▶▶

E X A M P L E 4   Practicing Length Conversions

Convert using the metric conversion line.

(a) 1.28 m to mm

Moving from **m** to **mm** is going three places to the right. In order to move the decimal point in 1.28 three places to the right, you must write a zero as a placeholder.

1.28**0**   Zero is written in as placeholder.

Move decimal point
three places to the right.

1.28 m = 1280 mm

(b) 60 cm to m

From **cm** to **m** is two places to the left. The decimal point in 60 starts at the *far right side* because 60 is a whole number. Then move it two places to the left.

60.   60.

Decimal point starts here.   Move decimal point two places
                              to the left.

60 cm = 0.60 m, which is equivalent to 0.6 m.

(c) 8 m to km

From **m** to **km** is three places to the left. The decimal point in 8 starts at the far right side. In order to move it three places to the left, you must write two zeros as placeholders.

                           ↓ ↓ ─── Two zeros written
8.   00**8**.   in as placeholders.

Decimal point starts here.   Move decimal point
                              three places to the left.

8 m = 0.008 km.

WORK PROBLEM 6 AT THE SIDE. ▶▶

5. Convert using the metric conversion line.

(a) 12.008 km to m

(b) 561.4 m to km

(c) 20.7 cm to m

(d) 20.7 cm to mm

(e) 4.66 m to cm

(f) 85.6 mm to cm

6. Convert using the metric conversion line.

(a) 9 m to mm

(b) 3 cm to m

(c) 14.6 km to m

(d) 5 mm to cm

(e) 70 m to km

(f) 0.8 m to cm

ANSWERS

5. (a) 12,008 m **(b)** 0.5614 km
 (c) 0.207 m **(d)** 207 mm
 (e) 466 cm **(f)** 8.56 cm
6. (a) 9000 mm **(b)** 0.03 m
 (c) 14,600 m **(d)** 0.5 cm
 (e) 0.07 km **(f)** 80 cm

NUMBERS IN THE
Real World *collaborative investigations*

HAIR AND NAIL GROWTH

Q How fast do hair and nails grow? Do they grow faster in the summer?

A Fingernails grow, on average, about one-tenth of a millimeter per day, although there is considerable variation among individuals. Fingernails grow faster than toenails, and nails on the longest fingers appear to grow the fastest.

Fingernails, as well as hair and skin, grow faster in the summer, presumably under the influence of sunlight, which expands blood vessels, bringing more oxygen and nutrients to the area and allowing for faster growth.

The rate the scalp hair grows is 0.3 to 0.4 millimeter per day, or about 6 inches a year.

Source: Minneapolis Star Tribune

▶ How much do nails grown in one week? One month? One year?
0.7 mm; 3 mm during 30-day month; 36.5 mm

▶ How much does scalp hair grow in one week? One month? One year? (Use metric units.)
2.1 to 2.8 mm; 9 to 12 mm; 109.5 to 146 mm

▶ When you have finished Section 7.5, come back to this article. Is the statement about hair growing 6 inches a year accurate? Explain your answer.
Converting 109.5 to 146 mm gives ≈ 4.3 to 5.7 inches

Measuring Up

New device measures distances within billionths of an inch

U.S. officials have unveiled "the ultimate ruler," a measuring device that can gauge distances to within billionths of an inch—the length of five individual atoms—and may help revolutionize high-tech manufacturing.

Developed at a cost of $8 million by the National Institute of Standards and Technology with other agencies, the Molecular Measuring Machine can measure distances to within 40 nanometers, or billionths of a meter. After further refinement it is expected to measure within 1 nanometer.

By way of comparison, the period at the end of this sentence is about 300,000 nanometers wide.

The machine is expected to prove a boon to U.S. manufacturers of computer chips and other tiny high-tech items that must meet exacting specifications. It can also help manufacturers better calibrate their own super-accurate equipment so that, for example, even more devices could be put onto silicon chips to increase their computing power.

"This is a key project where the United States is a long way in front of any other country," said Trevor Howe, a University of Connecticut professor of metallurgy and director of its Precision Manufacturing Center.

Source: Boston Globe

▶ The article states, "the Molecular Measuring Machine can measure distances to within 40 nanometers, or billionths of a meter." This suggests that the prefix *nano* means what?
one billionth, or $\dfrac{1}{1,000,000,000}$

▶ Write the two unit fractions you would use to convert between meters and nanometers. Use your unit fractions to change 40 nanometers to meters, and to change 300,000 nanometers to meters.

$$\frac{40 \text{ nm}}{1} \cdot \frac{1\,\text{m}}{1,000,000,000 \text{ nm}} = 0.00000004 \text{ m}$$

$$\frac{300,000 \text{ nm}}{1} \cdot \frac{1\,\text{m}}{1,000,000,000 \text{ nm}} = 0.0003 \text{ m}$$

▶ Why do you suppose the headline and first paragraph talk about "billionths of an inch" when the rest of the article specifies nanometers, which are billionths of meter? Would one–billionth of an inch be the same as one–billionth of a meter? How different would they be? **Headline may use inches because they are more familiar to U.S. readers.**

▶ Use the information in the article to find the length of one atom. Will you use inches or meters?
Use nanometers:
40 nm ÷ 5 atoms = 8 nm per atom

7.2 Exercises

Use your knowledge of the meaning of metric prefixes to fill in the blanks.

Example: *hecto* means _____ so 1 hm = _____ m

Solution: *hecto* means _____**100**_____ so 1 hm = _____**100**_____ m

---

1. *kilo* means _____**1000**_____ so

   1 km = _____**1000**_____ m

2. *deka* means _____**10**_____ so

   1 dam = _____**10**_____ m

3. *milli* means _____$\frac{1}{1000}$ **or 0.001**_____ so

   1 mm = _____$\frac{1}{1000}$ **or 0.001**_____ m

4. *deci* means _____$\frac{1}{10}$ **or 0.1**_____ so

   1 dm = _____$\frac{1}{10}$ **or 0.1**_____ m

5. *centi* means _____$\frac{1}{100}$ **or 0.01**_____ so

   1 cm = _____$\frac{1}{100}$ **or 0.01**_____ m

6. *hecto* means _____**100**_____ so

   1 hm = _____**100**_____ m

Use this ruler to measure the following.

7. the width of your hand in centimeters

   answer varies—about 8 cm

8. the width of your hand in millimeters

   10 times the number of cm measured in Exercise 7.

9. the width of your thumb in millimeters

   answer varies—about 20 mm

10. the width of your thumb in centimeters

    number of mm measured in Exercise 9 divided by 10

Write the most reasonable metric length unit in each blank. Choose from km, m, cm, and mm.

11. The child was 91 _____**cm**_____ tall.

12. The cardboard was 3 _____**mm**_____ thick.

13. Ming-Na swam in the 200 _____**m**_____ backstroke race.

14. The bookcase is 75 _____**cm**_____ wide.

15. Adriana drove 400 _____**km**_____ on her vacation.

16. The door is 2 _____**m**_____ high.

17. An aspirin tablet is 10 _____**mm**_____ across.

18. Lamard jogs 4 _____**km**_____ every morning.

19. A paper clip is about 3 _____**cm**_____ long.

20. My pen is 145 _____**mm**_____ long.

21. Dave's truck is 5 _____**m**_____ long.

22. Wheelchairs need doorways that are at least 80 _____**cm**_____ wide.

23. Describe at least three examples of metric length units that you have come across in your daily life.

    Examples include 35 mm film for cameras, track and field events, metric auto parts, and lead refills for mechanical pencils.

24. Explain one reason why the metric system would be easier for a child to learn than the English system.

    Conversions can be done using decimals instead of fractions; fewer conversion relationships to memorize.

✍ Writing   ◉ Conceptual   ▲ Challenging   ≈ Estimation

Convert each measurement. Use unit fractions or the metric conversion line.

Example: 16 mm to m	**Solution:** $\dfrac{16 \text{ mm}}{1} \cdot \dfrac{1 \text{ m}}{1000 \text{ mm}} = \dfrac{16}{1000}\text{m} = \textbf{0.016 m}$
	or: From mm to m is three places to the left $\underset{\smile\smile}{016.}$ mm = **0.016 m**

25. 7 m to cm

700 cm

26. 18 m to cm

1800 cm

27. 40 mm to m

0.040 m or 0.04 m

28. 6 mm to m

0.006 m

29. 9.4 km to m

9400 m

30. 0.7 km to m

700 m

31. 509 cm to m

5.09 m

32. 30 cm to m

0.3 m

33. 400 mm to cm

40 cm

34. 25 mm to cm

2.5 cm

35. 0.91 m to mm

910 mm

36. 4 m to mm

4000 mm

37. Is 82 cm greater than or less than 1 m? What is the difference in the lengths?

less; 18 cm or 0.18 m

38. Is 1022 m greater than or less than 1 km? What is the difference in the lengths?

greater; 22 m or 0.022 km

39. Many cameras use film that is 35 mm wide. Movie film may be 70 mm wide. Using the ruler on the previous page, draw a line that is 35 mm long and a line 70 mm long. Then convert each measurement to centimeters.

———————————— 35 mm = 3.5 cm

————————————————— 70 mm = 7 cm

40. Gold wedding bands may be very narrow or quite wide. Common widths are 3 mm, 5 mm, and 10 mm. Using the ruler on the previous page, draw lines that are 3 mm, 5 mm, and 10 mm long. Then convert each measurement to centimeters.

— 3 mm = 0.3 cm

—— 5 mm = 0.5 cm

———— 10 mm = 1 cm

▲ **41.** Convert 5.6 mm to km.   0.0000056 km

▲ **42.** Convert 16.5 km to mm.   16,500,000 mm

Review and Prepare

Write each decimal as a fraction in lowest terms. (For help, see Section 4.1.)

43. 0.875 $\dfrac{7}{8}$

44. 0.6 $\dfrac{3}{5}$

45. 0.08 $\dfrac{2}{25}$

46. 0.075 $\dfrac{3}{40}$

We use capacity units to measure liquids, such as the amount of milk in a recipe, the gasoline in our car tank, and the water in an aquarium. (The English capacity units we've been using are cups, pints, quarts, and gallons.) The basic metric unit for capacity is the **liter** (LEE-ter) (also spelled *litre*). The capital letter **L** is the symbol for liter, to avoid confusion with the numeral 1.

OBJECTIVE 1 The liter is related to metric length in this way: a box that measures 10 cm on every side holds exactly one liter. (The volume of the box is 1000 cubic centimeters. Volume is discussed in **Section 8.7**.) A liter is just a little more than 1 quart.

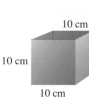
10 cm
10 cm
10 cm

Holds exactly
1 liter (L)

A liter is a little more
than one quart (just $\frac{1}{4}$ cup more).

In the metric system you use liters for things like buying milk at the store, filling a pail with water, and describing the size of your home aquarium.

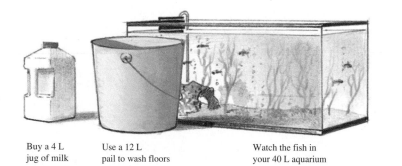

Buy a 4 L
jug of milk

Use a 12 L
pail to wash floors

Watch the fish in
your 40 L aquarium

WORK PROBLEM 1 AT THE SIDE. ▶▶

To make larger or smaller capacity units we use the same **prefixes** as we did with length units. For example, *kilo* means 1000 so a *kilo*meter is 1000 meters. In the same way, a *kilo*liter is 1000 liters.

Prefix	*kilo-* liter	*hecto-* liter	*deka-* liter	liter	*deci-* liter	*centi-* liter	*milli-* liter
Meaning	1000 liters	100 liters	10 liters	1 liter	$\frac{1}{10}$ of a liter	$\frac{1}{100}$ of a liter	$\frac{1}{1000}$ of a liter
Symbol	kL	hL	daL	L	dL	cL	mL

OBJECTIVES

1▶ Know the basic metric units of capacity.

2▶ Convert among metric capacity units.

3▶ Know the basic metric units of weight (mass).

4▶ Convert among metric weight (mass) units.

5▶ Distinguish among basic metric units of length, capacity, and weight (mass).

FOR EXTRA HELP

Tutorial   Tape 12   SSM, Sec. 7.3

1. Which things would you measure in liters?

   amount of water in the bathtub

   length of the bathtub

   width of your car

   amount of gasoline you buy for your car

   weight of your car

   height of a pail

   amount of water in a pail

ANSWERS

1. water in bathtub, gasoline, water in a pail

2. Write the most reasonable metric unit in each blank. Choose from L and mL.

(a) I bought 8 _____ of milk at the store.

(b) The nurse gave me 10 _____ of cough syrup.

(c) This is a 100 _____ garbage can.

(d) It took 10 _____ of paint to cover the bedroom walls.

(e) My car's gas tank holds 50 _____ .

(f) I added 15 _____ of oil to the pancake mix.

(g) The can of orange soda holds 350 _____ .

(h) My friend gave me a 30 _____ bottle of expensive perfume.

The capacity units you will use most often in daily life are liters (L) and *milli*liters (mL). A tiny box that measures 1 cm on every side holds exactly one milliliter. (In medicine, this small amount is also called 1 cubic centimeter, or 1 cc for short.) It takes 1000 mL to make 1 L. Here are some other useful comparisons.

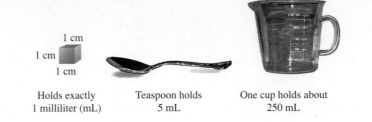

Holds exactly 1 milliliter (mL)   Teaspoon holds 5 mL   One cup holds about 250 mL

E X A M P L E 1   Using Metric Capacity Units

Write the most reasonable metric unit in each blank. Choose from L and mL.

(a) The bottle of shampoo held 500 _____ .

500 <u>mL</u> because 500 L would be about 500 quarts, which is too much.

(b) I bought a 2 _____ carton of orange juice.

2 <u>L</u> because 2 mL would be less than a teaspoon.

◀◀ **WORK PROBLEM 2 AT THE SIDE.**

OBJECTIVE 2 Just as with length units, you can convert between milliliters and liters using unit fractions. The units fractions you need are:

$$\frac{1000 \text{ mL}}{1 \text{ L}} \qquad \frac{1 \text{ L}}{1000 \text{ mL}}$$

Or you can use a metric conversion line to decide how to move the decimal point.

1000	100	10	1	$\frac{1}{10}$	$\frac{1}{100}$	$\frac{1}{1000}$
kL	hL	daL	L	dL	cL	mL

E X A M P L E 2   Conversions among Metric Capacity Units

Convert using the metric conversion line or unit fractions.

(a) 2.5 L to mL

Using the metric conversion line:
From **L** to **mL** is *three places* to the *right*.

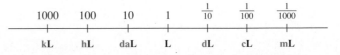

2.500   Write two zeros as placeholders.

2.5 L = 2500 mL

Using unit fractions:

Multiply by a unit fraction that allows you to cancel liters.

$$\frac{2.5 \text{ L̸}}{1} \cdot \frac{1000 \text{ mL}}{1 \text{ L̸}} = 2500 \text{ mL}$$

CONTINUED ON NEXT PAGE

ANSWERS

2. (a) L **(b)** mL **(c)** L
(d) L **(e)** L **(f)** mL
(g) mL **(h)** mL

(b) 80 mL to L

Using the metric conversion line:
From **mL** to **L** is *three places* to the *left*.

80.   080.

↑   ⌣⌣⌣
Decimal point   Move three
starts here.   places left.

80 mL = 0.080 L or 0.08 L

Using unit fractions:

Multiply by a unit fraction that allows you to cancel mL.

$$\frac{80 \text{ mL}}{1} \cdot \frac{1 \text{ L}}{1000 \text{ mL}}$$

$$= \frac{80}{1000} \text{L} = 0.08 \text{ L}$$

> **WORK PROBLEM 3 AT THE SIDE.** ▶▶

OBJECTIVE 3▶ The **gram** is the basic metric unit for mass. Although we often call it "weight," there is a difference. Weight is a measure of the pull of gravity; the farther you are from the center of the earth, the less you weigh. In outer space you become weightless, but your mass, the amount of matter in your body, stays the same regardless of where you are. We will use the word "weight" for everyday purposes.

The gram is related to metric length in this way: the weight of the water in a box measuring 1 cm on every side is 1 gram. This is a very tiny amount of water (1 mL) and a very small weight. One gram is also the weight of a dollar bill or a single raisin. A nickel weighs 5 grams. A regular hamburger weighs from 175 to 200 grams.

The 1 mL of water in this box weighs 1 gram.

A nickel weighs 5 grams.

A dollar bill weighs 1 gram.

A hamburger weighs 175 to 200 grams.

> **WORK PROBLEM 4 AT THE SIDE.** ▶▶

To make larger or smaller weight units, we use the same **prefixes** as we did with length and capacity units. For example, *kilo* means 1000 so a *kilo*meter is 1000 meters, a *kilo*liter is 1000 liters, and a *kilo*gram is 1000 grams.

3. Convert.

(a) 9 L to mL

(b) 0.75 L to mL

(c) 500 mL to L

(d) 5 mL to L

(e) 2.07 L to mL

(f) 3275 mL to L

4. Which things would weigh about 1 gram?

a small paperclip

a pair of scissors

one playing card from a deck of cards

a calculator

an average-size apple

the check you wrote at the grocery store

ANSWERS

3. (a) 9000 mL **(b)** 750 mL
(c) 0.5 L **(d)** 0.005 L
(e) 2070 mL **(f)** 3.275 L

4. paperclip, playing card, check

5. Write the most reasonable metric unit in each blank. Choose from kg, g, and mg.

(a) A thumbtack weighs 800 _____ .

(b) A teenager weighs 50 _____ .

(c) This large cast-iron frying pan weighs 1 _____ .

(d) Jerry's basketball weighed 600 _____ .

(e) Tamlyn takes a 500 _____ calcium tablet every morning.

(f) On his diet, Greg can eat 90 _____ of meat for lunch.

(g) One strand of hair weighs 2 _____ .

(h) One banana might weigh 150 _____ .

Prefix	*kilo-gram*	*hecto-gram*	*deka-gram*	*gram*	*deci-gram*	*centi-gram*	*milli-gram*
Meaning	1000 grams	100 grams	10 grams	1 gram	$\frac{1}{10}$ of a gram	$\frac{1}{100}$ of a gram	$\frac{1}{1000}$ of a gram
Symbol	**kg**	**hg**	**dag**	**g**	**dg**	**cg**	**mg**

The units you will use most often in daily life are kilograms (kg), grams (g), and milligrams (mg). *Kilo*grams are used instead of pounds. A kilogram is 1000 grams. It is about 2.2 pounds. This textbook weighs about 1.7 kg. An average newborn baby weighs 3 to 4 kg; a college football player might weigh 100 to 110 kg.

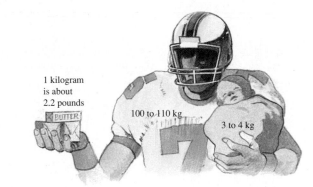

1 kilogram is about 2.2 pounds

100 to 110 kg

3 to 4 kg

Extremely small weights are measured in *milli*grams. It takes 1000 mg to make 1 g. Recall that a dollar bill weighs about 1 g. Imagine cutting it into 1000 pieces; the weight of one tiny piece would be 1 mg. Dosages of medicine and vitamins are given in milligrams. You will also use milligrams in science classes.

Cut a dollar bill into 1000 pieces. One tiny piece weighs 1 milligram.

E X A M P L E 3 Using Metric Weight Units

Write the most reasonable metric unit in each blank. Choose from kg, g, and mg.

(a) Ramon's suitcase weighed 20 _____ .
    20 <u>kg</u> because kilograms are used instead of pounds. 20 kg is about 44 pounds.

(b) LeTia took a 350 _____ aspirin tablet.
    350 <u>mg</u> because 350 g would be more than the weight of a hamburger, which is too much.

(c) Jenny mailed a letter that weighed 30 _____ .
    30 <u>g</u> because 30 kg would be much too heavy and 30 mg is less than the weight of a dollar bill.

◀◀ **WORK PROBLEM 5 AT THE SIDE.**

ANSWERS

5. (a) mg (b) kg (c) kg
    (d) g (e) mg (f) g
    (g) mg (h) g

OBJECTIVE 4 As with length and capacity, you can convert among metric weight units by using unit fractions. The unit fractions you need are shown here.

Converting between grams and kilograms	Converting between milligrams and grams
$\dfrac{1000 \text{ g}}{1 \text{ kg}}$ or $\dfrac{1 \text{ kg}}{1000 \text{ g}}$	$\dfrac{1000 \text{ mg}}{1 \text{ g}}$ or $\dfrac{1 \text{ g}}{1000 \text{ mg}}$

Or you can use a metric conversion line to decide how to move the decimal point.

1000	100	10	1	$\frac{1}{10}$	$\frac{1}{100}$	$\frac{1}{1000}$
kg	hg	dag	g	dg	cg	mg

E X A M P L E 4   Conversions among Metric Weight Units

Convert using the metric conversion line or unit fractions.

(a) 7 mg to g

Using the metric conversion line:
From **mg** to **g** is *three places* to the *left*.

7.   007.
↑
Decimal point   Move three
starts here.   places left.

7 mg = 0.007 g

Using unit fractions:

Multiply by a unit fraction that allows you to cancel mg.

$$\frac{7 \text{ mg}}{1} \cdot \frac{1 \text{ g}}{1000 \text{ mg}} = \frac{7}{1000} \text{ g}$$
$$= 0.007 \text{ g}$$

(b) 13.72 kg to g

Using the metric conversion line:
From **kg** to **g** is *three* places to the *right*.

13.720   Decimal point moves
          three places to
          the right.

13.72 kg = 13,720 g
                    ↑
              A comma
        (not a decimal point)

Using unit fractions:

Multiply by a unit fraction that allows you to cancel kg.

$$\frac{13.72 \text{ kg}}{1} \cdot \frac{1000 \text{ g}}{1 \text{ kg}} = 13,720 \text{ g}$$
                                              ↑
                                        A comma
                                  (not a decimal point)

WORK PROBLEM 6 AT THE SIDE. ▶▶

6. Convert.

(a) 10 kg to g

(b) 45 mg to g

(c) 6.3 kg to g

(d) 0.077 g to mg

(e) 5630 g to kg

(f) 90 g to kg

ANSWERS
6. **(a)** 10,000 g **(b)** 0.045 g
(c) 6300 g **(d)** 77 mg
(e) 5.63 kg **(f)** 0.09 kg

7. First decide which type of units are needed: length, capacity, or weight. Then write the most appropriate unit in the blank. Choose from km, m, cm, mm, L, mL, kg, g, and mg.

(a) Gail bought a 4 _____ can of paint.

Use _____ units.

(b) The bag of chips weighed 450 _____ .

Use _____ units.

(c) Give the child 5 _____ of liquid aspirin.

Use _____ units.

(d) The width of the window is 55 _____ .

Use _____ units.

(e) Akbar drives 18 _____ to work.

Use _____ units.

(f) Each computer weighs 5 _____ .

Use _____ units.

(g) A credit card is 55 _____ wide.

Use _____ units.

OBJECTIVE 5 As you encounter things to be measured at home, on the job, or in your classes at school, be careful to use the correct type of measurement unit.

Use *length units* (kilometers, meters, centimeters, millimeters) to measure:

how long	how high	how far away
how wide	how tall	how far around (perimeter)
how deep	distance	

Use *capacity units* (liters, milliliters) to measure liquids (things that can be poured) such as:

water	shampoo	gasoline
milk	perfume	oil
soft drinks	cough syrup	paint

Also use liters and milliliters to describe how much liquid something can hold, such as an eyedropper, measuring cup, pail, or bathtub.

Use *weight units* (kilogram, grams, milligrams) to measure:

the weight of something   how heavy something is

In Chapter 8 you will use square units (such as square meters) to measure area, and cubic units (such as cubic centimeters) to measure volume.

EXAMPLE 5   Using a Variety of Metric Units

First decide which type of units are needed: length, capacity, or weight. Then write the most appropriate metric unit in the blank. Choose from km, m, cm, mm, L, mL, kg, g, and mg.

(a) The letter needs another stamp because it weighs 40 _____ .
Use _____ units.

The letter weighs 40 **grams** because 40 mg is less than the weight of a dollar bill and 40 kg would be about 88 pounds.
Use **weight** units because of the word "weighs."

(b) The swimming pool is 3 _____ deep at the deep end.
Use _____ units.

The pool is 3 **meters** deep because 3 cm is only about an inch and 3 km is about 1.8 miles.
Use **length** units because of the word "deep."

(c) This is a 340 _____ can of juice.
Use _____ units.

It is a 340 **milliliter** can because 340 liters would be more than 340 quarts.
Use **capacity** units because juice is a liquid.

◀◀ **WORK PROBLEM 7 AT THE SIDE.**

ANSWERS

7. (a) L; capacity **(b)** g; weight
(c) mL; capacity **(d)** cm; length
(e) km; length **(f)** kg; weight
(g) mm; length

7.3 Exercises

Write the most reasonable metric unit in each blank. Choose from L, mL, kg, g, and mg.

1. The glass held 250 __mL__ of water.

2. Hiromi used 20 __L__ of water to wash the kitchen floor.

3. Dolores can make 10 __L__ of soup in that pot.

4. Jay gave 2 __mL__ of vitamin drops to the baby.

5. Our labrador dog grew up to weigh 40 __kg__ .

6. One dime weighs 2 __g__ .

7. Lori caught a small sunfish weighing 150 __g__ .

8. A small safety pin weighs 750 __mg__ .

9. Andre donated 500 __mL__ of blood today.

10. Barbara bought the large 2 __L__ bottle of cola.

11. The patient received 250 __mg__ of medication each hour.

12. The 8 people on the elevator weighed a total of 500 __kg__ .

13. The gas can for the lawn mower holds 4 __L__ .

14. Kevin poured 10 __mL__ of vanilla into the bowl.

15. Pam's backpack weighs 5 __kg__ when it is full of books.

16. One grain of salt weighs 2 __mg__ .

Today, medical measurements are usually given in the metric system. Since we convert among metric units of measure by moving the decimal point, it is possible that mistakes can be made. Examine the following dosages and indicate whether they are reasonable or unreasonable.

17. Drink 4.1 liters of Kaopectate after each meal.

unreasonable

18. Drop 1 mL of solution into the eye twice a day.

reasonable

19. Soak your feet in 5 kilograms of Epsom salts per liter of water.

unreasonable

20. Inject 0.5 liter of insulin each morning.

unreasonable

21. Take 15 milliliters of cough syrup every four hours.

reasonable

22. Take 200 milligrams of vitamin C each day.

reasonable

23. Take 350 milligrams of aspirin three times a day.

reasonable

24. Buy a tube of ointment weighing 0.002 gram.

unreasonable

✍ Writing   ⊙ Conceptual   ▲ Challenging   ≈ Estimation

25. Describe at least two examples of metric capacity units and two examples of metric weight units that you have come across in your daily life.

Some examples are 2 liter bottles of soda, shampoo bottles marked in mL, grams of fat listed on cereal boxes, vitamin doses in mg.

26. Explain in your own words how the meter, liter, and gram are related.

A box measuring 1 cm on each side holds 1 mL of water and the water weighs 1 g.

27. Describe how you decide which unit fraction to use when converting 6.5 kg to grams.

Unit for your answer (g) is in numerator; unit being changed (kg) is in denominator so it will cancel. The unit fraction is $\frac{1000 \text{ g}}{1 \text{ kg}}$.

28. Write out an explanation of each step you would use to convert 20 mg to grams using the metric conversion line.

From mg to g is three places to the left on the metric conversion line so move decimal point three places left.

020.   20 mg = 0.02 g

Convert each measurement. Use unit fractions or the metric conversion line.

Example: 9 g to kg

Solution:
$$\frac{9 \text{ g}}{1} \cdot \frac{1 \text{ kg}}{1000 \text{ g}} = \frac{9}{1000} \text{ kg} = \textbf{0.009 kg}$$

or: From g to kg is three places to the left.   **009.** g = **0.009 kg**

29. 15 L to mL   **15,000 mL**

30. 6 L to mL   **6000 mL**

31. 3000 mL to L   **3 L**

32. 18,000 mL to L   **18 L**

33. 925 mL to L   **0.925 L**

34. 200 mL to L   **0.2 L**

35. 8 mL to L   **0.008 L**

36. 25 mL to L   **0.025 L**

37. 4.15 L to mL   **4150 mL**

38. 11.7 L to mL   **11,700 mL**

39. 8000 g to kg   **8 kg**

40. 25,000 g to kg   **25 kg**

41. 5.2 kg to g   **5200 g**

42. 12.42 kg to g   **12,420 g**

43. 0.85 g to mg   **850 mg**

44. 0.2 g to mg   **200 mg**

45. 30,000 mg to g   **30 g**

46. 7500 mg to g   **7.5 g**

47. 598 mg to g   **0.598 g**

48. 900 mg to g   **0.9 g**

49. 60 mL to L   **0.06 L**

50. 6.007 kg to g   **6007 g**

51. 3 g to kg   **0.003 kg**

52. 12 mg to g   **0.012 g**

53. 0.99 L to mL   **990 mL**

54. 13,700 mL to L   **13.7 L**

Write the most appropriate metric unit in each blank. Choose from km, m, cm, mm, L, mL, kg, g, and mg.

55. The masking tape is 19 __mm__ wide.

56. The roll has 55 __m__ of tape on it.

57. Buy a 60 __mL__ jar of acrylic paint for art class.

58. One onion weighs 200 __g__ .

59. My waist measurement is 65 __cm__ .

60. Add 2 __L__ of windshield washer fluid to your car.

61. A single postage stamp weighs 90 __mg__ .

62. The hallway is 10 __m__ long.

Solve the following application problems.

63. The doctor told Sara to drink two liters of water each day. How many milliliters is that?

2000 mL

64. A juice can holds 1500 mL. How many liters of juice does it hold?

1.5 L

65. The premature infant weighed only 950 grams. How many kilograms did he weigh?

0.95 kg

66. Bill bought three kilograms of potatoes. How many grams did he buy?

3000 g

67. A healthy human heart pumps about 70 mL of blood per beat. How many liters of blood does it pump per beat?

0.07 L

68. In one sip, an elephant can suck 7.6 L of water into its trunk. How many milliliters does it suck into its trunk?

7600 mL

69. A small adult cat weighs about 3 kg. How many grams does it weigh?

3000 g

70. If the letter you are mailing weighs 29 g, you must put additional postage on it. How many kilograms does the letter weigh?

0.029 kg

71. Is 1005 mg greater than or less than 1 g? What is the difference in the weights?

greater; 5 mg or 0.005 g

72. Is 990 mL greater than or less than 1 L? What is the difference in the amounts?

less; 10 mL or 0.01 L

▲**73.** One nickel weighs 5 grams. How many nickels are in 1 kilogram of nickels?

200 nickels

▲**74.** Seawater contains about 3.5 grams of salt per 1000 milliliters of water. How many grams of salt would be in 1 liter of seawater?

3.5 grams

▲**75.** Helium weighs about 0.0002 grams per milliliter. How much would 1 liter of helium weigh?

0.2 gram

▲**76.** About 1500 grams of sugar can be dissolved in a liter of warm water. How much sugar could be dissolved in 1 milliliter of warm water?

1.5 grams

Review and Prepare

Name the digit that has the given place value in each of the following. (For help, see Section 4.1.)

77. 7250.6183
    7 thousands
    1 hundredths
    8 thousandths
    2 hundreds

78. 1358.0256
    6 ten-thousandths
    0 tenths
    5 tens
    8 ones

7.4 Applications of Metric Measurement

OBJECTIVE ▶ One advantage of the metric system is the ease of comparing measurements in application situations. Just be sure that you are comparing similar units: mg to mg, km to km, and so on.

E X A M P L E I Solving Metric Applications

(a) Cheddar cheese is on sale at $8.99 per kg. Jake bought 350 grams of the cheese. How much did he pay, to the nearest cent?

The price is $8.99 per *kilogram,* but the amount purchased is in *grams.* Convert grams to kilograms (the unit in the price). Then multiply the weight times the cost per kilogram.

$$350 \text{ g} = 0.35 \text{ kg}$$

$$\frac{\$8.99}{1 \text{ kg}} \cdot \frac{0.35 \text{ kg}}{1} = \$3.1465$$

Jake paid $3.15, to the nearest cent.

(b) Olivia has 2.5 meters of lace. How many centimeters of lace can she use to trim each of six hair ornaments? Round to the nearest tenth of a centimeter.

The given amount is in *meters,* but the answer must be in *centimeters,* so convert meters to centimeters. Then divide by the number of hair ornaments.

$$2.5 \text{ m} = 250 \text{ cm}$$

$$\frac{250 \text{ cm}}{6 \text{ ornaments}} = 41.6666 \text{ cm/ornament}$$

Olivia can use ≈41.7 cm of lace on each ornament (rounded to the nearest tenth).

WORK PROBLEM I AT THE SIDE. ▶▶

E X A M P L E 2 Measurement Applications

(a) Rubin measured a board and found that the length was 3 meters plus an additional 5 centimeters. He cut off a piece measuring 1 meter 40 centimeters for a shelf. Find the length of the remaining piece in meters.

The lengths involve two units, m and cm. To make the calculations easier, write each length in terms of meters (the unit called for in the answer). Then subtract to find the leftover length.

Board		Shelf	
3m →	3.00 m	1 m →	1.0 m
plus 5 cm →	+ 0.05 m	plus 40 cm →	+ 0.4 m
	3.05 m		1.4 m

Subtract to	3.05 m	← Board
find leftover	− 1.40 m	← Shelf
length.	1.65 m	← Leftover piece

The length of the leftover piece is 1.65 m.

— **CONTINUED ON NEXT PAGE**

— **CONTINUED ON NEXT PAGE**

OBJECTIVE

▶ Solve application problems involving metric measurements.

FOR EXTRA HELP

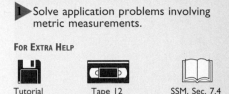

Tutorial   Tape 12   SSM, Sec. 7.4

1. Solve each problem.

(a) Satin ribbon is on sale at $0.89 per meter. How much will 75 cm cost, to the nearest cent?

(b) Lucinda's doctor wants her to take 1.2 grams of medication each day in three equal doses. How many milligrams should be in each dose?

ANSWERS

1. (a) $0.67 (rounded to the nearest cent)
   (b) 400 mg/dose

469

2. (a) Andrea has two pieces of fabric. One measures 2 meters 35 centimeters and the other measures 1 meter 85 centimeters. How many meters of fabric does she have in all?

(b) Amy put a basket of nuts on her scale and saw that they weighed 4 kilograms plus 140 grams. She plans to put the nuts into three gift packs of equal size. Find the number of kilograms in each pack.

Write 4 kg 140 g in terms of kg (the unit for the answer). Then divide by the number of gift packs.

$$
\begin{array}{r}
4 \text{ kg} \rightarrow \quad 4.00 \text{ kg} \\
\text{plus } 140 \text{ g} \rightarrow + \ 0.14 \text{ kg} \\
\hline
4.14 \text{ kg}
\end{array}
$$

$$\frac{4.14 \text{ kg}}{3 \text{ packs}} = 1.38 \text{ kg/pack}$$

Each gift pack will have 1.38 kg of nuts.

◀◀ **WORK PROBLEM 2 AT THE SIDE.**

(b) Mr. Green has 9 m 20 cm of rope. He is cutting it into eight pieces so his Scout troop can practice knot tying. How many meters of rope will each Scout get?

ANSWERS

2. (a) 4.2 m **(b)** 1.15 m/Scout

7.4 Exercises

Solve each application problem.

> **Example:** A basket of strawberries weighed 1 kg 80 g. Find the cost of the strawberries if they are priced at $2.19 per kg.
>
> **Solution:** Write 1 kg 80 g in terms of kg (the unit in the price).
>
> $$\begin{array}{rl} 1\text{ kg} \rightarrow & 1.00\text{ kg} \\ \text{plus } 80\text{ g} \rightarrow & +\ 0.08\text{ kg} \\ \hline & 1.08\text{ kg} \end{array} \qquad \frac{\$2.19}{1\text{ kg}} \cdot \frac{1.08\text{ kg}}{1} = \$2.3652$$
>
> In consumer situations, prices are rounded to the nearest cent, so the strawberries cost **$2.37**.

1. Bulk rice is on special at $0.65 per kilogram. Pam scooped some rice into a bag and put it on the scale. How much will she pay for 2 kg 50 g of rice?

$1.33 rounded to the nearest cent

2. Lanh is buying a piece of plastic tubing for the science lab that measures 3 m 15 cm. The price is $4.75 per meter. How much will Lanh pay?

$14.96 rounded to the nearest cent

3. Kendal works for a garden store. He put 15 grams of fertilizer on each of 650 tomato plants. How many kilograms of fertilizer did he use?

9.75 kg

4. The garden store ordered a 50 liter drum of liquid plant food. They repackaged the plant food into 125 mL bottles. How many bottles were filled?

400 bottles

5. An adult human body contains about 5 L of blood. If each beat of the heart pumps 70 mL of blood, how many times must the heart beat to pass all the blood through the heart? Round to the nearest whole number of beats.

≈71 beats

6. A floor tile measures 30 cm by 30 cm and weighs 185 g. How many kilograms would a carton of 24 tiles weigh?

4.44 kg

7. Rosa is building a bookcase. She has one board that is 2 m 8 cm long and another that is 2 m 95 cm long. How long are the two boards together in meters?

5.03 m

8. Eric's Scottie dog weighs 8 kg 600 g. Rob's Great Dane weighs 50 kg 50 g. The Great Dane weighs how much more than the Scottie, in kilograms?

41.45 kg

✍ Writing   ◉ Conceptual   ▲ Challenging   ≈ Estimation

9. The apartment building caretaker puts 750 mL of chlorine into the swimming pool every day. How many liters should he order to have a one-month (30-day) supply on hand?

22.5 L

10. Janet has 10 m 30 cm of fabric. She wants to make curtains for three windows that are all the same size. How much fabric is available for each window, to the nearest tenth of a meter?

≈3.4 m

It is difficult to weigh very light objects, such as a single sheet of paper or a single staple (unless you have a very expensive scientific scale). One way around this problem is to weigh a large number of the items and then divide to find the weight of one item. Of course, before dividing, you must subtract the weight of the box or wrapper that the items are packaged in to find the net weight. Complete this table.

Item	Total weight	Weight of packaging	Net weight	Weight of one item in grams	Weight of one item in milligrams
11. Box of 50 envelopes	255 g	40 g	215 g	4.3 g	4300 mg
12. Box of 1000 staples	350 g	20 g	330 g	0.33 g	330 mg
▲ **13.** Ream of paper (500 sheets)	1.55 kg	50 g	1500 g	3 g	3000 mg
▲ **14.** Box of 100 small paper clips	55 g	5 g	50 g	0.5 g	500 mg

▲ **15.** As a fund raiser, the PTA bought 40 kg of nuts for $113.50. They sold the nuts in 250 g bags for $2.95 each. Find the amount of profit.

$358.50

▲ **16.** In chemistry class, each of the 45 students needs 85 mL of acid. How many one-liter bottles of acid need to be ordered?

4 bottles

▲ **17.** Which case of shampoo is the better buy: a $16 case that holds 12 1-liter bottles or an $18 case that holds 36 400-mL bottles?

$18 case

▲ **18.** James needs 3 m 80 cm of wood molding to frame a picture. The price is $5.89 per meter plus a 7% sales tax. How much will James pay?

$23.95 rounded to the nearest cent

Review and Prepare

Multiply. (For help, see Section 4.5.)

19.   0.035
     $\times$ 18
     ─────
     0.63

20.   28.35
     $\times$ 12
     ─────
     340.2

21. $6.3 \cdot 0.91$

5.733

22. $14.7 \cdot 2.2$

32.34

7.5 Metric–English Conversions and Temperature

OBJECTIVES

1. ▶ Use unit fractions to convert from metric to English or English to metric units.

2. ▶ Know common temperatures on the Celsius scale.

3. ▶ Convert temperatures by using the order of operations.

FOR EXTRA HELP

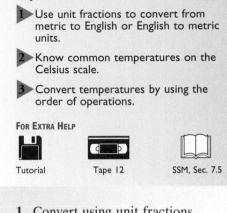

Tutorial   Tape 12   SSM, Sec. 7.5

OBJECTIVE ▶ Until the United States has switched completely from the English system to the metric system, it will be necessary to make conversions from one system to the other. *Approximate* conversions can be made with the help of the following table, in which the values have been rounded to the nearest hundredth or thousandth.

Metric to English		English to Metric	
1 kilometer	≈ 0.62 mile	1 mile	≈ 1.61 kilometers
1 meter	≈ 1.09 yards	1 yard	≈ 0.91 meter
1 meter	≈ 3.28 feet	1 foot	≈ 0.30 meter
1 centimeter	≈ 0.39 inch	1 inch	≈ 2.54 centimeters
1 liter	≈ 0.26 gallon	1 gallon	≈ 3.78 liters
1 liter	≈ 1.06 quarts	1 quart	≈ 0.95 liter
1 kilogram	≈ 2.20 pounds	1 pound	≈ 0.45 kilogram
1 gram	≈ 0.035 ounce	1 ounce	≈ 28.35 grams

EXAMPLE 1   Converting Metric and English (Length)

Convert using unit fractions. Round your answers to the nearest tenth, if necessary.

10 meters to yards

We're changing from a metric unit to an English unit. In the "Metric to English" part of the table, you see that 1 meter ≈ 1.09 yards. Two unit fractions can be written using that information:

$$\frac{1 \text{ meter}}{1.09 \text{ yards}} \quad \text{or} \quad \frac{1.09 \text{ yards}}{1 \text{ meter}}.$$

Multiply by the unit fraction that allows you to cancel meters (that is, meters is in the denominator).

$$10 \text{ meters} \cdot \frac{1.09 \text{ yards}}{1 \text{ meter}} = \frac{10 \text{ meters}}{1} \cdot \frac{1.09 \text{ yards}}{1 \text{ meter}} = 10.9 \text{ yards}$$

These units should match.

10 meters ≈ 10.9 yards

Note

You could also use the other numbers from the table involving meters and yards: 1 yard ≈ 0.91 meter.

$$\frac{10 \text{ meters}}{1} \cdot \frac{1 \text{ yard}}{0.91 \text{ meter}} = \frac{10}{0.91} \text{ yards} ≈ 10.99 \text{ yards}$$

The answer is slightly different because all the values in the table are approximate. Also, you have to divide instead of multiply, which is usually more difficult to do without a calculator. We will use the first method in this chapter.

1. Convert using unit fractions. Round your answers to the nearest tenth.

(a) 23 meters to yards

(b) 40 centimeters to inches

(c) 5 miles to kilometers (Look at the "English to Metric" side of the table.)

(d) 12 inches to centimeters

ANSWERS

1. **(a)** ≈25.1 yd
(b) ≈15.6 in.
(c) ≈8.1 km
(d) ≈30.5 cm

WORK PROBLEM 1 AT THE SIDE. ▶▶

2. Convert. Use the values from the table on the previous page to make unit fractions. Round answers to the nearest tenth.

(a) 17 kilograms to pounds

(b) 5 liters to quarts

(c) 90 grams to ounces

(d) 3.5 gallons to liters

(e) 145 pounds to kilograms

(f) 8 ounces to grams

EXAMPLE 2 Converting Metric and English (Weight and Capacity)

Convert using unit fractions. Round your answers to the nearest tenth.

(a) 3.5 kilograms to pounds

Look at the "Metric to English" side of the table on the previous page to see that 1 kilogram ≈ 2.20 pounds. Use this information to write a unit fraction that allows you to cancel kilograms.

$$\frac{3.5 \ \cancel{\text{kilograms}}}{1} \cdot \frac{2.20 \text{ pounds}}{1 \ \cancel{\text{kilogram}}} = \frac{3.5 \cdot 2.20 \text{ pounds}}{1} = 7.7 \text{ pounds}$$

3.5 kilograms ≈ 7.7 pounds

(b) 18 gallons to liters

Look at the "English to Metric" side of the table to see that 1 gallon ≈ 3.78 liters. Write a unit fraction that will allow you to cancel gallons.

$$\frac{18 \ \cancel{\text{gallons}}}{1} \cdot \frac{3.78 \text{ liters}}{1 \ \cancel{\text{gallon}}} = \frac{18 \cdot 3.78 \text{ liters}}{1} = 68.04 \text{ liters}$$

68.04 rounded to the nearest tenth is 68.0 so
18 gallons ≈ 68.0 liters.

(c) 300 grams to ounces

In the "Metric to English" side of the table, 1 gram ≈ 0.035 ounce.

$$\frac{300 \ \cancel{\text{grams}}}{1} \cdot \frac{0.035 \text{ ounce}}{1 \ \cancel{\text{gram}}} = 10.5 \text{ ounces}$$

300 grams ≈ 10.5 ounces

Note
Because the metric and English systems were developed independently, there are no exact comparisons. Your answers should be written with the "≈" symbol to show they are approximate.

◄◄ **WORK PROBLEM 2 AT THE SIDE.**

OBJECTIVE 2▶ In the metric system, temperature is measured on the **Celsius** (SELL-see-us) **scale.** On the Celsius scale, water freezes at 0°C and boils at 100°C. The small raised circle stands for "degrees" and capital **C** is for Celsius. Read the temperatures like this:

Water freezes at 0 degrees Celsius (0°C).

Water boils at 100 degrees Celsius (100°C).

The English temperature system that we now use is measured on the **Fahrenheit** (FAIR-en-hite) **scale.** On this scale:

Water freezes at 32 degrees Fahrenheit (32°F).

Water boils at 212 degrees Fahrenheit (212°F).

ANSWERS
2. (a) ≈37.4 pounds (b) ≈5.3 quarts
(c) ≈3.2 ounces (d) ≈13.2 L
(e) ≈65.3 kg (f) ≈226.8 g

The thermometer below shows some typical temperatures in both Celsius and Fahrenheit. For example, comfortable room temperature is about 20°C or 68°F, and normal body temperature is about 37°C or 98.6°F.

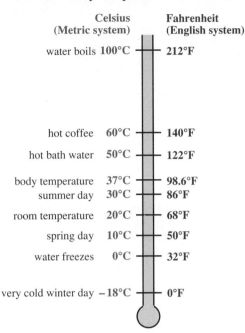

	Celsius (Metric system)	Fahrenheit (English system)
water boils	100°C	212°F
hot coffee	60°C	140°F
hot bath water	50°C	122°F
body temperature	37°C	98.6°F
summer day	30°C	86°F
room temperature	20°C	68°F
spring day	10°C	50°F
water freezes	0°C	32°F
very cold winter day	−18°C	0°F

Note
The freezing and boiling temperatures are exact. The other temperatures are approximate. Even normal body temperature varies slightly from person to person.

E X A M P L E 3 Using Celsius Temperatures

Circle the Celsius temperature that is most reasonable for each situation.

(a) warm summer day   29°C 64°C 90°C

29°C is reasonable. 64°C and 90°C are too hot; they're both above the temperature of hot bath water (above 122°F).

(b) inside a freezer   −10°C 3°C 25°C

−10°C is the reasonable temperature because it is the only one below the freezing point of water (0°C). Your frozen foods would start thawing at 3°C or 25°C.

WORK PROBLEM 3 AT THE SIDE. ▶▶

OBJECTIVE 3 You can use these formulas to convert temperatures.

Celsius-Fahrenheit Conversion Formulas

Converting from Fahrenheit (F) to Celsius (C)

$$C = \frac{5(F - 32)}{9}$$

Converting from Celsius (C) to Fahrenheit (F)

$$F = \frac{9 \cdot C}{5} + 32$$

3. Circle the Celsius temperature that is most reasonable for each situation.

(a) Set the living room thermostat at:
11°C 21°C 71°C

(b) The baby has a fever of:
29°C 39°C 49°C

(c) Wear a sweater outside because it's:
15°C 25°C 50°C

(d) My iced tea is:
−5°C 5°C 30°C

(e) Time to go swimming! It's:
95°C 65°C 35°C

(f) Inside a refrigerator (not the freezer) it's:
−15°C 0°C 3°C

(g) There's a blizzard outside. It's:
10°C 0°C −20°C

(h) I need hot water to get these clothes clean. It should be:
55°C 105°C 200°C

ANSWERS

3. (a) 21°C **(b)** 39°C **(c)** 15°C **(d)** 5°C
(e) 35°C **(f)** 3°C **(g)** −20°C **(h)** 55°C

4. Convert to Celsius.

(a) 59°F

(b) 41°F

(c) 212°F

(d) 98.6°F

As you use these formulas, be sure to follow the order of operations.

1. Do all operations inside parentheses.

2. Simplify any expressions with exponents and find any square roots.

3. Multiply or divide from left to right.

4. Add or subtract from left to right.

E X A M P L E 4 Converting Fahrenheit to Celsius

Convert 68°F to Celsius.

Use the formula and the order of operations.

$$C = \frac{5(F - 32)}{9}$$

$$= \frac{5(68 - 32)}{9} \qquad \text{Work inside parentheses first.}$$

$$= \frac{5(36)}{9}$$

$$= \frac{5(\overset{4}{\cancel{36}})}{\underset{1}{\cancel{9}}} = 20 \qquad \begin{array}{l}\text{Use cancellation, if possible.}\\ \text{Multiply.}\end{array}$$

Thus, 68°F = 20°C.

◀◀ **WORK PROBLEM 4 AT THE SIDE.**

5. Convert to Fahrenheit.

(a) 100°C

(b) 25°C

(c) 80°C

(d) 5°C

E X A M P L E 5 Converting Celsius to Fahrenheit

Convert 15°C to Fahrenheit.

Use the formula and the order of operations.

$$F = \frac{9 \cdot C}{5} + 32$$

$$= \frac{9 \cdot 15}{5} + 32$$

$$= \frac{9 \cdot \overset{3}{\cancel{15}}}{\underset{1}{\cancel{5}}} + 32 \qquad \begin{array}{l}\text{Use cancellation, if possible.}\\ \text{Multiply.}\end{array}$$

$$= 27 + 32 \qquad \text{Add.}$$

$$= 59$$

Thus, 15°C = 59°F.

◀◀ **WORK PROBLEM 5 AT THE SIDE.**

ANSWERS
4. (a) 15°C **(b)** 5°C **(c)** 100°C **(d)** 37°C
5. (a) 212°F **(b)** 77°F **(c)** 176°F **(d)** 41°F

7.5 Exercises

Use the table on page 473 and unit fractions to make approximate conversions from metric to English or English to metric. Round your answers to the nearest tenth.

> **Example:** 36 meters to yards
>
> **Solution:** In the "Metric to English" part of the table, 1 meter ≈ 1.09 yards.
>
> $$\frac{36 \ \cancel{meters}}{1} \cdot \frac{1.09 \ yards}{1 \ \cancel{meter}} = \frac{36 \cdot 1.09 \ yards}{1} = 39.24 \ yards$$
>
> 39.24 rounds to 39.2 so **36 meters ≈ 39.2 yards**

1. 20 meters to yards

≈**21.8 yards**

2. 8 kilometers to miles

≈**5.0 miles**

3. 80 meters to feet

≈**262.4 feet**

4. 85 centimeters to inches

≈**33.2 inches**

5. 16 feet to meters

≈**4.8 m**

6. 3.2 yards to meters

≈**2.9 m**

7. 150 grams to ounces

≈**5.3 ounces**

8. 2.5 ounces to grams

≈**70.9 g**

9. 248 pounds to kilograms

≈**111.6 kg**

10. 7.68 kilograms to pounds

≈**16.9 pounds**

11. 28.6 liters to quarts

≈**30.3 quarts**

12. 15.75 liters to gallons

≈**4.1 gallons**

13. Manuela's new sports car has a 16-gallon gas tank. How many liters does the tank hold?

≈**60.5 liters**

14. Jamal's foot is 11 inches long. How long is it in centimeters?

≈**27.9 cm**

Circle the more reasonable temperature for each of the following.

15. A snowy day

28°C   ⟨28°F⟩

16. Brewing coffee

⟨80°C⟩   80°F

17. A high fever

⟨40°C⟩   40°F

18. Swimming pool water

78°C   ⟨78°F⟩

19. Oven temperature

⟨150°C⟩   150°F

20. Light jacket weather

⟨10°C⟩   10°F

21. Would a drop of 20 Celsius degrees be more or less than a drop of 20 Fahrenheit degrees? Explain your answer.

More. There are 180 degrees between freezing and boiling on the Fahrenheit scale, but only 100 degrees on the Celsius scale, so each Celsius degree is a greater change in temperature.

22. Describe one advantage of switching from the Fahrenheit temperature scale to the Celsius scale. Describe one disadvantage.

Advantage: the rest of the world uses the Celsius scale.
Disadvantage: people would have to buy new thermometers and get used to a new system.

🗹 Writing   ◉ Conceptual   ▲ Challenging   ≈ Estimation

Use the conversion formulas on page 475 and the order of operations to convert Fahrenheit temperatures to Celsius and Celsius temperatures to Fahrenheit. Round your answers to the nearest degree, if necessary.

23. 60°F ≈16°C

24. 80°F ≈27°C

25. 104°F 40°C

26. 36°F ≈2°C

27. 8°C ≈46°F

28. 18°C ≈64°F

29. 35°C 95°F

30. 0°C 32°F

Solve the following application problems. Round your answers to the nearest degree, if necessary.

31. The highest temperature ever recorded on earth was 136°F at Aziza, Libya. Convert this temperature to Celsius.

≈58°C

32. A recipe for French pastry calls for an oven temperature of 175°C. Convert this to Fahrenheit.

347°F

33. Here is the tag on a pair of boots. In what kind of weather would you wear these boots?

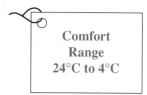

Comfort
Range
24°C to 4°C

Pleasant weather, above freezing but not hot.

For what Fahrenheit temperatures are the boots designed? ≈75°F to ≈39°F

What range of metric temperatures would you have in January where you live?

Varies; in Minnesota it's 0°C to −40°C; in California, 24°C to 0°C.

34. What are the picture directions on this tea bag package telling you to do?

for a
perfect
cup
of tea

100°C   4 MIN

Pour boiling water over the tea bag and leave it in the water for 4 minutes.
What Fahrenheit temperature would give the same result?   212°F

How many seconds should the tea bag be left in the water?   240 sec

35. Paint sells for $9.20 per gallon. Find the cost of 4 liters.

≈$9.57 (liters converted to gallons)

36. A 3-liter bottle of beverage sells for $2.80. A 1-gallon bottle of the same beverage sells for $3.50. What is the better value?

the 1-gallon bottle

Review and Prepare

*Work each problem. (For help, see **Section 1.8**.)*

37. 2 • 8 + 2 • 8   32

38. 2 • 12.2 + 2 • 5.6   35.6

39. 9^2 81

40. 7^2 49

41. $(5^2) + (4^2)$ 41

42. $(12^2) + (3^2)$ 153

CHAPTER 7 SUMMARY

KEY TERMS

7.1	**English system**	The English system of measurement (American system of units) is the system used for many daily activities in the United States. Common units in this system include quarts, pounds, feet, miles, and degrees Fahrenheit.
	metric system	The metric system of measurement is an international system of measurement used in manufacturing, science, medicine, sports, and other fields. The system uses meters, liters, grams, and degrees Celsius.
	unit fraction	A unit fraction involves measurement units and is equivalent to 1. Unit fractions are used to convert among different measurements.
7.2	**meter**	The meter is the basic unit of length in the metric system. The symbol **m** is used for meter. One meter is a little longer than a yard.
	prefixes	Attaching a prefix to meter, liter, or gram produces larger or smaller units. For example, the prefix *kilo* means 1000 so a *kilo*meter is 1000 meters.
	metric conversion line	The metric conversion line is a line showing the various metric measurement prefixes and their size relationship to each other. See pages 454, 460, and 463.
7.3	**liter**	The liter is the basic unit of capacity in the metric system. The symbol **L** is used for liter. One liter is a little more than one quart.
	gram	The gram is the basic unit of weight (mass) in the metric system. The symbol **g** is used for gram. One gram is the weight of 1 milliliter of water or one dollar bill.
7.5	**Celsius**	The Celsius scale is the scale used to measure temperature in the metric system. Water boils at 100°C and freezes at 0°C.
	Fahrenheit	The Fahrenheit scale is the scale used to measure temperature in the English system. Water boils at 212°F and freezes at 32°F.

QUICK REVIEW

Concepts	*Examples*
7.1 The English System of Measurement Memorize the basic measurement relationships. Then, to convert units, multiply when changing from a larger unit to a smaller unit; divide when changing from a smaller unit to a larger unit.	Convert each measurement. **(a)** 5 feet to inches $5 \text{ feet} = 5 \cdot 12 = 60 \text{ inches}$ **(b)** 3 pounds to ounces $3 \text{ pounds} = 3 \cdot 16 = 48 \text{ ounces}$ **(c)** 15 quarts to gallons $15 \text{ quarts} = \dfrac{15}{4} = 3\dfrac{3}{4} \text{ gallons}$
7.1 Using Unit Fractions Another, more useful, conversion method is multiplying by a unit fraction. The unit you want in the answer should be in the numerator. The unit you want to change should be in the denominator.	Convert 32 ounces to pounds. $\left.\dfrac{32 \text{ ounces}}{1} \cdot \dfrac{1 \text{ pound}}{16 \text{ ounces}}\right\}$ Unit fraction $= \dfrac{\overset{2}{\cancel{32 \text{ ounces}}}}{1} \cdot \dfrac{1 \text{ pound}}{\underset{1}{\cancel{16 \text{ ounces}}}}$ Cancel ounces. Cancel numbers. $= 2 \text{ pounds}$

Concepts	Examples
7.2 Knowing Basic Metric Length Units	

Concepts

7.2 Knowing Basic Metric Length Units

Use approximate comparisons to judge which units are appropriate:

    1 mm is the thickness of a dime.

    1 cm is about $\frac{1}{2}$ inch.

    1 m is a little more than 1 yard.

    1 km is about 0.6 mile.

7.2 and 7.3 Converting Within the Metric System

Using Unit Fractions

One conversion method is to multiply by a unit fraction. Use a fraction with the unit you want in the answer in the numerator and the unit you want to change in the denominator.

Using the Metric Conversion Line

Another conversion method is to find the unit you are given on the metric conversion line. Count the number of places to get from the unit you are given to the unit you want. Move the decimal point the same number of places and in the same direction.

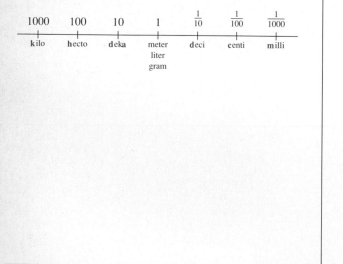

Examples

Write the most reasonable metric unit in each blank. Choose from km, m, cm, mm.

The room is 6 ___m___ long.

A paper clip is 30 ___mm___ long.

He drove 20 ___km___ to work.

Convert 9 g to kg

$$\frac{9 \not{g}}{1} \cdot \frac{1 \text{ kg}}{1000 \not{g}} = \frac{9}{1000} \text{ kg} = 0.009 \text{ kg}$$

Convert 3.6 m to cm

$$\frac{3.6 \not{m}}{1} \cdot \frac{100 \text{ cm}}{1 \not{m}} = 360 \text{ cm}$$

Convert each of the following.

(a) 68.2 kg to g

From kg to g is three places to the right.

    6 8.2 0 0   Decimal point is moved three places to the right.

68.2 kg = 68,200 g

(b) 300 mL to L

From mL to L is three places to the left.

    3 0 0.   Decimal point is moved three places to the left.

300 mL = 0.3 L

(c) 825 cm to m

From cm to m is two places to the left.

    8 2 5.   Decimal point is moved two places to the left.

825 cm = 8.25 m

Concepts	Examples
7.3 Knowing Basic Metric Capacity Units	Write the most appropriate metric unit in each blank. Choose from L or mL.
Use approximate comparisons to judge which units are appropriate:	
1 L is a little more than 1 quart.	The pail holds 12 __L__ .
1 mL is the amount of water in a cube 1 cm on each side.	The milk carton from the vending machine holds 250 __mL__ .
5 mL is about one teaspoon.	
250 mL is about one cup.	
7.3 Knowing Basic Metric Weight (Mass) Units	Write the most appropriate metric unit in each blank. Choose from kg, g, and mg.
Use approximate comparisons to judge which units are appropriate:	
1 kg is about 2.2 pounds.	The wrestler weighed 95 __kg__ .
1 g is the weight of 1 mL of water or one dollar bill.	She took a 500 __mg__ aspirin tablet.
1 mg is $\dfrac{1}{1000}$ of a gram; very tiny!	One banana weighs 150 __g__ .
7.4 Solving Metric Application Problems	**(a)** Grapes are \$3.95 per kg. How much will 400 g of grapes cost?
Convert units so you are comparing kg to kg, cm to cm, and so on. When a measurement involves two units, such as 6 m 20 cm, write it in terms of the unit called for in the answer (6.2 m or 620 cm).	$400 \text{ g} = 0.4 \text{ kg} \qquad \dfrac{0.4 \text{ kg}}{1} \cdot \dfrac{\$3.95}{1 \text{ kg}} = \$1.58$
	(b) How many meters are left if 1 m 35 cm is cut off a board measuring 3 m?
	$\begin{array}{lrll} 1 \text{ m} \rightarrow & 1.00 \text{ m} & 3.00 \text{ m} & \leftarrow \text{Board} \\ \text{plus } 35 \text{ cm} \rightarrow & +\ 0.35 \text{ m} & -\ 1.35 \text{ m} & \leftarrow \text{Cut off} \\ & \overline{1.35 \text{ m}} & \overline{1.65 \text{ m}} & \leftarrow \text{Left} \end{array}$
7.5 Converting from Metric to English and English to Metric	Convert. Round answers to the nearest tenth.
Use the values in the table of conversion factors to write a unit fraction. Because the values in the table are rounded, your answers will be approximate and should be written with the ≈ symbol.	**(a)** 23 meters to yards
	From the table, 1 meter ≈ 1.09 yards.
	$\dfrac{23 \text{ meters}}{1} \cdot \dfrac{1.09 \text{ yards}}{1 \text{ meter}} = 25.07 \text{ yards}$
	25.07 rounds to 25.1 so 23 meters ≈ 25.1 yards
	(b) 4 ounces to grams
	$\dfrac{4 \text{ ounces}}{1} \cdot \dfrac{28.35 \text{ grams}}{1 \text{ ounce}} = 113.4 \text{ grams}$
	So 4 ounces ≈ 113.4 grams.

Concepts	Examples
7.5 Knowing Common Celsius Temperatures Use approximate and exact comparisons to judge which temperatures are appropriate. Exact comparisons: 0°C is freezing point (32°F) 100°C is boiling point (212°F) Approximate comparisons: 10°C for a spring day (50°F) 20°C for room temperature (68°F) 30°C for summer day (86°F) 37°C for body temperature (98.6°F)	Circle the Celsius temperature that is most reasonable. **(a)** Hot summer day: (35°C) 90°C 110°C **(b)** The first snowy day in winter. −20°C (0°C) 15°C
7.5 Converting between Fahrenheit and Celsius Temperatures Use these formulas. $$C = \frac{5(F - 32)}{9}$$ $$F = \frac{9 \cdot C}{5} + 32$$	Convert 176°F to Celsius. $$C = \frac{5(176 - 32)}{9}$$ $$= \frac{5(\overset{16}{\cancel{144}})}{\underset{1}{\cancel{9}}} \quad \text{Cancel, if possible.}$$ $$\text{Then multiply.}$$ $$= 80$$ 176°F = 80°C Convert 80°C to Fahrenheit. $$F = \frac{9 \cdot 80}{5} + 32$$ $$= \frac{9 \cdot \overset{16}{\cancel{80}}}{\underset{1}{\cancel{5}}} + 32 \quad \begin{array}{l}\text{Cancel, if possible.}\\ \text{Then multiply.}\end{array}$$ $$= 144 + 32 \quad \text{Add.}$$ $$= 176$$ 80°C = 176°F

CHAPTER 7 REVIEW EXERCISES

[7.1] *Fill in the blanks with the measurement relationships you have memorized.*

1. 1 pound = ___16___ ounces **2.** ___3___ feet = 1 yard **3.** 1 ton = ___2000___ pounds

4. ___24___ hours = 1 day **5.** 1 hour = ___60___ minutes **6.** 1 cup = ___8___ fluid ounces

7. ___4___ quarts = 1 gallon **8.** ___5280___ feet = 1 mile **9.** ___12___ inches = 1 foot

10. 1 week = ___7___ days **11.** ___60___ seconds = 1 minute **12.** 1 pint = ___2___ cups

Convert using unit fractions.

13. 4 feet = ___48___ inches **14.** 15 yards = ___45___ feet **15.** 64 ounces = ___4___ pounds

16. 6000 pounds = ___3___ tons **17.** 150 minutes = ___$2\frac{1}{2}$ or 2.5___ hours **18.** 11 cups = ___$5\frac{1}{2}$ or 5.5___ pints

19. 18 hours = ___$\frac{3}{4}$ or 0.75___ day **20.** 9 quarts = ___$2\frac{1}{4}$ or 2.25___ gallons **21.** $6\frac{1}{2}$ feet = ___78___ inches

22. $1\frac{3}{4}$ pounds = ___28___ ounces **23.** 7 gallons = ___112___ cups **24.** 4 days = ___345,600___ seconds

[7.2] *Write the most reasonable metric length unit in each blank. Choose from km, m, cm, mm.*

25. My thumb is 20 ___mm___ wide. **26.** Her waist measurement is 66 ___cm___ .

27. The two towns are 40 ___km___ apart. **28.** A basketball court is 30 ___m___ long.

29. The height of the picnic bench is 45 ___cm___ . **30.** The eraser on the end of my pencil is 5 ___mm___ long.

Convert using unit fractions or the metric conversion line.

31. 5 m to cm **32.** 8.5 km to m **33.** 85 mm to cm

  500 cm **8500 m** **8.5 cm**

☑ Writing ◉ Conceptual ▲ Challenging ≈ Estimation

34. 370 cm to m

3.7 m

35. 70 m to km

0.07 km

36. 0.93 m to mm

930 mm

[7.3] *Write the most reasonable metric unit in each blank. Choose from L, mL, kg, g,*
and mg.

37. The eyedropper holds 1 ___mL___ .

38. I can heat 3 ___L___ of water in this pan.

39. Loretta's hammer weighed 650 ___g___ .

40. Yongshu's suitcase weighed 20 ___kg___ when it was packed.

41. My fish tank holds 80 ___L___ of water.

42. I'll buy the 500 ___mL___ bottle of mouthwash.

43. Mara took a 200-___mg___ antibiotic pill.

44. This piece of chicken weighs 100 ___g___ .

Convert using unit fractions or the metric conversion line.

45. 5000 mL to L

5 L

46. 8 L to mL

8000 mL

47. 4.58 g to mg

4580 mg

48. 0.7 kg to g

700 g

49. 6 mg to g

0.006 g

50. 35 mL to L

0.035 L

[7.4] *Solve each application problem.*

51. Each serving of punch at the wedding reception will be 180 mL. How many liters of punch are needed for 175 guests?

31.5 L

52. Jason is serving a 10-kg turkey to 28 people. How many grams of meat is he allowing for each person? Round to the nearest whole gram.

≈357 g

53. Yerald weighed 92 kg. Then he lost 4 kg 750 g. What is his weight now in kilograms?

87.25 kg

54. Young-Mi bought 2 kg 20 g of onions. The price was $1.49 per kilogram. How much did she pay, to the nearest cent?

≈$3.01

[7.5] *Use the table on page 473 and unit fractions to make approximate conversions. Round*
your answers to the nearest tenth, if necessary.

55. 6 m to yards

≈6.5 yards

56. 30 cm to inches

≈11.7 inches

57. 108 km to miles

≈**67.0 miles**

58. 800 miles to km

≈**1288 km**

59. 23 quarts to L

≈**21.9 L**

60. 41.5 L to quarts

≈**44.0 quarts**

Write the appropriate metric (Celsius) temperature in each blank.

61. Water freezes at __**0°C**__ .

62. Water boils at __**100°C**__ .

63. Normal body temperature is about __**37°C**__ .

64. Comfortable room temperature is about __**20°C**__ .

Use the conversion formulas on page 475 to convert each temperature to Fahrenheit or Celsius. Round to the nearest degree, if necessary.

65. 77°F **25°C**

66. 92°F ≈**33°C**

67. 6°C ≈**43°F**

68. 40°C **104°F**

──────────────── **MIXED REVIEW EXERCISES** ────────────────

Write the most reasonable metric unit in each blank. Choose from km, m, cm, mm, L, mL, kg, g, and mg.

69. I added 1 __**L**__ of oil to my car.

70. The box of books weighed 15 __**kg**__ .

71. Larry's shoe is 30 __**cm**__ long.

72. Jan used 15 __**mL**__ of shampoo on her hair.

73. My fingernail is 10 __**mm**__ wide.

74. I walked 2 __**km**__ to school.

75. The tiny bird weighed 15 __**g**__ .

76. The new library building is 18 __**m**__ wide.

77. The cookie recipe uses 250 __**mL**__ of milk.

78. Renee's pet mouse weighs 30 __**g**__ .

79. One postage stamp weighs 90 __**mg**__ .

80. I bought 30 __**L**__ of gas for my car.

Convert the following using unit fractions, the metric conversion line, or the temperature conversion formulas.

81. 10.5 cm to mm

105 mm

82. 45 minutes to hours

$\frac{3}{4}$ **hour or 0.75 hour**

83. 90 inches to feet

$7\frac{1}{2}$ **feet or 7.5 feet**

84. 1.3 m to cm

130 cm

85. 25°C to Fahrenheit
77°F

86. $3\frac{1}{2}$ gallons to quarts

14 quarts

87. 700 mg to g

0.7 g

88. 0.81 L to mL

810 mL

89. 5 pounds to ounces

80 ounces

90. 60 kg to g

60,000 g

91. 1.8 L to mL

1800 mL

92. 86°F to Celsius

30°C

93. 0.36 m to cm

36 cm

94. 55 mL to L

0.055 L

Solve the following application problems.

95. Peggy had a board measuring 2 m 4 cm. She cut off 78 cm. How long is the board now, in meters?

1.26 m

96. Imported wool fabric is $12.99 per meter. What is the cost, to the nearest cent, of a piece that measures 3 m 70 cm?

≈$48.06

97. Olivia is sending a recipe to her mother in Mexico. Among other things, the recipe calls for 4 ounces of rice and a baking temperature of 350°F. Convert these measurements to metric, rounding to the nearest gram and nearest degree.

≈113 g; ≈177°C

98. While on vacation in Canada, Jalo became ill and went to a health clinic. They said he weighed 80.9 kilograms and was 1.83 meters tall. Find his weight in pounds and height in feet. Round to the nearest tenth.

≈178.0 pounds; ≈6.0 feet

Convert the following measurements.

1. 9 gallons = _____ quarts

2. 45 feet = _____ yards

3. 135 minutes = _____ hours

4. 9 inches = _____ foot

5. $3\frac{1}{2}$ pounds = _____ ounces

6. 5 days = _____ minutes

Write the most reasonable metric unit in each blank. Choose from km, m, cm, mm, L, mL, kg, g, and mg.

7. My husband weighs 75 _____.

8. I hiked 5 _____ this morning.

9. She bought 125 _____ of cough syrup.

10. This apple weighs 180 _____.

11. This page is 21 _____ wide.

12. My watch band is 10 _____ wide.

13. I bought 10 _____ of soda for the picnic.

14. The bracelet is 16 _____ long.

Convert these measurements.

15. 250 cm to meters

16. 4.6 km to meters

17. 5 mm to centimeters

18. 325 mg to grams

19. 16 L to milliliters

20. 0.4 kg to grams

21. 10.55 m to centimeters

22. 95 mL to liters

1. 36 quarts

2. 15 yards

3. 2.25 or $2\frac{1}{4}$ hours

4. 0.75 or $\frac{3}{4}$ foot

5. 56 ounces

6. 7200 minutes

7. kg

8. km

9. mL

10. g

11. cm

12. mm

13. L

14. cm

15. 2.5 m

16. 4600 m

17. 0.5 cm

18. 0.325 g

19. 16,000 mL

20. 400 g

21. 1055 cm

22. 0.095 L

23. 6.32 kg _____

23. Stan's cat weighed 3 kg 740 g. His dog weighed 10 kg 60 g. How much heavier is the dog in kilograms?

24. 6.75 m _____

24. Denise is making five matching pillows. She needs 1 m 35 cm of braid to trim each pillow. How many meters of braid should she buy?

Pick the metric temperature that is most appropriate in each situation.

25. 95°C _____

25. The water is almost boiling.
    210°C   155°C   95°C

26. 0°C _____

26. The tomato plants may freeze tonight.
    30°C   20°C   0°C

Use the table on page 473 and unit fractions to convert the following measurements. Round your answers to the nearest tenth, if necessary.

27. ≈1.8 m _____

27. 6 feet to meters

28. 125 pounds to kilograms

28. ≈56.3 kg _____

29. ≈13 gallons _____

29. 50 liters to gallons

30. 8.1 kilometers to miles

30. ≈5.0 miles _____

Use the conversion formulas to convert each temperature. Round your answers to the nearest degree, if necessary.

31. ≈23°C _____

31. 74°F to Celsius

32. 2°C to Fahrenheit

32. ≈36°F _____

33. Possible answers: _____

Use same system as rest of the world; easier system for children to learn.

33. Describe two benefits the United States would achieve by switching entirely to the metric system.

34. Possible answers: _____

People like to use the system they're familiar with; cost of new signs, scales, etc.

34. Give two reasons why you think the United States has resisted changing to the metric system.

$\approx$ *Round the numbers in each problem so there is only one non-zero digit. Then add, subtract, multiply, or divide the rounded numbers, as indicated, to estimate the answer. Finally, solve for the exact answer.*

1. *estimate*   *exact*

$$\begin{array}{r} 100 \\ 3 \\ + \ 70 \\ \hline 173 \end{array} \qquad \begin{array}{r} 107.5 \\ 2.548 \\ + \ 68.79 \\ \hline 178.838 \end{array}$$

2. *estimate*   *exact*

$$\begin{array}{r} 30{,}000 \\ - \ 800 \\ \hline 29{,}200 \end{array} \qquad \begin{array}{r} 31{,}007 \\ - \ 829 \\ \hline 30{,}178 \end{array}$$

3. *estimate*   *exact*

$$\begin{array}{r} 90{,}000 \\ \times \ 200 \\ \hline 18{,}000{,}000 \end{array} \qquad \begin{array}{r} 92{,}075 \\ \times \ 183 \\ \hline 16{,}849{,}725 \end{array}$$

4. *estimate*   *exact*

$$\begin{array}{r} 60 \\ \times \ 5 \\ \hline 300 \end{array} \qquad \begin{array}{r} 56.52 \\ \times \ 4.7 \\ \hline 265.644 \end{array}$$

5. *estimate*   *exact*

$$40\overline{)20{,}000}^{\ 500} \qquad 37\overline{)19{,}610}^{\ 530}$$

6. *estimate*   *exact*

$$8\overline{)40}^{\ 5} \qquad 8.3\overline{)38.18}^{\ 4.6}$$

7. *estimate*   *exact*

$$\begin{array}{r} 2 \\ + \ 4 \\ \hline 6 \end{array} \qquad \begin{array}{r} 1\frac{7}{10} \\ + \ 3\frac{4}{5} \\ \hline 5\frac{1}{2} \end{array}$$

8. *estimate*   *exact*

$$\begin{array}{r} 6 \\ - \ 1 \\ \hline 5 \end{array} \qquad \begin{array}{r} 5\frac{1}{2} \\ - \ 1\frac{2}{7} \\ \hline 4\frac{3}{14} \end{array}$$

9. *exact*

$$3\frac{1}{6} \cdot 4\frac{2}{3} = 14\frac{7}{9}$$

estimate

$$\underline{\ 3\ } \cdot \underline{\ 5\ } = \underline{\ 15\ }$$

10. *exact*

$$2\frac{1}{4} \div \frac{9}{10} = 2\frac{1}{2}$$

estimate

$$\underline{\ 2\ } \div \underline{\ 1\ } = \underline{\ 2\ }$$

Add, subtract, multiply, or divide as indicated. Write answers to fraction problems in lowest terms and as whole or mixed numbers when possible.

11. $3 - 2\frac{5}{16}$

$\dfrac{11}{16}$

12. $7 + 484{,}099 + 3939$

$488{,}045$

13. $12 \cdot 2\frac{2}{9}$

$26\frac{2}{3}$

14. $0.86 \div 0.066$
Round to nearest tenth.

≈ 13.0

15. $\dfrac{3}{8} + \dfrac{5}{6}$

$1\dfrac{5}{24}$

16. $8 - 0.9207$

7.0793

17. $3\frac{3}{4} \div 6$

$\dfrac{5}{8}$

18. Write your answer using R for the remainder.

$$47\overline{)14{,}467}^{\ 307 \ \text{R}38}$$

19. $(2.54)(0.003)$

0.00762

Simplify by using the order of operations.

20. $24 - 12 \div 6 \cdot 8 + (25 - 25)$

8

21. $3^2 + 2^5 \cdot \sqrt{64}$

265

22. Write 307.19 in words.

three hundred seven and nineteen hundredths

23. Write eighty-two ten-thousandths in numbers.

0.0082

24. Arrange in order from smallest to largest.
0.67   0.067   0.6   0.6007

0.067, 0.6, 0.6007, 0.67

Complete this chart.

Fraction/Mixed Number	*Decimal*	*Percent*
$\dfrac{7}{8}$	**25.** ___0.875___	**26.** ___ 87.5% or $87\frac{1}{2}$%
27. ___ $\dfrac{1}{20}$ ___	0.05	**28.** ___5%___
29. ___ $3\dfrac{1}{2}$ ___	**30.** ___3.5___	350%

Write each rate or ratio in lowest terms. Change to the same units when necessary.

31. \$44 to \$4

$\dfrac{11}{1}$

32. 20 minutes to 3 hours.
Compare in minutes.

$\dfrac{1}{9}$

33. There are 200 students taking a math class and 350 students taking an English class. Find the ratio of English students to math students.

$\dfrac{7}{4}$

34. Find the best buy on extra large disposable diapers.
        package of 16 for \$3.87
        package of 22 for \$5.96
        package of 36 for \$11.69

16 diapers for \$3.87, ≈\$0.242 per diaper

Find the missing number in each proportion. Round your answers to hundredths, if necessary.

35. $\dfrac{x}{16} = \dfrac{3}{4}$

12

36. $\dfrac{0.9}{0.75} = \dfrac{2}{x}$

≈1.67

Solve each of the following.

37. $4 is what percent of $80?

5%

38. 36 hours is 120% of what number of hours?

30 hours

Convert the following measurements. Use the table on page 473 and the temperature conversion formulas when necessary.

39. $2\frac{1}{2}$ feet to inches

30 inches

40. 105 seconds to minutes

$1\frac{3}{4}$ **or 1.75 minutes**

41. 2.8 m to centimeters

280 cm

42. 65 mg to grams

0.065 g

43. 198 km to miles

≈122.76 miles

44. 50°F to Celsius

10°C

Write the most reasonable metric unit in each blank. Choose from km, m, cm, mm, L, mL, kg, g, and mg.

45. Ron bought the tube of toothpaste weighing 100 ___g___ .

46. The teacher's desk is 140 ___cm___ long.

47. The hallway is 3 ___m___ wide.

48. Joe's hammer weighed 1 ___kg___ .

49. Anne took a 500-___mg___ tablet of vitamin C.

50. Tia added 125 ___mL___ of milk to her cereal.

Circle the metric temperature that is most appropriate in each situation.

51. John has a slight fever.

(38°C)   70°C   99°C

52. You'll need a light jacket outside.

0°C   (12°C)   45°C

Solve the following application problems.

53. Calbert bought an automatic focus camera for $64.95. He paid $7\frac{1}{2}$% sales tax. Find the amount of tax, to the nearest cent, and the total cost of the camera.

≈$4.87; $69.82

54. Danielle had a roll of 35 mm film developed. She received 24 prints for $10.25. What is the cost per print, to the nearest cent?

≈$0.43

55. Mark bought 650 grams of maple sugar candy on his vacation in Montreal. The candy is priced at $14.98 per kilogram. How much did Mark pay, to the nearest cent?

≈$9.74

56. On the Illinois map, one centimeter represents 12 kilometers. The center of Springfield is 7.8 cm from the center of Bloomington on the map. What is the actual distance in kilometers?

93.6 km

57. Dimitri took out a $3\frac{1}{2}$ year car loan for $8750 at 9% simple interest. Find the interest and the total amount due on the loan.

$2756.25; $11,506.25

58. On a 35-problem math test, Juana solved 31 problems correctly. What percent of the problems were correct? Round to the nearest tenth of a percent.

≈88.6%

59. An elevator has a weight limit of 1500 pounds. On it are twin boys weighing 88 pounds each and 8 adults weighing this number of pounds each: 240, 189, 127, 165, 143, 219, 116, and 124. The total weight of the people is how much above or below the limit?

1 pound under the limit

60. The Jackson family is making three kinds of holiday cookies that require brown sugar. The recipes call for $2\frac{1}{4}$ cups, $1\frac{1}{2}$ cups, and $\frac{3}{4}$ cup, respectively. They bought two packages of brown sugar, each holding $2\frac{1}{3}$ cups. The amount bought is how much more or less than the amount needed?

$\frac{1}{6}$ cup more than the amount needed

61. Leather jackets are on sale at 30% off the regular price. Tracy likes a jacket with a regular price of $189. Find the amount of discount and the sale price she will pay.

$56.70; $132.30

62. Bags of slivered almonds weigh 115 g each. They are packed in a carton that weighs 450 g. How many kilograms would a carton containing 48 bags weigh?

5.97 kg

63. Akuba is knitting a scarf. Six rows of knitting result in 5 centimeters of scarf. At that rate, how many rows will she knit to make a 100-centimeter scarf?

120 rows

64. A survey of the 5600 students on our campus found that $\frac{3}{8}$ of the students work 20 hours or more per week. How many students work 20 hours or more?

2100 students

Geometry 8

Geometry developed centuries ago when people needed a way to measure land. The name geometry comes from the Greek words *ge,* meaning earth, and *metron,* meaning measure. Today we still use geometry to measure farmland. It is also important in architecture, construction, navigation, art and design, physics, chemistry, and astronomy. You can use it at home when you buy carpet or wallpaper, hang a picture, or build a fence. This chapter discusses the basic terms of geometry and the common geometric shapes that are all around us.

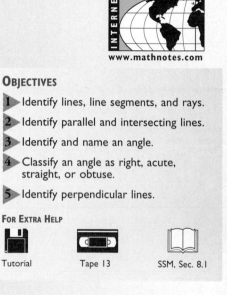

www.mathnotes.com

OBJECTIVES

1 ▶ Identify lines, line segments, and rays.
2 ▶ Identify parallel and intersecting lines.
3 ▶ Identify and name an angle.
4 ▶ Classify an angle as right, acute, straight, or obtuse.
5 ▶ Identify perpendicular lines.

FOR EXTRA HELP

Tutorial   Tape 13   SSM, Sec. 8.1

8.1 Basic Geometric Terms

Geometry starts with the idea of a point. A **point** is a location in space. It has no length or width. A point is represented by a dot and is named by writing a capital letter next to the dot.

•*P*

Point *P*

OBJECTIVE 1 ▶ A **line** is a straight row of points that goes on forever in both directions. A line is drawn by using arrowheads to show that it never ends. The line is named by using the letters of any two points on the line.

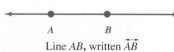

Line *AB*, written $\overleftrightarrow{AB}$

A piece of a line that has two endpoints is called a **line segment.** A line segment is named for its endpoints. The segment with endpoints *P* and *Q* is shown below. It can be named $\overline{PQ}$ or $\overline{QP}$.

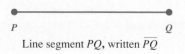

Line segment *PQ*, written $\overline{PQ}$

1. Identify each of the following as a line, line segment, or ray.

(a)

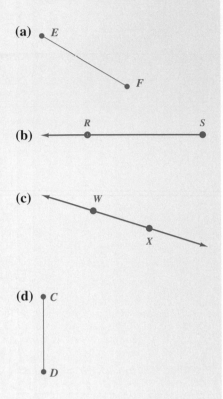

(b)

(c)

(d)

A **ray** is a part of a line that has only one endpoint and goes on forever in one direction. A ray is named by using the endpoint and some other point on the ray. The endpoint is always mentioned first.

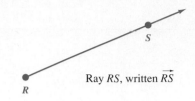

Ray *RS*, written $\overrightarrow{RS}$

E X A M P L E I Identifying Lines, Rays, and Line Segments

Identify each of the following as a line, line segment, or ray.

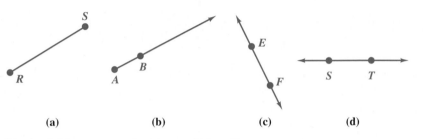

(a) (b) (c) (d)

Figure **(a)** has two endpoints, so it is a line segment.

Figure **(b)** starts at point *A* and goes on forever in one direction, so it is a ray.

Figures **(c)** and **(d)** go on forever in both directions, so they are lines.

◄◄ **WORK PROBLEM I AT THE SIDE.**

2. Label each pair of lines as parallel or intersecting.

(a)

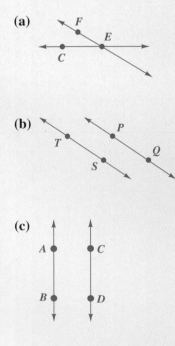

(b)

(c)

OBJECTIVE 2 A *plane* is a flat surface, like a floor or a wall. Lines that are in the same plane but that never intersect (never cross) are called **parallel** (PAIR-uh-lell) **lines,** while lines that cross or merge are called **intersecting** (in-tur-SEKT-ing) **lines.** (Think of an intersection, where two streets cross each other.)

E X A M P L E 2 Identifying Parallel and Intersecting Lines

Label each pair of lines as parallel or intersecting.

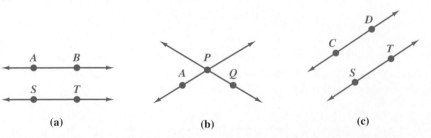

(a) (b) (c)

The lines in Figures **(a)** and **(c)** never intersect. They are parallel lines. The lines in Figure **(b)** cross at *P*, so they are intersecting lines.

◄◄ **WORK PROBLEM 2 AT THE SIDE.**

ANSWERS

1. (a) line segment **(b)** ray **(c)** line
  (d) line segment

2. (a) intersecting **(b)** parallel
  (c) parallel

OBJECTIVE 3 An **angle** (ANG-gul) is made up of two rays that start at a common endpoint. This common endpoint is called the *vertex*.

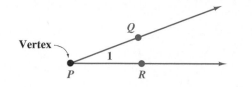

The rays *PQ* and *PR* are called *sides*. The angle can be named four ways:

∠1   ∠**P**   ∠*QPR*   ∠*RPQ*

Vertex   Vertex in
alone   the middle

Naming an Angle

When naming an angle, the vertex is written alone or it is written in the middle of two other points. If two or more angles have the same vertex, as in Example 3, do not use the vertex alone to name an angle.

E X A M P L E 3   Identifying and Naming an Angle

Name the highlighted angle.

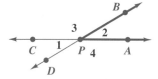

The angle can be named ∠*BPA*, ∠*APB*, or ∠2. It cannot be named ∠*P*, using the vertex alone, because four different angles have *P* as their vertex.

WORK PROBLEM 3 AT THE SIDE. ▶▶

OBJECTIVE 4 Angles can be measured in **degrees** (deh-GREEZ). The symbol for degrees is a small, raised circle °. Think of the minute hand on a clock as a ray of an angle. Suppose it is at 12:00. During one hour of time, the minute hand moves around in a complete circle. It moves 360 *degrees*, or 360°. In half an hour, at 12:30, the minute hand has moved half way around the circle, or 180°. An angle of 180° is called a **straight angle.** (Notice that the two rays in a straight angle form a straight line.)

Complete circle
360°

Straight angle
(half a circle)
180°

---

3. (a) Name the highlighted angle in three different ways.

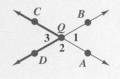

(b) Darken the lines that make up ∠*ZTW*.

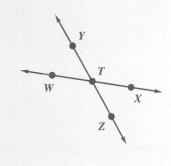

(c) Name this angle in four different ways.

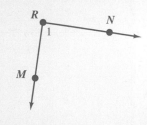

ANSWERS

3. (a) ∠3, ∠*CQD*, ∠*DQC*
  (b)

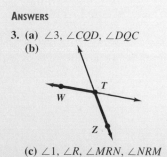

(c) ∠1, ∠*R*, ∠*MRN*, ∠*NRM*

In a quarter of an hour, at 12:15, the minute hand has moved $\frac{1}{4}$ of the way around the circle, or 90°. An angle of 90° is called a **right angle.** Sometimes you hear it called a *square angle*. The minute hands at 12:00 and 12:15 form one corner of a square. So, to show that an angle is a **right angle**, we draw a **small square** at the vertex.

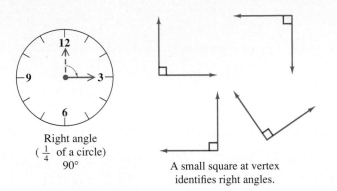

Right angle
($\frac{1}{4}$ of a circle)
90°

A small square at vertex
identifies right angles.

You can see that an angle of 1° is very small. To be precise, it is only the distance that the *minute hand* moves in ten *seconds*.

Some other terms used to describe angles are shown below.

Acute (uh-CUTE) **angles** measure between 0° and 90°.

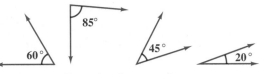

Examples of acute angles

Obtuse (ob-TOOS) **angles** measure between 90° and 180°.

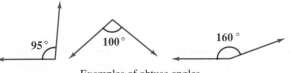

Examples of obtuse angles

Section 10.1 shows you how to use a tool called a *protractor* to measure the number of degrees in an angle.

Note

Angles can also be measured in radians, which you will learn about in a later math course.

E X A M P L E 4 Classifying an Angle

Label each of the following angles as acute, right, obtuse, or straight.

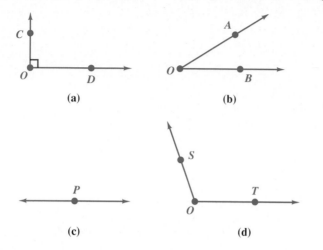

(a) (b)

(c) (d)

Figure **(a)** shows a right angle (exactly 90° and identified by a small square at the vertex).

Figure **(b)** shows an acute angle (between 0° and 90°).

Figure **(c)** shows a straight angle (exactly 180°).

Figure **(d)** shows an obtuse angle (between 90° and 180°).

WORK PROBLEM 4 AT THE SIDE. ▶▶

OBJECTIVE 5 Two lines are called **perpendicular** (per-pen-DIK-yoo-ler) **lines** if they intersect to form a right angle.

Lines *CB* and *ST* are **perpendicular**, because they intersect at right angles. This can be written in the following way: $\overleftrightarrow{CB} \perp \overleftrightarrow{ST}$.

E X A M P L E 5 Identifying Perpendicular Lines

Which of the following pairs of lines are perpendicular?

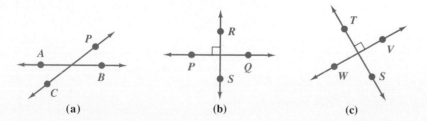

(a) (b) (c)

The lines in Figures **(b)** and **(c)** are perpendicular to each other, because they intersect at right angles. The lines in Figure **(a)** are intersecting lines, but are not perpendicular, because they do not form a right angle.

WORK PROBLEM 5 AT THE SIDE. ▶▶

4. Label each of the following as an acute, right, obtuse, or straight angle.

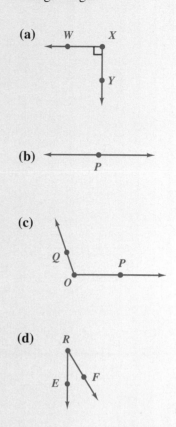

(a)

(b)

(c)

(d)

5. Which pair of lines is perpendicular? How can you describe the other pair of lines?

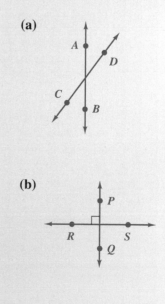

(a)

(b)

ANSWERS

4. (a) right **(b)** straight **(c)** obtuse
   (d) acute
5. (b); Figure **(a)** shows intersecting lines

NUMBERS IN THE
Real World *collaborative investigations*

Ratios may be used in the directions for mixing two items. This allows you to make as much of the mixture as you need. One example is the recipe in the article below for a sugar mixture to be used in hummingbird feeders. Complete the table below. Be sure you maintain the 1 to 4 ratio.

Sugar	Water
1 cup	4 cups
$1\frac{1}{4}$ cups	5 cups
$1\frac{1}{2}$ cups	6 cups
$1\frac{3}{4}$ cups	7 cups
2 cups	8 cups

Sugar	Water
1 cup	4 cups
$\frac{3}{4}$ cup	3 cups
$\frac{1}{2}$ cup	2 cups
$\frac{1}{4}$ cup	1 cup

Suppose you wanted to use 3 cups of *sugar*. How much water would you need? **12 cups**
How much water for 4 cups of sugar? **16 cups**

If you had only $\frac{1}{3}$ cup of sugar, how much water could you use? $1\frac{1}{3}$ **cups**

As you change the amounts of water and sugar, should you change the length of time that you boil the mixture? Explain your answer.
Probably not; once the water is boiling, 1 or 2 minutes is long enough to prevent fermentation in 1 cup or 4 cups.

Will the length of time it takes to get the water hot enough to start boiling change? Explain your answer.
Yes, the greater the amount of water, the longer it will take to get it up to boiling temperature.

The article says that the nectar wildflowers visited by hummingbirds has an average sugar concentration of 21 percent. For mixtures, a percentage can be calculated using the mass (weight) of the ingredients. One cup of sugar weighs about 200 grams and one cup of water weighs about 235 grams. If you mix 1 cup of sugar and 4 cups of water, what is the weight of the resulting mixture? **1140 grams**
What percent of the mixture's weight is sugar?
200 grams of sugar is ≈17.5% of 1140 grams of mixture.

If you wanted the mixture to be 21% sugar by weight, how many grams of sugar should be mixed with 4 cups of water? ≈**239 grams**

Feeding Hummingbirds

After getting a hummingbird feeder, the next step is to fill it! You have two choices at this point: you can either buy one of the commercial mixtures or you can make your own solution:

> **Recipe for Homemade Mixture:**
> **1 part sugar (not honey)**
> **4 parts water**
> **Boil for 1 to 2 minutes. Cool.**
> **Store extra in refrigerator.**

The concentration of the sugar is important. A 1 to 4 ratio of sugar to water is recommended because it approximates the ratio of sugar to water found in the nectar of many hummingbird flowers. A recent study of 21 native California wildflowers visited by hummingbirds showed that their nectar had an average sugar concentration of 21 percent. This is sweet enough to attract the hummers without being too sweet. If you increase the concentration of sugar, it may be harder for the birds to digest; if you decrease the concentration, they may lose interest.

Boiling the solution helps retard fermentation. Sugar-and-water solutions are subject to rapid spoiling, especially in hot weather.

Source: The Hummingbird Book

8.1 Exercises

Name each line, line segment, or ray using the appropriate symbol.

1.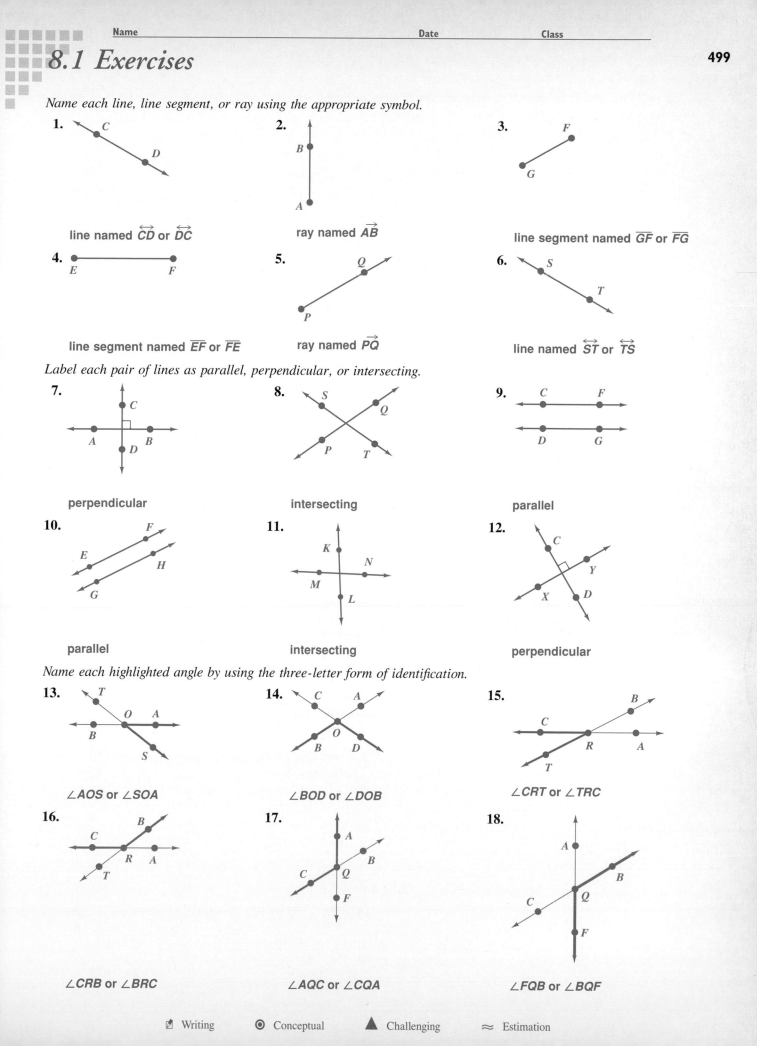

line named $\overleftrightarrow{CD}$ or $\overleftrightarrow{DC}$

2.

ray named $\overrightarrow{AB}$

3.

line segment named $\overline{GF}$ or $\overline{FG}$

4.

line segment named $\overline{EF}$ or $\overline{FE}$

5.

ray named $\overrightarrow{PQ}$

6.

line named $\overleftrightarrow{ST}$ or $\overleftrightarrow{TS}$

Label each pair of lines as parallel, perpendicular, or intersecting.

7.

perpendicular

8.

intersecting

9.

parallel

10.

parallel

11.

intersecting

12.

perpendicular

Name each highlighted angle by using the three-letter form of identification.

13.

$\angle AOS$ or $\angle SOA$

14.

$\angle BOD$ or $\angle DOB$

15.

$\angle CRT$ or $\angle TRC$

16.

$\angle CRB$ or $\angle BRC$

17.

$\angle AQC$ or $\angle CQA$

18.

$\angle FQB$ or $\angle BQF$

☑ Writing   ◉ Conceptual   ▲ Challenging   ≈ Estimation

Label each of the following as an acute, right, obtuse, or straight angle. For right angles and straight angles, indicate the number of degrees in the angle.

19.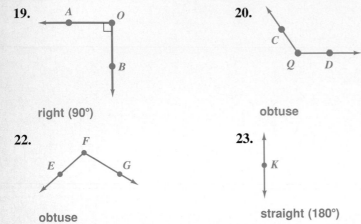

right (90°)

20.

obtuse

21.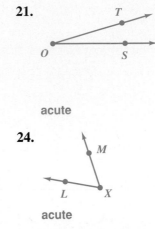

acute

22.

obtuse

23.

straight (180°)

24.

acute

25. Explain what is happening in each sentence.
  (a) The road was so slippery that my car did a 360.
  (b) After the election, the governor's view on taxes took a 180° turn.
  (a) The car turned around in a complete circle.
  (b) The governor took the opposite view, for example, having opposed taxes and now supporting them.

26. Find at least four examples of right angles in your home, at work, or on the street. Make a sketch of each example and label the right angle.
  There are many possibilities. Some examples: corner of a room, street corner, corners of a window, corner of a piece of paper.

▲ *Use the diagram to label each statement as true or false. If the statement is false, rewrite it to make a true statement.*

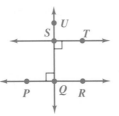

27. ∠UST is 90°.   true

28. $\overleftrightarrow{SQ}$ and $\overleftrightarrow{PQ}$ are perpendicular.   true

29. ∠USQ is smaller than ∠PQR.
  False; the angles have the same measure (both are 180°).

30. $\overleftrightarrow{ST}$ and $\overleftrightarrow{PR}$ are intersecting.
  False; $\overleftrightarrow{ST}$ and $\overleftrightarrow{PR}$ are parallel.

31. $\overleftrightarrow{QU}$ and $\overleftrightarrow{TS}$ are parallel.
  False; $\overleftrightarrow{QU}$ and $\overleftrightarrow{TS}$ are perpendicular.

32. ∠UST and ∠UQR measure the same number of degrees.
  true

Review and Prepare

*Evaluate the following expressions. (For help, see **Section 1.8**.)*

33. $(180 - 75) - 15$ 90

34. $180 - (75 - 15)$ 120

35. $(90 - 37) + 15$ 68

36. $90 - (37 + 15)$ 38

8.2 *Angles and Their Relationships*

OBJECTIVE ▸ Two angles are called **complementary** (kahm-pleh-MEN-tary) **angles** if their sum is 90°. If two angles are complementary, each angle is the *complement* of the other.

OBJECTIVES

1 ▸ Identify complementary angles and supplementary angles.

2 ▸ Identify congruent angles and vertical angles.

FOR EXTRA HELP

Tutorial   Tape 13   SSM, Sec. 8.2

EXAMPLE 1   Identifying Complementary Angles

Identify each pair of complementary angles.

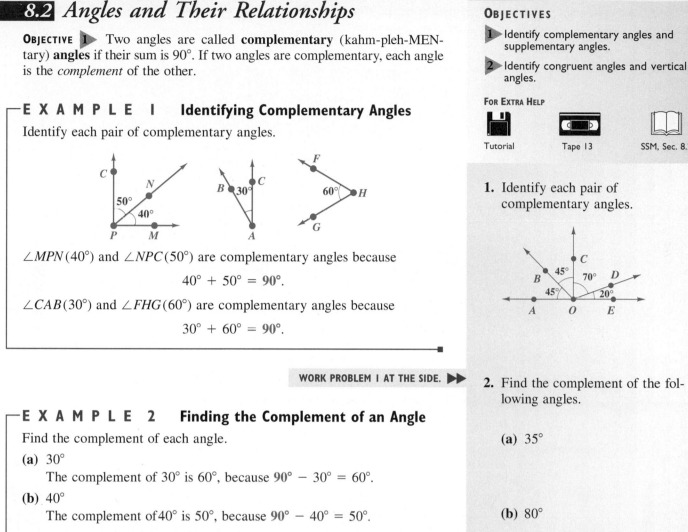

∠MPN (40°) and ∠NPC (50°) are complementary angles because

$$40° + 50° = 90°.$$

∠CAB (30°) and ∠FHG (60°) are complementary angles because

$$30° + 60° = 90°.$$

> **WORK PROBLEM 1 AT THE SIDE.** ▶▶

EXAMPLE 2   Finding the Complement of an Angle

Find the complement of each angle.

(a) 30°
  The complement of 30° is 60°, because **90°** − 30° = **60°**.

(b) 40°
  The complement of 40° is 50°, because **90°** − 40° = **50°**.

> **WORK PROBLEM 2 AT THE SIDE.** ▶▶

  Two angles are called **supplementary** (sup-luh-MEN-tary) **angles** if their sum is 180°. If two angles are supplementary, each angle is the *supplement* of the other.

EXAMPLE 3   Identifying Supplementary Angles

Identify each pair of supplementary angles.

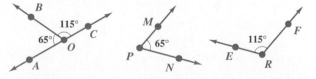

  ∠BOA and ∠BOC, because 65° + 115° = **180°**.

  ∠BOA and ∠ERF, because 65° + 115° = **180°**.

  ∠BOC and ∠MPN, because 115° + 65° = **180°**.

  ∠MPN and ∠ERF, because 65° + 115° = **180°**.

> **WORK PROBLEM 3 AT THE SIDE.** ▶▶

1. Identify each pair of complementary angles.

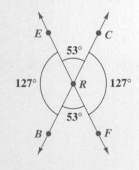

2. Find the complement of the following angles.

(a) 35°

(b) 80°

3. Identify each pair of supplementary angles.

ANSWERS

1. ∠AOB and ∠BOC; ∠COD and ∠DOE
2. (a) 55° (b) 10°
3. ∠CRF and ∠BRF
  ∠CRE and ∠ERB
  ∠BRF and ∠BRE
  ∠CRE and ∠CRF

501

4. Find the supplement of each angle.

(a) 175°

E X A M P L E 4 Finding the Supplement of an Angle

Find the supplement of the following angles.

(a) 70°

The supplement of 70° is 110°, because **180°** − 70° = 110°.

(b) 140°

The supplement of 140° is 40°, because **180°** − 140° = 40°.

◀◀ **WORK PROBLEM 4 AT THE SIDE.**

(b) 30°

OBJECTIVE 2 Two angles are called **congruent** (kuhn-GROO-ent) **angles** if they measure the same number of degrees. If two angles are congruent, this is written as

$$\angle A \cong \angle B$$

and read as, "angle A **is congruent to** angle B."

5. Identify the angles that are congruent.

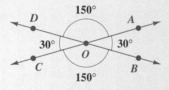

E X A M P L E 5 Identifying Congruent Angles

Identify the angles that are congruent.

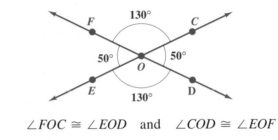

$$\angle FOC \cong \angle EOD \quad \text{and} \quad \angle COD \cong \angle EOF$$

◀◀ **WORK PROBLEM 5 AT THE SIDE.**

Angles that do not share a common side are called *nonadjacent* angles. Two nonadjacent angles formed by intersecting lines are called **vertical** (VUR-ti-kul) **angles.**

E X A M P L E 6 Identifying Vertical Angles

Identify the vertical angles in this figure.

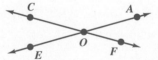

6. Identify the vertical angles.

∠AOF and ∠COE are vertical angles because they do not share a common side and they are formed by two intersecting lines ($\overleftrightarrow{CF}$ and $\overleftrightarrow{EA}$).

∠COA and ∠EOF are also vertical angles.

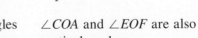

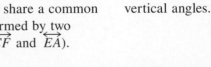 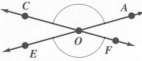

ANSWERS

4. (a) 5° (b) 150°

5. ∠BOC ≅ ∠AOD, ∠AOB ≅ ∠DOC

6. ∠SPB and ∠MPD, ∠BPD and ∠SPM

◀◀ **WORK PROBLEM 6 AT THE SIDE.**

Look back at Example 5 on the previous page. Notice that the two congruent angles that measure 130° are also vertical angles. Also, the two congruent angles that measure 50° are vertical angles. This illustrates the following property.

Congruent Angles

If two angles are vertical angles, they are congruent, that is, they measure the same number of degrees.

E X A M P L E 7   Finding the Measures of Vertical Angles

In the figure below, find the measure of the following angles.

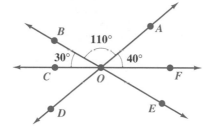

(a) $\angle COD$

$\angle COD$ and $\angle AOF$ are vertical angles so they are congruent. This means they measure the same number of degrees.

The measure of $\angle AOF$ is 40°   so   the measure of $\angle COD$ is 40° also.

(b) $\angle DOE$

$\angle DOE$ and $\angle BOA$ are vertical angles so they are congruent.

The measure of $\angle BOA$ is 110°   so   the measure of $\angle DOE$ is 110° also.

(c) $\angle EOF$

$\angle EOF$ and $\angle COB$ are vertical angles so they are congruent.

The measure of $\angle COB$ is 30°   so   the measure of $\angle EOF$ is 30° also.

WORK PROBLEM 7 AT THE SIDE. ▶▶

---

7. In the figure below, find the number of degrees in each of the following angles.

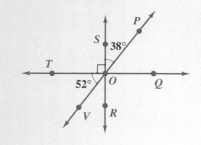

(a) $\angle VOR$

(b) $\angle POQ$

(c) $\angle QOR$

ANSWERS

7. (a) 38° **(b)** 52° **(c)** 90°

NUMBERS IN THE
Real World *collaborative investigations*

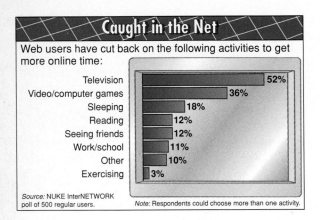

Caught in the Net

Web users have cut back on the following activities to get more online time:

Activity	Percent
Television	52%
Video/computer games	36%
Sleeping	18%
Reading	12%
Seeing friends	12%
Work/school	11%
Other	10%
Exercising	3%

Source: NUKE InterNETWORK poll of 500 regular users.

Note: Respondents could choose more than one activity.

Look at the "Source" information at the bottom of the graph. How were the numbers in the graph obtained?
A poll was taken of 500 regular users of the Web.

How many people in the poll said they cut back on television in order to find time to use the World Wide Web?
52% of 500 is 260 people

Find the number of people in the poll who cut back on each of the other activities shown in the graph. **Video/computer games, 180 people; sleeping, 90 people; reading, 60 people; seeing friends, 60 people; work/school, 55 people; other, 50 people; exercising, 15 people.**

Add up all the numbers of people you just calculated for all the different activities. Why is the total more than the 500 people that were in the poll? **Each person in the poll could choose more than one activity, according to the note at the bottom of the graph.**

Suppose you took a similar poll of 100 students at your school who regularly use the Web. Using the information in the graph, how many students would you expect to choose each of the activities? **TV 52; games 36; sleeping 18; reading 12; seeing friends 12; work/school 11; other 10; exercising 3.**

Why might your poll give different results than the ones shown in the graph? Give at least two possible explanations. **There are many possible explanations. The users in your school are all students, while the 500 who were polled for the graph probably are not. Or, 100 people may not be a large enough group to get reliable data.**

Suppose you were writing a paper about the Web and wanted to include some of the information from the graph, but without showing the whole graph and *without using percents*. You also decide that it would be all right to round the data up or down a little bit in order to make it easier for people to understand. Write a sentence that expresses the information about people who cut back on television. Also write sentences about the people who cut back on video/computer games, those who cut back on sleep, and those who cut back on work/school.
There are many possibilities. If data is rounded, use "about" or "approximately."
Some examples are:
   About half of the users cut back on watching TV.
   A little more than one-third of the users spent less time on video or computer games.
   Nearly one-fifth of users spent less time sleeping.
   About one in ten people cut back on work or school.

8.2 Exercises

Identify each pair of complementary angles.

1.

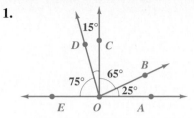

∠EOD and ∠COD; ∠AOB and ∠BOC

2.

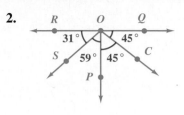

∠COQ and ∠COP; ∠SOR and ∠SOP

Identify each pair of supplementary angles.

3.

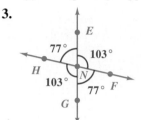

∠HNE and ∠ENF; ∠HNG and ∠GNF;
∠HNE and ∠HNG; ∠ENF and ∠GNF

4.

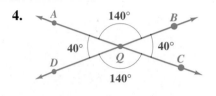

∠AQB and ∠AQD; ∠BQC and ∠CQD;
∠AQB and ∠BQC; ∠CQD and ∠AQD

Find the complement of each angle.

5. 40°   **6.** 35°   **7.** 86°   **8.** 59°

   50°   55°   4°   31°

Find the supplement of each angle.

9. 130°   **10.** 75°   **11.** 90°   **12.** 5°

   50°   105°   90°   175°

In each of the following, identify the angles that are congruent.

13.

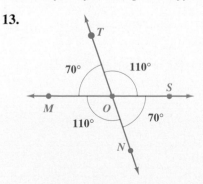

∠SON ≅ ∠TOM; ∠TOS ≅ ∠MON

14.

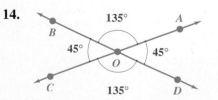

∠AOB ≅ ∠COD; ∠AOD ≅ ∠BOC

☑ Writing   ◉ Conceptual   ▲ Challenging   ≈ Estimation

In this figure, ∠AOH measures 37° and ∠COE measures 63°. Find the number of degrees in each angle.

15. ∠GOH

63°

16. ∠AOC

80°

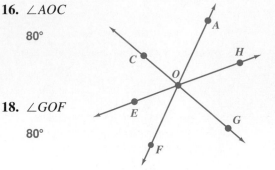

17. ∠EOF

37°

18. ∠GOF

80°

19. In your own words, write a definition of complementary angles and a definition of supplementary angles. Draw a picture to illustrate each definition.

Two angles are complementary if their sum is 90°. Two angles are supplementary if their sum is 180°. Drawings will vary; examples are:

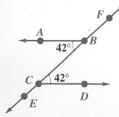

20. Make up a test problem in which a student has to use knowledge of vertical angles. Include a drawing with some angles labeled and ask the student to find the size of the remaining angles. Give the correct answer for your problem.

There are many possibilities. One example is:

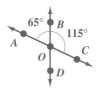

Find the number of degrees in ∠AOD and ∠DOC. ∠AOD measures 115° and ∠DOC measures 65°.

In each figure, ray AB is parallel to ray CD. Identify two pairs of congruent angles and the number of degrees in each congruent angle.

▲ **21.**

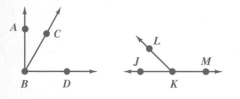

∠ABF ≅ ∠ECD **Both are 138°.**
∠ABC ≅ ∠BCD **Both are 42°.**

▲ **22.**

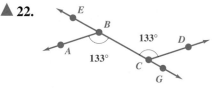

∠ABC ≅ ∠BCD **Both are 133°.**
∠DCG ≅ ∠EBA **Both are 47°.**

23. Can two obtuse angles be supplementary? Explain why or why not.

No, because obtuse angles are >90° so their sum would be >180°.

24. Can two acute angles be complementary? Explain why or why not.

Yes, because acute angles are <90° so their sum could equal 90°.

Review and Prepare

*Evaluate the following expressions. (For help, see **Section 1.8**.)*

25. $16 - (3 \cdot 3)$ **7**

26. $16 \cdot 4 \div 2$ **32**

27. $2 \cdot 3 + 2 \cdot 4$ **14**

28. $5 + 2 \cdot 6$ **17**

29. $6 \cdot 8 - 5^2$ **23**

30. $3^2 + 4^2$ **25**

8.3 Rectangles and Squares

A **rectangle** (REK-tang-gul) is a figure with four sides that meet to form 90° angles. Opposite sides are parallel and congruent (have the same length).

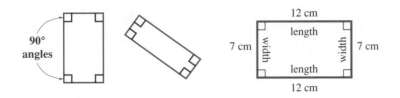

12 cm
length
7 cm width   width 7 cm
length
12 cm

Each longer side of a rectangle is called the length (*l*) and each shorter side is called the width (*w*).

> **WORK PROBLEM I AT THE SIDE.** ▶▶

OBJECTIVE 1 The distance around the outside edges of a figure is the **perimeter** (per-IM-it-er) of the figure. Think of how much fence you would need to put around the sides of a garden plot, or how far you would walk if you go around the outside edges of your backyard. In either case you would add up the lengths of the sides. In the rectangle on the right above,

$$\text{Perimeter} = 12 \text{ cm} + 7 \text{ cm} + 12 \text{ cm} + 7 \text{ cm} = 38 \text{ cm}.$$

Because the two long sides are the same, and the two short sides are the same, you can also use this formula.

Finding the Perimeter of a Rectangle

$$\text{Perimeter of a rectangle} = (2 \cdot \text{length}) + (2 \cdot \text{width})$$

$$P = 2 \cdot l + 2 \cdot w$$

E X A M P L E I   Finding the Perimeter of a Rectangle

Find the perimeter of each rectangle.

(a)

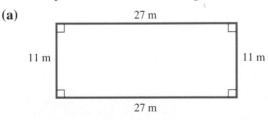

27 m
11 m   11 m
27 m

The length is 27 m and the width is 11 m.

$$P = 2 \cdot \quad l \quad + 2 \cdot \quad w$$
$$P = \underbrace{2 \cdot 27 \text{ m}} + \underbrace{2 \cdot 11 \text{ m}}$$
$$P = \quad 54 \text{ m} \quad + \quad 22 \text{ m}$$
$$P = 76 \text{ m}$$

The perimeter of the rectangle (the distance you would walk around the outside edges of the rectangle) is 76 m.

— **CONTINUED ON NEXT PAGE**

CONTINUED ON NEXT PAGE

OBJECTIVES

1 ▶ Find the perimeter and area of a rectangle.

2 ▶ Find the perimeter and area of a square.

3 ▶ Find the perimeter and area of a composite shape.

FOR EXTRA HELP

Tutorial   Tape 13   SSM, Sec. 8.3

1. Identify all the rectangles.

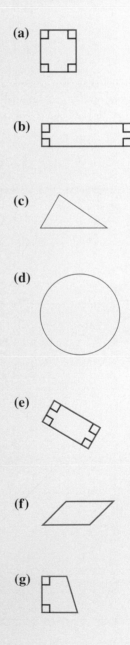

(a)

(b)

(c)

(d)

(e)

(f)

(g)

ANSWERS

1. (a), (b), and (e) are rectangles;
(c), (d), (f), and (g) are not.

507

2. Find the perimeter of each rectangle.

(a)

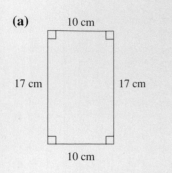

10 cm

17 cm   17 cm

10 cm

(b)

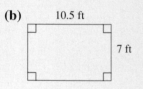

10.5 ft

7 ft

(c) 6 m wide and 11 m long

(d) 0.9 km by 2.8 km

(b) A rectangle 8.9 m by 12.3 m

You can use the formula:

$$P = 2 \cdot \quad l \quad + 2 \cdot \quad w$$
$$P = \underline{2 \cdot 12.3 \text{ m}} + \underline{2 \cdot 8.9 \text{ m}}$$
$$P = \quad 24.6 \text{ m} \quad + \quad 17.8 \text{ m}$$
$$P = 42.4 \text{ m}$$

or you can add up the lengths of the four sides:

$$P = 8.9 \text{ m} + 12.3 \text{ m} + 8.9 \text{ m} + 12.3 \text{ m}$$
$$P = 42.4 \text{ m}$$

◀◀ **WORK PROBLEM 2 AT THE SIDE.**

The perimeter of a rectangle is the distance around the outside edges. The **area** of a rectangle is the amount of surface *inside* the rectangle. We measure area by seeing how many squares of a certain size are needed to cover the surface inside the rectangle. Think of covering the floor of a rectangular living room with carpet. Carpet is measured in square yards, that is, square pieces that measure 1 yard along each side. Here is a drawing of a living room floor.

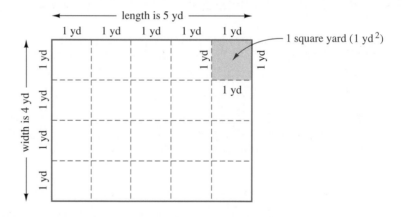

You can see from the drawing that it takes 20 squares to cover the floor. We say that the area of the floor is 20 *square yards*. A shorter way to write square yards is yd².

20 **square yards** can be written 20 **yd²**

To find the number of squares, you can count them, or you can multiply the number of squares in the length (5) times the number of squares in the width (4) to get 20. The formula is:

Finding the Area of a Rectangle

Area of a rectangle = length • width

$$A = l \cdot w$$

Remember to use square units when measuring area.

ANSWERS

2. (a) 54 cm **(b)** 35 ft **(c)** 34 m
   (d) 7.4 km

Squares of other sizes can be used to measure area. For smaller areas, you might use these:

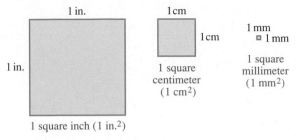

1 in.
1 in.
1 square inch (1 in.²)

1 cm
1 cm
1 square centimeter (1 cm²)

1 mm
1 mm
1 square millimeter (1 mm²)

Actual-size drawings

Other sizes of squares that are often used to measure area are listed here, but they are too large to draw on this page.

1 square meter (1 m²)   1 square foot (1 ft²)
1 square kilometer (1 km²)   1 square yard (1 yd²)
                                        1 square mile (1 mi²)

> **Note**
> The raised 2 in 4^2 means that you multiply $4 \cdot 4$ to get 16. The raised 2 in cm² or yd² is a short way to write the word "square." When you see 5 cm², say "five square centimeters." Do *not* multiply $5 \cdot 5$.

E X A M P L E 2   Finding the Area of a Rectangle

Find the area of each rectangle.

(a)

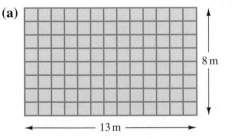

8 m

13 m

The length of this rectangle is 13 m and the width is 8 m.

$$A = \quad l \quad \cdot \quad w$$
$$A = \textbf{13 m} \cdot \textbf{8 m}$$
$$A = 104 \text{ square meters}$$

"Square meters" can be written as m², so the area is 104 m².

(b) A rectangle measuring 7 cm by 21 cm

The area is

$$A = 21 \text{ cm} \cdot 7 \text{ cm} = 147 \text{ cm}^2.$$

> **Note**
> The units for *area* will always be *square* units (cm², m², yd², mi², and so on). The units for *perimeter* will be cm, m, yd, mi, and so on (no square units).

WORK PROBLEM 3 AT THE SIDE. ▶▶

3. Find the area of each rectangle.

(a)
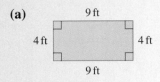
9 ft
4 ft
4 ft
9 ft

(b) A rectangle that is 6 meters long and 0.5 meter wide.

(c) 8.2 cm by 41.2 cm

ANSWERS

3. **(a)** 36 ft² **(b)** 3 m² **(c)** 337.84 cm²

4. Find the perimeter and area of each square.

(a)

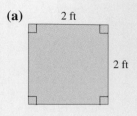

2 ft

2 ft

OBJECTIVE 2 A **square** is a rectangle with all sides the same length. Two squares are shown here. Notice the 90° angles.

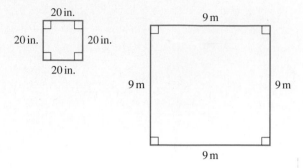

To find the *perimeter* (distance around) of the square on the right, you could add 9 m + 9 m + 9 m + 9 m to get 36 m. A shorter way is to multiply the length of one side times 4, because all 4 sides are the same length.

Finding the Perimeter of a Square

Perimeter of a square = 4 • side

$$P = 4 \cdot s$$

As with a rectangle, you can multiply length times width to find the area (surface inside) a square. Because the length and the width are the same in a square, the formula is written:

Finding the Area of a Square

Area of a square = side • side

$$A = s \cdot s$$
$$A = s^2$$

(b) 10.5 cm on each side

Remember to use square units when measuring area.

(c) 2.1 miles on a side

E X A M P L E 3 Finding the Perimeter and Area of a Square

(a) Find the perimeter of a square where each side measures 9 m.

$P = 4 \cdot s$	Or add up the four sides.
$P = 4 \cdot 9\text{ m}$	$P = 9\text{ m} + 9\text{ m} + 9\text{ m} + 9\text{ m}$
$P = 36\text{ m}$	$P = 36\text{ m}$

(b) Find the area of the same square.

$$A = s^2$$
$$A = s \cdot s$$
$$A = 9\text{ m} \cdot 9\text{ m}$$
$$A = 81\text{ m}^2 \text{ (square units for area)}$$

> **Note**
> s^2 does **not** mean 2 • s. In this example, s is 9 m so s^2 is 9 • 9 = 81, **not** 2 • 9 = 18.

ANSWERS

4. (a) $P = 8$ ft; $A = 4$ ft^2
   (b) $P = 42$ cm; $A = 110.25$ cm^2
   (c) $P = 8.4$ mi; $A = 4.41$ mi^2

◄◄ **WORK PROBLEM 4 AT THE SIDE.**

OBJECTIVE 3 As with any other shape, you can find the perimeter (distance around) an irregular shape by adding up the lengths of the sides. To find the area (surface inside the shape), try to break it up into pieces that are squares or rectangles. Find the area of each piece and then add them together.

E X A M P L E 4 **Finding Perimeter and Area of Composite Figures**

A room has the shape shown here.

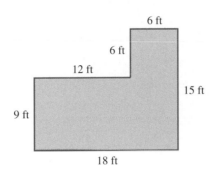

(a) Suppose you want to put a new baseboard (wooden strip) along the base of all the walls. How much material do you need?

Find the perimeter of the room by adding up the lengths of the sides.

$$P = 9 \text{ ft} + 12 \text{ ft} + 6 \text{ ft} + 6 \text{ ft} + 15 \text{ ft} + 18 \text{ ft} = 66 \text{ ft}$$

You need 66 feet of baseboard material.

(b) The carpet you like costs $20.50 per square yard. How much will it cost to carpet the room?

First change the measurements from feet to yards, because the carpet is sold in square yards. There are 3 feet in 1 yard, so multiply by the unit fraction that allows you to cancel feet. For example:

$$\frac{\overset{3}{\cancel{9} \text{ feet}}}{1} \cdot \frac{1 \text{ yard}}{\underset{1}{\cancel{3} \text{ feet}}} = 3 \text{ yards}$$

Cancel feet
Cancel numbers

Use the same unit fraction to change the other measurements from feet to yards.

$$\frac{\overset{4}{\cancel{12} \text{ feet}}}{1} \cdot \frac{1 \text{ yard}}{\underset{1}{\cancel{3} \text{ feet}}} = 4 \text{ yards} \qquad \frac{\overset{2}{\cancel{6} \text{ feet}}}{1} \cdot \frac{1 \text{ yard}}{\underset{1}{\cancel{3} \text{ feet}}} = 2 \text{ yards}$$

$$\frac{\overset{5}{\cancel{15} \text{ feet}}}{1} \cdot \frac{1 \text{ yard}}{\underset{1}{\cancel{3} \text{ feet}}} = 5 \text{ yards} \qquad \frac{\overset{6}{\cancel{18} \text{ feet}}}{1} \cdot \frac{1 \text{ yard}}{\underset{1}{\cancel{3} \text{ feet}}} = 6 \text{ yards}$$

CONTINUED ON NEXT PAGE

5. Carpet costs $19.95 per square yard. Find the cost of carpeting the following rooms. Round your answers to the nearest cent, if necessary.

(a)

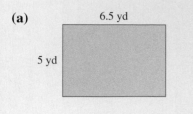

6.5 yd

5 yd

Next, break up the room into two pieces. Use just the measurements for the length and width of each piece.

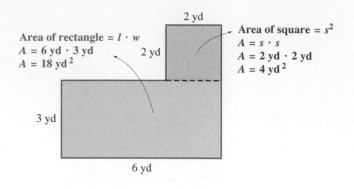

Area of rectangle = $l \cdot w$
$A = 6 \text{ yd} \cdot 3 \text{ yd}$
$A = 18 \text{ yd}^2$

Area of square = s^2
$A = s \cdot s$
$A = 2 \text{ yd} \cdot 2 \text{ yd}$
$A = 4 \text{ yd}^2$

2 yd

2 yd

3 yd

6 yd

Total area = **18 yd² + 4 yd²** = 22 yd²

so the cost of the carpet is

$$\frac{22 \text{ yd}^2}{1} \cdot \frac{\$20.50}{1 \text{ yd}^2} = \$451.00$$

You could have cut the room into two rectangles. The total area is the same.

(b)

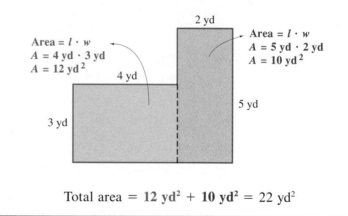

21 ft

21 ft

12 ft

12 ft

9 ft

9 ft

Area = $l \cdot w$
$A = 4 \text{ yd} \cdot 3 \text{ yd}$
$A = 12 \text{ yd}^2$

Area = $l \cdot w$
$A = 5 \text{ yd} \cdot 2 \text{ yd}$
$A = 10 \text{ yd}^2$

2 yd

4 yd

3 yd

5 yd

Total area = **12 yd² + 10 yd²** = 22 yd²

◀◀ **WORK PROBLEM 5 AT THE SIDE.**

(c) a classroom that is 24 ft long and 18 ft wide

ANSWERS

5. (a) 32.5 yd² costs ≈ $648.38
   (b) 37 yd² costs $738.15
   (c) 48 yd² costs $957.60

8.3 *Exercises*

Find the perimeter and area of each rectangle or square.

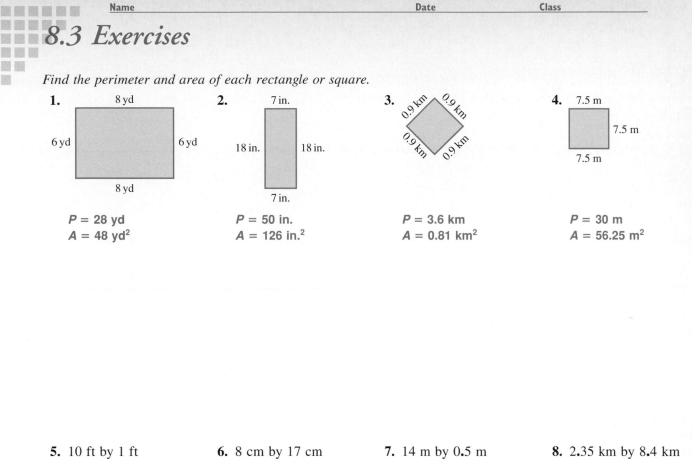

1. 8 yd   6 yd   6 yd   8 yd

P = 28 yd
A = 48 yd²

2. 7 in.   18 in.   18 in.   7 in.

P = 50 in.
A = 126 in.²

3. 0.9 km   0.9 km   0.9 km   0.9 km

P = 3.6 km
A = 0.81 km²

4. 7.5 m   7.5 m   7.5 m

P = 30 m
A = 56.25 m²

5. 10 ft by 1 ft

P = 22 ft
A = 10 ft²

6. 8 cm by 17 cm

P = 50 cm
A = 136 cm²

7. 14 m by 0.5 m

P = 29 m
A = 7 m²

8. 2.35 km by 8.4 km

P = 21.5 km
A = 19.74 km²

9. 76.1 ft by 22 ft

P = 196.2 ft
A = 1674.2 ft²

10. 12 m by 12 m

P = 48 m
A = 144 m²

11. a square 3 mi wide

P = 12 mi
A = 9 mi²

12. a square 20.3 cm on a side

P = 81.2 cm
A = 412.09 cm²

✎ Writing   ◉ Conceptual   ▲ Challenging   ≈ Estimation

Find the perimeter and area of each figure.

Example:

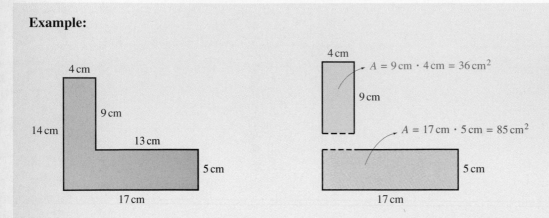

Solution:

perimeter = 14 cm + 4 cm + 9 cm + 13 cm   total area = **36 cm² + 85 cm²**
         + 5 cm + 17 cm = **62 cm**   = **121 cm²** (square units for area)

13.

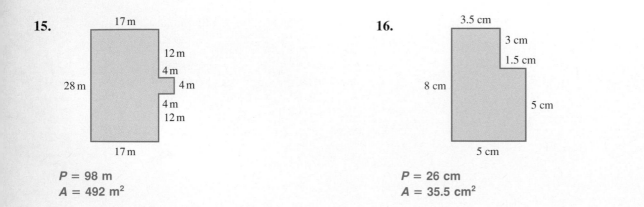

P = 38 m
A = 39 m²

14.

P = 48 ft
A = 72 ft²

15.

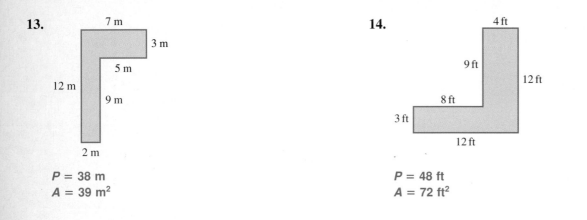

P = 98 m
A = 492 m²

16.

P = 26 cm
A = 35.5 cm²

▲ **17.**

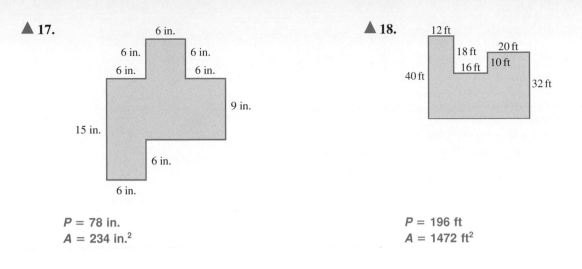

P = 78 in.
A = 234 in.²

▲ **18.**

P = 196 ft
A = 1472 ft²

Solve each application problem.

19. The Wang's family room measures 20 feet by 25 feet. They are covering the floor with square tiles that measure 1 foot on a side and cost $0.72 each. How much will they spend on tile?

$360

20. A page in this book measures 27.5 centimeters from top to bottom and 20 centimeters wide. Find the perimeter and the area of the page.

P = 95 cm
A = 550 cm²

21. Tyra's kitchen is 4.4 meters wide and 5.1 meters long. She is pasting a decorative strip that costs $4.99 per meter around the top edge of all the walls. How much will she spend?

$94.81

22. Mr. and Mrs. Gomez are buying carpet for their square-shaped bedroom that is 15 feet wide. The carpet is $23 per square yard and padding and installation is another $6 per square yard. How much will they spend in all?

$725

23. In your own words, describe the difference between perimeter and area. Then make a drawing of a square and a rectangle, label the sides with measurements, and show the steps in finding the perimeter and area of each figure.

Perimeter is the distance around the outside edges of a shape; area is the surface inside the shape measured in square units. Drawings will vary.

24. Suppose you had 16 feet of fencing. Draw three different square or rectangular garden plots that would use exactly 16 feet of fencing; label the lengths of the sides. What shape plot would have the greatest area?

A square plot has the greatest area. Examples:

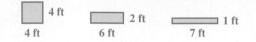

25. A lot is 124 feet by 172 feet. County rules require that nothing be built on land within 12 feet of any edge of the lot. Draw a sketch of the lot, showing the land that cannot be built on. What is the area of the land that cannot be built on?

A = 6528 ft²

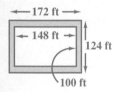

26. Find the cost of fencing needed for this rectangular field. Fencing along the country road costs $4.25 per foot. Fencing for the other three sides costs $2.75 per foot.

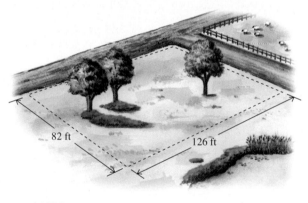

82 ft   126 ft

$1333

Review and Prepare

*Convert the following measurements. (For help, see **Sections 7.1 and 7.2**.)*

27. 50 cm to m   **0.5 m**

28. 8 km to m   **8000 m**

29. 12 ft to in.   **144 in.**

30. 75 ft to yd   **25 yd**

31. 2700 in. to yd   **75 yd**

32. 400 mm to cm   **40 cm**

8.4 *Parallelograms and Trapezoids*

A **parallelogram** (pair-uh-LELL-uh-gram) is a four-sided figure with opposite sides parallel, such as these. Notice that opposite sides have the same length.

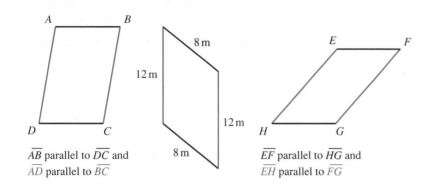

$\overline{AB}$ parallel to $\overline{DC}$ and
$\overline{AD}$ parallel to $\overline{BC}$

$\overline{EF}$ parallel to $\overline{HG}$ and
$\overline{EH}$ parallel to $\overline{FG}$

OBJECTIVE 1 Perimeter is the distance around a figure, so the easiest way to find the perimeter of a parallelogram is to add the lengths of the four sides.

E X A M P L E   I   Finding the Perimeter of a Parallelogram

Find the perimeter of the middle parallelogram above.

$$P = 12 \text{ m} + 8 \text{ m} + 12 \text{ m} + 8 \text{ m} = 40 \text{ m}$$

WORK PROBLEM I AT THE SIDE. ▶▶

To find the area of a parallelogram, first draw a dashed line as shown here.

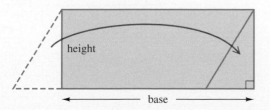

The length of the dashed line is the *height* of the parallelogram. It forms a *right angle* with the base. The height is the shortest distance between the base and the opposite side.

Now cut off the triangle created on the left side of the parallelogram and move it to the right side, as shown below.

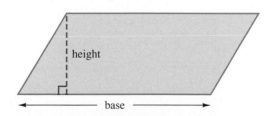

The parallelogram has been made into a rectangle. You can see that the area of the parallelogram and the rectangle are the same. The area of the rectangle is length times width. In the parallelogram, this translates into base times height.

OBJECTIVES

1 Find the perimeter and area of parallelograms.

2 Find the perimeter and area of trapezoids.

FOR EXTRA HELP

Tutorial   Tape 13   SSM, Sec. 8.4

1. Find the perimeter of each parallelogram.

(a)

27 m

15 m   15 m

27 m

(b)

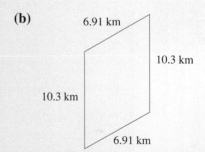

6.91 km

10.3 km

10.3 km

6.91 km

ANSWERS

1. (a) 84 m **(b)** 34.42 km

517

2. Find the area of each parallelo-gram.

(a)

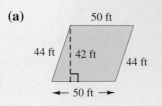

(b)

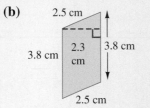

(c) a parallelogram with base $12\frac{1}{2}$ m and height $4\frac{3}{4}$ m (*Hint:* Write $12\frac{1}{2}$ as 12.5 and $4\frac{3}{4}$ as 4.75.)

3. Find the perimeter of each trapezoid.

(a)

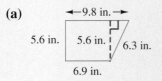

(b)

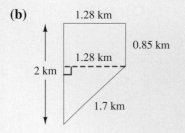

(c) a trapezoid with sides 39.7 cm, 29.2 cm, 74.9 cm, and 16.4 cm

Finding the Area of a Parallelogram

Area of parallelogram = base • height
$$A = b \cdot h$$

Remember to use square units when measuring area.

E X A M P L E 2   Finding the Area of a Parallelogram

Find the area of each parallelogram.

(a)

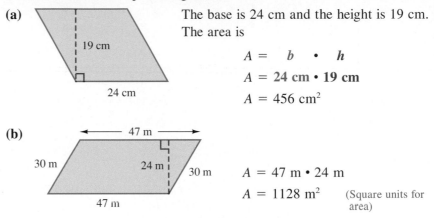

The base is 24 cm and the height is 19 cm. The area is

$$A = \quad b \quad \bullet \quad h$$
$$A = \textbf{24 cm} \bullet \textbf{19 cm}$$
$$A = 456 \text{ cm}^2$$

(b)

$$A = 47 \text{ m} \bullet 24 \text{ m}$$
$$A = 1128 \text{ m}^2 \quad \text{(Square units for area)}$$

Notice that the 30-m sides are *not* used in finding the area.

◀◀ **WORK PROBLEM 2 AT THE SIDE.**

OBJECTIVE 2 ▶ A **trapezoid** (TRAP-uh-zoyd) is a four-sided figure with one pair of parallel sides, such as the ones shown here. Opposite sides may *not* have the same length, as in parallelograms.

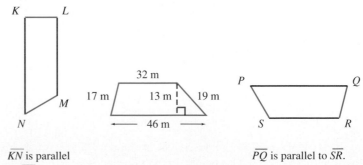

$\overline{KN}$ is parallel to $\overline{LM}$.

$\overline{PQ}$ is parallel to $\overline{SR}$.

E X A M P L E 3   Finding the Perimeter of a Trapezoid

Find the perimeter of the middle trapezoid above. Add the lengths of the sides.

$$P = 17 \text{ m} + 32 \text{ m} + 19 \text{ m} + 46 \text{ m} = 114 \text{ m}$$

Notice that the height (13 m) is *not* part of the perimeter.

◀◀ **WORK PROBLEM 3 AT THE SIDE.**

ANSWERS

2. **(a)** 2100 ft² **(b)** 8.74 cm²

(c) $59\frac{3}{8}$ m² or 59.375 m²

3. **(a)** 28.6 in. **(b)** 5.83 km **(c)** 160.2 cm

Use the following formula to find the area of a trapezoid.

Finding the Area of a Trapezoid

$$\text{Area} = \frac{1}{2} \cdot \text{height} \cdot (\text{short base} + \text{long base})$$

$$A = \frac{1}{2} \cdot h \cdot (b + B)$$

or $A = 0.5 \cdot h \cdot (b + B)$

Remember to use square units when measuring area.

E X A M P L E   4   Finding the Area of a Trapezoid

Find the area of this trapezoid. The short base and long base are the parallel sides.

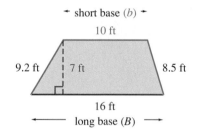

The height (h) is **7 ft**, the short base (b) is **10 ft**, and the long base (B) is **16 ft**. You do *not* need the 9.2 ft or 8.5 ft sides to find the area.

$$A = \frac{1}{2} \cdot h \cdot (b + B)$$

$$A = \frac{1}{2} \cdot 7 \text{ ft} \cdot (10 \text{ ft} + 16 \text{ ft})$$

$$A = \frac{1}{2} \cdot 7 \text{ ft} \cdot (\overset{13}{\underset{1}{26}} \text{ ft})$$

$$A = 91 \text{ ft}^2 \qquad \text{(Square units for area)}$$

You can also solve the problem by using 0.5, the decimal equivalent for $\frac{1}{2}$, in the formula.

$$A = 0.5 \cdot h \cdot (b + B)$$
$$A = 0.5 \cdot 7 \cdot (10 + 16)$$
$$A = 0.5 \cdot 7 \cdot 26$$
$$A = 91 \text{ ft}^2$$

Calculator Tip: Use the parentheses keys on your scientific calculator to work Example 4:

$$0.5 \boxed{\times} 7 \boxed{\times} \boxed{(} 10 \boxed{+} 16 \boxed{)} \boxed{=} 91$$

What happens if you do *not* use the parentheses keys? What order of operations will the calculator follow then?

WORK PROBLEM 4 AT THE SIDE. ▶▶

---

4. Find the area of each trapezoid.

(a)

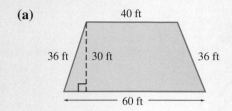

(b)

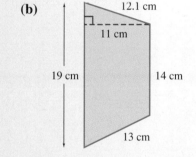

(c) a trapezoid with height 4.7 m, short base 9 m, and long base 10.5 m

ANSWERS

4. (a) 1500 ft² **(b)** 181.5 cm²
(c) 45.825 m²

5. Find the area of each floor.

(a)

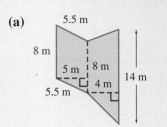

(b)

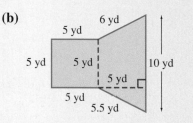

6. Find the cost of carpeting the floors in Problem 5. The cost of carpet is as follows:

(a) Floor (a), $18.50 per square meter.

(b) Floor (b), $28 per square yard.

EXAMPLE 5   Finding the Area of Composite Figures

Find the area of this figure.

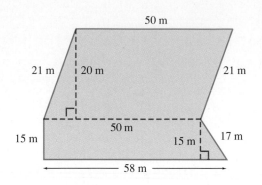

Break the figure into two pieces, a parallelogram and a trapezoid.

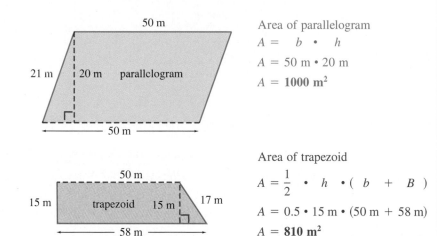

Area of parallelogram
$A = b \cdot h$
$A = 50 \text{ m} \cdot 20 \text{ m}$
$A = \mathbf{1000 \text{ m}^2}$

Area of trapezoid
$A = \dfrac{1}{2} \cdot h \cdot (b + B)$
$A = 0.5 \cdot 15 \text{ m} \cdot (50 \text{ m} + 58 \text{ m})$
$A = \mathbf{810 \text{ m}^2}$

Total area = **1000 m² + 810 m²** = 1810 m²

◀◀ **WORK PROBLEM 5 AT THE SIDE.**

EXAMPLE 6   Applying Knowledge of Area

Suppose the figure in Example 5 represents the floor plan of a hotel lobby. What is the cost of labor to install tile on the floor if the labor charge is $35.11 per square meter?

From Example 5, the floor area is 1810 m². To find the labor cost, multiply the number of square meters times the cost of labor per square meter.

$$\text{Cost} = \frac{1810 \text{ m}^2}{1} \cdot \frac{\$35.11}{1 \text{ m}^2}$$

$$\text{Cost} = \$63,549.10$$

The cost of the labor is $63,549.10.

◀◀ **WORK PROBLEM 6 AT THE SIDE.**

ANSWERS
5. (a) 84 m² (b) 62.5 yd²
6. (a) $1554 (b) $1750

8.4 Exercises

Find the perimeter of each figure.

1.
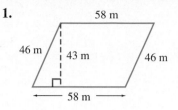
58 m
46 m 43 m 46 m
58 m

208 m

2.
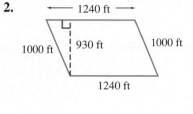
1240 ft
1000 ft 930 ft 1000 ft
1240 ft

4480 ft

3.
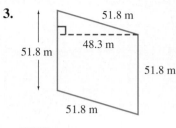
51.8 m
51.8 m 48.3 m
51.8 m
51.8 m

207.2 m

4.
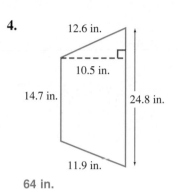
12.6 in.
10.5 in.
14.7 in. 24.8 in.
11.9 in.

64 in.

5.
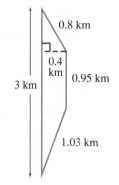
0.8 km
0.4 km
3 km 0.95 km
1.03 km

5.78 km

6.
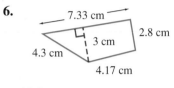
7.33 cm
2.8 cm
4.3 cm 3 cm
4.17 cm

18.6 cm

Find the area of each figure.

7.
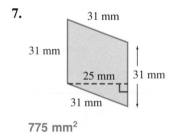
31 mm
31 mm
25 mm 31 mm
31 mm

775 mm²

8.
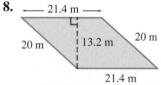
21.4 m
20 m 13.2 m 20 m
21.4 m

282.48 m²

9.
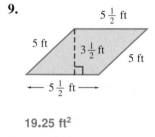
5 ½ ft
5 ft 3 ½ ft 5 ft
5 ½ ft

19.25 ft²

10.

9 ¼ in.
8 in. 8 in. 10 ¼ in.
15 ¾ in.

100 in.²

11.
42 cm
61.4 cm 86.2 cm
42 cm
48.8 cm

3099.6 cm²

12.
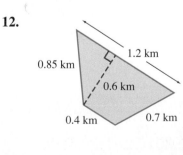
1.2 km
0.85 km
0.6 km
0.4 km 0.7 km

0.48 km²

Solve each application problem.

13. The backyard of a new home is shaped like a trapezoid with a height of 45 ft and bases of 80 ft and 110 ft. What is the cost of putting sod on the yard, if the landscaper charges $0.22 per square foot for sod?

$940.50

14. A swimming pool is in the shape of a parallelogram with a height of 9.6 m and base of 12.4 m. Find the labor cost to make a custom solar cover for the pool at a cost of $4.92 per square meter.

≈$585.68

✎ Writing   ◉ Conceptual   ▲ Challenging   ≈ Estimation

📝 *Find two errors in each student's solution below. Write a sentence explaining each error.*
◎ *Then show how to work the problem correctly.*

15.

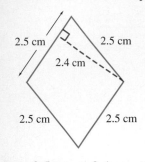

$P = 2.5 \text{ cm} + 2.4 \text{ cm} + 2.5 \text{ cm} + 2.5 \text{ cm}$
$\quad + 2.5 \text{ cm}$
$P = 12.4 \text{ cm}^2$

Height is not part of perimeter; square units are used for area, not perimeter.

$P = 2.5 \text{ cm} + 2.5 \text{ cm} + 2.5 \text{ cm} + 2.5 \text{ cm}$

$P = 10 \text{ cm}$

16.

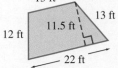

$A = 0.5 \cdot 11.5 \text{ ft} \cdot (12 \text{ ft} + 13 \text{ ft})$
$A = 143.75 \text{ ft}$

The bases are the parallel lines, 22 ft and 13 ft; area is measured in square units.

$A = 0.5 \cdot 11.5 \text{ ft} \cdot (22 \text{ ft} + 13 \text{ ft})$

$A = 201.25 \text{ ft}^2$

🖩 *Find the area of each figure.*

▲

17.

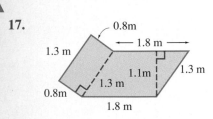

3.02 m²

18.

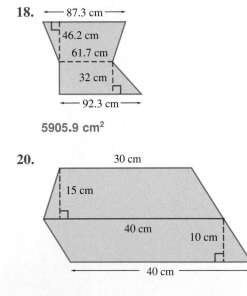

5905.9 cm²

19.

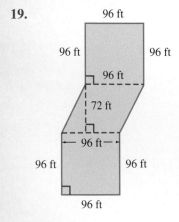

25,344 ft²

20.

30 cm

15 cm

40 cm

10 cm

40 cm

925 cm²

▪▪▪▪▪
▪▪▪▪
▪▪ ▪ *Review and Prepare*
▪ ▪ ▪
▪▪▪▪

*Multiply the following. Write all answers as whole numbers or mixed numbers in lowest terms. (For help, see **Sections 2.5 and 2.8**.)*

21. $6\frac{1}{2} \cdot 8$ **52**

22. $12 \cdot \frac{5}{6}$ **10**

23. $25 \cdot 1\frac{3}{7} \cdot \frac{28}{75}$ $13\frac{1}{3}$

24. $\frac{15}{34} \cdot 3\frac{2}{5} \cdot 1\frac{1}{3}$ **2**

8.5 Triangles

A **triangle** is a figure with exactly three sides, as shown below.

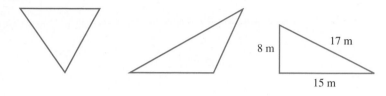

OBJECTIVE ▷ To find the perimeter of a triangle (the distance around the edges), add the lengths of the three sides.

EXAMPLE 1   Finding the Perimeter of a Triangle

The perimeter of the triangle above on the right is

$$P = 8 \text{ m} + 15 \text{ m} + 17 \text{ m} = 40 \text{ m}.$$

WORK PROBLEM 1 AT THE SIDE. ▶▶

As with parallelograms, you can find the *height* of a triangle by measuring the distance from one corner of the triangle to the opposite side (the base). The height line must be *perpendicular* to the base, that is, it must form a right angle with the base. Sometimes you have to extend the base line in order to draw the height perpendicular to it.

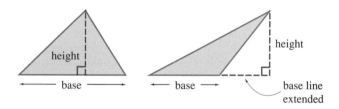

If you cut out two identical triangles and turn one upside down, you can fit them together to form a parallelogram.

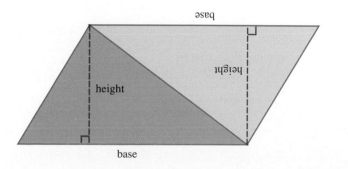

The area of the parallelogram is base times height. Because each triangle is *half* of the parallelogram, the area of one triangle is

$$\frac{1}{2} \text{ of base times height.}$$

OBJECTIVES

1 ▷ Find the perimeter of a triangle.

2 ▷ Find the area of a triangle.

3 ▷ Given two angles in a triangle, find the third angle.

FOR EXTRA HELP

Tutorial   Tape 14   SSM, Sec. 8.5

1. Find the perimeter of each triangle.

(a) 31 mm, 25 mm, 16 mm

(b) 25.9 m, 11.7 m, 16.2 m

(c) a triangle with sides $6\frac{1}{2}$ yd, $9\frac{3}{4}$ yd, and $11\frac{1}{4}$ yd

ANSWERS

1. (a) 72 mm **(b)** 53.8 m

(c) $27\frac{1}{2}$ yd or 27.5 yd

2. Find the area of each triangle.

(a)

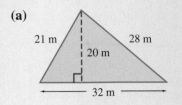

21 m   28 m
20 m
32 m

(b)

2.1 cm   1.9 cm
1.7 cm
2.1 cm

(c)

12 in.   4.5 in.
7.5 in.
6 in.

(d)

7 ft $11\frac{4}{5}$ ft
$9\frac{1}{2}$ ft

ANSWERS

2. (a) 320 m²   **(b)** 1.785 cm²
   (c) 13.5 in.²

   (d) 33.25 ft² or $33\frac{1}{4}$ ft²

OBJECTIVE 2   Use the following formula to find the area of a triangle.

Finding the Area of a Triangle

$$\text{Area of triangle} = \frac{1}{2} \cdot \text{base} \cdot \text{height}$$

$$A = \frac{1}{2} \cdot b \cdot h$$

$$\text{or} \quad A = 0.5 \cdot b \cdot h$$

Remember to use square units when measuring area.

E X A M P L E 2   **Finding the Area of a Triangle**

Find the area of each triangle.

(a)

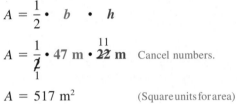

26 m   41 m
22 m
47 m

The base is 47 m and the height is 22 m. You do *not* need the 26 m or 41 m sides to find the area.

$$A = \frac{1}{2} \cdot \quad b \quad \cdot \quad h$$

$$A = \frac{1}{\overset{2}{1}} \cdot 47 \text{ m} \cdot \overset{11}{2\!\!\!/2} \text{ m} \quad \text{Cancel numbers.}$$

$$A = 517 \text{ m}^2 \quad \text{(Square units for area)}$$

(b)

58.6 cm
19.4 cm   21 cm
45.6 cm

$$A = 0.5 \cdot 45.6 \text{ cm} \cdot 19.4 \text{ cm}$$
$$A = 442.32 \text{ cm}^2$$

The base line must be extended to draw the height. However, still use 45.6 cm for b in the formula. Because the measurements are decimal numbers, it is easier to use 0.5 in the formula (the decimal equivalent of $\frac{1}{2}$).

(c)

$11\frac{1}{10}$ in.
$6\frac{1}{2}$ in.
9 in.

Because two sides of the triangle are perpendicular to each other, use those sides as the base and the height.

Equivalent

$$A = \frac{1}{2} \cdot 9 \text{ in.} \cdot 6\frac{1}{2} \text{ in.} \quad \text{or} \quad A = 0.5 \cdot 9 \text{ in.} \cdot 6.5 \text{ in.}$$

Equivalent

$$A = 29\frac{1}{4} \text{ in.}^2 \qquad \text{or} \quad A = 29.25 \text{ in.}^2$$

◄◄ **WORK PROBLEM 2 AT THE SIDE.**

E X A M P L E   3   Using the Concept of Area

Find the area of the shaded part in this figure.

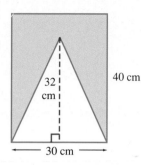

The *entire* figure is a rectangle.

$$A = l \cdot w$$
$$A = 30 \text{ cm} \cdot 40 \text{ cm} = 1200 \text{ cm}^2$$

The *un*shaded part is a triangle.

$$A = \frac{1}{2} \cdot \overset{15}{\cancel{30}} \text{ cm} \cdot 32 \text{ cm}$$
$$\underset{1}{}$$

$$A = 480 \text{ cm}^2$$

Subtract to find the area of the shaded part.

$$A = \overbrace{1200 \text{ cm}^2}^{\text{Entire area}} - \overbrace{480 \text{ cm}^2}^{\text{Unshaded part}} = \overbrace{720 \text{ cm}^2}^{\text{Shaded part}}$$

> **WORK PROBLEM 3 AT THE SIDE.** ▶▶

E X A M P L E   4   Applying the Concept of Area

The Department of Transportation cuts triangular signs out of rectangular pieces of metal using the measurements shown above in Example 3. If the metal costs $0.02 per square centimeter, how much does the metal cost for the sign? What is the cost of the metal that is *not* used?

From Example 3, the area of the triangle (the sign) is 480 cm². Multiply that times the cost per square centimeter.

$$\text{Cost of sign} = \frac{480 \text{ cm}^2}{1} \cdot \frac{\$0.02}{1 \text{ cm}^2} = \$9.60$$

The metal that is *not* used is the *shaded* part from Example 3. The area is 720 cm².

$$\text{Cost of unused metal} = \frac{720 \text{ cm}^2}{1} \cdot \frac{\$0.02}{1 \text{ cm}^2} = \$14.40$$

> **WORK PROBLEM 4 AT THE SIDE.** ▶

3. Find the area of the shaded part in this figure.

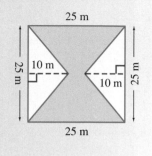

4. Suppose the figure directly above is an auditorium floor plan. The shaded part will be covered with carpet costing $27 per square meter. The rest will be covered with vinyl floor covering costing $18 per square meter. What is the total cost of covering the floor?

ANSWERS

3. 625 m² − 125 m² − 125 m² = 375 m²

4. $10,125 + $2250 + $2250 = $14,625

5. Find the number of degrees in the unlabeled angle.

(a)

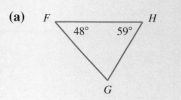

(b)

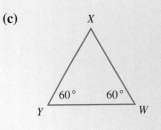

(c)

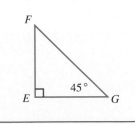

OBJECTIVE **3** The *tri* in *tri*angle means *three*. So the name tells you that a triangle has three angles. The sum of the three angles in any triangle is always 180° (a straight angle). You can see it by drawing a triangle, cutting off the three angles, and rearranging them to make a straight angle.

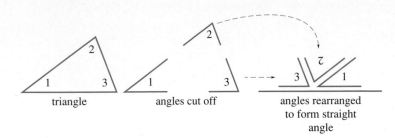

| triangle | angles cut off | angles rearranged to form straight angle |

Finding the Unknown Angle Measurement in a Triangle

Step 1 Add the number of degrees in the two angles you are given.

Step 2 Subtract the sum from 180°.

E X A M P L E   5 **Finding an Angle Measurement in a Triangle**

How many degrees are in:

(a) angle *R*

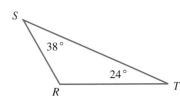

Step 1 Add the two angles you are given,

$$38° + 24° = 62°$$

Step 2 Subtract the sum from 180°.

$$180° - 62° = 118°$$

∠*R* measures 118°.

(b) angle *F*

∠*E* is a right angle, which is 90°.

Step 1 90° + 45° = 135°

Step 2 180° − 135° = 45°

∠*F* measures 45°.

◄◄ WORK PROBLEM 5 AT THE SIDE.

8.5 Exercises

Find the perimeter of each triangle.

1.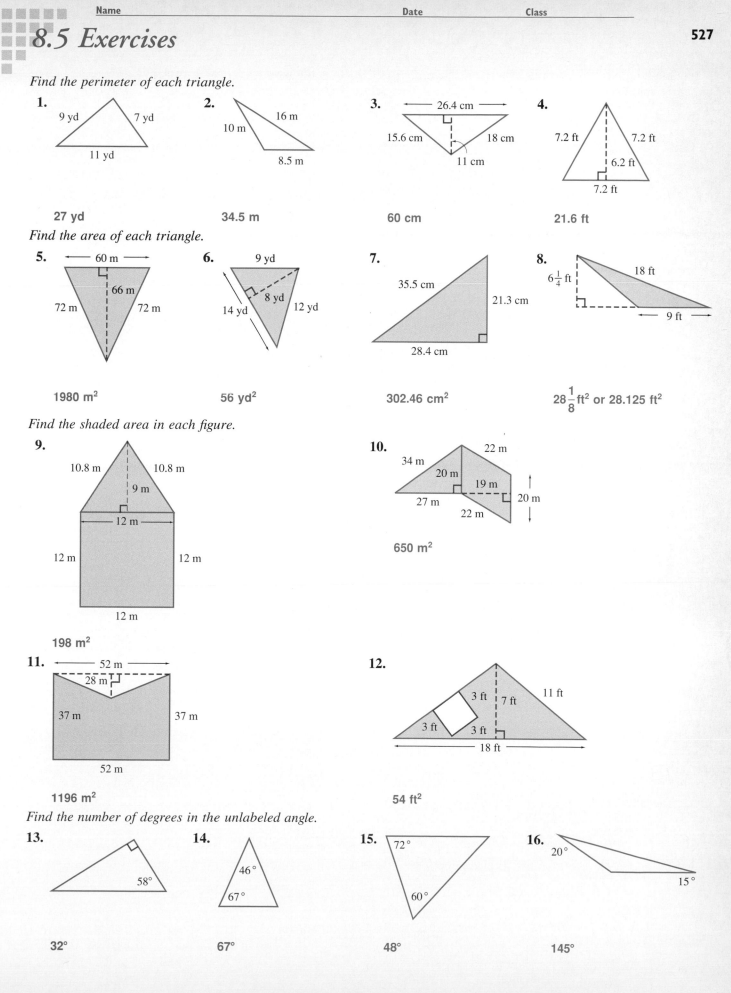
9 yd 7 yd
11 yd

27 yd

2.
16 m
10 m
8.5 m

34.5 m

3. ← 26.4 cm →
15.6 cm 18 cm
11 cm

60 cm

4.
7.2 ft 7.2 ft
6.2 ft
7.2 ft

21.6 ft

Find the area of each triangle.

5. ← 60 m →
66 m
72 m 72 m

1980 m²

6.
9 yd
8 yd
14 yd 12 yd

56 yd²

7.
35.5 cm 21.3 cm
28.4 cm

302.46 cm²

8.
18 ft
$6\frac{1}{4}$ ft
9 ft

$28\frac{1}{8}$ ft² or 28.125 ft²

Find the shaded area in each figure.

9.
10.8 m 10.8 m
9 m
← 12 m →
12 m 12 m
12 m

198 m²

10.
22 m
34 m
20 m 19 m
27 m 20 m
22 m

650 m²

11. ← 52 m →
28 m
37 m 37 m
52 m

1196 m²

12.
3 ft 7 ft 11 ft
3 ft 3 ft
← 18 ft →

54 ft²

Find the number of degrees in the unlabeled angle.

13.
58°

32°

14.
46°
67°

67°

15.
72°
60°

48°

16.
20°
15°

145°

✎ Writing   ◉ Conceptual   ▲ Challenging   ≈ Estimation

17. Can a triangle have two right angles? Explain your answer.

No. Right angles are 90° so two right angles is 180° and the sum of all *three* angles in a triangle equals 180°.

Solve each application problem.

19. A triangular tent flap measures $3\frac{1}{2}$ feet along the base and has a height of $4\frac{1}{2}$ feet. How much canvas is needed to make the flap?

$7\frac{7}{8}$ ft² or 7.875 ft²

21. A triangular space between three streets has the measurements shown. How much new curbing will be needed to go around the space? How much sod will be needed to cover the space?

126.8 m of curb; 672 m² of sod

18. In your own words, explain where the $\frac{1}{2}$ comes from in the formula for area of a triangle.

Two identical triangles form a parallelogram. The area of a parallelogram is base times height, so the area of one of the triangles is $\frac{1}{2} \cdot$ base $\cdot$ height.

20. A wooden sign in the shape of a right triangle has perpendicular sides measuring 1.5 m and 1.2 m. How much surface area does the sign have?

0.9 m²

22. Each gable end of a new house has a span of 36 ft and a rise of 9.5 ft. What is the total area of both gable ends of the house?

342 ft²

▲ **23.** (a) Find the area of one side of the house.
(b) Find the area of one roof section.
All sides of the house are congruent and all roof sections are congruent.

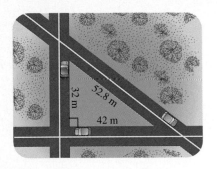

(a) 32 m² (b) 13.5 m²

▲ **24.** The sketch shows the plan for an office building. The shaded area will be a parking lot. What is the cost of building the parking lot, if the contractor charges $28.00 per square yard for materials and labor?

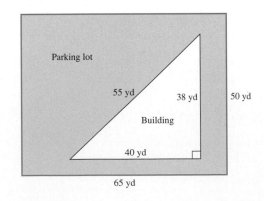

$69,720

Review and Prepare

*Multiply the following decimals. (For help, see **Section 4.5**.)*

25. 3.14 • 16   **50.24**   **26.** 2.13 • 4.65   **9.9045**   **27.** 0.8 • 0.8   **0.64**   **28.** 15 • 0.7   **10.5**

8.6 Circles

OBJECTIVES

 Find the radius and diameter of a circle.

 Find the circumference of a circle.

Find the area of a circle.

Become familiar with Latin and Greek prefixes used in math terminology.

OBJECTIVE ▶ Suppose you start with one dot on a piece of paper. Then you draw a bunch of dots that are each 2 cm away from the first dot. If you draw enough dots (points) you'll end up with a **circle.** Each point on the circle is exactly 2 cm away from the *center* of the circle. The 2 cm distance is called the **radius** (RAY-dee-us), *r*, of the circle. The distance across the circle (passing through the center) is called the **diameter** (dy-AH-meh-ter), *d*, of the circle.

FOR EXTRA HELP

Tutorial   Tape 14   SSM, Sec. 8.6

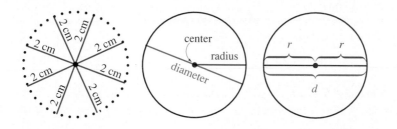

As the circle on the right shows,

> **Finding the Diameter and Radius of a Circle**
>
> $$\text{diameter} = 2 \cdot \text{radius}$$
> $$d = 2 \cdot r$$
> and $$r = \frac{d}{2}$$

1. Find the missing diameter or radius in each circle.

(a)
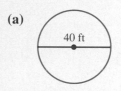
40 ft

(b)
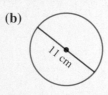
11 cm

E X A M P L E   1   **Finding the Diameter and Radius of a Circle**

Find the missing diameter or radius in each circle.

(a)
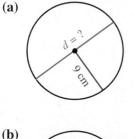
d = ?
9 cm

Because the radius is 9 cm, the diameter is twice as long.

$$d = 2 \cdot r$$
$$d = 2 \cdot \textbf{9 cm}$$
$$d = 18 \text{ cm}$$

(c)
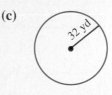
32 yd

(b)
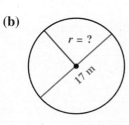
r = ?
17 m

The radius is half the diameter.

$$r = \frac{d}{2}$$
$$r = \frac{\textbf{17 m}}{2}$$
$$r = 8.5 \text{ m} \quad \text{or} \quad 8\frac{1}{2} \text{ m}$$

(d)

9.5 m

WORK PROBLEM 1 AT THE SIDE. ▶▶

ANSWERS

1. (a) $r = 20$ ft
  (b) $r = 5.5$ cm
  (c) $d = 64$ yd
  (d) $d = 19$ m

529

OBJECTIVE 2 The perimeter of a circle is called its **circumference** (sir-KUM-fer-ens). Circumference is the distance around the edge of a circle.

The diameter of the can in the drawing is about 10.6 cm, and the circumference of the can is about 33.3 cm. Dividing the circumference of the circle by the diameter gives an interesting result.

$$\frac{\text{circumference}}{\text{diameter}} = \frac{33.3}{10.6} \approx 3.14 \quad \text{(Rounded)}$$

Dividing the circumference of *any* circle by its diameter *always* gives an answer close to 3.14. This means that going around the edge of any circle is a little more than 3 times as far as going straight across the circle.

This ratio of circumference to diameter is called π (the Greek letter **pi**, pronounced PIE). There is no decimal that is exactly equal to π, but approximately:

$$\pi \approx 3.14159265359.$$

Rounding the Value of *Pi* (π)

We usually round π to 3.14. Therefore, calculations involving π will give approximate answers and should be written using the $\approx$ symbol.

Use the following formulas to find the circumference of a circle.

Finding the Distance around a Circle

$$\text{Circumference} = \pi \cdot \text{diameter}$$
$$C = \pi \cdot d$$

or, because $d = 2 \cdot r$ then $C = \pi \cdot 2 \cdot r$ usually written $2 \cdot \pi \cdot r$

E X A M P L E 2 Finding the Circumference of a Circle

Find the circumference of each circle. Use 3.14 as the approximate value for π. Round answers to the nearest tenth.

(a)

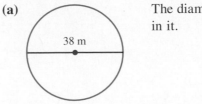

38 m

The diameter is 38 m, so use the formula with d in it.

$$C = \pi \cdot d$$
$$C \approx 3.14 \cdot 38 \text{ m}$$
$$C \approx 119.3 \text{ m} \quad \text{(Rounded)}$$

CONTINUED ON NEXT PAGE

(b)

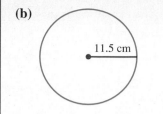

11.5 cm

In this example, r is known, so it is easier to use the formula

$$C = 2 \cdot \pi \cdot r$$
$$C \approx 2 \cdot 3.14 \cdot 11.5 \text{ cm}$$
$$C \approx 72.2 \text{ cm} \quad \text{(Rounded)}$$

▦ **Calculator Tip:** Scientific calculators have a $\boxed{\pi}$ key. Try pressing it. With a 10-digit display, you'll see the value of π to the nearest billionth.

$$\boxed{3.141592654}$$

But this is still an approximate value, although it is more precise than rounding π to 3.14. Try finding the circumference in Example 2(a) from the previous page using the $\boxed{\pi}$ key.

$$\boxed{\pi} \, \boxed{\times} \, 38 \, \boxed{=} \, 119.3805208$$

When you used 3.14 as the approximate value of π, the result was 119.32, so the answers are slightly different. In this book we will use 3.14 instead of the $\boxed{\pi}$ key. Our measurements of radius and diameter are given as whole numbers or with tenths, so it is acceptable to round π to hundredths. And, some students may be using standard calculators without a $\boxed{\pi}$ key, or doing the calculations by hand.

> **WORK PROBLEM 2 AT THE SIDE.** ▶▶

OBJECTIVE 3 To find the formula for the area of a circle, start by cutting two circles into many pie-shaped pieces.

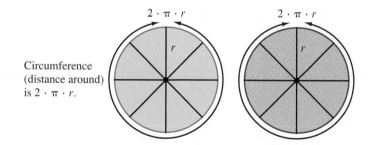

Circumference (distance around) is $2 \cdot \pi \cdot r$.

Unfold the circles, much as you might "unfold" a peeled orange, and put them together as shown here.

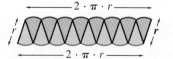

The figure is approximately a rectangle with width r (the radius of the original circle) and length $2 \cdot \pi \cdot r$ (the circumference of the original circle). The area of the "rectangle" is length times width.

$$\text{Area} = l \cdot w$$
$$\text{Area} = 2 \cdot \pi \cdot r \cdot r$$
$$\text{Area} = 2 \cdot \pi \cdot r^2$$

Because the "rectangle" was formed from *two* circles, the area of *one* circle is half as much.

$$\frac{1}{\cancel{2}} \cdot \cancel{2} \cdot \pi \cdot r^2 = 1 \cdot \pi \cdot r^2 \quad \text{or simply} \quad \pi \cdot r^2$$

2. Find the circumference of each circle. Use 3.14 as the approximate value for π. Round answers to the nearest tenth.

(a)

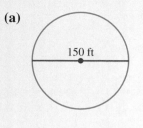

150 ft

(b)

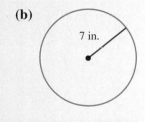

7 in.

(c) diameter 0.9 km

(d) radius 4.6 m

ANSWERS

2. (a) $C \approx 471$ ft (b) $C \approx 44.0$ in.
(c) $C \approx 2.8$ km (d) $C \approx 28.9$ m

3. Find the area of each circle. Use 3.14 for π. Round your answers to the nearest tenth.

(a)

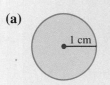

(b)

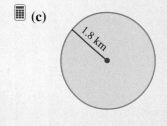

(*Hint:* The diameter is 12 m so $r =$ _____ m)

(c)

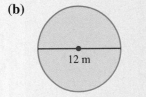

(d)

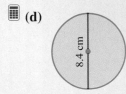

Finding the Area of a Circle

$$\text{Area of a circle} = \pi \cdot \text{radius} \cdot \text{radius}$$
$$A = \pi \cdot r^2$$

Remember to use square units when measuring area.

E X A M P L E 3   Finding the Area of a Circle

Find the area of each circle. Use 3.14 for π. Round your answers to the nearest tenth.

(a) A circle with radius 8.2 cm.

Use the formula $A = \pi \cdot r^2$, which means $\pi \cdot r \cdot r$.

$$A = \pi \cdot r \cdot r$$
$$A \approx 3.14 \cdot 8.2 \text{ cm} \cdot 8.2 \text{ cm}$$
$$A \approx 211.1 \text{ cm}^2 \text{ (square units for area)}$$

(b)

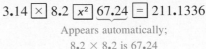

To use the formula, you need to know the radius (r). In this circle, the diameter is 10 ft. First find the radius.

$$r = \frac{d}{2}$$
$$r = \frac{10 \text{ ft}}{2} = 5 \text{ ft}$$

Now find the area.

$$A \approx 3.14 \cdot 5 \text{ ft} \cdot 5 \text{ ft}$$
$$A \approx 78.5 \text{ ft}^2$$

> **Note**
> When finding *circumference,* you can start with either the radius or the diameter. When finding *area,* you must use the *radius.* If you are given the diameter, divide it by 2 to find the radius. Then find the area.

Calculator Tip: You can find the area of the circle in Example 3(a) on your calculator. The first method works on both scientific and standard calculators:

$$3.14 \;\boxed{\times}\; 8.2 \;\boxed{\times}\; 8.2 \;\boxed{=}\; 211.1336$$

You round the answer to 211.1 (nearest tenth).

On a scientific calculator you can also use the $\boxed{x^2}$ key, which automatically squares the number you enter (that is, multiplies the number times itself):

$$3.14 \;\boxed{\times}\; 8.2 \;\boxed{x^2}\; \underline{67.24} \;\boxed{=}\; 211.1336$$

Appears automatically;
8.2 × 8.2 is 67.24

◀◀ **WORK PROBLEM 3 AT THE SIDE.**

ANSWERS

3. **(a)** $\approx 3.1 \text{ cm}^2$ **(b)** $\approx 113.0 \text{ m}^2$
   (c) $\approx 10.2 \text{ km}^2$ **(d)** $\approx 55.4 \text{ cm}^2$

E X A M P L E 4   Finding the Area of a Semicircle

Find the area of the semicircle. Use 3.14 for π. Round your answer to the nearest tenth.

First, find the area of an entire circle with a radius of 12 ft.

$$A = \pi \cdot r \cdot r$$
$$A \approx 3.14 \cdot 12 \text{ ft} \cdot 12 \text{ ft}$$
$$A \approx 452.16 \text{ ft}^2 \qquad \text{(Do not round yet.)}$$

Divide the area of the whole circle by 2 to find the area of the semicircle.

$$\frac{452.16 \text{ ft}^2}{2} = 226.08 \text{ ft}^2$$

The *last* step is rounding 226.08 to the nearest tenth.

$$\text{Area of semicircle} \approx 226.1 \text{ ft}^2$$

WORK PROBLEM 4 AT THE SIDE. ▶▶

E X A M P L E 5   Applying the Concept of Circumference

A circular rug is 8 feet in diameter. The cost of fringe for the edge is $2.25 per foot. What will it cost to add fringe to the rug? Use 3.14 for π.

$$\text{Circumference} = \pi \cdot d$$
$$C \approx 3.14 \cdot 8 \text{ ft}$$
$$C \approx 25.12 \text{ ft}$$

$$\text{Cost} = \text{Cost per foot} \cdot \text{Circumference}$$
$$\text{Cost} = \frac{\$2.25}{1 \text{ ft}} \cdot \frac{25.12 \text{ ft}}{1}$$
$$\text{Cost} = \$56.52$$

WORK PROBLEM 5 AT THE SIDE. ▶▶

E X A M P L E 6   Applying the Concept of Area

Find the cost of covering the rug in Example 5 with a plastic cover. The material for the cover costs $1.50 per square foot. Use 3.14 for π.

First find the radius.

$$r = \frac{d}{2} = \frac{8 \text{ ft}}{2} = 4 \text{ ft}$$

Then,
$$A = \pi \cdot r^2$$
$$A \approx 3.14 \cdot 4 \text{ ft} \cdot 4 \text{ ft}$$
$$A \approx 50.24 \text{ ft}^2$$

$$\text{Cost} = \frac{\$1.50}{1 \text{ ft}^2} \cdot \frac{50.24 \text{ ft}^2}{1} = \$75.36$$

WORK PROBLEM 6 AT THE SIDE. ▶▶

---

4. Find the area of each semicircle. Use 3.14 for π. Round your answers to the nearest tenth.

(a)
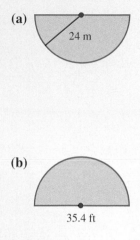
24 m

(b)
35.4 ft

(c)
9.8 m

5. Find the cost of binding around the edge of a circular rug that is 3 meters in diameter. The binder charges $4.50 per meter. Use 3.14 for π.

6. Find the cost of covering the rug in problem 5 above with a non-slip rubber backing. The rubber backing costs $2 per square meter.

ANSWERS

4. (a) $\approx 904.3 \text{ m}^2$ **(b)** $\approx 491.9 \text{ ft}^2$
    (c) $\approx 150.8 \text{ m}^2$
5. $42.39
6. $14.13

7. (a) Here are some more prefixes you have seen in this textbook. List at least one math term and one non-mathematical word that use each prefix.

dia (through):

fract (break):

par (beside):

per (divide):

peri (around):

rad (ray):

rect (right):

sub (below):

(b) How could you use your knowledge of prefixes to remember the difference between perimeter and area?

Objective 4 Many English words are built from Latin or Greek root words and prefixes. Knowing the meaning of the more common ones can help you figure out the meaning of terms in many subject areas, including math.

E X A M P L E   7   Using Prefixes to Understand Math Terms

(a) Listed below are some Latin and Greek root words and prefixes with their meanings in parentheses. You've already seen math terms in this textbook that use these prefixes. List at least one math term and one non-mathematical word that use each prefix or root word.

cent (100): ***cent*imeter; *cent*ury**
circum (around): ***circum*ference; *circum*vent**
de (down): ***de*nominator; *de*duction**
dec (10): ***dec*imal; *Dec*ember** (originally the 10th month in the old calendar)

There are many answers. These are some of the possibilities.

(b) Suppose you have trouble remembering which part of a fraction is the denominator. How could your knowledge of prefixes help in this situation?

The *de* prefix in *de*nominator means down so the denominator is the number down below the fraction bar.

◄◄ WORK PROBLEM 7 AT THE SIDE.

Note
Here are some additional prefixes and root words and their meanings that you will see in the rest of chapter 8, in chapter 9, and in other math classes. An example of a math term and a non-mathematical word are shown for each one.

equ (equal): ***equ*ation; *equ*inox**   *lateral* (side): **quadri*lateral*; bi*lateral***

hemi (half): ***hemi*sphere; *hemi*trope**   *re* (back or again): ***re*ciprocal; *re*duce***

Answers

7. (a) Some possibilities are:
diameter; diagonal
fraction; fracture
parallel; paramedic
percent; per capita
perimeter; periscope
radius; radiate
rectangle; rectify
subtract; submarine
(b) *Peri* in perimeter means around, so perimeter is the distance around the edges of a shape.

8.6 Exercises

Find the missing value in each circle.

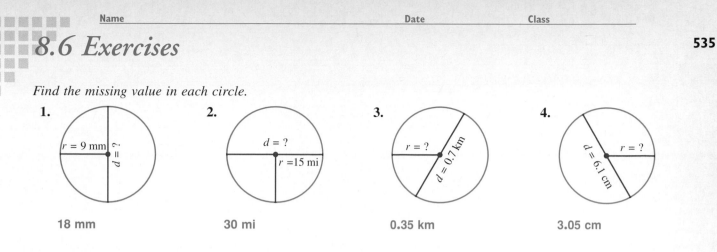

1. $r = 9\text{ mm}$ $d = ?$

   18 mm

2. $d = ?$ $r = 15\text{ mi}$

   30 mi

3. $r = ?$ $d = 0.7\text{ km}$

   0.35 km

4. $d = 6.1\text{ cm}$ $r = ?$

   3.05 cm

Find the circumference and area of each circle. Use 3.14 as the approximate value for π. Round your answers to the nearest tenth.

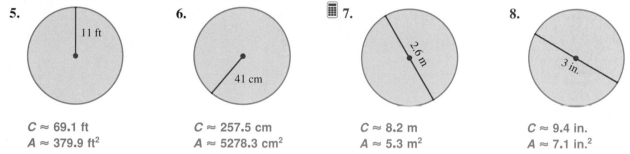

5. 11 ft

   $C \approx 69.1\text{ ft}$
   $A \approx 379.9\text{ ft}^2$

6. 41 cm

   $C \approx 257.5\text{ cm}$
   $A \approx 5278.3\text{ cm}^2$

▦ 7. 2.6 m

   $C \approx 8.2\text{ m}$
   $A \approx 5.3\text{ m}^2$

8. 3 in.

   $C \approx 9.4\text{ in.}$
   $A \approx 7.1\text{ in.}^2$

Find the circumference and area of circles having the following diameters. Use 3.14 for π. Round your answers to the nearest tenth.

9. $d = 15\text{ cm}$

   $C \approx 47.1\text{ cm}$
   $A \approx 176.6\text{ cm}^2$

10. $d = 39\text{ ft}$

   $C \approx 122.5\text{ ft}$
   $A \approx 1194.0\text{ ft}^2$

11. $d = 7\frac{1}{2}\text{ ft}$

   $C \approx 23.6\text{ ft}$
   $A \approx 44.2\text{ ft}^2$

12. $d = 4\frac{1}{2}\text{ yd}$

   $C \approx 14.1\text{ yd}$
   $A \approx 15.9\text{ yd}^2$

▦ 13. $d = 8.65\text{ km}$

   $C \approx 27.2\text{ km}$
   $A \approx 58.7\text{ km}^2$

14. $d = 19.5\text{ mm}$

   $C \approx 61.2\text{ mm}$
   $A \approx 298.5\text{ mm}^2$

Solve each application problem.

15. How far does a point on the tread of a tire move in one turn, if the diameter of the tire is 70 cm?

    ≈219.8 cm

16. If you swing a ball held at the end of a string 2 m long, how far will the ball travel on each turn?

    ≈12.6 m

17. A wave energy extraction device is a huge undersea dome used to harness the power of ocean waves. The base of the dome is 250 ft in diameter. Find its circumference.

    ≈785.0 ft

18. Find the area of the base of the dome in Exercise 17.

    ≈49,062.5 ft²

✎ Writing   ◉ Conceptual   ▲ Challenging   ≈ Estimation

Use the table below to solve Exercises 19–24.

Find the "best buy" for each type of pizza. The "best buy" is the lowest cost per square inch of pizza. All the pizzas are circular in shape, and the measurement given on the menu board is the diameter of the pizza in inches. Use 3.14 as the approximate value of π. Round the area to the nearest tenth. Round cost per square inch to the nearest thousandth.

PIZZA MENU	Small $7\frac{1}{2}$"	Medium 13"	Large 16"
Cheese only	$2.80	$6.50	$9.30
"The Works"	$3.70	$8.95	$14.30
Deep-dish combo	$4.35	$10.95	$15.65

19. Find the area of a small pizza.

≈44.2 in.²

20. Find the area of a medium pizza.

≈132.7 in.²

21. Find the area of a large pizza.

≈201.0 in.²

22. What is the cost per square inch for each size of cheese pizza? Which size is the "best buy"?

small ≈ $0.063
medium ≈ $0.049
large ≈ $0.046   Best Buy

23. What is the cost per square inch for each size of "the works" pizza? Which size is the "best buy"?

small ≈ $0.084
medium ≈ $0.067   Best Buy
large ≈ $0.071

24. What is the cost per square inch for each size of deep dish combo pizza? Which size is the "best buy"?

small ≈ $0.098
medium ≈ $0.083
large ≈ $0.078   Best Buy

25. How would you explain π to a friend who is not in your math class? Write an explanation. Then make up a test question which requires the use of π, and show how to solve it.

π is the ratio of circumference of a circle to its diameter. If you divide the circumference of any circle by its diameter, the answer is always a little more than 3. The approximate value is 3.14 which we call π (pi). Your test question could involve finding the circumference or the area of a circle.

26. Explain how circumference and perimeter are alike. How are they different? Make up two problems, one involving perimeter, the other circumference. Show how to solve your problems.

Circumference and perimeter are both the distance around a shape and are both measured in linear units like ft, yd, cm, or m. However, circumference applies *only* to circles. Perimeter can apply to many shapes such as squares, rectangles, triangles, and so on. Your circumference problem should use the formula $C = 2\pi r$ or $C = \pi d$. To find perimeter, add up the lengths of the sides of the shape.

*Find each shaded area in Exercises 27–30. Use 3.14 as the approximate value of π. Round
your answers to the nearest tenth, if necessary.*

Example:

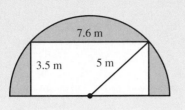

7.6 m

3.5 m   5 m

Solution:

First, find the area of the entire circle.

$$A = \pi \cdot r^2 \approx 3.14 \cdot 5 \text{ m} \cdot 5 \text{ m} \approx 78.5 \text{ m}^2$$

Next, find the area of the semicircle.

$$\frac{78.5 \text{ m}^2}{2} = 39.25 \text{ m}^2$$

Now, find the area of the white rectangle.

$$3.5 \text{ m} \cdot 7.6 \text{ m} = 26.6 \text{ m}^2$$

Finally, subtract to find the shaded area.

$$39.25 \text{ m}^2 - 26.6 \text{ m}^2 \approx \textbf{12.7 m}^2 \quad \text{(Rounded)}$$

27.

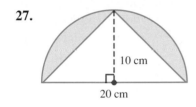

10 cm

20 cm

≈57 cm²

28.

8 ft

≈13.8 ft²

29.

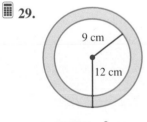

9 cm

12 cm

≈197.8 cm²

30.

20 in.

18 in.

≈232.8 in.²

Solve each application problem.

31. A radio station can be heard 150 miles in all
directions during evening hours. How many
square miles are in the station's broadcast
area?

≈70,650 mi²

32. An earthquake was felt by people 900 km
away in all directions from the epicenter (the
source of the earthquake). How much area was
affected by the quake?

≈2,543,400 km²

▲ 33. The circumference of a circular swimming
pool is 22 meters. What is the radius of the
pool, to the nearest tenth of a meter?

≈3.5 m

▲ 34. A forest ranger measured 56 ft around the
base of a giant sequoia tree. The ranger
wanted to find the diameter of the tree without
cutting it down. Find the diameter. Round
your answer to the nearest tenth.

≈17.8 ft

▲ **35.** Find the cost of sod, at $1.76 per square foot, for the following playing field.

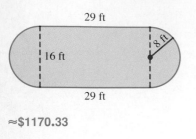

≈$1170.33

▲ **36.** Find the area of this skating rink.

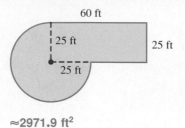

≈2971.9 ft²

📝 **37.** Explain how you could use the information
⊙ about prefixes on page 534 to remember the difference between radius, diameter, and circumference.

The *rad* prefix tells you that radius is a ray from the center of the circle. The *dia* prefix means the diameter goes through the circle, and *circum* means the circumference is the distance around.

📝 **38.** Explain how you could use the information
⊙ about prefixes on page 534 to avoid confusion between parallel and perpendicular lines.

The *par* prefix means beside, so parallel lines are beside each other. (Perpendicular lines cross so that they form a right angle.)

Review and Prepare

*Write each fraction or mixed number as a decimal. Round to the nearest thousandth, if necessary. (For help, see **Section 4.7**.)*

39. $\frac{1}{2}$ 0.5

40. $\frac{1}{3}$ ≈0.333

41. $2\frac{3}{4}$ 2.75

42. $4\frac{1}{4}$ 4.25

43. $\frac{2}{3}$ ≈0.667

44. $\frac{4}{3}$ ≈1.333

45. $10\frac{1}{2}$ 10.5

46. $18\frac{3}{4}$ 18.75

8.7 Volume

OBJECTIVE ▶1▶ A shoe box and a cereal box are examples of three-dimensional (or solid) figures. The three dimensions are length, width, and height. (A rectangle or square is a two-dimensional figure. The two dimensions are length and width.) If you want to know how much the shoe box will hold, you find its **volume.** We measure volume by seeing how many cubes of a certain size will fill the space inside the box. Three sizes of *cubic units* are shown here.

OBJECTIVES

Find the volume of a

▶1▶ rectangular solid;

▶2▶ sphere;

▶3▶ cylinder;

▶4▶ cone and pyramid.

FOR EXTRA HELP

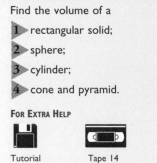

Tutorial   Tape 14   SSM, Sec. 8.7

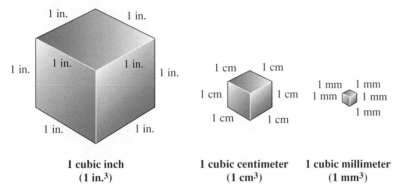

| 1 cubic inch (1 in.³) | 1 cubic centimeter (1 cm³) | 1 cubic millimeter (1 mm³) |

Notice that the edges of a cube all have the same length. Some other sizes of cubes that are used to measure volume are 1 cubic foot (1 ft^3), 1 cubic yard (1 yd^3), and 1 cubic meter (1 m^3).

> **Note**
> The raised 3 in 4^3 means that you multiply $4 \cdot 4 \cdot 4$ to get 64. The raised 3 in cm^3 or ft^3 is a short way to write the word "cubic." When you see 5 cm^3, say "five cubic centimeters." Do *not* multiply $5 \cdot 5 \cdot 5$.

Use the following formula for finding the volume of rectangular solids (box-like shapes).

Finding the Volume of Box-Like Shapes

Volume of rectangular solid = length • width • height

$$V = l \cdot w \cdot h$$

Remember to use cubic units when measuring volume.

E X A M P L E 1   Finding the Volume of a Rectangular Solid

Find the volume of each box.

(a)

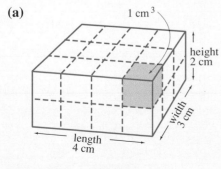

Each cube that fits in the box is 1 cubic centimeter (1 cm^3). To find the volume, you can count the number of cubes.

bottom layer has 12 cubes

top layer has 12 cubes

} total of 24 cubes (24 cm^3)

── **CONTINUED ON NEXT PAGE**

1. Find the volume of each box. Round your answers to the nearest tenth, if necessary.

(a)

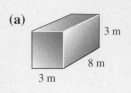

3 m
8 m
3 m

Or you can use the formula for rectangular solids.

$$V = l \cdot w \cdot h$$
$$V = 4 \text{ cm} \cdot 3 \text{ cm} \cdot 2 \text{ cm}$$
$$V = 24 \text{ cm}^3 \quad \text{Cubic units for volume}$$

(b)

10 in.

7 in.

$2\frac{1}{2}$ in.

Use the formula.

$$V = 7 \text{ in.} \cdot 2\frac{1}{2} \text{ in.} \cdot 10 \text{ in.}$$

$$V = \frac{7 \text{ in.}}{1} \cdot \frac{5 \text{ in.}}{\underset{1}{2}} \cdot \frac{\overset{5}{\cancel{10}} \text{ in.}}{1} = 175 \text{ in.}^3$$

If you like, use **2.5** inches, the decimal equivalent of $2\frac{1}{2}$ inches, for the width.

$$V = 7 \text{ in.} \cdot \textbf{2.5 in.} \cdot 10 \text{ in.} = 175 \text{ in.}^3$$

◄◄ **WORK PROBLEM I AT THE SIDE.**

🖩 **(b)** 23.4 cm

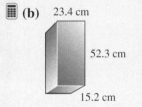

52.3 cm

15.2 cm

OBJECTIVE 2 A *sphere* is shown here. Examples of spheres include baseballs, oranges, and the earth. (The last two aren't perfect spheres, but they're close.)

r

As with circles, the *radius* of a sphere is the distance from the center to the edge of the sphere. Use the following formula to find the volume of a sphere.

> **Finding the Volume of a Sphere**
>
> $$\text{Volume of sphere} = \frac{4}{3} \cdot \pi \cdot r \cdot r \cdot r$$
>
> $$V = \frac{4}{3} \cdot \pi \cdot r^3 \quad \text{or} \quad \frac{4 \cdot \pi \cdot r^3}{3}$$
>
> Remember to use cubic units when measuring volume.

🖩 **(c)** length $6\frac{1}{4}$ ft, width $3\frac{1}{2}$ ft, height 2 ft

EXAMPLE 2 Finding the Volume of a Sphere

Find the volume of each sphere with the help of a calculator. Use **3.14** as the approximate value of π. Round your answers to the nearest tenth.

(a)

9 m

$$V = \frac{4}{3} \cdot \pi \cdot r^3$$

$$V \approx \frac{4 \cdot 3.14 \cdot 9 \text{ m} \cdot 9 \text{ m} \cdot 9 \text{ m}}{3}$$

$$V \approx 3052.08 \quad \text{Now round to tenths.}$$

$$V \approx 3052.1 \text{ m}^3$$

ANSWERS

1. **(a)** 72 m³ **(b)** ≈18,602.1 cm³

   (c) $43\frac{3}{4}$ ft³ or ≈43.8 ft³

CONTINUED ON NEXT PAGE

(b)

$$V \approx \frac{4 \cdot 3.14 \cdot 4.2 \text{ ft} \cdot 4.2 \text{ ft} \cdot 4.2 \text{ ft}}{3}$$

$$V \approx 310.18176 \qquad \text{Now round to tenths.}$$

$$V \approx 310.2 \text{ ft}^3$$

Calculator Tip: You can find the volume of the sphere in Example 2(b) on your calculator. The first method works on both scientific and standard calculators:

$$4 \boxed{\times} 3.14 \boxed{\times} 4.2 \boxed{\times} 4.2 \boxed{\times} 4.2 \boxed{\div} 3 \boxed{=} 310.18176$$

Round the answer to 310.2 ft³.

On a scientific calculator you can use the $\boxed{y^x}$ key to calculate r^3 (to multiply the radius times itself three times).

$$4 \boxed{\times} 3.14 \boxed{\times} \underbrace{4.2 \boxed{y^x} 3}_{r^3} \boxed{\div} 3 \boxed{=} 310.18176$$

Recall that we are using 3.14 as the approximate value for π instead of using the $\boxed{\pi}$ key.

You can also use the $\boxed{y^x}$ key with other exponents. For example:

To find 2^5 press $2 \boxed{y^x} 5 \boxed{=}$ Answer is 32.
To find 6^4 press $6 \boxed{y^x} 4 \boxed{=}$ Answer is 1296.

WORK PROBLEM 2 AT THE SIDE. ▶▶

Half a sphere is called a *hemisphere*. The volume of a hemisphere is *half* the volume of a sphere. Use the following formula to find the volume of a hemisphere.

Finding the Volume of a Hemisphere

$$\text{Volume of hemisphere} = \frac{1}{\overset{1}{\cancel{2}}} \cdot \frac{\overset{2}{\cancel{4}}}{3} \cdot \pi \cdot r \cdot r \cdot r$$

$$V = \frac{2}{3} \cdot \pi \cdot r^3 \quad \text{or} \quad \frac{2 \cdot \pi \cdot r^3}{3}$$

Remember to use cubic units when measuring volume.

E X A M P L E   3   Finding the Volume of a Hemisphere

Find the volume of the hemisphere with the help of a calculator. Use 3.14 for π. Round your answer to the nearest tenth.

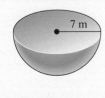

$$V = \frac{2 \cdot \pi \cdot r^3}{3}$$

$$V \approx \frac{2 \cdot 3.14 \cdot 7 \text{ m} \cdot 7 \text{ m} \cdot 7 \text{ m}}{3}$$

$$V \approx 718.0 \text{ m}^3 \qquad \text{(Rounded to nearest tenth)}$$

WORK PROBLEM 3 AT THE SIDE. ▶▶

---

2. Find the volume of each sphere. Use 3.14 for π. Round your answers to the nearest tenth.

(a)

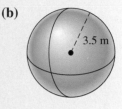

12 in.

(b)

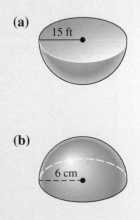

3.5 m

(c) radius 2.7 cm

3. Find the volume of each hemisphere. Use 3.14 for π. Round your answers to the nearest tenth.

(a)

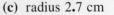

15 ft

(b)

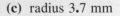

6 cm

(c) radius 3.7 mm

ANSWERS

2. (a) ≈7234.6 in.³ (b) ≈179.5 m³
   (c) ≈82.4 cm³
3. (a) ≈7065 ft³ (b) ≈452.2 cm³
   (c) ≈106.0 mm³

4. Find the volume of each cylinder. Use 3.14 for π. Round your answers to the nearest tenth. (A calculator is helpful on these problems.)

(a)

12 ft — 4 ft

(b)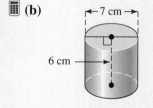

←7 cm→

6 cm

(c) radius 14.5 yd, height 3.2 yd

OBJECTIVE 3 Several *cylinders* are shown here.

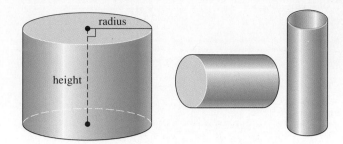

radius

height

These are called *right circular cylinders* because the top and bottom are circles, and the side makes a right angle with the top and bottom. Examples of cylinders are a soup can, a home water heater, and a piece of pipe.

Use the following formula to find the volume of a cylinder. Notice that the first part of the formula, $\pi \cdot r \cdot r$, is the area of the circular base.

> **Finding the Volume of a Cylinder**
>
> $$\text{Volume of cylinder} = \pi \cdot r \cdot r \cdot h$$
> $$V = \pi \cdot r^2 \cdot h$$
>
> Remember to use cubic units when measuring volume.

EXAMPLE 4 Finding the Volume of a Cylinder

Find the volume of each cylinder. Use 3.14 as the approximate value of π. Round your answers to the nearest tenth, if necessary.

(a)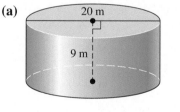

20 m

9 m

The diameter is 20 m so the radius is $\frac{20\ m}{2} = 10$ m. The height is 9 m. Use the formula to find the volume.

$$V = \pi \cdot r \cdot r \cdot h$$
$$V \approx 3.14 \cdot 10\ m \cdot 10\ m \cdot 9\ m$$
$$V \approx 2826\ m^3$$

(b)

6.2 cm

38.4 cm

$$V \approx 3.14 \cdot 6.2\ cm \cdot 6.2\ cm \cdot 38.4\ cm$$
$$V \approx 4634.94144 \quad \text{Now round to tenths.}$$
$$V \approx 4634.9\ cm^3$$

◀◀ **WORK PROBLEM 4 AT THE SIDE.**

OBJECTIVE 4 A cone and a pyramid are shown here.

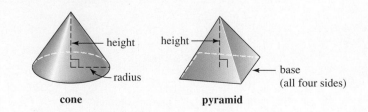

height — radius

cone

height — base (all four sides)

pyramid

ANSWERS

4. (a) ≈602.9 ft³ **(b)** ≈230.8 cm³
(c) ≈2112.6 yd³

Use the following formula to find the volume of a cone.

Finding the Volume of a Cone

$$\text{Volume of cone} = \frac{1}{3} \cdot B \cdot h$$

$$\text{or} \quad V = \frac{B \cdot h}{3}$$

where B is the area of the circular base of the cone and h is the height of the cone. Remember to use cubic units when measuring volume.

E X A M P L E 5   Finding the Volume of a Cone

Find the volume of the cone. Use 3.14 for π. Round your answer to the nearest tenth.

First find the area of the circular base. Recall that the formula for the area of a circle is πr^2.

$$B = \pi \cdot r \cdot r$$
$$B \approx 3.14 \cdot 4 \text{ cm} \cdot 4 \text{ cm}$$
$$B \approx 50.24 \text{ cm}^2 \quad \text{Do not round to tenths yet.}$$

Next, find the volume. The height is 9 cm.

$$V = \frac{B \cdot h}{3}$$
$$V \approx \frac{50.24 \text{ cm}^2 \cdot 9 \text{ cm}}{3}$$
$$V \approx 150.72 \text{ cm}^3 \quad \text{Now round to tenths.}$$
$$V \approx 150.7 \text{ cm}^3$$

WORK PROBLEM 5 AT THE SIDE. ▶▶

5. Find the volume of a cone with base radius 2 ft and height 11 ft. Use 3.14 for π. Round your answer to the nearest tenth.

ANSWERS

5. ≈ 46.1 ft³

6. Find the volume of a pyramid with base 10 m by 10 m and height 8 m. Round your answer to the nearest tenth.

Use the same formula to find the volume of a pyramid as you did to find the volume of a cone.

> **Finding the Volume of a Pyramid**
>
> $$\text{Volume of pyramid} = \frac{1}{3} \cdot B \cdot h$$
>
> $$\text{or} \quad V = \frac{B \cdot h}{3}$$
>
> where B is the area of the square or rectangular base of the pyramid and h is the height of the pyramid. Remember to use cubic units when measuring volume.

E X A M P L E 6   Finding the Volume of a Pyramid

Find the volume of the pyramid. Round your answer to the nearest tenth.

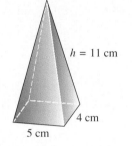

First find the area of the rectangular base by multiplying its length times its width.

$$B = 5 \text{ cm} \cdot 4 \text{ cm}$$

$$B = 20 \text{ cm}^2$$

Next, find the volume.

$$V = \frac{B \cdot h}{3}$$

$$V = \frac{20 \text{ cm}^2 \cdot 11 \text{ cm}}{3}$$

$$V \approx 73.3 \text{ cm}^3 \qquad \text{(Rounded to nearest tenth)}$$

◀◀ **WORK PROBLEM 6 AT THE SIDE.**

ANSWERS

6. $\approx 266.7 \text{ m}^3$

8.7 Exercises

Find the volume of each figure. Use 3.14 as the approximate value of π. Round your answers to the nearest tenth, if necessary.

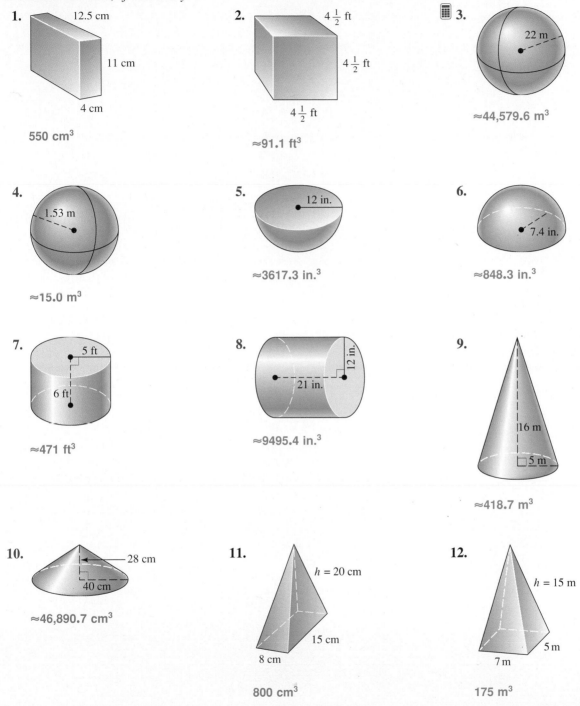

1.
12.5 cm
11 cm
4 cm

550 cm³

2.
4 ½ ft
4 ½ ft
4 ½ ft

≈91.1 ft³

▦ 3.
22 m

≈44,579.6 m³

4.
1.53 m

≈15.0 m³

5.
12 in.

≈3617.3 in.³

6.
7.4 in.

≈848.3 in.³

7.
5 ft
6 ft

≈471 ft³

8.
12 in.
21 in.

≈9495.4 in.³

9.
16 m
5 m

≈418.7 m³

10.
28 cm
40 cm

≈46,890.7 cm³

11.
h = 20 cm
15 cm
8 cm

800 cm³

12.
h = 15 m
5 m
7 m

175 m³

Solve each application problem. Use 3.14 as the approximate value of π. Round your final answers to the nearest tenth, if necessary.

13. A box to hold pencils measures 3 inches by 8 inches by ¾ inch high. Find the volume of the box.

18 in.³

14. A train is being loaded with shipping crates. Each one is 6 m long, 3.4 m wide, and 2 m high. How much space will each crate take?

40.8 m³

✍ Writing   ◉ Conceptual   ▲ Challenging   ≈ Estimation

15. An oil candle globe made of hand-blown glass has a diameter of 16.8 cm. What is the volume of the globe?

≈2481.5 cm³

16. A metal sphere used as part of a fountain has a diameter of $6\frac{1}{2}$ ft. Find its volume.

≈143.7 ft³

17. A city sewer pipe has a diameter of 5 ft and a length of 200 ft. Find the volume of the pipe.

≈3925 ft³

18. A cylindrical woven basket made by a Northwest Coast tribe is 8 cm high and has a diameter of 11 cm. What is the volume of the basket?

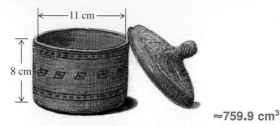

≈759.9 cm³

19. One of the ancient stone pyramids in Egypt has a square base that measures 145 m on each side. The height is 93 m. What is the volume of the pyramid?

651,775 m³

20. An ice cream cone has a diameter of 2 inches and a height of 4 inches. Find its volume.

≈4.2 in.³

21. Explain the two errors made by a student in finding the volume of a cylinder with a diameter of 7 cm and a height of 5 cm. Find the correct answer.

$$V \approx 3.14 \cdot 7 \cdot 7 \cdot 5$$
$$V \approx 769.3 \text{ cm}^2$$

Student used diameter of 7 cm; should use radius of 3.5 cm in formula. Units for volume are cm³ not cm². Correct answer is ≈192.3 cm³.

22. Compare the steps in finding the volume of a cylinder and a cone. How are they similar? Suppose you know the volume of a cylinder. How can you find the volume of a cone with the same radius and height by doing just a one-step calculation?

Both involve finding the area of a circular base and multiplying by the height. To find the volume of the cone, divide the volume of the cylinder by 3.

▲ 23. Find the volume.

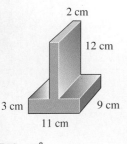

513 cm³

▲ 24. Find the volume of the shaded part. (*Hint:* Notice the hole that goes through the center of the shape.)

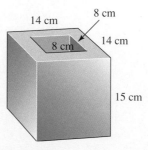

1980 cm³

Review and Prepare

Simplify each expression. (For help, see Section 1.8.)

25. 8^2 26. 14^2 27. $\sqrt{16}$ 28. $\sqrt{144}$ 29. $\sqrt{64}$ 30. $\sqrt{4}$

64   196   4   12   8   2

8.8 *Pythagorean Theorem*

Recall the formula for area of a square, $A = s^2$. The square on the left has an area of 25 cm².

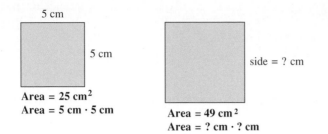

5 cm

5 cm

Area = 25 cm²
Area = 5 cm · 5 cm

side = ? cm

Area = 49 cm²
Area = ? cm · ? cm

The square on the right has an area of 49 cm². To find the length of a side, ask yourself, "What number can be multiplied by itself to give 49?" Because 7 • 7 = 49, the length of each side is 7 cm.

Remember: 7 • 7 = 49, so 7 is the **square root** of 49, or $\sqrt{49} = 7$. Also, $\sqrt{81} = 9$, since 9 • 9 = 81. See **Section 1.8** for further help.

WORK PROBLEM 1 AT THE SIDE. ▶▶

A number that has a whole number as its square root is called a *perfect square.* For example, 9 is a perfect square because $\sqrt{9} = 3$, and 3 is a whole number.

The first few perfect squares are listed here.

$\sqrt{1} = 1$	$\sqrt{16} = 4$	$\sqrt{49} = 7$	$\sqrt{100} = 10$
$\sqrt{4} = 2$	$\sqrt{25} = 5$	$\sqrt{64} = 8$	$\sqrt{121} = 11$
$\sqrt{9} = 3$	$\sqrt{36} = 6$	$\sqrt{81} = 9$	$\sqrt{144} = 12$

OBJECTIVE 1 ▶ If a number is not a perfect square, then you can find its approximate square root by using a calculator with a square root key.

🖩 *Calculator Tip:* To find a square root, use the $\boxed{\sqrt{}}$ key on a standard calculator or the $\boxed{\sqrt{x}}$ key on a scientific calculator. In either case, you do *not* need to use the $\boxed{=}$ key. Try these. Jot down your answers.

To find $\sqrt{16}$ press: 16 $\boxed{\sqrt{x}}$ Answer is 4.
To find $\sqrt{7}$ press: 7 $\boxed{\sqrt{x}}$

For $\sqrt{7}$, your calculator shows 2.645751311 which is an *approximate* answer. We will be rounding to the nearest thousandth so $\sqrt{7} \approx 2.646$. To check, multiply 2.646 times 2.646. Do you get 7 as the result? No, but 7.001316 is very close to 7. The difference is due to rounding.

EXAMPLE 1   Finding the Square Root of a Number

Use a calculator to find each square root. Round your answers to the nearest thousandth.

(a) $\sqrt{35}$ Calculator shows 5.916079783; round to 5.916

(b) $\sqrt{124}$ Calculator shows 11.13552873; round to 11.136

(c) $\sqrt{200}$ Calculator shows 14.14213562; round to 14.142

OBJECTIVES

1 ▶ Find square roots using the square root key on a calculator.

2 ▶ Find the unknown length in a right triangle.

3 ▶ Solve application problems involving right triangles.

FOR EXTRA HELP

Tutorial   Tape 14   SSM, Sec. 8.8

1. Find each square root.

(a) $\sqrt{36}$

(b) $\sqrt{25}$

(c) $\sqrt{9}$

(d) $\sqrt{100}$

(e) $\sqrt{121}$

ANSWERS

1. (a) 6 **(b)** 5 **(c)** 3 **(d)** 10 **(e)** 11

547

2. Use a calculator with a square root key to find each square root. Round to the nearest thousandth if necessary.

(a) $\sqrt{11}$

(b) $\sqrt{40}$

(c) $\sqrt{56}$

(d) $\sqrt{196}$

(e) $\sqrt{147}$

◀◀ **WORK PROBLEM 2 AT THE SIDE.**

OBJECTIVE 2 One place you will use square roots is when working with the *Pythagorean Theorem*. This theorem applies only to *right* triangles (triangles with a 90° angle). The longest side of a right triangle is called the **hypotenuse** (hy-POT-en-oos). It is opposite the right angle. The other two sides are called *legs*. The legs form the right angle.

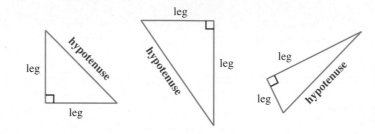

Examples of right triangles

Pythagorean Theorem

$$(\text{hypotenuse})^2 = (\text{leg})^2 + (\text{leg})^2$$

In other words, square the length of each side. After you have squared all the sides, the sum of the squares of the two legs will equal the square of the hypotenuse.

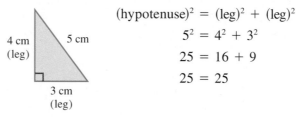

$$(\text{hypotenuse})^2 = (\text{leg})^2 + (\text{leg})^2$$
$$5^2 = 4^2 + 3^2$$
$$25 = 16 + 9$$
$$25 = 25$$

The theorem is named after Pythagoras, a Greek mathematician who lived about 2500 years ago. He and his followers may have used floor tiles to prove the theorem, as shown here.

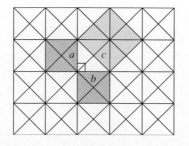

The right triangle in the center of the drawing has sides a, b, and c. The square drawn on side a contains four triangles. The square on side b contains four triangles. The square on side c contains eight triangles. The number of triangles in the square on side c equals the sum of the number of triangles in the squares on sides a and b, that is, 8 triangles = 4 triangles + 4 triangles. As a result, you often see the Pythagorean Theorem written as $c^2 = a^2 + b^2$.

ANSWERS

2. (a) ≈3.317 (b) ≈6.325
   (c) ≈7.483 (d) 14
   (e) ≈12.124

If you know the lengths of any two sides in a right triangle, you can use the Pythagorean Theorem to find the length of the third side.

Using the Pythagorean Theorem

To find the hypotenuse, use this formula:

$$\text{hypotenuse} = \sqrt{(\text{leg})^2 + (\text{leg})^2}$$

To find a leg, use this formula:

$$\text{leg} = \sqrt{(\text{hypotenuse})^2 - (\text{leg})^2}$$

Note

Remember: A small square drawn in one angle of a triangle indicates a right angle. You can use the Pythagorean Theorem *only* on triangles that have a right angle.

E X A M P L E   2   **Finding the Unknown Length in a Right Triangle**

Find the unknown length in each right triangle.

(a)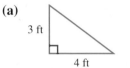

The length of the side opposite the right angle is unknown. That side is the hypotenuse, so use this formula.

$$\text{hypotenuse} = \sqrt{(\text{leg})^2 + (\text{leg})^2} \qquad \text{Find the hypotenuse.}$$
$$\text{hypotenuse} = \sqrt{(3)^2 + (4)^2} \qquad \text{Legs are 3 and 4.}$$
$$= \sqrt{9 + 16} \qquad 3 \cdot 3 \text{ is } 9 \quad \text{and} \quad 4 \cdot 4 \text{ is } 16$$
$$= \sqrt{25}$$
$$= 5$$

The hypotenuse is 5 ft long.

(b)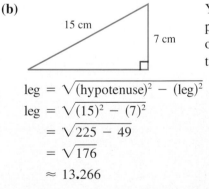

You *do* know the length of the hypotenuse (15 cm), so it is the length of one of the legs that is unknown. Use this formula.

$$\text{leg} = \sqrt{(\text{hypotenuse})^2 - (\text{leg})^2} \qquad \text{Find a leg.}$$
$$\text{leg} = \sqrt{(15)^2 - (7)^2} \qquad \text{Hypotenuse is 15, one leg is 7.}$$
$$= \sqrt{225 - 49} \qquad 15 \cdot 15 \text{ is } 225 \quad \text{and} \quad 7 \cdot 7 \text{ is } 49$$
$$= \sqrt{176} \qquad \text{Use calculator to find } \sqrt{176}.$$
$$\approx 13.266 \qquad \text{Round } 13.26649916 \text{ to } 13.266.$$

The length of the leg is approximately 13.266 cm.

Note

You use the Pythagorean Theorem to find the *length* of one side, *not* the area of the triangle. Your answer will be in linear units, such as ft, yd, cm, m, and so on (*not* ft², cm², m²).

WORK PROBLEM 3 AT THE SIDE. ▶▶

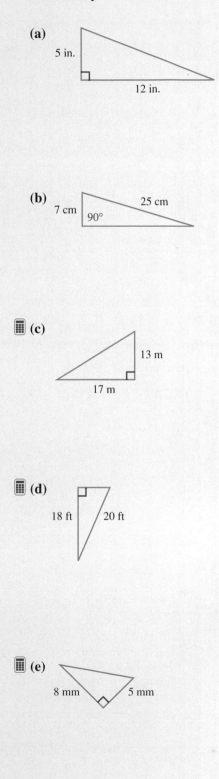

3. Find the unknown length in each right triangle. Round your answers to the nearest thousandth, if necessary.

(a)

5 in.   12 in.

(b)

7 cm   25 cm   90°

(c)

13 m   17 m

(d)

18 ft   20 ft

(e)

8 mm   5 mm

ANSWERS

3. **(a)** 13 in. **(b)** 24 cm
   (c) ≈21.401 m **(d)** ≈8.718 ft
   (e) ≈9.434 mm

4. These problems show ladders leaning against buildings. Find the unknown lengths. Round to the nearest thousandth of a foot, if necessary.

(a)

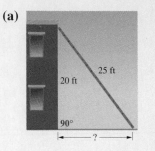

How far away from the building is the bottom of the ladder?

(b)

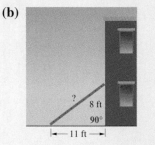

How long is the ladder?

(c) A 17-foot ladder is leaning against a building. The bottom of the ladder is 10 ft from the building. How high up on the building will the ladder reach? (*Hint:* Start by drawing the building and the ladder.)

OBJECTIVE 3 The next example shows an application of the Pythagorean Theorem.

E X A M P L E 3   Using the Pythagorean Theorem

A television antenna is on the roof of a house, as shown. Find the length of the support wire.

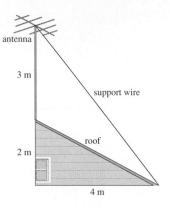

A right triangle is formed. The total length of the side at the left is
3 m + 2 m = 5 m.

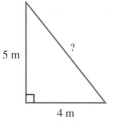

The support wire is the hypotenuse of the right triangle.

$$\text{hypotenuse} = \sqrt{(\text{leg})^2 + (\text{leg})^2}$$ Find the hypotenuse.

$$\text{hypotenuse} = \sqrt{(5)^2 + (4)^2}$$ Legs are 5 and 4.

$$= \sqrt{25 + 16}$$ 5^2 is 25 and 4^2 is 16.

$$= \sqrt{41}$$ Use $\sqrt{x}$ key on a calculator.

$$\approx 6.403$$ Round 6.403124237 to 6.403.

The length of the support wire is ≈ 6.403 m.

◀◀ **WORK PROBLEM 4 AT THE SIDE.**

ANSWERS

4. (a) $\sqrt{225} = 15$ ft
   (b) $\sqrt{185} \approx 13.601$ ft
   (c) $\sqrt{189} \approx 13.748$ ft

8.8 Exercises

Find each square root. Starting with Exercise 5, use the square root key on a calculator.
Round your answers to the nearest thousandth, when necessary.

1. $\sqrt{16}$ 4

2. $\sqrt{4}$ 2

3. $\sqrt{64}$ 8

4. $\sqrt{81}$ 9

5. $\sqrt{11}$ ≈3.317

6. $\sqrt{23}$ ≈4.796

7. $\sqrt{5}$ ≈2.236

8. $\sqrt{2}$ ≈1.414

9. $\sqrt{73}$ ≈8.544

10. $\sqrt{80}$ ≈8.944

11. $\sqrt{101}$ ≈10.050

12. $\sqrt{125}$ ≈11.180

13. $\sqrt{190}$ ≈13.784

14. $\sqrt{160}$ ≈12.649

15. $\sqrt{1000}$ ≈31.623

16. $\sqrt{2000}$ ≈44.721

Find the areas of the squares on the sides of the right triangles in Exercises 17 and 18. Check
to see if the Pythagorean Theorem holds true.

17.

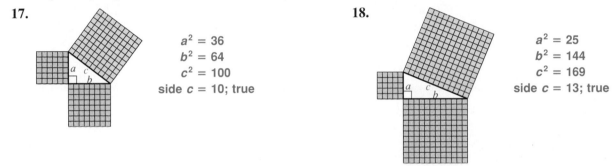

$a^2 = 36$
$b^2 = 64$
$c^2 = 100$
side $c = 10$; true

18.

$a^2 = 25$
$b^2 = 144$
$c^2 = 169$
side $c = 13$; true

Find the unknown length in each right triangle. Use a calculator to find square roots. Round
your answers to the nearest thousandth, if necessary.

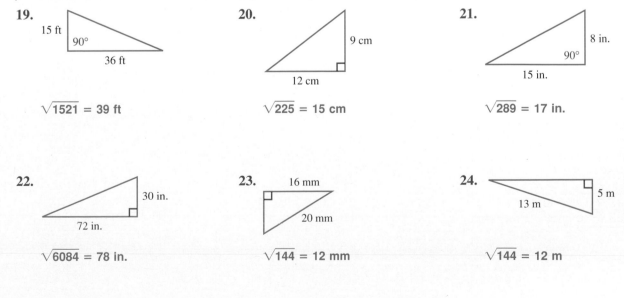

19.

15 ft
90°
36 ft

$\sqrt{1521} = 39$ ft

20.

9 cm
12 cm

$\sqrt{225} = 15$ cm

21.

8 in.
90°
15 in.

$\sqrt{289} = 17$ in.

22.

30 in.
72 in.

$\sqrt{6084} = 78$ in.

23.

16 mm
20 mm

$\sqrt{144} = 12$ mm

24.

5 m
13 m

$\sqrt{144} = 12$ m

✍ Writing   ⊙ Conceptual   ▲ Challenging   ≈ Estimation

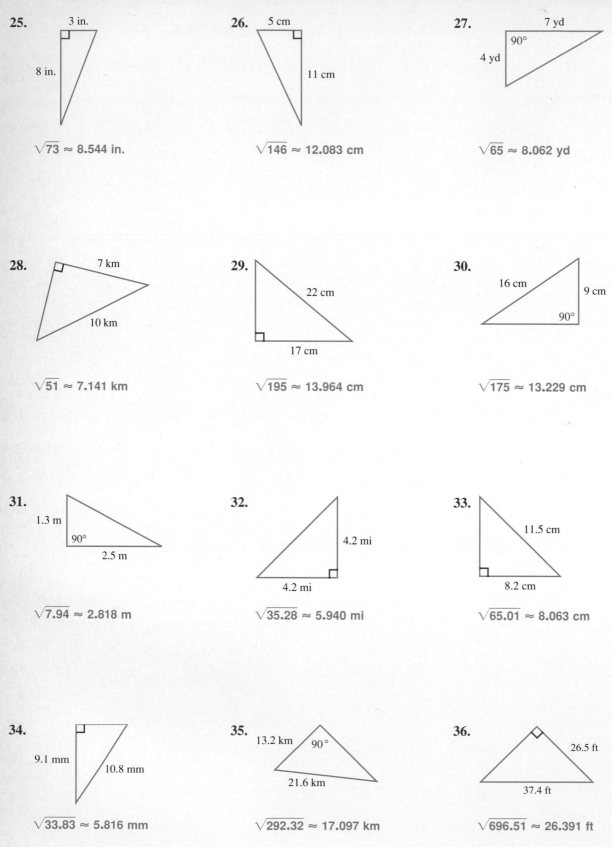

25. 3 in.

8 in.

$\sqrt{73} \approx 8.544$ in.

26. 5 cm

11 cm

$\sqrt{146} \approx 12.083$ cm

27. 7 yd

90°

4 yd

$\sqrt{65} \approx 8.062$ yd

28. 7 km

10 km

$\sqrt{51} \approx 7.141$ km

29. 22 cm

17 cm

$\sqrt{195} \approx 13.964$ cm

30. 16 cm

9 cm

90°

$\sqrt{175} \approx 13.229$ cm

31. 1.3 m

90°

2.5 m

$\sqrt{7.94} \approx 2.818$ m

32. 4.2 mi

4.2 mi

$\sqrt{35.28} \approx 5.940$ mi

33. 11.5 cm

8.2 cm

$\sqrt{65.01} \approx 8.063$ cm

34. 9.1 mm

10.8 mm

$\sqrt{33.83} \approx 5.816$ mm

35. 13.2 km 90°

21.6 km

$\sqrt{292.32} \approx 17.097$ km

36. 26.5 ft

37.4 ft

$\sqrt{696.51} \approx 26.391$ ft

Solve each application problem. Round your answers to the nearest tenth when necessary.

37. Find the length of this loading ramp.

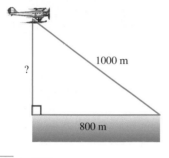

4 ft

?

7 ft

$\sqrt{65} \approx 8.1$ ft

38. Find the unknown length in this roof plan.

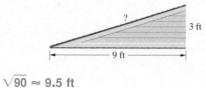

?

3 ft

9 ft

$\sqrt{90} \approx 9.5$ ft

39. How high is the airplane above the ground?

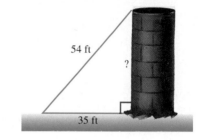

1000 m

?

800 m

$\sqrt{360,000} = 600$ m

40. Find the height of this farm silo.

54 ft

?

35 ft

$\sqrt{1691} \approx 41.1$ ft

41. To reach his lady-love, a knight placed a 12-foot ladder against the castle wall. If the base of the ladder is 3 feet from the building, how high on the castle will the top of the ladder reach? Draw a sketch of the castle and ladder and solve the problem.

ladder 12 ft

building

?

3 ft

$\sqrt{135} \approx 11.6$ ft

42. William drove his car 15 miles north, then made a right turn and drove 7 miles east. How far is he, in a straight line, from his starting point? Draw a sketch to illustrate the problem and solve it.

7 mi

15 mi

?

$\sqrt{274} \approx 16.6$ miles

43. You know that $\sqrt{25} = 5$ and $\sqrt{36} = 6$. Using just that information (no calculator), describe how you could estimate $\sqrt{30}$. How would you estimate $\sqrt{26}$ or $\sqrt{35}$? Now check your estimates using a calculator.

30 is about halfway between 25 and 36 so $\sqrt{30}$ should be about halfway between 5 and 6, or $\approx$5.5. Using a calculator, $\sqrt{30} \approx$ 5.477. Similarly $\sqrt{26}$ should be a little more than $\sqrt{25}$; by calculator it is $\approx$5.099. And $\sqrt{35}$ should be a little less than $\sqrt{36}$; by calculator it is $\approx$5.916.

44. Describe the two errors made by a student in solving this problem. Also find the correct answer. Round to the nearest tenth.

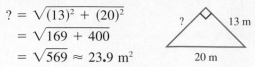

$$? = \sqrt{(13)^2 + (20)^2}$$
$$= \sqrt{169 + 400}$$
$$= \sqrt{569} \approx 23.9 \text{ m}^2$$

The student used the formula for finding the hypotenuse but the unknown side is a leg, so $? = \sqrt{(20)^2 - (13)^2}$. Also, the final answer should be m, not m². Correct answer is $\sqrt{231} \approx$ 15.2 m.

45. Find the lengths of $\overline{BC}$ and $\overline{BD}$.

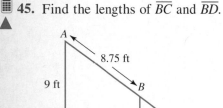

$\overline{BC}$ = 6.25 ft
$\overline{BD}$ = 3.75 ft

46. Find the lengths of $\overline{CD}$ and $\overline{DB}$. Round your answers to the nearest tenth.

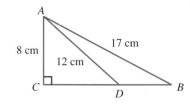

$\overline{CD} \approx$ 8.9 cm
$\overline{BD} \approx$ 6.1 cm

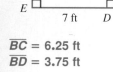
Review and Prepare

*Find the missing number in each of the following proportions. (For help, see **Section 5.4**.)*

47. $\dfrac{2}{9} = \dfrac{x}{36}$ 8

48. $\dfrac{7}{x} = \dfrac{21}{24}$ 8

49. $\dfrac{x}{9.2} = \dfrac{15.6}{7.8}$ 18.4

50. $\dfrac{0.8}{5} = \dfrac{12.4}{x}$ 77.5

8.9 Similar Triangles

Two triangles with the same shape (but not necessarily the same size) are called **similar triangles.** Three pairs of similar triangles are shown here.

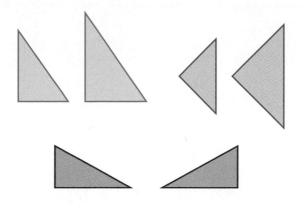

OBJECTIVE 1 The two triangles shown below are different sizes but have the same shape, so they are similar triangles. Angles A and P measure the same number of degrees and are called *corresponding angles.* Angles B and Q are corresponding angles, as are angles C and R.

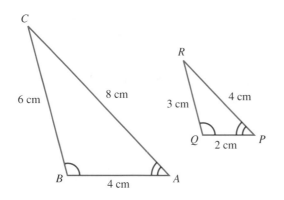

$\overline{PR}$ and $\overline{AC}$ are called *corresponding sides,* since they are *opposite* corresponding angles. Also, $\overline{QR}$ and $\overline{BC}$ are corresponding sides, as are $\overline{PQ}$ and $\overline{AB}$. Although corresponding angles measure the same number of degrees, corresponding sides do *not* need to be the same in length. In the triangles here, each side in the smaller triangle is half the length of the corresponding side in the larger triangle.

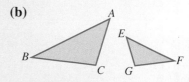
WORK PROBLEM I AT THE SIDE.

OBJECTIVE 2 Similar triangles are useful because of the following property.

Similar Triangles

In similar triangles, the ratios of the lengths of corresponding sides are equal.

OBJECTIVES

1 Identify corresponding parts in similar triangles.

2 Find the unknown lengths of sides in similar triangles.

3 Solve problems with similar triangles.

FOR EXTRA HELP

Tutorial

Tape 15

SSM, Sec. 8.9

1. Identify corresponding angles and sides in these similar triangles.

(a)

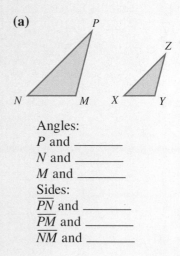

Angles:
P and _____
N and _____
M and _____
Sides:
$\overline{PN}$ and _____
$\overline{PM}$ and _____
$\overline{NM}$ and _____

(b)

Angles:
A and _____
B and _____
C and _____
Sides:
$\overline{AB}$ and _____
$\overline{BC}$ and _____
$\overline{AC}$ and _____

ANSWERS

1. (a) Z; X; Y; $\overline{ZX}$; $\overline{ZY}$; $\overline{XY}$
   (b) F; F; G; $\overline{EF}$; $\overline{GF}$; $\overline{EG}$

555

2. Find the length of $\overline{EF}$ in Example 1.

EXAMPLE 1 Finding the Unknown Lengths of Sides in Similar Triangles

Find the length of y in the smaller triangle. Assume the triangles are similar.

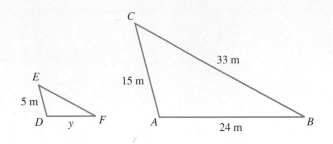

$\overline{ED}$ and $\overline{CA}$ are corresponding sides. The ratio of the lengths of these sides can be written as a fraction in lowest terms.

$$\frac{ED}{CA} = \frac{5 \text{ m}}{15 \text{ m}} = \frac{1}{3} \qquad \text{Lowest terms}$$

As mentioned earlier, the ratios of the lengths of corresponding sides are equal. $\overline{DF}$ in the smaller triangle corresponds to $\overline{AB}$ in the larger triangle. Since the ratios of corresponding sides are equal,

$$\frac{DF}{AB} = \frac{1}{3}$$

Replace DF with y and AB with 24 to get the proportion

$$\frac{y}{24} = \frac{1}{3}.$$

Find cross products.

$$24 \cdot 1 = 24$$

$$\frac{y}{24} = \frac{1}{3}$$

$$y \cdot 3$$

Show that cross products are equivalent.

$$y \cdot 3 = 24$$

Divide both sides by 3.

$$\frac{y \cdot \overset{1}{\cancel{3}}}{\underset{1}{\cancel{3}}} = \frac{24}{3}$$

$$y = 8$$

$\overline{DF}$ has a length of 8 m.

◀◀ **WORK PROBLEM 2 AT THE SIDE.**

EXAMPLE 2 Finding an Unknown Length and the Perimeter

Find the perimeter of the smaller triangle. Assume the triangles are similar.

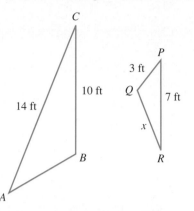

First find the length of $\overline{RQ}$, then add up the sides to find the perimeter. The smaller triangle is turned "upside down" compared to the larger triangle, so be careful when identifying corresponding sides. $\overline{AC}$ is the longest side in the larger triangle, and $\overline{PR}$ is the longest side in the smaller triangle. So $\overline{PR}$ and $\overline{AC}$ are corresponding sides. The ratio of their lengths can be written as a fraction in lowest terms.

$$\frac{7 \text{ ft}}{14 \text{ ft}} = \frac{1}{2} \qquad \text{Lowest terms}$$

The two triangles are similar, so the ratio of any pair of corresponding sides will also equal $\frac{1}{2}$. Because $\overline{RQ}$ and $\overline{CB}$ are corresponding sides,

$$\frac{RQ}{CB} = \frac{1}{2}.$$

Replace RQ with x and CB with 10 to make a proportion.

$$\frac{x}{10} = \frac{1}{2}$$

Find cross products.

$$10 \cdot 1 = 10$$
$$\frac{x}{10} = \frac{1}{2}$$
$$x \cdot 2$$

Show that cross products are equivalent.

$$x \cdot 2 = 10$$

Divide both sides by 2.

$$\frac{x \cdot \overset{1}{\cancel{2}}}{\underset{1}{\cancel{2}}} = \frac{10}{2}$$

$$x = 5$$

$\overline{RQ}$ has a length of 5 ft. Now add the lengths of all three sides to find the perimeter.

$$\text{Perimeter} = 5 \text{ ft} + 3 \text{ ft} + 7 \text{ ft} = 15 \text{ ft}$$

3. (a) Find the perimeter of triangle ABC in Example 2.

(b) Find the perimeter of each triangle. Assume the triangles are similar.

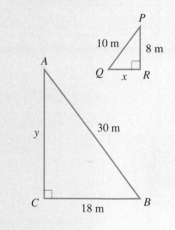

ANSWERS

3. (a) $\overline{AB}$ is 6 ft; perimeter $=$ 14 ft $+$ 10 ft $+$ 6 ft $=$ 30 ft
(b) $x = 6$ m, perimeter $= 24$ m; $y = 24$ m, perimeter $= 72$ m

WORK PROBLEM 3 AT THE SIDE. ▶▶

4. Find the height of each flagpole.

(a)

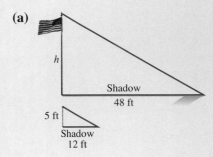

(b)

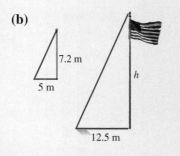

OBJECTIVE ▶ **3** The next example shows an application of similar triangles.

E X A M P L E 3   Using Similar Triangles in Applications

A flagpole casts a shadow 99 m long at the same time that a pole 10 m tall casts a shadow 18 m long. Find the height of the flagpole.

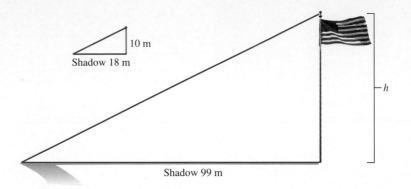

The triangles shown are similar, so write a proportion to find h.

Height in larger triangle → $\dfrac{h}{10} = \dfrac{99}{18}$ ← Shadow in larger triangle

Height in smaller triangle → $\phantom{\dfrac{h}{10} = \dfrac{99}{18}}$ ← Shadow in smaller triangle

Find cross products and show that they are equivalent.

$$h \cdot 18 = 10 \cdot 99$$
$$h \cdot 18 = 990$$

Divide both sides by 18.

$$\frac{h \cdot \overset{1}{\cancel{18}}}{\underset{1}{\cancel{18}}} = \frac{990}{18}$$

$$h = 55$$

The flagpole is 55 m high.

◀◀ **WORK PROBLEM 4 AT THE SIDE.**

8.9 Exercises

Write similar *or* not similar *for each pair of triangles.*

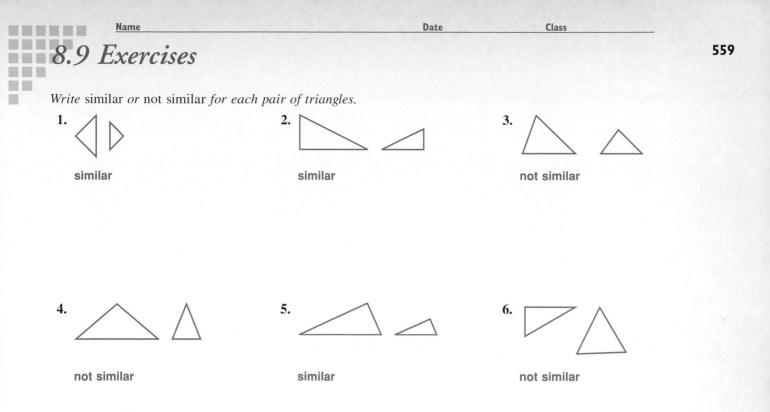

1. similar

2. similar

3. not similar

4. not similar

5. similar

6. not similar

Name the corresponding angles and the corresponding sides in each pair of similar triangles.

7.

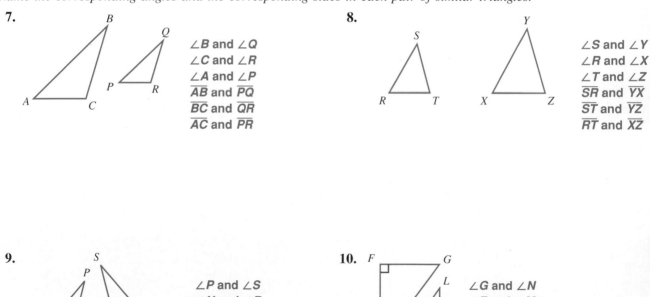

∠B and ∠Q
∠C and ∠R
∠A and ∠P
$\overline{AB}$ and $\overline{PQ}$
$\overline{BC}$ and $\overline{QR}$
$\overline{AC}$ and $\overline{PR}$

8.

∠S and ∠Y
∠R and ∠X
∠T and ∠Z
$\overline{SR}$ and $\overline{YX}$
$\overline{ST}$ and $\overline{YZ}$
$\overline{RT}$ and $\overline{XZ}$

9.

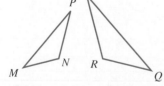

∠P and ∠S
∠N and ∠R
∠M and ∠Q
$\overline{MP}$ and $\overline{QS}$
$\overline{MN}$ and $\overline{QR}$
$\overline{NP}$ and $\overline{RS}$

10.

∠G and ∠N
∠F and ∠M
∠E and ∠L
$\overline{FG}$ and $\overline{NM}$
$\overline{FE}$ and $\overline{LM}$
$\overline{EG}$ and $\overline{LN}$

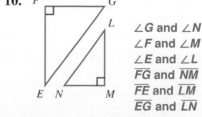

✎ Writing   ◉ Conceptual   ▲ Challenging   ≈ Estimation

Find all the ratios for the triangles shown below. Write the ratios as fractions in lowest terms.

11. $\dfrac{AB}{PQ}; \dfrac{AC}{PR}; \dfrac{BC}{QR}$

$\dfrac{3}{2}; \dfrac{3}{2}; \dfrac{3}{2}$

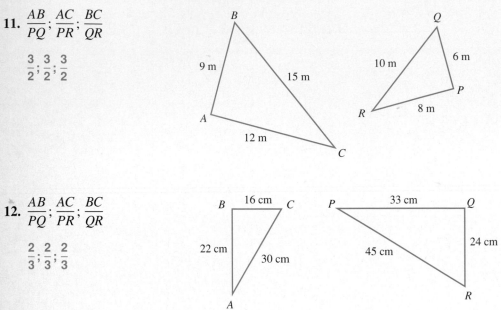

12. $\dfrac{AB}{PQ}; \dfrac{AC}{PR}; \dfrac{BC}{QR}$

$\dfrac{2}{3}; \dfrac{2}{3}; \dfrac{2}{3}$

Find the unknown lengths in each pair of similar triangles.

13.

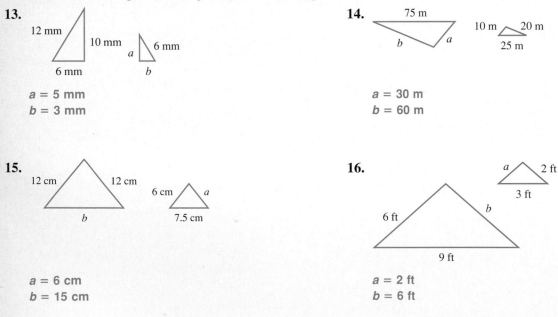

$a = 5$ mm
$b = 3$ mm

14.

$a = 30$ m
$b = 60$ m

15.

$a = 6$ cm
$b = 15$ cm

16.

$a = 2$ ft
$b = 6$ ft

Find the perimeter of each triangle. Assume the triangles are similar.

17.

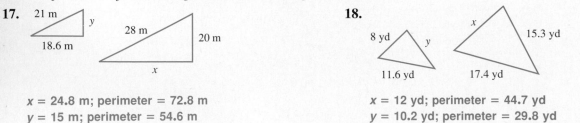

$x = 24.8$ m; perimeter = 72.8 m
$y = 15$ m; perimeter = 54.6 m

18.

$x = 12$ yd; perimeter = 44.7 yd
$y = 10.2$ yd; perimeter = 29.8 yd

Solve the following application problems.

19. The height of the house shown here can be found by comparing its shadow to the shadow cast by a 3 foot stick. Find the height of the house by writing a proportion and solving it.

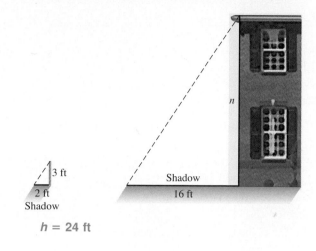

3 ft

2 ft
Shadow

Shadow

16 ft

n

h = 24 ft

20. Refer to the building in Exercise 19. Later in the day, the same building had a shadow 6 ft long. How long would the stick's shadow be at that time?

$\frac{3}{4}$ **ft or 0.75 ft**

21. A sailor on the USS *Ramapo* saw one of the highest waves ever recorded. He used the height of the ship's mast, the width of the deck, and similar triangles to find the height of the wave. Using the information in the figure, write a proportion and then find the height of the wave.

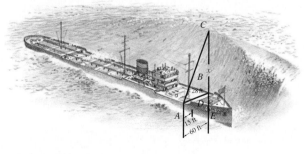

C

B

x

A

E

15 ft

60 ft

x = 112 ft

22. A fire lookout tower provides an excellent view of the surrounding countryside. The height of the tower can be found by lining up the top of the tower with the top of a 2-meter stick. Use similar triangles to find the height of the tower.

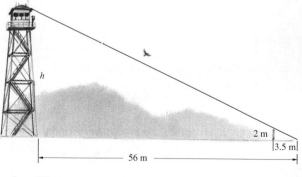

h

2 m

3.5 m

56 m

h = 32 m

▲ **23.** Triangles *CDE* and *FGH* are similar. Find the perimeter and area of triangle *FGH*.

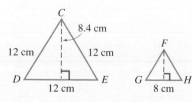

C

8.4 cm

12 cm

12 cm

D

12 cm

E

F

G

8 cm

H

Perimeter = 8 cm + 8 cm + 8 cm = 24 cm
Area = 0.5 • 8 cm • 5.6 cm = 22.4 cm²

▲ **24.** Triangles *JKL* and *MNO* are similar. Find the perimeter and area of triangle *MNO*.

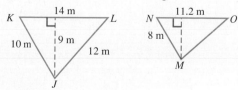

14 m

K

L

10 m

9 m

12 m

J

11.2 m

N

O

8 m

M

Perimeter = 11.2 m + 8 m + 9.6 m = 28.8 m
Area = 0.5 • 11.2 m • 7.2 m = 40.32 m²

25. Look up the word *similar* in a dictionary. What is the non-mathematical definition of this word? Find two examples of similar objects at home or school.

One dictionary definition is, "Resembling, but not identical." Examples of similar objects are sets of different size pots or measuring cups; small and large size cans of beans; child's tennis shoe and adult tennis shoe.

26. *Congruent* objects have the same shape and the same size. Sketch a pair of congruent triangles. Find two examples of congruent objects at home or school.

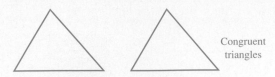

Congruent triangles

Examples of congruent objects include two matching chairs, two contact lenses, and two pieces of notebook paper.

27. Use similar triangles and a proportion to find the length of the lake shown here. (*Hint:* The side 100 m long in the smaller triangle corresponds to a side of 100 + 120 = 220 m in the larger triangle.)

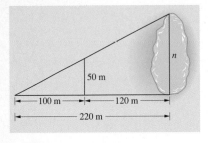

50 m

100 m | 120 m

220 m

n = 110 m

28. To find the height of the tree, find *y* and then add $5\frac{1}{2}$ feet for the distance from the ground to eye level.

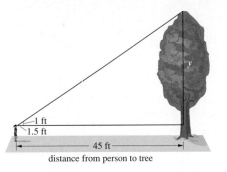

1 ft
1.5 ft

45 ft

distance from person to tree

35.5 ft or $35\frac{1}{2}$ ft

Find the unknown length. Round your answers to the nearest tenth.
Note: When a line is drawn parallel to one side of a triangle, the smaller triangle that is formed will be similar to the original triangle.

29.

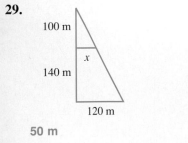

100 m

x

140 m

120 m

50 m

30.

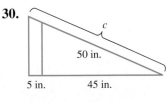

c

50 in.

5 in. | 45 in.

≈55.6 in.

Review and Prepare

*Work each problem by using the order of operations. (For help, see **Section 1.8**.)*

31. $8 \div 4 + 3 \cdot 2$ 8

32. $16 + 4 \div 2$ 18

33. $13 - 2 \cdot 5 + 7$ 10

34. $18 + 7 \cdot 2 - 33 \div 3$ 21

35. $16 - 5 + 2(6 - 4)$ 15

36. $3(2 + 2) \div 12 \cdot 4$ 4

8.1	**point**	A point is a location in space.
	line	A line is a straight row of points that goes on forever in both directions.
	line segment	A line segment is a piece of a line with two endpoints.
	ray	A ray is a part of a line that has one endpoint and extends forever in one direction.
	angle	An angle is made up of two rays that have a common endpoint called the vertex.
	degrees	A system used to measure angles in which a complete circle is 360 degrees, written 360°.
	right angle	A right angle is an angle that measures 90°; it is also called a square angle.
	acute angle	An acute angle is an angle that measures between 0° and 90°.
	obtuse angle	An obtuse angle is an angle that measures between 90° and 180°.
	straight angle	A straight angle is an angle that measures 180°; its sides form a straight line.
	intersecting lines	Intersecting lines cross or merge.
	perpendicular lines	Perpendicular lines are two lines that intersect to form a right angle.
	parallel lines	Parallel lines are two lines in the same plane that never intersect and are equidistant from each other.
8.2	**complementary angles**	Complementary angles are two angles with a sum of 90°.
	supplementary angles	Supplementary angles are two angles with a sum of 180°.
	congruent angles	Congruent angles are angles that measure the same number of degrees.
	vertical angles	Vertical angles are two nonadjacent congruent angles formed by intersecting lines.
8.3, 8.4, 8.5	**perimeter**	Perimeter is the distance around the outside edges of a figure. It is measured in linear units such as ft, yd, cm, m, km, and so on.
8.3, 8.4, 8.5, 8.6	**area**	Area is the surface inside a two-dimensional (flat) figure. It is measured by seeing how many squares of a certain size are needed to cover the surface inside the figure. Some of the commonly used units are square inches (in.^{2}); square feet (ft^2); square yards (yd^2); square centimeters (cm^2); and square meters (m^2).
8.3	**rectangle**	A rectangle is a four-sided figure with all sides meeting at 90° angles.
	square	A square is a rectangle with all four sides the same length.
8.4	**parallelogram**	A parallelogram is a four-sided figure with both pairs of opposite sides parallel.
	trapezoid	A trapezoid is a four-sided figure with one pair of parallel sides.
8.5	**triangle**	A triangle is a figure with exactly three sides.
8.6	**circle**	A circle is a figure with all points the same distance from a fixed center point.
	radius	Radius is the distance from the center of a circle to any point on the circle.
	diameter	Diameter is the distance across a circle, passing through the center.
	circumference	Circumference is the distance around a circle.
	π (pi)	π is the ratio of the circumference to the diameter of any circle. It is approximately equal to 3.14.
8.7	**volume**	Volume is the space inside a three-dimensional (solid) figure.
8.8	**square root**	A square root is one of two equal factors of a number.
	hypotenuse	The hypotenuse is the side of a right triangle opposite the 90° angle; it is the longest side.
8.9	**similar triangles**	Similar triangles are triangles with the same shape but not necessarily the same size; corresponding angles measure the same number of degrees.

QUICK REVIEW

Concepts	Examples

8.1 Lines

If a line has one endpoint, it is a ray. If it has two endpoints, it is a line segment.

Identify each of the following as a line, line segment, or ray.

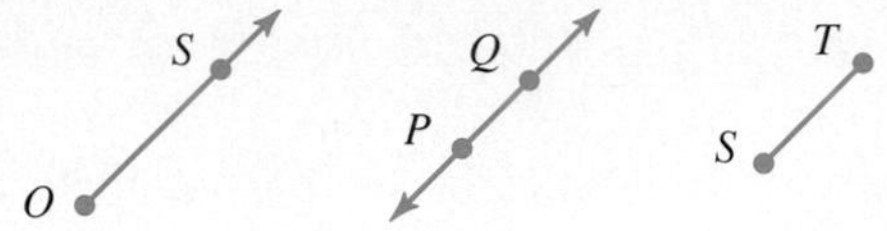

(a) (b) (c)

(a) shows a ray, (b) shows a line, and (c) shows a line segment.

If two lines intersect at right angles, they are perpendicular. If two lines in the same plane never intersect, they are parallel.

Label each pair of lines as parallel or perpendicular.

(a) (b)

(a) shows two perpendicular lines (they intersect at 90°).
(b) shows two parallel lines (they never intersect).

8.2 Angles

If the sum of two angles is 90°, they are complementary.
If the sum of two angles is 180°, they are supplementary.

Find the complement and supplement of a 35° angle.

$$90° - 35° = 55° \text{ (the complement)}$$
$$180° - 35° = 145° \text{ (the supplement)}$$

If two angles measure the same number of degrees, the angles are congruent. The symbol for congruent is ≅.

Two nonadjacent angles formed by intersecting lines are called vertical angles. Vertical angles are congruent.

Identify the congruent and vertical angles in the following figure.

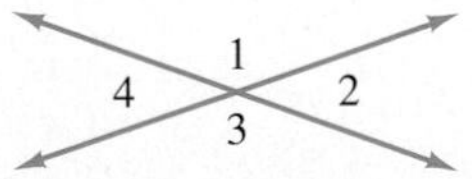

$$\angle 1 \cong \angle 3 \quad \text{and} \quad \angle 2 \cong \angle 4$$

$\angle 1$ and $\angle 3$ are vertical angles.
$\angle 2$ and $\angle 4$ are vertical angles.

[8.1] *Name each line, line segment, or ray.*

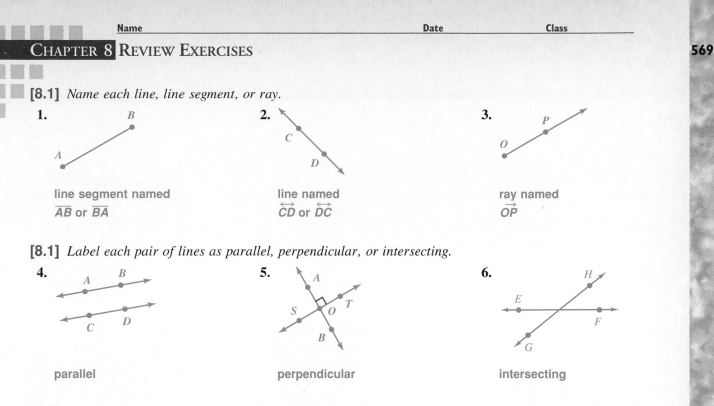

1.

line segment named
$\overline{AB}$ or $\overline{BA}$

2.

line named
$\overleftrightarrow{CD}$ or $\overleftrightarrow{DC}$

3.

ray named
$\overrightarrow{OP}$

[8.1] *Label each pair of lines as parallel, perpendicular, or intersecting.*

4.

parallel

5.

perpendicular

6.

intersecting

[8.1] *Label each angle as an acute, right, obtuse, or straight angle. For right and straight angles, indicate the number of degrees in the angle.*

7.

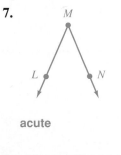

acute

8.

obtuse

9.

straight; 180°

10.

right; 90°

[8.2] *First identify the congruent angles in each figure. Then find all pairs of complementary angles.*

11.

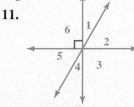

∠1 ≅ ∠4; ∠2 ≅ ∠5; ∠3 ≅ ∠6
∠1 and ∠2; ∠4 and ∠5
∠5 and ∠1; ∠4 and ∠2

12.

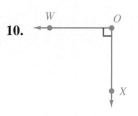

∠4 ≅ ∠7; ∠3 ≅ ∠6
∠1 and ∠2; ∠3 and ∠4
∠6 and ∠7; ∠6 and ∠4
∠7 and ∠3

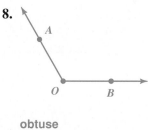

✎ Writing   ◉ Conceptual   ▲ Challenging   ≈ Estimation

[8.2] *Name the pairs of supplementary angles in each of the following.*

13.

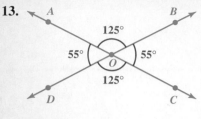

∠AOB and ∠BOC; ∠BOC and ∠COD
∠COD and ∠DOA; ∠DOA and ∠AOB

14.

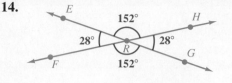

∠ERH and ∠HRG; ∠HRG and ∠GRF
∠FRG and ∠FRE; ∠FRE and ∠ERH

[8.2] *Find the complement or supplement of each angle.*

15. Find the complement of:
  (a) 80°   10°
  (b) 45°   45°
  (c) 7°   83°

16. Find the supplement of:
  (a) 155°   25°
  (b) 90°   90°
  (c) 33°   147°

[8.3] *Find each perimeter.*

17.

1.5 m

0.92 m   0.92 m

1.5 m

4.84 m

18.

32 in.

32 in.

32 in.

128 in.

19. A square-shaped pillow measures 38 cm along each side. How much lace is needed to trim all the edges?

152 cm

20. A rectangular garden plot is $8\frac{1}{2}$ feet wide and 12 feet long. How much fencing is needed to surround the garden?

41 feet

Find the area of each rectangle or square. Round your answers to the nearest tenth when necessary.

21.

27 mm

18 mm   18 mm

27 mm

486 mm²

22.

3 ft

$5\frac{1}{2}$ ft $5\frac{1}{2}$ ft

3 ft

16.5 ft² or $16\frac{1}{2}$ ft²

23.

6.3 m

6.3 m   6.3 m

6.3 m

≈39.7 m²

[8.4] *Find the perimeter and area of each parallelogram or trapezoid. Round your answers to the nearest tenth when necessary.*

24.

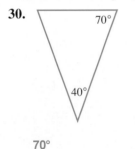

$P = 50$ cm
$A = 140$ cm^2

25.

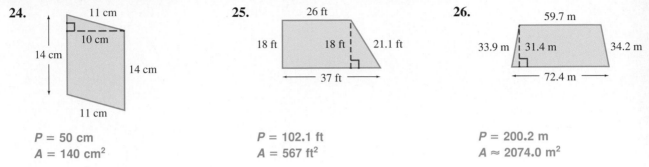

$P = 102.1$ ft
$A = 567$ ft^2

26.

$P = 200.2$ m
$A \approx 2074.0$ m^2

[8.5] *Find the perimeter and area of each triangle.*

27.

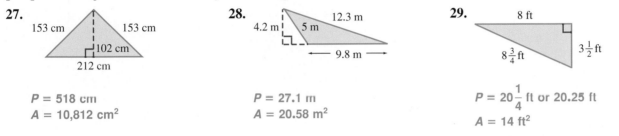

$P = 518$ cm
$A = 10,812$ cm^2

28.

$P = 27.1$ m
$A = 20.58$ m^2

29.

$P = 20\frac{1}{4}$ ft or 20.25 ft
$A = 14$ ft^2

Find the number of degrees in the unlabeled angle.

30.

70°
40°

70°

31.

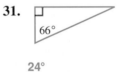

66°

24°

[8.6] *Find the missing value.*

32. The radius of a circular irrigation field is 68.9 m. What is the diameter of the field?

137.8 m

33. The diameter of a juice can is 3 inches. What is the radius of the can?

$1\frac{1}{2}$ in. or 1.5 in.

Find the circumference and area of each circle. Use 3.14 as the approximate value for π. Round your answers to the nearest tenth.

34.

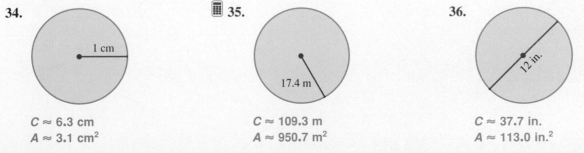

1 cm

$C \approx 6.3$ cm
$A \approx 3.1$ cm^2

🔳 **35.**

17.4 m

$C \approx 109.3$ m
$A \approx 950.7$ m^2

36.

12 in.

$C \approx 37.7$ in.
$A \approx 113.0$ in.^{2}

[8.3–8.6] *Find each shaded area. Use 3.14 as the approximate value for π. Round your answers to the nearest tenth when necessary.*

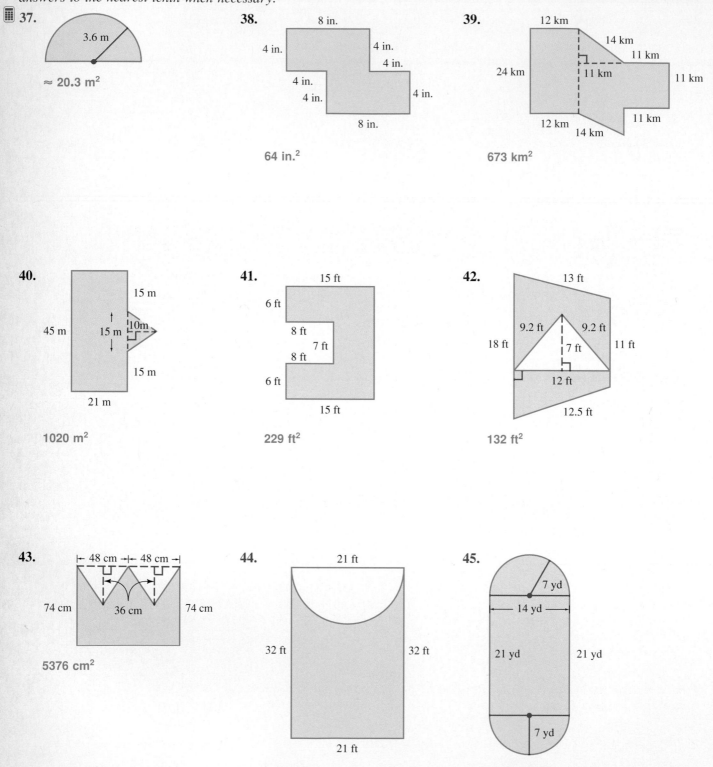

37.

3.6 m

≈ 20.3 m²

38.

8 in.

4 in.

4 in.

4 in.

4 in.

4 in.

4 in.

8 in.

64 in.²

39.

12 km

14 km

11 km

24 km

11 km

11 km

12 km

14 km

11 km

673 km²

40.

15 m

45 m

15 m

10 m

15 m

21 m

1020 m²

41.

15 ft

6 ft

8 ft

7 ft

8 ft

6 ft

15 ft

229 ft²

42.

13 ft

9.2 ft

9.2 ft

18 ft

7 ft

11 ft

12 ft

12.5 ft

132 ft²

43.

48 cm

48 cm

74 cm

36 cm

74 cm

5376 cm²

44.

21 ft

32 ft

32 ft

21 ft

≈498.9 ft²

45.

7 yd

14 yd

21 yd

21 yd

7 yd

≈447.9 yd²

[8.7] *Find each volume. Use 3.14 as the approximate value for π. Round your answers to the nearest tenth when necessary.*

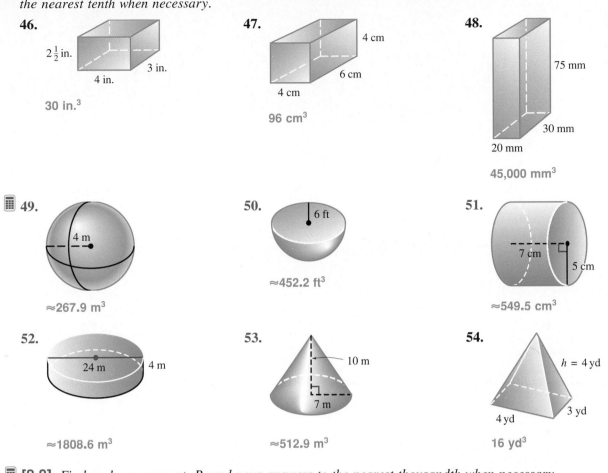

46.

$2\frac{1}{2}$ in. 3 in. 4 in.

30 in.³

47.

4 cm 6 cm 4 cm

96 cm³

48.

75 mm 30 mm 20 mm

45,000 mm³

49.

4 m

≈267.9 m³

50.

6 ft

≈452.2 ft³

51.

7 cm 5 cm

≈549.5 cm³

52.

24 m 4 m

≈1808.6 m³

53.

10 m 7 m

≈512.9 m³

54.

h = 4 yd 3 yd 4 yd

16 yd³

[8.8] *Find each square root. Round your answers to the nearest thousandth when necessary.*

55. $\sqrt{49}$

7

56. $\sqrt{8}$

≈2.828

57. $\sqrt{3000}$

≈54.772

58. $\sqrt{144}$

12

59. $\sqrt{58}$

≈7.616

60. $\sqrt{625}$

25

61. $\sqrt{105}$

≈10.247

62. $\sqrt{80}$

≈8.944

Find the unknown length in each right triangle. Use a calculator to find square roots. Round your answers to the nearest thousandth when necessary.

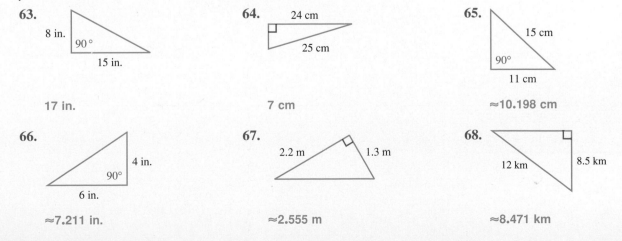

63.

8 in. 90° 15 in.

17 in.

64.

24 cm 25 cm

7 cm

65.

15 cm 90° 11 cm

≈10.198 cm

66.

4 in. 90° 6 in.

≈7.211 in.

67.

2.2 m 1.3 m

≈2.555 m

68.

12 km 8.5 km

≈8.471 km

[8.9] *Find the unknown lengths in each pair of similar triangles.*

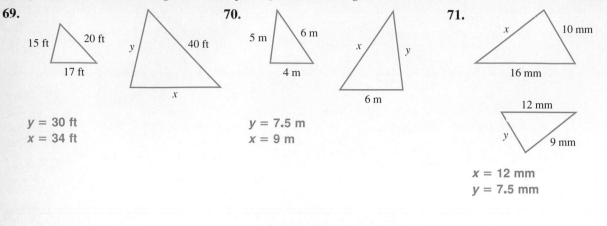

69.

y = 30 ft
x = 34 ft

70.

y = 7.5 m
x = 9 m

71.

x = 12 mm
y = 7.5 mm

MIXED REVIEW EXERCISES

Find the perimeter (or circumference) and area of each figure. Use 3.14 as the approximate value for π.

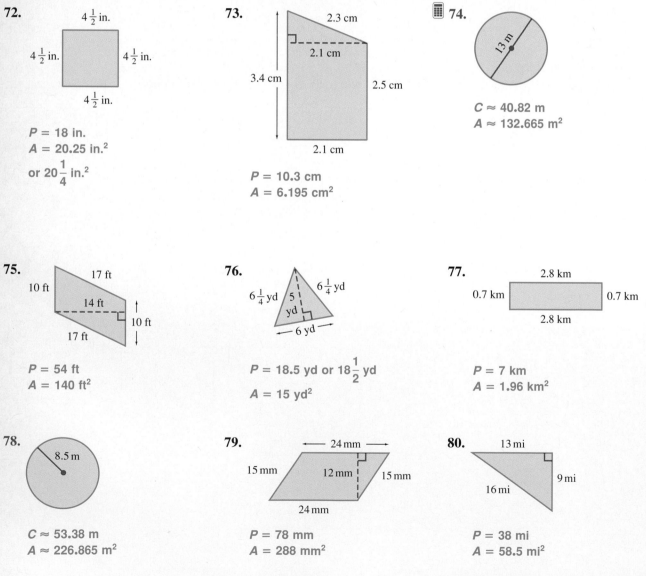

72.

$4\frac{1}{2}$ in.

$4\frac{1}{2}$ in. $4\frac{1}{2}$ in.

$4\frac{1}{2}$ in.

P = 18 in.
A = 20.25 in.²
or $20\frac{1}{4}$ in.²

73.

2.3 cm
2.1 cm
3.4 cm
2.5 cm
2.1 cm

P = 10.3 cm
A = 6.195 cm²

74.

13 m

C ≈ 40.82 m
A ≈ 132.665 m²

75.

17 ft
10 ft
14 ft
10 ft
17 ft

P = 54 ft
A = 140 ft²

76.

$6\frac{1}{4}$ yd   5 yd $6\frac{1}{4}$ yd
6 yd

P = 18.5 yd or $18\frac{1}{2}$ yd
A = 15 yd²

77.

2.8 km
0.7 km   0.7 km
2.8 km

P = 7 km
A = 1.96 km²

78.

8.5 m

C ≈ 53.38 m
A ≈ 226.865 m²

79.

24 mm
15 mm   12 mm   15 mm
24 mm

P = 78 mm
A = 288 mm²

80.

13 mi
9 mi
16 mi

P = 38 mi
A = 58.5 mi²

Label each figure. Choose from these labels: line, line segment, ray, parallel lines, perpendicular lines, intersecting lines, acute angle, right angle, straight angle, obtuse angle. Indicate the number of degrees in the right angle and the straight angle.

81.

parallel lines

82.

line segment

83.

acute angle

84.

intersecting lines

85.

right angle; 90°

86.

ray

87.

straight angle; 180°

88.

obtuse angle

89.

perpendicular lines

90. What is the complement of an angle measuring 9°?

81°

91. What is the supplement of an angle measuring 42°?

138°

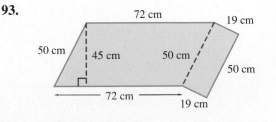

Find the perimeter and area of the following figures.

92.

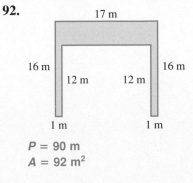

P = 90 m
A = 92 m²

93.

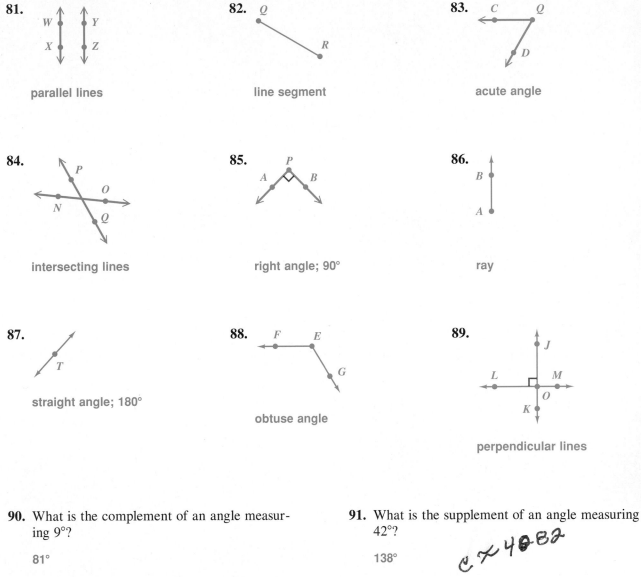

P = 282 cm
A = 4190 cm²

Find the volume of each figure. Use 3.14 as the approximate value for π. Round your answers to the nearest tenth when necessary.

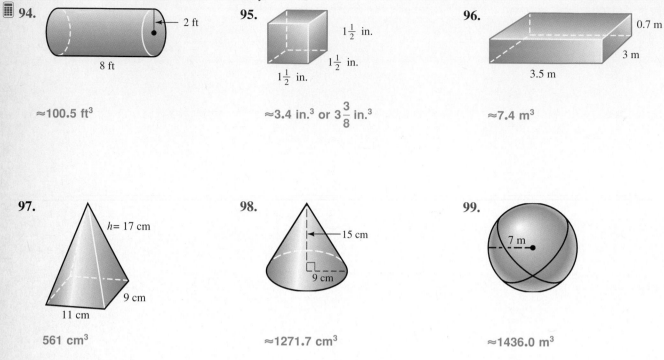

94.   — 2 ft

8 ft

≈100.5 ft³

95. $1\frac{1}{2}$ in.

$1\frac{1}{2}$ in.

$1\frac{1}{2}$ in.

≈3.4 in.³ or $3\frac{3}{8}$ in.³

96.   0.7 m

3 m

3.5 m

≈7.4 m³

97.   h= 17 cm

9 cm

11 cm

561 cm³

98.   15 cm

9 cm

≈1271.7 cm³

99.   7 m

≈1436.0 m³

Find the unknown angle or side measurement. Round your answers to the nearest thousandth when necessary.

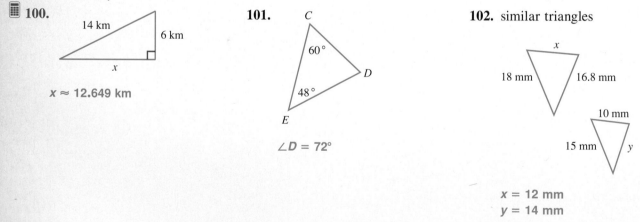

100.   14 km

6 km

x

x ≈ 12.649 km

101.   C

60°

D

48°

E

∠D = 72°

102. similar triangles

x

18 mm   16.8 mm

10 mm

15 mm   y

x = 12 mm
y = 14 mm

103. Explain how you could use the information about prefixes on page 534 to solve a problem that asks, "How many decades are in two centuries?"

The *dec* prefix in decade means 10 and the *cent* prefix in century means 100, so divide 200 (two centuries) by 10. The answer is 20 decades.

Choose the figure that matches each label.

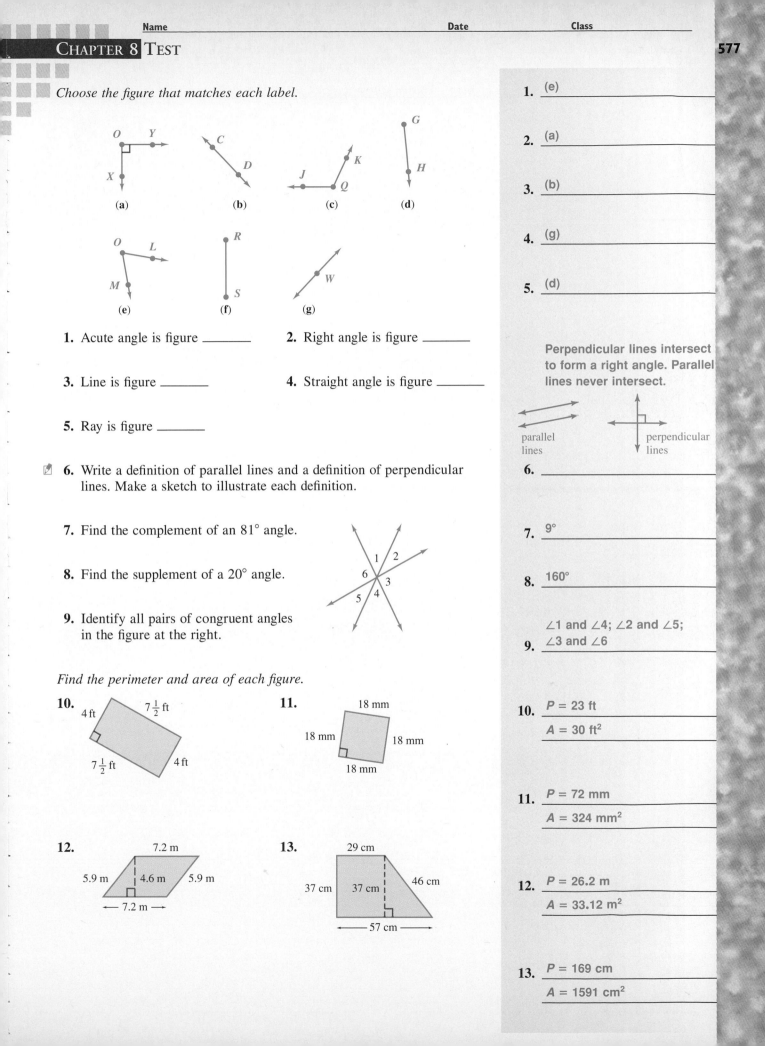

(a)   **(b)**   **(c)**   **(d)**

(e)   **(f)**   **(g)**

1. Acute angle is figure _____   2. Right angle is figure _____

3. Line is figure _____   4. Straight angle is figure _____

5. Ray is figure _____

✏ 6. Write a definition of parallel lines and a definition of perpendicular lines. Make a sketch to illustrate each definition.

7. Find the complement of an 81° angle.

8. Find the supplement of a 20° angle.

9. Identify all pairs of congruent angles in the figure at the right.

Find the perimeter and area of each figure.

10.
4 ft   7½ ft
7½ ft   4 ft

11.
18 mm
18 mm   18 mm
18 mm

12.
7.2 m
5.9 m   4.6 m   5.9 m
← 7.2 m →

13.
29 cm
37 cm   37 cm   46 cm
← 57 cm →

1. (e)

2. (a)

3. (b)

4. (g)

5. (d)

Perpendicular lines intersect to form a right angle. Parallel lines never intersect.

parallel lines   perpendicular lines

6. _____

7. 9°

8. 160°

9. ∠1 and ∠4; ∠2 and ∠5; ∠3 and ∠6

10. P = 23 ft
A = 30 ft²

11. P = 72 mm
A = 324 mm²

12. P = 26.2 m
A = 33.12 m²

13. P = 169 cm
A = 1591 cm²

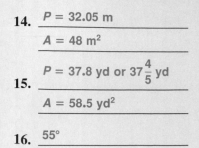

14. P = 32.05 m
_____

A = 48 m²
_____

15. P = 37.8 yd or 37$\frac{4}{5}$ yd
_____

A = 58.5 yd²
_____

16. 55°
_____

17. 12.5 in. or 12$\frac{1}{2}$ in.
_____

18. ≈5.7 km
_____

19. ≈206.0 cm²
_____

20. ≈39.3 m²
_____

21. 6480 m³
_____

22. ≈33.5 ft³
_____

23. ≈5086.8 ft³
_____

24. ≈9.220 cm
_____

25. y = 12 cm; z = 6 cm
_____

Linear units like cm are used to measure perimeter, radius, diameter, and circumference. Area is measured in square units like cm² (squares that measure 1 cm on each side). Volume is measured in cubic units like cm³.

26. _____

Find the perimeter and area of each triangle.

14.

11.4 m   8 m   8.65 m

⟵ 12 m ⟶

15.   9 yd

13 yd   15$\frac{4}{5}$ yd

16. A triangle has angles that measure 90° and 35°. What does the third angle measure?

In problems 17–23, use 3.14 as the approximate value for π. Round your answers to the nearest tenth when necessary.

17. Find the radius.

25 in.

18. Find the circumference.

0.9 km

📗 *Find the area of each figure.*

19.

16.2 cm

20.

5 m

Find the volume of each figure.

21.

12 m   30 m   18 m

📗 **22.**

2 ft

23.   18 ft
5 ft

Find the unknown lengths. Round your answers to the nearest thousandth when necessary.

📗 **24.**

?   6 cm

7 cm

25. similar triangles

18 cm   15 cm   10 cm   z   y

9 cm

📝 **26.** Explain the difference between cm, cm², and cm³. In what types of geometry problems might you use each of these units?

$\approx$ *Round the numbers in each problem so there is only one non-zero digit. Then add, subtract, multiply, or divide the rounded numbers, as indicated, to estimate the answer. Finally, solve for the exact answer.*

1. *estimate*   *exact*

$$\begin{array}{r} 300 \\ 60{,}000 \\ +\ \ 6000 \\ \hline 66{,}300 \end{array} \qquad \begin{array}{r} 319 \\ 58{,}028 \\ +\ \ 6{,}227 \\ \hline 64{,}574 \end{array}$$

2. *estimate*   *exact*

$$\begin{array}{r} 20 \\ -\ 10 \\ \hline 10 \end{array} \qquad \begin{array}{r} 20.07 \\ -\ 9.828 \\ \hline 10.242 \end{array}$$

3. *estimate*   *exact*

$$\begin{array}{r} 4 \\ \times\ 7 \\ \hline 28 \end{array} \qquad \begin{array}{r} 3.664 \\ \times\ 7.3 \\ \hline 26.7472 \end{array}$$

4. *estimate*   *exact*

$$\begin{array}{r} 30{,}000 \\ \times\ \ \ \ \ 70 \\ \hline 2{,}100{,}000 \end{array} \qquad \begin{array}{r} 28{,}419 \\ \times\ \ \ \ 73 \\ \hline 2{,}074{,}587 \end{array}$$

5. *estimate*   *exact*

$$3\overline{)600} = 200 \qquad 2.8\overline{)562.24} = 200.8$$

6. *estimate*   *exact*

$$50\overline{)5000} = 100 \qquad 52\overline{)4888} = 94$$

7. *estimate*   *exact*

$$\begin{array}{r} 5 \\ +\ 5 \\ \hline 10 \end{array} \qquad \begin{array}{r} 4\frac{1}{2} \\ +\ 4\frac{9}{10} \\ \hline 9\frac{2}{5} \end{array}$$

8. *estimate*   *exact*

$$\begin{array}{r} 3 \\ -\ 2 \\ \hline 1 \end{array} \qquad \begin{array}{r} 3\frac{1}{6} \\ -\ 1\frac{7}{8} \\ \hline 1\frac{7}{24} \end{array}$$

9. *exact*

$$3\frac{1}{9} \cdot 1\frac{5}{7} = \underline{\ 5\frac{1}{3}\ }$$

estimate

$$\underline{\ 3\ } \cdot \underline{\ 2\ } = \underline{\ 6\ }$$

Add, subtract, multiply, or divide as indicated. Write answers to fraction problems in lowest terms and as whole or mixed numbers when possible.

10. $3\frac{3}{5} \div 8$

$$\frac{9}{20}$$

11. $1 - 0.0868$

$$0.9132$$

12. Write your answer using R for the remainder

$$81\overline{)5749} = 70\ \text{R}79$$

13. $10 \div \frac{5}{16}$

$$32$$

14. $(0.006)(0.013)$

$$0.000078$$

15. $40{,}020 - 915$

$$39{,}105$$

16. $0.7 \div 0.036$ Round your answer to the nearest hundredth.

$$\approx 19.44$$

17. $6\frac{1}{6} - 1\frac{3}{4}$

$$4\frac{5}{12}$$

18. $752.6 + 83 + 0.485$

$$836.085$$

Simplify by using order of operations.

19. $16 - (10 - 2) \div 2 \cdot 3 + 5$

$$9$$

20. $2^4 \div \sqrt{64} + 6^2$

$$38$$

21. Write 0.0208 in words.

two hundred eight ten-thousandths

22. Write six hundred sixty and five hundredths in numbers.

660.05

23. Arrange in order from smallest to largest.

2.55   2.505   2.055   2.5005

2.055;   2.5005;   2.505;   2.55

24. Explain how you could use the information on prefixes and root words on page 534 to remember the way to change a percent to a decimal.

Per means divide and *cent* means 100, so divide by 100 to change a percent to a decimal.

Complete this chart.

	Fraction/Mixed Number		Decimal		Percent
25.	$\dfrac{1}{50}$		0.02	**26.**	2%
	$1\dfrac{3}{4}$	**27.**	1.75	**28.**	175%
29.	$\dfrac{2}{5}$	**30.**	0.4		40%

Write each rate or ratio in lowest terms. Change to the same units when necessary.

31. 4 feet to 6 inches; compare in inches.

$\dfrac{8}{1}$

32. Last month there were 9 cloudy days and 21 sunny days. What was the ratio of sunny days to cloudy days?

$\dfrac{7}{3}$

Find the unknown number in each proportion. Round your answer to hundredths, if necessary.

33. $\dfrac{5}{13} = \dfrac{x}{91}$

35

34. $\dfrac{207}{69} = \dfrac{300}{x}$

100

35. $\dfrac{4.5}{x} = \dfrac{6.7}{3}$

≈ 2.01

Solve each of the following.

36. 72 patients is what percent of 45 patients?

160%

37. $18 is 3% of what number of dollars?

$600

Convert the following measurements.

38. $2\frac{1}{4}$ hours to minutes

135 minutes

39. 40 ounces to pounds

$2\dfrac{1}{2}$ or 2.5 pounds

40. 8 cm to meters

0.08 m

41. 1.8 L to mL

1800 mL

Write the most reasonable metric unit in each blank. Choose from km, m, cm, mm, L, mL, kg, g, and mg.

42. Her wristwatch strap is 15 __mm__ wide.

43. Jon added 2 __L__ of oil to his car.

44. The child weighs 15 __kg__ .

45. The bookcase is 90 __cm__ high.

46. List the metric temperatures at which water freezes and water boils.   0°C and 100°C

Find the perimeter or circumference and area of each figure. Use 3.14 as the approximate value of π.

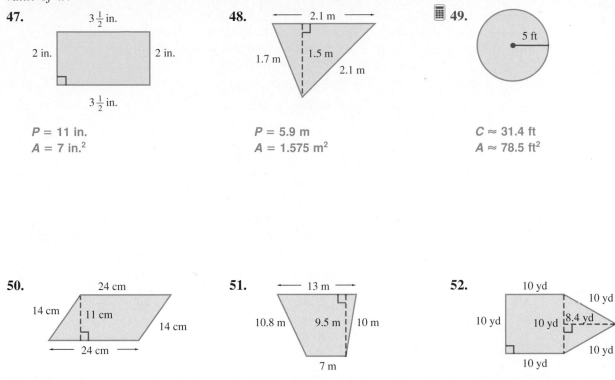

47.

$3\frac{1}{2}$ in.

2 in.   2 in.

$3\frac{1}{2}$ in.

P = 11 in.
A = 7 in.²

48.   2.1 m

1.7 m   1.5 m

2.1 m

P = 5.9 m
A = 1.575 m²

🖩 **49.**

5 ft

C ≈ 31.4 ft
A ≈ 78.5 ft²

50.   24 cm

14 cm   11 cm

14 cm

24 cm

P = 76 cm
A = 264 cm²

51.   13 m

10.8 m   9.5 m   10 m

7 m

P = 40.8 m
A = 95 m²

52.   10 yd

10 yd

10 yd   10 yd   8.4 yd

10 yd

10 yd

P = 50 yd
A = 142 yd²

Find the unknown length in each figure. Round your answers to the nearest tenth.

🖩 **53.**   19 mm

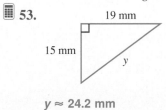

15 mm

y

y ≈ 24.2 mm

54. similar triangles

15 ft

5.5 ft

x

5.5 ft

22 ft

15 ft

x ≈ 8.1 ft

Solve the following application problems.

55. Mei Ling must earn 90 credits to receive an associate of arts degree. She has 53 credits. What percent of the necessary credits does she have? Round to the nearest whole percent.

≈59%

56. Which bag of chips is the best buy: Brand T is $15\frac{1}{2}$ ounces for $2.99, Brand F is 14 ounces for $2.49, and Brand H is 18 ounces for $3.89. You have a coupon for 40¢ off Brand H and another for 30¢ off Brand T.

Brand T at 15.5 ounces for
$2.99 − $0.30 coupon

57. A coffee can has a diameter of 13 cm and a height of 17 cm. Find the volume of the can. Use 3.14 for π and round your answer to the nearest tenth.

≈2255.3 cm³

58. Swimsuits are on sale in August at 65% off the regular price. How much will Lanece pay for a suit that has a regular price of $44?

$15.40

59. Steven bought $4\frac{1}{2}$ yards of canvas material to repair the tents used by the scout troop. He used $1\frac{2}{3}$ yards on one tent and $1\frac{3}{4}$ yards on another. How much material is left?

$1\frac{1}{12}$ yard

60. The cooks at a homeless shelter used 30 pounds of meat to make stew for 140 people. At that rate, how much meat is needed for stew to feed 200 people? Round to the nearest tenth.

≈42.9 pounds

61. Graciela needs 85 cm of yarn to make a tassel for one corner of a pillow. How many meters of yarn does she need to put a tassel on each corner of a square-shaped pillow?

3.4 m

62. A photograph measures 8 inches by 10 inches. Earl put it in a frame that is 2 inches wide. Find the perimeter of the frame. Draw a sketch to help solve the problem.

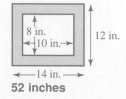

52 inches

Basic Algebra

9.1 Signed Numbers

All the numbers you have studied so far in this book have been either 0 or greater than 0. Numbers greater than 0 are called *positive numbers*. For example, you have worked with these positive numbers:

    salary of $800

    temperature of 98.6°F

    length of $3\frac{1}{2}$ feet.

OBJECTIVE 1 Not all numbers are positive. For example, "15 degrees below 0" or "a loss of $500" is expressed with a number less than 0. Numbers less than 0 are called **negative** (NEG-uh-tiv) **numbers.** Zero is neither positive nor negative.

Writing Negative Numbers

Write negative numbers with a *negative sign, −*.

For example, "15 degrees below 0" is written with a negative sign, as −15°. And "a loss of $500" is written −$500.

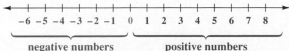
WORK PROBLEM 1 AT THE SIDE.

OBJECTIVE 2 In **Section 3.5** you graphed positive numbers on a number line. Negative numbers can also be shown on a number line. Zero separates the positive numbers from the negative numbers on the number line. The number −5 is read "negative five."

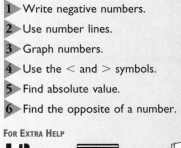

negative numbers   positive numbers

Note
For every positive number on a number line, there is a corresponding negative number on the opposite side of 0.

When you work with both positive and negative numbers (and zero) we say you are working with **signed numbers.**

www.mathnotes.com

OBJECTIVES

1. Write negative numbers.
2. Use number lines.
3. Graph numbers.
4. Use the < and > symbols.
5. Find absolute value.
6. Find the opposite of a number.

FOR EXTRA HELP

Tutorial   Tape 15   SSM, Sec. 9.1

---

1. Write each number.

    (a) A temperature at the North Pole of 70 degrees below 0.

    (b) Your checking account is overdrawn by 15 dollars.

    (c) The altitude of a place 284 feet below sea level.

ANSWERS

1. (a) −70° (b) −$15 (c) −284 ft

583

2. Write *positive, negative,* or *neither* for each number.

(a) −8

(b) −$\frac{3}{4}$

(c) 1

(d) 0

Writing Positive Numbers

Positive numbers can be written in two ways:

1. Use a "+" sign. For example, +2 is "positive two."
2. Do not write any sign. For example, 3 is assumed to be "positive three."

◀◀ WORK PROBLEM 2 AT THE SIDE.

OBJECTIVE 3 The next example shows you how to graph signed numbers.

E X A M P L E 1   Graphing Signed Numbers

Graph **(a)** −4   **(b)** 3   **(c)** −1   **(d)** 0   **(e)** $1\frac{1}{4}$

Place a dot at the correct location for each number.

◀◀ WORK PROBLEM 3 AT THE SIDE.

OBJECTIVE 4 As shown on the following number line, 3 is to the left of 5.

Also, 3 is *less than* 5.

Recall the following symbols for comparing two numbers.

$<$ means "is less than"

$>$ means "is greater than"

Use these symbols to write "3 is less than 5" as follows.

$$3 \quad < \quad 5$$
$$\downarrow \quad \downarrow \quad \downarrow$$

3 is less than 5

As this example suggests,

The lesser of two numbers is the one farther to the *left* on a number line.

E X A M P L E 2   Using the Symbols $<$ and $>$

Use this number line and $>$ or $<$ to make true statements.

CONTINUED ON NEXT PAGE

---

3. Graph each list of numbers.

(a) −1, 1, −3, 3

(b) −2, 4, 0, −1, −4

ANSWERS

2. (a) negative (b) negative (c) positive
(d) neither

3. (a)

(b)

(a) 2 < 6 (read "2 is less than 6") because 2 is to the *left* of 6 on the number line.

(b) −9 < −4 because −9 is to the *left* of −4.

(c) 2 > −1 because 2 is to the *right* of −1.

(d) −4 < 0 because −4 is to the *left* of 0.

Note

When using > and <, the *small* pointed end of the symbol points to the *smaller* (lesser) number.

WORK PROBLEM 4 AT THE SIDE. ▶▶

OBJECTIVE 5 In order to graph a number on the number line, you need to know two things:

1. Which *direction* it is from 0. It can be in a *positive* direction or a *negative* direction. You can tell the direction by looking for a positive or negative sign.
2. How *far* it is from zero. The *distance* from zero is the **absolute** (ab-soh-LOOT) **value** of a number.

Absolute value is indicated by two vertical bars. For example, | 6 | is read "the **absolute value** of 6."

Note

Absolute value is never negative, because it is a distance and distance is never negative.

E X A M P L E 3 Finding Absolute Value

Find each of the following.

(a) |8| The distance from 0 to 8 is 8, so |8| = 8.

distance is 8, direction is positive

(b) |−8| The distance from 0 to −8 is also 8, so |−8| = 8.

distance is 8, direction is negative

(c) |0| = 0

(d) −|−3| First, |−3| is 3. But there is also a negative sign outside the absolute value bars. So, −3 is the solution.

Note

A negative sign *outside* the absolute value bars is *not* affected by the absolute value bars. Therefore, your final answer is negative, as in Example 3(d) above.

WORK PROBLEM 5 AT THE SIDE. ▶▶

4. Write < or > in each blank to make a true statement.

(a) 4 _____ 0

(b) −1 _____ 0

(c) −3 _____ −1

(d) −8 _____ −9

(e) 0 _____ −3

5. Find each of the following.

(a) |5|

(b) |−5|

(c) |−17|

(d) −|−9|

(e) −|2|

ANSWERS

4. (a) > **(b)** < **(c)** < **(d)** > **(e)** >
5. (a) 5 **(b)** 5 **(c)** 17 **(d)** −9 **(e)** −2

6. Find the opposite of each number.

(a) 4

(b) 10

(c) 49

(d) $\frac{2}{5}$

7. Find the opposite of each number.

(a) −4

(b) −10

(c) −25

(d) −1.9

(e) −0.85

(f) $-\frac{3}{4}$

ANSWERS
6. (a) −4 (b) −10 (c) −49 (d) $-\frac{2}{5}$
7. (a) 4 (b) 10 (c) 25 (d) 1.9 (e) 0.85
    (f) $\frac{3}{4}$

OBJECTIVE 6 Two numbers that are the same distance from 0 on a number line, but on opposite sides of 0, are called **opposites** of each other. As this number line shows, −3 and 3 are opposites of each other.

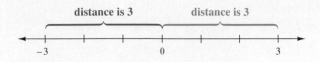

To indicate the opposite of a number, write a negative sign in front of the number.

E X A M P L E 4   Finding Opposites

Find the opposite of each number.

Number	Opposite
5	$-(5) = {}^-5$ Write a negative sign.
9	$-(9) = {}.-9$
$\frac{4}{5}$	$-\left(\frac{4}{5}\right) = -\frac{4}{5}$
0	$-(0) = \quad 0$ No negative sign.

The opposite of 0 is 0. Zero is neither positive nor negative.

◄◄ WORK PROBLEM 6 AT THE SIDE.

Some numbers have two negative signs, such as

$$-(-3).$$

The negative sign in front of −3 means the *opposite* of −3. The opposite of −3 is 3, so

$$-(-3) = 3.$$

Use the following rule to find the opposite of a negative number.

Double Negative Rule

$$-(-x) = x$$

The opposite of a negative number is positive.

E X A M P L E 5   Finding Opposites

Find the opposite of each number.

Number	Opposite	
−2	$-(-2) = 2$	By double negative rule
−9	$-(-9) = 9$	
$-\frac{1}{2}$	$-\left(-\frac{1}{2}\right) = \frac{1}{2}$	

◄◄ WORK PROBLEM 7 AT THE SIDE.

9.1 Exercises

Write a signed number for each of the following.

1. Water freezes at 32 degrees above zero on the Fahrenheit temperature scale.

+32

2. She made a profit of $920.

+920

3. The price of the stock fell $12.

−12

4. His checking account is overdrawn by $30.

−30

5. The river is 20 feet above flood stage.

+20

6. The team lost 6 yards on that play.

−6

Write positive, negative, *or* neither *for each of the following numbers.*

7. 24

positive

8. −8

negative

9. $-\dfrac{7}{10}$

negative

10. $2\dfrac{1}{3}$

positive

11. 0

neither

12. +6

positive

13. −6.3

negative

14. −0.25

negative

Graph each of the following lists of numbers.

Example: $-3, -\dfrac{2}{3}, -5, 1, 2, \dfrac{3}{4}$

Solution: Place a dot on a number line for each number.

15. $4, -1, 2, 3, 0, -2$

16. $-5, -3, 1, 4, 0$

17. $-\dfrac{1}{2}, -3, -5, \dfrac{1}{2}, 1\dfrac{3}{4}, 3$

18. $-4, -\dfrac{3}{4}, -2, 4, 1, 2\dfrac{1}{2}$

✎ Writing   ⊙ Conceptual   ▲ Challenging   ≈ Estimation

19. $-2, -4, -3\frac{1}{5}, -\frac{5}{8}, 1, 2$

$-5\ -4\ -3\ -2\ -1\ \ 0\ \ 1\ \ 2\ \ 3\ \ 4\ \ 5$

20. $-5, -3, -2, -4\frac{2}{3}, -1\frac{1}{2}, 0, 1$

$-5\ -4\ -3\ -2\ -1\ \ 0\ \ 1\ \ 2\ \ 3\ \ 4\ \ 5$

21. $3, 4.5, -1.5, 2.2, 0.1$

$-5\ -4\ -3\ -2\ -1\ \ 0\ \ 1\ \ 2\ \ 3\ \ 4\ \ 5$

22. $3.25, -1, 4.5, 1.25, 2$

$-5\ -4\ -3\ -2\ -1\ \ 0\ \ 1\ \ 2\ \ 3\ \ 4\ \ 5$

23. $-14, -11, -10.5, -13, -7.3$

$-14\ \ -12\ \ -10\ \ -8\ \ -6$

24. $-10, -7, -14.8, -9.25, -13.75$

$-14\ \ -12\ \ -10\ \ -8\ \ -6$

Write $<$ or $>$ in each blank to make a true statement.

Examples: -4 _____ 2	-5 _____ -9
Solutions: Because -4 is to the *left* of 2 on a number line, -4 **is less than** 2.	Because -5 is to the *right* of -9 on a number line, -5 **is greater than** -9.
$-4 < 2$	$-5 > -9$

25. $9 \underline{\ <\ } 14$ **26.** $6 \underline{\ <\ } 11$ **27.** $0 \underline{\ >\ } -2$ **28.** $0 \underline{\ <\ } 2$

29. $-6 \underline{\ <\ } 3$ **30.** $-4 \underline{\ <\ } 7$ **31.** $1 \underline{\ >\ } 0$ **32.** $-1 \underline{\ <\ } 0$

33. $-11 \underline{\ <\ } -2$ **34.** $-5 \underline{\ <\ } -1$ **35.** $-75 \underline{\ <\ } -72$ **36.** $-50 \underline{\ >\ } -60$

37. $2 \underline{\ >\ } -1$ **38.** $4 \underline{\ >\ } -9$ **39.** $-115 \underline{\ >\ } -120$ **40.** $-205 \underline{\ >\ } -210$

Find the absolute value of the following.

> **Examples:** $|8|$ $|-7|$ $-|-2|$
>
> **Solutions:** $|8| = 8$ $|-7| = 7$ $-|-2| = -2$

41. $|5|$

5

42. $|12|$

12

43. $|-5|$

5

44. $|-16|$

16

45. $|-1|$

1

46. $|-14|$

14

47. $|251|$

251

48. $|397|$

397

49. $|0|$

0

50. $|-199|$

199

51. $\left|-\dfrac{1}{2}\right|$

$\dfrac{1}{2}$

52. $\left|-\dfrac{9}{5}\right|$

$\dfrac{9}{5}$

53. $|-9.5|$

9.5

54. $|-0.72|$

0.72

55. $|0.618|$

0.618

56. $|4.7|$

4.7

57. $-|-10|$

−10

58. $-|-8|$

−8

59. $-\left|-\dfrac{5}{2}\right|$

$-\dfrac{5}{2}$

60. $-\left|-\dfrac{1}{3}\right|$

$-\dfrac{1}{3}$

61. $-|-7.6|$

−7.6

62. $-|-1.03|$

−1.03

63. $-|4|$

−4

64. $-|20|$

−20

Find the opposite of each number.

> **Examples:**   **Solutions:**
>
> Number   Opposite
> 8 $-(8) = -8$
> −5 $-(-5) = 5$
> 0 $-(0) = 0$

65. 2   −2

66. 7   −7

67. −54   54

68. −75   75

69. −11   11

70. −24   24

71. 163   −163

72. 502   −502

73. 0   0

74. $\dfrac{5}{8}$ $-\dfrac{5}{8}$

75. $-4\dfrac{1}{2}$ $4\dfrac{1}{2}$

76. $-1\dfrac{2}{3}$ $1\dfrac{2}{3}$

77. 5.2   −5.2

78. 3.7   −3.7

79. −1.4   1.4

80. −0.65   0.65

81. In your own words, explain opposite numbers. Include an example of opposite numbers and draw a number line to illustrate your example.

Opposite numbers are the same distance from 0 but on opposite sides of it.

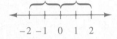

−2 and +2 are opposites.

82. Explain why the opposite of zero is zero.

Zero is in the "middle" of the number line and is neither positive nor negative, so its opposite can be neither positive nor negative.

83. Describe three different situations at home, work, or school where you have used negative numbers.

Some possible answers are temperatures below zero, loss of points in a game, and overdrawn checking account.

84. Explain in your own words why absolute value is never negative.

The absolute value of a number is its *distance* from zero and distance is never negative.

▲ *Write* true *or* false *for each statement.*

85. $|-5| > 0$

true

86. $|-12| > |-15|$

false

87. $0 < -(-6)$

true

88. $-9 < -(-6)$

true

89. $-|-4| < -|-7|$

false

90. $-|-0| > 0$

false

Review and Prepare

*Add or subtract the following numbers. (For help, see **Section 3.3**.)*

91. $\dfrac{3}{4} + \dfrac{1}{5}$ $\dfrac{19}{20}$

92. $\dfrac{3}{10} + \dfrac{3}{8}$ $\dfrac{27}{40}$

93. $\dfrac{5}{6} - \dfrac{1}{4}$ $\dfrac{7}{12}$

94. $\dfrac{2}{3} - \dfrac{5}{9}$ $\dfrac{1}{9}$

9.2 Addition and Subtraction of Signed Numbers

You can show a positive number on a number line by drawing an arrow pointing to the right. In the following examples both arrows represent positive 4 units.

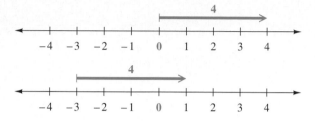

Draw arrows pointing to the left to show negative numbers. Both of the following arrows represent −3 units.

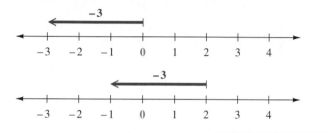

WORK PROBLEM I AT THE SIDE. ▶▶

OBJECTIVE ▶ You can use a number line to add signed numbers. For example, the next number line shows how to add 2 and 3.

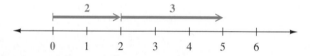

Add 2 and 3 by starting at zero and drawing an arrow 2 units to the right. From the end of this arrow, draw another arrow 3 units to the right. This second arrow ends at 5, showing that

$$2 + 3 = 5.$$

E X A M P L E I   Adding Signed Numbers by Using a Number Line

Add by using a number line.

(a) $4 + (-1)$

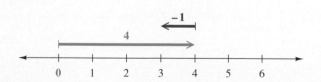

Start at zero and draw an arrow 4 units to the right. From the end of this arrow, draw an arrow 1 unit to the *left*. (Remember to go to the left for a negative number.) This second arrow ends at 3, so

$$4 + (-1) = 3.$$

Note
Always start at 0 when adding on the number line.

CONTINUED ON NEXT PAGE

OBJECTIVES

1 ▶ Add signed numbers by using a number line.

2 ▶ Add signed numbers without using a number line.

3 ▶ Find the additive inverse of a number.

4 ▶ Subtract signed numbers.

5 ▶ Add or subtract a series of signed numbers.

FOR EXTRA HELP

Tutorial   Tape 15   SSM, Sec. 9.2

1. Complete each arrow so it represents the indicated number of units.

(a)

(b)
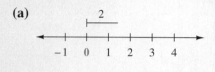

(c)
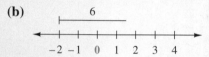

(d)
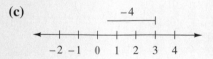

ANSWERS

1. (a)

(b)

(c)

(d)

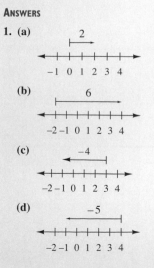

2. Draw arrows to find each of the following.

(a) $3 + (-2)$

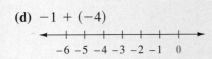

(b) $-4 + 1$

(c) $-3 + 7$

(d) $-1 + (-4)$

ANSWERS

2. (a) $3 + (-2) = 1$

(b) $-4 + 1 = -3$

(c) $-3 + 7 = 4$

(d) $-1 + (-4) = -5$

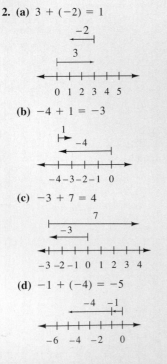

(b) $-6 + 2$

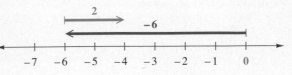

Draw an arrow from zero going 6 units to the left. From the end of this arrow, draw an arrow 2 units to the right. This second arrow ends at -4, so

$$-6 + 2 = -4.$$

(c) $-3 + (-5)$

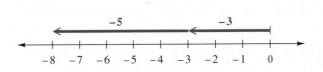

As the arrows along the number line show,

$$-3 + (-5) = -8.$$

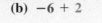 **WORK PROBLEM 2 AT THE SIDE.**

OBJECTIVE 2 ▶ After working with number lines for awhile, you will see ways to add signed numbers without drawing arrows. You already know how to add two positive numbers (from Chapter 1). Here are the steps for adding two negative numbers.

Adding Negative Numbers

Step 1 Add the absolute values of the numbers.

Step 2 Write a negative sign in front of the sum.

E X A M P L E 2   Adding Negative Numbers

Add (without number lines).

(a) $-4 + (-12)$

The absolute value of -4 is **4**.
The absolute value of -12 is **12**.
Add the absolute values.

$$4 + 12 = 16$$

Write a negative sign in front of the sum.

$$-4 + (-12) = -16 \qquad \text{Write a negative sign in front of 16.}$$

(b) $-5 + (-25) = -30$ ← Sum of absolute values, with a negative sign written in front of 30.

(c) $-11 + (-46) = -57$

CONTINUED ON NEXT PAGE

(d) $-\dfrac{3}{4} + \left(-\dfrac{1}{2}\right)$

The absolute value of $-\dfrac{3}{4}$ is $\dfrac{3}{4}$, and the absolute value of $-\dfrac{1}{2}$ is $\dfrac{1}{2}$. Add the absolute values. Check that the answer is in lowest terms.

$$\frac{3}{4} + \frac{1}{2} = \frac{3}{4} + \frac{2}{4} = \frac{5}{4} \quad \leftarrow \text{Lowest terms.}$$

Write a negative sign in front of the sum.

$$-\frac{3}{4} + \left(-\frac{1}{2}\right) = -\frac{5}{4} \quad \text{Write a negative sign.}$$

Note

In algebra we always write fractions in lowest terms, but usually do *not* change improper fractions to mixed numbers because improper fractions are easier to work with. In Example 2(d) above, we checked that $-\dfrac{5}{4}$ was in lowest terms but did *not* rewrite it as $-1\dfrac{1}{4}$.

WORK PROBLEM 3 AT THE SIDE. ▶▶

Use the following steps to add two numbers with *different* signs.

Adding Two Numbers with Different Signs

Step 1 *Subtract* the smaller absolute value from the larger absolute value.

Step 2 Write the sign of the number with the *larger* absolute value in front of the answer.

┌ **E X A M P L E   3**   **Adding Two Numbers with Different Signs**

Add.

(a) $8 + (-3)$

Find this sum with a number line as follows.

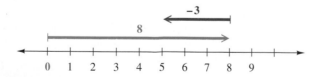

Because the top arrow ends at 5,

$$8 + (-3) = 5.$$

Find the sum by using the rule as follows. First, find the absolute value of each number.

$$|8| = 8 \qquad |-3| = 3$$

Subtract the absolute value of the smaller number from the larger number.

$$8 - 3 = 5$$

Here the positive number 8 has the larger absolute value, so the answer is positive.

$$8 + (-3) = 5 \leftarrow \text{Positive answer}$$

CONTINUED ON NEXT PAGE

---

3. Add.

(a) $-4 + (-4)$

(b) $-3 + (-20)$

(c) $-31 + (-5)$

(d) $-10 + (-8)$

(e) $-\dfrac{9}{10} + \left(-\dfrac{3}{5}\right)$

Answers

3. **(a)** -8 **(b)** -23 **(c)** -36 **(d)** -18

   (e) $-\dfrac{3}{2}$

4. Add.

(a) $10 + (-2)$

(b) $-7 + 8$

(c) $-11 + 11$

(d) $23 + (-32)$

(e) $-\dfrac{7}{8} + \dfrac{1}{4}$

(b) $-12 + 4$

First, find absolute values.

$$|-12| = 12 \qquad |4| = 4$$

Subtract:

$$12 - 4 = 8$$

The negative number -12 has the larger absolute value, so the answer is negative.

$$-12 + 4 = -8$$

↑ —— Write negative sign in front of the answer because -12 has the larger absolute value.

(c) $15 + (-21) = -6$

↑ —— Write negative sign in front of the answer because -21 has the larger absolute value.

(d) $13 + (-9) = 4$ ← Positive answer because the positive number 13 has the larger absolute value.

(e) $-\dfrac{1}{2} + \dfrac{2}{3}$

The absolute value of $-\frac{1}{2}$ is $\frac{1}{2}$, and the absolute value of $\frac{2}{3}$ is $\frac{2}{3}$. Subtract the absolute value of the smaller number from the larger.

$$\frac{2}{3} - \frac{1}{2} = \frac{4}{6} - \frac{3}{6} = \frac{1}{6}$$

Because the positive number $\frac{2}{3}$ has the larger absolute value, the answer is positive.

$$-\frac{1}{2} + \frac{2}{3} = \frac{1}{6} \leftarrow \text{Positive answer}$$

◀◀ **WORK PROBLEM 4 AT THE SIDE.**

OBJECTIVE 3 Recall that the opposite of 9 is -9, and the opposite of -4 is $-(-4)$, or 4. Add these opposites as follows.

$$9 + (-9) = 0 \quad and \quad -4 + 4 = 0$$

The sum of a number and its opposite is always 0. For this reason, opposites are also called **additive** (ADD-ih-tiv) **inverses** of each other.

> **Additive Inverse**
>
> The opposite of a number is called its **additive inverse.** The sum of a number and its opposite is zero.

EXAMPLE 4   Finding the Additive Inverse

This chart shows you several numbers and the additive inverse of each.

Number	Additive Inverse	Sum of Number and Inverse
6	−6	$6 + (−6) = 0$
−8	−(−8) or 8	$(−8) + 8 = 0$
4	−4	$4 + (−4) = 0$
−3	−(−3) or 3	$−3 + 3 = 0$
$\frac{5}{8}$	$−\frac{5}{8}$	$\frac{5}{8} + \left(−\frac{5}{8}\right) = 0$
0	0	$0 + 0 = 0$

WORK PROBLEM 5 AT THE SIDE. ▶▶

OBJECTIVE 4 ▶ You may have noticed that negative numbers are often written with parentheses, like $(−8)$. This is especially helpful when subtracting because the − sign is used both to indicate a **negative number** and to indicate **subtraction**.

Example	How to Say It
$(−5)$	**negative five**
8	positive eight
$−3 − 2$	**negative three** minus positive two
$6 − (−4)$	positive six **minus negative four**
$−7 − (−1)$	**negative seven** minus **negative one**
$8 − 3$	positive eight **minus** positive three

Note
Be sure you understand when the "−" sign means "subtract," and when it means "negative number."

WORK PROBLEM 6 AT THE SIDE. ▶▶

When working with signed numbers, it is helpful to write a subtraction problem as an addition problem. For example, you know that $6 − 4 = 2$. But you get the same result by *adding* 6 and the *opposite* of 4, that is, $6 + (−4)$.

$$6 − 4 = 2$$
$$6 + (−4) = 2$$
Same result

This suggests the following definition of subtraction.

Defining Subtraction

The difference of two numbers, a and b, is
$$a − b = a + (−b).$$

Subtract two numbers by adding the first number and the opposite (additive inverse) of the second.

5. Give the additive inverse of each number. Then find the sum of the number and its inverse.

(a) 12

(b) −9

(c) 3.5

(d) $−\frac{7}{10}$

(e) 0

6. Write each example in words.

(a) $−7 − 2$

(b) $−10$

(c) $3 − (−5)$

(d) 4

(e) $−8 − (−6)$

(f) $2 − 9$

ANSWERS
5. (a) −12; 12 + (−12) = 0
   (b) 9; −9 + 9 = 0
   (c) −3.5; 3.5 + (−3.5) = 0
   (d) $\frac{7}{10}; −\frac{7}{10} + \frac{7}{10} = 0$
   (e) 0; 0 + 0 = 0
6. (a) negative seven minus positive two
   (b) negative ten
   (c) positive three minus negative five
   (d) positive four
   (e) negative eight minus negative six
   (f) positive two minus positive nine

7. Subtract.

(a) $-6 - 5$

(b) $3 - (-10)$

(c) $-8 - (-2)$

(d) $4 - 9$

(e) $-7 - (-15)$

(f) $-\dfrac{2}{3} - \left(-\dfrac{5}{12}\right)$

Subtracting Signed Numbers

To subtract two signed numbers, add the opposite of the second number to the first number. Use these steps.

Step 1 Change the subtraction sign to addition.

Step 2 Change the sign of the second number to its opposite.

Step 3 Proceed as in addition.

Note

The pattern in a subtraction problem is:

$$\begin{array}{c} \text{1st} \\ \text{number} \end{array} - \begin{array}{c} \text{2nd} \\ \text{number} \end{array} = \begin{array}{c} \text{1st} \\ \text{number} \end{array} + \begin{array}{c} \text{opposite of} \\ \text{2nd number} \end{array}$$

E X A M P L E 5   Subtracting Signed Numbers

Subtract.

(a) $8 - 11$

The first number, 8, stays the same. Change the subtraction sign to addition. Change the sign of the second number to its opposite.

Positive 8 stays the same $8 - 11$ Positive 11 changed to
                              $\downarrow \quad \downarrow$ its opposite -11
                              $8 + (-11)$

Subtraction changed to addition

Now add.

$$8 + (-11) = -3$$
$$\text{So} \quad 8 - 11 = -3 \quad \text{also.}$$

(b) $-9 - 15$
       $\quad\downarrow \quad \downarrow$ Positive 15 is changed to its opposite (-15).
$-9 + (-15) = -24$

(c) $-5 - (-7)$
       $\quad\downarrow \quad \downarrow$ Negative 7 is changed to its opposite $(+7)$.
$-5 + (+7) = 2$

(d) $7.6 - (-8.3)$
       $\quad\downarrow \quad \downarrow$ Negative 8.3 is changed to its opposite $(+8.3)$.
$7.6 + (+8.3) = 15.9$

(e) $\dfrac{5}{8} - \left(-\dfrac{1}{2}\right)$
       $\quad\downarrow \quad \downarrow$
$\dfrac{5}{8} + \left(+\dfrac{1}{2}\right) = \dfrac{5}{8} + \left(+\dfrac{4}{8}\right) = \dfrac{9}{8}$

◀◀ **WORK PROBLEM 7 AT THE SIDE.**

Answers

7. (a) -11 **(b)** 13 **(c)** -6 **(d)** -5 **(e)** 8
  (f) $-\dfrac{1}{4}$

OBJECTIVE 5 If a problem involves both addition and subtraction, use the order of operations and work from left to right.

EXAMPLE 6 Combining Addition and Subtraction of Signed Numbers

Perform the addition and subtraction from left to right.

(a) $-6 + (-11) - 5$

$-17 - 5$ Change subtraction to addition; change positive 5 to its opposite (-5)

$-17 + (-5)$

-22

(b) $4 - (-3) + (-9)$ Change subtraction to addition; change -3 to its opposite $(+3)$.

$4 + (+3) + (-9)$

$7 + (-9)$

-2

WORK PROBLEM 8 AT THE SIDE. ▶▶

EXAMPLE 7 Using the Order of Operations to Combine More than Two Numbers

Find each sum.

(a) $-7 + 12 + (-3)$

$5 + (-3)$

2

(b) $14 + (-9) - (-8) + 10$

$5 - (-8) + 10$ Change subtraction to addition; change -8 to its opposite $(+8)$.

$5 + (+8) + 10$

$13 + 10$

23

(c) -6.3
   -14.9
     8.5
   -7.4
     5.2

Start at the top.

-6.3, -14.9 → -21.2, 8.5 → -12.7, -7.4 → -20.1, 5.2

-14.9

WORK PROBLEM 9 AT THE SIDE. ▶▶

8. Perform the addition and subtraction from left to right.

(a) $6 - 7 + (-3)$

(b) $-2 + (-3) - (-5)$

(c) $-3 - (-9) - (-5)$

(d) $8 - (-2) + (-6)$

9. Add.

(a) $-1 - 2 + 3 - 4$

(b) $7 - 6 - 5 + (-4)$

(c) $-6 + (-15) - (-19) + (-25)$

(d) -19.2
   -6.7
   15.8
   17.1
   -5.4

ANSWERS
8. (a) -4 (b) 0 (c) 11 (d) 4
9. (a) -4 (b) 8 (c) 27 (d) 1.6

A Census Bureau report gives the information at the right about family composition in 1970 and in 1995. There were approximately 99 million households in the United States in 1995.

Write 99 million using digits. **99,000,000**

In 1995, 1 of 10 households had five or more people living in it. 1 of 10 means the same as $\frac{1}{10}$. To find the number of households with five or more people, multiply $\frac{1}{10}$ times the number of households.
9,900,000 households

Then and Now

	1970	1995
People per household	3.14	2.65
Households with five or more people	1 of 5	1 of 10
Households with people living alone	One-sixth	One-fourth
Families maintained by women with no husband present	5.6 million	12.2 million
Families maintained by men with no wife present	1.2 million	3.2 million
Households in metropolitan areas	2 of 3	4 of 5

Source: Commerce Department's Census Bureau

What shortcut involving division could you use to find $\frac{1}{10}$ of the number of households?
Divide by 10 (move decimal point one place to left).

Why is multiplying by $\frac{1}{10}$ the same as dividing by 10?
The 10 in $\frac{1}{10}$ means the whole is divided into 10 equal parts. The 1 in the numerator means we are interested in only one of the 10 parts, so dividing by 10 gives the same result.

Show how to use both multiplication and a division shortcut to find the number of households with people living alone in 1995.
Multiply by $\frac{1}{4}$ or divide by 4 to get 24,750,000 households.

Use a signed number to express the change from 1970 to 1995 in the number of people per household.
–0.49 people per household.

If $\frac{1}{4}$ of the households in 1995 have people living alone, what fraction of the households have more than one person? $\frac{3}{4}$

If $\frac{4}{5}$ of the households in 1995 are in metropolitan areas, what fraction of the households are in rural areas? $\frac{1}{5}$

Use multiplication to find the number of households in metropolitan areas in 1995 and the number of households in rural areas. Can you also use a shortcut? If so, explain your shortcut.
79,200,000 metropolitan households; 19,800,000 rural households. One possible shortcut is to divide by 5 to get rural households, then multiply rural households by 4 to get metropolitan households.

Married couples with children made up 40% of all households in 1970 and 25% of all households in 1995. Suppose you wanted to add this information to the table. None of the other figures are given as percents. How could you change this information so it is more consistent with the rest of the table?
Express 40% as $\frac{2}{5}$ or 2 of 5 households. Express 25% as $\frac{1}{4}$ or 1 of 4 households.

9.2 Exercises

Add by using the number line.

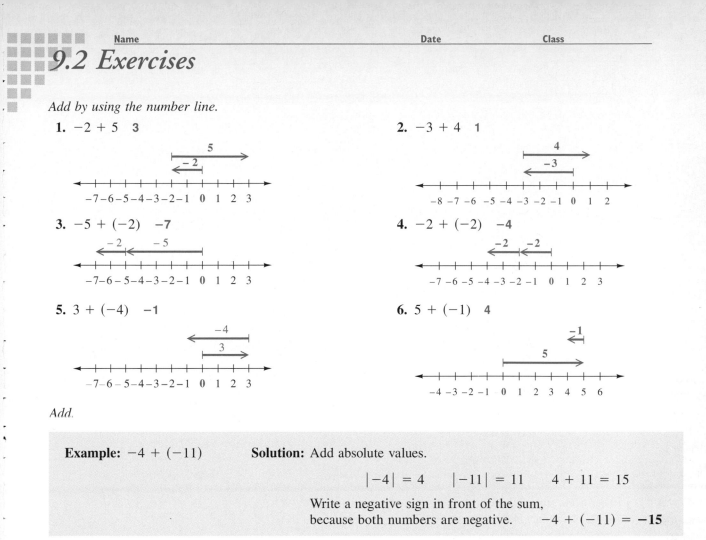

1. −2 + 5   **3**

2. −3 + 4   **1**

3. −5 + (−2)   **−7**

4. −2 + (−2)   **−4**

5. 3 + (−4)   **−1**

6. 5 + (−1)   **4**

Add.

Example: −4 + (−11)   **Solution:** Add absolute values.

$$|-4| = 4 \qquad |-11| = 11 \qquad 4 + 11 = 15$$

Write a negative sign in front of the sum, because both numbers are negative.   −4 + (−11) = **−15**

7. −8 + 5   **−3**

8. −3 + 2   **−1**

9. −1 + 8   **7**

10. −4 + 10   **6**

11. −2 + (−5)   **−7**

12. −7 + (−3)   **−10**

13. 6 + (−5)   **1**

14. 11 + (−3)   **8**

15. 4 + (−12)   **−8**

16. 9 + (−10)   **−1**

17. −10 + (−10)   **−20**

18. −5 + (−20)   **−25**

Write and solve an addition problem for each situation.

19. The football team gained 13 yards on the first play and lost 17 yards on the second play.

   13 + (−17) = −4

20. At penguin breeding grounds on Antarctic islands, winter temperatures routinely drop to −15°C; but at the interior of the continent, temperatures may easily drop another 60° below that.

   −15 + (−60) = −75

✐ Writing   ◉ Conceptual   ▲ Challenging   ≈ Estimation

21. Nicole's checking account was overdrawn by $52.50. She deposited $50 in the account.

$$-\$52.50 + \$50 = -\$2.50$$

22. $48.40 was stolen from Jay's car. He got $30 of it back.

$$-\$48.40 + \$30 = -\$18.40$$

Add.

23. $7.8 + (-14.6)$
-6.8

24. $4.9 + (-8.1)$
-3.2

25. $-\dfrac{1}{2} + \dfrac{3}{4}$ $\dfrac{1}{4}$

26. $-\dfrac{2}{3} + \dfrac{5}{6}$ $\dfrac{1}{6}$

27. $-\dfrac{7}{10} + \dfrac{2}{5}$ $-\dfrac{3}{10}$

28. $-\dfrac{3}{4} + \dfrac{3}{8}$ $-\dfrac{3}{8}$

29. $-\dfrac{7}{3} + \left(-\dfrac{5}{9}\right)$ $-\dfrac{26}{9}$

30. $-\dfrac{8}{5} + \left(-\dfrac{3}{10}\right)$ $-\dfrac{19}{10}$

31. Explain in your own words how to add two numbers with different signs. Include two examples in your explanation, one that has a positive answer and one that has a negative answer.

Subtract the smaller absolute value from the larger absolute value. The answer has the same sign as the addend with the larger absolute value. Examples: $-6 + 2 = -4$ and $6 + (-2) = 4$

32. Work these two examples:
  (a) $-5 + 3$ (b) $3 + (-5)$
Explain why the answers are the same.

Addition is commutative, so the order of the numbers does not affect the sum. Both answers are -2.

Give the additive inverse of each number.

	Number	Additive Inverse		Number	Additive Inverse		Number	Additive Inverse		Number	Additive Inverse
33.	3	-3	**34.**	4	-4	**35.**	-9	9	**36.**	-14	14
37.	$\dfrac{1}{2}$	$-\dfrac{1}{2}$	**38.**	$\dfrac{7}{8}$	$-\dfrac{7}{8}$	**39.**	-6.2	6.2	**40.**	-0.5	0.5

Subtract by changing subtraction to addition.

Examples: $-8 - (-2)$
             $7 - 11$

Solutions:
Change subtraction to addition. Change the sign of the second number to its opposite.
    $-8 - (-2)$ $7 - 11$
     $\downarrow$ $\downarrow$ $\downarrow$ $\downarrow$
  $-8 + (+2) = -6$ $7 + (-11) = -4$

41. $19 - 5$ 14

42. $24 - 11$ 13

43. $10 - 12$ -2

44. $1 - 8$ -7

45. $7 - 19$ -12

46. $2 - 17$ -15

47. $-15 - 10$ -25

48. $-10 - 4$ -14

49. $-9 - 14$ **−23**   **50.** $-3 - 11$ **−14**   **51.** $-3 - (-8)$ **5**   **52.** $-1 - (-4)$ **3**

53. $6 - (-14)$ **20**   **54.** $8 - (-1)$ **9**   **55.** $1 - (-10)$ **11**   **56.** $6 - (-1)$ **7**

57. $-30 - 30$ **−60**   **58.** $-25 - 25$ **−50**   **59.** $-16 - (-16)$ **0**   **60.** $-20 - (-20)$ **0**

61. $-\dfrac{7}{10} - \dfrac{4}{5}$

$-\dfrac{3}{2}$

62. $-\dfrac{8}{15} - \dfrac{3}{10}$

$-\dfrac{5}{6}$

63. $\dfrac{1}{2} - \dfrac{9}{10}$

$-\dfrac{2}{5}$

64. $\dfrac{2}{3} - \dfrac{11}{12}$

$-\dfrac{1}{4}$

65. $-8.3 - (-9)$

0.7

66. $-2 - (-3.9)$

1.9

📝 **67.** Explain the purpose of the "−" sign in each of
◉   these examples.
   (a) $6 - 9$ **(b)** (-9) **(c)** $-(-2)$

(a) 6 *minus* 9 (b) *negative* 9
(c) the *opposite* of *negative* 2

📝 **68.** Solve these two examples.
◉   **(a)** $8 - 3$ **(b)** $3 - 8$
How are the answers similar? How are they
different? Write a rule that explains what happens when you switch the order of the numbers in a subtraction problem.

(a) 5 (b) −5
The answers have the same absolute value but
opposite signs. When the numbers in a
subtraction problem are switched, change the sign
on the answer to its opposite.

Follow the order of operations to work each problem.

Example: $\underbrace{-6 + (-9)} - (-10)$	Add and subtract from left to right.
Solution: $\quad -15 \qquad - (-10)$ $\qquad\qquad\quad \downarrow \qquad \downarrow$ $\qquad -15 \qquad + (+10)$ $\qquad \underbrace{\qquad\qquad\qquad}$ $\qquad\qquad -5$	Change subtraction to addition; change -10 to its opposite $(+10)$.

69. $-2 + (-11) - (-3)$

   −10

70. $-5 - (-2) + (-6)$

   −9

71. $4 - (-13) + (-5)$

   12

72. $6 - (-1) + (-10)$

   −3

73. $-12 - (-3) - (-2)$

   −7

74. $-1 - (-7) - (-4)$

   10

75. $4 - (-4) - 3$ **5**

76. $5 - (-2) - 8$ **−1**

77. $\dfrac{1}{2} - \dfrac{2}{3} + \left(-\dfrac{5}{6}\right)$

   −1

78. $\dfrac{2}{5} - \dfrac{7}{10} + \left(-\dfrac{3}{2}\right)$

   $-\dfrac{9}{5}$

79. $-5.7 - (-9.4) - 8.1$

   −4.4

80. $-6.5 - (-11.2) - 1.4$

   3.3

▲ *Add or subtract the following.*

81. $-2 + (-11) + \left|-2\right|$ **−11**

82. $\left|-7 + 2\right| + (-2) + 4$ **7**

83. $-3 - (-2 + 4) + (-5)$ **−10**
(*Hint:* Work inside parentheses first.)

84. $5 - 8 - (6 - 7) + 1$ **−1**

85. $2\dfrac{1}{2} + 3\dfrac{1}{4} - \left(-1\dfrac{3}{8}\right) - 2\dfrac{3}{8}$ $\dfrac{19}{4}$ or $4\dfrac{3}{4}$

86. $\dfrac{5}{8} - \left(-\dfrac{1}{2} - \dfrac{3}{4}\right)$ $\dfrac{15}{8}$

Review and Prepare

*Multiply or divide the following. (For help, see **Sections 1.4, 1.5, 2.5, 2.7, 2.8, or 4.5.**)*

87. $23 \cdot 46$

   1058

88. $\dfrac{8}{11} \cdot \dfrac{3}{5}$

   $\dfrac{24}{55}$

89. $71.20 \cdot 21.25$

   1513

90. $2\dfrac{2}{3} \cdot 2\dfrac{3}{4}$

   $\dfrac{22}{3}$ or $7\dfrac{1}{3}$

91. $1235 \div 5$ **247**

92. $1\dfrac{2}{3} \div 2\dfrac{7}{9}$ $\dfrac{3}{5}$

93. $\dfrac{7}{9} \div \dfrac{14}{27}$ $\dfrac{3}{2}$ or $1\dfrac{1}{2}$

9.3 Multiplication and Division of Signed Numbers

How do you multiply two numbers with different signs? Look for a pattern in the following list of products.

$$4 \cdot 2 = 8$$
$$3 \cdot 2 = 6$$
$$2 \cdot 2 = 4$$
$$1 \cdot 2 = 2$$
$$0 \cdot 2 = 0$$
$$-1 \cdot 2 = ?$$

As the numbers in blue decrease by 1, the numbers in red decrease by 2. You can continue the pattern by replacing **?** with a number 2 *less than* 0, which is **−2**. Therefore:

$$-1 \cdot 2 = -2$$

OBJECTIVE 1 The pattern above suggests a rule for multiplying numbers with different signs.

Multiplying Numbers with Different Signs

The product of two numbers with *different* signs is *negative*.

EXAMPLE 1   Multiplying Numbers with Different Signs

Multiply.

(a) $-8 \cdot 4 = -32$ ← The product is negative.

(b) $6 \cdot (-3) = -18$

(c) $-5 \cdot (11) = -55$

(d) $12 \cdot (-7) = -84$

WORK PROBLEM 1 AT THE SIDE. ▶▶

For two numbers with the same signs, look at this pattern.

$$4 \cdot (-2) = -8$$
$$3 \cdot (-2) = -6$$
$$2 \cdot (-2) = -4$$
$$1 \cdot (-2) = -2$$
$$0 \cdot (-2) = 0$$
$$-1 \cdot (-2) = ?$$

This time, as the numbers in blue decrease by 1, the products increase by 2. You can continue the pattern by replacing **?** with a number 2 *greater than* 0, which is positive 2. Therefore:

$$-1 \cdot (-2) = 2$$

OBJECTIVE 2 In the pattern above, a negative number times a negative number gave a positive result.

OBJECTIVES

1 ▶ Multiply or divide two numbers with opposite signs.

2 ▶ Multiply or divide two numbers with the same sign.

FOR EXTRA HELP

Tutorial   Tape 15   SSM, Sec. 9.3

1. Multiply.

(a) $5 \cdot (-4)$

(b) $-9 \cdot (15)$

(c) $12 \cdot (-1)$

(d) $-6 \cdot (6)$

(e) $\left(-\dfrac{7}{8}\right)\left(\dfrac{4}{3}\right)$

ANSWERS

1. (a) -20 **(b)** -135 **(c)** -12 **(d)** -36
   (e) $-\dfrac{7}{6}$

603

2. Multiply.

(a) $(-5) \cdot (-5)$

(b) $(-14)(-1)$

(c) $-7 \cdot (-8)$

(d) $3 \cdot 12$

(e) $\left(-\dfrac{2}{3}\right)\left(-\dfrac{6}{5}\right)$

3. Divide.

(a) $\dfrac{-20}{4}$

(b) $\dfrac{-50}{-5}$

(c) $\dfrac{44}{2}$

(d) $\dfrac{6}{-6}$

(e) $\dfrac{-15}{-1}$

(f) $\dfrac{-\dfrac{3}{5}}{\dfrac{9}{10}}$

(g) $\dfrac{-35}{0}$

ANSWERS

2. (a) 25 (b) 14 (c) 56 (d) 36 (e) $\dfrac{4}{5}$

3. (a) -5 (b) 10 (c) 22 (d) -1 (e) 15

(f) $-\dfrac{2}{3}$ (g) undefined

Multiplying Numbers with the Same Sign

The product of two numbers with the *same* sign is *positive*.

E X A M P L E 2   Multiplying Two Numbers with the Same Sign

Multiply.

(a) $(-9)(-2)$ The numbers have the same sign (both are negative).

$(-9)(-2) = 18 \leftarrow$ The product is positive.

(b) $-7 \cdot (-4) = 28$ (c) $(-6)(-2) = 12$

(d) $(-10)(-5) = 50$ (e) $7 \cdot 5 = 35$

◀◀ **WORK PROBLEM 2 AT THE SIDE.**

You can use the same rules for dividing signed numbers as you use for multiplying signed numbers.

Dividing Signed Numbers

When two nonzero numbers with *different* signs are divided, the result is *negative*.

When two nonzero numbers with the *same* sign are divided, the result is *positive*.

Division with zero works the same as it did for whole numbers (see **Section 1.5**). Division by zero cannot be done. We say it is undefined. Zero divided by any other number is zero.

E X A M P L E 3   Dividing Signed Numbers

Divide.

(a) $\dfrac{-15}{5}$ ⟵ Numbers have different signs so the answer is negative. $\dfrac{-15}{5} = -3$

(b) $\dfrac{-8}{-4}$ ⟵ Numbers have same sign (both negative) so answer is positive. $\dfrac{-8}{-4} = 2$

(c) $\dfrac{-75}{-25} = 3$ (d) $\dfrac{-6}{0}$ is undefined   Division by zero cannot be done.

(e) $\dfrac{0}{-5} = 0$ (f) $\dfrac{90}{-9} = -10$

(g) $\dfrac{-\dfrac{2}{3}}{-\dfrac{5}{9}} = -\dfrac{2}{3} \cdot \left(-\dfrac{9}{5}\right)$ Invert the divisor: $-\dfrac{5}{9}$ becomes $-\dfrac{9}{5}$.

$= -\dfrac{2}{\underset{1}{3}} \cdot \left(-\dfrac{\overset{3}{9}}{5}\right)$ Cancel when possible. Then multiply.

$= \dfrac{6}{5}$ Both numbers were negative so answer is positive.

◀◀ **WORK PROBLEM 3 AT THE SIDE.**

9.3 Exercises

Multiply.

Examples: $-6 \cdot 5$ $\qquad$ $(-4)(-3)$ $\qquad$ $-\dfrac{3}{4} \cdot \left(-\dfrac{2}{3}\right)$

Solutions: $-6 \cdot 5 = -30$ $\qquad$ $(-4)(-3) = 12$ $\qquad$ $-\dfrac{\overset{1}{\cancel{3}}}{\underset{2}{\cancel{4}}} \cdot \left(-\dfrac{\overset{1}{\cancel{2}}}{\underset{1}{\cancel{3}}}\right) = \dfrac{1}{2}$

1. $-5 \cdot 7$

-35

2. $-10 \cdot 2$

-20

3. $(-5)(9)$

-45

4. $(-9)(4)$

-36

5. $3 \cdot (-6)$

-18

6. $8 \cdot (-6)$

-48

7. $10 \cdot (-5)$

-50

8. $5 \cdot (-11)$

-55

9. $(-1)(40)$

-40

10. $(75)(-1)$

-75

11. $-8 \cdot (-4)$

32

12. $-3 \cdot (-9)$

27

13. $11 \cdot 7$

77

14. $4 \cdot 25$

100

15. $-19 \cdot (-7)$

133

16. $-21 \cdot (-3)$

63

17. $-13 \cdot (-1)$

13

18. $-1 \cdot (-31)$

31

19. $(0)(-25)$

0

20. $(-50)(0)$

0

21. $-\dfrac{1}{2} \cdot (-8)$

4

22. $\dfrac{1}{3} \cdot (-15)$

-5

23. $-10 \cdot \left(\dfrac{2}{5}\right)$

-4

24. $-25 \cdot \left(-\dfrac{7}{10}\right)$

$\dfrac{35}{2}$

25. $\left(\dfrac{3}{5}\right)\left(-\dfrac{1}{6}\right)$

$-\dfrac{1}{10}$

26. $\left(-\dfrac{7}{9}\right)\left(-\dfrac{3}{4}\right)$

$\dfrac{7}{12}$

27. $-\dfrac{7}{5} \cdot \left(-\dfrac{10}{3}\right)$

$\dfrac{14}{3}$

28. $-\dfrac{9}{10} \cdot \dfrac{5}{4}$

$-\dfrac{9}{8}$

✎ Writing $\qquad$ ◉ Conceptual $\qquad$ ▲ Challenging $\qquad$ ≈ Estimation

29. $-\dfrac{7}{15} \cdot \dfrac{25}{14}$

$-\dfrac{5}{6}$

30. $-\dfrac{5}{9} \cdot \dfrac{18}{25}$

$-\dfrac{2}{5}$

31. $-\dfrac{5}{2} \cdot \left(-\dfrac{7}{10}\right)$

$\dfrac{7}{4}$

32. $-\dfrac{8}{5} \cdot \left(-\dfrac{15}{16}\right)$

$\dfrac{3}{2}$

33. $9 \cdot (-4.7)$

−42.3

34. $15 \cdot (-6.3)$

−94.5

35. $(-0.5)(-12)$

6

36. $(-3.15)(-5)$

15.75

37. $-6.2 \cdot (5.1)$

−31.62

38. $-4.3 \cdot (9.7)$

−41.71

39. $-1.25 \cdot (-3.6)$

4.5

40. $6.33 \cdot 0.2$

1.266

41. $(-8.23)(-1)$

8.23

42. $(-1)(-0.69)$

0.69

43. $0 \cdot (-58.6)$

0

44. $-91.3 \cdot 0$

0

Divide.

Examples: $\dfrac{-10}{2}$ $\qquad$ $\dfrac{32}{-8}$ $\qquad$ $\dfrac{-21}{-3}$ $\qquad$ $\dfrac{-8}{0}$

Solutions: $\dfrac{-10}{2} = \mathbf{-5}$ $\qquad$ $\dfrac{32}{-8} = \mathbf{-4}$ $\qquad$ $\dfrac{-21}{-3} = \mathbf{7}$ $\qquad$ Division by zero is undefined.

45. $\dfrac{-14}{7}$ −2

46. $\dfrac{-8}{2}$ −4

47. $\dfrac{30}{-6}$ −5

48. $\dfrac{21}{-7}$ −3

49. $\dfrac{-28}{0}$ undefined

50. $\dfrac{-40}{0}$ undefined

51. $\dfrac{14}{-1}$ −14

52. $\dfrac{25}{-1}$ −25

53. $\dfrac{-20}{-2}$ 10

54. $\dfrac{-80}{-4}$ 20

55. $\dfrac{-48}{-12}$ 4

56. $\dfrac{-30}{-15}$ 2

57. $\dfrac{-18}{18}$ -1

58. $\dfrac{50}{-50}$ -1

59. $\dfrac{-573}{-3}$ 191

60. $\dfrac{-580}{-5}$ 116

61. $\dfrac{0}{-9}$ 0

62. $\dfrac{0}{-4}$ 0

63. $\dfrac{-30}{-30}$ 1

64. $\dfrac{-25}{-25}$ 1

65. $\dfrac{-\dfrac{5}{7}}{-\dfrac{15}{14}}$ $\dfrac{2}{3}$

66. $\dfrac{-\dfrac{3}{4}}{-\dfrac{9}{16}}$ $\dfrac{4}{3}$

67. $-\dfrac{2}{3} \div (-2)$ $\dfrac{1}{3}$

68. $-\dfrac{3}{4} \div (-9)$ $\dfrac{1}{12}$

69. $5 \div \left(-\dfrac{5}{8}\right)$ -8

70. $7 \div \left(-\dfrac{14}{15}\right)$ $-\dfrac{15}{2}$

71. $-\dfrac{7}{5} \div \dfrac{3}{10}$ $-\dfrac{14}{3}$

72. $-\dfrac{4}{9} \div \dfrac{8}{3}$ $-\dfrac{1}{6}$

73. $\dfrac{-18.92}{-4}$ 4.73

74. $\dfrac{-22.75}{-7}$ 3.25

75. $\dfrac{-7.05}{1.5}$ -4.7

76. $\dfrac{-17.02}{7.4}$ -2.3

77. $\dfrac{45.58}{-8.6}$ -5.3

78. $\dfrac{6.27}{-0.3}$ -20.9

Following the order of operations, work from left to right in each of the following.

79. $(-4) \cdot (-6) \cdot \dfrac{1}{2}$

12

80. $(-9) \cdot (-3) \cdot \dfrac{2}{3}$

18

81. $(-0.6)(-0.2)(-3)$

-0.36

82. $(-4)(-1.2)(-0.7)$

-3.36

83. $\left(-\dfrac{1}{2}\right) \cdot \left(\dfrac{2}{5}\right) \cdot \left(\dfrac{7}{8}\right)$

$-\dfrac{7}{40}$

84. $\left(\dfrac{3}{4}\right) \cdot \left(-\dfrac{5}{6}\right) \cdot \left(\dfrac{2}{3}\right)$

$-\dfrac{5}{12}$

85. Write three examples for each of these situations:
(a) a positive number multiplied by -1
(b) a negative number multiplied by -1
Now write a rule that explains what happens when you multiply a signed number by -1.

(a) $6 \cdot (-1) = -6$; $2 \cdot (-1) = -2$; $15 \cdot (-1) = -15$
(b) $-6 \cdot (-1) = 6$; $-2 \cdot (-1) = 2$; $-15 \cdot (-1) = 15$

The result of multiplying any nonzero number times -1 is the number with the opposite sign.

86. Write three examples for each of these situations:
(a) a negative number divided by -1
(b) a positive number divided by -1
(c) a negative number divided by itself
Now write a rule that explains what happens when you divide a signed number by -1. Write another rule for a negative number divided by itself.

(a) $\frac{-6}{-1} = 6$; $\frac{-2}{-1} = 2$; $\frac{-15}{-1} = 15$

(b) $\frac{6}{-1} = -6$; $\frac{2}{-1} = -2$; $\frac{15}{-1} = -15$

(c) $\frac{-6}{-6} = 1$; $\frac{-2}{-2} = 1$; $\frac{-15}{-15} = 1$

When dividing by -1, the sign of the number changes to its opposite. A negative number divided by itself gives a result of 1.

87. Explain what is different and what is similar between multiplying and dividing signed numbers.

Similar: If the signs match, the result is positive. If the signs are different, the result is negative.
Different: Multiplication is commutative, division is not. You can multiply by zero, but dividing by zero is not allowed.

88. Explain why $\frac{0}{-3}$ and $\frac{-3}{0}$ do **not** give the same result.

Division is not commutative. $\frac{0}{-3} = 0$ because $0 \cdot (-3) = 0$. But $\frac{-3}{0}$ is undefined because when $\frac{-3}{0} = ?$ is rewritten as $? \cdot 0 = -3$, no number can replace ? and make a true statement.

▲ *Simplify the following.*

89. $-36 \div (-2) \div (-3) \div (-3) \div (-1)$

-2

90. $-48 \div (-8) \cdot (-4) \div (-4) \div (-3)$

-2

91. $|-8| \div (-4) \cdot |-5|$

-10

92. $-6 \cdot |-3| \div |9| \cdot (-2)$

4

Review and Prepare

Simplify the following. (For help, see Section 1.8.)

93. $8 + 4 \cdot 2 \div 8$ 9

94. $16 \div 2 \cdot 4 + (15 - 2 \cdot 3)$ 41

95. $7 + 6 \div 2 \cdot 4 - 9$ 10

96. $9 \div 3 + 8 \div 4 \cdot 2 - 7$ 0

9.4 Order of Operations

In the last two sections you worked examples that mixed either addition and subtraction or multiplication and division. In those situations you worked from left to right. Here are two more examples.

Work additions and subtractions from left to right.

$$-8 - (-6) + (-11)$$
$$\underbrace{-8 - (-6)}_{-2} + (-11)$$
$$\underbrace{-2 + (-11)}_{-13}$$

Work multiplications and divisions from left to right.

$$(-15) \div (-3) \cdot 6$$
$$\underbrace{(-15) \div (-3)}_{5} \cdot 6$$
$$\underbrace{5 \cdot 6}_{30}$$

WORK PROBLEM I AT THE SIDE. ▶▶

OBJECTIVE 1 Before working examples that mix division with addition or include parentheses, let's review the order of operations from **Section 1.8**.

Order of Operations

1. Work inside **parentheses.**
2. Simplify expressions with **exponents,** and find any **square roots.**
3. Multiply or divide from **left to right.**
4. Add or subtract from **left to right.**

E X A M P L E I   Using the Order of Operations

Use the order of operations to simplify the following.

$4 - 10 \div 2 + 7$ Check for parentheses: none.
Check for exponents and square roots: none.
Move from left to right, checking for multiplying and dividing.

$4 - \underbrace{10 \div 2}_{} + 7$ Yes, here is dividing. Use the number on either side of the ÷ sign.
$10 \div 2$ is 5. Bring down the other numbers and signs you haven't used.

$4 - 5 + 7$

Move from left to right, checking for adding and subtracting.
Yes, here is subtracting.

$4 - 5 + 7$ Change subtraction to addition; change 5 to its opposite (-5).

$\underbrace{4 + (-5)}_{-1} + 7$ Add $4 + (-5)$ to get -1.

$\underbrace{-1 + 7}_{6}$ Add $-1 + 7$ to get 6.

WORK PROBLEM 2 AT THE SIDE. ▶▶

OBJECTIVES

1 ▶ Use the order of operations.
2 ▶ Use the order of operations with exponents.
3 ▶ Use the order of operations with fraction bars.

FOR EXTRA HELP

Tutorial   Tape 16   SSM, Sec. 9.4

1. Simplify.

   (a) $-9 + (-15) + (-3)$

   (b) $-8 - (-2) + (-6)$

   (c) $-2 - (-7) - (-4)$

   (d) $3 \cdot (-4) \div (-6)$

   (e) $-18 \div 9 \cdot (-4)$

2. Use the order of operations to simplify each of the following.

   (a) $10 + 8 \div 2$

   (b) $4 - 6 \cdot (-2)$

   (c) $-3 + (-5) \cdot 2 - 1$

   (d) $-6 \div 2 + 3 \cdot (-2)$

   (e) $7 - 6 \cdot 2 \div (-3)$

ANSWERS

1. (a) -27 (b) -12 (c) 9 (d) 2 (e) 8
2. (a) 14 (b) 16 (c) -14 (d) -9 (e) 11

3. Simplify each of the following.

(a) $2 + 40 \div (-5 + 3)$

(b) $-5 \cdot 5 - (15 + 5)$

(c) $(-24 \div 2) + (15 - 3)$

(d) $-3 \cdot (2 - 8) - 5 \cdot (4 - 3)$

(e) $3 \cdot 3 - (10 \cdot 3) \div 5$

(f) $6 - (2 + 7) \div (-4 + 1)$

E X A M P L E 2 Parentheses and Order of Operations

Use the order of operations to simplify each of the following.

(a) $-8 \cdot \underbrace{(7 - 5)}_{} - 9$ Check for parentheses first. Yes; so do whatever you can inside the parentheses.

$-8 \cdot \quad 2 \quad - 9$ Bring down the other numbers and signs you haven't used yet. Check for exponents and square roots: none.

$\underbrace{-8 \cdot \quad 2}_{} \quad - 9$ Move from left to right, checking for multiplying and dividing. Yes, multiplying.

$-16 \quad - 9$ Change subtraction to addition.

$\underbrace{-16 \quad + (-9)}_{}$ Add $-16 + (-9)$ to get -25.

-25

(b) $3 + 2 \cdot \underbrace{(6 - 8)}_{} \cdot (15 \div 3)$ Work inside first set of parentheses; change $6 - 8$ to $6 + (-8)$ to get (-2).

$3 + 2 \cdot \quad (-2) \quad \cdot \underbrace{(15 \div 3)}_{}$ Work inside second set of parentheses.

$3 + \underbrace{2 \cdot (-2)}_{} \quad \cdot \quad 5$ Multiply from left to right; first multiply $2 \cdot (-2)$ to get -4.

$3 + \underbrace{\quad -4 \quad \cdot \quad 5}_{}$ Then multiply $-4 \cdot 5$ to get -20.

$\underbrace{3 + \quad\quad -20}_{}$ Add last. $3 + (-20)$ gives -17.

-17

◄◄ **WORK PROBLEM 3 AT THE SIDE.**

OBJECTIVE **2**▶ Remember that 2^3 means 2 is used as a factor 3 times:

$$2^3 = 2 \cdot 2 \cdot 2 = 8.$$

The 3 is called an *exponent*. Exponents are also used with signed numbers. For example:

$$(-3)^2 = (-3) \cdot (-3) = 9$$

$$(-4)^3 = \underbrace{(-4) \cdot (-4)}_{} \cdot (-4)$$ Be careful! Multiply two numbers at a time. Watch the signs.
$$= \underbrace{16 \quad \cdot (-4)}_{}$$
$$= -64$$

$$\left(-\frac{1}{2}\right)^4 = \underbrace{\left(-\frac{1}{2}\right) \cdot \left(-\frac{1}{2}\right)}_{} \cdot \left(-\frac{1}{2}\right) \cdot \left(-\frac{1}{2}\right)$$
$$= \underbrace{\frac{1}{4} \quad \cdot \left(-\frac{1}{2}\right)}_{} \cdot \left(-\frac{1}{2}\right)$$
$$= \underbrace{-\frac{1}{8} \quad \cdot \left(-\frac{1}{2}\right)}_{}$$
$$= \frac{1}{16}$$

ANSWERS

3. (a) -18 **(b)** -45 **(c)** 0 **(d)** 13 **(e)** 3 **(f)** 9

Be very careful with exponents and signed numbers. For example:

$$(-3)^2 = (-3) \cdot (-3) = 9. \leftarrow \text{Positive 9}$$

But the expression -3^2, with *no parentheses*, is different.

$$-3^2 = -(3 \cdot 3) = -9 \leftarrow \text{Negative 9}$$

Note

$(-3)^2$ is **not** the same as -3^2.

$$(-3)^2 = (-3) \cdot (-3) = 9 \quad but \quad -3^2 = -(3 \cdot 3) = -9$$

You will need this information as you take more algebra classes.

E X A M P L E 3 **Using Exponents and Order of Operations**

Simplify each of the following.

(a) $4^2 - (-3)^2$ There are parentheses around (-3) but no work can be done inside these parentheses.

$16 \overset{\downarrow}{-} 9$ Work with the exponents: $4^2 = 4 \cdot 4 = 16$ and $(-3)^2 = (-3)(-3) = 9$

7 Subtract: $16 - 9 = 7$

(b) $(-5)^2 - (4 - 6)^2 \cdot (-3)$ Work inside parentheses first.

$(-5)^2 - (-2)^2 \cdot (-3)$ Use the exponents next.

$25 - 4 \cdot (-3)$ Multiply.

$25 - (-12)$ Change subtraction to addition.

$25 + (+12)$ Add.

37

(c) $\left(\dfrac{2}{3} - \dfrac{1}{6}\right)^2 \div \left(-\dfrac{3}{8}\right)$ Inside parentheses: $\frac{2}{3} - \frac{1}{6} = \frac{4}{6} - \frac{1}{6} = \frac{3}{6} = \frac{1}{2}$

$\left(\dfrac{1}{2}\right)^2 \div \left(-\dfrac{3}{8}\right)$ Use the exponent: $(\frac{1}{2})^2 = \frac{1}{2} \cdot \frac{1}{2} = \frac{1}{4}$

$\dfrac{1}{4} \div \left(-\dfrac{3}{8}\right)$ Invert the divisor: $-\frac{3}{8}$ becomes $-\frac{8}{3}$

$\dfrac{1}{4} \cdot \left(-\dfrac{8}{3}\right)$ Cancel and multiply: $\frac{1}{4} \cdot -\frac{\overset{2}{8}}{3} = -\frac{2}{3}$

$-\dfrac{2}{3}$

WORK PROBLEM 4 AT THE SIDE. ▶▶

Note

Parentheses can be used in several different ways:

To indicate multiplication. $(4)(-3) = -12$

To separate a negative number from a minus sign. $8 - (-2) = 10$

To indicate which operation to do first. $35 + (6 - 2)$
$35 + 4$
39

4. Simplify each of the following.

(a) $2^3 - 3^2$

(b) $4^2 - 3^2 \cdot (5 - 2)$

(c) $-18 \div (-3) \cdot 2^3$

(d) $(-3)^3 + (3 - 8)^2$

(e) $\dfrac{3}{8} + \left(-\dfrac{1}{2}\right)^2 \div \dfrac{1}{4}$

ANSWERS

4. (a) -1 **(b)** -11 **(c)** 48
(d) -2 **(e)** $\dfrac{11}{8}$

5. Simplify each of the following.

(a) $\dfrac{-3 \cdot 2^3}{-10 - 6 + 8}$

(b) $\dfrac{(-10)(-5)}{-6 \div 3 \cdot 5}$

(c) $\dfrac{6 + 18 \div (-2)}{(1 - 10) \div 3}$

(d) $\dfrac{6^2 - 3^2 \cdot 4}{5 + (3 - 7)^2}$

OBJECTIVE ▶3 A fraction bar indicates division, as in $\frac{-6}{2}$ which means $-6 \div 2$. In an example like the one below

$$\frac{-5 + 3^2}{16 - 7 \cdot 2}$$

the fraction bar also tells us to do the work in the numerator, then the work in the denominator. The last step is to divide the results.

$$\frac{-5 + 3^2}{16 - 7 \cdot 2} \rightarrow \frac{-5 + 9}{16 - 14} \rightarrow \frac{4}{2} \rightarrow \text{Now divide.} \quad 4 \div 2 = 2$$

The final result is 2.

E X A M P L E 4   Fraction Bars and Order of Operations

Simplify the following.

$$\frac{-8 + (4 - 6) \cdot 5}{4 - 4^2 \div 8}$$

First do the work in the numerator.

$$-8 + \underbrace{(4 - 6)} \cdot 5 \qquad \text{Work inside parentheses.}$$
$$-8 + \underbrace{-2 \cdot 5} \qquad \text{Multiply.}$$
$$\underbrace{-8 + (-10)} \qquad \text{Add.}$$
$$\text{Numerator} \rightarrow -18$$

Now do the work in the denominator.

$$4 - \underbrace{4^2} \div 8 \qquad \text{No parentheses; use exponent.}$$
$$4 - \underbrace{16 \div 8} \qquad \text{Divide.}$$
$$\underbrace{4 - 2} \qquad \text{Subtract.}$$
$$\text{Denominator} \rightarrow 2$$

The last step is the division.

$$\frac{\text{Numerator} \rightarrow -18}{\text{Denominator} \rightarrow 2} = -9$$

◀◀ **WORK PROBLEM 5 AT THE SIDE.**

ANSWERS

5. (a) $\dfrac{-24}{-8} = 3$ **(b)** $\dfrac{50}{-10} = -5$

(c) $\dfrac{-3}{-3} = 1$ **(d)** $\dfrac{0}{21} = 0$

9.4 Exercises

Simplify each of the following.

Examples: $-6 - \underbrace{3^2}$ $4 \cdot \underbrace{(6 - 11)}^2 - (-8)$

Solutions: $-6 - \quad 9$ $4 \cdot \underbrace{(-5)^2} \quad - (-8)$

$$\underbrace{-6 + (-9)}$$
$$\mathbf{-15}$$

$4 \cdot \underbrace{\quad 25 \quad} - (-8)$

$100 \qquad - (-8)$

$100 \qquad + (+8)$

$$\underbrace{\qquad\qquad}$$
$$\mathbf{108}$$

1. $6 + 3 \cdot (-4)$

   -6

2. $10 - 30 \div 2$

   -5

3. $-1 + 15 + (-7) \cdot 2$

   0

4. $9 + (-5) + 2 \cdot (-2)$

   0

5. $6^2 + 4^2$

   52

6. $3^2 + 8^2$

   73

7. $10 - 7^2$

   -39

8. $5 - 5^2$

   -20

9. $(-2)^5 + 2$

   -30

10. $(-2)^4 - 7$

   9

11. $4^2 + 3^2 + (-8)$

   17

12. $5^2 + 2^2 + (-12)$

   17

13. $2 - (-5) + 3^2$

   16

14. $6 - (-9) + 2^3$

   23

15. $(-4)^2 + (-3)^2 + 5$

   30

16. $(-5)^2 + (-6)^2 + 12$

   73

17. $3 + 5 \cdot (6 - 2)$

   23

18. $4 + 3 \cdot (8 - 3)$

   19

19. $-7 + 6 \cdot (8 - 14)$

   -43

20. $-3 + 5 \cdot (9 - 12)$

   -18

21. $-6 + (-5) \cdot (9 - 14)$

   19

   ✐ Writing   ⊙ Conceptual   ▲ Challenging   ≈ Estimation

22. $-5 + (-3) \cdot (6 - 7)$

-2

23. $(-5) \cdot (7 - 13) \div (-10)$

-3

24. $(-4) \cdot (9 - 17) \div (-8)$

-4

25. $9 \div (-3)^2 + (-1)$

0

26. $-48 \div (-4)^2 + 3$

0

27. $2 - (-5) \cdot (-3)^2$

47

28. $1 - (-10) \cdot (-2)^3$

-79

29. $(-2) \cdot (-7) + 3 \cdot 9$

41

30. $4 \cdot (-2) + (-3) \cdot (-5)$

7

31. $30 \div (-5) - 36 \div (-9)$

-2

32. $8 \div (-4) - 42 \div (-7)$

4

33. $2 \cdot 5 - 3 \cdot 4 + 5 \cdot 3$

13

34. $9 \cdot 3 - 6 \cdot 4 + 3 \cdot 7$

24

35. $4 \cdot 3^2 + 7 \cdot (3 + 9) - (-6)$

126

36. $5 \cdot 4^2 - 6 \cdot (1 + 4) - (-3)$

53

37. $\dfrac{-1 + 5^2 - (-3)}{-6 - 9 + 12}$ $\dfrac{27}{-3} = -9$

38. $\dfrac{-6 + 3^2 - (-7)}{7 - 9 - 3}$ $\dfrac{10}{-5} = -2$

39. $\dfrac{-2 \cdot 4^2 - 4 \cdot (6 - 2)}{-4 \cdot (8 - 13) \div (-5)}$ $\dfrac{-48}{-4} = 12$

40. $\dfrac{3 \cdot 3^2 - 5 \cdot (9 - 2)}{8 \cdot (6 - 9) \div (-3)}$ $\dfrac{-8}{8} = -1$

41. $\dfrac{2^3 \cdot (-2 - 5) + 4 \cdot (-1)}{4 + 5 \cdot (-6 \cdot 2) + (5 \cdot 11)}$ $\dfrac{-60}{-1} = 60$

42. $\dfrac{3^3 + (-1 - 2) \cdot 4 - 25}{-4 + 4 \cdot (3 \cdot 5) + (-6 \cdot 9)}$ $\dfrac{-10}{2} = -5$

43. $(-4)^2 \cdot (7-9)^2 \div 2^3$

    8

44. $(-5)^2 \cdot (9-17)^2 \div (-10)^2$

    16

45. $(-0.3)^2 + (-0.5)^2 + 0.9$

    1.24

46. $(0.2)^3 - (-0.4)^2 + 3.02$

    2.868

47. $(-0.75) \cdot (3.6-5)^2$

    −1.47

48. $(-0.3) \cdot (4-6.8)^2$

    −2.352

49. $(0.5)^2 \cdot (-8) - (0.31)$

    −2.31

50. $(0.3)^3 \cdot (-5) - (-2.8)$

    2.665

51. $\dfrac{2}{3} \div \left(-\dfrac{5}{6}\right) - \dfrac{1}{2}$

    $-\dfrac{13}{10}$

52. $\dfrac{5}{8} \div \left(-\dfrac{10}{3}\right) - \dfrac{3}{4}$

    $-\dfrac{15}{16}$

53. $\left(-\dfrac{1}{2}\right)^2 - \left(\dfrac{3}{4} - \dfrac{7}{4}\right)$

    $\dfrac{5}{4}$

54. $\left(-\dfrac{2}{3}\right)^2 - \left(\dfrac{1}{6} - \dfrac{11}{6}\right)$

    $\dfrac{19}{9}$

55. $\dfrac{3}{5} \cdot \left(-\dfrac{7}{6}\right) - \left(\dfrac{1}{6} - \dfrac{5}{3}\right)$

    $\dfrac{4}{5}$

56. $\dfrac{2}{7} \cdot \left(-\dfrac{14}{5}\right) - \left(\dfrac{4}{3} - \dfrac{13}{9}\right)$

    $-\dfrac{31}{45}$

57. $5^2 \cdot (9-11) \cdot (-3) \cdot (-2)^3$

    −1200

58. $4^2 \cdot (13-17) \cdot (-2) \cdot (-3)^2$

    1152

59. $1.6 \cdot (-0.8) \div (-0.32) \div 2^2$

    1

60. $6.5 \cdot (-4.8) \div (-0.3) \div (-2)^3$

    −13

61. Solve this series of examples.

$$(-2)^2 = \qquad (-2)^6 =$$
$$(-2)^3 = \qquad (-2)^7 =$$
$$(-2)^4 = \qquad (-2)^8 =$$
$$(-2)^5 = \qquad (-2)^9 =$$

What pattern do you see in the sign of the answers?

The answers are 4, −8, 16, −32, 64, −128, 256, −512. When a negative number is raised to an even power, the answer is positive; when raised to an odd power, the answer is negative.

62. Explain the difference between -5^2 and $(-5)^2$.

-5^2 is the opposite of 5^2 or $-(5 \cdot 5) = -25$. The parentheses in $(-5)^2$ mean that -5 is multiplied times itself. So $(-5)^2 = (-5) \cdot (-5) = 25$. One answer is negative and the other is positive.

▲ *Simplify each of the following.*

63. $\dfrac{-9 + 18 \div (-3) \cdot (-6)}{5 - 4 \cdot 12 \div 3 \cdot 2}$

$$\dfrac{27}{-27} = -1$$

64. $\dfrac{-20 - 15 \cdot (-4) - (-40)}{4 + 27 \div 3 \cdot (-2) - 6}$

$$\dfrac{80}{-20} = -4$$

65. $-7 \cdot \left(6 - \dfrac{5}{8} \cdot 24 + 3 \cdot \dfrac{8}{3}\right)$

7

66. $(-0.3)^2 \cdot (-5 \cdot 3) + (6 \div 2 \cdot 0.4)$

−0.15

67. $|-12| \div 4 + 2 \cdot 3^2 \div 6$

6

68. $6 - (2 - 3 \cdot 4) + 5^2 \div \left(-2 \cdot \dfrac{5}{2}\right) + (2)^2$

15

Review and Prepare

*Evaluate the following. (For help, see **Section 3.5**.)*

69. $\dfrac{6}{7} \cdot \dfrac{14}{9}$

$\dfrac{4}{3}$ or $1\dfrac{1}{3}$

70. $\dfrac{17}{15} \div \dfrac{34}{5}$

$\dfrac{1}{6}$

71. $\dfrac{5}{3} - \dfrac{3}{8} \cdot \dfrac{4}{9}$

$\dfrac{3}{2}$ or $1\dfrac{1}{2}$

72. $\dfrac{7}{12} \div \dfrac{21}{10} \cdot \dfrac{4}{5}$

$\dfrac{2}{9}$

9.5 *Evaluating Expressions and Formulas*

In formulas you have seen that numbers can be represented by letters. For example, you used this formula for finding simple interest in **Section 6.8**.

$$I = p \cdot r \cdot t$$

In this formula, p (principal) represents the amount of money borrowed, r is the rate of interest, and t is the time in years. In algebra, we often write multiplication without the multiplication dots. If there is no operation sign written between two letters, or between a letter and a number, you assume it is multiplication.

> **Showing Multiplication in Algebra**
>
> If there is no operation sign, it is understood to be multiplication. Here are some examples.
>
> | | | |
> |---|---|---|
> | $I = p \cdot r \cdot t$ | is written | $I = prt$ |
> | $2 \cdot r$ | is written | $2r$ |
> | $3 \cdot x + 4 \cdot y$ | is written | $3x + 4y$ |

OBJECTIVE 1 Letters (such as the I, p, r, or t used above) that represent numbers are called **variables** (VAIR-ee-uh-buls). A combination of letters and numbers is an **expression** (eks-PRESH-un). Three examples of expressions are shown here.

$$9 + p \qquad 8r \qquad 7k - 2m$$

OBJECTIVE 2 The value of an expression changes depending upon the value of each variable. To find the value of an expression, replace the variables with their values. It is helpful to write each value inside parentheses when multiplication is involved.

EXAMPLE 1   Finding the Value of an Expression

Find the value of $5x - 3y$, if $x = 2$ and $y = 7$.

Replace x with 2. Replace y with 7.

Using the order of operations, multiply first.

$$5x - 3y$$
$$5(2) - 3(7)$$
$$10 - 21$$
$$-11$$

WORK PROBLEM 1 AT THE SIDE. ▶▶

EXAMPLE 2   Finding the Value of an Expression

Find the value of $7m - 8n + p$, if $m = -2$, $n = 4$, and $p = 3$.

Replace m with -2, n with 4, and p with 3.

$$7m - 8n + p$$
$$7(-2) - 8(4) + 3 \qquad \text{Multiply from left to right.}$$
$$-14 - 32 + 3 \qquad \text{Add and subtract from left to right.}$$
$$-46 + 3$$
$$-43$$

WORK PROBLEMS 2 AND 3 AT THE SIDE. ▶▶

OBJECTIVES

1 ▶ Define variable and expression.

2 ▶ Find the value of an expression when values of the variables are given.

FOR EXTRA HELP

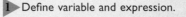

Tutorial   Tape 16   SSM, Sec. 9.5

1. Find the value of $5x - 3y$ if

   (a) $x = 1$, $y = 2$.

   (b) $x = 3$, $y = -4$.

   (c) $x = 0$, $y = 6$.

2. Find the value of $7m - 8n + p$ if

   (a) $m = 1$, $n = 2$, $p = 5$.

   (b) $m = -4$, $n = -3$, $p = -7$.

   (c) $m = -5$, $n = 0$, $p = -1$.

3. Find the value of $x + 6y$ if

   (a) $x = 9$, $y = -3$.

   (b) $x = -2$, $y = 1$.

   (c) $x = 6$, $y = -1$.

ANSWERS

1. (a) -1 (b) 27 (c) -18
2. (a) -4 (b) -11 (c) -36
3. (a) -9 (b) 4 (c) 0

4. Find the value of $\dfrac{3k + r}{2s}$ if

(a) $k = 1, r = 1, s = 2.$

(b) $k = 8, r = -2,$
$s = -4.$

(c) $k = -3, r = 1,$
$s = -2.$

5. Find the value of A, P, d, and C in these formulas.

(a) $A = \dfrac{1}{2}bh; b = 6$ yd,
$h = 12$ yd

(b) $P = 2l + 2w; l = 10,$
$w = 8$

(c) $d = rt; r = 4,$
$t = 80$

(d) $C = 2\pi r; \pi \approx 3.14,$
$r = 6$

ANSWERS

4. (a) $\dfrac{4}{4} = 1$ (b) $\dfrac{22}{-8} = -\dfrac{11}{4}$ (c) $\dfrac{-8}{-4} = 2$

5. (a) $A = 36$ yd^2 (b) $P = 36$
(c) $d = 32$ (d) $C \approx 37.68$

E X A M P L E 3 Finding the Value of an Expression

Find the value of $5x - y$, if $x = 2$ and $y = -3$.

Replace x with 2. $5\ x\ -\ y$ Replace y with -3.

$$5(2) - (-3)$$

$$\underbrace{10\ +\ (+3)}$$

$$13$$

E X A M P L E 4 Finding the Value of an Expression

What is the value of $\dfrac{6k + 2r}{5s}$, if $k = -2$, $r = 5$, and $s = -1$?

Replace k with -2, r with 5, and s with -1.

$$\frac{6k + 2r}{5s} = \frac{6(-2) + 2(5)}{5(-1)}$$

$$= \frac{-12 + 10}{-5} \qquad \text{Multiply.}$$

$$= \frac{-2}{-5} \qquad \text{Add in numerator.}$$

$$= \frac{2}{5} \qquad \begin{array}{l}\text{Dividing two numbers with the}\\ \text{same sign gives a positive answer.}\end{array}$$

◀◀ **WORK PROBLEM 4 AT THE SIDE.**

E X A M P L E 5 Evaluating a Formula

The formula you used in Chapter 8 for the area of a triangle can now be written without the multiplication dots.

$$A = \frac{1}{2} \cdot b \cdot h \qquad \text{is written} \qquad A = \frac{1}{2}bh$$

In this formula, b is the length of the base and h is the height. What is the area if $b = 9$ cm and $h = 24$ cm?

$$A = \frac{1}{2}\ \ b\ \ \ \ h \qquad \text{Replace } b \text{ with 9 cm and } h \text{ with 24 cm.}$$

$$A = \frac{1}{2}(9\ \text{cm})(24\ \text{cm})$$

$$A = \frac{1}{2}(9\ \text{cm})(\overset{12}{24}\ \text{cm}) \qquad \text{Cancel if possible.}$$

$$A = 108\ \text{cm}^2$$

The area of the triangle is 108 cm^2.

Note
Area is measured in square units. The short way to write square centimeters is cm^2.

◀◀ **WORK PROBLEM 5 AT THE SIDE.**

9.5 Exercises

Find the value of the expression $2r + 4s$ for each of the following values of r and s.

> **Example:** $r = 3$, $s = -5$
>
> **Solution:**
> Replace r with 3.
> Replace s with -5.
> Using the order of operations,
> multiply first, then add.
>
> $$\underbrace{\underset{\downarrow\ \downarrow}{2\ r}}_{\underbrace{2(3)}_{6}}\ +\ \underbrace{\underset{\downarrow\ \downarrow}{4\ s}}_{\underbrace{4(-5)}_{+\ (-20)}}$$
>
> $$-14$$

1. $r = 2$, $s = 6$

28

2. $r = 6$, $s = 1$

16

3. $r = 1$, $s = -3$

-10

4. $r = 7$, $s = -2$

6

5. $r = -4$, $s = 4$

8

6. $r = -3$, $s = 5$

14

7. $r = -1$, $s = -7$

-30

8. $r = -3$, $s = -5$

-26

9. $r = 0$, $s = -2$

-8

10. $r = -7$, $s = 0$

-14

Use the given values of the variables to find the value of each expression.

11. $8x - y$; $x = 1$, $y = 8$

0

12. $a - 5b$; $a = 10$, $b = 2$

0

13. $6k + 2s$; $k = 1$, $s = -2$

2

14. $7p + 7q$; $p = -4$, $q = 1$

-21

✍ Writing　　◉ Conceptual　　▲ Challenging　　≈ Estimation

15. $\dfrac{-m + 5n}{2s + 2}$; $m = 4$, $n = -8$, $s = 0$

-22

16. $\dfrac{2y - z}{x - 2}$; $y = 0$, $z = 5$, $x = 1$

5

17. $-m - 3n$; $m = \dfrac{1}{2}$, $n = \dfrac{3}{8}$

$-\dfrac{13}{8}$

18. $7k - 3r$; $k = \dfrac{2}{3}$, $r = -\dfrac{1}{3}$

$\dfrac{17}{3}$

Be careful when an expression has a negative sign and the value of the variable is also negative. Use the given values to find the value of each expression.

Example: Find the value of $-c - 5b$ when $c = -2$ and $b = -3$

Solution: Replace c with -2.
Replace b with -3.
$-(-2)$ is the opposite
of (-2) which is $(+2)$.

$$\begin{array}{cccc} - & c & - 5 & b \\ \downarrow & \downarrow & \downarrow & \downarrow \\ -(-2) & - & 5(-3) & \\ \hline +2 & - & (-15) & \\ & \downarrow & \downarrow & \\ 2 & + & (+15) & \end{array}$$

Multiply $5 \cdot (-3)$

Change subtraction to addition.

$$\underbrace{}$$

17

19. $-c - 5b$; $c = -8$, $b = -4$

28

20. $-c - 5b$; $c = -1$, $b = -2$

11

21. $-4x - y$; $x = 5$, $y = -15$

-5

22. $-4x - y$; $x = 3$, $y = -8$

-4

23. $-k - m - 8n$; $k = 6$, $m = -9$, $n = 0$

3

24. $-k - m - 8n$; $k = 0$, $m = -7$, $n = -1$

15

25. $\dfrac{-3s - t - 4}{-s + 6 + t}$; $s = -1$, $t = -13$

$\dfrac{12}{-6} = -2$

26. $\dfrac{-3s - t - 4}{-s - 20 - t}$; $s = -3$, $t = -6$

$\dfrac{11}{-11} = -1$

In each of the following, use the given formula and values of the variables to find the value of the remaining variable.

Example: $P = 2l + 2w$; $l = 33$, $w = 16$ **Solution:** Replace l with 33 and w with 16.

$$P = 2 \quad l \quad + 2 \quad w$$
$$P = \underbrace{2 \cdot (33)} + \underbrace{2 \cdot (16)}$$
$$P = \quad 66 \quad + \quad 32$$
$$P = \mathbf{98}$$

27. $P = 4s$; $s = 7.5$

30

28. $P = 4s$; $s = 0.8$

3.2

29. $P = 2l + 2w$; $l = 9$, $w = 5$

28

30. $P = 2l + 2w$; $l = 12$, $w = 2$

28

31. $A = \pi r^2$; $\pi \approx 3.14$, $r = 5$

≈ 78.5

32. $A = \pi r^2$; $\pi \approx 3.14$, $r = 10$

≈ 314

33. $A = \dfrac{1}{2}bh$; $b = 15$, $h = 3$

$\dfrac{45}{2}$ or $22\dfrac{1}{2}$

34. $A = \dfrac{1}{2}bh$; $b = 5$, $h = 11$

$\dfrac{55}{2}$ or $27\dfrac{1}{2}$

35. $V = \dfrac{1}{3}Bh$; $B = 30$, $h = 60$

600

36. $V = \dfrac{1}{3}Bh$; $B = 105$, $h = 5$

175

37. $d = rt$; $r = 53$, $t = 6$

318

38. $d = rt$; $r = 180$, $t = 5$

900

39. $C = 2\pi r$; $\pi \approx 3.14$, $r = 4$

≈ 25.12

40. $C = 2\pi r$; $\pi \approx 3.14$, $r = 18$

≈ 113.04

41. Find and correct the error made by the student who solved this example:

Find the value of $-x - 4y$ if $x = -3$ and $y = -1$.

$$-x - 4y$$
$$-3 - 4(-1)$$
$$-3 - (-4)$$
$$-3 + (+4)$$
$$1$$

Also write a sentence next to each step, explaining what is being done in that step.

The error was made when replacing x with -3; should be $-(-3)$, not -3.

$-x - 4y$	Replace x with (-3) and y with (-1).
$-(-3) - 4(-1)$	Opposite of (-3) is $+3$
$(+3) - (-4)$	Change subtraction to addition.
$3 + (+4)$	Add.
7	

42. Go back to Chapter 8 and find one of the formulas listed below. Pick values for the variables indicated and find the value of A or V. Then pick different values for the variables and again find the value of A or V.

Area of a trapezoid: pick values for h, a, and b.

Volume of a rectangular solid: pick values for l, w, and h.

Area of trapezoid $= \dfrac{1}{2} h(b + B)$

If h is 12, b is 6, and B is 10, then

$$A = \frac{1}{2}(12)(6 + 10)$$

$$A = \frac{1}{\cancel{2}}\left(\frac{\overset{6}{\cancel{12}}}{1}\right)(16)$$

$$A = 96$$

Volume of rectangular solid $= lwh$
If $l = 5$, $w = 3$, and $h = 4$, then

$$V = (5)(3)(4)$$
$$V = 60$$

▲ *Use the given formula and values of the variables to find the value of the remaining variable. If you studied Chapters 7 and 8, write a sentence telling when you would use each formula.*

43. $F = \dfrac{9C}{5} + 32$; $C = -40$

−40; convert a Celsius temperature to Fahrenheit

44. $C = \dfrac{5(F - 32)}{9}$; $F = -4$

−20; convert a Fahrenheit temperature to Celsius

45. $V = \dfrac{4\pi r^3}{3}$; $\pi \approx 3.14$, $r = 3$

≈113.04; find the volume of a sphere

46. $c^2 = a^2 + b^2$; $a = 3$, $b = 4$

5; Pythagorean theorem for finding the hypotenuse or a leg in a right triangle.

47. $A = \dfrac{1}{2}h(b + B)$; $h = 7$, $b = 4$, $B = 12$

56; find the area of a trapezoid

48. $V = \dfrac{\pi r^2 h}{3}$; $\pi \approx 3.14$, $r = 6$, $h = 10$

≈376.8; find the volume of a cone

Review and Prepare

*Simplify the following. (For help, see **Sections 2.5 and 9.3**.)*

49. $-\dfrac{1}{9} \cdot 9$

−1

50. $4 \cdot \left(-\dfrac{3}{2}\right)$

−6

51. $-\dfrac{4}{3} \cdot \left(-\dfrac{3}{4}\right)$

1

52. $-\dfrac{2}{5} \cdot \left(-\dfrac{5}{2}\right)$

1

9.6 Solving Equations

An **equation** (ee-KWAY-zhuhn) is a statement that says two expressions are equal. Examples of equations are shown here.

$$x + 1 = 9 \qquad 20 = 5k \qquad 6r - 1 = 17$$

The **equal sign** in an equation divides the equation into two parts, the *left side* and the *right side*. In $6r - 1 = 17$, the left side is $6r - 1$, and the right side is 17.

$$\underbrace{6r - 1}_{\text{Left side}} = \underbrace{17}_{\text{Right side}}$$

You solve an equation by finding all numbers that can be substituted for the variable to make the equation true. These numbers are called **solutions** of the equation.

OBJECTIVE 1 To tell whether a number is a solution of the equation, substitute the number in the equation to see whether the result is true.

E X A M P L E I   Determining If a Number Is a Solution of an Equation

Is 7 a solution of the following equations?

(a) $12 = x + 5$

Replace x with 7.

$$
\begin{aligned}
12 &= x + 5 \\
12 &= 7 + 5 \qquad \text{Replace } x \text{ with 7.} \\
12 &= 12 \qquad \text{True}
\end{aligned}
$$

Because the statement is true, 7 is a solution of $12 = x + 5$.

(b) $2y + 1 = 16$

Replace y with 7.

$$
\begin{aligned}
2y + 1 &= 16 \\
2(7) + 1 &= 16 \\
14 + 1 &= 16 \\
15 &= 16 \qquad \text{False}
\end{aligned}
$$

The false statement shows that 7 is *not* a solution of $2y + 1 = 16$.

WORK PROBLEM I AT THE SIDE. ▶▶

OBJECTIVE 2 If the equation $a = b$ is true, and if a number c is added to both a and b, the new equation is also true. This rule is called the **addition property of equations.** It means that you can add the same number to both sides of an equation and still have a true equation.

Addition Property of Equations

If $a = b$, then $a + c = b + c$. In other words, you may add the same number to both sides of an equation.

You can use the addition property to solve equations. The idea is to get the variable (the letter) by itself on one side of the equal sign and a number by itself on the other side.

OBJECTIVES

1. ▶ Tell whether a number is a solution of an equation.

2. ▶ Solve equations using the addition property of equations.

3. ▶ Solve equations using the multiplication property of equations.

FOR EXTRA HELP

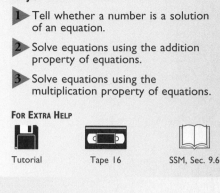

Tutorial   Tape 16   SSM, Sec. 9.6

1. Decide if the given number is a solution of the equation.

(a) $p + 1 = 8$; 7

(b) $30 = 5r$; 6

(c) $3k - 2 = 4$; 3

(d) $23 = 4y + 3$; 5

ANSWERS

1. **(a)** solution **(b)** solution
   (c) not a solution **(d)** solution

623

2. Solve each equation. Check each solution.

(a) $n - 5 = 8$

(b) $5 = r - 10$

(c) $3 = z + 1$

(d) $k + 9 = 0$

(e) $-2 = y + 9$

(f) $x - 2 = -6$

E X A M P L E   2   **Solving an Equation by Using the Addition Property**

Solve each equation.

(a) $k - 4 = 6$

To get k by itself on the left side, add 4 to the left side, because $k - 4 + 4$ gives $k + 0$. You must then add 4 to the right side also.

$$k - 4 = 6 \qquad \leftarrow \text{Original equation}$$
$$k \underbrace{- 4 + 4}_{} = 6 + 4 \qquad \text{Add 4 to both sides.}$$
$$k + 0 = 10$$
$$k = 10$$

The solution is 10. Check by replacing k with 10 in the original equation.

$$k - 4 = 6$$
$$10 - 4 = 6 \qquad \text{Replace } k \text{ with 10.}$$
$$6 = 6 \qquad \text{True}$$

This result is true, so 10 is the solution.

(b) $2 = z + 8$

To get z by itself on the right side, add -8 to both sides.

$$2 = z + 8 \qquad \leftarrow \text{Original equation}$$
$$2 + (-8) = z + 8 + (-8) \qquad \text{Add } (-8) \text{ to both sides.}$$
$$-6 = z + 0$$
$$-6 = z$$

Check the solution by replacing z with -6 in the original equation.

$$2 = z + 8$$
$$2 = -6 + 8$$
$$2 = 2 \qquad \text{True, so } -6 \text{ is the solution}$$

Notice that we *added* -8 to both sides to get z by itself. We can accomplish the same thing by *subtracting* 8 from both sides. Recall from **Section 9.2** that subtraction is defined in terms of addition. On the left side of the equation above

$$2 - 8 \text{ gives the same result as } 2 + (-8).$$

Note
You may add *or subtract* the same number on both sides of an equation.

◀◀ **WORK PROBLEM 2 AT THE SIDE.**

OBJECTIVE 3▶ Here is a summary of the rules you can use to solve equations using the addition property. In these rules, x is the variable and a and b are numbers.

Solving Equations Using the Addition Property

Solve $x - a = b$ or $b = x - a$ by adding a to both sides.

Solve $x + a = b$ or $b = x + a$ by subtracting a from both sides.

ANSWERS
2. (a) 13 **(b)** 15 **(c)** 2 **(d)** -9 **(e)** -11
 (f) -4

Multiplication Property of Equations

If $a = b$ and c does not equal 0, then

$$a \cdot c = b \cdot c \quad \text{and} \quad \frac{a}{c} = \frac{b}{c}.$$

In other words, you can multiply or divide both sides of an equation by the same number. (The only exception is you cannot divide by zero.)

E X A M P L E   3   Solving an Equation by Using the Multiplication Property

Solve each equation.

(a) $9p = 63$

You want to get the variable, p, by itself on the left side. The expression $9p$ means $9 \cdot p$. To get p by itself, *divide* both sides by 9.

$$9p = 63$$

$$\frac{\overset{1}{\cancel{9}} \cdot p}{\underset{1}{\cancel{9}}} = \frac{63}{9} \qquad \text{Divide both sides by 9.}$$

$$p = 7$$

Check.

$$9p = 63$$
$$9 \cdot 7 = 63 \qquad \text{Replace } p \text{ with 7.}$$
$$63 = 63 \qquad \text{True}$$

The result is true, so 7 is the solution.

(b) $-4r = 24$

Divide both sides by -4 to get r by itself on the left.

$$\frac{\overset{1}{\cancel{-4}} \cdot r}{\underset{1}{\cancel{-4}}} = \frac{24}{-4} \qquad \text{Divide both sides by } -4.$$

$$r = -6$$

Check this solution: $-4 \cdot (-6) = 24$ is true.

(c) $-55 = -11m$

Divide both sides by -11 to get m by itself on the right.

$$\frac{-55}{-11} = \frac{\overset{1}{\cancel{-11}} \cdot m}{\underset{1}{\cancel{-11}}}$$

$$5 = m$$

Check this solution: $-55 = -11 \cdot (5)$ is true, so 5 is the solution.

WORK PROBLEM 3 AT THE SIDE. ▶▶

---

3. Solve each equation. Check each solution.

(a) $2y = 14$

(b) $42 = 7p$

(c) $-8a = 32$

(d) $-3r = -15$

(e) $-60 = -6k$

(f) $10x = 0$

Answers

3. (a) 7 (b) 6 (c) -4 (d) 5 (e) 10 (f) 0

4. Solve each equation. Check each solution.

(a) $\dfrac{a}{4} = 2$

(b) $\dfrac{y}{7} = -3$

(c) $-8 = \dfrac{k}{6}$

(d) $8 = -\dfrac{4}{5}z$

(e) $-\dfrac{5}{8}p = -10$

E X A M P L E   4   **Solving an Equation by Using the Multiplication Property**

Solve each equation.

(a) $\dfrac{x}{2} = 9$

Replace $\frac{x}{2}$ with $\frac{1}{2}x$ because dividing by 2 is the same as multiplying by $\frac{1}{2}$. Then, to get x by itself, multiply both sides by $\frac{2}{1}$ (because $\frac{1}{2}$ times $\frac{2}{1}$ is 1).

$$\frac{1}{2}x = 9$$

$$\overset{1}{\underset{1}{\frac{2}{1}}} \cdot \underset{1}{\frac{1}{2}}x = \mathbf{2 \cdot 9} \qquad \text{Multiply both sides by } \tfrac{2}{1} \text{ (which equals 2)}$$

$$1x = 18$$

$$x = 18$$

Check. $\qquad\qquad \dfrac{x}{2} = 9$

$$\dfrac{18}{2} = 9 \qquad \text{Replace } x \text{ with 18.}$$

$$9 = 9 \qquad \text{True}$$

18 is the correct solution.

(b) $-\dfrac{2}{3}r = 4$

Multiply both sides by $-\frac{3}{2}$ (because the product of $-\frac{3}{2}$ and $-\frac{2}{3}$ is 1).

$$-\frac{2}{3}r = 4$$

$$-\underset{1}{\overset{1}{\frac{3}{2}}} \cdot \left(-\underset{1}{\frac{2}{3}}r\right) = -\underset{1}{\frac{3}{2}} \cdot \overset{2}{\underset{1}{\frac{4}{1}}}$$

$$r = -6$$

Check by replacing r with -6. Write -6 as $\frac{-6}{1}$.

$$-\frac{2}{\underset{1}{3}} \cdot \frac{\overset{-2}{\cancel{-6}}}{1} = 4$$

$$4 = 4 \qquad \text{True, so } -6 \text{ is the solution.}$$

◀◀ **WORK PROBLEM 4 AT THE SIDE.**

Here is a summary of the rules for using the multiplication property. In these rules, x is the variable and a, b, and c are numbers.

Solving Equations Using the Multiplication Property
Solve the equation $ax = b$ by dividing both sides by a.
Solve the equation $\frac{a}{b}x = c$ by multiplying both sides by $\frac{b}{a}$.

ANSWERS

4. (a) 8 **(b)** -21 **(c)** -48 **(d)** -10 **(e)** 16

Decide whether the given number is a solution of the equation.

1. $x + 7 = 11$; 4

yes

2. $k - 2 = 7$; 9

yes

3. $4y = 28$; 7

yes

4. $5p = 30$; 6

yes

5. $2z - 1 = -15$; -8

no

6. $6r - 3 = -14$; -2

no

Solve each equation by using the addition property. Check each solution.

Example: $m - 2 = 7$	**Solution:**	**Check:**
	Add 2 to both sides.	$m - 2 = 7$
	$m - 2 + 2 = 7 + 2$	$9 - 2 = 7$ Replace m with 9.
	$m = 9$	$7 = 7$ True
		The solution is 9.

7. $p + 5 = 9$

4

8. $a + 3 = 12$

9

9. $k + 15 = 0$

-15

10. $y + 6 = 0$

-6

11. $z - 5 = 3$

8

12. $x - 9 = 4$

13

13. $8 = r - 2$

10

14. $3 = b - 5$

8

15. $-5 = n + 3$

-8

16. $-1 = a + 8$

-9

17. $7 = r + 13$

-6

18. $12 = z + 7$

5

19. $-4 + k = 14$

18

20. $-9 + y = 7$

16

21. $-12 + x = -1$

11

22. $-3 + m = -9$

-6

23. $-5 = -2 + r$

-3

24. $-1 = -10 + y$

9

✏ Writing   ◉ Conceptual   ▲ Challenging   ≈ Estimation

25. $d + \dfrac{2}{3} = 3$

$\dfrac{7}{3}$ or $2\dfrac{1}{3}$

26. $x + \dfrac{1}{2} = 4$

$\dfrac{7}{2}$ or $3\dfrac{1}{2}$

27. $z - \dfrac{7}{8} = 10$

$\dfrac{87}{8}$ or $10\dfrac{7}{8}$

28. $m - \dfrac{3}{4} = 6$

$\dfrac{27}{4}$ or $6\dfrac{3}{4}$

29. $\dfrac{1}{2} = k - 2$

$\dfrac{5}{2}$ or $2\dfrac{1}{2}$

30. $\dfrac{3}{5} = t - 1$

$\dfrac{8}{5}$ or $1\dfrac{3}{5}$

31. $m - \dfrac{7}{5} = \dfrac{11}{4}$

$\dfrac{83}{20}$

32. $z - \dfrac{7}{3} = \dfrac{32}{9}$

$\dfrac{53}{9}$

33. $x - 0.8 = 5.07$

5.87

34. $a - 3.82 = 7.9$

11.72

35. $3.25 = 4.76 + r$

−1.51

36. $8.9 = 10.5 + b$

−1.6

Solve each equation. Check each solution.

Example: $5k = 60$

Solution:
Divide both sides by 5.

$$\dfrac{\overset{1}{\cancel{5}} \cdot k}{\underset{1}{\cancel{5}}} = \dfrac{60}{5}$$

$k = \mathbf{12}$

Check:

$5k = 60$

$5(12) = 60$

$60 = 60$ \quad True

The solution is 12.

37. $6z = 12$

2

38. $8k = 24$

3

39. $48 = 12r$

4

40. $99 = 11m$

9

41. $3y = 0$

0

42. $5a = 0$

0

43. $-6k = 36$

−6

44. $-7y = 70$

−10

45. $-36 = -4p$

9

46. $-54 = -9r$

6

47. $-1.2m = 8.4$

−7

48. $-5.4z = 27$

−5

49. $-8.4p = -9.24$

1.1

50. $-3.2y = -16.64$

5.2

Solve each equation. Check each solution.

Example: $\dfrac{p}{4} = -3$

Solution:

Replace $\dfrac{p}{4}$ with $\dfrac{1}{4}p$.

Multiply both sides by 4.

$$\dfrac{\overset{1}{\cancel{4}}}{1} \cdot \dfrac{1}{\underset{1}{\cancel{4}}}p = 4 \cdot (-3)$$

$$p = \mathbf{-12}$$

Check:

$$\dfrac{p}{4} = -3$$

$$\dfrac{-12}{4} = -3$$

$$-3 = -3 \qquad \text{True}$$

The solution is -12.

51. $\dfrac{k}{2} = 17$ 34

52. $\dfrac{y}{3} = 5$ 15

53. $11 = \dfrac{a}{6}$ 66

54. $5 = \dfrac{m}{8}$ 40

55. $\dfrac{r}{3} = -12$ −36

56. $\dfrac{z}{9} = -3$ −27

57. $-\dfrac{2}{5}p = 8$ −20

58. $-\dfrac{5}{6}k = 15$ −18

59. $-\dfrac{3}{4}m = -3$ 4

60. $-\dfrac{9}{10}b = -18$ 20

61. $6 = \dfrac{3}{8}x$ 16

62. $4 = \dfrac{2}{3}a$ 6

63. $\dfrac{y}{2.6} = 0.5$

1.3

64. $\dfrac{k}{0.7} = 3.2$

2.24

65. $\dfrac{z}{-3.8} = 1.3$

−4.94

66. $\dfrac{m}{-5.2} = 2.1$

−10.92

67. Explain the addition property of equations. Then show an example of an equation where you would use the addition property to solve it. Make the equation so it has -3 as the solution.

You may add or subtract the same number on both sides of an equation. Many different equations could have -3 as the solution. One possibility is:

$$x + 5 = 2$$
$$x + 5 - 5 = 2 - 5$$
$$x = -3$$

68. Explain the multiplication property of equations. Then show an example of an equation where you would use the multiplication property to solve it. Make the equation so it has $+6$ as the solution.

You may multiply or divide both sides of an equation by the same number (except you cannot divide by zero). Many different equations could have $+6$ as the solution. One possibility is:

$$8r = 48$$
$$\frac{\overset{1}{\cancel{8}} \cdot r}{\underset{1}{\cancel{8}}} = \frac{48}{8}$$
$$r = 6$$

▲ *Solve the following equations.*

69. $x - 17 = 5 - 3$

19

70. $y + 4 = 10 - 9$

-3

71. $3 = x + 9 - 15$

9

72. $-1 = y + 7 - 9$

1

73. $\dfrac{7}{2}x = \dfrac{4}{3}$

$\dfrac{8}{21}$

74. $\dfrac{3}{4}x = \dfrac{5}{3}$

$\dfrac{20}{9}$

Review and Prepare

*Use the order of operations to simplify the following. (For help, see **Sections 3.5 and 9.4**.)*

75. $2\dfrac{1}{5} \div \left(3\dfrac{1}{3} - 4\dfrac{1}{5}\right)$

$-\dfrac{33}{13}$ or $-2\dfrac{7}{13}$

76. $\dfrac{2}{3} + \dfrac{5}{6} \cdot \left(-\dfrac{3}{4}\right)$

$\dfrac{1}{24}$

77. $\dfrac{1}{2} + \dfrac{3}{4} \cdot \dfrac{8}{9} - \dfrac{1}{6}$

1

78. $2 - \left(3 \cdot \dfrac{1}{2} \div \dfrac{2}{3}\right) - \dfrac{7}{4}$

-2

9.7 *Solving Equations with Several Steps*

You cannot solve the equation $5m + 1 = 16$ by just adding the same number to both sides, nor by just dividing both sides by the same number.

OBJECTIVE 1 Instead, you will use a combination of both operations. Here are the steps.

Solving Equations Using the Addition and Multiplication Properties

Step 1 Add or subtract the same amount on both sides of the equation so that the variable term ends up by itself on one side.

Step 2 Multiply or divide both sides by the same number to find the solution.

Step 3 Check the solution.

E X A M P L E I **Solving Equations with Several Steps**

Solve $5m + 1 = 16$.

First subtract 1 from both sides so that $5m$ will be by itself on the left side.

$$5m + 1 - 1 = 16 - 1$$
$$5m = 15$$

Next, divide both sides by 5.

$$\frac{\overset{1}{\cancel{5}} \cdot m}{\underset{1}{\cancel{5}}} = \frac{15}{5}$$

$$m = 3$$

Check.

$$5m + 1 = 16$$
$$5\,(3) + 1 = 16 \qquad \text{Replace } m \text{ with 3.}$$
$$15 + 1 = 16$$
$$16 = 16 \qquad \text{True}$$

The solution is 3.

WORK PROBLEM I AT THE SIDE. ▶▶

OBJECTIVE 2 We can use the order of operations to simplify these two expressions:

$$2\underbrace{(6 + 8)} \qquad \text{and} \qquad \underbrace{2 \cdot 6} + \underbrace{2 \cdot 8}$$
$$\underbrace{2(14)} \qquad\qquad\qquad \underbrace{12 \;+\; 16}$$
$$28 \qquad\qquad\qquad\qquad 28$$

Because both answers are the same, the two expressions are equivalent.

$$2(6 + 8) = 2 \cdot 6 + 2 \cdot 8$$

This is an example of the **distributive** (dis-TRIB-yoo-tiv) **property.**

Distributive Property

$$a\,(b + c) = ab + ac$$

OBJECTIVES

1 Solve equations with several steps.

2 Use the distributive property.

3 Combine like terms.

4 Solve more difficult equations.

FOR EXTRA HELP

Tutorial   Tape 16   SSM, Sec. 9.7

1. Solve each equation. Check each solution.

   (a) $2r + 7 = 13$

   (b) $20 = 6y - 4$

   (c) $7m + 9 = 9$

   (d) $-2 = 4p + 10$

   (e) $-10z - 9 = 11$

ANSWERS

1. **(a)** 3 **(b)** 4 **(c)** 0 **(d)** -3 **(e)** -2

631

2. Use the distributive property.

(a) $3(2 + 6)$

(b) $8(k - 3)$

(c) $-6(r + 5)$

(d) $-9(s - 8)$

E X A M P L E 2   Using the Distributive Property

Simplify each expression by using the distributive property.

(a) $9(4 + 2) = 9 \cdot 4 + 9 \cdot 2 = 36 + 18 = 54$

The 9 on the outside of the parentheses is *distributed* over the 4 and the 2 on the inside of the parentheses. That means that each number inside the parentheses is multiplied by 9.

(b) $-3(k + 9) = -3 \cdot k + (-3) \cdot 9 = -3k + (-27) = -3k - 27$

(c) $6(y - 5) = 6 \cdot y - 6 \cdot 5 = 6y - 30$

(d) $-2(x - 3) = -2 \cdot x - (-2) \cdot 3 = -2x - (-6) = -2x + 6$

◀◀ WORK PROBLEM 2 AT THE SIDE.

3. Combine like terms.

(a) $5y + 11y$

(b) $10a - 28a$

(c) $3x + 3x - 9x$

(d) $k + k$

(e) $6b - b - 7b$

OBJECTIVE **3**▶ A single letter or number, or the product of a variable and a number, makes up a *term*. Here are six examples of terms.

$$3y \qquad 5 \qquad -9 \qquad 8r \qquad 10r^2 \qquad x$$

*Terms with exactly the same variable and the same exponent are called **like terms**.*

$5x$	and	$3x$	like terms
$5x$	and	$3m$	not like terms; variables are different
$5x^2$	and	$5x^3$	not like terms; exponents are different
$5x^4$	and	$3x^4$	like terms

The distributive property can be used to simplify a sum of like terms such as $6r + 3r$.

$$6r + 3r = (6 + 3)r = 9r$$

This process is called *combining like terms*.

E X A M P L E 3   Combining Like Terms

Use the distributive property to combine like terms.

(a) $5k + 11k = (5 + 11)k = 16k$

(b) $10m - 14m + 2m = (10 - 14 + 2)m = -2m$

(c) $-5x + x$ can be written $-5x + 1x = (-5 + 1)x = -4x$

◀◀ WORK PROBLEM 3 AT THE SIDE.

ANSWERS

2. (a) $3 \cdot 2 + 3 \cdot 6 = 24$ (b) $8k - 24$
   (c) $-6r + (-6) \cdot 5 = -6r + (-30)$
       $= -6r - 30$
   (d) $-9s - (-9) \cdot 8$
       $= -9s - (-72) = -9s + 72$
3. (a) $16y$ (b) $-18a$ (c) $-3x$ (d) $2k$
   (e) $-2b$

OBJECTIVE ▶4 The next examples show you how to solve more difficult equations using the addition, multiplication, and distributive properties.

E X A M P L E 4   Solving Equations

Solve each equation.

(a) $6r + 3r = 36$

You can combine $6r$ and $3r$ because they are like terms.

$6r + 3r$ is $9r$, so the equation becomes:

$$9r = 36$$

Next, divide both sides by 9.

$$\frac{\overset{1}{\cancel{9}} \cdot r}{\underset{1}{\cancel{9}}} = \frac{36}{9}$$

$$r = 4$$

Check.

$$6r + 3r = 36$$
$$6(4) + 3(4) = 36 \qquad \text{Replace } r \text{ with 4.}$$
$$24 + 12 = 36$$
$$36 = 36 \qquad \text{True}$$

The solution is 4.

(b) $2k - 2 = 5k - 11$

First, to get the variable term on one side, subtract $5k$ from both sides.

$$2k - 2 - \mathbf{5k} = 5k - 11 - \mathbf{5k}$$
$$2k - 5k - 2 = 5k - 5k - 11$$
$$-3k - 2 = -11$$

Next, add 2 to both sides.

$$-3k - 2 + 2 = -11 + 2$$
$$-3k = -9$$

Finally, divide both sides by -3.

$$\frac{\overset{1}{-\cancel{3}} \cdot k}{\underset{1}{-\cancel{3}}} = \frac{-9}{-3}$$

$$k = 3$$

Check.

$$2k - 2 = 5k - 11$$
$$2(3) - 2 = 5(3) - 11 \qquad \text{Replace } k \text{ with 3.}$$
$$6 - 2 = 15 - 11$$
$$4 = 4 \qquad\qquad \text{True}$$

The solution is 3.

WORK PROBLEM 4 AT THE SIDE. ▶▶

4. Solve each equation. Check each solution.

(a) $3y - 1 = 2y + 7$

(b) $5a + 7 = 3a - 9$

(c) $3p - 2 = p - 6$

ANSWERS

4. (a) $y = 8$ **(b)** $a = -8$ **(c)** $p = -2$

5. Solve each equation. Check each solution.

(a) $-12 = 4(y - 1)$

(b) $5(m + 4) = 20$

(c) $6(t - 2) = 18$

Now that you know about the distributive property and combining like terms, here is a summary of all the steps you can use to solve an equation.

> **Solving Equations**
>
> *Step 1* If possible, use the **distributive property** to remove parentheses.
>
> *Step 2* **Combine** any like terms on the left side of the equation. Combine any like terms on the right side of the equation.
>
> *Step 3* **Add or subtract** the same amount on both sides of the equation so that the variable term ends up by itself on one side.
>
> *Step 4* **Multiply or divide** both sides by the same number to find the solution.
>
> *Step 5* **Check** your solution by going back to the original equation. Replace the variable with your solution. Follow the order of operations to complete the calculations. If the two sides of the equation are equal, your solution is correct.

E X A M P L E   5   Solving Equations Using the Distributive Property

Solve $-6 = 3(y - 2)$

Step 1 Use the distributive property on the right side of the equation.

$$3(y - 2) \text{ becomes } 3 \cdot y - 3 \cdot 2 \quad \text{or} \quad 3y - 6$$

Now the equation looks like this.

$$-6 = 3y - 6$$

Step 2 Combine like terms. Check the left side of the equation. There are no like terms. Check the right side. No like terms there either, so go on to Step 3.

Step 3 Add 6 to both sides in order to get the variable term by itself on the right side.

$$-6 + 6 = 3y - 6 + 6$$
$$0 = 3y$$

Step 4 Divide both sides by 3.

$$\frac{0}{3} = \frac{\overset{1}{\cancel{3}} \cdot y}{\underset{1}{\cancel{3}}}$$

$$0 = y$$

Step 5 Check. Go back to the original equation.

$$-6 = 3(y - 2)$$
$$-6 = 3(0 - 2) \qquad \text{Replace } y \text{ with 0.}$$
$$-6 = 3(-2)$$
$$-6 = -6 \qquad \text{True}$$

The solution is 0.

◀◀ **WORK PROBLEM 5 AT THE SIDE.**

ANSWERS

5. (a) $y = -2$ **(b)** $m = 0$ **(c)** $t = 5$

9.7 Exercises

Solve each equation. Check each solution.

Examples:

$9p - 7 = 11$ $-3m + 2 = 8$

Solutions:

Add 7 to both sides.   Subtract 2 from both sides.

$9p - 7 + 7 = 11 + 7$ $-3m + 2 - 2 = 8 - 2$

$\qquad 9p = 18$ $\qquad -3m = 6$

Divide both sides by 9.   Divide both sides by -3.

$$\frac{\overset{1}{\cancel{9}} \cdot p}{\underset{1}{\cancel{9}}} = \frac{18}{9}$$ $$\frac{\overset{1}{\cancel{-3}} \cdot m}{\underset{1}{\cancel{-3}}} = \frac{6}{-3}$$

$\qquad p = 2$ $\qquad m = -2$

Check: Replace p with 2.   Check: Replace m with -2.

$\qquad 9(2) - 7 = 11$ $\qquad -3(-2) + 2 = 8$

$\qquad 18 - 7 = 11$ $\qquad 6 + 2 = 8$

$\qquad\qquad 11 = 11$ True $\qquad\qquad 8 = 8$ True

The solution is **2.**   The solution is **−2.**

1. $7p + 5 = 12$

1

2. $6k + 3 = 15$

2

3. $2 = 8y - 6$

1

4. $10 = 11p - 12$

2

5. $-3m + 1 = 1$

0

6. $-4k + 5 = 5$

0

7. $28 = -9a + 10$

−2

8. $5 = -10p + 25$

2

9. $-5x - 4 = 16$

−4

10. $-12a - 3 = 21$

−2

11. $-\frac{1}{2}z + 2 = -1$

6

12. $-\frac{5}{8}r + 4 = -6$

16

Use the distributive property to simplify.

13. $6(x + 4)$

$6x + 24$

14. $8(k + 5)$

$8k + 40$

15. $7(p - 8)$

$7p - 56$

16. $9(t - 4)$

$9t - 36$

17. $-3(m + 6)$

$-3m - 18$

18. $-5(a + 2)$

$-5a - 10$

19. $-2(y - 3)$

$-2y + 6$

20. $-4(r - 7)$

$-4r + 28$

21. $-5(z - 9)$

$-5z + 45$

Combine like terms.

22. $11r + 6r$

$17r$

23. $2m + 5m$

$7m$

24. $8z + 7z$

$15z$

25. $10x - 2x$

$8x$

26. $9y - 3y$

$6y$

27. $-10a + a$

$-9a$

28. $-4t + t$

$-3t$

29. $3y - y - 4y$

$-2y$

30. $7p - 9p - p$

$-3p$

Solve each equation. Check each solution.

31. $4k + 6k = 50$

5

32. $3a + 2a = 15$

3

33. $54 = 10m - m$

6

34. $28 = x + 6x$

4

35. $2b - 6b = 24$

-6

36. $3r - 9r = 18$

-3

37. $-12 = 6y - 18y$

1

38. $-5 = 10z - 15z$

1

39. $6p - 2 = 4p + 6$

4

40. $5y - 5 = 2y + 10$

    5

41. $9 + 7z = 9z + 13$

    -2

42. $8 + 4a = 2a + 2$

    -3

43. $-2y + 6 = 6y - 10$

    2

44. $5x - 4 = -3x + 4$

    1

45. $b + 3.05 = 2$

    -1.05

46. $t + 0.8 = -1.7$

    -2.5

47. $2.5r + 9 = -1$

    -4

48. $0.5x - 6 = 2$

    16

49. $-10 = 2(y + 4)$

    -9

50. $-3 = 3(x + 6)$

    -7

51. $-4(t + 2) = 12$

    -5

52. $-5(k + 3) = 25$

    -8

53. $6(x - 5) = -30$

    0

54. $7(r - 5) = -35$

    0

55. Solve $-2t - 10 = 3t + 5$. Show each step you take in solving it. Next to each step, write a sentence that explains what you did in that step. Be sure to tell when you used the addition property of equations and when you used the multiplication property of equations.

$$-2t - 10 = 3t + 5$$

$-2t - 3t - 10 = 3t - 3t + 5$ Subtract 3t from both sides (addition property).

$-5t - 10 = 5$

$-5t - 10 + 10 = 5 + 10$ Add 10 to both sides (addition property).

$-5t = 15$

$\dfrac{-5 \cdot t}{-5} = \dfrac{15}{-5}$ Divide both sides by -5 (multiplication property).

$t = -3$

56. Here is one student's solution to an equation.

$$3(2x + 5) = -7$$
$$6x + 5 = -7$$
$$6x + 5 - 5 = -7 - 5$$
$$6x = -12$$
$$x = -2$$

Show how to check the solution. If the solution does not check, find and correct the error.

Check: $3(2(-2) + 5) = -7$

       $3(-4 + 5) = -7$

          $3(1) = -7$

            $3 = -7$ Not true

The first step on the left side should have been $3(2x + 5)$ is $6x + 15$.

▲ *Solve each equation.*

57. $30 - 40 = -2x + 7x - 4x$

-10

58. $-6 - 5 + 14 = -50a + 51a$

3

59. $0 = -2(y - 2)$

2

60. $0 = -9(b - 1)$

1

61. $\dfrac{y}{2} - 2 = \dfrac{y}{4} + 3$

20

62. $\dfrac{z}{3} + 1 = \dfrac{z}{2} - 3$

24

Review and Prepare

Solve each word problem. (For help, see Section 6.7.)

63. A chip and dip set is on sale after Christmas at 60% off the regular price. How much will Beyanjeru pay for a set regularly priced at $28?

$11.20

64. A "going out of business" sale at the Great Goods store promises 75% off on all items. Gel-pack batteries for electric wheelchairs are regularly priced at $99. Find the amount that Will paid for a battery during the sale.

$24.75

65. If the sales tax rate is 6% and a refrigerator costs $420, what is the amount of sales tax?

$25.20

66. A VCR sells for $450 plus 4% sales tax. Find the price of the VCR including sales tax.

$468

9.8 Applications

It is rare for an application problem to be presented as an equation. Usually, the problem is given in words. You need to *translate* these words into an equation that you can solve.

OBJECTIVE 1 The following examples show you how to translate word phrases into algebra.

E X A M P L E 1 Translating Word Phrases by Using Variables

Write in symbols by using x as the variable.

Words	Algebra
a number **plus** 2	$x + 2$ or $2 + x$
the **sum** of 8 and a number	$8 + x$ or $x + 8$
5 **more than** a number	$x + 5$ or $5 + x$
-35 **added to** a number	$-35 + x$ or $x + (-35)$
a number **increased by** 6	$x + 6$ or $6 + x$
9 **less than** a number	$x - 9$
a number **subtracted from** 3	$3 - x$
a number **decreased by** 4	$x - 4$
10 **minus** a number	$10 - x$

Note
Recall that addition can be done in any order, so $x + 2$ gives the same result as $2 + x$. This is *not* true in subtraction, so be careful. $10 - x$ does *not* give the same result as $x - 10$.

WORK PROBLEM 1 AT THE SIDE. ▶▶

E X A M P L E 2 Translating Word Phrases by Using Variables

Write in symbols by using x as the variable.

Words	Algebra
8 **times** a number	$8x$
the **product** of 12 and a number	$12x$
double a number (meaning "2 times")	$2x$
the **quotient** of 6 and a number	$\dfrac{6}{x}$
a number **divided by** 10	$\dfrac{x}{10}$
one-third of a number	$\dfrac{1}{3}x$ or $\dfrac{x}{3}$
the result **is**	$=$

WORK PROBLEM 2 AT THE SIDE. ▶▶

OBJECTIVE 2 The next examples show you how to solve application problems. Notice that you begin each solution by selecting a variable to represent the unknown.

OBJECTIVES

1 ▶ Translate word phrases by using variables.

2 ▶ Solve application problems.

FOR EXTRA HELP

Tutorial   Tape 17   SSM, Sec. 9.8

1. Write in symbols by using x as the variable.

   (a) 15 less than a number

   (b) 12 more than a number

   (c) a number increased by 13

   (d) a number minus 8

   (e) -10 plus a number

   (f) 6 minus a number

2. Write in symbols by using x as the variable.

   (a) double a number

   (b) the product of -8 and a number

   (c) the quotient of 15 and a number

   (d) one-half of a number

ANSWERS

1. (a) $x - 15$ (b) $x + 12$ or $12 + x$
   (c) $x + 13$ or $13 + x$ (d) $x - 8$
   (e) $-10 + x$ or $x + (-10)$ (f) $6 - x$

2. (a) $2x$ (b) $-8x$ (c) $\dfrac{15}{x}$ (d) $\dfrac{1}{2}x$ or $\dfrac{x}{2}$

3. Solve each application problem. Check your solution by going back to the words in the original problem.

(a) If 3 times a number is added to 4, the result is 19. Find the number.

(b) If −6 times a number is added to 5, the result is −13. Find the number.

(c) Susan donated $10 more than twice what LuAnn donated. If Susan donated $22, how much did LuAnn donate?

E X A M P L E 3   Solving Application Problems

If 5 times a number is added to 11, the result is 26. Find the number.

Let x represent the unknown number.
    Use the information in the problem to write an equation.

$$\underbrace{5 \text{ times a number}} \quad \underbrace{\text{added to} \quad 11} \quad \text{is} \quad 26.$$
$$5x \qquad\qquad + \qquad 11 \;=\; 26$$

Note
The phrase "the result is" translates to "=."

Next, solve the equation. First subtract 11 from both sides.

$$5x + 11 - \mathbf{11} = 26 - \mathbf{11}$$
$$5x = 15$$
$$\frac{\overset{1}{\cancel{5}}x}{\underset{1}{\cancel{5}}} = \frac{15}{5} \qquad \text{Divide both sides by 5.}$$
$$x = 3$$

The number is 3.
To check the solution, go back to the words of the original problem.

$$\text{If 5 times} \quad \underbrace{\text{a number}} \quad \underbrace{\text{is added to}} \quad 11, \quad \underbrace{\text{the result is}} \quad 26.$$
$$5 \quad \cdot \quad 3 \qquad + \qquad 11 \qquad = \qquad 26$$

Does $5 \cdot 3 + 11$ really equal 26? Yes it does. So 3 is the correct solution because it "works" when you put it back into the original problem.

E X A M P L E 4   Solving Application Problems

Michael has 5 less than three times as many lab experiments completed as David. If Michael has completed 13 experiments, how many lab experiments has David completed?

Let x represent the number of experiments David has completed.

$$\underbrace{3 \text{ times David's number}} \quad \underbrace{\text{minus} \quad 5} \quad \text{is} \quad \underbrace{\text{Michael's number.}}$$
$$3x \qquad\qquad - \quad 5 \;=\; \qquad 13$$

Next, solve the equation. First add 5 to both sides.

$$3x - 5 + \mathbf{5} = 13 + \mathbf{5}$$
$$3x = 18$$
$$\frac{\overset{1}{\cancel{3}}x}{\underset{1}{\cancel{3}}} = \frac{18}{3} \qquad \text{Divide both sides by 3.}$$
$$x = 6$$

David has completed 6 lab experiments. Check by using the words of the original problem. First take three times the solution $(3 \cdot 6)$, which is 18. Then, 18 decreased by 5 is 13, which matches the 13 experiments completed by Michael. So 6 is the correct solution.

ANSWERS

3. (a) $3x + 4 = 19$
        $x = 5$
(b) $-6x + 5 = -13$
        $x = 3$
(c) $2x + 10 = 22$
        $x = 6$
    LuAnn donated $6.

 **WORK PROBLEM 3 AT THE SIDE.**

The steps in solving an application problem are summarized for you here.

Solving Algebra Application Problems

Step 1 Choose a variable to represent the unknown. If there are several unknowns, let the variable represent the one you know the least about.

Step 2 Use the information in the problem to write an equation that relates known information to the unknown. Make a sketch or drawing, if possible.

Step 3 Solve the equation.

Step 4 Answer the question raised in the problem.

Step 5 Check the solution with the original words of the problem.

E X A M P L E 5   Solving Application Problems

During the day, Sheila drove 72 km more than Russell. The total distance traveled by them both was 232 km. Find the distance traveled by each person.

There are two unknowns: Sheila's distance and Russell's distance.

Let x be the distance traveled by Russell, because you know less about his distance than Sheila's distance.

Since Sheila drove 72 km more than Russell, the distance she traveled is $x + 72$ km, that is, Russell's distance (x) plus 72 km.

Now write an equation and solve it.

$$\underbrace{\text{Distance for Russell}}_{x} \quad \underset{\downarrow}{\text{plus}} \quad \underbrace{\text{distance for Sheila}}_{x + 72} \quad \underset{\downarrow}{\text{is}} \quad \underbrace{\text{total distance.}}_{232}$$

Recall that x is really $1x$, so the sum $x + x$ is $1x + 1x$ which is $2x$. The equation becomes:

$$2x + 72 = 232$$
$$2x + 72 - 72 = 232 - 72 \qquad \text{Subtract 72 from both sides.}$$
$$2x = 160 \qquad\qquad \text{Divide both sides by 2.}$$
$$x = 80$$

Russell's distance is x, so Russell traveled 80 km.
Sheila's distance is $x + 72$ so Sheila traveled $80 + 72 = 152$ km.
Check using the words of the original problem.

Sheila drove 72 km more than Russell.

Sheila's 152 km is 72 km more than Russell's 80 km, so that checks.

The total distance traveled by them both was 232 km.

Sheila's 152 km + Russell's 80 km = 232 km, so that checks. ∎

Note

Check the solution to an application problem by putting the numbers back into the original problem. If they do *not* work, solve the problem a different way.

WORK PROBLEM 4 AT THE SIDE. ▶▶

---

4. (a) In a day of work, Keonda made $12 more than her daughter. Together they made $182. Find the amount made by each person. (*Hint:* Which amount do you know the least about, Keonda's, or her daughter's? Let x be that amount.)

(b) A rope is 21 m long. Marcos cut it into two pieces, so that one piece is 3 m longer than the other. Find the length of each piece.

ANSWERS

4. (a) daughter made x
Keonda made $x + 12$
$x + x + 12 = 182$
daughter made $85
Keonda made $97
(b) shorter piece is x
longer piece is $x + 3$
$x + x + 3 = 21$
shorter piece is 9 m
longer piece is 12 m

5. Make a drawing to help solve this problem. The length of Ann's rectangular garden plot is 3 m more than the width. She used 22 m of fencing around the edge. Find the length and the width of the garden.

E X A M P L E   6   Solving a Geometry Problem

The length of a rectangle is 2 cm more than the width. The perimeter is 68 cm. Find the length and width.

Let x be the width of the rectangle because you know the least about the width.

Since the length is 2 cm more than the width, the length is $x + 2$.

A drawing of the rectangle will help you see these relationships.

width $= x$
length $= x + 2$
perimeter $= 68$

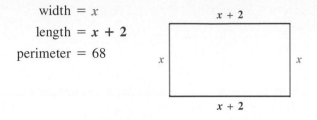

Use the formula for perimeter of a rectangle, $P = 2 \cdot l + 2 \cdot w$.

$$P = 2 \cdot l \quad + 2 \cdot w$$
$$68 = \underbrace{2(x + 2)} + \underbrace{2 \cdot x} \quad \text{Use distributive property.}$$
$$68 = 2x + 4 \quad + \quad 2x \quad \text{Combine like terms.}$$
$$68 = 4x + 4$$
$$68 - 4 = 4x + 4 - 4 \quad \text{Subtract 4 from both sides.}$$
$$64 = 4x$$
$$\frac{64}{4} = \frac{\overset{1}{4} \cdot x}{\underset{1}{4}} \quad \text{Divide both sides by 4.}$$
$$16 = x$$

This does **not** answer the entire question in the problem, because x represents the *width* of the rectangle and the problem asks for the *length* and width.

width $= x$, so width is 16 cm

length $= x + 2$, so length is $16 + 2$ or 18 cm

Check using the words of the original problem. It says the length is 2 cm more than the width. 18 cm is 2 cm more than 16 cm, so that part checks. The original problem also says the perimeter is 68 cm. Use 18 cm and 16 cm to find the perimeter.

$$68 = 2 \cdot l \quad + 2 \cdot w \quad \text{Let } l = 18 \text{ and } w = 16.$$
$$68 = 2 \cdot 18 + 2 \cdot 16$$
$$68 = 36 + 32$$
$$68 = 68 \quad \text{True}$$

The complete solution is width $= 16$ cm and length $= 18$ cm.

◀◀ **WORK PROBLEM 5 AT THE SIDE.**

ANSWERS

5.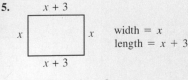

width $= x$
length $= x + 3$

$22 = 2(x + 3) + 2 \cdot x$
width is 4 m
length is 7 m

9.8 Exercises

Write in symbols by using x as the variable.

Examples:	**Solutions:**
the sum of 7 and a number	$7 + x$ or $x + 7$
12 added to a number	$x + 12$ or $12 + x$
a number subtracted from 57	$57 - x$
three times a number	$3x$
the sum of 15 and twice a number	$15 + 2x$ or $2x + 15$

1. 14 plus a number

$14 + x$ or $x + 14$

2. the sum of a number and -8

$x + (-8)$ or $-8 + x$

3. -5 added to a number

$-5 + x$ or $x + (-5)$

4. 16 more than a number

$x + 16$ or $16 + x$

5. 20 minus a number

$20 - x$

6. a number decreased by 25

$x - 25$

7. 9 less than a number

$x - 9$

8. a number subtracted from -7

$-7 - x$

9. subtract 4 from a number

$x - 4$

10. 3 fewer than a number

$x - 3$

11. six times a number

$6x$

12. the product of -3 and a number

$-3x$

13. double a number

$2x$

14. half a number

$\dfrac{x}{2}$ or $\dfrac{1}{2}x$

15. a number divided by 2

$\dfrac{x}{2}$

16. 4 divided by a number

$\dfrac{4}{x}$

17. twice a number added to 8

$8 + 2x$ or $2x + 8$

18. five times a number plus 5

$5x + 5$ or $5 + 5x$

✍ Writing   ⊙ Conceptual   ▲ Challenging   ≈ Estimation

19. 10 fewer than seven times a number

$7x - 10$

20. 12 less than six times a number

$6x - 12$

21. the sum of twice a number and the number

$2x + x$ or $x + 2x$

22. triple a number subtracted from the number

$x - 3x$

23. In your own words, write a definition for each of these words: variable, expression, equation. Give three examples to illustrate each definition.

A variable is a letter that represents an unknown quantity. Examples: x, w, p.
An expression is a combination of letters and numbers. Examples: $6x$, $w - 5$, $2p + 3x$.
An equation has an $=$ sign and shows that two expressions are equal. Examples: $2y = 14$; $x + 5 = 2x$; $8p - 10 = 54$.

24. "You can use any letter to represent the unknown in an application problem." Is this statement true or false? Explain your answer.

True. Choose any letter you like, although it may help to choose a letter which reminds you of what it represents, such as w for an unknown width.

Solve each application problem. Use the five steps for solving algebra application problems. Be sure to show your equation and the steps you use to solve the equation.

Example: If a number is multiplied by 5 and the product is added to 2, the result is -13. Find the number.

Solution:

Let x represent the unknown number.

$$\underbrace{\text{a number multiplied by 5}}_{5x} \quad \underbrace{\text{added to}}_{+} \quad \underset{\downarrow}{2} \quad \underbrace{\text{result is}}_{=} \quad \underset{\downarrow}{-13}$$

Solve the equation.

$5x + 2 - 2 = -13 - 2$ Subtract 2 from both sides.

$5x = -15$ Divide both sides by 5.

$x = -3$

The number is -3.

Check with the words of the original problem: $-3 \cdot 5 + 2 = -13$

$-13 = -13$ True

25. If four times a number is decreased by 2, the result is 26. Find the number.

$4n - 2 = 26$
$n = 7$

26. The sum of 8 and five times a number is 53. Find the number.

$8 + 5n = 53$
$n = 9$

27. If twice a number is added to the number, the result is −15. What is the number?

$2n + n = -15$
$n = -5$

28. If a number is subtracted from three times the number, the result is −8. What is the number?

$3n − n = -8$
$n = -4$

29. If half a number is added to twice the number, the answer is 50. Find the number.

$\frac{1}{2}n + 2n = 50$
$n = 20$

30. If one-third of a number is added to three times the number, the result is 30. Find the number.

$\frac{1}{3}n + 3n = 30$
$n = 9$

31. A board is 78 cm long. Rosa cut the board into two pieces, with one piece 10 cm longer than the other. Find the length of both pieces.

$p + p + 10 = 78$
34 cm and 44 cm

32. Ed and Marge were candidates for city council. Marge won, with 93 more votes than Ed. The total number of votes cast in the election was 587. Find the number of votes received by each candidate.

$v + v + 93 = 587$
247 votes for Ed; 340 votes for Marge

33. Kerwin rented a chain saw for a one-time $9 sharpening fee plus $16 a day rental. His total bill was $89. For how many days did Kerwin rent the saw?

$16d + 9 = 89$
5 days

34. Mrs. Chao made a $50 down payment on a sofa. Her monthly payments were $158. She paid $998 in all. For how many months did she make payments?

$158m + 50 = 998$
6 months

In the next exercises, use the formula for the perimeter of a rectangle, $P = 2l + 2w$. Make a drawing to help you solve each problem.

35. The perimeter of a rectangle is 48 m. The width is 5 m. Find the length.

19 m

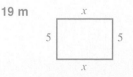

36. The length of a rectangle is 27 cm, and the perimeter is 74 cm. Find the width of the rectangle.

10 cm

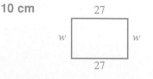

37. A rectangular dog pen is twice as long as it is wide. The perimeter of the pen is 36 ft. Find the length and the width of the pen.

12 ft, 6 ft

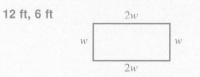

38. A new city park is a rectangular shape. The length is triple the width. It will take 240 meters of fencing to go around the park. Find the length and width of the park.

90 m, 30 m

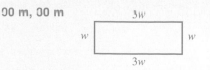

▲ *Solve each application problem. Show your equation and the steps you use to solve the equation.*

39. When 75 is subtracted from four times Tamu's age, the result is Tamu's age. How old is Tamu?

$4a - 75 = a$
25 years old

40. If three times Linda's age is decreased by 36, the result is twice Linda's age. How old is Linda?

$3a - 36 = 2a$
36 years old

41. A fence is 706 m long. It is to be cut into three parts. Two parts are the same length, and the third part is 25 m longer than the other two. Find the length of each part.

$x + x + x + 25 = 706$
227 m, 227 m, 252 m

42. A wooden railing is 82 m long. It is to be divided into four pieces. Three pieces will be the same length, and the fourth piece will be 2 m longer than each of the other three. Find the length of each piece.

$x + x + x + x + 2 = 82$
20 m, 20 m, 20 m, 22 m

▲ *In the following exercises, use the formula for interest, $I = prt$.*

43. For how long must $800 be deposited at 12% per year to earn $480 interest?

$480 = 800 \cdot 0.12 \cdot t$
5 years

44. How much money must be deposited at 12% per year for 7 years to earn $1008 interest?

$1008 = p \cdot 0.12 \cdot 7$
$1200

Review and Prepare

Solve the following problems. (For help, see Section 6.5 or Section 6.6.)

45. 80% of $2900 is how much money? $2320

46. What is 15% of $360? $54

47. Jeffrey puts 5% of his $1830 monthly salary into a retirement plan. How much goes into his plan each month?

$91.50

48. Marshall College expects 9% of incoming students to need emergency loans to buy books. How many of the 2200 incoming students are expected to need loans?

198 students

KEY TERMS

9.1	**negative numbers**	Negative numbers are numbers that are less than zero.
	signed numbers	Signed numbers are positive numbers, negative numbers, and zero.
	absolute value	Absolute value is the distance of a number from zero on a number line. Absolute value is never negative.
	opposite of a number	The opposite of a number is a number the same distance from zero on a number line as the original number but on the opposite side of zero.
9.2	**additive inverse**	The additive inverse is the opposite of a number. The sum of a number and its additive inverse is always 0.
9.5	**variables**	Variables are letters that represent numbers.
	expression	An expression is a combination of letters and numbers.
9.6	**equation**	An equation is a statement that says two expressions are equal.
	solution	The solution is a number that can be substituted for the variable in an equation, so that the equation is true.
	addition property of equations	The addition property of equations states that the same number can be added or subtracted on both sides of an equation.
	multiplication property of equations	The multiplication property of equations states that both sides of an equation can be multiplied or divided by the same number, except division by zero is not allowed.
9.7	**distributive property**	If a, b, and c are three numbers, the distributive property says that $a(b + c) = ab + ac$.
	like terms	Like terms are terms with exactly the same variable and the same exponent.

QUICK REVIEW

Concepts	*Examples*
9.1 Graphing Signed Numbers Place a dot at the correct location on the number line.	Graph -2, 1, 0, and $2\frac{1}{2}$. (number line from -2 to 3 with dots at -2, 0, 1, and between 2 and 3)
9.1 Identifying the Smaller of Two Numbers Place the symbols $<$ (less than) or $>$ (greater than) between two numbers to make the statement true. The small pointed end of the symbol points to the smaller number.	Use the symbol $<$ or $>$ to make the following statements true. $2 \underline{\quad > \quad} 1$ $-3 \underline{\quad > \quad} -5$ $-6 \underline{\quad < \quad} 2$
9.1 Finding the Absolute Value of a Number Determine the distance from 0 to the given number on the number line.	Find each of the following. **(a)** $|8|$ **(b)** $|-7|$ **(c)** $-|-5|$ $\quad |8| = 8$ $\quad |-7| = 7$ $\quad \to -5$

Concepts	Examples
9.1 Finding the Opposite of a Number Determine the number that is the same distance from 0 as the given number, but on the opposite side of 0 on a number line.	Find the opposite of each of the following. **(a)** -6 **(b)** $+9$ $-(-6) = 6$ $-(+9) = -9$

9.2 Adding Two Signed Numbers
 	
Case 1: *Two positive numbers*
Add the numerical values.
 	
Case 2: *Two negative numbers*
Add the absolute values and write a negative sign in front of the sum.
 	
Case 3: *Two numbers with different signs*
Subtract the absolute values and write the sign of the number with the larger absolute value in front of the answer.
	Add the following.
	
	(a) $8 + 6 = 14$
	
	(b) $-8 + (-6)$
	    Find absolute values.
	$$\lvert -8 \rvert = 8 \quad \lvert -6 \rvert = 6$$
	    Add absolute values.
	$$8 + 6 = 14$$
	    Write a negative sign: -14.
	    So, $-8 + (-6) = -14$
	
	(c) $5 + (-7)$
	    Find absolute values.
	$$\lvert 5 \rvert = 5 \quad \lvert -7 \rvert = 7$$
	    Subtract the smaller absolute value from the larger.
	$$7 - 5 = 2$$
	    The number with the larger absolute value is -7. Its sign is negative, so write a negative sign in front of the answer.
	$$5 + (-7) = -2$$

9.2 Subtracting Two Signed Numbers Follow these steps: Change the subtraction sign to addition. Change the sign of the second number to its opposite. Proceed as in addition.	Subtract. **(a)** $-6 - 5$ $\quad \downarrow \quad \downarrow$ $-6 + (-5) = -11$ **(b)** $5 - (-8)$ $\quad \downarrow \quad \downarrow$ $5 + (+8) = 13$
9.3 Multiplying Signed Numbers Use these rules: The product of two numbers with the same sign is positive. The product of two numbers with different signs is negative.	Multiply the following. **(a)** $7 \cdot 3 = 21$ **(b)** $(-3) \cdot 4 = -12$ **(c)** $(-9) \cdot (-6) = 54$
9.3 Dividing Signed Numbers Use the same rules as for multiplying signed numbers: When two numbers have the same sign, the quotient is positive. When two numbers have different signs, the quotient is negative.	Divide. **(a)** $\dfrac{8}{4} = 2$ **(b)** $\dfrac{-20}{5} = -4$ **(c)** $\dfrac{50}{-5} = -10$ **(d)** $\dfrac{-12}{-6} = 2$

Concepts	Examples
9.4 Using the Order of Operations to Evaluate Numerical Expressions Use the following order of operations to evaluate numerical expressions: Work inside parentheses first. Simplify expressions with exponents and find any square roots. Multiply or divide from left to right. Add or subtract from left to right.	Simplify the following. **(a)** $-4 + \underbrace{6 \div (-2)}$ **(b)** $3^2 \cdot 4 + 3 \cdot \underbrace{(8 \div 2)}$ $\underbrace{-4 + \quad (-3)}$ $3^2 \cdot 4 + 3 \cdot \quad 4$ -7 $\mathbf{9} \cdot 4 + 3 \cdot \quad 4$ $\underbrace{36 \quad + \quad 12}$ 48
9.5 Evaluating Expressions Replace the variables in the expression with the numerical values. Use the order of operations to evaluate.	What is the value of $6p - 5s$, if $p = -3$ and $s = -4$? $6 \; p \quad - \; 5 \; s$ $\downarrow \; \downarrow \quad\quad \downarrow \; \downarrow$ $\underline{6(-3)} \quad \underline{5(\mathbf{-4})}$ $\underbrace{-18 \; - \; (-20)}$ 2
9.6 Determining if a Number Is a Solution of an Equation Substitute the number for the variable in the equation. If the equation is true, the number is a solution.	Is the number 4 a solution of the following equation? $$3x - 5 = 7$$ Replace x with 4. $$3(4) - 5 = 7$$ $$12 - 5 = 7$$ $$7 = 7 \quad \text{True}$$ 4 is the solution.

9.6 Using the Addition Property of Equations to Solve an Equation
 	
 Add or subtract the same number on both sides of the equation, so that you get the variable by itself on one side.
	Solve each equation.
	
	 (a) $x - 6 = 9$
	      $x - 6 \mathbf{\,+\,6} = 9 \mathbf{\,+\,6}$ Add 6 to both sides.
	         $x + 0 = 15$
	            $x = 15$
	
	 (b) $-7 = x + 9$
	    $-7 \mathbf{\,-\,9} = x + 9 \mathbf{\,-\,9}$ Subtract 9 from both sides.
	       $-16 = x + 0$
	       $-16 = x$

Concepts	*Examples*
9.6 Using the Multiplication Property of Equations to Solve an Equation Multiply or divide both sides of the original equation by the same number so that you get the variable by itself on one side. (Do not divide by zero.)	Solve each equation. **(a)** $-54 = 6x$ **(b)** $\frac{1}{3}x = 8$ $$\frac{-54}{6} = \frac{\overset{1}{\cancel{6}} \cdot x}{\underset{1}{\cancel{6}}} \qquad \frac{\overset{1}{\cancel{3}}}{1} \cdot \frac{1}{\underset{1}{\cancel{3}}}x = 3 \cdot 8$$ $$-9 = x \qquad\qquad 1x = 24$$ $$\qquad\qquad\qquad\qquad x = 24$$
9.7 Solving Equations with Several Steps Use the following steps: *Step 1* Add or subtract the same amount on both sides of the equation so that the variable ends up by itself on one side. *Step 2* Multiply or divide both sides by the same number to find the solution. *Step 3* Check the solution.	Solve: $2p - 3 = 9$ $$2p - 3 + 3 = 9 + 3 \qquad \text{Add 3 to both sides.}$$ $$2p = 12$$ $$\frac{\overset{1}{\cancel{2}} \cdot p}{\underset{1}{\cancel{2}}} = \frac{12}{2} \qquad \begin{array}{l}\text{Divide both}\\\text{sides by 2.}\end{array}$$ $$p = 6$$ Check: $2p - 3 = 9$ $$2(6) - 3 = 9 \qquad \text{Replace } p \text{ with 6.}$$ $$12 - 3 = 9$$ $$9 = 9 \qquad \text{True}$$ The solution is 6.
9.7 Using the Distributive Property To simplify expressions, use the distributive property: $$a(b + c) = ab + ac.$$	Simplify: $-2(x + 4)$ $$= -2 \cdot x + (-2) \cdot 4$$ $$= -2x - 8$$
9.7 Combining Like Terms If terms are like, combine the numbers that multiply each variable.	Combine like terms in the following. **(a)** $6p + 7p$ $$6p + 7p = (6 + 7)p = 13p$$ **(b)** $8m - 11m$ $$8m - 11m = (8 - 11)m = -3m$$
9.8 Translating Word Phrases by Using Variables Use x as a variable and symbolize the operations described by the words of the problem.	Write the following word phrases in symbols using x as the variable. **(a)** Two more than a number $x + 2$ or $2 + x$ **(b)** A number decreased by 8 $x - 8$ **(c)** The product of a number and 15 $15x$ **(d)** A number divided by 9 $\dfrac{x}{9}$

CHAPTER 9 REVIEW EXERCISES

[9.1] *Graph the following lists of numbers.*

1. $2, -3, 4, 1, 0, -5$

2. $-2, 5, -4, -1, 3, -6$

3. $-1\frac{1}{4}, -\frac{5}{8}, -3\frac{3}{4}, 2\frac{1}{8}, 1\frac{1}{2}, -2\frac{1}{8}$

4. $0, -\frac{3}{4}, 1\frac{1}{4}, -4\frac{1}{2}, \frac{7}{8}, -7\frac{2}{3}$

Place $<$ or $>$ in each of the following to get a true statement.

5. $0 \underline{\quad > \quad} -2$

6. $-5 \underline{\quad < \quad} 0$

7. $-1 \underline{\quad > \quad} -4$

8. $-9 \underline{\quad < \quad} -6$

Find each of the following.

9. $|8|$ 8

10. $|-19|$ 19

11. $-|-7|$ -7

12. $-|15|$ -15

[9.2] *Add.*

13. $-4 + 6$ 2

14. $-10 + 3$ -7

15. $-11 + (-8)$ -19

16. $-9 + (-24)$ -33

17. $12 + (-11)$ 1

18. $1 + (-20)$ -19

19. $\frac{9}{10} + \left(-\frac{3}{5}\right)$ $\frac{3}{10}$

20. $-\frac{7}{8} + \frac{1}{2}$ $-\frac{3}{8}$

21. $-6.7 + 1.5$ -5.2

22. $-0.8 + (-0.7)$ -1.5

✏ Writing   ◉ Conceptual   ▲ Challenging   ≈ Estimation

[9.2] *Give the additive inverse (opposite) of each number.*

23. 6

−6

24. −14

14

25. $-\dfrac{5}{8}$

$\dfrac{5}{8}$

26. 3.75

−3.75

Subtract.

27. 4 − 10

−6

28. 7 − 15

−8

29. −6 − 1

−7

30. −12 − 5

−17

31. 8 − (−3)

11

32. 2 − (−9)

11

33. −1 − (−14)

13

34. −10 − (−4)

−6

35. −40 − 40

−80

36. −15 − (−15)

0

37. $\dfrac{1}{3} - \dfrac{5}{6}$

$-\dfrac{1}{2}$

38. 2.8 − (−6.2)

9

[9.3] *Multiply or divide.*

39. −4 • 6

−24

40. 5 • (−4)

−20

41. −3 • (−5)

15

42. −8 • (−8)

64

43. $\dfrac{80}{-10}$

−8

44. $\dfrac{-9}{3}$

−3

45. $\dfrac{-25}{-5}$

5

46. $\dfrac{-120}{-6}$

20

47. (−37)(0)

0

48. (−1)(81)

−81

49. $\dfrac{0}{-10}$

0

50. $\dfrac{-20}{0}$

undefined

51. $\dfrac{2}{3} \cdot \left(-\dfrac{6}{7}\right)$

$-\dfrac{4}{7}$

52. $-\dfrac{4}{5} \div \left(-\dfrac{2}{15}\right)$

6

53. −0.5 • (−2.8)

1.4

54. $\dfrac{-5.28}{0.8}$

−6.6

[9.4] *Use the order of operations to simplify each of the following.*

55. $2 - 11 \cdot (-5)$

57

56. $(-4) \cdot (-8) - 9$

23

57. $48 \div (-2)^3 - (-5)$

−1

58. $-36 \div (-3)^2 - (-2)$

−2

59. $5 \cdot 4 - 7 \cdot 6 + 3 \cdot (-4)$

−34

60. $2 \cdot 8 - 4 \cdot 9 + 2 \cdot (-6)$

−32

61. $-4 \cdot 3^3 - 2 \cdot (5 - 9)$

−100

62. $6 \cdot (-4)^2 - 3 \cdot (7 - 14)$

117

63. $\dfrac{3 - (5^2 - 4^2)}{14 + 24 \div (-3)}$

−1

64. $(-0.8)^2 \cdot (0.2) - (-1.2)$

1.328

65. $\left(-\dfrac{1}{3}\right)^2 + \dfrac{1}{4} \cdot \left(-\dfrac{4}{9}\right)$

0

66. $\dfrac{12 \div (2 - 5) + 12 \cdot (-1)}{2^3 - (-4)^2}$

2

[9.5] *Find the value of each expression using the given values of the variables.*

67. $3k + 5m$
   $k = 4, \quad m = 3$

27

68. $3k + 5m$
   $k = -6, \quad m = 2$

−8

69. $2p - q$
   $p = -5, \quad q = -10$

0

70. $2p - q$
   $p = 6, \quad q = -7$

19

71. $\dfrac{5a - 7y}{2 + m}$
   $a = 1, \quad y = 4, \quad m = -3$

23

72. $\dfrac{5a - 7y}{2 + m}$
   $a = 2, \quad y = -2, \quad m = -26$

−1

In each of the following, use the formula and the values of the variables to find the value of the remaining variable.

73. $P = a + b + c;$ $a = 9, \quad b = 12, \quad c = 14$

35

74. $A = \dfrac{1}{2}bh;$ $b = 6, \quad h = 9$

27

[9.6–9.7] *Solve each equation. Check each solution.*

75. $y + 3 = 0$

-3

76. $a - 8 = 8$

16

77. $-5 = z - 6$

1

78. $-8 = -9 + r$

1

79. $-\dfrac{3}{4} + x = -2$

$-\dfrac{5}{4}$ or $-1\dfrac{1}{4}$

80. $12.92 + k = 4.87$

-8.05

81. $-8r = 56$

-7

82. $3p = 24$

8

83. $\dfrac{z}{4} = 5$

20

84. $\dfrac{a}{5} = -11$

-55

85. $20 = 3y - 7$

9

86. $-5 = 2b + 3$

-4

Use the distributive property to simplify.

87. $6(r - 5)$

$6r - 30$

88. $11(p + 7)$

$11p + 77$

89. $-9(z - 3)$

$-9z + 27$

90. $-8(x + 4)$

$-8x - 32$

Combine like terms.

91. $3r + 8r$

$11r$

92. $10z - 15z$

$-5z$

93. $3p - 12p + p$

$-8p$

94. $-6x - x + 9x$

$2x$

Solve each equation. Check each solution.

95. $-4z + 2z = 18$

-9

96. $-35 = 9k - 2k$

-5

97. $4y - 3 = 7y + 12$

-5

98. $b + 6 = 3b - 8$

7

99. $-14 = 2(a - 3)$

-4

100. $42 = 7(t + 6)$

0

[9.8] *Write in symbols by using x to represent the variable.*

101. 18 plus a number

$18 + x$ or $x + 18$

102. half a number

$\frac{1}{2}x$ or $\frac{x}{2}$

103. the sum of four times a number and 6

$4x + 6$ or $6 + 4x$

104. five times a number decreased by 10

$5x - 10$

Solve each application problem. Show your equation and the steps you use to solve it.

105. If eight times a number is subtracted from eleven times the number, the result is −9. Find the number.

$11n - 8n = -9$
$n = -3$

106. In Cicely, Alaska, Ruth Anne rents snowmobiles for $45 for the first day and $35 for each additional day. The bill for Joel's rental was $255. For how many days did he rent a snowmobile?

$45 + 35d = 255$
7 days

107. The perimeter of a rectangle is 124 cm. The width is 25 cm. Find the length. (Use the formula for the perimeter of a rectangle, $P = 2l + 2w$.)

$124 = 2l + 2(25)$
37 cm

108. My sister is 9 years older than I am. The sum of our ages is 51. Find our ages.

$a + a + 9 = 51$
21 and 30 years old

MIXED REVIEW EXERCISES

Add, subtract, multiply, or divide as indicated.

109. $-6 - (-9)$

3

110. $-8 \cdot (-5)$

40

111. $-12 + 11$

−1

112. $\frac{-70}{10}$

−7

113. $-4 \cdot 4$

−16

114. $5 - 14$

−9

115. $\frac{-42}{-7}$

6

116. $16 + (-11)$

5

117. $-10 - 10$

−20

118. $\frac{-5}{0}$

undefined

119. $-\frac{2}{3} + \frac{1}{9}$

$-\frac{5}{9}$

120. $0.7(-0.5)$

−0.35

121. $|-6| + 2 - 3 \cdot (-8) - 5^2$

7

122. $9 \div |-3| + 6 \cdot (-5) + 2^3$

−19

Solve each equation.

123. $-45 = -5y$

9

124. $b - 8 = -12$

−4

125. $6z - 3 = 3z + 9$

4

126. $-5 = r + 5$

−10

127. $-3x = 33$

−11

128. $2z - 7z = -15$

3

129. $3(k - 6) = 6 - 12$

4

130. $6(t + 3) = -2 + 20$

0

131. $-10 = \dfrac{a}{5} - 2$

−40

132. $4 + 8p = 4p + 16$

3

Solve each application problem. Show your equation and the steps you use to solve it.

133. When twice a number is decreased by 8, the result is the number increased by 7. Find the number.

$2n - 8 = n + 7$
$n = 15$

134. The length of a rectangle is 3 inches more than twice the width. The perimeter is 36 inches. Find the length and the width. (Use $P = 2l + 2w$.)

$36 = 2(2w + 3) + 2w$
13 inches; 5 inches

135. A cheetah's sprinting speed is 25 miles per hour faster than a zebra can run. The sum of their running speeds is 111 miles per hour. How fast can each animal run?

$s + s + 25 = 111$
Cheetah sprints 68 miles per hour; zebra runs 43 miles per hour.

136. A 90-centimeter pipe is cut into two pieces so that one piece is 6 centimeters shorter than the other. Find the length of each piece.

$l + l - 6 = 90$
48 cm and 42 cm

CHAPTER 9 TEST

Work each of the following problems.

1. Graph the numbers $-4, -1, 1\frac{1}{2}, 3, 0$

1. (number line showing points at $-4, -1, 0, 1\frac{1}{2}, 3$)
$-5\ -4\ -3\ -2\ -1\ \ 0\ \ 1\ \ 2\ \ 3\ \ 4$

2. Place $<$ or $>$ in the blanks to make true statements:
-3 _____ 0 -4 _____ -8

2. $<, >$

3. Find $|-7|$ and $|15|$

3. 7, 15

Add, subtract, multiply, or divide.

4. $-8 + 7$

5. $-11 + (-2)$

4. -1

5. -13

6. $6.7 + (-1.4)$

7. $8 - 15$

6. 5.3

7. -7

8. $4 - (-12)$

9. $-\frac{1}{2} - \left(-\frac{3}{4}\right)$

8. 16

9. $\frac{1}{4}$

10. $8(-4)$

11. $-7(-12)$

10. -32

11. 84

12. $-16 \cdot 0$

13. $\frac{-100}{4}$

12. 0

13. -25

14. $\frac{-24}{-3}$

15. $-\frac{1}{4} \div \frac{5}{12}$

14. 8

15. $-\frac{3}{5}$

Use the order of operations to simplify each of the following.

16. $-5 + 3 \cdot (-2) - (-12)$

17. $2 - (6 - 8) - (-5)^2$

16. 1

17. -21

Find the value of $8k - 3m$, given the following.

18. $k = -4, \quad m = 2$

19. $k = 3, \quad m = -1$

18. -38

19. 27

20. _____

When evaluating, you are given specific values to replace each variable. When solving an equation, you are not given the value of the variable. You must find a value that "works"; that is, when your solution is substituted for the variable, the two sides of the equation are equal.

📝 **20.** In Exercises 18 and 19, you were evaluating an expression. Explain ◎ the difference between evaluating an expression and solving an equation.

21. 110 _____

21. The formula for the area of a triangle is $A = \frac{1}{2}bh$. Find A, if $b = 20$ and $h = 11$.

Solve each equation.

22. 5 _____

22. $x - 9 = -4$

23. $30 = -1 + r$

23. 31 _____

24. 0 _____

24. $0 = -7y$

25. $\dfrac{p}{5} = -3$

25. −15 _____

26. 5 _____

26. $3t - 8t = -25$

27. $3m - 5 = 7m - 13$

27. 2 _____

28. −3 _____

28. $-15 = 3(a - 2)$

Solve each application problem. Show your equation and the steps you use to solve it.

29. $7n - 23 = 47$
$n = 10$ _____

29. If seven times a number is decreased by 23, the result is 47. Find the number.

30. $l + l + 4 = 118$
57 cm; 61 cm _____

30. A board is 118 cm long. Karin cut it into two pieces, with one piece 4 cm longer than the other. Find the length of both pieces.

31. $420 = 2(4w) + 2w$
168 feet; 42 feet

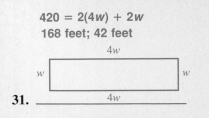

31. _____

31. The perimeter of a rectangular building is 420 feet. The length is four times as long as the width. Find the length and the width. Make a drawing to help solve this problem.

32. $h + h - 3 = 19$
Tim spent 8 hours;
Marcella spent 11 hours _____

32. Marcella and her husband Tim spent a total of 19 hours redecorating their living room. Tim spent 3 hours less time than Marcella. How long did each person work on the room?

≈ *Round the numbers in each problem so there is only one non-zero digit. Then add, subtract, multiply, or divide the rounded numbers, as indicated, to estimate the answer. Finally, solve for the exact answer. Write answers to fractions problems in lowest terms and as whole or mixed numbers when possible.*

1. estimate   exact
$$\begin{array}{r} 9 \\ 1 \\ +\ 40 \\ \hline 50 \end{array} \qquad \begin{array}{r} 8.7 \\ 0.902 \\ +\ 41 \\ \hline 50.602 \end{array}$$

2. estimate   exact
$$\begin{array}{r} 6 \\ \times\ 50 \\ \hline 300 \end{array} \qquad \begin{array}{r} 6.27 \\ \times\ 49.2 \\ \hline 308.484 \end{array}$$

3. estimate   exact
$$\begin{array}{r} 500 \\ 80\overline{)40{,}000} \end{array} \qquad \begin{array}{r} 503 \\ 78\overline{)39{,}234} \end{array}$$

4. estimate   exact
$$\begin{array}{r} 4 \\ -\ 3 \\ \hline 1 \end{array} \qquad \begin{array}{r} 3\frac{3}{5} \\ -\ 2\frac{3}{4} \\ \hline \frac{17}{20} \end{array}$$

5. *exact*

$$5\frac{5}{6} \cdot \frac{9}{10} = \underline{\ 5\frac{1}{4}\ }$$

estimate

$$\underline{\ 6\ } \cdot \underline{\ 1\ } = \underline{\ 6\ }$$

6. *exact*

$$4\frac{1}{6} \div 1\frac{2}{3} = \underline{\ 2\frac{1}{2}\ }$$

estimate

$$\underline{\ 4\ } \div \underline{\ 2\ } = \underline{\ 2\ }$$

Add, subtract, multiply, or divide as indicated.

7. $17 - 8.094$

8.906

8. $(1309)(408)$

534,072

9. $4.06 \div 0.072$ Round to nearest tenth.

≈56.4

10. $-12 + 7$ −5

11. $-5(-8)$ 40

12. $-3 - (-7)$ 4

13. $3.2 + (-4.5)$ −1.3

14. $\dfrac{30}{-6}$ −5

15. $\dfrac{1}{4} - \dfrac{3}{4}$ $-\dfrac{1}{2}$

Simplify by using the order of operations.

16. $45 \div \sqrt{25} - 2 \cdot 3 + (10 \div 5)$

5

17. $-6 - (4 - 5) + (-3)^2$

4

Write < or > in the blanks to make true statements.

18. $\dfrac{3}{10} \underline{\ >\ } \dfrac{4}{15}$

19. $0.7072 \underline{\ <\ } 0.72$

20. $-5 \underline{\ <\ } -2$

21. Write 8% as a decimal and as a fraction in lowest terms.

$0.08; \dfrac{2}{25}$

22. Write $4\frac{1}{2}$ as a decimal and as a percent.

4.5; 450%

23. There are 12 infants and 48 toddlers in the daycare center. Write the ratio of toddlers to infants as a fraction in lowest terms.

$\dfrac{4}{1}$

Find the unknown number in each proportion. Round your answer to hundredths, if necessary.

24. $\dfrac{x}{12} = \dfrac{1.5}{45}$

0.4

25. $\dfrac{350}{x} = \dfrac{3}{2}$

≈233.33

26. $\dfrac{38}{190} = \dfrac{9}{x}$

45

Solve each of the following.

27. 0.5% of 3000 students is how many students?

15 students

28. What percent of 12.5 miles is 6.8 miles?

54.4%

29. 90 cars is 180% of what number of cars?

50 cars

30. $5.80 is what percent of $145?

4%

Convert the following measurements.

31. $3\frac{1}{2}$ gallons to quarts

14 quarts

32. 72 hours to days

3 days

33. 3.7 L to milliliters

3700 mL

34. 40 cm to meters

0.4 m

Write the most reasonable metric unit in each blank. Choose from km, m, cm, mm, L, mL, kg, g, and mg.

35. The building is 15 __m__ high.

36. Rita took 15 __mL__ of cough syrup.

37. Bruce walked 2 __km__ to work.

38. The robin weighs 100 __g__ .

39. Identify which of these temperatures is normal body temperature and which is comfortable room temperature: 52°C, 70°C, 20°C, 10°C, 37°C, 98°C.   37°C; 20°C

Find the perimeter or circumference and the area of each figure. Use 3.14 as the approximate value of π.

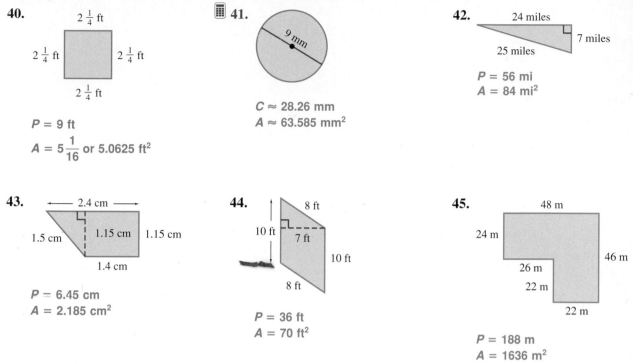

40.

$2\frac{1}{4}$ ft

$2\frac{1}{4}$ ft $2\frac{1}{4}$ ft

$2\frac{1}{4}$ ft

P = 9 ft

A = 5$\frac{1}{16}$ or 5.0625 ft²

41.

9 mm

C ≈ 28.26 mm

A ≈ 63.585 mm²

42.

24 miles

7 miles

25 miles

P = 56 mi

A = 84 mi²

43.

2.4 cm

1.5 cm   1.15 cm   1.15 cm

1.4 cm

P = 6.45 cm

A = 2.185 cm²

44.

8 ft

10 ft   7 ft

10 ft

8 ft

P = 36 ft

A = 70 ft²

45.

48 m

24 m

26 m   46 m

22 m

22 m

P = 188 m

A = 1636 m²

Find the unknown length in each figure. Round your answers to the nearest tenth, if necessary.

46.

20 yd   y

15 yd

y ≈ 13.2 yd

47. similar triangles

18 in.

7 in.   8.5 in.   17 in.   x

9 in.

x = 14 in.

Solve each equation.

48. $-20 = 6 + y$

    −26

49. $-2t - 6t = 40$

    −5

50. $3x + 5 = 5x - 11$

    8

51. $6(p + 3) = -6$

    −4

Chapter 9   Basic Algebra

Solve each application problem. Show your equation and the steps you use to solve it.

52. If 40 is added to four times a number, the result is zero. Find the number.

$4n + 40 = 0$
-10

53. $1000 in prize money is being split between Reggie and Donald. Donald should get $300 more than Reggie. How much will each man receive?

$p + p + 300 = 1000$
$350 for Reggie, $650 for Donald

54. Make a drawing to help solve this problem. The length of a photograph is 5 cm more than the width. The perimeter of the photograph is 82 cm. Find the length and the width.

23 cm, 18 cm

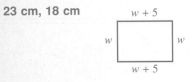

Solve the following application problems.

55. Portia bought two CDs at $11.98 each. The sales tax rate is $6\frac{1}{2}\%$. Find the total amount charged to Portia's credit card, to the nearest cent.

≈$25.52

56. Brian's spaghetti sauce recipe calls for $3\frac{1}{3}$ cups of tomato sauce. He wants to make $2\frac{1}{2}$ times the usual amount. How much tomato sauce does he need?

$8\frac{1}{3}$ **cups**

57. The local food shelf received 2480 pounds of food this month. Their goal was 2000 pounds. What percent of their goal was received?

124%

58. Sayoko bought 720 g of chicken priced at $5.97 per kilogram. How much did she pay for the chicken, to the nearest cent?

≈$4.30

59. A packing crate measures 2.4 m long, 1.2 m wide, and 1.2 m high. A trucking company wants crates that hold 4 m³. The crate's volume is how much more or less than 4 m³?

0.544 m³ less

60. The Mercado family has 35 feet of fencing to put around a garden plot. The plot is rectangular in shape. If it is $6\frac{1}{2}$ feet wide, find the length of the plot.

11 feet

61. Rich spent 25 minutes reading 14 pages in his sociology textbook. At that rate, how long will it take him to read 30 pages? Round to the nearest whole number of minutes.

≈54 minutes

62. Jackie drove her car 364 miles on 14.5 gallons of gas. Maya used 16.3 gallons to drive 406 miles. Naomi drove 300 miles on 11.9 gallons. Which car had the highest number of miles per gallon? How many miles per gallon did that car get, rounded to the nearest tenth?

Naomi's car; ≈25.2 miles per gallon

Statistics 10

The word *statistics* originally came from words that mean *state numbers*. State numbers refer to numerical information, or **data,** gathered by the government such as the number of births, deaths, or marriages in a population. Today the word *statistics* has a much broader application; data from the fields of economics, social science, science, and business can all be organized and studied under the branch of mathematics called **statistics.**

10.1 *Circle Graphs*

OBJECTIVE 1 It can be hard to understand a large collection of data. The graphs described in this section can be used to help you make sense of such data. The **circle graph** is used to show how a total amount is divided into parts. The circle graph below shows you how 24 hours in the life of a college student are divided among different activities.

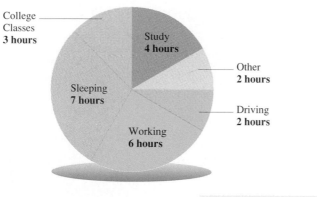

WORK PROBLEM 1 AT THE SIDE. ▶▶

OBJECTIVE 2 This circle graph uses pie-shaped pieces called **sectors** to show the amount of time spent on each activity (the total must be 24 hours); a circle graph can therefore be used to compare the time spent on one activity to the total number of hours in the day.

E X A M P L E 1 Using a Circle Graph

Find the ratio of time spent in college classes to the total number of hours in a day. Write the ratio as a fraction in lowest terms. (See **Section 5.1.**)

The circle graph shows that 3 of the 24 hours in a day are spent in class. The ratio of class time to the hours in a day is

$$\frac{3 \text{ hours (college classes)}}{24 \text{ hours (whole day)}} = \frac{3 \text{ hours}}{24 \text{ hours}} = \frac{1}{8} \leftarrow \text{Lowest terms}$$

www.mathnotes.com

OBJECTIVES

1 ▶ Read and understand a circle graph.
2 ▶ Use a circle graph.
3 ▶ Draw a circle graph.

FOR EXTRA HELP

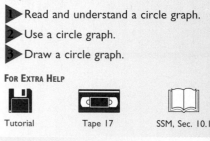

Tutorial   Tape 17   SSM, Sec. 10.1

1. Use the circle graph to answer each of the following.

   (a) The greatest number of hours is spent in which activity?

   (b) How many more hours are spent working than studying?

   (c) Find the total number of hours spent studying, working, and attending classes.

ANSWERS

1. (a) sleeping (b) 2 hours (c) 13 hours

2. Use the circle graph to find the following ratios. Write the ratios as fractions in lowest terms.

(a) hours spent driving to whole day

(b) hours spent studying to whole day

(c) hours spent sleeping and doing other to all day

(d) hours spent working and studying to whole day

3. Use the circle graph to find the following ratios. Write the ratios as fractions in lowest terms.

(a) hours spent studying to hours spent working

(b) hours spent working to hours spent sleeping

(c) hours spent studying to hours spent driving

(d) hours spent in class to hours spent for other

◄◄ **WORK PROBLEM 2 AT THE SIDE.**

This circle graph can also be used to find the ratio of the time spent on one activity to the time spent on any other activity.

EXAMPLE 2   Finding a Ratio from a Circle Graph

Find the ratio of working time to class time.

The circle graph shows 6 hours spent working and 3 hours spent in class. The ratio of working time to class time is

$$\frac{6 \text{ hours (working)}}{3 \text{ hours (class)}} = \frac{6 \text{ hours}}{3 \text{ hours}} = \frac{2}{1}$$

◄◄ **WORK PROBLEM 3 AT THE SIDE.**

A circle graph often shows data as percents. For example, the recorded music industry had total annual sales of $12 billion in 1995. The next circle graph shows how the sales were divided among the various types of recording formats. The entire circle represents the $12 billion in sales. Each sector represents the sales of one format as a percent of the total sales (the total must be 100%).

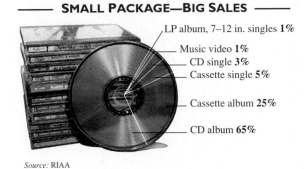

——— **SMALL PACKAGE—BIG SALES** ———

LP album, 7–12 in. singles **1%**
Music video **1%**
CD single **3%**
Cassette single **5%**
Cassette album **25%**
CD album **65%**

Source: RIAA

EXAMPLE 3   Calculating Amounts Using a Circle Graph

Use the circle graph on recorded music sales to find the amount spent on CD albums for the year.

Recall the percent equation:

$$\text{amount} = \text{percent} \cdot \text{base}$$

or $a = p \cdot b$

The total sales are $12 billion, so $b = \$12$ billion. The percent is 65, so $p = 0.65$. Find a.

$$\text{amount} = \text{percent} \cdot \text{base}$$
$$a = 0.65 \cdot 12 \text{ billion} = 7.8 \text{ billion}$$

The amount spent on CD albums was $7.8 billion or $7,800,000,000.

ANSWERS

2. (a) $\frac{1}{12}$ **(b)** $\frac{1}{6}$ **(c)** $\frac{3}{8}$ **(d)** $\frac{5}{12}$

3. (a) $\frac{2}{3}$ **(b)** $\frac{6}{7}$ **(c)** $\frac{2}{1}$ **(d)** $\frac{3}{2}$

WORK PROBLEM 4 AT THE SIDE. ▶▶

OBJECTIVE 3 The coordinator of the Fair Oaks Youth Soccer League organizes teams in five age groups. She places the players in various age groups as follows.

Age Group	Percent of Total
Under 8 years	20%
Under 10 years	15%
Under 12 years	25%
Under 14 years	25%
Under 16 years	15%
Total	100%

You can show these percents by using a circle graph. A circle has 360 degrees (written 360°). The 360° represents the entire league, or 100% of the soccer league.

EXAMPLE 4 Drawing a Circle Graph

Using the data on *age groups*, find the number of degrees in the sector that would represent the "Under 8" group, and begin constructing a circle graph.

Recall that a complete circle has 360° (**Section 8.1**). Because the "Under 8" group makes up 20% of the total number of players, the number of degrees needed for the "Under 8" sector of the circle graph is 20% of 360°.

$$360° \times 20\% = 360° \times 0.2 = 72°$$

Use a tool called a **protractor** (PRO-trak-ter) to make a circle graph. First, using a straight edge, draw a line from the center of a circle to the left edge. Place the hole in the protractor over the center of the circle, making sure that zero on the protractor lines up with the line that was drawn. Find 72° and make a mark as shown in the illustration. Then remove the protractor and use the straight edge to draw a line from the center of the circle to the 72° mark at the edge of the circle.

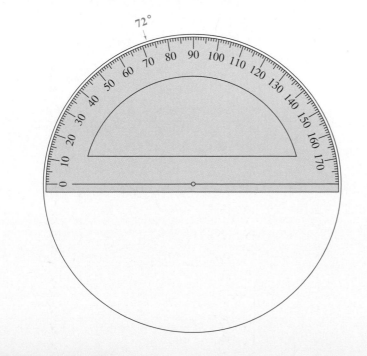

CONTINUED ON NEXT PAGE

4. Use the circle graph on recorded music sales to find the following:

(a) the amount spent on CD singles

(b) the amount spent on LP albums, 7–12 inch singles

(c) the amount spent on cassette albums

(d) the amount spent on cassette singles

ANSWERS
4. (a) $0.36 billion or $360,000,000
   (b) $0.12 billion or $120,000,000
   (c) $3 billion or $3,000,000,000
   (d) $0.6 billion or $600,000,000

5. Using the information on the soccer age groups in the table, find the number of degrees needed for each of the following and complete the circle graph.

(a) Under 12 group

To draw the "Under 10" sector, begin by finding the number of degrees in the sector.

$$360° \times 15\% = 360° \times 0.15 = 54°$$

Again, place the hole of the protractor at the center of the circle, but this time align zero on the second line that was drawn. Make a mark at 54° and draw a line as before. This sector is 54° and represents the "Under 10" group.

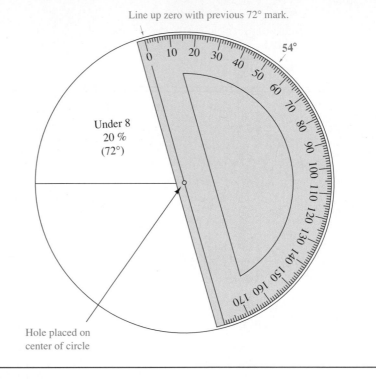

Line up zero with previous 72° mark.

Under 8
20 %
(72°)

Hole placed on center of circle

(b) Under 14 group

(c) Under 16 group

Note
You must be certain that the hole in the protractor is placed on the exact center of the circle each time you measure the size of a sector.

◀◀ **WORK PROBLEM 5 AT THE SIDE.**

ANSWERS

5. **(a)** 90° **(b)** 90° **(c)** 54°

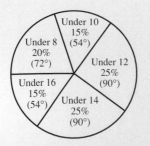

10.1 *Exercises*

Use the circle graph below, which shows the cost of adding an art studio to an existing building, to find each of the following. Write ratios as fractions in lowest terms.

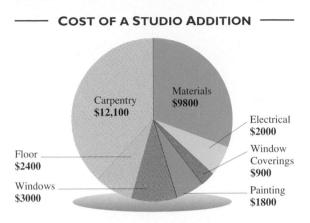

—— **COST OF A STUDIO ADDITION** ——

Materials $9800
Carpentry $12,100
Electrical $2000
Window Coverings $900
Floor $2400
Windows $3000
Painting $1800

1. Find the total cost of adding the art studio.

   $32,000

2. What is the largest single expense in adding the studio?

   carpentry

3. Find the ratio of the cost of materials to the total remodeling cost.

   $$\frac{9800}{32,000} = \frac{49}{160}$$

4. Find the ratio of the cost of painting to the total remodeling cost.

   $$\frac{1800}{32,000} = \frac{9}{160}$$

5. Find the ratio of the cost of carpentry to the cost of window coverings.

   $$\frac{12,100}{900} = \frac{121}{9}$$

6. Find the ratio of the cost of windows to the cost of the floor.

   $$\frac{3000}{2400} = \frac{5}{4}$$

7. The circle graph at the right shows the number of students at Rockfield College who are enrolled in various majors. In which major are the least number of students enrolled?

   history

STUDENTS AT ROCKFIELD COLLEGE

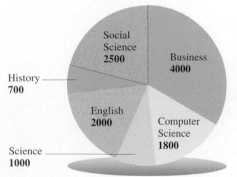

Social Science 2500
Business 4000
History 700
English 2000
Computer Science 1800
Science 1000

8. Using the circle graph at the right, which major has the second highest number of students enrolled?

   social science

Use the circle graph from Exercises 7 and 8 to find each ratio. Write the ratios as fractions in lowest terms.

9. business majors to the total number of students

$$\frac{4000}{12,000} = \frac{1}{3}$$

10. English majors to the total number of students

$$\frac{2000}{12,000} = \frac{1}{6}$$

11. computer science majors to the number of English majors

$$\frac{1800}{2000} = \frac{9}{10}$$

12. history majors to the number of social science majors

$$\frac{700}{2500} = \frac{7}{25}$$

13. business majors to the number of science majors

$$\frac{4000}{1000} = \frac{4}{1}$$

14. science majors to the number of history majors

$$\frac{1000}{700} = \frac{10}{7}$$

The following circle graph shows the costs necessary to comply with the Americans with Disabilities Act (ADA) at the Dos Pueblos College. Each cost item is expressed as a percent of the total cost of $1,740,000. Use the graph to find the dollar amount spent for each of the following.

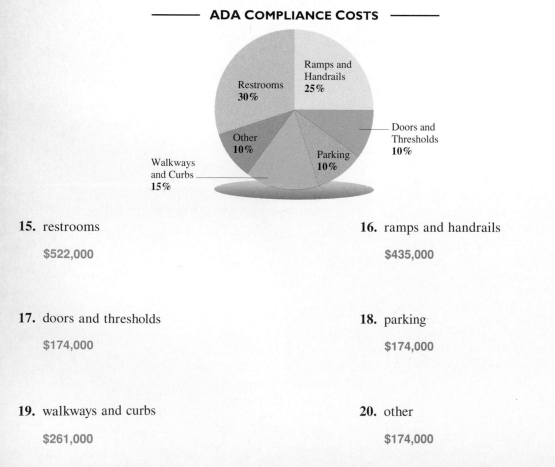

ADA COMPLIANCE COSTS

Ramps and Handrails 25%
Restrooms 30%
Other 10%
Walkways and Curbs 15%
Parking 10%
Doors and Thresholds 10%

15. restrooms

$522,000

16. ramps and handrails

$435,000

17. doors and thresholds

$174,000

18. parking

$174,000

19. walkways and curbs

$261,000

20. other

$174,000

The circle graph below is adapted from USA TODAY *and shows the percent of employees who take time off from work during one year because of colds. If Folsum Electronics has 8740 employees, use the graph to find the number of employees who miss work for various lengths of time because of colds. Round to the nearest whole number.*

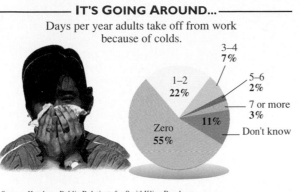

IT'S GOING AROUND...

Days per year adults take off from work
because of colds.

3–4
7%

1–2
22%

5–6
2%

7 or more
3%

11%

Zero
55%

Don't know

Source: Ketchum Public Relations for SmithKline Beecham

21. 1–2 days

≈1923 people

22. don't know

≈961 people

23. 5–6 days

≈175 people

24. 3–4 days

≈612 people

25. 7 or more days

≈262 people

26. never take off for colds

4807 people

27. Describe the procedure for determining how large each sector must be to represent each of the items in a circle graph.

First find the percent of the total that is to be represented by each item. Next, multiply the percent by 360° to find the size of each sector. Finally, use a protractor to draw each sector.

28. A protractor is the tool used to draw a circle graph. Give a brief explanation of what the protractor does and how you would use it to measure and draw each sector in the circle graph.

A protractor is used to measure the number of degrees in a sector. First, you must draw a line from the center of the circle to the left edge. Next, place the hole of the protractor at the center of the circle, making sure that the zero on the protractor is on the line. Finally, make a mark at the desired number of degrees. This gives you the size of the sector.

During one semester Zoë Werner spent $4200 for school expenses as shown in the following chart. Find all numbers missing from the chart.

Item	Dollar Amount	Percent of Total	Degrees of a Circle
29. rent	$1050	25%	90°
30. food	$840	20%	72°
31. clothing	$420	10%	36°
32. books	$420	10%	36°
33. entertainment	$630	15%	54°
34. savings	$210	5%	18°
35. other	$630	15%	54°

36. Draw a circle graph by using the above information.

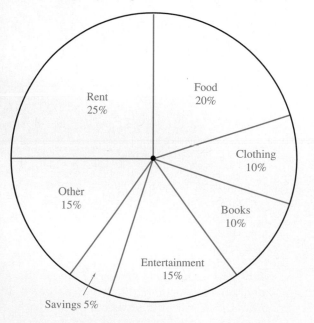

37. White Water Rafting Company divides its annual sales into five categories, as follows.

Category	Annual Sales
Adventure classes	$12,500
Grocery and provision sales	$40,000
Equipment rentals	$60,000
Rafting tours	$50,000
Equipment sales	$37,500

(a) Find the total sales for the year.

$200,000

(b) Find the number of degrees in a circle graph for each item.

22.5°; 72°; 108°; 90°; 67.5°

(c) Make a circle graph showing this information.

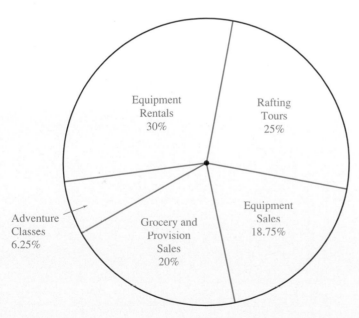

38. A book publisher had 25% of total sales in mysteries, 10% in biographies, 15% in cookbooks, 15% in romantic novels, 20% in science, and the rest in business books.
(a) Find the number of degrees in a circle graph for each type of book.

90°; 36°; 54°; 54°; 72°; 54°

(b) Draw a circle graph.

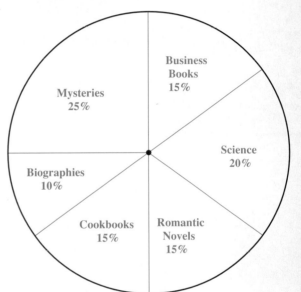

▲ **39.** A family kept track of its expenses for a year and recorded the following results. Complete the chart and draw a circle graph.

Item	Amount	Percent of Total	Number of Degrees
Housing	$9600	30%	108°
Food	$6400	20%	72°
Automobile	$4800	15%	54°
Clothing	$3200	10%	36°
Medical	$1600	5%	18°
Savings	$1600	5%	18°
Other	$4800	15%	54°
Total	$32,000		

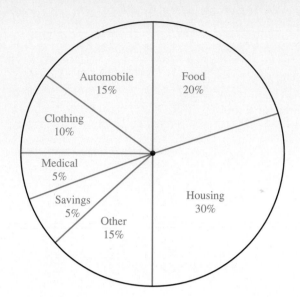

Review and Prepare

Write < *or* > *to make a true statement. (For help, see* **Sections 3.5, 4.7,** *and* **6.1.***)*

40. $\dfrac{1}{4}$ ___<___ $\dfrac{3}{8}$

41. $\dfrac{4}{5}$ ___<___ $\dfrac{7}{8}$

42. 0.4219 ___<___ 0.422

43. 0.0118 ___>___ 0.01

44. 38.25% ___<___ 38.29%

45. 25.9% ___<___ 26.01%

46. 60,500 ___>___ 60,498

47. 44,272.68 ___>___ 44,272.098

48. 799,802 ___<___ 799,899

10.2 Bar Graphs and Line Graphs

OBJECTIVE 1 **Bar graphs** are useful when showing comparisons. For example, the bar graph below compares the number of college graduates who continued taking advanced courses in their major field during each of five years.

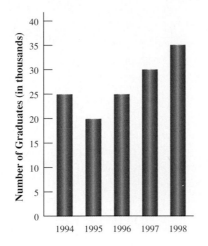

EXAMPLE 1 Using a Bar Graph

How many college graduates took advanced classes in their major field in 1996?

The bar for 1996 rises to 25. Notice the label along the left side of the graph that says "Number of Graduates (in thousands)." The phrase *in thousands* means you have to multiply 25 by 1000 to get 25,000. So, 25,000 (not 25) graduates took advanced classes in their major field in 1996.

WORK PROBLEM 1 AT THE SIDE. ▶▶

OBJECTIVE 2 A **double-bar graph** can be used to compare two sets of data. This graph shows the number of new cable television installations each quarter for two different years.

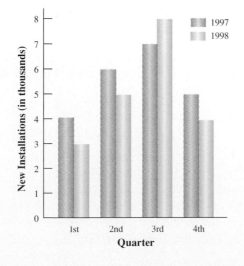

OBJECTIVES

Read and understand

1 ▶ a bar graph;

2 ▶ a double-bar graph;

3 ▶ a line graph;

4 ▶ a comparison line graph.

FOR EXTRA HELP

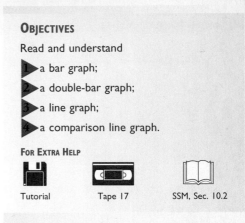

Tutorial   Tape 17   SSM, Sec. 10.2

1. Use the bar graph in the text to find the number of college graduates who took advanced classes in their major field in each of these years.

   (a) 1994

   (b) 1995

   (c) 1997

   (d) 1998

ANSWERS

1. (a) 25,000 (b) 20,000 (c) 30,000
   (d) 35,000

673

2. Use the double-bar graph to find the number of new cable television installations in 1997 and 1998 for each of the following quarters.

(a) 1st quarter

(b) 3rd quarter

(c) 4th quarter

(d) Find the greatest number of installations. Identify the quarter and the year in which they occurred.

3. Use the line graph in the text to find the number of trout stocked in each of the following months.

(a) June

(b) May

(c) April

(d) July

ANSWERS

2. (a) 4000; 3000 **(b)** 7000; 8000
   (c) 5000; 4000
   (d) 8000; 3rd quarter of 1998
3. (a) 55,000 **(b)** 30,000 **(c)** 40,000
   (d) 60,000

E X A M P L E 2   Reading a Double-Bar Graph

Use the double-bar graph to find each of the following.

(a) the number of new cable television installations in the second quarter of 1997

There are two bars for the second quarter. The color code in the upper right hand corner of the graph tells you that the **red bars** represent 1997. So the **red bar** on the *left* is for the 2nd quarter of 1997. It rises to 6. Multiply 6 by 1000 because the label on the left side of the graph says *in thousands*. So there were 6000 new installations for the second quarter in 1997.

(b) the number of new cable television installations in the second quarter of 1998

The **green bar** for the second quarter rises to 5 and 5 times 1000 is 5000. So, in the second quarter of 1998, there were 5000 new installations.

> **Note**
> Use a ruler or straight edge to line up the top of the bar with the number on the left side of the graph.

◄◄ **WORK PROBLEM 2 AT THE SIDE.**

OBJECTIVE ▶ A **line graph** is often useful for showing a trend. The line graph that follows shows the number of trout stocked along the Feather River over a five-month period. Each dot indicates the number of trout stocked during the month directly below that dot.

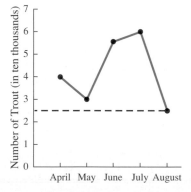

E X A M P L E 3   Understanding a Line Graph

Use the line graph to find the following.

(a) In which month were the least number of trout stocked?

The lowest point on the graph is the dot directly over August, so the least number of trout were stocked in August.

(b) How many trout were stocked in August?

Use a ruler or straight edge to line up the August dot with the numbers along the left edge of the graph. The August dot is halfway between the 2 and 3. Notice the label on the left side says "in ten thousands." So August is halfway between 2 • 10,000 and 3 • 10,000. It is halfway between 20,000 and 30,000. That means 25,000 trout were stocked in August.

◄◄ **WORK PROBLEM 3 AT THE SIDE.**

OBJECTIVE ▶ Two sets of data can also be compared by drawing two line graphs together as a **comparison line graph.** For example, the following line graph compares the number of thermal-paper fax machines and the number of plain-paper fax machines sold during each of five years.

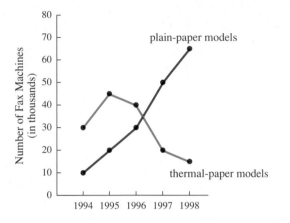

E X A M P L E   4   Interpreting a Comparison Line Graph

Use the comparison line graph above to find the following.

(a) the number of thermal-paper machines sold in 1995

Find the dot on the blue line above 1995. Use a ruler or straight edge to line up the dot with the numbers along the left edge. The dot is halfway between 40 and 50, which is 45. Then, 45 times 1000 is 45,000 thermal-paper machines that were sold in 1995.

(b) the number of plain-paper machines sold in 1998

The **red line** on the graph shows that 65,000 plain-paper machines were sold in 1998. ■

> **Note**
> Both the double-bar graph and the comparison line graph are used to compare two or more sets of data.

WORK PROBLEM 4 AT THE SIDE. ▶▶

4. Use the comparison line graph in the text to find the following.

   (a) the number of thermal-paper machines sold in 1994, 1996, 1997, and 1998

   (b) the number of plain-paper machines sold in 1994, 1995, 1996, and 1997

   (c) the first full year in which the number of plain-paper machines sold was greater than the number of thermal-paper machines sold

ANSWERS

4. **(a)** 30,000; 40,000; 20,000; 15,000
   (b) 10,000; 20,000; 30,000; 50,000
   (c) 1997

NUMBERS IN THE
Real World collaborative investigations

The graph at the right is a double-bar graph. What is it about? Write several sentences describing the general purpose of the graph. **The graph shows what percent of men and what percent of women use various strategies for saving money on groceries.**

Which two savings strategies do women use most often? Which two do men use most often?
Women: check ads for specials; mail/newspaper coupons. Men: stock-up on bargains; stick to list

Which two strategies show the greatest difference in use between men and women? Which two strategies show the least difference?
Greatest: check ads for specials; mail/newspaper coupons. Least: buy store brands; stock-up on bargains

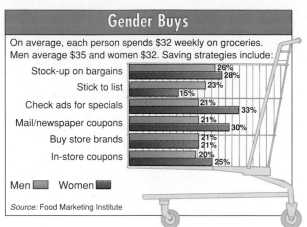

Gender Buys

On average, each person spends $32 weekly on groceries. Men average $35 and women $32. Saving strategies include:

Strategy	Men	Women
Stock-up on bargains	26%	28%
Stick to list	23%	15%
Check ads for specials	21%	33%
Mail/newspaper coupons	21%	30%
Buy store brands	21%	21%
In-store coupons	20%	25%

Men ▭ Women ▬

Source: Food Marketing Institute

Sometimes people "jump to conclusions" without enough evidence. Which of these conclusions are appropriate, *based on the information in the graph*?

Men and women use a variety of savings strategies when grocery shopping. **appropriate**
Men spend more for groceries than women. **appropriate**
Men eat more groceries than women do. **not appropriate**
There are some differences in the grocery shopping strategies used by men and women. **appropriate**
Women are better grocery shoppers than men. **not appropriate**

Write two additional conclusions that you can make, based on the information in the graph. Share your conclusions with other class members. Be prepared to defend your conclusions. **There are many possible answers, but all must be clearly based on data in the graph, not on personal assumptions.**

Conduct a survey of your class members. Find out how many of them regularly use each of the savings strategies shown in the graph. Complete the table below. **Results will vary.**

Total number of women in survey _____			Total number of men in survey _____	
	Number of Women Using Strategy	Percent of Women Using Strategy	Number of Men Using Strategy	Percent of Men Using Strategy
Stock-up on bargains				
Stick to list				
Check ads for specials				
Mail/newspaper coupons				
Buy store brands				
In-store coupons				

Now make a double-bar graph showing your survey data. How is your data similar to the graph above? How is it different? **Results will vary.**

10.2 Exercises

The National Insurance Crime Bureau says that "thieves are stealing fewer cars but more motorcycles." There has been a 46% increase in motorcycle thefts since 1992. The bar graph below shows the number of motorcycles stolen in a large eastern city over a five-year period.

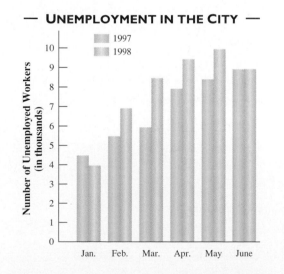

— CITY MOTORCYCLE THEFTS —

1. Find the number of motorcycles stolen in 1994.

   8000 motorcycles

2. Find the number of motorcycles stolen in 1993.

   7000 motorcycles

3. Which year had the greatest number of motorcycles stolen? How many were stolen that year?

   1996; 10,500 motorcycles

4. Which year had the least number of motorcycles stolen? How many were stolen that year?

   1992; 6000 motorcycles

5. How many more motorcycles were stolen in 1995 than in 1993?

   2500 motorcycles

6. How many more motorcycles were stolen in 1996 than in 1992?

   4500 motorcycles

This double-bar graph shows the number of workers who were unemployed in a city during the first six months of 1997 and 1998.

— UNEMPLOYMENT IN THE CITY —

■ 1997
■ 1998

7. In which month in 1998 were the greatest number of workers unemployed? What was the total number unemployed in that month?

   May; 10,000 unemployed

8. How many workers were unemployed in January of 1997?

   4500 workers

9. How many more workers were unemployed in February of 1998 than in February of 1997?

   1500 workers

10. How many fewer workers were unemployed in March of 1997 than in March of 1998?

    2500 workers

📝 Writing   ◉ Conceptual   ▲ Challenging   ≈ Estimation

11. Find the increase in the number of unemployed workers from February 1997 to April 1997.

2500 workers

12. Find the increase in the number of unemployed workers from January 1998 to June 1998.

5000 workers

This double-bar graph shows sales of super unleaded and supreme unleaded gasoline at a service station for each of five years.

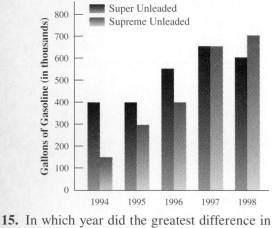

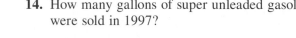

13. How many gallons of supreme unleaded gasoline were sold in 1994?

150,000 gallons

14. How many gallons of super unleaded gasoline were sold in 1997?

650,000 gallons

15. In which year did the greatest difference in sales between super unleaded and supreme unleaded gasoline occur? Find the difference.

1994; 250,000 gallons

16. In which year did the sales of supreme unleaded gasoline surpass the sales of super unleaded gasoline?

1998

17. Find the increase in supreme unleaded gasoline sales from 1994 to 1998.

550,000 gallons

18. Find the increase in super unleaded gasoline sales from 1994 to 1998.

200,000 gallons

This line graph shows the number of burglaries in a community during the first six months of last year.

19. In which month did the greatest number of burglaries occur? How many were there?

April; 600 burglaries

20. In which month did the least number of burglaries occur? How many were there?

February; 200 burglaries

21. Find the increase in the number of burglaries from March to April.

200 burglaries

22. Find the decrease in the number of burglaries from April to May.

200 burglaries

23. Give two possible explanations for the decrease in burglaries during February.

Extremely cold or snowy weather; greater police activity.

24. Give two possible explanations for the increase in burglaries in March and April.

Improved weather; decrease in police activity.

This comparison line graph below shows the number of compact discs (CDs) sold by two different chain stores during each of five years. Find the annual number of CDs sold in each of the following years.

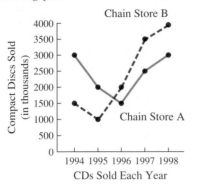

25. Chain Store A in 1998

3,000,000 CDs

26. Chain Store A in 1997

2,500,000 CDs

27. Chain Store A in 1996

1,500,000 CDs

28. Chain Store B in 1998

4,000,000 CDs

29. Chain Store B in 1997

3,500,000 CDs

30. Chain Store B in 1996

2,000,000 CDs

31. Looking at the comparison line graph above, which store would you like to own? Explain why. Based on the graph, what amount of sales would you predict for your store in 1999?

Probably Store B with greater sales. Predicted sales might be 4,500,000 CDs to 5,000,000 CDs in 1999.

32. In the comparison line graph above, Store B used to have lower sales than Store A. What might have happened to cause this change? Give two possible explanations.

Some possible answers are that Store B may have started to: do more advertising; keep longer store hours; give better training to their staff; employ more help; give better service than Store A.

33. Explain in your own words why a bar graph or a line graph (not a double-bar graph or comparison line graph) can be used to show only one set of data.

A single bar or a single line must be used for each set of data. To show multiple sets of data, multiple sets of bars or lines must be used.

34. The double-bar graph and the comparison line graph are both useful for comparing two sets of data. Explain how this works and give your own example.

You would use a set of bars or a set of lines for each set of data. One example is the number of miles driven by two salespeople over a three year period.

▲ *This comparison line graph shows the sales and profits of Tacos-To-Go for each of four years. Use the graph to answer the following questions.*

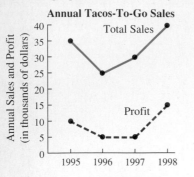

Annual Tacos-To-Go Sales

35. total sales in 1998   **$40,000**

36. total sales in 1997   **$30,000**

37. total sales in 1996   **$25,000**

38. profit in 1998   **$15,000**

39. profit in 1997   **$5000**

40. profit in 1996   **$5000**

41. Give two possible explanations for the decrease in sales from 1995 to 1996 and two possible explanations for the increase in sales from 1996 to 1998.

The decrease in sales may have resulted from poor service or greater competition. The increase in sales may have been a result of more advertising or better service.

42. Based on the graph, what conclusion can you make about the relationship between sales and profits?

As sales increase or decrease, so do profits.

Review and Prepare

 *The circle graph below shows how the city of Santa Barbara spent their annual budget. Use the graph to find the amount spent on each budget item. Round to the nearest thousand dollars. (For help, see **Section 10.1.**)*

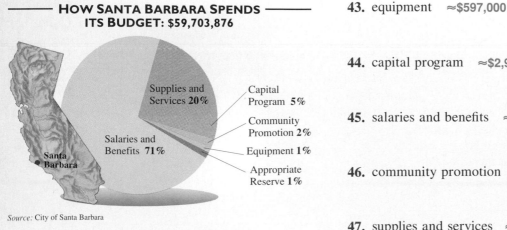

HOW SANTA BARBARA SPENDS ITS BUDGET: $59,703,876

Supplies and Services **20%**

Capital Program **5%**

Community Promotion **2%**

Salaries and Benefits **71%**

Equipment **1%**

Appropriate Reserve **1%**

Santa Barbara

Source: City of Santa Barbara

43. equipment   ≈**$597,000**

44. capital program   ≈**$2,985,000**

45. salaries and benefits   ≈**$42,390,000**

46. community promotion   ≈**$1,194,000**

47. supplies and services   ≈**$11,941,000**

48. appropriate reserve   ≈**$597,000**

10.3 Frequency Distributions and Histograms

OBJECTIVES

1. Understand a frequency distribution.
2. Arrange data in class intervals.
3. Read and understand a histogram.

FOR EXTRA HELP

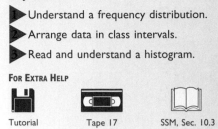

Tutorial   Tape 17   SSM, Sec. 10.3

The owner of a small insurance agency has kept track of her personal phone sales call activity over the past 50 weeks. The number of sales calls made for each of the weeks is given below. Read down the columns, beginning with the left column, for successive weeks of the year.

75	65	40	50	45	30	30	35	45	25
75	70	60	55	30	25	44	30	35	30
75	70	50	30	50	20	30	30	20	25
60	62	45	45	48	40	35	25	20	25
75	45	50	40	35	40	40	30	27	40

OBJECTIVE 1 A long list of numbers can be confusing. You can make the data easier to read by putting it in a table. This type of table is called a **frequency distribution** (FREE-kwen-see dis-trih-BYOO-shun).

E X A M P L E   1   Preparing a Frequency Distribution

Using the data above, construct a table that shows each possible number of sales calls. Then go through the original data and place a **tally** mark (|) in the tally column next to each corresponding value. The result is a frequency distribution table.

Number of Sales Calls	Tally	Frequency	Number of Sales Calls	Tally	Frequency
20					3	48			1
25	NJ	5	50	||||	4
27			1	55			1
30	NJ ||||	9	60	||	2
35						4	62			1
40	NJ		6	65			1
44			1	70				2
45	NJ	5	75	||||	4

WORK PROBLEM 1 AT THE SIDE. ▶▶

OBJECTIVE 2 The frequency distribution given in Example 1 contains a great deal of information—perhaps too much to digest. It can be simplified by combining the number of sales calls into groups, forming the class intervals shown here.

GROUPED DATA

Class Intervals (Number of Sales Calls)	Class Frequency (Number of Weeks)
20–29	9
30–39	13
40–49	13
50–59	5
60–69	4
70–79	6

1. Use the frequency distribution table to find the following.

   (a) the least number of sales calls made in a week

   (b) the most common number of sales calls made in a week

   (c) the number of weeks in which 50 calls were made

   (d) the number of weeks in which 40 calls were made

ANSWERS

1. (a) 20 calls (b) 30 calls (c) 4 weeks
   (d) 6 weeks

2. Use the grouped data for the insurance agency to answer the following questions.

(a) During how many weeks were less than 50 calls made?

(b) During how many weeks were 50 or more calls made?

Note

The number of class intervals in the left column of the table is arbitrary. Grouped data usually has between 5 and 15 class intervals.

E X A M P L E 2   Analyzing a Frequency Distribution

Use the grouped data for the insurance agency (on the preceeding page) to answer the following questions.

(a) During how many weeks were fewer than 30 calls made?

The first class in the grouped data table above (20–29) is the number of weeks during which fewer than 30 calls were made. Therefore, the owner made fewer than 30 calls during 9 weeks out of the 50 weeks shown.

(b) During how many weeks were 40 or more calls made?

The last four classes in the grouped data table are the number of weeks during which 40 or more calls were made.

$$13 + 5 + 4 + 6 = 28 \text{ weeks}$$

◀◀ **WORK PROBLEM 2 AT THE SIDE.**

OBJECTIVE 3 The results in the grouped data table have been used to draw this special bar graph, called a **histogram** (HIST-oh-gram). In a histogram, the width of each bar represents a range of numbers (*class interval*). The height of each bar in a histogram gives the *class frequency,* that is, the number of occurrences in each class interval.

3. Use the histogram to find the following.

(a) During how many weeks were less than 60 calls made?

(b) During how many weeks were 60 or more calls made?

SALES CALL DATA FOR THE PAST 50 WEEKS (GROUPED DATA)

These are the 9 weeks in which 20–29 calls were made.

Number of Weeks (vertical axis)
Number of Sales Calls per Week (horizontal axis): 20–29, 30–39, 40–49, 50–59, 60–69, 70–79

E X A M P L E 3   Using a Histogram

Use the histogram to find the number of weeks in which less than 40 calls were made.

Because 20–29 calls were made during 9 of the weeks and 30–39 calls were made during 13 of the weeks, the number of weeks in which less than 40 calls were made was $9 + 13 = 22$ weeks.

◀◀ **WORK PROBLEM 3 AT THE SIDE.**

ANSWERS

2. (a) 35 weeks **(b)** 15 weeks

3. (a) 40 weeks **(b)** 10 weeks

10.3 Exercises

The Toy Train Collectors' Club recorded the ages of its members and used the results to construct the histogram below. Use the histogram to find each of the following.

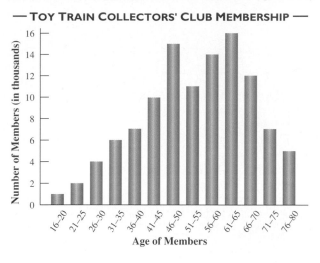

— TOY TRAIN COLLECTORS' CLUB MEMBERSHIP —

1. The greatest number of members are in which age group? How many members are in that group?

   61–65 years; 16,000 members

2. The least number of members are in which age group? How many members are in that group?

   16–20 years; 1000 members

3. Find the number of members 35 years of age and under.

   13,000 members

4. Find the number of members 61 years of age and older.

   40,000 members

5. How many members are 41 to 60 years of age?

   50,000 members

6. How many members are 46 to 55 years of age?

   26,000 members

This histogram shows the annual salaries for part-time and full-time employees of the Tickle Time Pickle Factory.

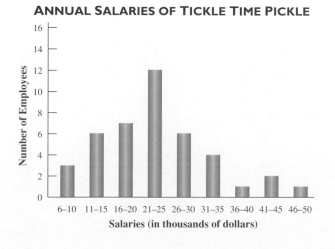

ANNUAL SALARIES OF TICKLE TIME PICKLE

7. The greatest number of employees are in which salary group? How many are in that group?

   $21,000 to $25,000; 12 employees

8. The least number of employees are in which salary groups? How many are in each group?

   $36,000 to $40,000, one employee; $46,000 to $50,000, one employee

---

✎ Writing   ◎ Conceptual   ▲ Challenging   ≈ Estimation

9. Find the number of employees who earn $16,000 to $20,000.

   7 employees

10. Find the number of employees who earn $6000 to $10,000.

    3 employees

11. How many employees earn $25,000 or less?

    28 employees

12. How many employees earn $31,000 or more?

    8 employees

13. Describe class interval and class frequency. How are these both used when preparing a histogram?

    Class intervals are the result of combining data into groupings. Class frequency is the number of data items that fit in each class frequency. These are used to group data and to have multiple responses (frequency) in a class interval—this makes the data easier to interpret.

14. What might be a problem of using too few or too many class intervals?

    If too few class intervals were used, the class frequencies would be high and any differences in the data might not be observable. If too many class intervals were used, interpretation might become impossible because class frequencies would be very low or nonexistent.

The list shows the number of sets of encyclopedias sold annually by the members of the local sales staff. Use it to complete the table.

120	130	144	132	147	158	174
135	142	155	174	162	151	178
145	151	139	128	147	134	146

	Class Intervals (Number of Sets)	*Tally*	*Class Frequency (Number of Staff)*
15.	120–129	II	2
16.	130–139	⊎Ⱶ	5
17.	140–149	⊎Ⱶ I	6
18.	150–159	IIII	4
19.	160–169	I	1
20.	170–179	III	3

The manager of Rocklin Building Supply asked her 30 employees how many college credits each had completed. The list of responses is shown below. Use these responses to complete the following table.

74	133	4	127	20	30
103	27	158	118	138	121
149	132	64	141	130	76
42	50	95	56	65	104
4	140	12	88	119	64

	Class Intervals (Number of Credits)	*Tally*	*Class Frequency (Number of Employees)*
21.	0–24	IIII	4
22.	25–49	III	3
23.	50–74	IIII I	6
24.	75–99	III	3
25.	100–124	IIII	5
26.	125–149	IIII III	8
27.	150–174	I	1

28. Construct a histogram by using the data in Exercises 21–27.

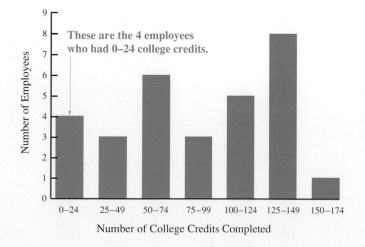

These are the 4 employees who had 0–24 college credits.

Number of Employees

Number of College Credits Completed

▲ *Southside Real Estate has 60 salespeople spread over its five offices. The number of new homes sold by each of these salespeople during the past year is shown below. Use these numbers to complete the following table.*

9	33	14	8	17	10	25	11	4	16	3	9
15	24	19	30	16	31	21	20	30	2	6	6
3	8	5	11	15	26	7	18	29	10	7	3
11	6	10	4	2	35	10	25	5	19	34	2
8	13	25	15	23	26	12	4	22	12	21	12

	Class Interval (New Homes Sold)	Tally	Class Frequency (Number of Salespeople)
29.	1–5	⩕ ⩕ I	11
30.	6–10	⩕ ⩕ IIII	14
31.	11–15	⩕ ⩕ I	11
32.	16–20	⩕ II	7
33.	21–25	⩕ III	8
34.	26–30	⩕	5
35.	31–35	IIII	4

36. Make a histogram showing the results from Exercises 29–35.

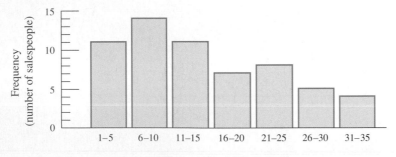

Number of New Homes Sold

Review and Prepare

*Solve each problem by using the order of operations. (For help, see **Section 1.8**.)*

37. $(18 + 53 + 31 + 26) \div 5$

25.6

38. $(17 + 23 + 46 + 36 + 29) \div 4$

37.75

39. $(8 \cdot 6) + (3 \cdot 8) \div 5$

52.8

40. $(9 \cdot 2) + (12 \cdot 4) \div 6$

26

41. $(4 \cdot 3) + (3 \cdot 9) + (2 \cdot 3) \div 15$

39.4

42. $(4 \cdot 3) + (3 \cdot 6) + (2 \cdot 6) + (3 \cdot 3) \div 18$

42.5

10.4 *Mean, Median, and Mode*

OBJECTIVE 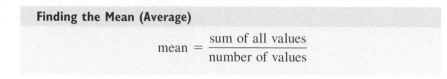 When analyzing data, one of the first things to look for is a *measure of central tendency*—a single number that we can use to represent the entire list of numbers. One such measure is the *average* or **mean.** The mean can be found with the following formula.

OBJECTIVES

 Find the mean of a list of numbers.

Find a weighted mean.

Find the median.

Find the mode.

FOR EXTRA HELP

Tutorial   Tape 17   SSM, Sec. 10.4

Finding the Mean (Average)

$$\text{mean} = \frac{\text{sum of all values}}{\text{number of values}}$$

E X A M P L E 1   Finding the Mean

David had test scores of 84, 90, 95, 98, and 88. Find his average or mean of these scores.

Use the formula for finding mean. Add up all the test scores and then divide by the number of tests.

$$\text{mean} = \frac{84 + 90 + 95 + 98 + 88}{5} \quad \begin{array}{l} \leftarrow \text{Sum of test scores} \\ \\ \leftarrow \text{Number of tests} \end{array}$$

$$= \frac{455}{5} \quad \text{Divide.}$$

$$= 91$$

David has a mean score of 91.

WORK PROBLEM 1 AT THE SIDE. ▶▶

1. Tanya had test scores of 96, 98, 84, 88, 82, and 92. Find her average or mean score.

E X A M P L E 2   Applying the Average or Mean

The sales of photo albums at Sarah's Card Shop for each day last week were

$86, $103, $118, $117, $126, $158, and $149.

For Sarah's Card Shop, the mean (rounded to the nearest cent) is

$$\text{mean} = \frac{\$86 + \$103 + \$118 + \$117 + \$126 + \$158 + \$149}{7}$$

$$= \frac{\$857}{7}$$

$$\approx \$122.43.$$

2. Find the mean for each list of numbers.

(a) Monthly phone bills of $25.12, $42.58, $76.19, $32, $81.11, $26.41, $19.76, $59.32, $71.18, and $21.03.

(b) A list of the sales for one year at eight different office supply stores: $749,820; $765,480 $643,744; $824,222 $485,886; $668,178 $702,294; $525,800

WORK PROBLEM 2 AT THE SIDE. ▶▶

ANSWERS

1. 90

2. (a) $\dfrac{\$454.70}{10} = \$45.47$

   (b) $\dfrac{\$5,365,424}{8} = \$670,678$

687

3. The numbers below show the amount that Alison Nakano spent for lottery tickets and the number of days that she spent that amount. Find the weighted mean.

Value	Frequency
$ 6	2
$ 7	3
$ 8	3
$ 9	4
$10	6

OBJECTIVE 2 Some items in a list might appear more than once. In this case, we find a **weighted mean,** in which each value is "weighted" by multiplying it by the number of times it occurs.

E X A M P L E 3   Understanding the Weighted Mean

The following table shows the amount of contribution and the number of times the amount was given (frequency) to a food pantry. Find the weighted mean.

Contribution Value	Frequency
$ 3	4
$ 5	2
$ 7	1
$ 8	5
$ 9	3
$10	2
$12	1
$13	2

The same amount was given by more than one person: for example, $5 was given twice and $8 was given five times. Other amounts, such as $12, were given once. To find the mean, multiply each contribution value by its frequency. Then add the products. Next, add the numbers in the *frequency* column to find the total number of values.

Value	Frequency	Product
$ 3	4	$(3 \cdot 4) = \$12$
$ 5	2	$(5 \cdot 2) = \$10$
$ 7	1	$(7 \cdot 1) = \$ 7$
$ 8	5	$(8 \cdot 5) = \$40$
$ 9	3	$(9 \cdot 3) = \$27$
$10	2	$(10 \cdot 2) = \$20$
$12	1	$(12 \cdot 1) = \$12$
$13	2	$(13 \cdot 2) = \$26$
Totals	**20**	**$154**

Finally, divide the totals.

$$\text{mean} = \frac{\$154}{20} = \$7.70.$$

The mean contribution to the food pantry was $7.70.

◄◄ WORK PROBLEM 3 AT THE SIDE.

ANSWERS

3. $8.50

A common use of the weighted mean is to find a student's *grade point average,* as shown by the next example.

E X A M P L E 4   Applying the Weighted Mean

Find the grade point average for a student earning the following grades. Assume A = 4, B = 3, C = 2, D = 1, and F = 0. The number of credits determines how many times the grade is counted (the frequency).

Course	Credits	Grade	Credits • Grade
Mathematics	3	A (= 4)	3 • 4 = 12
Speech	3	C (= 2)	3 • 2 = 6
English	3	B (= 3)	3 • 3 = 9
Computer Science	3	A (= 4)	3 • 4 = 12
Lab for Computer Science	2	D (= 1)	2 • 1 = 2
Totals	14		41

$\approx$ It is common to round grade point averages to the nearest hundredth. So the grade point average for this student is

$$\frac{41}{14} \approx 2.93.$$

WORK PROBLEM 4 AT THE SIDE. ▶▶

OBJECTIVE ▶ Because it can be affected by extremely high or low numbers, the mean is often a poor indicator of central tendency for a list of numbers. In cases like this, another measure of central tendency, called the **median** (MEE-dee-un), can be used. The *median* divides a group of numbers in half; half the numbers lie above the median, and half lie below the median.

Find the median by listing the numbers *in order* from *smallest* to *largest.* If the list contains an *odd* number of items, the median is the *middle number.*

E X A M P L E 5   Using the Median

Find the median for the following list of prices.

$7, $23, $15, $6, $18, $12, $24

First arrange the numbers in numerical order from smallest to largest.

Smallest → 6, 7, 12, 15, 18, 23, 24 ← Largest

Next, find the middle number in the list.

6, 7, 12, 15, 18, 23, 24

Three are below Three are above
Middle number

The median price is $15.

WORK PROBLEM 5 AT THE SIDE. ▶▶

If a list contains an *even* number of items, there is no single middle number. In this case, the median is defined as the mean (average) of the *middle two* numbers.

---

4. Find the grade point average for a student earning the following grades. Round to the nearest hundredth.

Course	Credits	Grade
Mathematics	3	A (= 4)
P.E.	1	C (= 2)
English	3	C (= 2)
Keyboarding	3	B (= 3)
Recreation	3	B (= 3)

5. Find the median for the following number of customers helped each hour at the order desk. 35, 33, 27, 31, 39, 50, 59, 25, 30

ANSWERS

4. ≈ 2.92

5. 33 customers (the middle number when the numbers are arranged from smallest to largest)

6. Find the median for the following list of measurements. 178 ft, 261 ft, 126 ft, 189 ft, 121 ft, 195 ft

7. Find the mode for each list of numbers.

(a) Ages of part-time employees (in years): 28, 16, 22, 28, 34

(b) Total points on a screening exam of 312, 219, 782, 312, 219, 426

(c) Monthly commissions of sales people: $1706, $1289, $1653, $1892, $1301, $1782

ANSWERS
6. 183.5 ft
7. (a) 28 years
   (b) bimodal, 219 points and 312 points
       (this list has two modes)
   (c) no mode (no number occurs more
       than once)

E X A M P L E   6   Finding the Median

Find the median for the following list of ages.

$$74, 7, 15, 13, 25, 28, 47, 59, 32, 68$$

First arrange the numbers in numerical order. Then find the middle two numbers.

Smallest → 7, 13, 15, 25, 28, 32, 47, 59, 68, 74 ← Largest

Middle two numbers

The median age is the mean of these two numbers.

$$\text{median} = \frac{28 + 32}{2} = \frac{60}{2} = 30 \text{ years}$$

◀◀ **WORK PROBLEM 6 AT THE SIDE.**

OBJECTIVE 4 The last important statistical measure is the **mode,** the number that occurs most often in a list of numbers. For example, if the test scores for 10 students were

$$74, 81, 39, 74, 82, 80, 100, 92, 74, \text{ and } 85$$

then the mode is 74. Three students earned a score of 74, so 74 appears more times on the list than any other score.

A list can have two modes; such a list is sometimes called **bimodal.** If no number occurs more frequently than any other number in a list, the list has *no mode.*

E X A M P L E   7   Finding the Mode

Find the mode for each list of numbers.

(a) 51, 32, 49, 73, 49, 90

The number 49 occurs more often than any other number; therefore, 49 is the mode. (It is not necessary to place the numbers in numerical order when looking for the mode.)

(b) 482, 485, 483, 485, 487, 487, 489

Because both 485 and 487 occur twice, each is a mode. This list is *bimodal.*

(c) 10,708; 11,519; 10,972; 12,546; 13,905; 12,182

No number occurs more than once. This list has *no mode.*

Measures of Central Tendency

The **mean** is the sum of all the values divided by the number of values. It is the mathematical average.

The **median** is the middle number in a group of values that are listed from smallest to largest. It divides a group of numbers in half.

The **mode** is the value that occurs most often in a group of values.

◀◀ **WORK PROBLEM 7 AT THE SIDE.**

10.4 Exercises

Find the mean for each list of numbers. Round answers to the nearest tenth, if necessary.

1. Earth moving equipment ages (in years) of
7, 18, 3, 5,27, 12

12 years

2. Monthly phone bills of $53, $77, $38, $29,
$49, $48

$49

3. Final exam scores of 92, 51, 59, 86, 68, 73,
49, 80

≈69.8

4. Quiz scores of 18, 25, 21, 8, 16, 13, 23, 19

≈17.9

5. Annual salaries of $21,900, $22,850, $24,930,
$29,710, $28,340, $40,000

$27,955

6. Numbers of people attending baseball games:
27,500; 18,250; 17,357; 14,298; 33,110

22,103 people

Solve the following application problems.

7. The Athletic Shoe Store sold shoes at the following prices: $75.52, $36.15, $58.24, $21.86,
$47.68, $106.57, $82.72, $52.14, $28.60,
$72.92. Find the average (mean) shoe sales
amount.

$58.24

8. In one evening, a waitress collected the following checks from her dinner customers: $30.10,
$42.80, $91.60, $51.20, $88.30, $21.90,
$43.70, $51.20. Find the average (mean) dinner check amount.

$52.60

🖉 Writing   ◉ Conceptual   ▲ Challenging   ≈ Estimation

9. The table below shows the face value (policy amount) of life insurance policies sold and the number of policies sold for each amount by the New World Life Company during one week. Find the weighted mean amount for the policies sold.

Policy Amount	Number of Policies Sold
$10,000	6
$20,000	24
$25,000	12
$30,000	8
$50,000	5
$100,000	3
$250,000	2

$35,500

10. Detroit Metro-Sales Company prepares the following table showing the gasoline mileage obtained by each of the cars in their automobile fleet. Find the weighted mean to determine the miles per gallon for the fleet of cars.

Miles per Gallon	Number of Autos
15	5
20	6
24	10
30	14
32	5
35	6
40	4

27.7 miles per gallon

Find the weighted mean. Round answers to the nearest tenth, if necessary.

11. Quiz

Scores	Frequency
3	4
5	2
9	1
12	3

6.7

12.

Value	Frequency
9	3
12	5
15	1
18	1

12

13.

Value	Frequency
12	4
13	2
15	5
19	3
22	1
23	5

≈17.2

14. Students

per Class	Frequency
25	1
26	2
29	5
30	4
32	3
33	5

≈30.2 students

Find the median for the following lists of numbers.

15. Number of voice mail messages received:
9, 12, 14, 15, 23, 24, 28

15 messages

16. Deliveries by a newspaper distributor: 99, 108, 109, 123, 126, 129, 146, 168, 170

126 deliveries

17. Customers served each day:
328, 549, 420, 592, 715, 483

516 customers

18. Number of cars in the parking lot each day:
520, 523, 513, 1283, 338, 509, 290, 420

511 cars

Find the mode or modes for each list of numbers.

19. Number of samples taken each hour:
3, 8, 5, 1, 7, 6, 8, 4

8 samples

20. Water bills of
$21, $32, $46, $32, $49, $32, $49

$32

21. Ages of retirees (in years) at the village:
74, 68, 68, 68, 75, 75, 74, 74, 70

68 and 74 years

22. Tires balanced by different employees:
30, 19, 25, 78, 36, 20, 45, 85, 38

no mode

23. When can the mean be a poor indicator of the central tendency? What might you do in such a situation?

When the data contains a few very low or a few very high values, the mean will give a poor indication of the average. Consider using the median or mode instead.

24. What is the purpose of the weighted mean? Give an example of where it is used.

When the same value occurs more than once, the value must be multiplied by the number of times it occurs (weighted).
A good example of using the weighted mean is when finding grade point averages. It must be used because you have several *credits* with the same *grade value*.

25. When is the median a better average to use than the mean to describe a set of data? Make up a list of numbers to illustrate your explanation. Calculate both the mean and the median.

The median is a better average when the list contains one or more extreme values.
Find the mean and the median of the following home values. $82,000; $64,000; $91,000; $115,000; $982,000
mean home value = $266,800
median home value = $91,000

26. Suppose you own a hat shop and can order a certain hat in only one size. You look at last year's sales to decide on the size to order. Should you find the mean, median, or mode for these sales? Explain your answer.

The size to order is the mode.
The mode is the size most worn by customers and it would be wise to order most hats in this size.

▲ *Find the grade point average for students earning the following grades. Assume* A = 4, B = 3, C = 2, D = 1, *and* F = 0. *Round answers to the nearest hundredth.*

27.

Credits	Grade	
4	B	
2	A	
5	C	2.60
1	F	
3	B	

28.

Credits	Grade	
3	A	
3	B	
4	B	≈2.87
2	C	
3	C	

Find the median for the following lists of numbers.

29. The number of computer service calls taken each day:
51, 48, 96, 40, 47, 23, 95, 56, 34

48 calls

30. Number of gallons of paint sold per week:
1072, 1068, 1093, 1042, 1056, 205, 1009

1056 gallons

Find the mode or modes for each list of numbers.

31. The number of boxes of candy sold by each child:
5, 9, 17, 3, 2, 8, 19, 1, 4, 20

no mode

32. The weights of soccer players (in pounds):
158, 161, 165, 162, 165, 157, 163

165 pounds

10.1	**circle graph**	A circle graph shows how a total amount is divided into parts or sectors. It is based on percents of 360°.
	protractor	A protractor is a device (usually in the shape of a half-circle) used to measure the number of degrees in an angle or parts of a circle.
10.2	**bar graph**	A bar graph uses bars of various heights to show quantity or frequency.
	double-bar graph	A double-bar graph compares two sets of data by showing two sets of bars.
	line graph	A line graph uses dots connected by lines to show trends.
	comparison line graph	A comparison line graph shows how several different items relate to each other by showing a line graph for each item.
10.3	**frequency distribution**	A table that includes a column showing each possible number in the data collected. The original data is then entered in another column using a tally mark for each corresponding value. The tally marks are totaled and placed in a third column. The result is a frequency distribution.
	histogram	A histogram is a bar graph in which the width of each bar represents a range of numbers (class interval) and the height represents the quantity or frequency of items that fall within the interval.
10.4	**mean**	The mean is the sum of all the values divided by the number of values. It is often called the *average*.
	weighted mean	The weighted mean is a mean calculated so that each value is multiplied by its frequency.
	median	The median is the middle number in a group of values that are listed from smallest to largest. It divides a group of values in half. If there are an even number of values, the median is the mean (average) of the two middle values.
	mode	The mode is the value that occurs most often in a group of values.

Concepts	*Examples*

10.1 Constructing a Circle Graph

1. Determine the percent of the total for each item.
2. Find the number of degrees out of 360° that each percent represents.
3. Use a protractor to measure the number of degrees for each item in the circle.

Construct a circle graph for the following table, which lists expenses for a business trip.

Item	*Amount*
Transportation	$200
Lodging	$300
Food	$250
Entertainment	$150
Other	$100
Total	**$1000**

Item	*Amount*	*Percent of Total*	*Sector Size*
Transportation	$200	$\frac{\$200}{\$1000} = \frac{1}{5} = 20\%$ so $360° \cdot 20\%$ $= 360 \cdot 0.20$	$= 72°$
Lodging	$300	$\frac{\$300}{\$1000} = \frac{3}{10} = 30\%$ so $360° \cdot 30\%$ $= 360 \cdot 0.30$	$= 108°$
Food	$250	$\frac{\$250}{\$1000} = \frac{1}{4} = 25\%$ so $360° \cdot 25\%$ $= 360 \cdot 0.25$	$= 90°$
Entertainment	$150	$\frac{\$150}{\$1000} = \frac{3}{20} = 15\%$ so $360° \cdot 15\%$ $= 360 \cdot 0.15$	$= 54°$
Other	$100	$\frac{\$100}{\$1000} = \frac{1}{10} = 10\%$ so $360° \cdot 10\%$ $= 360 \cdot 0.10$	$= 36°$

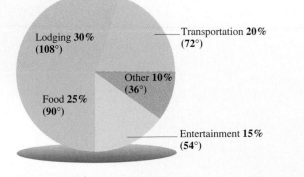

Lodging **30%** (108°)

Transportation **20%** (72°)

Other **10%** (36°)

Food **25%** (90°)

Entertainment **15%** (54°)

Concepts	Examples

10.2 Reading a Bar Graph

The height of the bar is used to show the quantity or frequency (number) in a specific category. Use a ruler or straight edge to line up the top of each bar with the numbers on the left side of the graph.

Use the bar graph below to determine the number of students who earned each letter grade.

	Number of
Grade	Students
A	3
B	7
C	4
D	2

10.2 Reading a Line Graph

A dot is used to show the number or quantity in a specific class. The dots are connected with lines. This kind of graph is used to show a trend.

The line graph below shows the sales volume for each of four years.

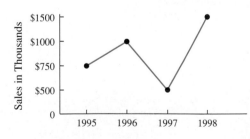

Find the sales in each year.

Year	Total Sales
1995	$750 · 1000 = $750,000
1996	$1000 · 1000 = $1,000,000
1997	$500 · 1000 = $500,000
1998	$1500 · 1000 = $1,500,000

Concepts	Examples

Concepts

10.3 Prepare a Frequency Distribution and a Histogram from Raw Data

1. Construct a table listing each value, and the number of times this value occurs.

2. Divide the data into groups, categories, or classes.

3. Draw bars representing these groups to make a histogram.

Examples

Draw a histogram for the following list of student quiz scores.

12	15	15	14
13	20	10	12
11	9	10	12
17	20	16	17
14	18	19	13

Quiz Score	Tally	Frequency	
9	I	1	1st
10	II	2	class
11	I	1	interval
12	III	3	2nd
13	II	2	class
14	II	2	interval
15	II	2	3rd
16	I	1	class
17	II	2	interval
18	I	1	4th
19	I	1	class
20	II	2	interval

Class Interval (Quiz Scores)	Frequency (Number of Students)
9–11	4
12–14	7
15–17	5
18–20	4

Concepts	Examples
10.4 Finding the Mean (average) of a Set of Numbers **1.** Add all values to obtain a total. **2.** Divide the total by the number of values.	The test scores for Heather Hall in her business math course were as follows: $$85 \quad 76 \quad 93 \quad 91$$ $$78 \quad 82 \quad 87 \quad 85$$ Find Heather's test score average (mean) to the nearest tenth. Mean $$= \frac{85 + 76 + 93 + 91 + 78 + 82 + 87 + 85}{8}$$ $$= \frac{677}{8} \approx 84.6$$
10.4 Finding the Weighted Mean **1.** Multiply frequency by value. **2.** Add all the products from Step 1. **3.** Divide the sum in Step 2 by the total number of pieces of data.	This table shows the distribution of the number of school-age children in a survey of 30 families. 	Number of School-Age Children	Frequency (Number of Families)
:---:	:---:
0	12
1	6
2	7
3	3
4	2 | **Total of 30 Families** Find the mean number of school-age children per family. Round to the nearest hundredth. | Value	Frequency	Product
:---:	:---:	:---:
0	12	$(0 \cdot 12) = 0$
1	6	$(1 \cdot 6) = 6$
2	7	$(2 \cdot 7) = 14$
3	3	$(3 \cdot 3) = 9$
4	2	$(4 \cdot 2) = 8$
Totals	30	37	 $$\text{Mean} = \frac{37}{30} \approx 1.23$$ The mean number of school-age children per family is 1.23.

Concepts	*Examples*
10.4 Finding the Median of a Set of Numbers	Find the median for Heather Hall's grades from the previous page.
1. Arrange the data from smallest to largest. 2. Select the middle value or the average of the two middle values, if there is an even number of values.	The data arranged from smallest to largest is as follows: 76 78 82 **85 85** 87 91 93 Middle values The middle two values are 85 and 85. The average of these two values is $$\frac{85 + 85}{2} = 85$$
10.4 Determining the Mode of a Set of Values Find the value that appears most often in the list of values. If no value appears more than once, there is no mode. If two different values appear the same number of times, the list is bimodal.	Find the mode for Heather's grades in the previous example. The most frequently occurring score is 85 (it occurs twice). Therefore, the mode is 85.

[10.1]

1. This circle graph shows the cost of a family vacation. What is the largest single expense of the vacation? How much is that item?

—— **COST OF A FAMILY VACATION** ——

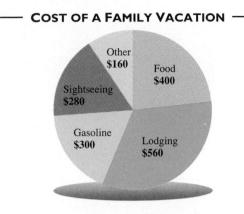

Lodging; $560

Using the circle graph in Exercise 1, find each of the following ratios. Write ratios as fractions in lowest terms.

2. cost of the food to the total cost of the vacation

$$\frac{400}{1700} = \frac{4}{17}$$

3. cost of the gasoline to the total cost of the vacation

$$\frac{300}{1700} = \frac{3}{17}$$

4. cost of sightseeing to the total cost of the vacation

$$\frac{280}{1700} = \frac{14}{85}$$

5. cost of gasoline to the cost of the *other* category

$$\frac{300}{160} = \frac{15}{8}$$

6. cost of the lodging to the cost of the food

$$\frac{560}{400} = \frac{7}{5}$$

✎ Writing ◉ Conceptual ▲ Challenging ≈ Estimation

[10.2] *The bar graph below is adapted from* USA TODAY. *If there were 16,500,000 cellular phones sold last year, find the number of cellular phones sold in each of the following categories.*

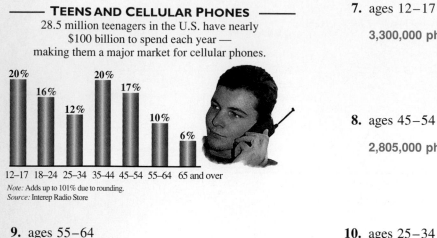

—— **TEENS AND CELLULAR PHONES** ——
28.5 million teenagers in the U.S. have nearly
$100 billion to spend each year —
making them a major market for cellular phones.

20% 16% 12% 20% 17% 10% 6%

12–17 18–24 25–34 35–44 45–54 55–64 65 and over

Note: Adds up to 101% due to rounding.
Source: Interep Radio Store

7. ages 12–17

3,300,000 phones

8. ages 45–54

2,805,000 phones

9. ages 55–64

1,650,000 phones

10. ages 25–34

1,980,000 phones

11. What two age groups buy the most cellular phones? Give one possible explanation why these age groups buy more phones than others.

Ages 12–17 and 35–44 buy more cellular phones.
Perhaps they use a phone more often and feel the need to be reached by phone at all times. Or, it may be a status symbol among these age groups.

12. What two age groups buy the fewest cellular phones? Give one possible explanation why these age groups buy fewer phones than others.

Ages 55–64 and 65 and over buy the fewest phones.
Perhaps they use the phone less often and feel that they do not need to be reached by phone at all times and in different places. Or, those over 65 may not have enough income to afford cellular phones.

This double-bar graph shows the number of acre-feet of water in Lake Natoma for each of the first six months of 1997 and 1998.

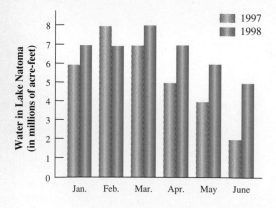

1997
1998

Water in Lake Natoma
(in millions of acre-feet)

8
7
6
5
4
3
2
1
0

Jan. Feb. Mar. Apr. May June

13. During which month in 1998 was the greatest amount of water in the lake? How much was there?

March; 8 million acre-feet

14. During which month in 1997 was the least amount of water in the lake? How much was there?

June; 2 million acre-feet

15. How many acre-feet of water were in the lake in June of 1998?

5 million acre-feet

16. How many acre-feet of water were in the lake in May of 1997?

4 million acre-feet

17. Find the decrease in the amount of water in the lake from March 1997 to June 1997.

5 million acre-feet

18. Find the decrease in the amount of water in the lake from April 1998 to June 1998.

2 million acre-feet

This comparison line graph shows the annual grocery purchases of two different childcare centers during each of five years. Find the amount of annual grocery purchases in each of the following years.

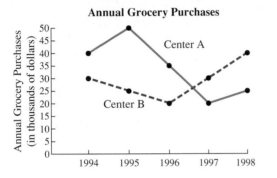

19. Center A in 1995

$50,000

20. Center A in 1997

$20,000

21. Center B in 1996

$20,000

22. Center B in 1998

$40,000

23. What trend do you see in Center A's purchases? Why might this have happened?

The grocery purchases decreased for two years and then moved up slightly.
Less children are attending the center or fewer children are eating at the childcare center.

24. What trend do you see in Center B's purchases? Why might this have happened?

The grocery purchases are increasing. A greater number of children are attending the center or a greater number are eating at the childcare center.

[10.4] *Find the mean for each list of numbers.*

25. Digital cameras sold:
18, 12, 15, 24, 9, 42, 54, 87, 21, 3

28.5 digital cameras

26. Number of harassment complaints filed:
31, 9, 8, 22, 46, 51, 48, 42, 53, 42

35.2 complaints

Find the weighted mean for each of the following. Round to the nearest tenth if necessary.

27. Dollar

Value	Frequency
$42	3
$47	7
$53	2
$55	3
$59	5

≈$51.10

28. Total

Points	Frequency
243	1
247	3
251	5
255	7
263	4
271	2
279	2

≈257.3 points

Find the median for each list of numbers.

29. Number of insurance claims processed:
54, 28, 35, 43, 13, 37, 68, 75, 39

39 claims

30. Commissions of $576, $578, $542, $151,
$559, $565, $525, $590

$562

Find the mode or modes for each list of numbers.

31. Hiking boots priced at $80, $72, $64, $64,
$72, $53, $64

$64

32. Boat launchings: 18, 25, 63, 32, 28, 37, 32,
26, 18

18 and 32 launchings (bimodal)

MIXED REVIEW EXERCISES

The Broadway Hair Salon spent $22,400 to open a new shop. This amount was spent as shown in the following chart. Find all the missing numbers in Exercises 33–37.

Item	Dollar Amount	Percent of Total	Degrees of Circle
33. plumbing and electrical changes	$2240	10%	36°
34. work stations	$7840	35%	126°
35. small appliances	$4480	20%	72°
36. interior decoration	$5600	25%	90°
37. supplies	$2240	10%	36°

38. Draw a circle graph by using the information in Exercises 33–37.

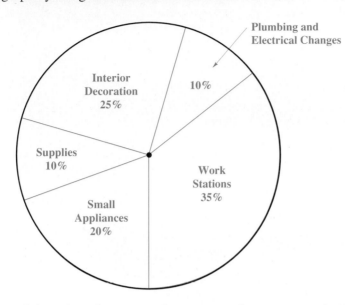

Find the mean for each list of numbers. Round answers to the nearest tenth, if necessary.

39. Contestant ages (in years) of 24, 36, 26, 74, 90

50 years

40. Number of tacks in a handful: 122, 135, 146, 159, 128, 147, 168, 139, 158

≈144.7 tacks

Find the mode or modes for each list of numbers.

41. Units meeting quality standards: 97, 95, 94, 95, 94, 97, 97

97

42. Number of 2-bedroom apartments in each building: 26, 31, 31, 37, 43, 51, 31, 43, 43

31 and 43 2-bedroom apartments (bimodal)

Find the median for each list of numbers.

43. Hours worked: 4.7, 3.2, 2.9, 5.3, 7.1, 8.2, 9.4, 1.0

5.0 hours

44. Number of yard sales each Saturday: 7, 15, 28, 3, 14, 18, 46, 59, 1, 2, 9, 21

14.5 yard sales

Here are the scores of 40 students on a computer science exam. Complete the table.

78	89	36	59	78	99	92	86
73	78	85	57	99	95	82	76
63	93	53	76	92	79	72	62
74	81	77	76	59	84	76	94
58	37	76	54	80	30	45	38

	Class Intervals *(Scores)*	**Tally**	**Class Frequency** *(Number of Students)*
45.	30–39	IIII	4
46.	40–49	I	1
47.	50–59	LHI I	6
48.	60–69	II	2
49.	70–79	LHI LHI III	13
50.	80–89	LHI II	7
51.	90–99	LHI II	7

52. Construct a histogram by using the data in Exercises 45–51.

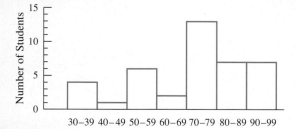

Find the weighted mean for each of the following. Round answers to the nearest tenth, if necessary.

53.

Test Score	Frequency	
23	2	
27	5	
31	4	≈32.3
35	6	
39	5	

54.

Dollar Value	Frequency	
$104	6	
$112	14	
$115	21	
$119	13	≈$118.80
$123	22	
$127	6	
$132	9	

CHAPTER 10 TEST

This circle graph shows the development costs for the Shady Brook Public Housing Subdivision. Find the dollar amount budgeted for each category. The total budget of the subdivision is $5,600,000.

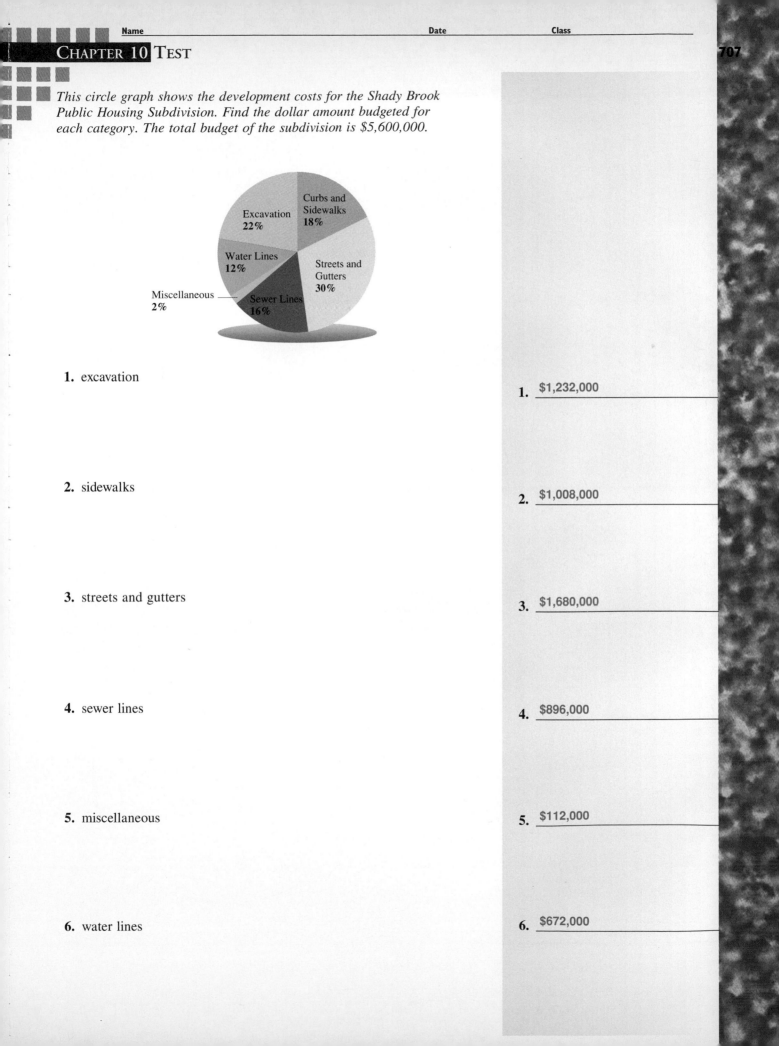

Excavation 22%

Curbs and Sidewalks 18%

Water Lines 12%

Streets and Gutters 30%

Miscellaneous 2%

Sewer Lines 16%

1. excavation

1. $1,232,000

2. sidewalks

2. $1,008,000

3. streets and gutters

3. $1,680,000

4. sewer lines

4. $896,000

5. miscellaneous

5. $112,000

6. water lines

6. $672,000

7. 108°

8. 36°

9. 72°

10. 108°

11. 10%

12.

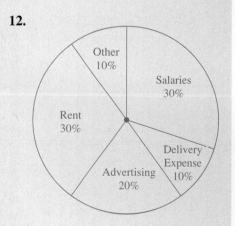

During a one-year period, Oak Mill Furniture Sales had the following expenses. Find all numbers missing from the chart.

Item	Dollar Amount	Percent of Total	Degrees of a Circle
7. salaries	$36,000	30%	_____
8. delivery expense	$12,000	10%	_____
9. advertising	$24,000	20%	_____
10. rent	$36,000	30%	_____
11. other	$12,000	_____	36°

12. Draw a circle graph using the information in Exercises 7–11.

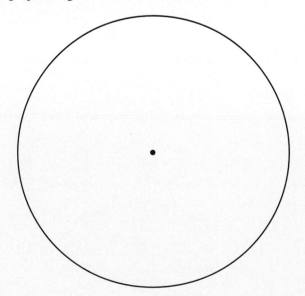

Here are the profits for each of the past 20 weeks from the snack bar vending machines. Complete the following table.

$142 $137 $125 $132 $147 $129 $151 $172 $175 $129
$159 $148 $173 $160 $152 $174 $169 $163 $149 $173

Profit	*Number of Weeks*
13. $120–129	_____
14. $130–139	_____
15. $140–149	_____
16. $150–159	_____
17. $160–169	_____
18. $170–179	_____

19. Use the numbers found in Exercises 13–18 to draw a histogram.

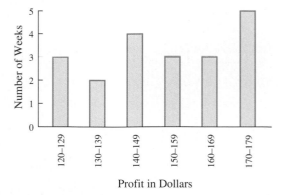

Find the mean for each of the following. Round answers to the nearest tenth if necessary.

20. Number of books loaned: 52, 61, 68, 69, 73, 75, 79, 84, 91, 98

21. Number of miles per gallon by older cars: 22, 28, 24, 27, 29, 32, 33, 35

22. Airplane speeds in miles per hour: 458, 432, 496, 491, 500, 508, 512, 396, 492, 504

13. 3
14. 2
15. 4
16. 3
17. 3
18. 5
19.
20. 75 books
21. ≈28.8 miles per gallon
22. 478.9 miles per hour

Credits	Grade	
3	A	$3 \times 4 = 12$
2	C	$2 \times 2 = 4$
4	B	$4 \times 3 = 12$
9		28

23. $28 \div 9 \approx 3.11$

24. 8, 17, 23, 32, 64 ⌐ Median

25. $\approx\$11.30$

26. ≈ 173.7

27. 31.5 degrees

28. 9.3 liters

29. 57 meters

30. 103° and 104°

23. Explain why a weighted mean must be used to determine a student grade point average. Calculate your own grade point average for last semester or quarter. If you are a new student, make up a grade point average problem of your own and solve it.

The weighted mean must be used because different classes are worth different numbers of credits.

24. Explain in your own words the procedure for finding the median when there are an odd number of numbers in a list. Make up a problem with a list of five numbers and solve for the median.

Arrange the numbers in order from smallest to largest. When there is an odd number of numbers in a list, the median is the middle number.

Find the weighted mean for each of the following. Round answers to the nearest tenth, if necessary.

25. Cost	Frequency		26. Value	Frequency
$ 6	7		150	15
$10	3		160	17
$11	4		170	21
$14	2		180	28
$19	3		190	19
$24	1		200	7

Find the median for each list of numbers.

27. Low daily temperatures in degrees Fahrenheit:
32, 41, 28, 28, 37, 35, 16, 31

28. Number of liters of water lost in evaporation:
9.3, 10.0, 8.1, 6.3, 1.2, 11.4, 22.8, 10.3, 8.6

Find the mode or modes for each list of numbers.

29. Drilling depth (in meters) of 61, 57, 58, 42, 81, 92, 57

30. Hot tub temperatures (Fahrenheit) of 96°, 104°, 103°, 104°, 103°, 104°, 91°, 74°, 103°

≈*Round each number to the place shown.*

1. $72.648 to the nearest cent. ≈$72.65

2. $926.499 to the nearest dollar. ≈$926

3. 75,696 to the nearest ten. ≈75,700

4. 983,168 to the nearest ten thousand. ≈980,000

Simplify each of the following by using the order of operations.

5. $3 + 8 \div 4 + 6 \cdot 2$ 17

6. $\sqrt{81} - 4 \cdot 2 + 9$ 10

Solve each of the following.

7. $3^2 \cdot 2^4$ 144

8. $6^2 \cdot 3^2$ 324

≈*Round the numbers in each problem so that there is only one non-zero digit. Then add, subtract, multiply, or divide the rounded numbers as indicated to estimate the answer. Finally, solve for the exact answer.*

9. *estimate* *exact*

60,000	62,318
200,000	159,680
90	89
+ 20,000	+ 22,308
280,090	244,395

10. *estimate* *exact*

3	2.607
800	796.2
40	37.96
50	53.72
+ 8	+ 8.06
901	898.547

11. *estimate* *exact*

300,000	321,508
− 100,000	− 147,725
200,000	173,783

12. *estimate* *exact*

900	875.62
− 60	− 63.757
840	811.863

13. *estimate* *exact*

7000	7064
× 600	× 635
4,200,000	4,485,640

14. *estimate* *exact*

60	62.75
× 3	× 2.644
180	165.911

15. *estimate* *exact*

$40\overline{)20,000}$ = 500 $36\overline{)23,112}$ = 642

16. *estimate* *exact*

$4\overline{)60}$ = 15 $4.25\overline{)62.56}$ = 14.72

Add, subtract, multiply, or divide as indicated. Write answers in lowest terms and as whole or mixed numbers when possible.

17. $\frac{3}{4} + \frac{3}{8}$ $1\frac{1}{8}$

18. $\frac{3}{4} + \frac{5}{8} + \frac{1}{2}$ $1\frac{7}{8}$

19. $3\frac{2}{3}$
$+ \ 4\frac{4}{5}$
_____
$8\frac{7}{15}$

20. $\frac{5}{6} - \frac{2}{3}$ $\frac{1}{6}$

21. $5\frac{1}{3}$
$- \ 2\frac{3}{4}$
_____
$2\frac{7}{12}$

22. $46\frac{3}{4}$
$- \ 15\frac{4}{5}$
_____
$30\frac{19}{20}$

23. $\frac{7}{8} \cdot \frac{4}{5}$ $\frac{7}{10}$

24. $9\frac{3}{5} \cdot 4\frac{5}{8}$ $44\frac{2}{5}$

25. $22 \cdot \frac{2}{5}$ $8\frac{4}{5}$

26. $\frac{5}{6} \div \frac{5}{8}$ $1\frac{1}{3}$

27. $12 \div \frac{2}{3}$ 18

28. $3\frac{1}{3} \div 8\frac{3}{4}$ $\frac{8}{21}$

Simplify each of the following. Use the order of operations.

29. $\left(\frac{7}{8} - \frac{3}{4}\right) \cdot \frac{2}{3}$ $\frac{1}{12}$

30. $\left(\frac{5}{6} - \frac{1}{3}\right) + \left(\frac{1}{2}\right)^2 \cdot \frac{3}{4}$ $\frac{11}{16}$

Write each fraction in decimal form. Round to the nearest thousandth, if necessary.

31. $\frac{2}{5}$ 0.4

32. $\frac{3}{8}$ 0.375

33. $\frac{3}{4}$ 0.75

34. $\frac{13}{20}$ 0.65

Write in order, from smallest to largest.

35. $0.218, 0.22, 0.199, 0.207, 0.2215$

36. $0.6319, \frac{5}{8}, 0.608, \frac{13}{20}, 0.58$

$0.199, 0.207, 0.218, 0.22, 0.2215$

$0.58, 0.608, \frac{5}{8}, 0.6319, \frac{13}{20}$

Write each of the following ratios in lowest terms. Be sure to make all necessary conversions.

37. $2\frac{1}{2}$ inches to 20 inches $\frac{1}{8}$

38. 2 hours to 20 minutes $\frac{6}{1}$

Use cross-multiplication to decide whether the following proportions are true or false.

39. $\frac{6}{15} = \frac{18}{45}$

(True)   False

40. $\frac{52}{180} = \frac{36}{120}$

True   (False)

Find the missing numbers in each proportion.

41. $\frac{1}{4} = \frac{x}{12}$ 3

42. $\frac{14}{x} = \frac{364}{104}$ 4

43. $\frac{200}{135} = \frac{24}{x}$ 16.2

44. $\frac{x}{208} = \frac{6.5}{26}$ 52

Write the percents as decimals. Write the decimals as percents.

45. 35%   0.35

46. 0.025   2.5%

47. 250%   2.50 or 2.5

48. 4.35%   0.0435

Write each percent as a fraction or mixed number in lowest terms. Write each fraction as a percent.

49. 2% $\frac{1}{50}$

50. $62\frac{1}{2}$% $\frac{5}{8}$

51. $\frac{3}{20}$ 15%

52. $3\frac{1}{4}$ 325%

Solve these percent problems.

53. 75% of $640 is how much?

$480

54. Find 2.7% of 3000 chairs.

81 chairs

55. $8\frac{1}{2}$% of what number of people is 238 people?

2800 people

56. 48 is 15% of what number?

320

57. What percent of 520 is 182?

35%

58. 13 weeks is what percent of 52 weeks?

25%

Convert the following.

59. _____3_____ feet = 1 yard

60. 16 quarts = _____4_____ gallons

61. 5 days = _____120_____ hours

62. _____12,000_____ pounds = 6 tons

Convert each of the following measures as indicated.

63. 5 km to m   5000 m

64. 3815 mm to m   3.815 m

65. 8.3 g to mg   8300 mg

66. 230 g to kg   0.23 kg

67. 6 mL to L   0.006 L

68. 0.28 L to mL   280 mL

Write the most reasonable metric unit in each blank. Choose from L, mL, kg, g, mg, km, cm, m, and mm.

69. The fuel tank on the chain saw has a capacity of 750 _____mL_____ of fuel.

70. A nickel weighs 5 _____g_____.

71. The distance of the run this Saturday is 10 _____km_____.

72. The heaviest player on the team weighs 108 _____kg_____.

Find the area of each figure. Use 3.14 as the approximate value of π. Round answers to the nearest tenth.

73. a rectangle 2.8 m by 4.35 m.   ≈12.2 m²

74. a trapezoid with bases of 6.2 cm and 8.4 cm and height 5.3 cm.   ≈38.7 cm²

75. a triangle with base 8.5 ft and height 9 ft.   ≈38.3 ft²

76. a circle with diameter of 13 cm.   ≈132.7 cm²

Solve the following. Use 3.14 as the approximate value for π. Round answers to the nearest tenth, if necessary.

77. Find the volume of a cylinder with radius 8.6 cm and height 3.8 cm.   ≈**882.5 cm³**

78. Find the volume of a rectangular solid with length 5.5 m, width 2 m, and height 9 m.   **99 m³**

Find the unknown length in each right triangle.

79.

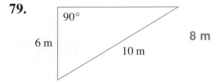

80.

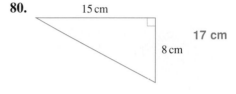

Add, subtract, multiply, or divide as indicated.

81. $-10 + (-6)$ **−16**

82. $-5.7 - (-12.6)$ **6.9**

83. $7 \cdot (-6)$ **−42**

84. $-14.6 \cdot (-5.7)$ **83.22**

85. $\dfrac{-36}{-6}$ **6**

86. $\dfrac{-34.04}{14.8}$ **−2.3**

Solve each equation.

87. $4x - 3 = 17$ **x = 5**

88. $-12 = 3(x + 2)$ **x = −6**

89. $19x - 12x = 14$ **x = 2**

90. $3.4x + 6 = 1.4x - 8$ **x = −7**

Find the mean, the median, and the mode for each of the following.

91. Cable hookups per installer: 16, 37, 27, 31, 19, 25, 15, 38, 43, 19   27 hookups; 26 hookups; 19 hookups

92. Number of tons of beets processed each hour: 20.6, 8.6, 3.3, 5.7, 10.6, 11.4, 4.6, 8.7, 5.7

8.8 tons; 8.6 tons; 5.7 tons

Solve the following application problems.

93. Trader Joes sold 1840 pints of cottage cheese in a recent week. If 690 of these cartons were non-fat cottage cheese, find the percent that were non-fat.

37.5%   or $37\frac{1}{2}\%$

94. In one state the sales tax is 7%. On a recent purchase the amount of sales tax was $78.68. Find the cost of the item purchased.

$1124

95. A gasoline additive is used at the rate of $2\frac{3}{4}$ liters for each storage tank. If $280\frac{1}{2}$ liters of additive are available, how many storage tanks can receive the additive?

102 tanks

96. A survey found that 19 out of every 25 adults are nonsmokers. If the Food Club has 2850 employees, how many would be expected to be nonsmokers?

2166 employees

97. Maya Chow had sales of $48,250 in hardware and fasteners last month. If her commission rate is 8.5%, find the amount of her commission.

$4101.25

98. The sketch below shows the plans for a lobby in a large commercial complex. What is the cost of carpeting the lobby, excluding the atrium, if the contractor charges $43.50 per square yard? Use 3.14 for π.

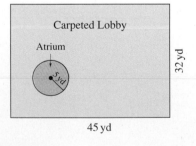

$59,225.25

99. The Spa Service Center services 140 spas. If each spa will need 125 mL of muriatic acid, how many liters of acid will be needed to service the spas?

17.5 L

100. A loan of $3500 will be paid back with $7\frac{1}{2}\%$ interest at the end of 6 months. Find the total amount due.

$3631.25

Appendix A

Scientific Calculators

Calculators are among the more popular inventions of the last three decades. Each year better calculators are developed and costs drop. The first all-transistor desktop calculator was introduced to the market in 1966; it weighed 55 pounds, cost $2500, and was slow. Today, these same calculations are performed quite well on a calculator costing less than $10. And today's $200 calculators have more ability to solve problems than some of the early computers.

Many colleges allow students to use calculators in basic mathematics courses. There are many types of calculators available, from the inexpensive basic calculator to the more complex **financial** and **graphing** calculators. The discussion here is confined to the common scientific calculator with the percent key, reciprocal key, exponent key, square root key, memory function, order of operations, and parentheses keys.

> **Note**
>
> Any explanation needed for specific calculator models or special function keys is best gained by referring to the booklet supplied with your calculator.

OBJECTIVE 1 Most calculators use **algebraic logic.** Some problems can be solved by entering number and function keys in the same order as you would solve problems by hand. Other problems require a knowledge of the order of operations when entering the problem. The problem $14 + 28$ would be entered as

$$14 \boxed{+} 28 \boxed{=}$$

and 42 would appear as the answer. Enter $387 - 62$ as

$$387 \boxed{-} 62 \boxed{=}$$

and 325 appears as the answer. If your calculator does not work problems in this way, check its instruction book to see how to proceed.

OBJECTIVE 2 All calculators have a

$$\boxed{C}, \qquad \boxed{ON/C}, \qquad \text{or} \qquad \boxed{ON/AC}$$

key. Pressing this key erases everything in the calculator and prepares the calculator to begin a new problem. Some calculators also have a

$$\boxed{CE}$$

key. Pressing this key erases *only* the number displayed and allows the person using the calculator to correct a mistake without having to start the problem over.

OBJECTIVES

1. Learn the basic calculator keys.
2. Understand the $\boxed{C}$, $\boxed{CE}$, and $\boxed{ON/C}$ or $\boxed{ON/AC}$ keys.
3. Understand the floating decimal point.
4. Use the $\boxed{\%}$ key.
5. Use the $\boxed{x^2}$ and the $\boxed{x^3}$ keys.
6. Use the $\boxed{y^x}$ and $\boxed{\sqrt{x}}$ keys.
7. Use the $\boxed{a^{b/c}}$ key.
8. Solve problems with negative numbers.
9. Use the calculator memory function.
10. Solve chain calculations using the order of operations.
11. Use the parentheses keys.

Many calculators combine the $\boxed{C}$ key and the $\boxed{CE}$ key and use an $\boxed{ON/C}$ key. This key turns the calculator on and is also used to erase the calculator display. If the $\boxed{ON/C}$ key is pressed after the $\boxed{=}$ or one of the operation keys ($\boxed{+}$, $\boxed{-}$, $\boxed{\times}$, $\boxed{\div}$), everything in the calculator is erased. If the wrong operation key is pressed, you press the correct key and the error is corrected. For example, $7 \boxed{+} \boxed{-} 3 \boxed{=} 4$. Pressing the $\boxed{-}$ key cancels out the previous $\boxed{+}$ key entry.

OBJECTIVE 3▶ Most calculators have a **floating decimal** which locates the decimal point in the final result. For example, to buy 55.75 square yards of vinyl floor covering at $18.99 per square yard, proceed as follows.

$$55.75 \boxed{\times} 18.99 \boxed{=} 1058.6925$$

The decimal point is automatically placed in the answer. You should **round** money answers to the nearest cent. Draw a cut-off line after the hundredths place.

Look only at the first digit being cut off.

$$1058.69 | 25$$

Cent position (hundredths)

Because the first digit being cut off is less than 5 (0, 1, 2, 3, or 4), the part you are keeping remains the same. The answer is rounded to $1058.69. If the first digit being cut off had been 5 or greater (5, 6, 7, 8, or 9), we would have rounded up by adding 1 to the cent position.

When using a calculator with a floating decimal, enter the decimal point as needed. For example, enter $47 as

$$47$$

with no decimal point, but enter 95¢ as

$$\boxed{\cdot} 95$$

One problem using a floating decimal is shown by the following example (adding $21.38 and $1.22).

$$21.38 \boxed{+} 1.22 \boxed{=} 22.6$$

The calculator does not show the final 0. You must remember that the problem dealt with money and write the final 0 making the answer $22.60.

OBJECTIVE 4▶ The $\boxed{\%}$ key moves the decimal point two places to the left when pressed following multiplication or division. The problem, 8% of $4205 is solved as follows.

$$4205 \boxed{\times} 8 \boxed{\%} \boxed{=} 336.4$$

Because the problem involved money, write the answer as $336.40.

OBJECTIVE 5▶ The squaring key, $\boxed{x^2}$, allows you to square the number in the display (multiply the number by itself). The square of 7 is found as follows.

$$7 \boxed{x^2} 49$$

The cubing key, $\boxed{x^3}$, allows you to find the cube of a number (the number is multiplied by itself three times). To find the cube of 6.8, (that is, $6.8 \cdot 6.8 \cdot 6.8$), follow these keystrokes.

$$6.8 \boxed{x^3} 314.432$$

OBJECTIVE 6 The product of $3 \times 3 \times 3 \times 3 \times 3$ can be written as

$$3^5 \quad \text{Exponent}$$
$$\text{Base}$$

The exponent (5) shows how many times the base is multiplied by itself (multiply 3 by itself 5 times). The $\boxed{y^x}$ key raises a base to any desired power. Find 3^5 as

$$3 \;\boxed{y^x}\; 5 = 243.$$

Since $3^2 = 9$, the number 3 is called the square root of 9. Square roots are written with the symbol $\sqrt{}$. Use the $\boxed{\sqrt{x}}$ key to find the square root of 144 ($\sqrt{144}$) as follows.

$$144 \;\boxed{\sqrt{x}}\; 12$$

Find $\sqrt{20}$ as

$$20 \;\boxed{\sqrt{x}}\; 4.472135955$$

which may be rounded to the desired position.

OBJECTIVE 7 The $\boxed{a b/c}$ key is used when solving problems containing fractions and mixed numbers.

Solve $\dfrac{3}{4} + \dfrac{6}{11}$ as

$$3 \;\boxed{a b/c}\; 4 \;\boxed{+}\; 6 \;\boxed{a b/c}\; 11 \;\boxed{=}\; 1 \lrcorner 13 \lrcorner 44$$

The answer is $1\dfrac{13}{44}$.

Solve the mixed number problem $4\dfrac{7}{8} \div 3\dfrac{4}{7}$ as

$$4 \;\boxed{a b/c}\; 7 \;\boxed{a b/c}\; 8 \;\boxed{\div}\; 3 \;\boxed{a b/c}\; 4 \;\boxed{a b/c}\; 7 \;\boxed{=}\; 1 \lrcorner 73 \lrcorner 200$$

The answer is $1\dfrac{73}{200}$.

> **Note**
> The calculator automatically shows fractions in lowest terms and as mixed numbers when possible.

OBJECTIVE 8 Negative numbers may be entered by first entering the number and then using the $\boxed{+/-}$ key. This changes the number entered to a negative number. For example, solve $-10 + 6 - 8$ as follows.

$$10 \;\boxed{+/-}\; \boxed{+}\; 6 \;\boxed{-}\; 8 \;\boxed{=}\; -12$$

OBJECTIVE 9 Many calculators feature memory keys, which are a sort of electronic scratch paper. These memory keys are used to store intermediate steps in a calculation. On some basic calculators, a key labeled $\boxed{M}$ is used to store the numbers in the display, with $\boxed{MR}$ used to recall the numbers from memory.

Other basic calculators have $\boxed{\text{M+}}$ and $\boxed{\text{M-}}$ keys. The $\boxed{\text{M+}}$ key adds the number displayed to the number already in memory. For example, if the memory contains the number 0 at the beginning of a problem, and the calculator display contains the number 29.4, then pushing $\boxed{\text{M+}}$ will cause 29.4 to be stored in the memory (the result of adding 0 and 29.4). If 57.8 is then entered into the display, pushing $\boxed{\text{M+}}$ will cause

$$29.4 + 57.8 = 87.2$$

to be stored. If 11.9 is then entered into the display, with $\boxed{\text{M-}}$ pushed, the memory will contain

$$87.2 - 11.9 = 75.3.$$

The $\boxed{\text{MR}}$ key is used to recall the number in memory as needed, with $\boxed{\text{MC}}$ used to clear the memory.

Scientific calculators typically have one or more storage registers in which to store numbers. These memory keys are usually labeled as $\boxed{\text{STO}}$ for store and $\boxed{\text{RCL}}$ for recall. For example, you can store 25.6 in register 1 by pressing

$$25.6 \ \boxed{\text{STO}} \ 1$$

or you can store it in register 2 by pressing 25.6 $\boxed{\text{STO}}$ 2 and so forth for other registers. Values are retrieved from a particular memory register by using the $\boxed{\text{RCL}}$ key followed by the number of the register, for example, $\boxed{\text{RCL}}$ 2 recalls the contents of memory in register 2.

With a scientific calculator, a number stays in memory until it is replaced by another number or until the memory is cleared. With some calculators, the contents of the memory is saved even when the calculator is turned off.

Here is an example of a problem that uses the memory keys. Suppose an elevator technician wants to find the average weight of a person using an elevator. To do this, she counts the number of people entering an elevator and also measures the weight of each group of people.

Number of People	Total Weight
6	839 pounds
8	1184 pounds
4	640 pounds

First, find the total weight of all three groups and store this in memory register 1.

$$839 \ \boxed{+} \ 1184 \ \boxed{+} \ 640 \ \boxed{=} \ 2663 \ \boxed{\text{STO}} \ 1$$

Then, find the total number of people.

$$6 \ \boxed{+} \ 8 \ \boxed{+} \ 4 \ \boxed{=} \ 18 \ \boxed{\text{STO}} \ 2$$

Finally, divide the contents of memory register 1 (total weight) by the contents of memory register 2 (18 people).

$$\boxed{\text{RCL}} \ 1 \ \boxed{\div} \ \boxed{\text{RCL}} \ 2 \ \boxed{=} \ 147.9444444 \ \text{pounds}$$

This answer can be rounded as needed.

OBJECTIVE 10 ► Long calculations involving several different operations must be done in a specific sequence called the **order of operations**, and they are called chain calculations. The logic of the following order of operations is built into most scientific calculators and can help you work problems without having to store or write down a lot of intermediate steps.

1. Do all operations inside parentheses first.
2. Simplify any expressions with exponents and find any square roots.
3. Multiply and divide from left to right.
4. Add and subtract from left to right.

Your scientific calculator can be used to solve the problem $3 + 7 \times 9\frac{3}{4}$.

$$3 \boxed{+} 7 \boxed{\times} 9 \boxed{a^{b}/_{c}} 3 \boxed{a^{b}/_{c}} 4 \boxed{=} 71\frac{1}{4}$$

The calculator automatically multiplies $7 \times 9\frac{3}{4}$ *before* adding 3.

The problem $42.1 \times 5 - 90 \div 4$ is solved as

$$42.1 \boxed{\times} 5 \boxed{-} 90 \boxed{\div} 4 \boxed{=} 188.$$

The calculator automatically multiplies 42.1×5 and divides $90 \div 4$ *before* doing the subtraction.

OBJECTIVE 11 ► The parentheses keys allow you to group numbers in a complex chain calculation. For example, $\dfrac{4}{5 + 7}$ can be written as $\dfrac{4}{(5 + 7)}$ which can be solved as follows.

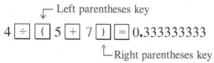

Left parentheses key

$$4 \boxed{\div} \boxed{(} 5 \boxed{+} 7 \boxed{)} \boxed{=} 0.333333333$$

Right parentheses key

Without the parentheses the calculator would have automatically divided 4 by 5 before adding 7, giving an *incorrect* answer of 7.8.

To solve the problem

$$\frac{16 - 2.5}{39.2 - 29.8 \times 0.6}$$

you must think of the problem as

$$\frac{(16 - 2.5)}{(39.2 - 29.8 \times 0.6)}.$$

Use parentheses to set off the numerator and the denominator.

$$\boxed{(} 16 \boxed{-} 2.5 \boxed{)} \boxed{\div} \boxed{(} 39.2 \boxed{-} 29.8 \boxed{\times} 0.6 \boxed{)} \boxed{=} 0.633208255$$

Appendix

Inductive and Deductive Reasoning

OBJECTIVE 1 In many scientific experiments, conclusions are drawn from specific outcomes. After many repetitions and similar outcomes, the findings are generalized into statements that appear to be true. When general conclusions are drawn from specific observations, we are using a type of reasoning called **inductive reasoning.** In the next several examples, this type of reasoning will be illustrated.

E X A M P L E I   Using Inductive Reasoning

Find the next number in the sequence 3, 7, 11, 15, ...

To discover a pattern, calculate the difference between each pair of successive numbers.

$$7 - 3 = 4$$
$$11 - 7 = 4$$
$$15 - 11 = 4$$

As shown, the difference is 4. Each number is 4 greater than the previous one. Thus, the next number in the pattern is 15 + 4, or 19.

WORK PROBLEM I AT THE SIDE. ▶▶

E X A M P L E 2   Using Inductive Reasoning

Find the number that comes next in the sequence.

$$7, 11, 8, 12, 9, 13, ...$$

The pattern in this example can be determined as follows.

$$7 + 4 = 11$$
$$11 - 3 = 8$$
$$8 + 4 = 12$$
$$12 - 3 = 9$$
$$9 + 4 = 13$$

To get the second number, we add 4 to the first number. To get the third number, we subtract 3 from the second number. To obtain subsequent numbers, this pattern is continued. The next number is 13 − 3, or 10.

WORK PROBLEM 2 AT THE SIDE. ▶▶

OBJECTIVES

1 ▶ Use inductive reasoning to analyze patterns.

2 ▶ Use deductive reasoning to analyze arguments.

3 ▶ Use deductive reasoning to solve problems.

---

1. Find the next number in the sequence 2, 8, 14, 20, ...

2. Find the next number in the sequence 6, 11, 7, 12, 8, 13, ...

ANSWERS

1. 26

2. 9

A-7

3. Find the next number in the sequence 2, 6, 18, 54, ...

E X A M P L E   3   Using Inductive Reasoning

Find the next number in the sequence 1, 2, 4, 8, 16, ...

Each number after the first is obtained by multiplying the previous number by 2. So the next number would be $16 \cdot 2 = 32$.

◀◀ **WORK PROBLEM 3 AT THE SIDE.**

E X A M P L E   4   Using Inductive Reasoning

Find the next geometric shape in the following sequence.

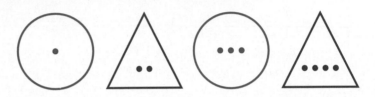

In this sequence, the figures alternate between a circle and a triangle. In addition, the number of dots increases by 1 in each subsequent figure. Thus, the next figure should be a circle with five dots contained in it, or

4. Find the next shape in the following sequence.

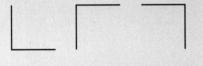

E X A M P L E   5   Using Inductive Reasoning

Find the next geometric shape in the following sequence.

The first two shapes consist of vertical lines with horizontal lines at the bottom facing left and right. The third shape is a vertical line with a horizontal line at the top facing to the left. The fourth shape should be a vertical line with a horizontal line at the top facing to the right, or

◀◀ **WORK PROBLEM 4 AT THE SIDE.**

ANSWERS

3. 162

4.

OBJECTIVE 2 In the previous discussion, specific cases were used to find patterns and predict the next event. There is another type of reasoning called **deductive reasoning,** which moves from general cases to specific conclusions.

E X A M P L E   6   Using Deductive Reasoning

Does the conclusion follow from the premises in this argument?

> All Buicks are automobiles.
> All automobiles have horns.
> ∴   All Buicks have horns.

In this example, the first two statements are called *premises* and the third statement (below the line) is called a conclusion. The symbol ∴ is a mathematical symbol meaning "therefore." The entire set of statements is called an *argument*. The focus of deductive reasoning is to determine if the conclusion follows (is valid) from the premises. A series of circles called **Euler** (OI-ler) **circles** is used to analyze the argument. In Example 6, the statement "All Buicks are automobiles" can be represented by two circles, one for Buicks and one for automobiles.

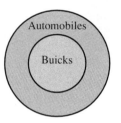

Note that the circle representing Buicks is totally inside the circle representing automobiles.

If a circle representing the second statement is added, a circle representing vehicles with horns must surround the circle representing automobiles.

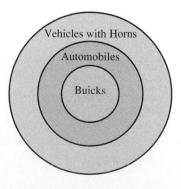

Notice that the circle representing Buicks is completely inside the circle representing vehicles with horns. It must follow that

> all Buicks have horns.

WORK PROBLEM 5 AT THE SIDE. ▶▶

5. Does the conclusion follow from the premises in the following argument?

> All cars have four wheels.
> All Fords are cars.
> ∴   All Fords have four wheels.

ANSWERS

5. The conclusion follows from the premises.

6. Does each conclusion follow from the premises?

  (a) All animals are wild.
     All cats are animals.
     ∴   All cats are wild.

  (b) All students use math.
     All adults use math.
     ∴   All adults are students.

E X A M P L E 7   Using Deductive Reasoning

Does the conclusion follow from the premises in this argument?

All tables are round.
All glasses are round.
∴   All glasses are tables.

Using Euler circles, a circle representing tables is drawn inside a circle representing round objects.

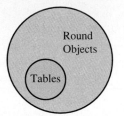

The second statement requires that a circle representing glasses must now be drawn inside the circle representing round objects but not necessarily inside the circle representing tables.

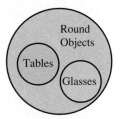

The conclusion does not follow from the premises. This means that the conclusion is invalid or untrue.

◀◀ **WORK PROBLEM 6 AT THE SIDE.**

OBJECTIVE 3   Another type of deductive reasoning problem occurs when a set of facts is given in a problem and a conclusion must be drawn by using these facts.

E X A M P L E 8   Using Deductive Reasoning

There were 25 students enrolled in a ceramics class. During the class, 10 of the students made a bowl and 8 students made a birdbath. Three students made both a bowl and a birdbath. How many students did not make either a bowl or a birdbath?

This type of problem is best solved by organizing the data using a device called a *Venn diagram.* Two overlapping circles are drawn, with each circle representing one item made by students.

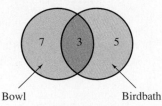

CONTINUED ON NEXT PAGE

ANSWERS

6. (a) The conclusion follows from the premises.
  (b) The conclusion does not follow from the premises.

In the region where the circles overlap, place the number that represents the number of students who made both items, namely 3. In the remaining portion of the birdbath circle, write the number 5, which when added to 3 will give the total number of students who made a birdbath, namely 8. In a similar manner, write 7 in the remaining portion of the bowl circle, since $7 + 3 = 10$, the total number of students who made a bowl. The three numbers that have been written in the regions total 15. Since there are 25 students in the class, this means $25 - 15$ or 10 students did not make either a birdbath or a bowl.

WORK PROBLEM 7 AT THE SIDE. ▶▶

E X A M P L E   9   Using Deductive Reasoning

Four cars in a race finish first, second, third, and fourth. The following facts are known.

(a) Car A beat Car C.

(b) Car D finished between Cars C and B.

(c) Car C beat Car B.

In which order did the cars finish?

To solve this type of problem, it is helpful to use a line diagram.

1. *Write A before C,* since Car A beat Car C (fact a).

$$A \qquad C$$

2. *Write B after C,* since Car C beat Car B (fact c).

$$A \qquad C \qquad B$$

3. *Write D between C and B,* since Car D finished between Cars C and B (fact b).

So

$$A \qquad C \qquad D \qquad B$$

is the correct order of finish.

WORK PROBLEM 8 AT THE SIDE. ▶▶

---

7. In a college class of 100 students, 35 take both math and history, 50 take history, and 40 take math. How many take neither math nor history?

8. A Chevy, BMW, Cadillac, and Oldsmobile are parked side by side.

(a) The Oldsmobile is on the right end.

(b) The BMW is next to the Cadillac.

(c) The Chevy is between the Oldsmobile and the Cadillac.

Which car is parked on the left end?

ANSWERS

7. 45
8. BMW

Numbers in the *Real World*

Read the information at the top of the graph. What does the phrase "on average" mean? How is an average calculated?

The average, or mean, represents a group of numbers. Find the average by adding all the values in the group, then dividing by the number of values.

Is the 43 percent figure correct? Show how you would check it.

Solve the percent equation: $p \cdot 1,420,850 = 608,810$

$p \approx 0.428$ **or** $\approx 43\%$ **(It was rounded to the nearest whole percent.)**

Find at least two other calculations you could do to compare "no high-school diploma" earnings with "bachelor's degree" earnings.

Subtract $1,420,850 – $608,810 to see that bachelor's degree earns $812,040 more than "no diploma."

Then, $812,040 is what percent of $608,810? Earning a bachelor's degree increases earnings by ≈133% over "no diploma."

All of the earning amounts in the graph end in zero. This suggests that the amounts may have been rounded. If so, to what place were they rounded? **To nearest ten.**

If you wanted to round the amounts even more, would it make sense to round them all to the nearest million? Explain why or why not. **No. First five amounts would all round to $1,000,000 and you could no longer see the differences.**

To what place might you choose to round the earnings? Explain your choice

Answers will vary, but nearest thousand or ten thousand will still reveal major differences.

Round each amount to the place you chose. Now look at the sequence of numbers from "no high-school diploma" to "master's degree." Suppose you wanted to add the next level to the graph, lifetime earnings of people with doctorate degrees. Can you use inductive reasoning to find the next number in the sequence of lifetime earnings? Explain why or why not.

There is no apparent pattern in the sequence. The Census Bureau lists average lifetime earnings for people with doctorate degrees as $2,142,440.

Here is the deductive reasoning used by one student. Do you agree with the reasoning? If not, what would you change?

> A bachelor's degree guarantees a lifetime earnings of exactly $1,420,850.
> I have a bachelor's degree.
> _____
> ∴ I will have lifetime earnings of $1,420,850.

The first premise is false; the earnings figure is an average, so some people with bachelor's degrees earn more than $1,420,850 and some earn less.

If you wanted to verify that the earnings shown in the graph are correct for your city or state, how might you do that? Discuss some possibilities with several classmates, then share your ideas with the whole class. **Ideas include checking for local earning averages on the Internet, at a library, or at a city or state employment agency. A more ambitious plan would be to conduct a survey and compute average earnings.**

Lifetime Earnings

People who don't graduate from high school will earn, on average, only 43 percent of what someone with a bachelor's degree will earn, as seen below:

Education	Work-life earnings
No high-school diploma	$608,810
High-school graduate	$820,870
Some college	$992,890
Associate's degree	$1,062,130
Bachelor's degree	$1,420,850
Master's degree	$1,618,970

Source: U.S. Census Bureau

Appendix B Exercises

Find the next number in each of the following sequences.

1. 2, 9, 16, 23, 30, ...   **37**

2. 5, 8, 11, 14, 17, ...   **20**

3. 1, 6, 11, 16, 21, ...   **26**

4. 3, 5, 7, 9, 11, ...   **13**

5. 1, 2, 4, 8, ...   **16**

6. 1, 8, 27, 64, ...   **125**

7. 1, 3, 9, 27, 81, ...   **243**

8. 3, 6, 12, 24, 48, ...   **96**

9. 1, 4, 9, 16, 25, ...   **36**

10. 6, 7, 9, 12, 16, ...   **21**

Find the next shape in each of the following sequences.

11.

12.

13.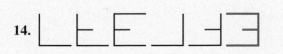

14.

In each of the following, state whether or not the conclusion follows from the premises.

15. All animals are wild.
All lions are animals.
∴   All lions are wild.

Conclusion follows.

16. All students are hard workers.
All business majors are students.
∴   All business majors are hard workers.

Conclusion follows.

17. All teachers are serious.
All mathematicians are serious.
∴   All mathematicians are teachers.

Conclusion does not follow.

18. All boys ride bikes.
All Americans ride bikes.
∴   All Americans are boys.

Conclusion does not follow.

Solve the following application problems.

19. In a given 30-day period, a man watched television 20 days and his wife watched television 25 days. If they watched television together 18 days, how many days did neither watch television?

3 days

20. In a class of 40 students, 21 students take both calculus and physics. If 30 students take calculus and 25 students take physics, how many do not take either calculus or physics?

6 students

21. Tom, Dick, Mary, and Joan all work for the same company. One is a secretary, one is a computer operator, one is a receptionist, and one is a mail clerk.
   (a) Tom and Joan eat dinner with the computer operator.
   (b) Dick and Mary carpool with the secretary.
   (c) Mary works on the same floor as the computer operator and the mail clerk.
Who is the computer operator?   **Dick**

22. Four cars—a Ford, a Buick, a Mercedes, and an Audi—are parked in a garage in four spaces.
   (a) The Ford is in the last space.
   (b) The Buick and Mercedes are next to each other.
   (c) The Audi is next to the Ford but not next to the Buick.
Which car is in the first space?   **Buick**

The solutions to selected odd-numbered exercises are given in the section beginning on page A-39.

In this section we provide the answers that we think most students will obtain when they work the exercises using the methods explained in the text. If your answer does not look exactly like the one given here, it is not necessarily wrong. In many cases there are equivalent forms of the answer that are correct. For example, if the answer section shows $\frac{3}{4}$ and your answer is 0.75, you have obtained the right answer but written it in a different (yet equivalent) form. Unless the directions specify otherwise, 0.75 is just as valid an answer as $\frac{3}{4}$.

In general, if your answer does not agree with the one given in the text, see whether it can be transformed into the other form. If it can, then it is the correct answer. If you still have doubts, talk with your instructor.

CHAPTER 1

SECTION 1.1 (page 5)

1. 5, 3   **3.** 1, 0   **5.** 8, 2
7. 7, 536, 175   **9.** 60, 0, 502, 109
11. Evidence suggests that this is true. It is common to count using fingers.   **13.** sixty-four thousand, two hundred fifteen   **15.** seven hundred twenty-five thousand, nine   **17.** twenty-five million, seven hundred fifty-six thousand, six hundred sixty-five   **19.** 32,526
21. 10,000,223   **23.** 4020   **25.** 2,000,000,000
27. 280,489,000   **29.** 800,000,621,020,215

SECTION 1.2 (page 13)

1. 59   **3.** 97   **5.** 889   **7.** 999   **9.** 7785
11. 997   **13.** 7676   **15.** 78,446   **17.** 8928
19. 59,224   **21.** 150   **23.** 154   **25.** 121
27. 162   **29.** 102   **31.** 1651   **33.** 1154
35. 413   **37.** 1771   **39.** 1410   **41.** 9253
43. 11,624   **45.** 17,611   **47.** 15,954   **49.** 10,648
51. 16,858   **53.** 11,557   **55.** 12,078   **57.** 4250
59. 12,268   **61.** correct   **63.** incorrect; should be 769   **65.** correct   **67.** incorrect; should be 11,577   **69.** correct   **71.** Changing the order in which numbers are added does not change the sum. You can add from bottom to top when checking addition.
73. 37 miles   **75.** 38 miles   **77.** $79
79. 699 people   **81.** 13,051 books   **83.** 294 inches
85. 708 feet

SECTION 1.3 (page 23)

1. 22   **3.** 33   **5.** 17   **7.** 213   **9.** 101
11. 6211   **13.** 3412   **15.** 2111   **17.** 13,160
19. 41,110   **21.** correct   **23.** incorrect; should be 62
25. incorrect; should be 121   **27.** correct
29. incorrect; should be 7222   **31.** 8   **33.** 25
35. 16   **37.** 61   **39.** 519   **41.** 9177   **43.** 7589
45. 8859   **47.** 3   **49.** 23   **51.** 8   **53.** 2833
55. 7775   **57.** 503   **59.** 156   **61.** 1942
63. 5687   **65.** 19,038   **67.** 31,556   **69.** 6584
71. correct   **73.** correct   **75.** correct   **77.** correct
79. Possible answers are:
$3 + 2 = 5$ could be changed to $5 - 2 = 3$ or $5 - 3 = 2$
$6 - 4 = 2$ could be changed to $2 + 4 = 6$ or $4 + 2 = 6$
81. 15 calories   **83.** 121 passengers   **85.** $8700
87. 367 feet   **89.** $940   **91.** 1329 students
93. 9539 flags   **95.** $57,500   **97.** $263
99. 758 people

SECTION 1.4 (page 35)

1. 9   **3.** 63   **5.** 0   **7.** 24   **9.** 36   **11.** 0
13. Factors may be multiplied in any order to get the same answer. They are the same; you may add or multiply numbers in any order.   **15.** 245   **17.** 168
19. 2048   **21.** 1872   **23.** 8612   **25.** 10,084
27. 19,092   **29.** 258,447   **31.** 120   **33.** 150
35. 2220   **37.** 2000   **39.** 3750   **41.** 44,550
43. 270,000   **45.** 86,000,000   **47.** 48,500
49. 320,000   **51.** 1,940,000   **53.** 405   **55.** 1496
57. 3735   **59.** 2378   **61.** 6164   **63.** 15,200
65. 32,805   **67.** 14,564   **69.** 82,320   **71.** 183,996
73. 1,616,076   **75.** 66,005   **77.** 86,028
79. 19,422,180   **81.** 2,278,410
83. To multiply by 10, 100, or by 1000 just add the number of zeros to the number you are multiplying and that's your answer.   **85.** 24,000 pages
87. 216 plants   **89.** 418 miles   **91.** $288
93. $1560   **95.** $3996   **97.** 50,568
99. 175 joggers   **101.** 1058 calories   **103.** $556

SECTION 1.5 (page 47)

1. $4\overline{)12}^{\,3}$; $\dfrac{12}{4} = 3$ **3.** $9\overline{)45}^{\,5}$; $45 \div 9 = 5$

5. $16 \div 2 = 8$; $\dfrac{16}{2} = 8$

7. 1   **9.** 6   **11.** undefined   **13.** 0   **15.** 0

ANSWERS

17. undefined   **19.** 0   **21.** 8   **23.** 27   **25.** 36
27. 608   **29.** 627 R1   **31.** 1522 R5   **33.** 309
35. 1006   **37.** 5006   **39.** 811 R1   **41.** 2589 R2
43. 7324 R2   **45.** 3157 R2   **47.** 6671
49. 12,458 R3   **51.** 10,253 R5   **53.** 18,377 R6
55. correct   **57.** incorrect; should be 1908 R1
59. incorrect; should be 670 R2
61. incorrect; should be 3568 R1   **63.** correct
65. correct   **67.** incorrect; should be 9628 R3
69. correct   **71.** Multiply the quotient by the divisor
and add any remainder. The result shuld be the dividend.
73. 156 cartons   **75.** $16,600   **77.** $18,200
79. 205 acres   **81.** $225,000

	2	3	5	10
83.	√	√	√	√
85.	√	X	X	X
87.	X	X	√	X
89.	X	√	X	X
91.	√	√	X	X
93.	X	X	X	X

95. $9135

Section 1.6 (page 57)

1. 32   **3.** 250   **5.** 120 R7   **7.** 1308 R9
9. 7134 R12   **11.** 900 R100   **13.** 207 R5
15. 236 R29   **17.** 2407 R1   **19.** 1239 R15
21. 3331 R82   **23.** 850   **25.** incorrect; should be
106 R7   **27.** incorrect; should be 658
29. incorrect; should be 62   **31.** When dividing by 10,
100, or 1000 drop the same number of zeros from the
dividend to get the quotient. One example is

$$2500 \div 100 = 25.$$

33. 25 hours   **35.** 630 bronze medals   **37.** $108
39. 1680 circuits   **41.** $375

Section 1.7 (page 65)

1. ≈620   **3.** ≈1090   **5.** ≈7900
7. ≈86,800   **9.** ≈42,500   **11.** ≈6000
13. ≈15,800   **15.** ≈78,000   **17.** ≈6000
19. ≈53,000   **21.** ≈600,000   **23.** ≈9,000,000
25. ≈1480; ≈1500; ≈1000
27. ≈4480; ≈4500; ≈4000
29. ≈5050; ≈5000; ≈5000
31. ≈3130; ≈3100; ≈3000
33. ≈19,540; ≈19,500; ≈20,000
35. ≈26,290; ≈26,300; ≈26,000
37. ≈23,500; ≈23,500; ≈24,000
39. 1. Locate the place to be rounded and underline it.
2. Look only at the next digit to the right. If this digit is 5
or more, increase the underlined digit by 1.
3. Change all digits to the right of the underlined place to
zeros.   **41.** 70 + 40 + 90 + 90 = 290; 292
43. 100 − 30 = 70; 71   **45.** 80 × 20 = 1600; 1672
47. 800 + 800 + 300 + 700 = 2600; 2635
49. 900 − 500 = 400; 416
51. 400 × 400 = 160,000; 160,448

53. 8000 + 60 + 700 + 4000 = 12,760; 12,605
55. 700 − 300 = 400; 365
57. 800 × 40 = 32,000; 31,958   **59.** Perhaps the best
explanation is that 648 is closer to 650 than 640, but 648 is
closer to 600 than to 700.   **61.** ≈3,025,940,000 pesos;
≈3,026,000,000 pesos; ≈3,000,000,000 pesos
63. ≈$5,465,485,400,000; ≈$5,465,500,000,000;
≈$5,465,000,000,000

Section 1.8 (page 71)

1. 3   **3.** 4   **5.** 12   **7.** 11   **9.** 2; 5; 25
11. 2; 6; 36   **13.** 2; 12; 144   **15.** 2; 15; 225
17. 100; $\sqrt{100}$ **19.** 225; $\sqrt{225}$
21. 1225; $\sqrt{1225}$ **23.** 1600; $\sqrt{1600}$
25. 2916; $\sqrt{2916}$ **27.** A perfect square is the square
of a whole number. The number 25 is the square of 5
because $5 \cdot 5 = 25$. The number 50 is not a perfect square.
There is no whole number that can be squared to get 50.
29. 31   **31.** 26   **33.** 12   **35.** 20
37. 24   **39.** 63   **41.** 118   **43.** 36   **45.** 49
47. 102   **49.** 10   **51.** 63   **53.** 33   **55.** 58
57. 7   **59.** 26   **61.** 36   **63.** 108   **65.** 19
67. 9   **69.** 23   **71.** 16   **73.** 8   **75.** 3   **77.** 7
79. 20   **81.** 7   **83.** 25   **85.** 16   **87.** 20   **89.** 209

Section 1.9 (page 79)

1. $400 + $500 = $900; $880   **3.** 200 − 60 = 140
types; 138 types   **5.** 200 × 20 = 4000 kits; 5664 kits
7. 3000 ÷ 700 ≈ 4 toys; 4 toys
9. 8000 − 4000 = 4000 people; 4174 people
11. $10 × 5 = $50; $70
13. 3000 ÷ 20 = 150 miles; 175 miles
15. $2000 − $500 − $300 − $300 − $200 − $200
= $500; $193   **17.** 40,000 × 100 = 4,000,000 square
feet; 6,011,280 square feet   **19.** $500 + $800 +
$100 + $100 + $100 = $1600; $1635
21. $500 + $800 + $200 + $1000 = $2500;
$2500 − $1800 = $700; $725
23. ($1000 × 6) + ($900 × 20) = $24,000; $20,961
25. Possible answers are: Addition: more; total; gain of;
Subtraction: less; less of; decreased by;
Multiplication: twice; of; product; Division: divided by;
goes into; per; Equals: is; are   **27.** Estimating the
answer can help you avoid careless mistakes like decimal
errors and calculation errors. Examples of reasonable
answers in daily life might be a $20 bag of groceries, $15
to fill the gas tank, or $45 for a phone bill.   **29.** $165
31. 2477 pounds   **33.** $125   **35.** 352 machines
37. 20 seats

Chapter 1 Review Exercises (page 87)

1. 4; 621   **2.** 87; 328   **3.** 105; 724
4. 1; 768; 710; 618   **5.** seven hundred twenty-five
6. twelve thousand, four hundred twelve
7. three hundred nineteen thousand, two hundred fifteen
8. sixty-two million, five hundred thousand, five

9. 4004   **10.** 200,000,455   **11.** 92   **12.** 113
13. 5464   **14.** 15,657   **15.** 9179   **16.** 6979
17. 40,602   **18.** 49,855   **19.** 22   **20.** 21
21. 39   **22.** 184   **23.** 3803   **24.** 4327
25. 224   **26.** 25,866   **27.** 25   **28.** 0
29. 21   **30.** 64   **31.** 42   **32.** 36   **33.** 56
34. 81   **35.** 48   **36.** 45   **37.** 48   **38.** 8
39. 0   **40.** 64   **41.** 48   **42.** 0   **43.** 172
44. 434   **45.** 522   **46.** 98   **47.** 3834
48. 5467   **49.** 5396   **50.** 45,815   **51.** 14,518
52. 32,640   **53.** 465,525   **54.** 174,984
55. 612   **56.** 1872   **57.** 1176   **58.** 5100
59. 13,755   **60.** 30,184   **61.** 887,169
62. 500,856   **63.** $300   **64.** $672   **65.** $13,344
66. $1512   **67.** 19,200   **68.** 25,200   **69.** 206,800
70. 300,800   **71.** 128,000,000   **72.** 90,300,000
73. 3   **74.** 6   **75.** 6   **76.** 2   **77.** 9   **78.** 4
79. 9   **80.** 0   **81.** undefined   **82.** 0   **83.** 8
84. 9   **85.** 108   **86.** 24   **87.** 6251   **88.** 352
89. 150 R4   **90.** 124 R25   **91.** ≈320
92. ≈14,300   **93.** ≈20,000   **94.** ≈70,000
95. ≈2400; ≈2400; ≈2000
96. ≈20,070; ≈20,100; ≈20,000
97. ≈98,200; ≈98,200; ≈98,000
98. ≈352,120; ≈352,100; ≈352,000   **99.** 6
100. 7   **101.** 12   **102.** 14   **103.** 2; 3; 9
104. 3; 2; 8   **105.** 3; 5; 125   **106.** 5; 4; 1024
107. 72   **108.** 4   **109.** 9   **110.** 4
111. 9   **112.** 6   **113.** 50 × $10 = $500; $528
114. 1000 × 60 = 60,000 revolutions; 84,000 revolutions
115. 100 × 6 = 600 cups; 720 cups
116. 6000 × 30 = 180,000 brackets; 180,000 brackets
117. 2000 × 10 = 20,000 hours; 24,000 hours
118. 80 × 5 = 400 miles; 400 miles
119. ($20 × 20) + ($10 × 30) = $700; $582
120. (60 × $20) + (20 × $10) = $1400; $1024
121. $400 − $200 = $200; $180
122. $400 − $100 = $300; $247
123. 9000 ÷ 200 = 45 pounds; 50 pounds
124. 30,000 ÷ 1000 ≈ 30 hours; 33 hours
125. 6000 ÷ 300 = 20 acres; 23 acres
126. 6000 ÷ 200 = 30 homes; 32 homes
127. 282   **128.** 546   **129.** 107   **130.** 563
131. 1041   **132.** 1030   **133.** 32,062   **134.** 24,947
135. 3   **136.** 7   **137.** 93,635   **138.** 83,178
139. undefined   **140.** 6   **141.** 13,800   **142.** 2305
143. 1,079,040   **144.** 115,713   **145.** 108   **146.** 207
147. two hundred eighty-six thousand, seven hundred fifty-three   **148.** one hundred eight thousand, two hundred ten   **149.** ≈3300   **150.** ≈200,000
151. 5   **152.** 10   **153.** $5940   **154.** $33,800
155. $2220   **156.** $15,782   **157.** 468 cards
158. 2856 textbooks   **159.** $280   **160.** $27,940
161. $1905   **162.** $12,420

Chapter 1 Test (page 95)

1. eight thousand, two hundred eight
2. seventy-five thousand, sixty-five
3. 138,008   **4.** 12,893   **5.** 112,630
6. 3048   **7.** 3084   **8.** 140   **9.** 171,000
10. 1785   **11.** 4,450,743   **12.** 7747

13. undefined   **14.** 458   **15.** 160   **16.** ≈4760
17. ≈68,000   **18.** 35   **19.** 28
20. $500 + $500 + $500 + $400 − $800 = $1100; $1140   **21.** 60,000 ÷ 500 = 120 days; 118 days
22. $1000 − $700 − $200 − $70 = $30; $165
23. (100 × 4) + (100 × 4) = 800 ovens; 1028 ovens
24. 1. Locate the place to be rounded and underline it. 2. Look only at the next digit to the right. If this digit is a 4 or less, do not change the underlined digit. If the digit is a 5 or more, increase the underlined digit by 1. 3. Change all digits to the right of the underlined place to zeros. Each person's example will vary.   **25.** 1. Read the problem carefully.   2. Work out a plan.   3. Estimate a reasonable answer.   4. Solve the problem being certain to check your work.

CHAPTER 2
SECTION 2.1 (page 101)

1. $\frac{5}{8}; \frac{3}{8}$ **3.** $\frac{2}{3}; \frac{1}{3}$ **5.** $\frac{7}{5}; \frac{3}{5}$ **7.** $\frac{2}{11}$ **9.** $\frac{8}{25}$

11. $\frac{13}{71}$ **13.** 3; 4   **15.** 12; 7

17. proper: $\frac{1}{3}, \frac{5}{8}, \frac{7}{16}$ improper: $\frac{8}{5}, \frac{6}{6}, \frac{12}{2}$

19. proper: $\frac{3}{4}, \frac{9}{11}, \frac{7}{15}$ improper: $\frac{3}{2}, \frac{5}{5}, \frac{19}{18}$

21. One possibility is

$\frac{3}{4}$ ← Numerator, ← Denominator

The denominator shows the number of equal parts in the whole and the numerator shows how many of the parts are being considered.   **23.** 9; 16
25. 100   **27.** 0   **29.** 4   **31.** 19

SECTION 2.2 (page 107)

1. $\frac{5}{3}$ **3.** $\frac{11}{4}$ **5.** $\frac{19}{4}$ **7.** $\frac{27}{4}$ **9.** $\frac{18}{11}$

11. $\frac{19}{3}$ **13.** $\frac{34}{3}$ **15.** $\frac{43}{4}$ **17.** $\frac{27}{8}$ **19.** $\frac{44}{5}$

21. $\frac{54}{11}$ **23.** $\frac{183}{8}$ **25.** $\frac{233}{13}$ **27.** $\frac{269}{15}$

29. $\frac{115}{18}$ **31.** $2\frac{2}{3}$ **33.** $1\frac{4}{5}$ **35.** 5

37. $3\frac{3}{8}$ **39.** $4\frac{3}{4}$ **41.** 9   **43.** $11\frac{3}{5}$ **45.** $5\frac{2}{9}$

47. $7\frac{1}{7}$ **49.** $16\frac{4}{5}$ **51.** $30\frac{3}{4}$ **53.** $26\frac{1}{7}$

55. Multiply the denominator by the whole number and add the numerator. The result becomes the new numerator which is placed over the original denominator.

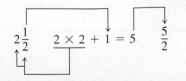

ANSWERS

57. $\dfrac{203}{2}$ **59.** $\dfrac{1000}{3}$ **61.** 171

63. 20   **65.** 66   **67.** 49

Section 2.3 (page 115)

1. 1, 2, 4, 8   **3.** 1, 2, 3, 6   **5.** 1, 5, 25
7. 1, 2, 3, 6, 9, 18   **9.** 1, 2, 4, 5, 8, 10, 20, 40
11. 1, 2, 4, 8, 16, 32, 64   **13.** composite   **15.** prime
17. composite   **19.** prime   **21.** prime
23. composite   **25.** composite   **27.** composite
29. $2 \cdot 3$ **31.** $2^2 \cdot 5$ **33.** 5^2 **35.** $2^2 \cdot 3^2$
37. $2^2 \cdot 11$ **39.** $2^3 \cdot 11$ **41.** $3 \cdot 5^2$ **43.** $2^2 \cdot 5^2$
45. $2^3 \cdot 3 \cdot 5$ **47.** $3^2 \cdot 5^2$ **49.** $2^6 \cdot 5$
51. $2^3 \cdot 3^2 \cdot 5$ **53.** A composite number has a factor(s) other than itself or 1. Examples include 4, 6, 8, 9, 10. A prime number is a whole number that has exactly two *different* factors, itself and 1. Examples include 2, 3, 5, 7, 11.   **55.** 8   **57.** 125   **59.** 81   **61.** 108
63. 1125   **65.** 972   **67.** $2^3 \cdot 5 \cdot 7$ **69.** $2^6 \cdot 3 \cdot 5$
71. $2^6 \cdot 5^2$ **73.** 24   **75.** 48   **77.** 9   **79.** 27

Section 2.4 (page 123)

1. $\dfrac{1}{2}$ **3.** $\dfrac{2}{3}$ **5.** $\dfrac{5}{8}$ **7.** $\dfrac{6}{7}$ **9.** $\dfrac{9}{10}$ **11.** $\dfrac{6}{7}$

13. $\dfrac{4}{7}$ **15.** $\dfrac{1}{50}$ **17.** $\dfrac{8}{11}$ **19.** $\dfrac{5}{9}$

21. $\dfrac{\cancel{2} \cdot \cancel{2} \cdot \cancel{3}}{\cancel{2} \cdot \cancel{3} \cdot 3} = \dfrac{2}{3}$ **23.** $\dfrac{\cancel{5} \cdot 7}{2 \cdot 2 \cdot 2 \cdot \cancel{5}} = \dfrac{7}{8}$

25. $\dfrac{\cancel{2} \cdot \cancel{3} \cdot \cancel{3} \cdot \cancel{5}}{\cancel{2} \cdot 2 \cdot \cancel{3} \cdot \cancel{3} \cdot \cancel{5}} = \dfrac{1}{2}$ **27.** $\dfrac{\cancel{2} \cdot 2 \cdot \cancel{3} \cdot 3}{\cancel{2} \cdot 2 \cdot \cancel{3}} = 3$

29. $\dfrac{7 \cdot \cancel{11}}{2 \cdot 2 \cdot 2 \cdot 3 \cdot \cancel{11}} = \dfrac{7}{24}$ **31.** equivalent

33. not equivalent   **35.** not equivalent
37. equivalent   **39.** not equivalent   **41.** equivalent
43. A fraction is in lowest terms when the numerator and the denominator have no common factors other than 1.

Some examples: $\dfrac{1}{2}, \dfrac{3}{8}, \dfrac{2}{3}$

45. $\dfrac{7}{8}$ **47.** $\dfrac{2}{1} = 2$ **49.** 1, 3, 5, 15
51. 1, 2, 4, 8, 16, 32, 64

Section 2.5 (page 131)

1. $\dfrac{3}{16}$ **3.** $\dfrac{4}{15}$ **5.** $\dfrac{9}{10}$ **7.** $\dfrac{3}{10}$ **9.** $\dfrac{5}{12}$

11. $\dfrac{9}{32}$ **13.** $\dfrac{1}{2}$ **15.** $\dfrac{13}{32}$ **17.** $\dfrac{21}{128}$ **19.** 3

21. 36   **23.** 12   **25.** 21   **27.** $31\dfrac{1}{2}$ **29.** 240

31. $272\dfrac{1}{4}$ **33.** 400   **35.** 810   **37.** $\dfrac{7}{16}$ square foot

39. 9 square yards   **41.** $\dfrac{3}{10}$ square inch

43. Multiply the numerators and multiply the denominators. An example is

$$\dfrac{3}{4} \cdot \dfrac{1}{2} = \dfrac{3 \cdot 1}{4 \cdot 2} = \dfrac{3}{8}.$$

45. $1\dfrac{1}{3}$ square yards   **47.** 1 square mile

49. They are both the same size: $\dfrac{3}{64}$ square mile

51. 290,175 cars

Section 2.6 (page 137)

1. $\dfrac{1}{2}$ square yard   **3.** $\dfrac{2}{3}$ square foot   **5.** 10 of them
7. $1750   **9.** 375 students   **11.** 325 women
13. $38,000   **15.** $7600   **17.** $2375
19. Solution is

$$\dfrac{9}{10} \times \dfrac{20}{21} = \dfrac{\overset{3}{\cancel{9}}}{\underset{1}{\cancel{10}}} \times \dfrac{\overset{2}{\cancel{20}}}{\underset{7}{\cancel{21}}} = \dfrac{6}{7}$$

21. $42   **23.** 9000 votes   **25.** $\dfrac{1}{32}$ of the estate

27. 74 cartons

Section 2.7 (page 145)

1. $\dfrac{1}{2}$ **3.** $2\dfrac{5}{8}$ **5.** $\dfrac{9}{20}$ **7.** $\dfrac{5}{8}$ **9.** 4   **11.** $\dfrac{24}{25}$

13. 9   **15.** $22\dfrac{1}{2}$ **17.** $\dfrac{1}{14}$ **19.** $\dfrac{2}{9}$ acre

21. 16 measuring cups   **23.** 88 dispensers
25. 60 trips   **27.** 12 batches   **29.** You can divide two fractions by inverting the second fraction (divisor) and multiplying.   **31.** 108 miles   **33.** $120,000
35. $\dfrac{19}{8}$ **37.** $\dfrac{25}{2}$ **39.** $\dfrac{604}{5}$

Section 2.8 (page 155)

1. $7\dfrac{7}{8}$; 2, 4, 8   **3.** $4\dfrac{1}{2}$; 2, 3, 6   **5.** 4; 3, 1, 3

7. $72\dfrac{1}{2}$; 10, 7, 70   **9.** $49\dfrac{1}{2}$; 5, 2, 5, 50

11. 12; 3, 2, 3, 18   **13.** $1\dfrac{5}{21}$; 3, 3, 1   **15.** $\dfrac{5}{6}$; 3, 3, 1

17. $4\dfrac{4}{5}$; 6, 1, 6   **19.** $\dfrac{2}{9}$; 1, 2, $\dfrac{1}{2}$ **21.** $\dfrac{3}{10}$; 2, 6, $\dfrac{1}{3}$

23. $\frac{17}{18}$; 6, 6, 1   **25.** $16 \cdot 2 = 32$ yards; 36 yards

27. $1314 \div 110 = 12$ homes; 12 homes

29. $2 \cdot 13 = 26$ ounces; $21\frac{7}{8}$ ounces

31. The answer should include

Step 1 Change mixed numbers to improper fractions.

Step 2 Multiply the fractions.

Step 3 Write the answer in lowest terms changing to mixed
or whole numbers where possible.

33. $25{,}730 \div 10 = 2573$ anchors; 2480 anchors

35. $10 \div 1 = 10$ spacers; 13 spacers

37. $13 \cdot 28 = \quad 364$

$\, 7 \cdot 16 = + 112$

$\,\overline{476 \text{ rolls; 471 rolls}}$

39. $5025 \div 8 \approx 628$ shares; 600 shares

41. $\frac{3}{4}$ **43.** $\frac{7}{10}$ **45.** $\frac{7}{8}$

CHAPTER 2 REVIEW EXERCISES (page 161)

1. $\frac{3}{4}$ **2.** $\frac{5}{8}$ **3.** $\frac{1}{4}$

4. *proper:* $\frac{1}{4}, \frac{5}{8}, \frac{2}{3}$; *improper:* $\frac{3}{2}, \frac{4}{4}$

5. *proper:* $\frac{15}{16}, \frac{1}{8}$;   *improper:* $\frac{6}{5}, \frac{16}{13}, \frac{5}{3}$

6. $\frac{35}{8}$ **7.** $\frac{54}{5}$ **8.** $5\frac{1}{4}$ **9.** $12\frac{3}{5}$

10. 1, 2, 4, 8   **11.** 1, 2, 3, 6, 9, 18   **12.** 1, 5, 11, 55

13. 1, 2, 3, 5, 6, 9, 10, 15, 18, 30, 45, 90   **14.** 2^4

15. $2 \cdot 3 \cdot 5^2$ **16.** $3^2 \cdot 5^2$ **17.** 25   **18.** 72

19. 1728   **20.** 2048   **21.** $\frac{3}{4}$ **22.** $\frac{7}{8}$ **23.** $\frac{15}{16}$

24. $\dfrac{\overset{1}{\cancel{3}} \cdot 5}{2 \cdot 2 \cdot 3 \cdot \underset{1}{\cancel{3}}}$; $\frac{5}{12}$ **25.** $\dfrac{\overset{1}{\cancel{2}} \cdot \overset{1}{\cancel{2}} \cdot \overset{1}{\cancel{2}} \cdot \overset{1}{\cancel{2}} \cdot \overset{1}{\cancel{2}} \cdot 2 \cdot 2 \cdot \overset{1}{\cancel{3}}}{\underset{1}{\cancel{2}} \cdot \underset{1}{\cancel{2}} \cdot \underset{1}{\cancel{2}} \cdot \underset{1}{\cancel{2}} \cdot \underset{1}{\cancel{2}} \cdot \underset{1}{\cancel{3}}}$; 4

26. equivalent   **27.** not equivalent   **28.** $\frac{1}{2}$ **29.** $\frac{1}{3}$

30. $\frac{1}{7}$ **31.** $\frac{4}{21}$ **32.** 15   **33.** 625   **34.** $\frac{1}{2}$

35. $\frac{5}{3} = 1\frac{2}{3}$ **36.** $\frac{5}{2} = 2\frac{1}{2}$ **37.** 2   **38.** 8

39. 24   **40.** $\frac{7}{16}$ **41.** $\frac{2}{15}$ **42.** $\frac{4}{13}$

43. $\frac{15}{32}$ square yard   **44.** $\frac{7}{12}$ square inch

45. 10 square feet   **46.** 36 square yards

47. $3\frac{9}{16}$; 2, 2, 4   **48.** $21\frac{3}{8}$; 2, 7, 1, 14

49. $5\frac{1}{6}$; 16, 3, $5\frac{1}{3}$ **50.** $\frac{35}{64}$; 3, 6, $\frac{1}{2}$ **51.** 300 bags

52. $\frac{2}{15}$ of the estate   **53.** $158 \div 4 \approx 40$ pull cords;

36 pull cords   **54.** $\$9 \cdot 38 = \$342$; $\$323$

55. 30 pounds   **56.** $\$375$

57. $\frac{5}{32}$ of the budget   **58.** $\frac{1}{12}$ of the total

59. $\frac{1}{3}$ **60.** $\frac{1}{6}$ **61.** $25\frac{5}{8}$ **62.** $28\frac{1}{8}$ **63.** $\frac{7}{48}$

64. $\frac{5}{32}$ **65.** 30   **66.** $2\frac{1}{6}$ **67.** $1\frac{3}{5}$ **68.** $45\frac{2}{3}$

69. $\frac{17}{3}$ **70.** $\frac{307}{8}$ **71.** $\dfrac{\overset{1}{\cancel{2}} \cdot \overset{1}{\cancel{2}} \cdot 2}{\underset{1}{\cancel{2}} \cdot \underset{1}{\cancel{2}} \cdot 3} = \frac{2}{3}$

72. $\dfrac{\overset{1}{\cancel{2}} \cdot 2 \cdot \overset{1}{\cancel{3}} \cdot 3 \cdot 3}{\underset{1}{\cancel{2}} \cdot \underset{1}{\cancel{3}} \cdot 5 \cdot 7} = \frac{18}{35}$ **73.** $\frac{3}{4}$ **74.** $\frac{1}{3}$ **75.** $\frac{2}{5}$

76. $\frac{1}{3}$ **77.** $4 \cdot 44 = 176$ ounces; $152\frac{4}{9}$ ounces

78. $7 \cdot 26 = 182$ quarts; $184\frac{7}{8}$ quarts

79. $\frac{1}{2}$ square inch   **80.** $\frac{7}{16}$ square meter

CHAPTER 2 TEST (page 167)

1. $\frac{3}{8}$ **2.** $\frac{5}{6}$ **3.** $\frac{3}{4}, \frac{7}{8}, \frac{1}{6}, \frac{2}{9}$ **4.** $\frac{35}{8}$ **5.** $20\frac{5}{6}$

6. 1, 2, 3, 6, 9, 18   **7.** $2^2 \cdot 3^2$ **8.** $2^5 \cdot 3$

9. $2^2 \cdot 5^3$ **10.** $\frac{5}{6}$ **11.** $\frac{2}{3}$ **12.** Write the prime

factorization of both numerator and denominator. Use
cancellation to divide numerator and denominator by any
common factors. Multiply the remaining factors in
numerator and denominator.

$$\frac{56}{84} = \frac{\overset{1}{\cancel{2}} \cdot \overset{1}{\cancel{2}} \cdot 2 \cdot \overset{1}{\cancel{7}}}{\underset{1}{\cancel{2}} \cdot \underset{1}{\cancel{2}} \cdot 3 \cdot \underset{1}{\cancel{7}}} = \frac{2}{3}$$

13. Multiply fractions by multiplying the numerators and
multiplying the denominators. Divide two fractions by
inverting the second fraction (divisor) and multiplying.

14. $\frac{1}{2}$ **15.** 18   **16.** $\frac{3}{8}$ square meter

17. 7392 students   **18.** $\frac{5}{6}$ **19.** $15\frac{3}{4}$

20. 100 vehicles   **21.** $5 \cdot 3 = 15$; $17\frac{23}{32}$

22. $2 \cdot 4 = 8$; $7\frac{17}{18}$ **23.** $5 \div 1 = 5$; $4\frac{4}{15}$

24. $9 \div 2 = 4\frac{1}{2}$; $5\frac{1}{10}$ **25.** $3 \cdot 12 = 36$; $30\frac{5}{8}$ grams

ANSWERS

CUMULATIVE REVIEW 1–2 (page 169)

1. 7; 1 **2.** 6; 4 **3.** 166 **4.** 149,199 **5.** 1572
6. 3,221,821 **7.** 768 **8.** 90 **9.** 2,168,232
10. 440,300 **11.** 9 **12.** 7581 **13.** 4235 R2
14. 22 R26 **15.** 8630; 8600; 9000 **16.** 85,460; 85,500; 85,000 **17.** 7 **18.** 5
19. $615 **20.** $130 **21.** 47,000 hairs **22.** $175
23. $\frac{7}{16}$ square foot **24.** 6¢ **25.** proper

26. improper **27.** proper **28.** $\frac{5}{2}$ **29.** $\frac{22}{3}$

30. $1\frac{5}{7}$ **31.** $12\frac{7}{8}$ **32.** $2 \cdot 5^2$ **33.** $2^4 \cdot 5$
34. $2 \cdot 5^2 \cdot 7$ **35.** 36 **36.** 675 **37.** 640
38. $\frac{7}{8}$ **39.** $\frac{2}{3}$ **40.** $\frac{5}{9}$ **41.** $\frac{1}{4}$ **42.** 12 **43.** 25

44. $\frac{9}{16}$ **45.** $\frac{3}{10}$ **46.** $2\frac{2}{5}$

CHAPTER 3
SECTION 3.1 (page 175)

1. $\frac{5}{8}$ **3.** $\frac{9}{10}$ **5.** $\frac{1}{2}$ **7.** $1\frac{1}{4}$ **9.** $\frac{1}{3}$

11. $\frac{13}{20}$ **13.** $\frac{10}{17}$ **15.** $1\frac{1}{2}$ **17.** $\frac{11}{27}$ **19.** $\frac{3}{5}$

21. $\frac{7}{15}$ **23.** $\frac{3}{5}$ **25.** $1\frac{1}{7}$ **27.** $\frac{1}{5}$ **29.** $1\frac{1}{3}$

31. $\frac{1}{9}$ **33.** $\frac{3}{5}$

35. Three steps to add like fractions are:
1. Add the numerators of the fractions to find the answer numerator (sum).
2. Use the denominator of the fractions as the denominator of the sum.
3. Write the answer in lowest terms.

37. $\frac{1}{2}$ mile **39.** $\frac{3}{8}$ mile **41.** $\frac{3}{4}$ acre
43. $2 \cdot 5$ **45.** $2 \cdot 2 \cdot 5 \cdot 5$ **47.** $3 \cdot 5 \cdot 5$
49. $5 \cdot 5 \cdot 5$

SECTION 3.2 (page 183)

1. 10 **3.** 12 **5.** 36 **7.** 14 **9.** 20
11. 100 **13.** 18 **15.** 72 **17.** 360 **19.** 120
21. 180 **23.** 120 **25.** $\frac{12}{24}$ **27.** $\frac{18}{24}$ **29.** $\frac{9}{24}$
31. 2 **33.** 36 **35.** 28 **37.** 55 **39.** 45
41. 72 **43.** 136 **45.** 96 **47.** 27
49. It probably depends on how large the numbers are. If the numbers are small, the method using multiples of the largest number seems best. If numbers are larger, or there are more of them, then the factorization method will be better.
51. 7200 **53.** 10,584

55. $1\frac{3}{5}$ **57.** $1\frac{7}{8}$ **59.** $3\frac{6}{7}$

SECTION 3.3 (page 191)

1. $\frac{7}{8}$ **3.** $\frac{1}{2}$ **5.** $\frac{3}{4}$ **7.** $\frac{39}{40}$ **9.** $\frac{23}{36}$

11. $\frac{14}{15}$ **13.** $\frac{29}{36}$ **15.** $\frac{17}{20}$ **17.** $\frac{23}{30}$

19. $\frac{7}{12}$ **21.** $\frac{23}{48}$ **23.** $\frac{5}{8}$ **25.** $\frac{1}{2}$

27. $\frac{1}{6}$ **29.** $\frac{1}{6}$ **31.** $\frac{19}{45}$ **33.** $\frac{2}{15}$

35. $\frac{17}{48}$ **37.** $\frac{23}{24}$ cubic yard **39.** $\frac{7}{12}$ acre

41. $\frac{41}{48}$ mile **43.** $\frac{3}{8}$ gallon **45.** You cannot add or subtract until all the fractional pieces are the same size. For example, halves are larger than fourths, so you cannot add $\frac{1}{2} + \frac{1}{4}$ until you rewrite $\frac{1}{2}$ as $\frac{2}{4}$. **47.** $\frac{1}{4}$

49. work and travel; 8 hours **51.** $\frac{1}{12}$ mile

53. $4\frac{3}{8}$ **55.** $13\frac{1}{2}$ **57.** $\frac{2}{5}$

SECTION 3.4 (page 199)

1. $6 + 3 = 9; 8\frac{5}{6}$ **3.** $10 + 5 = 15; 15\frac{1}{2}$

5. $27 + 9 = 36; 35\frac{17}{24}$ **7.** $25 + 19 = 44; 43\frac{2}{3}$

9. $34 + 19 = 53; 52\frac{1}{10}$ **11.** $23 + 15 = 38; 38\frac{5}{28}$

13. $19 + 48 + 26 = 93; 91\frac{5}{6}$

15. $33 + 6 + 15 = 54; 53\frac{17}{24}$ **17.** $19 - 16 = 3; 3\frac{1}{8}$

19. $11 - 5 = 6; 6\frac{17}{20}$ **21.** $15 - 6 = 9; 9\frac{9}{40}$

23. $35 - 17 = 18; 17\frac{5}{8}$ **25.** $47 - 7 = 40; 40\frac{19}{24}$

27. $26 - 12 = 14; 13\frac{59}{72}$

29. $374 - 212 = 162; 162\frac{1}{6}$ **31.** $3\frac{5}{6}$ **33.** $6\frac{11}{12}$

35. $8\frac{1}{8}$ **37.** $\frac{5}{6}$ **39.** $2\frac{7}{8}$ **41.** $2\frac{7}{12}$

43. Find the least common denominator. Change the fraction parts so that they have the same denominator. Add the fraction parts. Add the whole number parts. Write the answer as a mixed number.

45. $15 - 7 = 8$ hours; $8\frac{5}{6}$ hours

47. $13 + 9 = 22$ ft; $21\frac{1}{6}$ ft

49. $5 + 6 + 5 + 4 + 7 = 27$ hours; $26\frac{7}{8}$ hours

51. $24 + 35 + 24 + 35 = 118$ inches; $116\frac{1}{2}$ inches

53. $10 - 2 - 3 - 3 = 2$ cubic yards; $2\frac{3}{8}$ cubic yards

55. $527 - 108 - 151 - 139 = 129$ ft; 130 ft

57. $3 + 8 + 2 + 2 = 15$ tons; $14\frac{23}{24}$ tons

59. $4\frac{11}{16}$ in.   **61.** $21\frac{3}{8}$ in.

63. 30   **65.** 63   **67.** 27   **69.** 53

SECTION 3.5 (page 209)

1.–12.

2. 1.   10.   4.   3.   12.   7. 5.   6.   11. 9.   8.

13. $<$ **15.** $<$ **17.** $<$ **19.** $<$

21. $<$ **23.** $>$ **25.** $\frac{9}{64}$ **27.** $\frac{25}{49}$ **29.** $\frac{8}{27}$

31. $\frac{125}{216}$ **33.** $\frac{81}{16} = 5\frac{1}{16}$ **35.** $\frac{1}{32}$

37. A number line is a horizontal line with a range of numbers placed on it. The lowest number is on the left, and the highest number is on the right. It can be used to compare the size or value of numbers.

39. 2   **41.** 16   **43.** 1   **45.** $\frac{3}{16}$ **47.** $\frac{1}{4}$ **49.** $\frac{1}{3}$

51. $\frac{11}{16}$ **53.** $\frac{3}{8}$ **55.** $\frac{1}{4}$ **57.** $1\frac{1}{2}$ **59.** $\frac{1}{5}$ **61.** 3

63. $\frac{23}{112}$ **65.** $\frac{1}{4}$ **67.** $\frac{1}{32}$

69. five thousand, seven hundred twenty-eight
71. four million, seventy-one thousand, two hundred eighty

CHAPTER 3 REVIEW EXERCISES (page 217)

1. $\frac{7}{8}$ **2.** $\frac{3}{5}$ **3.** $\frac{3}{4}$ **4.** $\frac{5}{12}$ **5.** $\frac{1}{5}$

6. $\frac{1}{4}$ **7.** $\frac{13}{31}$ **8.** $\frac{8}{27}$ **9.** $\frac{1}{2}$ of the lumber

10. $\frac{3}{5}$ less   **11.** 12   **12.** 40   **13.** 60   **14.** 180

15. 120   **16.** 240   **17.** 12   **18.** 10   **19.** 10

20. 45   **21.** 63   **22.** 12   **23.** $\frac{7}{12}$ **24.** $\frac{7}{8}$

25. $\frac{31}{48}$ **26.** $\frac{11}{20}$ **27.** $\frac{5}{12}$ **28.** $\frac{17}{36}$

29. $\frac{23}{24}$ cubic yard   **30.** $\frac{59}{60}$ of the amount needed

31. $26 + 16 = 42$; $42\frac{1}{8}$ **32.** $78 + 18 = 96$; $96\frac{2}{7}$

33. $13 + 9 + 10 = 32$; $31\frac{43}{80}$ **34.** $18 - 13 = 5$; $5\frac{7}{12}$

35. $74 - 56 = 18$; $17\frac{5}{6}$ **36.** $215 - 136 = 79$; $79\frac{7}{16}$

37. $5\frac{3}{4}$ **38.** $6\frac{1}{12}$ **39.** $6\frac{4}{15}$ **40.** $2\frac{5}{6}$ **41.** $5\frac{1}{2}$

42. $2\frac{19}{24}$ **43.** $15 - 6 - 7 = 2$ gallons; $2\frac{5}{12}$ gallons

44. $15 + 19 = 34$ tons; $33\frac{5}{12}$ tons

45. $6 + 5 + 5 = 16$ pounds; $15\frac{23}{24}$ pounds

46. $9 - 2 - 3 = 4$ acres; $4\frac{1}{16}$ acres

47.–50.

51. $<$ **52.** $<$ **53.** $>$ **54.** $>$ **55.** $<$

56. $>$ **57.** $<$ **58.** $>$ **59.** $\frac{1}{9}$ **60.** $\frac{9}{16}$

61. $\frac{27}{125}$ **62.** $\frac{81}{4096}$ **63.** $\frac{5}{16}$ **64.** $11\frac{1}{4}$ **65.** $\frac{4}{9}$

66. 2   **67.** $\frac{3}{16}$ **68.** $1\frac{25}{64}$ **69.** $\frac{1}{2}$

70. $\frac{5}{12}$ **71.** $\frac{67}{86}$ **72.** $\frac{11}{16}$ **73.** $3\frac{1}{6}$

74. $26\frac{1}{4}$ **75.** $5\frac{3}{8}$ **76.** $11\frac{43}{80}$ **77.** $65\frac{5}{16}$

78. $\frac{8}{11}$ **79.** $\frac{1}{250}$ **80.** $\frac{1}{2}$ **81.** $\frac{2}{9}$ **82.** $\frac{11}{27}$

83. $>$ **84.** $>$ **85.** $<$ **86.** $>$

87. 72   **88.** 300   **89.** 72   **90.** 18   **91.** 108

92. 180

93. $93 - 14 - 22 = 57$ feet; $56\frac{7}{8}$ feet

94. $10 - 2 - 5 = 3$ liters; $3\frac{1}{8}$ liters

CHAPTER 3 TEST (page 223)

1. $\frac{1}{2}$ **2.** $\frac{3}{5}$ **3.** $\frac{1}{2}$ **4.** $\frac{1}{6}$ **5.** 12   **6.** 105

7. 108   **8.** $\frac{11}{12}$ **9.** $\frac{23}{36}$ **10.** $\frac{5}{24}$ **11.** $\frac{7}{40}$

12. $5 + 7 = 12$; $11\frac{5}{6}$ **13.** $16 - 12 = 4$; $4\frac{11}{15}$

14. $19 + 9 + 12 = 40$; $40\frac{29}{60}$ **15.** $24 - 18 = 6$; $5\frac{5}{8}$

16. Probably addition and subtraction of fractions is more difficult because you have to find the least common denominator and then change the fractions to the same denominator.

17. Round mixed numbers to the nearest whole number. Then add, subtract, multiply, or divide to estimate the answer. The estimate may vary from the exact answer but it lets you know if your answer is reasonable.

ANSWERS

18. $5 + 7 + 3 + 5 + 7 = 27$ hours; $27\frac{1}{4}$ hours

19. $148 - 69 - 37 - 6 = 36$ gallons; $35\frac{7}{8}$ gallons

20. $>$ **21.** $>$ **22.** 3   **23.** $\frac{13}{48}$ **24.** $1\frac{1}{4}$

25. $1\frac{1}{2}$

Cumulative Review 1–3 (page 225)

1. 5; 3   **2.** 2; 5   **3.** 1750; 1700; 2000
4. 59,800; 59,800; 60,000
5. $10,000 + 300 + 50,000 + 50,000 = 110,300$; 113,321
6. $20,000 - 10,000 = 10,000$; 14,389
7. $2000 \times 400 = 800,000$; 878,750
8. $100,000 \div 40 = 2500$; 3211   **9.** 24
10. 1,255,609   **11.** 292   **12.** 2,801,695   **13.** 135
14. 112   **15.** 216   **16.** 476   **17.** 369,408
18. 27,000   **19.** 158   **20.** 2693 R2   **21.** 32 R166
22. $20 + 9 + 5 + 20 + 9 + 5 = 68$ feet; 64 feet
23. $20 \cdot 10 = 200$ square feet; 252 square feet
24. $20,000 \div 30 \approx 667$ cans; 630 cans
25. $2000 \times 50 = 100,000$ revolutions;
90,000 revolutions

26. $2 \cdot 3 = 6$ square yards; $4\frac{2}{3}$ square yards

27. $4 \cdot 5 = 20$ square miles; $18\frac{13}{16}$ square miles

28. $5 \cdot 4 = 20$ cords; $18\frac{3}{8}$ cords

29. $1537 - 83 = 1454$ feet; $1454\frac{3}{8}$ feet

30. $2^2 \cdot 5$ **31.** $2^4 \cdot 3^2$ **32.** $2 \cdot 5^3$ **33.** 128
34. 144   **35.** 432   **36.** 5   **37.** 7   **38.** 12

39. 15   **40.** 44   **41.** $\frac{1}{48}$ **42.** $\frac{9}{10}$ **43.** $1\frac{35}{192}$

44. proper   **45.** improper   **46.** improper   **47.** $\frac{7}{10}$

48. $\frac{19}{25}$ **49.** $\frac{7}{20}$ **50.** $\frac{1}{2}$ **51.** $\frac{5}{22}$ **52.** $36\frac{3}{4}$

53. $1\frac{1}{5}$ **54.** $2\frac{3}{16}$ **55.** $13\frac{1}{2}$ **56.** $\frac{23}{24}$ **57.** $\frac{15}{16}$

58. $\frac{7}{36}$ **59.** $2 + 4 = 6$; $5\frac{7}{8}$

60. $22 + 4 = 26$; $26\frac{7}{24}$ **61.** $5 - 2 = 3$; $2\frac{5}{8}$

62. 36   **63.** 300   **64.** 144   **65.** 35   **66.** 77
67. 81   **68.** 60

69.–72.

70.69.   71.   72.

73. $<$ **74.** $<$ **75.** $<$

CHAPTER 4

Section 4.1 (page 235)

1. 7; 6; 3   **3.** 5; 1; 8   **5.** 4; 7; 0   **7.** 1; 6; 3
9. 1; 8; 9   **11.** 6; 2; 1   **13.** 410.25   **15.** 6.5432
17. 5406.045   **19.** $\frac{7}{10}$ **21.** $13\frac{2}{5}$ **23.** $\frac{7}{20}$
25. $\frac{33}{50}$ **27.** $10\frac{17}{100}$ **29.** $\frac{3}{50}$ **31.** $\frac{41}{200}$
33. $5\frac{1}{500}$ **35.** $\frac{343}{500}$ **37.** five tenths
39. seventy-eight hundredths   **41.** one hundred five thousandths   **43.** twelve and four hundredths
45. one and seventy-five thousandths   **47.** 6.7
49. 0.32   **51.** 420.008   **53.** 0.0703
55. 75.030   **57.** Anne should not say "and" because that denotes a decimal point.   **59.** 3-C   **61.** 4-A
63. One and six hundred two thousandths centimeters
65. millionths, ten-millionths, hundred-millionths, billionths; these match the words on the left side of the chart with "ths" added.   **67.** Seventy-two million four hundred thirty-six thousand nine hundred fifty-five hundred-millionths   **69.** eight thousand six and five hundred thousand one millionths
71. 8240; 8200; 8000
73. 19,710; 19,700; 20,000

Section 4.2 (page 243)

1. ≈ 16.9 **3.** ≈ 0.956 **5.** ≈ 0.80 **7.** ≈ 3.661
9. ≈ 794.0 **11.** ≈ 0.0980 **13.** ≈ 49 **15.** ≈ 9.09
17. ≈ 82.0002 **19.** 0.82 **21.** 1.22 **23.** 0.50
25. $\approx \$17,250$ **27.** $\approx \$310$ **29.** $\approx \$379$
31. Rounds to $0 (zero dollars) because $0.499 is closer to $0 than to $1.   **33.** Round amounts less than $1.00 to nearest cent instead of nearest dollar.   **35.** $\approx \$500$
37. $\approx \$1.00$ **39.** $\approx \$1000$
41. $8000 + 6000 + 8000 = 22,000$; 22,223
43. $80,000 + 100 + 800 = 80,900$; 82,859

Section 4.3 (page 247)

1. 17.72   **3.** 8321.412   **5.** 348.513   **7.** 11.98
9. 115.861   **11.** 59.323   **13.** 330.86895
15. 6 should be written 6.00; sum is 46.22.
17. $40 + 20 + 8 = 68$; 63.65
19. $400 + 1 + 20 = 421$; 414.645
21. $60 + 500 + 6 = 566$; 608.4363
23. $400 + 600 + 600 = 1600$; 1586.308
25. $\$300 + \$1 = \$301$; $311.09
27. $5 + 6 + 4 = 15$ days; 14.49 days
29. $\$7 + \$10 + \$1 = \$18$; $18.20
31. $8000 + 200 + 200 = 8400$ miles; 8251.7 miles
33. $10 + 5 = 15$ hours; 14.3 hours
35. $5 + 6 + 5 + 10 + 5 = 31$ hours; 30 hours

37. $5000 + $700 = $5700; $5651.64

39. $0.3000 = \dfrac{3000 \div 1000}{10,000 \div 1000} = \dfrac{3}{10} = 0.3$

41. 90.3 miles **43.** 19.486
45. 20 + 6 + 20 + 6 = 52 inches; 52.1 inches
47. 300 − 100 = 200; 197
49. 7000 − 100 = 6900; 6569

Section 4.4 (page 253)

1. 54.3 **3.** 32.566 **5.** 38.554 **7.** 20.104
9. 12.848 **11.** 89.7 **13.** 0.109 **15.** 0.91
17. 6.661 **19.** 15.32 should be on top; correct answer
is 7.87. **21.** $20 − $7 = $13; $13.16
23. 9 − 4 = 5; 4.849 **25.** 2 − 2 = 0; 0.019 inch
27. 400 − 9 = 391; 375.194 liters **29.** 0.275
31. 6.493 **33.** 1.81 **35.** 5951.2858
37. 40 − 20 = 20 hours; 26.15 hours
39. $20 − $9 = $11; $10.88 **41.** 11 − 10 = 1 ounce;
0.65 ounce **43.** $1105.75 **45.** $94.97
47. $219.22 **49.** $75.53
51. 20 + 1 = 21; 21 − 9 = 12 gallons; 9.372 gallons
53. $b = 1.39$ cm **55.** $k = 2.812$ inches
57. 80 × 30 = 2400; 2324
59. 4000 × 200 = 800,000; 776,745

Section 4.5 (page 259)

1. 0.1344 **3.** 159.10 **5.** 15.5844 **7.** $34,500.20
9. 43.2 **11.** 0.432 **13.** 0.0432 **15.** 0.00432
17. 0.0000312 **19.** 0.000006 **21.** Multiplying by
10, decimal point moves one place to the right; by 100, two
places to the right; by 1000, three places to the right.
23. 40 × 5 = 200; 190.08 **25.** 40 × 40 = 1600;
1558.2 **27.** 7 × 5 = 35; 30.038
29. 3 × 7 = 21; 19.24165
31. unreasonable; $189.00 **33.** reasonable
35. unreasonable; $3.19 **37.** unreasonable; 9.5 pounds
39. ≈$592.37 **41.** ≈$2.45 **43.** ≈$27.04
45. $8771.00 **47.** $347.52 **49.** $76.50
51. $121.30 **53.** 46.3 gallons **55.** $388.34
57. ≈$4.09
59. 1000 ÷ 5 = 200; 190 R4
61. 20,000 ÷ 20 = 1000; 905 R15

Section 4.6 (page 269)

1. 3.9 **3.** 0.47 **5.** 400.2 **7.** 36
9. 0.06 **11.** 6000 **13.** 25.3 **15.** ≈516.67
17. ≈24.291 **19.** ≈10,082.647 **21.** Dividing by
10, decimal point moves one place to the left; by 100, two
places to the left; by 1000, three places to the left.
23. unreasonable; 40 ÷ 8 = 5; $8\overline{)37.8}$ (4.725)
25. reasonable; 50 ÷ 50 = 1
27. unreasonable; 300 ÷ 5 = 60; $5.1\overline{)307.02}$ (60.2)

29. unreasonable; 9 ÷ 1 = 9; $1.25\overline{)9.3}$ (7.44)
31. ≈$4.00 **33.** $19.46 **35.** $0.30
37. $5.89 per hour **39.** ≈21.2 miles per gallon
41. ≈8.30 meters **43.** 0.05 meter
45. 25.03 meters **47.** 14.25 **49.** 73.4
51. 1.205 **53.** 0.334 **55.** ≈$0.03 **57.** $237.25
59. < **61.** > **63.** <

Section 4.7 (page 277)

1. 0.5 **3.** 0.75 **5.** 0.3 **7.** 0.9 **9.** 0.6
11. 0.875 **13.** 2.25 **15.** 14.7 **17.** 3.625
19. ≈0.333 **21.** ≈0.833 **23.** ≈1.889
25. $\dfrac{5}{9}$ means 5 ÷ 9 or $9\overline{)5}$ so correct answer is ≈0.556.
27. Just add the whole number part to 0.375.
So $1\dfrac{3}{8} = 1.375$; $3\dfrac{3}{8} = 3.375$; $295\dfrac{3}{8} = 295.375$.
29. $\dfrac{2}{5}$ **31.** $\dfrac{5}{8}$ **33.** $\dfrac{7}{20}$ **35.** 0.35 **37.** $\dfrac{1}{25}$
39. $\dfrac{3}{20}$ **41.** 0.2 **43.** $\dfrac{9}{100}$
45. Too much; 0.005 gram **47.** Shorter; 0.72 inch
49. 0.9991 cm, 1.0007 cm **51.** More; 0.05 inch
53. 0.5399, 0.54, 0.5455 **55.** 5.0079, 5.79, 5.8, 5.804
57. 0.6009, 0.609, 0.628, 0.62812
59. 2.8902, 3.88, 4.876, 5.8751
61. 0.006, 0.043, $\dfrac{1}{20}$, 0.051 **63.** 0.37, $\dfrac{3}{8}$, $\dfrac{2}{5}$, 0.4001
65. ≈1.4 in. **67.** ≈0.3 in. **69.** ≈0.4 in.
71. $\dfrac{6}{11}$, $\dfrac{5}{9}$, 0.571, $\dfrac{4}{7}$ **73.** 0.25, $\dfrac{4}{15}$, $\dfrac{3}{11}$, $\dfrac{1}{3}$
75. $\dfrac{1}{6}$, $\dfrac{3}{16}$, 0.188, $\dfrac{1}{5}$
77. $\dfrac{3}{4}$ **79.** $\dfrac{3}{4}$ **81.** $\dfrac{8}{11}$

Chapter 4 Review Exercises (page 287)

1. 0; 5 **2.** 0; 6 **3.** 8; 9 **4.** 5; 9 **5.** 7; 6
6. $\dfrac{1}{2}$ **7.** $\dfrac{3}{4}$ **8.** $4\dfrac{1}{20}$ **9.** $\dfrac{7}{8}$ **10.** $\dfrac{27}{1000}$
11. $27\dfrac{4}{5}$ **12.** eight tenths
13. four hundred and twenty-nine hundredths
14. twelve and seven thousandths
15. three hundred six ten-thousandths
16. 8.3 **17.** 0.205 **18.** 70.0066 **19.** 0.30
20. ≈275.6 **21.** ≈72.79 **22.** ≈0.160
23. ≈0.091 **24.** ≈1.0 **25.** ≈$15.83
26. ≈$0.70 **27.** ≈$17,625.79 **28.** ≈$350
29. ≈$130 **30.** ≈$100 **31.** ≈$29
32. 6 + 400 + 20 = 426; 444.86
33. 80 + 1 + 100 + 1 + 30 = 212; 233.515
34. 300 − 20 = 280; 290.7 **35.** 9 − 8 = 1; 1.2684
36. 13 − 10 = 3 hours; 2.75 hours

ANSWERS

37. $200 + $40 = $240; $260.00
38. $2 + $5 + $20 = $27; $30 − $27 = $3; $4.14
39. 2 + 4 + 5 = 11 kilometers; 11.55 kilometers
40. 6 × 4 = 24; 22.7106   41. 40 × 3 = 120; 141.57
42. 0.0112   43. 0.000355
44. reasonable; 700 ÷ 10 = 70

45. unreasonable; 30 ÷ 3 = 10; $2.8\overline{)26.6}^{\,9.5}$
46. ≈14.467   47. 1200   48. 0.4   49. ≈$350
50. ≈$1.99   51. ≈133 shares   52. ≈$3.12
53. 29.215   54. 10.15   55. 3.8   56. 0.64
57. 1.875   58. ≈0.111   59. 3.6008, 3.68, 3.806

60. 0.209, 0.2102, 0.215, 0.22   61. $\frac{1}{8}, \frac{3}{20}$, 0.159, 0.17

62. 404.865   63. 254.8   64. ≈3583.261
65. 29.0898   66. 0.03066   67. 9.4   68. 175.675
69. 9.04   70. 19.50   71. 8.19   72. 0.928
73. 35   74. 0.259   75. 0.3   76. ≈$3.00
77. ≈$2.17   78. $35.96   79. $199.71
80. $78.50

CHAPTER 4 TEST   (page 291)

1. $18\frac{2}{5}$ 2. $\frac{3}{40}$ 3. sixty and seven thousandths
4. two hundred eight ten-thousandths   5. ≈725.6
6. ≈0.630   7. ≈$1.49   8. ≈$7860
9. 8 + 80 + 40 = 128; 129.2028   10. 80 − 4 = 76; 75.498   11. 6 • 1 = 6; 6.948
12. 20 ÷ 5 = 4; 4.175   13. 839.762   14. 669.004
15. 0.0000483   16. 480   17. 2.625
18. 0.44, $\frac{9}{20}$, 0.4506, 0.451   19. 35.49
20. $446.87   21. Davida, by 0.441 minute
22. ≈$5.35   23. 2.8 degrees   24. $4.55 per meter
25. Answer varies.

CUMULATIVE REVIEW 1–4   (page 293)

1. 5, 9, 2   2. 0, 5, 8   3. ≈500,000   4. ≈602.49
5. ≈$710   6. ≈$0.05
7. 4000 + 600 + 9000 = 13,600; 13,339
8. 4 + 20 + 1 = 25; 20.683
9. 5000 − 2000 = 3000; 3209
10. 50 − 7 = 43; 44.506
11. 3000 × 200 = 600,000; 550,622
12. 7 × 7 = 49; 49.786
13. 100,000 ÷ 50 = 2000; 2690
14. 40 ÷ 8 = 5; 4.5   15. 2, 4, 8; $7\frac{1}{8}$
16. 2, 1, 2; $2\frac{4}{5}$ 17. 2, 2, 4; $3\frac{7}{15}$ 18. 5, 2, 3; $2\frac{5}{8}$
19. 9.671   20. $1\frac{4}{9}$ 21. 73,225   22. $1\frac{2}{5}$
23. 4914   24. 93.603   25. 404 R3   26. $1\frac{17}{24}$
27. 233,728   28. 0.03264   29. 8   30. 45
31. $\frac{4}{31}$ 32. $\frac{2}{3}$ 33. ≈0.51   34. 4   35. 14

36. $\frac{1}{4}$ 37. 20.81   38. 576   39. 14
40. $2^3 \cdot 5^2$ 41. forty and thirty-five thousandths
42. 0.0306   43. $\frac{1}{8}$ 44. $3\frac{2}{25}$ 45. 2.6
46. ≈0.636   47. >   48. 7.005, 7.5, 7.5005, 7.505
49. 0.8, 0.8015, $\frac{21}{25}, \frac{7}{8}$ 50. Too short, by 0.045 meter
51. 84 children   52. $20 − $8 − $1 = $11; $11.17
53. 50 − 47 = 3 inches; $3\frac{3}{8}$ inches
54. $9 × 17 = $153; ≈$144.05
55. (8 × 20) + (10 × 30) = 160 + 300 = 460 students; 488 students   56. 2 + 4 = 6 yards; $6\frac{5}{24}$ yards
57. $30 + $200 − $40 − $20 = $170; $191.50
58. $3 ÷ 3 = $1 per pound; ≈$0.95 per pound
59. $80,000 ÷ 100 = $800; ≈$729

CHAPTER 5

SECTION 5.1 (page 303)

1. $\frac{8}{9}$ 3. $\frac{2}{1}$ 5. $\frac{1}{3}$ 7. $\frac{8}{5}$ 9. $\frac{3}{8}$
11. $\frac{9}{7}$ 13. $\frac{6}{1}$ 15. $\frac{5}{6}$ 17. $\frac{8}{5}$ 19. $\frac{1}{12}$
21. $\frac{5}{16}$ 23. $\frac{4}{1}$ 25. $\frac{22}{3}$ 27. $\frac{5}{4}$ 29. $\frac{1}{100}$
31. $\frac{11}{1}$ 33. $\frac{3}{10}$ 35. A ratio of 3 to 1 means your income is 3 times your friend's income.   37. $\frac{2}{1}$
39. $\frac{3}{8}$ 41. $\frac{7}{5}$ 43. $\frac{6}{1}$ 45. $\frac{38}{17}$
47. $\frac{1}{2}$ 49. $\frac{34}{35}$
51. Answer varies. Some possibilities are:
$$\frac{4}{5} = \frac{8}{10} = \frac{12}{15} = \frac{16}{20} = \frac{20}{25} = \frac{24}{30} = \frac{28}{35}$$
53. ≈0.093   55. 1.025   57. ≈9.465

SECTION 5.2 (page 311)

1. $\frac{5 \text{ cups}}{3 \text{ people}}$ 3. $\frac{3 \text{ feet}}{7 \text{ seconds}}$ 5. $\frac{1 \text{ person}}{2 \text{ dresses}}$
7. $\frac{5 \text{ letters}}{1 \text{ minute}}$ 9. $\frac{\$21}{2 \text{ visits}}$ 11. $\frac{18 \text{ miles}}{1 \text{ gallon}}$
13. $12 per hour or $12/hour
15. 5 eggs per chicken or 5 eggs/chicken
17. 1.25 pounds/person   19. $103.30/day
21. 325.9; ≈21.0   23. 338.6; ≈20.9
25. 4 oz for $0.89   27. 17 ounces for $2.89
29. 18 ounces for $1.41   31. You might choose Brand B because you like more chicken, so the cost per chicken chunk may actually be the same or less than Brand A.   33. 1.75 pounds/week   35. $12.26/hour

37. $11.50/share　**39.** 0.1 second/meter or $\frac{1}{10}$ second/meter; 10 meters/second　**41.** $11.25/yard

43. One battery for $1.79; like getting 3 batteries so $1.79 \div 3 \approx \$0.597$ per battery

45. Brand P with the 50¢ coupon is the best buy. $3.39 - \$0.50 = \$2.89 \div 16.5$ ounces $\approx \$0.175$/ounce

47. 11　**49.** 108　**51.** $3\frac{1}{2}$

SECTION 5.3 (page 317)

1. $\dfrac{\$9}{12 \text{ cans}} = \dfrac{\$18}{24 \text{ cans}}$　**3.** $\dfrac{200 \text{ adults}}{450 \text{ children}} = \dfrac{4 \text{ adults}}{9 \text{ children}}$

5. $\dfrac{120}{150} = \dfrac{8}{10}$　**7.** $\dfrac{2.2}{3.3} = \dfrac{3.2}{4.8}$　**9.** $\dfrac{1\frac{1}{2}}{4\frac{1}{2}} = \dfrac{6}{18}$

11. true　**13.** true　**15.** false　**17.** true
19. true　**21.** false　**23.** True　**25.** False
27. False　**29.** True　**31.** False　**33.** True
35. True

37.　$\dfrac{16 \text{ hits}}{50 \text{ at bats}} = \dfrac{128 \text{ hits}}{400 \text{ at bats}}$　$\begin{array}{l} 50 \cdot 128 = 6400 \\ 16 \cdot 400 = 6400 \end{array}$

Cross products are equal so the proportion is true; they hit equally well.
39. False

41. $\dfrac{4}{5}$　**43.** $1\frac{1}{4}$　**45.** $\dfrac{4}{7}$　**47.** $6\frac{1}{2}$

SECTION 5.4 (page 325)

1. 4　**3.** 2　**5.** 88　**7.** 91　**9.** 5
11. 10　**13.** ≈ 24.44　**15.** 50.4　**17.** ≈ 17.64
19. $\dfrac{6.67}{4} = \dfrac{5}{3}$ or $\dfrac{10}{6} = \dfrac{5}{3}$ or $\dfrac{10}{4} = \dfrac{7.5}{3}$ or $\dfrac{10}{4} = \dfrac{5}{2}$

21. 1　**23.** $3\frac{1}{2}$　**25.** 0.2 or $\dfrac{1}{5}$

27. 0.005 or $\dfrac{1}{200}$

29. $\dfrac{25 \text{ feet}}{18 \text{ sec}} = \dfrac{15 \text{ feet}}{10 \text{ sec}}$; False

31. $\dfrac{170 \text{ miles}}{6.8 \text{ gallons}} = \dfrac{330 \text{ miles}}{13.2 \text{ gallons}}$; True

SECTION 5.5 (page 329)

1. 22.5 hours　**3.** $7.20　**5.** 7 pounds
7. $153.45　**9.** 14 feet, 10 feet　**11.** 14 feet, 8 feet
13. 3 runs　**15.** 665 students (reasonable); ≈ 1357 students with incorrect setup (only 950 students in the group).　**17.** ≈ 190 people (reasonable); ≈ 298 people with incorrect setup (only 238 people attended).　**19.** $5.50　**21.** 625 stocks
23. ≈ 4.6 hours　**25.** ≈ 4.06 meters
27. ≈ 10.53 meters

29. You cannot solve this problem using a proportion because the ratio of age to weight is not constant. As Jim's age increases, his weight may decrease, stay the same, or increase.

31. $4\frac{3}{8}$ cups　**33.** $7\frac{1}{2}$ tbsp　**35.** 3800 students
37. 6　**39.** 2870　**41.** 0.0193

CHAPTER 5 REVIEW EXERCISES (page 337)

1. $\dfrac{3}{11}$　**2.** $\dfrac{19}{7}$　**3.** $\dfrac{3}{2}$　**4.** $\dfrac{9}{5}$　**5.** $\dfrac{2}{1}$
6. $\dfrac{2}{3}$　**7.** $\dfrac{5}{2}$　**8.** $\dfrac{1}{6}$　**9.** $\dfrac{3}{1}$　**10.** $\dfrac{3}{8}$
11. $\dfrac{4}{3}$　**12.** $\dfrac{1}{9}$　**13.** $\dfrac{10}{7}$　**14.** $\dfrac{7}{5}$　**15.** $\dfrac{5}{6}$
16. $\dfrac{5}{9}$　**17.** $\dfrac{\$11}{1 \text{ dozen}}$　**18.** $\dfrac{12 \text{ children}}{5 \text{ families}}$

19. Both compare two things. In a ratio the common units cancel, but in a rate the units are different and must be written. Examples are:

(ratio) $\dfrac{5 \text{ feet}}{10 \text{ feet}} = \dfrac{1}{2}$　$\dfrac{55 \text{ miles}}{1 \text{ hour}}$ (rate)

20. A unit rate has 1 in the denominator. Examples are 55 miles in 1 hour, $440 in 1 week, or 30 miles on 1 gallon of gas. We usually write them using "per" or a slash mark: 55 miles per hour, etc.

21. 0.2 page/minute or $\dfrac{1}{5}$ page/minute; 5 minutes/page

22. $8/hour; 0.125 hour/dollar or $\dfrac{1}{8}$ hour/dollar

23. 13 ounces for $2.29

24. 25 pounds for $10.40 − $1 coupon

25. $\dfrac{5}{10} = \dfrac{20}{40}$　**26.** $\dfrac{7}{2} = \dfrac{35}{10}$　**27.** $\dfrac{1\frac{1}{2}}{6} = \dfrac{2\frac{1}{4}}{9}$
28. true　**29.** false　**30.** false　**31.** true
32. true　**33.** true　**34.** 1575　**35.** 20
36. 400　**37.** 12.5　**38.** ≈ 14.67　**39.** ≈ 8.17
40. 50.4　**41.** ≈ 0.57　**42.** ≈ 2.47　**43.** 27 cats
44. 46 hits　**45.** $\approx \$22.06$　**46.** ≈ 3299 students
47. 68 feet　**48.** 14.7 milligrams　**49.** 80 hours
50. $27\frac{1}{2}$ hours or 27.5 hours
51. 105　**52.** 0　**53.** 128　**54.** ≈ 23.08
55. 6.5　**56.** ≈ 117.36　**57.** False　**58.** False
59. True　**60.** $\dfrac{8}{5}$　**61.** $\dfrac{33}{80}$　**62.** $\dfrac{15}{4}$　**63.** $\dfrac{4}{1}$
64. $\dfrac{4}{5}$　**65.** $\dfrac{37}{7}$　**66.** $\dfrac{3}{8}$　**67.** $\dfrac{1}{12}$　**68.** $\dfrac{45}{13}$
69. $\approx 24,900$ fans　**70.** $\dfrac{8}{3}$
71. 75 feet for $1.99 − $0.50 coupon　**72.** 16.5 feet
73. $\dfrac{1}{2}$ teaspoon or 0.5 teaspoon　**74.** ≈ 21 points

75. Set up the proportion to compare teaspoons to pounds on both sides.

$$\frac{1.5 \text{ tsp}}{24 \text{ pounds}} = \frac{x \text{ tsp}}{8 \text{ pounds}}$$

Show that cross products are equal.

$$24 \cdot x = 1.5 \cdot 8$$

Divide both sides by 24.

$$\frac{\overset{1}{\cancel{24}} \cdot x}{\underset{1}{\cancel{24}}} = \frac{12}{24} \quad \text{so } x = \frac{1}{2} \text{ tsp or } 0.5 \text{ tsp}$$

76. 7.5 hours or $7\frac{1}{2}$ hours

CHAPTER 5 TEST (page 341)

1. $\frac{4}{5}$ **2.** $\frac{20 \text{ miles}}{1 \text{ gallon}}$ **3.** $\frac{\$1}{5 \text{ minutes}}$ **4.** $\frac{15}{4}$

5. $\frac{1}{80}$ **6.** $\frac{9}{2}$ **7.** 18 ounces for $1.89 − $0.25 coupon

8. You earned less this year. An example is:

$$\frac{\text{Last year} \to \$15,000}{\text{This year} \to \$10,000} = \frac{3}{2}$$

9. False **10.** True **11.** 25 **12.** ≈2.67

13. 325 **14.** $10\frac{1}{2}$ **15.** 576 words

16. 6.4 hours **17.** ≈87 students
18. No, 4875 cannot be correct because there are only 650 students in the whole school.
19. ≈23.8 grams **20.** 60 feet

CUMULATIVE REVIEW 1–5 (page 343)

1. 5; 3; 6 **2.** 9; 0; 5 **3.** ≈9900 **4.** ≈617.1
5. ≈$100 **6.** ≈$3.06
7. 30 + 5000 + 400 = 5430; 5585
8. 60 − 6 = 54; 57.408
9. 5000 × 800 = 4,000,000; 3,791,664
10. 1 × 18 = 18; 17.4796
11. 50,000 ÷ 50 = 1000; 907
12. 2000 ÷ 5 = 400; 364
13. 2 · 4 = 8; $6\frac{3}{5}$ **14.** 5 ÷ 1 = 5; 6
15. 3 − 2 = 1; $\frac{29}{30}$ **16.** 3 + 11 = 14; $13\frac{2}{5}$
17. 374,416 **18.** 29.34 **19.** 610 R27 **20.** 0.0076
21. 2312 **22.** 68.381 **23.** 55.6 **24.** 35,852,728
25. 39 **26.** 18 **27.** 64 **28.** 0.95
29. one hundred five ten-thousandths **30.** 60.071
31. 0.313 **32.** 4.778 **33.** 0.07, 0.0711, 0.7, 0.707
34. 0.305, $\frac{1}{3}$, $\frac{7}{20}$, $\frac{3}{8}$ **35.** $\frac{4}{1}$ **36.** $\frac{\$13}{2 \text{ hours}}$ **37.** $\frac{1}{12}$
38. $\frac{1}{3}$ **39.** $\frac{11}{5}$

40. 36 servings for $3.24 − $0.50 coupon **41.** 21
42. ≈17.14 **43.** $11\frac{1}{4}$ **44.** ≈0.98
45. 250 pounds **46.** ≈26.7 centimeters
47. 2000 + 2000 + 2000 + 2000 = 8000 students; 8400 students **48.** $200,000 ÷ 2000 = $100; ≈$78
49. $\frac{2}{7}$ **50.** $4 · 8000 = $32,000; $31,500

51. 4 + 3 = 7; 7 · 1 = 7 miles; $7\frac{3}{20}$ miles

52. 900 miles ÷ 50 gallons = 18 miles per gallon; ≈18.0 miles per gallon **53.** 200 residents
54. $1\frac{1}{4}$ teaspoons

CHAPTER 6

SECTION 6.1 (page 351)

1. 0.25 **3.** 0.30 or 0.3 **5.** 0.55 **7.** 1.40 or 1.4
9. 0.078 **11.** 1.00 or 1 **13.** 0.005 **15.** 0.0035
17. 50% **19.** 62% **21.** 3% **23.** 12.5%
25. 62.9% **27.** 200% **29.** 260% **31.** 3.12%
33. 416.2% **35.** 0.17%
37. Possible answers:
No common denominators are needed with percents. The denominator is always 100 with percent which makes comparisons easier to understand.
39. 0.08 **41.** 0.65 **43.** 3.5% **45.** 200%
47. 0.5% **49.** 1.536 **51.** $78 **53.** 20 children
55. 90 miles **57.** $142.50 **59.** 4100 students
61. Since 100% means 100 parts out of 100 parts, 100% is all of the number.
63. 95%; 5% **65.** 30%; 70% **67.** 75%; 25%
69. 55%; 45% **71.** 0.4 **73.** 0.75 **75.** 0.875
77. 0.8

SECTION 6.2 (page 361)

1. $\frac{1}{5}$ **3.** $\frac{1}{2}$ **5.** $\frac{11}{20}$ **7.** $\frac{3}{8}$ **9.** $\frac{1}{16}$ **11.** $\frac{1}{6}$

13. $\frac{1}{15}$ **15.** $\frac{1}{200}$ **17.** $1\frac{3}{10}$ **19.** $2\frac{1}{2}$

21. 25% **23.** 30% **25.** 60% **27.** 37%
29. 62.5% **31.** 87.5% **33.** 44% **35.** 46%
37. 5% **39.** ≈83.3% **41.** ≈55.6%

43. ≈14.3% **45.** $\frac{3}{4}$, 75% **47.** $\frac{7}{8}$, 0.875

49. $\frac{4}{5}$, 80% **51.** ≈0.167, ≈16.7%

53. $\frac{1}{4}$, 25% **55.** $\frac{1}{8}$, 0.125

57. ≈0.667, ≈66.7% **59.** 0.1, 10%
61. 0.01, 1% **63.** 0.005, 0.5%

65. $2\frac{1}{2}$, 250% **67.** 3.25, 325%

69. There are many possible answers. Examples 2 and 3 show the steps that students should include in their answers.

71. $\frac{19}{25}$; 0.76; 76%　　**73.** $\frac{1}{5}$; 0.2; 20%

75. $\frac{3}{5}$; 0.6; 60%　　**77.** $\frac{1}{5}$; 0.20; 20%

79. $\frac{1}{10}$; 0.1; 10%　　**81.** $\frac{1}{4}$; 0.25; 25%

83. $\frac{2}{5}$; 0.4; 40%　　**85.** 30　　**87.** 5　　**89.** 10

SECTION 6.3 (page 369)

1. 50　　**3.** 120　　**5.** 80　　**7.** ≈416.7　　**9.** 50%
11. 300%　　**13.** ≈33.3%　　**15.** 26　　**17.** 21.6
19. ≈13.2
21. One answer is:

```
            5 · 8 = 40 ←              3 · 12 = 36 ←
  2     8                   2     12
  — =  —                    — =  —
  5     20                  3     15
            2 · 20 = 40 ←              2 · 15 = 30 ←
            True                       False
```

23. 356　　**25.** 0.4%　　**27.** 2.5%　　**29.** 0.3%
31. 50%　　**33.** 57%　　**35.** 87.5%　　**37.** ≈66.7%

SECTION 6.4 (page 373)

1. 20, unknown, 80　　**3.** 80, 950, 760
5. 15, 75, unknown　　**7.** 36, unknown, 9
9. 50, 68, 34　　**11.** unknown, 296, 177.6
13. 6.5, unknown, 27.17　　**15.** 0.68, 487, unknown
17. p (percent)—the ratio of the part to the whole. It appears with the word "percent" or "%" after it. b (base)—the entire quantity or total. Often appears after the word "of." a (amount)—the part being compared with the whole.
19. p is unknown; 810; 640
21. p is unknown; 15; 0.90　　**23.** 23; 610; a is unknown
25. 25; b is unknown; 240　　**27.** 78; 650; a is unknown
29. 32; 272; a is unknown　　**31.** 16.8; b is unknown; 504

33. $\frac{8}{x} = \frac{15}{30}$, 16　　**35.** $\frac{r}{36} = \frac{\frac{4}{3}}{12}$, 4

SECTION 6.5 (page 383)

1. 52 bowlers　　**3.** 819 volunteers　　**5.** 4.8 feet
7. 135 folders　　**9.** 247.5 boxes
11. $3.28　　**13.** 1850 sales　　**15.** 42.625 pounds

17. $21.60　　**19.** 160 printers　　**21.** 170
23. 550 students　　**25.** 680 books　　**27.** 2800
29. 50%　　**31.** 52%　　**33.** 8%　　**35.** 1.5%
37. ≈18.6%　　**39.** 9.2%　　**41.** 150% of $30 cannot be less than $30 because 150% is greater than 1 (100%). The answer must be greater than $30. 25% of $16 cannot be greater than $16 because 25% is less than 1 (100%). The answer must be less than $16.
43. $19.80　　**45.** 836 drivers　　**47.** 220 students
49. $22,620.40　　**51.** ≈15.5%　　**53. (a)** 39% female **(b)** ≈2.3 million male workers
55. 110%　　**57.** 248.0 million　　**59.** March
61. 52.5 million cans　　**63.** $645
65. 2156 products　　**67.** 19.7808　　**69.** 9.1854
71. 3600　　**73.** 0.25

SECTION 6.6 (page 393)

1. 231 programs　　**3.** 845 species
5. 83.2 quarts　　**7.** 700 tablets　　**9.** 1029.2 meters
11. $4.16　　**13.** 284 employees
15. 325 salads　　**17.** 680 circuits
19. 1080 people　　**21.** 300 gallons
23. 50%　　**25.** 76%　　**27.** 125%　　**29.** 1.5%
31. 135%　　**33.** You must first change the fraction in the percent to a decimal, then divide the percent by 100 to change it to a decimal.

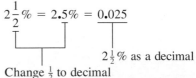

35. ≈82.3 million homes　　**37.** 924 employers
39. ≈4.9%　　**41.** ≈$1817.2 million
43. ≈478,175 Mustangs　　**45.** $510,390
47. ≈$477.32　　**49.** 23, 500, 115
51. 18, unknown, 72　　**53.** unknown, 830, 128.65

SECTION 6.7 (page 401)

1. $6, $106　　**3.** 3%, $70.04　　**5.** $21.90, $386.90　　**7.** $12.10, $232.10　　**9.** $21
11. 25%　　**13.** $173.49　　**15.** $4525　　**17.** $15, $85　　**19.** 30%, $126　　**21.** $4.38, $13.12
23. $3.75, $33.75　　**25.** On the basis of commission alone I would choose Company A. Other considerations might be: reputation of the company; expense allowances; other fringe benefits; travel; promotion and training, to name a few.　　**27.** $52.50　　**29.** $2310
31. 5%　　**33.** ≈18.0%　　**35.** ≈22.5%
37. $74.25　　**39.** 6%　　**41.** $106.20; $483.80
43. ≈5.6%　　**45.** ≈$18.43　　**47.** ≈$4276.97
49. ≈$11,943.70　　**51.** 78 brackets　　**53.** $3.84
55. 800　　**57.** 32%

ANSWERS

SECTION 6.8 (page 409)

1. $8   **3.** $144   **5.** $28.80   **7.** $488.75
9. $763.75   **11.** $6   **13.** $37.50   **15.** $131.20
17. ≈$17.19   **19.** ≈$634.38
21. $312   **23.** $773.30   **25.** $1725   **27.** $2535.75
29. $18,801.25
31. The answer should include:
Amount of principal—This is the amount of money borrowed or loaned.
Interest rate—This is the percent used to calculate the interest.
Time of loan—The length of time that money is loaned or borrowed is an important factor in determining interest.
33. $291   **35.** $6750   **37.** $1025   **39.** $323.40
41. $159.50   **43.** ≈$12,254.69   **45.** 8   **47.** 48
49. 25   **51.** 60

SECTION 6.9 (page 419)

1. $1102.50   **3.** $562.43   **5.** $4587.79
7. ≈$899.89   **9.** ≈$1391.13   **11.** ≈$2027.46
13. ≈$16,044.61   **15.** $1262.50; $262.50
17. $6205.20; $2205.20
19. ≈$10,976.01; $2547.84   **21.** Interest paid on past interest as well as on the principal.
Many people describe compound interest as "interest on interest."
23. ≈$7757.90   **25.** (a) $11,901.75   (b) $4401.75
27. (a) ≈$37,202.08   (b) ≈$7202.08   **29.** 9.3
31. 106.3   **33.** 0.14

CHAPTER 6 REVIEW EXERCISES (page 427)

1. 0.25   **2.** 1.8   **3.** 0.125   **4.** 0.00085
5. 265%   **6.** 2%   **7.** 87.5%   **8.** 0.2%   **9.** $\frac{3}{25}$
10. $\frac{3}{8}$ **11.** $2\frac{1}{2}$ **12.** $\frac{1}{400}$ **13.** 75%   **14.** 62.5%
or $62\frac{1}{2}$%   **15.** 325%   **16.** 0.5%   **17.** 0.125
18. 12.5%   **19.** $\frac{3}{20}$ **20.** 15%   **21.** $1\frac{4}{5}$
22. 1.8   **23.** 1000   **24.** 96   **25.** 40; 150; 60
26. unknown; 90; 73   **27.** 28; 320; unknown
28. 32; unknown; 209   **29.** unknown; 8; 3
30. 88; 1280; unknown   **31.** 54 telephones
32. 870 reference books   **33.** 43.2 miles
34. 2.8 kilograms   **35.** 500 athletes
36. 1160 capsules   **37.** 242 meters
38. 17,000 cases   **39.** 50%   **40.** ≈4.1%
41. ≈9.5%   **42.** ≈30.8%
43. 0.6 million or 600,000 patients   **44.** ≈10.4%
45. $25.96   **46.** 80 dumpsters   **47.** 0.4%
48. 175%   **49.** 120 miles   **50.** $575
51. $8.40, $218.40   **52.** $7\frac{1}{2}$%, $838.50
53. $308   **54.** 5%   **55.** $3.75, $33.75
56. 25%, $189   **57.** $5   **58.** $72   **59.** $12
60. $144   **61.** $397.25   **62.** $1598.85

63. $3257.80, $1257.80   **64.** ≈$1861.55, $331.55
65. $4534.92, $934.92   **66.** $16,170.90, $4770.90
67. 12   **68.** 1640   **69.** 23.28 meters
70. 150%   **71.** $0.51   **72.** 495 students
73. 40%   **74.** 249.4 liters   **75.** 0.75
76. 2   **77.** 400%   **78.** 471%   **79.** 0.062
80. 62.1%   **81.** 0.00375   **82.** 0.06%   **83.** 25%
84. $\frac{19}{50}$ **85.** $\frac{5}{8}$ **86.** 37.5% or $37\frac{1}{2}$%
87. $\frac{13}{40}$ **88.** 20%   **89.** $\frac{1}{200}$
90. 225%   **91.** $351.45   **92.** $1960.20
93. ≈32.4%   **94.** (a) ≈$15,780.96   (b) $3280.96
95. $3750   **96.** ≈8.1%   **97.** ≈13.1%
98. 5,000,000 cars   **99.** $8823   **100.** ≈$396.05

CHAPTER 6 TEST (page 433)

1. 0.75   **2.** 60%   **3.** 180%   **4.** 87.5% or $87\frac{1}{2}$%
5. 3.00 or 3   **6.** 0.0005   **7.** $\frac{5}{8}$ **8.** $\frac{1}{400}$
9. 25%   **10.** 62.5% or $62\frac{1}{2}$%   **11.** 175%
12. 320 files   **13.** 20%   **14.** $19,500
15. $2854.20   **16.** $249.60   **17.** 35%
18. A possible answer is:
Amount is the increase in salary.

$$\text{this year} - \text{last year} = \text{increase}$$

Base is last year's salary. Percent of increase is unknown.

$$\frac{\text{amount of increase}}{\text{last year's salary}} = \frac{p}{100}$$

19. The interest formula is $I = p \cdot r \cdot t$.
If time is in months it is expressed as a fraction with 12 as the denominator.
If time is expressed in years it is placed over 1 or shown as a decimal number.

$$I = \quad p \quad \cdot \quad r \quad \cdot \quad t$$

9 months $1000 \cdot 0.05 \cdot \frac{9}{12} = \$37.50$

$2\frac{1}{2}$ years $1000 \cdot 0.05 \cdot 2.5 = \$125$

20. $3.84; $44.16   **21.** $68.25; $113.75
22. $262.50   **23.** $52   **24.** $4876
25. (a) $10,668   (b) $1668

CUMULATIVE REVIEW 1–6 (page 435)

1. $9000 + 80 + 500 = 9580$; 9334
2. $1 + 40 + 5 = 46$; 42.511
3. $60,000 - 50,000 = 10,000$; 9993
4. $6 - 3 = 3$; 3.4945
5. $7000 \times 700 = 4,900,000$; 4,628,904
6. $70 \times 5 = 350$; 322.2
7. $40,000 \div 40 = 1000$; 902
8. $2000 \div 8 = 250$; 320

9. $7 \div 1 = 7$; 8.45
10. 44   **11.** 31   **12.** 39
13. ≈ 2360 **14.** $\approx 5{,}700{,}000$
15. $\approx \$718$ **16.** $\approx \$451.83$
17. $1\frac{1}{8}$ **18.** $1\frac{1}{6}$ **19.** $13\frac{3}{8}$
20. $\frac{1}{8}$ **21.** $2\frac{5}{6}$ **22.** $8\frac{8}{15}$
23. $\frac{7}{12}$ **24.** $26\frac{5}{32}$ **25.** $28\frac{4}{5}$
26. $\frac{8}{9}$ **27.** 25   **28.** $\frac{11}{30}$
29. <   **30.** <   **31.** >
32. $\frac{1}{12}$ **33.** $1\frac{1}{4}$ **34.** $\frac{1}{4}$
35. 0.625   **36.** 0.8   **37.** 0.85
38. ≈ 0.857 **39.** $\frac{3}{1}$ **40.** $\frac{5}{4}$
41. $\frac{1}{8}$ **42.** True   **43.** True
44. 5   **45.** 3   **46.** 14   **47.** 30
48. 0.25   **49.** 0.07   **50.** 3.00 or 3
51. 0.005   **52.** 56%   **53.** 270%
54. 2.3%   **55.** $\frac{3}{50}$ **56.** $\frac{5}{8}$
57. $1\frac{3}{4}$ **58.** 87.5% or $87\frac{1}{2}\%$ **59.** 5%
60. 350%   **61.** 540 officers   **62.** \$72.25
63. 180 cans   **64.** 1700 miles   **65.** 50%
66. 40%   **67.** $\approx \$2.16$; \$56.15
68. 6.5% or $6\frac{1}{2}\%$; \$489.90
69. \$731.10   **70.** 2.5% or $2\frac{1}{2}\%$
71. \$53.20; \$98.80
72. 15%; \$922.25   **73.** \$842.52
74. $\approx \$19{,}863.88$ **75.** 28 watches
76. 59.5 ounces   **77.** 12.5%   **78.** 2%
79. $\approx 5.5\%$ **80.** $\approx 8.1\%$
81. \$48,000   **82.** (a) $\approx \$7293.04$ (b) \$1293.04

CHAPTER 7
SECTION 7.1 (page 447)

1. 3   **3.** 8   **5.** 5280   **7.** 2000   **9.** 60
11. 2   **13.** 2   **15.** 76,000 to 80,000 pounds
17. 4   **19.** 112   **21.** 10   **23.** $1\frac{1}{2}$ or 1.5
25. $\frac{1}{4}$ or 0.25   **27.** $1\frac{1}{2}$ or 1.5   **29.** $2\frac{1}{2}$ or 2.5
31. $\frac{1}{2}$ day or 0.5 day   **33.** 5000   **35.** 17
37. 44 ounces   **39.** 216   **41.** 28
43. 518,400   **45.** 48,000
47. (a) pound/ounces   (b) quarts/pints or pints/cups
(c) min/hr or sec/min   (d) feet/inches   (e) pounds/tons
(f) days/weeks

49. 174,240   **51.** 800   **53.** 0.75 or $\frac{3}{4}$
55. 1.5 or $1\frac{1}{2}$ **57.** <   **59.** >   **61.** >

SECTION 7.2 (page 457)

1. 1000; 1000   **3.** $\frac{1}{1000}$ or 0.001; $\frac{1}{1000}$ or 0.0001
5. $\frac{1}{100}$ or 0.01; $\frac{1}{100}$ or 0.01
7. answer varies—about 8 cm
9. answer varies—about 20 mm   **11.** cm
13. m   **15.** km   **17.** mm   **19.** cm   **21.** m
23. Examples include 35 mm film for cameras, track and field events, metric auto parts, and lead refills for mechanical pencils.
25. 700 cm   **27.** 0.040 m or 0.04 m   **29.** 9400 m
31. 5.09 m   **33.** 40 cm   **35.** 910 mm
37. less; 18 cm or 0.18 m
39. ─────────── 35 mm = 3.5 cm

──────────────────────
70 mm = 7 cm

41. 0.0000056 km   **43.** $\frac{7}{8}$ **45.** $\frac{2}{25}$

SECTION 7.3 (page 465)

1. mL   **3.** L   **5.** kg   **7.** g   **9.** mL
11. mg   **13.** L   **15.** kg   **17.** unreasonable
19. unreasonable   **21.** reasonable   **23.** reasonable
25. Some examples are 2 liter bottles of soda, shampoo bottles marked in mL, grams of fat listed on cereal boxes, vitamin doses in mg.
27. Unit for your answer (g) is in numerator; unit being changed (kg) is in denominator so it will cancel. The unit fraction is $\frac{1000 \text{ g}}{1 \text{ kg}}$.
29. 15,000 mL   **31.** 3 L   **33.** 0.925 L
35. 0.008 L   **37.** 4150 mL   **39.** 8 kg
41. 5200 g   **43.** 850 mg   **45.** 30 g   **47.** 0.598 g
49. 0.06 L   **51.** 0.003 kg   **53.** 990 mL
55. mm   **57.** mL   **59.** cm   **61.** mg
63. 2000 mL   **65.** 0.95 kg   **67.** 0.07 L
69. 3000 g   **71.** greater; 5 mg or 0.005 g
73. 200 nickels   **75.** 0.2 gram   **77.** 7, 1, 8, 2

SECTION 7.4 (page 471)

1. \$1.33 rounded to the nearest cent   **3.** 9.75 kg
5. ≈ 71 beats   **7.** 5.03 m   **9.** 22.5 L
11. 215 g; 4.3 g; 4300 mg   **13.** 1.55; 1500 g; 3 g
15. \$358.50   **17.** \$18 case   **19.** 0.63   **21.** 5.733

Section 7.5 (page 477)

1. ≈21.8 yards   **3.** ≈262.4 feet   **5.** ≈4.8 m
7. ≈5.3 ounces   **9.** ≈111.6 kg   **11.** ≈30.3 quarts
13. ≈60.5 liters   **15.** 28°F   **17.** 40°C   **19.** 150°C
21. More. There are 180 degrees between freezing and
boiling on the Fahrenheit scale, but only 100 degrees on
the Celsius scale, so each Celsius degree is a greater change
in temperature.
23. ≈16°C   **25.** 40°C   **27.** ≈46°F   **29.** 95°F
31. ≈58°C
33. Pleasant weather, above freezing but not hot.
≈75°F to ≈39°F
Varies; in Minnesota it's 0°C to −40°C; in California, 24°C
to 0°C.
35. ≈$9.57 (liters converted to gallons)
37. 32   **39.** 81   **41.** 41

Chapter 7 Review Exercises (page 483)

1. 16   **2.** 3   **3.** 2000   **4.** 24   **5.** 60   **6.** 8
7. 4   **8.** 5280   **9.** 12   **10.** 7   **11.** 60
12. 2   **13.** 48   **14.** 45   **15.** 4   **16.** 3
17. $2\frac{1}{2}$ or 2.5   **18.** $5\frac{1}{2}$ or 5.5   **19.** $\frac{3}{4}$ or 0.75
20. $2\frac{1}{4}$ or 2.25   **21.** 78   **22.** 28   **23.** 112
24. 345,600   **25.** mm   **26.** cm   **27.** km
28. m   **29.** cm   **30.** mm   **31.** 500 cm
32. 8500 m   **33.** 8.5 cm   **34.** 3.7 m
35. 0.07 km   **36.** 930 mm   **37.** mL   **38.** L
39. g   **40.** kg   **41.** L   **42.** mL   **43.** mg
44. g   **45.** 5L   **46.** 8000 mL   **47.** 4580 mg
48. 700 g   **49.** 0.006 g   **50.** 0.035 L
51. 31.5 L   **52.** ≈357 g   **53.** 87.25 kg
54. ≈$3.01   **55.** ≈6.5 yards   **56.** ≈11.7 inches
57. ≈67.0 miles   **58.** ≈1288 km
59. ≈21.9 L   **60.** ≈44.0 quarts   **61.** 0°C
62. 100°C   **63.** 37°C   **64.** 20°C
65. 25°C   **66.** ≈33°C   **67.** ≈43°F
68. 104°F   **69.** L   **70.** kg   **71.** cm   **72.** mL
73. mm   **74.** km   **75.** g   **76.** m
77. mL   **78.** g   **79.** mg   **80.** L
81. 105 mm   **82.** $\frac{3}{4}$ hour or 0.75 hour
83. $7\frac{1}{2}$ feet or 7.5 feet   **84.** 130 cm
85. 77°F   **86.** 14 quarts   **87.** 0.7 g
88. 810 mL   **89.** 80 ounces   **90.** 60,000 g
91. 1800 mL   **92.** 30°C   **93.** 36 cm
94. 0.055 L   **95.** 1.26 m   **96.** ≈$48.06
97. ≈113 g; ≈177°C   **98.** ≈178.0 pounds; ≈6.0 feet

Chapter 7 Test (page 487)

1. 36 quarts   **2.** 15 yards   **3.** 2.25 or $2\frac{1}{4}$ hours
4. 0.75 or $\frac{3}{4}$ foot   **5.** 56 ounces   **6.** 7200 minutes
7. kg   **8.** km   **9.** mL   **10.** g   **11.** cm
12. mm   **13.** L   **14.** cm   **15.** 2.5 m

16. 4600 m   **17.** 0.5 cm   **18.** 0.325 g
19. 16,000 mL   **20.** 400 g   **21.** 1055 cm
22. 0.095 L   **23.** 6.32 kg   **24.** 6.75 m
25. 95°C   **26.** 0°C   **27.** ≈1.8 m
28. ≈56.3 kg   **29.** ≈13 gallons   **30.** ≈5.0 miles
31. ≈23°C   **32.** ≈36°F
33. Possible answers:
Use same system as rest of the world; easier system for
children to learn.
34. Possible answers:
People like to use the system they're familiar with; cost of
new signs, scales, etc.

Cumulative Review 1–7 (page 489)

1. 100 + 3 + 70 = 173; 178.838
2. 30,000 − 800 = 29,200; 30,178
3. 90,000 × 200 = 18,000,000; 16,849,725
4. 60 × 5 = 300; 265.644
5. 20,000 ÷ 40 = 500; 530   **6.** 40 ÷ 8 = 5; 4.6
7. 2 + 4 = 6; $5\frac{1}{2}$ **8.** 6 − 1 = 5; $4\frac{3}{14}$
9. 3 • 5 = 15; $14\frac{7}{9}$ **10.** 2 ÷ 1 = 2; $2\frac{1}{2}$
11. $\frac{11}{16}$ **12.** 488,045   **13.** $26\frac{2}{3}$
14. ≈13.0   **15.** $1\frac{5}{24}$ **16.** 7.0793
17. $\frac{5}{8}$ **18.** 307 R38   **19.** 0.00762
20. 8   **21.** 265
22. three hundred seven and nineteen hundredths
23. 0.0082   **24.** 0.067, 0.6, 0.6007, 0.67
25. 0.875   **26.** 87.5% or $87\frac{1}{2}$%   **27.** $\frac{1}{20}$
28. 5%   **29.** $3\frac{1}{2}$ **30.** 3.5   **31.** $\frac{11}{1}$
32. $\frac{1}{9}$ **33.** $\frac{7}{4}$
34. 16 diapers for $3.87, ≈$0.242 per diaper
35. 12   **36.** ≈1.67   **37.** 5%
38. 30 hours   **39.** 30 inches
40. $1\frac{3}{4}$ or 1.75 minutes   **41.** 280 cm
42. 0.065 g   **43.** ≈122.76 miles   **44.** 10°C
45. g   **46.** cm   **47.** m   **48.** kg
49. mg   **50.** mL   **51.** 38°C   **52.** 12°C
53. ≈$4.87; $69.82   **54.** ≈$0.43
55. ≈$9.74   **56.** 93.6 km
57. $2756.25; $11,506.25   **58.** ≈88.6%
59. 1 pound under the limit
60. $\frac{1}{6}$ cup more than the amount needed
61. $56.70; $132.30   **62.** 5.97 kg
63. 120 rows   **64.** 2100 students

Chapter 8

Section 8.1 (page 499)

1. line named $\overleftrightarrow{CD}$ or $\overleftrightarrow{DC}$ **3.** line segment named $\overline{GF}$
or $\overline{FG}$ **5.** ray named $\overrightarrow{PQ}$ **7.** perpendicular
9. parallel   **11.** intersecting   **13.** ∠AOS or ∠SOA
15. ∠CRT or ∠TRC   **17.** ∠AQC or ∠CQA

19. right (90°)   **21.** acute   **23.** straight (180°)
25. (a) The car turned around in a complete circle.
(b) The governor took the opposite view, for example, having opposed taxes and now supporting them.
27. true   **29.** False; the angles have the same measure (both are 180°).   **31.** False; $\overleftrightarrow{QU}$ and $\overrightarrow{TS}$ are perpendicular.   **33.** 90   **35.** 68

Section 8.2 (page 505)

1. $\angle EOD$ and $\angle COD$; $\angle AOB$ and $\angle BOC$ **3.** $\angle HNE$ and $\angle ENF$; $\angle HNG$ and $\angle GNF$; $\angle HNE$ and $\angle HNG$; $\angle ENF$ and $\angle GNF$ **5.** 50°   **7.** 4°   **9.** 50°
11. 90°   **13.** $\angle SON \cong \angle TOM$; $\angle TOS \cong \angle MON$
15. 63°   **17.** 37°
19. Two angles are complementary if their sum is 90°. Two angles are supplementary if their sum is 180°. Drawings will vary; examples are:

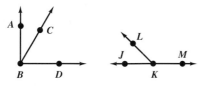

21. $\angle ABF \cong \angle ECD$ Both are 138°.
$\angle ABC \cong \angle BCD$ Both are 42°.   **23.** No, because obtuse angles are > 90° so their sum would be >180°.
25. 7   **27.** 14   **29.** 23

Section 8.3 (page 513)

1. $P = 28$ yd, $A = 48$ yd^2 **3.** $P = 3.6$ km, $A = 0.81$ km^2 **5.** $P = 22$ ft, $A = 10$ ft^2
7. $P = 29$ m, $A = 7$ m^2 **9.** $P = 196.2$ ft, $A = 1674.2$ ft^2 **11.** $P = 12$ mi, $A = 9$ mi^2
13. $P = 38$ m, $A = 39$ m^2 **15.** $P = 98$ m, $A = 492$ m^2 **17.** $P = 78$ in., $A = 234$ in.^{2}
19. \$360   **21.** \$94.81   **23.** Perimeter is the distance around the outside edges of a shape; area is the surface inside the shape measured in square units. Drawings will vary.
25. $A = 6528$ ft^2 ← 172 ft →
← 148 ft →
124 ft
100 ft

27. 0.5 m   **29.** 144 in.   **31.** 75 yd

Section 8.4 (page 521)

1. 208 m   **3.** 207.2 m   **5.** 5.78 km
7. 775 mm^2 **9.** 19.25 ft^2 **11.** 3099.6 cm^2
13. \$940.50   **15.** Height is not part of perimeter; square units are used for area, not perimeter.

$$P = 2.5 \text{ cm} + 2.5 \text{ cm} + 2.5 \text{ cm} + 2.5 \text{ cm}$$
$$P = 10 \text{ cm}$$

17. 3.02 m^2 **19.** 25,344 ft^2 **21.** 52   **23.** $13\frac{1}{3}$

Section 8.5 (page 527)

1. 27 yd   **3.** 60 cm   **5.** 1980 m^2
7. 302.46 cm^2 **9.** 198 m^2 **11.** 1196 m^2
13. 32°   **15.** 48°   **17.** No. Right angles are 90° so two right angles are 180° and the sum of all *three* angles in a triangle equals 180°.   **19.** $7\frac{7}{8}$ ft^2 or 7.875 ft^2
21. 126.8 m of curb; 672 m^2 of sod
23. (a) 32 m^2 **(b)** 13.5 m^2 **25.** 50.24   **27.** 0.64

Section 8.6 (page 535)

1. 18 mm   **3.** 0.35 km   **5.** $C \approx 69.1$ ft, $A \approx 379.9$ ft^2 **7.** $C \approx 8.2$ m, $A \approx 5.3$ m^2
9. $C \approx 47.1$ cm, $A \approx 176.6$ cm^2
11. $C \approx 23.6$ ft, $A \approx 44.2$ ft^2
13. $C \approx 27.2$ km, $A \approx 58.7$ km^2 **15.** ≈ 219.8 cm
17. ≈ 785.0 ft   **19.** ≈ 44.2 in.^{2} **21.** ≈ 201.0 in.^{2}
23. small $\approx$ \$0.084
medium $\approx$ \$0.067   Best Buy
large $\approx$ \$0.071
25. π is the ratio of circumference of a circle to its diameter. If you divide the circumference of any circle by its diameter, the answer is always a little more than 3. The approximate value is 3.14 which we call π (pi). Your test question could involve finding the circumference or the area of a circle.   **27.** ≈ 57 cm^2 **29.** ≈ 197.8 cm^2
31. $\approx 70,650$ mi^2 **33.** ≈ 3.5 m   **35.** $\approx$\$1170.33
37. The *rad* prefix tells you that radius is a ray from the center of the circle. The *dia* prefix means the diameter goes through the circle, and *circum* means the circumference is the distance around.   **39.** 0.5   **41.** 2.75
43. ≈ 0.667 **45.** 10.5

Section 8.7 (page 545)

1. 550 cm^3 **3.** $\approx 44,579.6$ m^3 **5.** ≈ 3617.3 in.^{3}
7. ≈ 471 ft^3 **9.** ≈ 418.7 m^3 **11.** 800 cm^3
13. 18 in.^{3} **15.** ≈ 2481.5 cm^3 **17.** ≈ 3925 ft^3
19. 651,775 m^3 **21.** Student used diameter of 7 cm; should use radius of 3.5 cm in formula. Units for volume are cm^3 not cm^2. Correct answer is ≈ 192.3 cm^3.
23. 513 cm^3 **25.** 64   **27.** 4   **29.** 8

Section 8.8 (page 551)

1. 4   **3.** 8   **5.** ≈ 3.317 **7.** ≈ 2.236
9. ≈ 8.544 **11.** ≈ 10.050 **13.** ≈ 13.784
15. ≈ 31.623
17. $a^2 = 36$, $b^2 = 64$, $c^2 = 100$, side $c = 10$; true
19. $\sqrt{1521} = 39$ ft   **21.** $\sqrt{289} = 17$ in.
23. $\sqrt{144} = 12$ mm   **25.** $\sqrt{73} \approx 8.544$ in.
27. $\sqrt{65} \approx 8.062$ yd   **29.** $\sqrt{195} \approx 13.964$ cm
31. $\sqrt{7.94} \approx 2.818$ m   **33.** $\sqrt{65.01} \approx 8.063$ cm

35. $\sqrt{292.32} \approx 17.097$ km   **37.** $\sqrt{65} \approx 8.1$ ft
39. $\sqrt{360,000} = 600$ m
41. $\sqrt{135} \approx 11.6$ ft

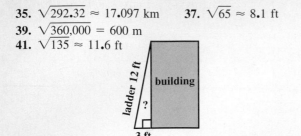

43. 30 is about halfway between 25 and 36 so $\sqrt{30}$ should be about halfway between 5 and 6, or ≈ 5.5. Using a calculator, $\sqrt{30} \approx 5.477$. Similarly $\sqrt{26}$ should be a little more than $\sqrt{25}$; by calculator it is ≈ 5.099. And $\sqrt{35}$ should be a little less than $\sqrt{36}$; by calculator it is ≈ 5.916.   **45.** $BC = 6.25$ ft; $BD = 3.75$ ft
47. 8   **49.** 18.4

SECTION 8.9 (page 559)

1. similar   **3.** not similar   **5.** similar
7. $\angle B$ and $\angle Q$, $\angle C$ and $\angle R$, $\angle A$ and $\angle P$, $\overline{AB}$ and $\overline{PQ}$, $\overline{BC}$ and $\overline{QR}$, $\overline{AC}$ and $\overline{PR}$ **9.** $\angle P$ and $\angle S$, $\angle N$ and $\angle R$, $\angle M$ and $\angle Q$, $\overline{MP}$ and $\overline{QS}$, $\overline{MN}$ and $\overline{QR}$, $\overline{NP}$ and $\overline{RS}$
11. $\frac{3}{2}$; $\frac{3}{2}$; $\frac{3}{2}$ **13.** $a = 5$ mm; $b = 3$ mm
15. $a = 6$ cm; $b = 15$ cm   **17.** $x = 24.8$ m, perimeter = 72.8 m; $y = 15$ m, perimeter = 54.6 m
19. $h = 24$ ft   **21.** $x = 112$ ft
23. Perimeter = 8 cm + 8 cm + 8 cm = 24 cm; Area = $0.5 \cdot 8$ cm $\cdot 5.6$ cm = 22.4 cm^2
25. One dictionary definition is, "Resembling, but not identical." Examples of similar objects are sets of different size pots or measuring cups; small and large size cans of beans; child's tennis shoe and adult tennis shoe.
27. $n = 110$ m   **29.** 50 m   **31.** 8
33. 10   **35.** 15

CHAPTER 8 REVIEW EXERCISES (page 569)

1. line segment named $\overline{AB}$ or $\overline{BA}$ **2.** line named $\overleftrightarrow{CD}$ or $\overleftrightarrow{DC}$ **3.** ray named $\overrightarrow{OP}$ **4.** parallel
5. perpendicular   **6.** intersecting   **7.** acute
8. obtuse   **9.** straight; 180°   **10.** right; 90°
11. $\angle 1 \cong \angle 4$; $\angle 2 \cong \angle 5$; $\angle 3 \cong \angle 6$; $\angle 1$ and $\angle 2$; $\angle 4$ and $\angle 5$; $\angle 5$ and $\angle 1$; $\angle 4$ and $\angle 2$ **12.** $\angle 4 \cong \angle 7$; $\angle 3 \cong \angle 6$; $\angle 1$ and $\angle 2$; $\angle 3$ and $\angle 4$; $\angle 6$ and $\angle 7$; $\angle 6$ and $\angle 4$; $\angle 7$ and $\angle 3$ **13.** $\angle AOB$ and $\angle BOC$; $\angle BOC$ and $\angle COD$; $\angle COD$ and $\angle DOA$; $\angle DOA$ and $\angle AOB$
14. $\angle ERH$ and $\angle HRG$; $\angle HRG$ and $\angle GRF$; $\angle FRG$ and $\angle FRE$; $\angle FRE$ and $\angle ERH$
15. (a) 10°   (b) 45°   (c) 83°   **16.** (a) 25°   (b) 90° (c) 147°   **17.** 4.84 m   **18.** 128 in.   **19.** 152 cm
20. 41 feet   **21.** 486 mm^2
22. 16.5 ft^2 or $16\frac{1}{2}$ ft^2 **23.** ≈ 39.7 m^2
24. $P = 50$ cm, $A = 140$ cm^2 **25.** $P = 102.1$ ft, $A = 567$ ft^2 **26.** $P = 200.2$ m, $A \approx 2074.0$ m^2
27. $P = 518$ cm, $A = 10,812$ cm^2 **28.** $P = 27.1$ m,

$A = 20.58$ m^2 **29.** $P = 20\frac{1}{4}$ ft   or   20.25 ft, $A = 14$ ft^2
30. 70°   **31.** 24°   **32.** 137.8 m
33. $1\frac{1}{2}$ in.   or   1.5 in.   **34.** $C \approx 6.3$ cm, $A \approx 3.1$ cm^2
35. $C \approx 109.3$ m, $A \approx 950.7$ m^2 **36.** $C \approx 37.7$ in., $A \approx 113.0$ in.^{2} **37.** ≈ 20.3 m^2 **38.** 64 in.^{2}
39. 673 km^2 **40.** 1020 m^2 **41.** 229 ft^2
42. 132 ft^2 **43.** 5376 cm^2 **44.** ≈ 498.9 ft^2
45. ≈ 447.9 yd^2 **46.** 30 in.^{3} **47.** 96 cm^3
48. 45,000 mm^3 **49.** ≈ 267.9 m^3 **50.** ≈ 452.2 ft^3
51. ≈ 549.5 cm^3 **52.** ≈ 1808.6 m^3 **53.** ≈ 512.9 m^3
54. 16 yd^3 **55.** 7   **56.** ≈ 2.828 **57.** ≈ 54.772
58. 12   **59.** ≈ 7.616 **60.** 25   **61.** ≈ 10.247
62. ≈ 8.944 **63.** 17 in.   **64.** 7 cm
65. ≈ 10.198 cm   **66.** ≈ 7.211 in.   **67.** ≈ 2.555 m
68. ≈ 8.471 km   **69.** $y = 30$ ft; $x = 34$ ft
70. $y = 7.5$ m; $x = 9$ m   **71.** $x = 12$ mm; $y = 7.5$ mm
72. $P = 18$ in.; $A = 20.25$ in.^{2} or $20\frac{1}{4}$ in.^{2}
73. $P = 10.3$ cm; $A = 6.195$ cm^2
74. $C \approx 40.82$ m; $A \approx 132.665$ m^2
75. $P = 54$ ft; $A = 140$ ft^2
76. $P = 18.5$ yd   or $18\frac{1}{2}$ yd; $A = 15$ yd^2
77. $P = 7$ km; $A = 1.96$ km^2 **78.** $C \approx 53.38$ m; $A \approx 226.865$ m^2 **79.** $P = 78$ mm; $A = 288$ mm^2
80. $P = 38$ mi; $A = 58.5$ mi^2 **81.** parallel lines
82. line segment   **83.** acute angle   **84.** intersecting lines   **85.** right angle; 90°   **86.** ray   **87.** straight angle; 180°   **88.** obtuse angle   **89.** perpendicular lines   **90.** 81°   **91.** 138°   **92.** $P = 90$ m; $A = 92$ m^2 **93.** $P = 282$ cm; $A = 4190$ cm^2
94. ≈ 100.5 ft^3 **95.** ≈ 3.4 in.^{3} or $3\frac{3}{8}$ in.^{3}
96. ≈ 7.4 m^3 **97.** 561 cm^3 **98.** ≈ 1271.7 cm^3
99. ≈ 1436.0 m^3 **100.** $x \approx 12.649$ km
101. $\angle D = 72°$ **102.** $x = 12$ mm; $y = 14$ mm
103. The *dec* prefix in decade means 10 and the *cent* prefix in century means 100, so divide 200 (two centuries) by 10. The answer is 20 decades.

CHAPTER 8 TEST (page 577)

1. (e)   **2.** (a)   **3.** (b)   **4.** (g)   **5.** (d)
6. Perpendicular lines intersect to form a right angle. Parallel lines never intersect.

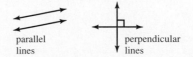

parallel lines   perpendicular lines

7. 9°   **8.** 160°
9. $\angle 1$ and $\angle 4$; $\angle 2$ and $\angle 5$; $\angle 3$ and $\angle 6$
10. $P = 23$ ft, $A = 30$ ft^2
11. $P = 72$ mm, $A = 324$ mm^2 **12.** $P = 26.2$ m, $A = 33.12$ m^2 **13.** $P = 169$ cm, $A = 1591$ cm^2
14. $P = 32.05$ m, $A = 48$ m^2
15. $P = 37.8$ yd or $37\frac{4}{5}$ yd, $A = 58.5$ yd^2 **16.** 55°

17. 12.5 in. or $12\frac{1}{2}$ in. **18.** ≈ 5.7 km

19. ≈ 206.0 cm^2 **20.** ≈ 39.3 m^2 **21.** ≈ 6480 m^3
22. ≈ 33.5 ft^3 **23.** ≈ 5086.8 ft^3 **24.** ≈ 9.220 cm
25. $y = 12$ cm; $z = 6$ cm **26.** Linear units like cm are
used to measure perimeter, radius, diameter, and
circumference. Area is measured in square units like cm^2
(squares that measure 1 cm on each side). Volume is
measured in cubic units like cm^3.

CUMULATIVE REVIEW 1–8 (page 579)

1. $300 + 60{,}000 + 6000 = 66{,}300$; 64,574
2. $20 - 10 = 10$; 10.242 **3.** $4 \times 7 = 28$; 26.7472
4. $30{,}000 \times 70 = 2{,}100{,}000$; 2,074,587
5. $600 \div 3 = 200$; 200.8
6. $5000 \div 50 = 100$; 94 **7.** $5 + 5 = 10$; $9\frac{2}{5}$

8. $3 - 2 = 1$; $1\frac{7}{24}$ **9.** $3 \cdot 2 = 6$; $5\frac{1}{3}$ **10.** $\frac{9}{20}$

11. 0.9132 **12.** 70 R79 **13.** 32 **14.** 0.000078

15. 39,105 **16.** ≈ 19.44 **17.** $4\frac{5}{12}$ **18.** 836.085

19. 9 **20.** 38
21. two hundred eight ten-thousandths **22.** 660.05
23. 2.055; 2.5005; 2.505; 2.55
24. *Per* means divide and *cent* means 100, so divide by

100 to change a percent to a decimal. **25.** $\frac{1}{50}$

26. 2% **27.** 1.75 **28.** 175% **29.** $\frac{2}{5}$ **30.** 0.4

31. $\frac{8}{1}$ **32.** $\frac{7}{3}$ **33.** 35 **34.** 100 **35.** ≈ 2.01

36. 160% **37.** \$600 **38.** 135 minutes

39. $2\frac{1}{2}$ or 2.5 pounds **40.** 0.08 m **41.** 1800 mL

42. mm **43.** L **44.** kg **45.** cm
46. 0°C and 100°C
47. $P = 11$ in., $A = 7$ in.^{2}
48. $P = 5.9$ m, $A = 1.575$ m^2
49. $C \approx 31.4$ ft, $A \approx 78.5$ ft^2
50. $P = 76$ cm, $A = 264$ cm^2
51. $P = 40.8$ m, $A = 95$ m^2
52. $P = 50$ yd, $A = 142$ yd^2 **53.** $y \approx 24.2$ mm
54. $x \approx 8.1$ ft **55.** $\approx 59\%$
56. Brand T at 15.5 ounces for \$2.99 − \$0.30 coupon
57. ≈ 2255.3 cm^3 **58.** \$15.40

59. $1\frac{1}{12}$ yard **60.** ≈ 42.9 pounds **61.** 3.4 m

62. 52 inches

CHAPTER 9
SECTION 9.1 (page 587)

1. $+32$ **3.** -12 **5.** $+20$ **7.** positive
9. negative **11.** neither **13.** negative
15.

17.

19.

21.

23.

25. $<$ **27.** $>$ **29.** $<$ **31.** $>$ **33.** $<$
35. $<$ **37.** $>$ **39.** $>$ **41.** 5 **43.** 5 **45.** 1

47. 251 **49.** 0 **51.** $\frac{1}{2}$ **53.** 9.5 **55.** 0.618

57. -10 **59.** $-\frac{5}{2}$ **61.** -7.6 **63.** -4

65. -2 **67.** 54 **69.** 11 **71.** -163 **73.** 0

75. $4\frac{1}{2}$ **77.** -5.2 **79.** 1.4

81. Opposite numbers are the same distance from 0 but on
opposite sides of it.

-2 and $+2$ are opposites.
83. Some possible answers are temperatures below zero,
loss of points in a game, and overdrawn checking account.

85. true **87.** true **89.** false **91.** $\frac{19}{20}$ **93.** $\frac{7}{12}$

SECTION 9.2 (page 599)

1. 3

3. -7

5. -1

7. -3 **9.** 7 **11.** -7 **13.** 1 **15.** -8
17. -20 **19.** $13 + (-17) = -4$
21. $-\$52.50 + \$50 = -\$2.50$ **23.** -6.8

25. $\frac{1}{4}$ **27.** $-\frac{3}{10}$ **29.** $-\frac{26}{9}$

31. Subtract the smaller absolute value from the larger absolute value. The answer has the same sign as the addend with the larger absolute value. Examples: $-6 + 2 = -4$ and $6 + (-2) = 4$

33. -3 **35.** 9   **37.** $-\dfrac{1}{2}$ **39.** 6.2   **41.** 14

43. -2 **45.** -12 **47.** -25 **49.** -23

51. 5   **53.** 20   **55.** 11   **57.** -60 **59.** 0

61. $-\dfrac{3}{2}$ **63.** $-\dfrac{2}{5}$ **65.** 0.7

67. **(a)** 6 *minus* 9   **(b)** *negative* 9   **(c)** the *opposite* of *negative* 2   **69.** -10 **71.** 12   **73.** -7

75. 5   **77.** -1 **79.** -4.4 **81.** -11 **83.** -10

85. $\dfrac{19}{4}$ or $4\dfrac{3}{4}$ **87.** 1058   **89.** 1513   **91.** 247

93. $\dfrac{3}{2}$ or $1\dfrac{1}{2}$

SECTION 9.3 (page 605)

1. -35 **3.** -45 **5.** -18 **7.** -50 **9.** -40

11. 32   **13.** 77   **15.** 133   **17.** 13   **19.** 0

21. 4   **23.** -4 **25.** $-\dfrac{1}{10}$ **27.** $\dfrac{14}{3}$

29. $-\dfrac{5}{6}$ **31.** $\dfrac{7}{4}$ **33.** -42.3 **35.** 6

37. -31.62 **39.** 4.5   **41.** 8.23   **43.** 0

45. -2 **47.** -5 **49.** undefined

51. -14 **53.** 10   **55.** 4   **57.** -1

59. 191   **61.** 0   **63.** 1   **65.** $\dfrac{2}{3}$ **67.** $\dfrac{1}{3}$

69. -8 **71.** $-\dfrac{14}{3}$ **73.** 4.73   **75.** -4.7

77. -5.3 **79.** 12   **81.** -0.36 **83.** $-\dfrac{7}{40}$

85. **(a)** $6 \cdot (-1) = -6; 2 \cdot (-1) = -2; 15 \cdot (-1) = -15$ **(b)** $-6 \cdot (-1) = 6; -2 \cdot (-1) = 2; -15 \cdot (-1) = 15$ The result of multiplying any nonzero number times -1 is the number with the opposite sign.

87. Similar: If the signs match, the result is positive. If the signs are different, the result is negative. Different: Multiplication is commutative, division is not. You can multiply by zero, but dividing by zero is not allowed. **89.** -2 **91.** -10 **93.** 9   **95.** 10

SECTION 9.4 (page 613)

1. -6 **3.** 0   **5.** 52   **7.** -39 **9.** -30

11. 17   **13.** 16   **15.** 30   **17.** 23   **19.** -43

21. 19   **23.** -3 **25.** 0   **27.** 47   **29.** 41

31. -2 **33.** 13   **35.** 126   **37.** $\dfrac{27}{-3} = -9$

39. $\dfrac{-48}{-4} = 12$ **41.** $\dfrac{-60}{-1} = 60$ **43.** 8   **45.** 1.24

47. -1.47 **49.** -2.31 **51.** $-\dfrac{13}{10}$ **53.** $\dfrac{5}{4}$

55. $\dfrac{4}{5}$ **57.** -1200 **59.** 1

61. The answers are 4, -8, 16, -32, 64, -128, 256, -512. When a negative number is raised to an even power,

the answer is positive; when raised to an odd power, the answer is negative.

63. $\dfrac{27}{-27} = -1$ **65.** 7   **67.** 6

69. $\dfrac{4}{3}$ or $1\dfrac{1}{3}$ **71.** $\dfrac{3}{2}$ or $1\dfrac{1}{2}$

SECTION 9.5 (page 619)

1. 28   **3.** -10 **5.** 8   **7.** -30 **9.** -8

11. 0   **13.** 2   **15.** -22 **17.** $-\dfrac{13}{8}$ **19.** 28

21. -5 **23.** 3   **25.** $\dfrac{12}{-6} = -2$ **27.** 30

29. 28   **31.** ≈78.5 **33.** $\dfrac{45}{2}$ or $22\dfrac{1}{2}$ **35.** 600

37. 318   **39.** ≈25.12

41. The error was made when replacing x with -3; should be $-(-3)$, not -3.

$$-x - 4y \quad \text{Replace } x \text{ with } (-3) \text{ and } y \text{ with } (-1).$$
$$-(-3) - 4(-1) \quad \text{Opposite of } (-3) \text{ is } +3$$
$$(+3) - (-4) \quad \text{Change subtraction to addition.}$$
$$3 + (+4) \quad \text{Add.}$$
$$7$$

43. -40; convert a Celsius temperature to Fahrenheit

45. ≈113.04; find the volume of a sphere

47. 56; find the area of a trapezoid

49. -1 **51.** 1

SECTION 9.6 (page 627)

1. yes   **3.** yes   **5.** no   **7.** 4   **9.** -15 **11.** 8

13. 10   **15.** -8 **17.** -6 **19.** 18   **21.** 11

23. -3 **25.** $\dfrac{7}{3}$ or $2\dfrac{1}{3}$ **27.** $\dfrac{87}{8}$ or $10\dfrac{7}{8}$

29. $\dfrac{5}{2}$ or $2\dfrac{1}{2}$ **31.** $\dfrac{83}{20}$ **33.** 5.87   **35.** -1.51

37. 2   **39.** 4   **41.** 0   **43.** -6 **45.** 9

47. -7 **49.** 1.1   **51.** 34   **53.** 66   **55.** -36

57. -20 **59.** 4   **61.** 16   **63.** 1.3   **65.** -4.94

67. You may add or subtract the same number on both sides of an equation. Many different equations could have -3 as the solution. One possibility is:

$$x + 5 = 2$$
$$x + 5 - 5 = 2 - 5$$
$$x = -3$$

69. 19   **71.** 9   **73.** $\dfrac{8}{21}$

75. $-\dfrac{33}{13}$ or $-2\dfrac{7}{13}$ **77.** 1

SECTION 9.7 (page 635)

1. 1   **3.** 1   **5.** 0   **7.** -2 **9.** -4

11. 6   **13.** $6x + 24$ **15.** $7p - 56$

17. $-3m - 18$ **19.** $-2y + 6$

21. $-5z + 45$ **23.** $7m$ **25.** $8x$ **27.** $-9a$
29. $-2y$ **31.** 5   **33.** 6   **35.** -6 **37.** 1
39. 4   **41.** -2 **43.** 2   **45.** -1.05 **47.** -4
49. -9 **51.** -5 **53.** 0
55. $-2t - 10 = 3t + 5$

$$-2t - 3t - 10 = 3t - 3t + 5$$ Subtract $3t$ from both
$$-5t - 10 = 5$$ sides (addition property).

$$-5t - 10 + 10 = 5 + 10$$ Add 10 to both sides
$$-5t = 15$$ (addition property).

$$\frac{-5 \cdot t}{-5} = \frac{15}{-5}$$ Divide both sides by -5
   (multiplication
$$t = -3$$ property).

57. -10 **59.** 2   **61.** 20   **63.** \$11.20
65. \$25.20

Section 9.8 (page 643)

1. $14 + x$ or $x + 14$
3. $-5 + x$ or $x + (-5)$ **5.** $20 - x$
7. $x - 9$ **9.** $x - 4$ **11.** $6x$ **13.** $2x$
15. $\dfrac{x}{2}$ **17.** $8 + 2x$ or $2x + 8$ **19.** $7x - 10$
21. $2x + x$ or $x + 2x$
23. A variable is a letter that represents an unknown quantity. Examples: x, w, p.
An expression is a combination of letters and numbers. Examples: $6x$, $w - 5$, $2p + 3x$.
An equation has an $=$ sign and shows that two expressions are equal. Examples: $2y = 14$; $x + 5 = 2x$; $8p - 10 = 54$
25. $4n - 2 = 26$; $n = 7$ **27.** $2n + n = -15$; $n = -5$
29. $\dfrac{1}{2}n + 2n = 50$; $n = 20$ **31.** $p + p + 10 = 78$;
34 cm and 44 cm   **33.** $16d + 9 = 89$; 5 days
35. 19 m   **37.** 12 ft, 6 ft

39. $4a - 75 = a$; 25 years old
41. $x + x + x + 25 = 706$; 227 m, 227 m, 252 m
43. $480 = 800 \cdot 0.12 \cdot t$; 5 years   **45.** \$2320
47. \$91.50

Chapter 9 Review Exercises   (page 651)

1.
$$-7\,-6\,-5\,-4\,-3\,-2\,-1\ \ 0\ \ 1\ \ 2\ \ 3\ \ 4$$

2.
$$-6\,-5\,-4\,-3\,-2\,-1\ \ 0\ \ 1\ \ 2\ \ 3\ \ 4\ \ 5$$

3.
$$-7\,-6\,-5\,-4\,-3\,-2\,-1\ \ 0\ \ 1\ \ 2\ \ 3$$

4.
$$-8\,-7\,-6\,-5\,-4\,-3\,-2\,-1\ \ 0\ \ 1\ \ 2$$

5. $>$ **6.** $<$ **7.** $>$
8. $<$ **9.** 8   **10.** 19   **11.** -7 **12.** -15
13. 2   **14.** -7 **15.** -19
16. -33 **17.** 1   **18.** -19
19. $\dfrac{3}{10}$ **20.** $-\dfrac{3}{8}$ **21.** -5.2 **22.** -1.5
23. -6 **24.** 14   **25.** $\dfrac{5}{8}$ **26.** -3.75 **27.** -6
28. -8 **29.** -7 **30.** -17 **31.** 11   **32.** 11
33. 13   **34.** -6 **35.** -80 **36.** 0   **37.** $-\dfrac{1}{2}$
38. 9   **39.** -24 **40.** -20 **41.** 15   **42.** 64
43. -8 **44.** -3 **45.** 5   **46.** 20   **47.** 0
48. -81 **49.** 0   **50.** undefined   **51.** $-\dfrac{4}{7}$
52. 6   **53.** 1.4   **54.** -6.6 **55.** 57   **56.** 23
57. -1 **58.** -2 **59.** -34 **60.** -32
61. -100 **62.** 117   **63.** -1 **64.** 1.328
65. 0   **66.** 2   **67.** 27   **68.** -8 **69.** 0
70. 19   **71.** 23   **72.** -1 **73.** 35   **74.** 27
75. -3 **76.** 16   **77.** 1   **78.** 1
79. $-\dfrac{5}{4}$ or $-1\dfrac{1}{4}$ **80.** -8.05 **81.** -7 **82.** 8
83. 20   **84.** -55 **85.** 9   **86.** -4
87. $6r - 30$ **88.** $11p + 77$ **89.** $-9z + 27$
90. $-8x - 32$ **91.** $11r$ **92.** $-5z$ **93.** $-8p$
94. $2x$ **95.** -9 **96.** -5 **97.** -5 **98.** 7
99. -4 **100.** 0   **101.** $18 + x$ or $x + 18$
102. $\dfrac{1}{2}x$ or $\dfrac{x}{2}$ **103.** $4x + 6$ or $6 + 4x$
104. $5x - 10$ **105.** $11n - 8n = -9$; $n = -3$
106. $45 + 35d = 255$; 7 days
107. $124 = 2l + 2(25)$; 37 cm
108. $a + a + 9 = 51$; 21 and 30 years old
109. 3   **110.** 40   **111.** -1 **112.** -7
113. -16 **114.** -9 **115.** 6   **116.** 5
117. -20 **118.** undefined   **119.** $-\dfrac{5}{9}$
120. -0.35 **121.** 7   **122.** -19 **123.** 9
124. -4 **125.** 4   **126.** -10 **127.** -11
128. 3   **129.** 4   **130.** 0   **131.** -40 **132.** 3
133. $2n - 8 = n + 7$; $n = 15$
134. $36 = 2(2w + 3) + 2w$; 13 inches; 5 inches
135. $s + s + 25 = 111$; Cheetah sprints 68 miles per hour; zebra runs 43 miles per hour.
136. $l + l - 6 = 90$; 48 cm and 42 cm

Chapter 9 Test   (page 657)

1.
$$-5\,-4\,-3\,-2\,-1\ \ 0\ \ 1\ \ 2\ \ 3\ \ 4\ \ 5$$

2. $<, >$ **3.** 7,15   **4.** -1 **5.** -13
6. 5.3   **7.** -7 **8.** 16   **9.** $\dfrac{1}{4}$ **10.** -32
11. 84   **12.** 0   **13.** -25 **14.** 8
15. $-\dfrac{3}{5}$ **16.** 1   **17.** -21 **18.** -38 **19.** 27

20. When evaluating, you are given specific values to replace each variable. When solving an equation, you are not given the value of the variable. You must find a value that "works;" that is, when your solution is substituted for the variable, the two sides of the equation are equal.
21. 110   **22.** 5   **23.** 31   **24.** 0   **25.** −15
26. 5   **27.** 2   **28.** −3
29. $7n - 23 = 47$; $n = 10$
30. $l + l + 4 = 118$; 57 cm; 61 cm
31. $420 = 2(4w) + 2w$; 168 feet; 42 feet

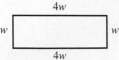

32. $h + h - 3 = 19$; Tim spent 8 hours; Marcella spent 11 hours

Cumulative Review 1–9   (page 659)

1. $9 + 1 + 40 = 50$; 50.602
2. $6 \times 50 = 300$; 308.484
3. $40,000 \div 80 = 500$; 503
4. $4 - 3 = 1$; $\frac{17}{20}$ **5.** $6 \cdot 1 = 6$; $5\frac{1}{4}$
6. $4 \div 2 = 2$; $2\frac{1}{2}$ **7.** 8.906
8. 534,072   **9.** ≈56.4   **10.** −5
11. 40   **12.** 4   **13.** −1.3   **14.** −5
15. $-\frac{1}{2}$ **16.** 5   **17.** 4   **18.** >   **19.** <
20. <   **21.** 0.08; $\frac{2}{25}$
22. 4.5; 450%   **23.** $\frac{4}{1}$ **24.** 0.4
25. ≈233.33   **26.** 45   **27.** 15 students
28. 54.4%   **29.** 50 cars   **30.** 4%
31. 14 quarts   **32.** 3 days   **33.** 3700 mL
34. 0.4 m   **35.** m   **36.** mL   **37.** km
38. g   **39.** 37°C; 20°C
40. $P = 9$ ft; $A = 5\frac{1}{16}$ or 5.0625 ft²
41. $C \approx 28.26$ mm; $A \approx 63.585$ mm²
42. $P = 56$ mi; $A = 84$ mi²
43. $P = 6.45$ cm; $A = 2.185$ cm²
44. $P = 36$ ft; $A = 70$ ft²
45. $P = 188$ m; $A = 1636$ m²   **46.** $y \approx 13.2$ yd
47. $x = 14$ in.   **48.** −26   **49.** −5   **50.** 8
51. −4   **52.** $4n + 40 = 0$; −10
53. $p + p + 300 = 1000$; $350 for Reggie, $650 for Donald
54. 23 cm, 18 cm

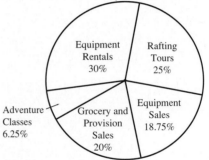

55. ≈$25.52   **56.** $8\frac{1}{3}$ cups   **57.** 124%

58. ≈$4.30   **59.** 0.544 m³ less
60. 11 feet   **61.** ≈54 minutes
62. Naomi's car; ≈25.2 miles per gallon

CHAPTER 10

SECTION 10.1 (page 667)

1. $32,000   **3.** $\frac{9800}{32,000} = \frac{49}{160}$ **5.** $\frac{12,100}{900} = \frac{121}{9}$
7. history   **9.** $\frac{4000}{12,000} = \frac{1}{3}$ **11.** $\frac{1800}{2000} = \frac{9}{10}$
13. $\frac{4000}{1000} = \frac{4}{1}$ **15.** $522,000   **17.** $174,000
19. $261,000   **21.** ≈1923 people   **23.** ≈175 people
25. ≈262 people
27. First find the percent of the total that is to be represented by each item. Next, multiply the percent by 360° to find the size of each sector. Finally, use a protractor to draw each sector.
29. 90°   **31.** 10%   **33.** 15%   **35.** $630; 15%
37. (a) $200,000   **(b)** 22.5°; 72°; 108°; 90°; 67.5°
(c)

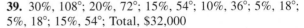

39. 30%, 108°; 20%, 72°; 15%, 54°; 10%, 36°; 5%, 18°; 5%, 18°; 15%, 54°; Total, $32,000

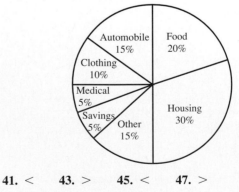

41. <   **43.** >   **45.** <   **47.** >

SECTION 10.2 (page 677)

1. 8000 motorcycles   **3.** 1996; 10,500 motorcycles
5. 2500 motorcycles   **7.** May; 10,000 unemployed
9. 1500 workers   **11.** 2500 workers

13. 150,000 gallons   **15.** 1994; 250,000 gallons
17. 550,000 gallons   **19.** April; 600 burglaries
21. 200 burglaries   **23.** Extremely cold or snowy
weather; greater police activity.   **25.** 3,000,000 CDs
27. 1,500,000 CDs   **29.** 3,500,000 CDs
31. Probably Store B with greater sales. Predicted sales
might be 4,500,000 CDs to 5,000,000 CDs in 1999.
33. A single bar or a single line must be used for each set of
data. To show multiple sets of data, multiple sets of bars or
lines must be used.
35. $40,000   **37.** $25,000   **39.** $5000
41. The decrease in sales may have resulted from poor
service or greater competition. The increase in sales may
have been a result of more advertising or better service.
43. ≈$597,000   **45.** ≈$42,390,000
47. ≈$11,941,000

SECTION 10.3 (page 683)

1. 61–65 years; 16,000 members
3. 13,000 members
5. 50,000 members
7. $21,000 to $25,000; 12 employees
9. 7 employees
11. 28 employees
13. Class intervals are the result of combining data into
groupings. Class frequency is the number of data items
that fit in each class frequency. These are used to group
data and to have multiple responses (frequency) in a class
interval—this makes the data easier to interpret.
15. ||; 2   **17.** ⊞⊞ |; 6   **19.** |; 1   **21.** ||||; 4
23. ⊞⊞ |; 6   **25.** ⊞⊞; 5   **27.** |; 1
29. ⊞⊞ ⊞⊞ |; 11   **31.** ⊞⊞ ⊞⊞ |; 11
33. ⊞⊞ |||; 8   **35.** ||||; 4
37. 25.6   **39.** 52.8   **41.** 39.4

SECTION 10.4 (page 691)

1. 12 years   **3.** ≈69.8   **5.** $27,955   **7.** $58.24
9. $35,500   **11.** 6.7   **13.** ≈17.2
15. 15 messages   **17.** 516 customers   **19.** 8 samples
21. 68 and 74 years
23. When the data contains a few very low or a few very
high values, the mean will give a poor indication of the
average. Consider using the median or mode instead.
25. The median is a better average when the list contains
one or more extreme values.
Find the mean and the median of the following home
values. $82,000; $64,000; $91,000; $115,000; $982,000
mean home value = $266,800
median home value = $91,000
27. 2.60   **29.** 48 calls   **31.** no mode

CHAPTER 10 REVIEW EXERCISES (page 701)

1. Lodging; $560   **2.** $\frac{400}{1700} = \frac{4}{17}$ **3.** $\frac{300}{1700} = \frac{3}{17}$
4. $\frac{280}{1700} = \frac{14}{85}$ **5.** $\frac{300}{160} = \frac{15}{8}$ **6.** $\frac{560}{400} = \frac{7}{5}$

7. 3,300,000 phones   **8.** 2,805,000 phones
9. 1,650,000 phones   **10.** 1,980,000 phones
11. Ages 12–17 and 35–44 buy more cellular phones.
Perhaps they use a phone more often and feel the need to
be reached by phone at all times. Or, it may be a status
symbol among these age groups.
12. Ages 55–64 and 65 and over buy the fewest phones.
Perhaps they use the phone less often and feel that they do
not need to be reached by phone at all times and in
different places. Or, those over 65 may not have enough
income to afford cellular phones.
13. March; 8 million acre-feet
14. June; 2 million acre-feet   **15.** 5 million acre-feet
16. 4 million acre-feet   **17.** 5 million acre-feet
18. 2 million acre-feet   **19.** $50,000   **20.** $20,000
21. $20,000   **22.** $40,000
23. The grocery purchases decreased for two years and
then moved up slightly. Less children are attending the
center or fewer children are eating at the childcare
center.
24. The grocery purchases are increasing. A greater
number of children are attending the center or a greater
number are eating at the childcare center.
25. 28.5 digital cameras   **26.** 35.2 complaints
27. ≈$51.10   **28.** ≈257.3 points
29. 39 claims   **30.** $562   **31.** $64
32. 18 and 32 launchings (bimodal)
33. 36°   **34.** 35%   **35.** 20%   **36.** 25%
37. 36°
38.

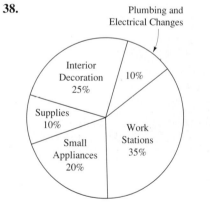

39. 50 years
40. ≈144.7 tacks   **41.** 97
42. 31 and 43 2-bedroom apartments (bimodal)
43. 5.0 hours   **44.** 14.5 yard sales
45. ||||; 4   **46.** |; 1   **47.** ⊞⊞ |; 6
48. ||; 2   **49.** ⊞⊞ ⊞⊞ |||; 13   **50.** ⊞⊞ ||; 7
51. ⊞⊞ ||; 7
52.

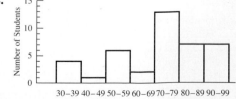

53. ≈32.3   **54.** ≈$118.80

CHAPTER 10 TEST (page 707)

1. $1,232,000 **2.** $1,008,000 **3.** $1,680,000
4. $896,000 **5.** $112,000 **6.** $672,000
7. 108° **8.** 36° **9.** 72° **10.** 108° **11.** 10%
12.

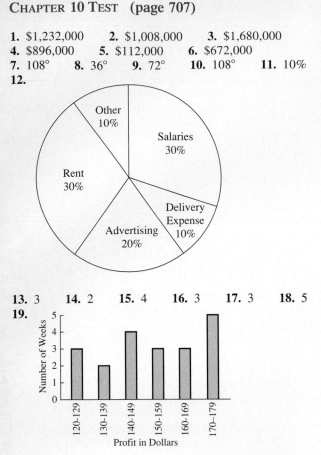

13. 3 **14.** 2 **15.** 4 **16.** 3 **17.** 3 **18.** 5
19.

20. 75 books
21. ≈28.8 miles per gallon **22.** 478.9 miles per hour
23. The weighted mean must be used because different
classes are worth different numbers of credits.

Credits	Grade	
3	A	$3 \times 4 = 12$
2	C	$2 \times 2 = 4$
4	B	$4 \times 3 = 12$
9		28

$$28 \div 9 \approx 3.11$$

24. Arrange the numbers in order from smallest to largest.
When there is an odd number of numbers in a list, the
median is the middle number,

8, 17, 23, 32, 64

Median

25. $11.30 **26.** ≈173.7 **27.** 31.5 degrees
28. 9.3 liters **29.** 57 meters **30.** 103° and 104°

CUMULATIVE REVIEW 1–10 (page 711)

1. ≈$72.65 **2.** ≈$926 **3.** ≈75,700
4. ≈980,000 **5.** 17 **6.** 10
7. 144 **8.** 324
9. $60,000 + 200,000 + 90 + 20,000 = 280,090$; 244,395
10. $3 + 800 + 40 + 50 + 8 = 901$; 898.547
11. $300,000 - 100,000 = 200,000$; 173,783
12. $900 - 60 = 840$; 811.863

13. $7000 \times 600 = 4,200,000$; 4,485,640
14. $60 \times 3 = 180$; 165.911
15. $20,000 \div 40 = 500$; 642
16. $60 \div 4 = 15$; 14.72
17. $1\frac{1}{8}$ **18.** $1\frac{7}{8}$ **19.** $8\frac{7}{15}$
20. $\frac{1}{6}$ **21.** $2\frac{7}{12}$ **22.** $30\frac{19}{20}$
23. $\frac{7}{10}$ **24.** $44\frac{2}{5}$ **25.** $8\frac{4}{5}$
26. $1\frac{1}{3}$ **27.** 18 **28.** $\frac{8}{21}$
29. $\frac{1}{12}$ **30.** $\frac{11}{16}$ **31.** 0.4
32. 0.375 **33.** 0.75 **34.** 0.65
35. 0.199, 0.207, 0.218, 0.22, 0.2215
36. $0.58, 0.608, \frac{5}{8}, 0.6319, \frac{13}{20}$
37. $\frac{1}{8}$ **38.** $\frac{6}{1}$ **39.** True **40.** False **41.** 3
42. 4 **43.** 16.2 **44.** 52 **45.** 0.35
46. 2.5% **47.** 2.50 or 2.5 **48.** 0.0435
49. $\frac{1}{50}$ **50.** $\frac{5}{8}$ **51.** 15% **52.** 325%
53. $480 **54.** 81 chairs **55.** 2800 people
56. 320 **57.** 35% **58.** 25%
59. 3 **60.** 4 **61.** 120 **62.** 12,000
63. 5000 m **64.** 3.815 m **65.** 8300 mg
66. 0.23 kg **67.** 0.006 L **68.** 280 mL
69. mL **70.** g **71.** km **72.** kg
73. ≈12.2 m² **74.** ≈38.7 cm² **75.** ≈38.3 ft²
76. ≈132.7 cm² **77.** ≈882.5 cm³ **78.** 99 m³
79. 8 m **80.** 17 cm **81.** −16 **82.** 6.9
83. −42 **84.** 83.22 **85.** 6 **86.** −2.3
87. $x = 5$ **88.** $x = -6$ **89.** $x = 2$
90. $x = -7$ **91.** 27 hookups; 26 hookups; 19 hookups
92. 8.8 tons; 8.6 tons; 5.7 tons **93.** 37.5% or $37\frac{1}{2}\%$
94. $1124 **95.** 102 tanks **96.** 2166 employees
97. $4101.25 **98.** $59,225.25 **99.** 17.5 L
100. $3631.25

APPENDIX B EXERCISES (page A-13)

1. 37 **3.** 26 **5.** 16 **7.** 243 **9.** 36
11.

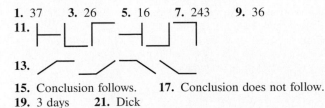

13.

15. Conclusion follows. **17.** Conclusion does not follow.
19. 3 days **21.** Dick

Solutions to Selected Exercises

CHAPTER 1

SECTION 1.1 (page 5)

1. 5: thousands; 3: tens **5.** 8: millions; 2: thousands
9. 60: billions; 0: millions; 502: thousands; 109: ones
13. sixty-four thousand, two hundred fifteen
17. twenty-five million, seven hundred fifty-six thousand, six hundred sixty-five **21.** 10,000,223
25. In digits, the number is 2,000,000,000.
29. In digits, the number is 800,000,621,020,215.

SECTION 1.2 (page 13)

1.
```
   26
 + 33
   59
```
↑↑── Ones added
└── Tens added

5.
```
   317
 + 572
   889
```
↑↑↑── Ones added
└── Tens added
└── Hundreds added

9.
```
   6310
    252
 + 1223
   7785
```
↑↑↑↑── Ones added
└── Tens added
└── Hundreds added
└── Thousands added

13.
```
   1251
   4311
 + 2114
   7676
```

17.
```
   3213
 + 5715
   8928
```

21.
```
    1
   67
 + 83
  150
```

25.
```
    1
   47
 + 74
  121
```

29.
```
    1
   73
 + 29
  102
```

33.
```
    1
   306
 + 848
  1154
```

37.
```
    1
   928
 + 843
  1771
```

41.
```
   111
   7968
 + 1285
   9253
```

45.
```
    111
   9625
 + 7986
  17,611
```

49.
```
      2
    ₁18
   ₁708
   9286
 +  636
  10,648
```

53.
```
      1
    ₁321
    9603
       8
      21
 +  1604
   11,557
```

57.
```
     33
    553
   ₂ 97
   2772
    437
     63
 +  328
   4250
```

61.
```
   1134    Correct
    537
    382
 +  215
   1134
```

65.
```
   5420    Correct
   4713
     28
    615
 +  64
   5420
```

69.
```
   14,332    Correct
    4 714
       27
       77
    8 878
 +    636
   14,332
```

73. Southtown and Rena

Southtown to Oakton	14 miles
Oakton to Rena	+ 23 miles
Southtown to Rena	37 miles

Any other route is longer.

77.
```
   $65    Cost of tune-up
 + $14    Cost of tire rotation
   $79    Total cost for both services
```

81.
```
   9 792      Books from library
 + 3 259      Books from book dealer
  13,051      Total books for sale
```

85.
```
    286
    308
 +  114
    708 feet
```

SECTION 1.3 (page 23)

1.
```
   46    Check:    24
 - 24            + 22
   22              46    Match
```

5.
```
   77    Check:    60
 - 60            + 17
   17              77    Match
```

9.
```
   552    Check:    451
 - 451            + 101
   101             552    Match
```

13.
```
   5546    Check:    2134
 - 2134            + 3412
   3412             5546    Match
```

17.
```
   24,392    Check:    11,232
 - 11,232            + 13,160
   13,160             24,392    Match
```

21.
```
   37       25
 - 25     + 12
   12       37    Correct
```

25.
```
   382      261
 - 261    + 131
   131      392    Incorrect; answer should be 121
```

SOLUTIONS

29.
```
 8643        1421
-1421       +7212
 7212        8633    Incorrect; answer should be 7222
```

33.
```
  7 13
  8 3
- 5 8
-----
  2 5
```
37.
```
 6 11
 7 1 9
-6 5 8
------
   6 1
```
41.
```
   7 15 11
 9 8 6 1
-  6 8 4
--------
 9 1 7 7
```

45.
```
 2 17 12 12 15
 3 8,3 3 5
-2 9,4 7 6
----------
   8 8 5 9
```
49.
```
  5 10
  6 0
- 3 7
-----
  2 3
```
53.
```
   3 10 3 11
 4 0 4 1
-1 2 0 8
--------
 2 8 3 3
```

57.
```
    7 10
 1 5 8 0
-1 0 7 7
--------
   5 0 3
```
61.
```
       9
    5 10 11 10
 6 0 2 0
-4 0 7 8
--------
 1 9 4 2
```

65.
```
       9
    7 10 6 10 15
 8 0,7 0 5
-6 1,6 6 7
----------
 1 9,0 3 8
```
69.
```
       9 9
    1 10 10 17 10
 2 0,0 8 0
-1 3,9 4 6
----------
   6 5 8 4
```

73.
```
 1439        1169
-1169       + 270
  270        1439    Correct
```

77.
```
 27,689       22,306
-22,306      + 5,383
  5 383       27,689    Correct
```

81.
```
     9
   0 10 13
 1 0 3    Calories burned by man
-  8 8    Calories burned by woman
-----
   1 5
```
The woman burned 15 fewer calories than the man during 30 minutes of bowling.

85.
```
        12
      4 2 13
 $2 5 3,3 0 0    Radiologist's yearly income
-$2 4 4,6 0 0    Surgeon's yearly income
 $      8 7 0 0
```
A radiologist earns $8700 more per year than a surgeon.

89.
```
    0 17
 $1 7 8 0    Average balance today
-$  8 4 0    Average balance 10 years ago
 $   9 4 0    Increase in average credit
                 card balance
```

93.
```
 0 14 5 10 18
 1 4,6 0 8    Flags manufactured
- 5 0 6 9    Flags sold
 9 5 3 9    Flags remaining
```

97.
```
  8 11
 $9 1 3    House payment per month
-$6 5 0    Rent payment per month
 $2 6 3    The Jordanos' monthly housing expense
              would increase by $263.
```

SECTION 1.4 (page 35)

1. $(3 \times 1) \times 3$
$3 \times 3 = 9$

5. $(9 \cdot 5) \cdot 0$
$45 \cdot 0 = 0$

9. $(2)(3)(6)$
$(6)(6) = 36$

13. Answers will vary. A sample answer follows:
The commutative property of multiplication means that factors may be multiplied in any order to get the same answer. The commutative properties of addition and multiplication are the same; you may add or multiply numbers in any order.

17.
```
   4
  28
×  6
-----
 168
```
21.
```
    1
  624
×   3
------
 1872
```
25.
```
     2
  2521
×    4
-------
 10,084
```
29.
```
    46 1
 36,921
×     7
--------
 258,447
```

33.
```
 30   First    3      30
× 5          × 5     × 5
            ---     ---
             15     150    Attach 1 zero.
```

37.
```
 500   First    5      500
×  4          × 4     × 4
             ---     ----
              20     2000   Attach 2 zeros.
```

41.
```
 1485   First   1485     1485
×  30          ×   3    ×  30
              ------    ------
                4455    44,550   Attach 1 zero.
```

45.
```
 43,000   First    43      43,000
× 2 000          ×  2    × 2 000
                ----    --------
                  86    86,000,000   Attach 6 zeros.
```

49. $400 \cdot 800$ First $4 \cdot 8 = 32$
$400 \cdot 800 = 320,000$ Attach 4 zeros.

53.
```
   27
×  15
-----
  135
   27
-----
  405
```
57.
```
    83
×   45
------
   415
   332
------
  3735
```
61.
```
    67
×   92
------
   134
   603
------
  6164
```

65.
```
    729
×    45
-------
  3 645
 29 16
-------
 32,805
```
69.
```
    735
×   112
-------
  1 470
  7 35
 73 5
-------
 82,320
```
73.
```
    8162
×    198
--------
   65296
   73458
  8162
---------
 1,616,076
```

77.
```
    428
×   201
-------
    428
 85 60
-------
 86,028
```

81.
```
     2195
×    1038
---------
   17 560
   65 85
  2 195 0
---------
 2,278,410
```

85.
```
   800    Pages in each volume
×   30    Volumes
------
 24,000   Pages
```

89.
```
   38    Miles per gallon
×  11    Gallons
----
  418    Miles
```

93.
```
   65    Rebuilt alternators
×  24    Cost per alternator
----
 1560
```
The total cost is $1560.

97. $21 \cdot 43 \cdot 56 = 50,568$

101.
$$\begin{array}{r} 1406 \\ -\ 348 \\ \hline 1058 \end{array}$$ Calories in large meal
Calories in small meal
More calories in large meal

SECTION 1.5 (page 47)

1. $4\overline{)12}^{\,3} \quad \dfrac{12}{14} = 3$

5. $16 \div 2 = 8 \quad \dfrac{16}{2} = 8$

9. $\dfrac{12}{2} = 6$ **13.** $\dfrac{0}{4} = 0$

17. $0\overline{)21}$ Undefined   **21.** $\dfrac{8}{1} = 8$

25. $9\overline{)32^54}^{\,36}$
Check: $9 \times 36 = 324$

29. $4\overline{)25^10^29}^{\,6\ 2\ 7\ R1}$
Check: $4 \times 627 + 1 = 2508 + 1 = 2509$

33. $6\overline{)1854}^{\,309}$
Check: $6 \times 309 = 1854$

37. $3\overline{)15,018}^{\,5\ 006}$
Check: $3 \times 5006 = 15,018$

41. $5\overline{)12,^29^44^47}^{\,2\ 5\ 8\ 9\ R2}$
Check: $5 \times 2589 + 2 = 12,945 + 2 = 12,947$

45. $4\overline{)12,6^23^30}^{\,3\ 1\ 5\ 7\ R2}$
Check: $4 \times 3157 + 2 = 12,628 + 2 = 12,630$

49. $6\overline{)7^14,^27^35^51}^{\,1\ 2,\ 4\ 5\ 8\ R3}$
Check: $6 \times 12,458 + 3 = 74,748 + 3 = 74,751$

53. $7\overline{)12^58,^26^54^55}^{\,1\ 8,\ 3\ 7\ 7\ R6}$
Check: $7 \times 18,377 + 6 = 128,639 + 6 = 128,645$

57. $3 \times 1908 + 2 = 5724 + 2 = 5726$ Incorrect
$3\overline{)5^27\ 2^25}^{\,1\ 9\ 0\ 8\ R1}$

61. $6 \times 3568 + 2 = 21,408 + 2 = 21,410$ Incorrect
$6\overline{)21,^34^40^49}^{\,3\ 5\ 6\ 8\ R1}$

65. $6 \times 11,523 + 2 = 69,138 + 2 = 69,140$ Correct

69. $8 \times 27,822 = 222,576$ Correct

73. $4\overline{)6^22^24}^{\,1\ 5\ 6}$ Cartons are needed.

77. $7\overline{)12^57^1400}^{\,1\ 8\ 200}$ Each family member
received $18,200.

81. $8,100,000 \div 36 = 225,000$
Each person received $225,000.

85. The number 184 is
divisible by 2 since it ends in 4
not divisible by 3 since $1 + 8 + 4 = 13$
not divisible by 5 since it does not end in 0 or 5
not divisible by 10 since it does not end in 0

	2	3	5	10
184	√	×	×	×

89. The number 903 is
not divisible by 2 since it does not end in 0, 2, 4, 6, or 8
divisible by 3 since $9 + 3 = 12$
not divisible by 5 since it does not end in 0 or 5
not divisible by 10 since it does not end in 0

	2	3	5	10
903	×	√	×	×

93. The number 21,763 is
not divisible by 2 since it does not end in 0, 2, 4, 6, or 8
not divisible by 3 since $2 + 1 + 7 + 6 + 3 = 19$
not divisible by 5 since it does not end in 0 or 5
not divisible by 10 since it does not end in 0

	2	3	5	10
21,763	×	×	×	×

SECTION 1.6 (page 57)

1. $24\overline{)768}^{\,3}$
↑ 3 goes over the 6 because $\frac{76}{24}$ is about 3. The answer must then be a two digit number or 32.

5. $86\overline{)10,327}^{\,1}$
↑ 1 goes over the 3 because $\frac{103}{86}$ is about 1. The answer must then be a three digit number, or 120 R7.

9. $21\overline{)149,826}^{\,7}$
↑ 7 goes over the 9 because $\frac{149}{21}$ is about 7. The answer must then be a four digit number or 7134 R12.

13.
$$\begin{array}{r} 207 \ R5 \\ 42\overline{)8699} \\ 84 \\ \hline 299 \\ 294 \\ \hline 5 \end{array}$$
Check:
$$\begin{array}{r} 207 \\ \times\ \ 42 \\ \hline 414 \\ 828 \\ \hline 8694 \\ +\ \ \ \ 5 \\ \hline 8699 \ \text{Match} \end{array}$$

17.
$$\begin{array}{r} 2\ 407 \ R1 \\ 26\overline{)62,583} \\ 52 \\ \hline 10\ 5 \\ 10\ 4 \\ \hline 183 \\ 182 \\ \hline 1 \end{array}$$
Check:
$$\begin{array}{r} 2407 \\ \times\ \ 26 \\ \hline 14442 \\ 4814 \\ \hline 62582 \\ +\ \ \ \ 1 \\ \hline 62583 \ \text{Match} \end{array}$$

21.
$$\begin{array}{r} 3\ 331 \ R82 \\ 153\overline{)509,725} \\ 459 \\ \hline 507 \\ 459 \\ \hline 482 \\ 459 \\ \hline 235 \\ 153 \\ \hline 82 \end{array}$$
Check:
$$\begin{array}{r} 3331 \\ \times\ \ 153 \\ \hline 9\ 993 \\ 166\ 55 \\ 333\ 1 \\ \hline 509,643 \\ +\ \ \ \ 82 \\ \hline 509,725 \ \text{Match} \end{array}$$

25. $56 \times 106 + 17 = 5936 + 17 = 5953$ Incorrect

$$
\begin{array}{r}
106 \text{ R7} \\
56\overline{)5943} \\
56 \\
\hline
343 \\
336 \\
\hline
7
\end{array}
$$

29. $614 \times 62 + 3 = 38{,}068 + 3 = 38{,}071$ Incorrect

$$
\begin{array}{r}
62 \\
614\overline{)38{,}068} \\
36\ 84 \\
\hline
1\ 228 \\
1\ 228 \\
\hline
0
\end{array}
$$

33. Divide.

$$
\begin{array}{r}
25 \quad \leftarrow \text{Hours traveled} \\
54\overline{)1350} \quad \leftarrow \text{Miles traveled} \\
108 \\
\hline
270 \\
270 \\
\hline
0
\end{array}
$$

Miles per hour ↗

37. Divide. $3888 \div 36 = 108$
Judy's monthly payment is $108.

41. First subtract.

$$
\begin{array}{rl}
\$7588 & \text{Money raised} \\
-\ \$\ 838 & \text{Expenses} \\
\hline
\$6750 & \text{Remaining money}
\end{array}
$$

Then divide.

$$
\begin{array}{r}
375 \\
18\overline{)6750} \quad \text{Remaining money} \\
54 \\
\hline
135 \\
126 \\
\hline
90 \\
90 \\
\hline
0
\end{array}
$$

Number of teams ↗

Each team received $375.

Section 1.7 (page 65)

1. ≈620   623
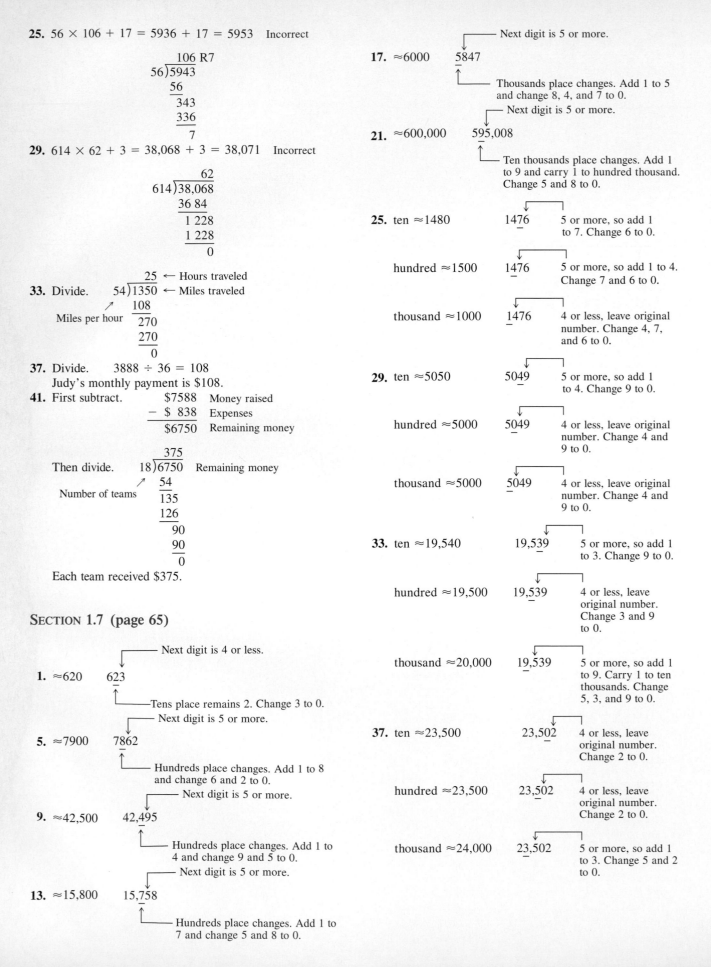
Next digit is 4 or less.
Tens place remains 2. Change 3 to 0.

5. ≈7900   7862
Next digit is 5 or more.
Hundreds place changes. Add 1 to 8 and change 6 and 2 to 0.

9. ≈42,500   42,495
Next digit is 5 or more.
Hundreds place changes. Add 1 to 4 and change 9 and 5 to 0.

13. ≈15,800   15,758
Next digit is 5 or more.
Hundreds place changes. Add 1 to 7 and change 5 and 8 to 0.

17. ≈6000   5847
Next digit is 5 or more.
Thousands place changes. Add 1 to 5 and change 8, 4, and 7 to 0.

21. ≈600,000   595,008
Next digit is 5 or more.
Ten thousands place changes. Add 1 to 9 and carry 1 to hundred thousand. Change 5 and 8 to 0.

25. ten ≈1480   1476   5 or more, so add 1 to 7. Change 6 to 0.

hundred ≈1500   1476   5 or more, so add 1 to 4. Change 7 and 6 to 0.

thousand ≈1000   1476   4 or less, leave original number. Change 4, 7, and 6 to 0.

29. ten ≈5050   5049   5 or more, so add 1 to 4. Change 9 to 0.

hundred ≈5000   5049   4 or less, leave original number. Change 4 and 9 to 0.

thousand ≈5000   5049   4 or less, leave original number. Change 4 and 9 to 0.

33. ten ≈19,540   19,539   5 or more, so add 1 to 3. Change 9 to 0.

hundred ≈19,500   19,539   4 or less, leave original number. Change 3 and 9 to 0.

thousand ≈20,000   19,539   5 or more, so add 1 to 9. Carry 1 to ten thousands. Change 5, 3, and 9 to 0.

37. ten ≈23,500   23,502   4 or less, leave original number. Change 2 to 0.

hundred ≈23,500   23,502   4 or less, leave original number. Change 2 to 0.

thousand ≈24,000   23,502   5 or more, so add 1 to 3. Change 5 and 2 to 0.

41.

estimate		exact
70	Rounds to ←	66
40	←	43
90	←	89
+ 90	←	+ 94
290	←	292

45.

estimate		exact
80	Rounds to ←	76
× 20	←	× 22
1600		1672

49.

estimate		exact
900	Rounds to ←	874
− 500	←	− 458
400		416

53.

estimate		exact
8000	Rounds to ←	8215
60	←	56
700	←	729
+ 4000	←	+ 3605
12,760		12,605

57.

estimate		exact
800	Rounds to ←	841
× 40	←	× 38
32,000		6 728
		25 23
		31,958

61. ten thousand: 3,025,940,000 pesos

3,025,935,000   5 or more, so add 1 to 3.
Change 5 to 0.

million: 3,026,000,000 pesos

3,025,935,000   5 or more, so add 1 to 5.
Change 9, 3, and 5 to 0.

billion: 3,000,000,000 pesos

3,025,935,000   4 or less, leave original number.
Change 2, 5, 9, 3, and 5 to 0.

SECTION 1.8 (page 71)

1. $3^2 = 9$, so $\sqrt{9} = 3$ **5.** $12^2 = 144$, so $\sqrt{144} = 12$

9. Exponent is 2, base is 5. $5^2 = 5 \times 5 = 25$

13. Exponent is 2, base is 12. $12^2 = 12 \times 12 = 144$

17. $10^2 = 10 \cdot 10 = 100$, so $\sqrt{100} = 10$

21. $35^2 = 35 \cdot 35 = 1225$, so $\sqrt{1225} = 35$

25. $54^2 = 54 \cdot 54 = 2916$, so $\sqrt{2916} = 54$

29. $5^2 + 8 - 2 = 25 + 8 - 2$
$= 33 - 2$
$= 31$

33. $20 \cdot 3 \div 5 = 60 \div 5$
$= 12$

37. $6 \cdot 2^2 + \dfrac{0}{6} = 6 \cdot 4 + \dfrac{0}{6}$
$= 24 + 0$
$= 24$

41. $3^3 \cdot 2^2 + (10 - 5) \cdot 2 = 3^3 \cdot 2^2 + 5 \cdot 2$
$= 27 \cdot 4 + 5 \cdot 2$
$= 108 + 10$
$= 118$

45. $6 \cdot 4 + 5 \cdot 7 - 10 = 24 + 35 - 10$
$= 59 - 10$
$= 49$

49. $8 + 10 \div 5 + \dfrac{0}{3} = 8 + 2 + 0$
$= 10 + 0$
$= 10$

53. $7 \cdot \sqrt{81} - 5 \cdot 6 = 7 \cdot 9 - 5 \cdot 6$
$= 63 - 30$
$= 33$

57. $4 \cdot \sqrt{49} - 7(5 - 2) = 4 \cdot \sqrt{49} - 7(3)$
$= 4 \cdot 7 - 7(3)$
$= 28 - 21$
$= 7$

61. $6^2 + 2^2 - 6 + 2 = 36 + 4 - 6 + 2$
$= 40 - 6 + 2$
$= 34 + 2$
$= 36$

65. $7 + 6 \div 3 + 5 \cdot 2 = 7 + 2 + 10$
$= 9 + 10$
$= 19$

69. $4^2 - 2^3 + 5 \cdot 3 = 16 - 8 + 5 \cdot 3$
$= 16 - 8 + 15$
$= 8 + 15$
$= 23$

73. $4 \cdot \sqrt{25} - 6 \cdot 2 = 4 \cdot 5 - 6 \cdot 2$
$= 20 - 12$
$= 8$

77. $7 \div 1 \cdot 8 \cdot 2 \div (21 - 5) = 7 \div 1 \cdot 8 \cdot 2 \div 16$
$= 7 \cdot 8 \cdot 2 \div 16$
$= 56 \cdot 2 \div 16$
$= 112 \div 16$
$= 7$

81. $4 \cdot \sqrt{16} - 3 \cdot \sqrt{9} = 4 \cdot 4 - 3 \cdot 3$
$= 16 - 9$
$= 7$

85. $8 \cdot 9 \div \sqrt{36} - 4 \div 2 + (14 - 8)$
$= 8 \cdot 9 \div \sqrt{36} - 4 \div 2 + 6$
$= 8 \cdot 9 \div 6 - 4 \div 2 + 6$
$= 72 \div 6 - 2 + 6$
$= 12 - 2 + 6$
$= 10 + 6$
$= 16$

89. $6 \cdot \sqrt{25} \cdot \sqrt{100} \div 3 \cdot \sqrt{4} + 9$
$= 6 \cdot 5 \cdot 10 \div 3 \cdot 2 + 9$
$= 30 \cdot 10 \div 3 \cdot 2 + 9$
$= 300 \div 3 \cdot 2 + 9$
$= 100 \cdot 2 + 9$
$= 200 + 9$
$= 209$

SOLUTIONS

SECTION 1.9 (page 79)

1. *Step 1*   The cost of a self-propelled lawn mower is given. We must find the cost of a "riding type" mower which costs $500 more than a self-propelled mower.

Step 2   Add $500 to the cost of a self-propelled lawn mower, which is $380.

Step 3   A reasonable answer would be $500 + $400 = $900. (estimate)

Step 4   Add.

$$\begin{array}{r} \$380 \\ +\ \$500 \\ \hline exact: \quad \$880 \end{array}$$

A "riding type" mower cost $880. The answer is reasonable. Check:

$$\begin{array}{r} \$880 \\ -\ \$500 \\ \hline \$380 \end{array}$$

5. *Step 1*   The number of kits packaged in one hour is given and the total number of kits packaged in 24 hours must be found.

Step 2   Multiply the number of kits packaged in 1 hour by 24.

Step 3   A reasonable answer would be 4000 (estimate) (200 × 20)

Step 4   Multiply.

$$\begin{array}{r} 236 \\ \times\ 24 \\ \hline exact: \quad 5664 \end{array}$$ Kits packaged in 24 hours

The answer is reasonable. Check:

$$\begin{array}{r} 236 \\ 24\overline{)5664} \end{array}$$

9. *Step 1*   The number of people at the lake on Friday is higher than that on Wednesday. The number on Friday and the amount higher is given. The number on Wednesday must be found.

Step 2   Subtract the number higher from the number on Friday.

Step 3   A reasonable answer would be 4000 (estimate) (8000 − 4000)

Step 4   Subtract.

$$\begin{array}{r} 8392 \\ -\ 4218 \\ \hline exact: \quad 4174 \end{array}$$ People at the lake on Wednesday

The answer is reasonable. Check:

$$\begin{array}{r} 4218 \\ +\ 4174 \\ \hline 8392 \end{array}$$

13. *Step 1*   The total miles traveled was divided among 18 days. The miles traveled each day must be found.

Step 2   Divide the total miles by the number of days.

Step 3   A reasonable answer is 150 (estimate) (3000 ÷ 20)

Step 4   Divide.

$$\begin{array}{r} 175 \\ exact: \quad 18\overline{)3150} \end{array}$$ 175 miles were traveled each day.

The answer is reasonable. Check:

$$\begin{array}{r} 175 \\ \times\ 18 \\ \hline 3150 \end{array}$$

17. *Step 1*   The square feet in one acre is given and the square feet in the given number of acres must be found.

Step 2   Multiply the square feet in one acre by the total acres.

Step 3   A reasonable answer is 4,000,000 (estimate) (40,000 × 100)

Step 4   Multiply.

$$\begin{array}{r} 43,560 \\ \times\ 138 \\ \hline exact: \quad 6,011,280 \end{array}$$ Square feet in 138 acres

The answer is reasonable. Check:

$$\begin{array}{r} 43,560 \\ 138\overline{)6,011,280} \end{array}$$

21. *Step 1*   A value package is offered. We must find how much can be saved if the customer buys the value pack of options instead of paying for each item separately.

Step 2   First, add the cost of air bags, antilock brakes, a central locking system, and air-conditioning. Then subtract the cost of the value package from this total.

Step 3   *Estimate*:
$500 + $800 + $200 + $1000 = $2500

$2500 − $1800 = $700

A reasonable answer is $700.

Step 4   Add.

$$\begin{array}{rl} \$475 & \text{Air bags} \\ 780 & \text{Antilock brakes} \\ 245 & \text{Central locking system} \\ +\ 975 & \text{Air-conditioning} \\ \hline \$2475 & \text{Total cost} \end{array}$$

Subtract.

$$\begin{array}{rl} \$2475 & \text{Total cost} \\ -\ 1750 & \text{Option value package} \\ \hline exact: \quad \$725 & \text{Amount saved} \end{array}$$

The answer is reasonable. Check:

$$\begin{array}{r} \$1750 \\ +\ \$725 \\ \hline \$2475 \end{array}$$

25. Answers will vary. Possible answers are:
Add: more; total; gain of
Subtract: less; loss of; decreased by
Multiply: twice; of; product
Divide: divided by; goes into; per
Equals: is; are

29. *Step 1* The cost of packages of undershirts and socks is given and the total cost of a different number of these items must be found.

Step 2 First, find the number of packages of undershirts and socks required. Then find the cost of each item. Finally, add these costs to get the total cost.

Step 3 Packages of undershirts is 10 (30 ÷ 3)
Packages of socks is 3 (18 ÷ 6)
Cost of undershirts is about $100 ($10 × 10)
Cost of socks is about $60 ($20 × 3)
A reasonable answer is $160 (estimate)
($100 + $60)

Step 4 Cost of undershirts   Cost of socks

$$
\begin{array}{r}
10 \\
\times\ \$12 \\
\hline
\$120
\end{array}
\qquad
\begin{array}{r}
3 \\
\times\ \$15 \\
\hline
\$45
\end{array}
$$

exact: Total cost = $120 + $45 = $165
The answer is reasonable. Check:

$$
\begin{array}{r}
\$165 \\
-\ \$45 \\
\hline
\$120
\end{array}
$$

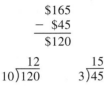

33. *Step 1* The cost of the least expensive and the most expensive hotels are given. We must find the amount saved on a five night stay at the least expensive hotel as opposed to the most expensive hotel.

Step 2 First subtract the cost of the least expensive hotel from the most expensive hotel and then multiply this savings by 5 nights.

Step 3 Estimate: $90 − $70 = $20

$$
\begin{array}{r}
\$20 \\
\times\ \quad 5 \\
\hline
\$100
\end{array}
$$

A reasonable answer is $100.

Step 4 Subtract.

$$
\begin{array}{rl}
\$90 & \text{Most expensive hotel} \\
-\ \$65 & \text{Least expensive hotel} \\
\hline
\$25 & \text{Savings for 1 night}
\end{array}
$$

Multiply.

$$
\begin{array}{rl}
\$25 & \text{Savings for 1 night} \\
\times\ \quad 5 & \text{Number of nights} \\
\hline
\textit{exact:}\quad \$125 & \text{Savings for 5 nights}
\end{array}
$$

The answer is reasonable. Check:

$$
\begin{array}{r}
\$25 \\
\$25 \quad 5\overline{)\$125} \\
+\ \$65 \\
\hline
\$90
\end{array}
$$

37. *Step 1* The total seating is given along with the information to find the number of seats on the main floor. The number of rows of seats in the balcony is given and the number of seats in each row of the balcony must be found.

Step 2 First the number of seats on the main floor must be found. Next, the number of seats on the main floor must be subtracted from the total number of seats to find the number of seats in the balcony. Finally the number of seats in the balcony must be divided by the number of rows of seats in the balcony.

Step 3 Number of seats on the main floor is 750 (30 × 25). Number of seats in balcony is about 500 (1300 − 800). A reasonable answer is about 20 (500 ÷ 25).

Step 4 Number of seats in balcony

$$
\begin{array}{r}
1250 \\
-\ 750 \\
\hline
500
\end{array}
$$

Number of seats in each row

$$
\textit{exact:}\quad 25\overline{)500}^{\ 20}
$$

Number of seats in each row of the balcony is 20.
Answer is reasonable. Check:

$$
\begin{array}{r}
25 \qquad\qquad 1250 \qquad 30\overline{)750}^{\ 25} \\
\times\ 20 \qquad -\ 500 \\
\hline
500 \qquad\qquad 750
\end{array}
$$

CHAPTER 2

SECTION 2.1 (page 101)

1. Shaded: $\dfrac{5}{8}$ (There are 8 parts and 5 are shaded.)

Unshaded: $\dfrac{3}{8}$ (There are 8 parts and 3 are unshaded.)

5. Shaded: $\dfrac{7}{5}$ (The object is divided into 5 parts, and 7 of these parts are shaded.)

Unshaded: $\dfrac{3}{5}$ (The object is divided into 5 parts, and 3 of these parts are unshaded.

9. $\dfrac{8}{25}$ (There are 25 students, and 8 are hearing-impaired.)

13. $\dfrac{3}{4}$ $\leftarrow$ Numerator
$\phantom{\dfrac{3}{4}}$ $\leftarrow$ Denominator

17. Proper   Improper
$\dfrac{1}{3}, \dfrac{5}{8}, \dfrac{7}{16}$ $\dfrac{8}{5}, \dfrac{6}{6}, \dfrac{12}{2}$

21. Answers will vary. A sample answer follows:

$\dfrac{3}{4}$ ← Numerator
← Denominator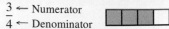

The denominator shows the number of equal parts in the whole and the numerator shows how many of the parts are being considered.

25. $4 \times 5 \times 5 = 20 \times 5$
$= 100$

29. $32 \div 8 = 4$

SECTION 2.2 (page 107)

1. $1\dfrac{2}{3}$ $1 \cdot 3 = 3$ $3 + 2 = 5$ $1\dfrac{2}{3} = \dfrac{5}{3}$

5. $4\dfrac{3}{4}$ $4 \cdot 4 = 16$ $16 + 3 = 19$ $4\dfrac{3}{4} = \dfrac{19}{4}$

9. $1\dfrac{7}{11}$ $1 \cdot 11 = 11$ $11 + 7 = 18$ $1\dfrac{7}{11} = \dfrac{18}{11}$

13. $11\dfrac{1}{3}$ $11 \cdot 3 = 33$ $33 + 1 = 34$ $11\dfrac{1}{3} = \dfrac{34}{3}$

17. $3\dfrac{3}{8}$ $3 \cdot 8 = 24$ $24 + 3 = 27$ $3\dfrac{3}{8} = \dfrac{27}{8}$

21. $4\dfrac{10}{11}$ $4 \cdot 11 = 44$ $44 + 10 = 54$ $4\dfrac{10}{11} = \dfrac{54}{11}$

25. $17\dfrac{12}{13}$ $17 \cdot 13 = 221$ $221 + 12 = 233$

$17\dfrac{12}{13} = \dfrac{233}{13}$

29. $6\dfrac{7}{18}$ $6 \cdot 18 = 108$ $108 + 7 = 115$ $6\dfrac{7}{18} = \dfrac{115}{18}$

33. $\begin{array}{r} 1 \text{ R4} \\ 5\overline{)9} \\ 5 \\ \hline 4 \end{array}$ $\dfrac{9}{5} = 1\dfrac{4}{5}$

37. $\begin{array}{r} 3 \text{ R3} \\ 8\overline{)27} \\ 24 \\ \hline 3 \end{array}$ $\dfrac{27}{8} = 3\dfrac{3}{8}$

41. $\begin{array}{r} 3 \\ 9\overline{)27} \\ 27 \\ \hline 0 \end{array}$ $\dfrac{27}{3} = 9$

45. $\begin{array}{r} 5 \text{ R2} \\ 9\overline{)47} \\ 45 \\ \hline 2 \end{array}$ $\dfrac{47}{9} = 5\dfrac{2}{9}$

49. $\begin{array}{r} 16 \text{ R4} \\ 5\overline{)84} \\ 5 \\ \hline 34 \\ 30 \\ \hline 4 \end{array}$ $\dfrac{84}{5} = 16\dfrac{4}{5}$

53. $\begin{array}{r} 26 \text{ R1} \\ 7\overline{)183} \\ 14 \\ \hline 43 \\ 42 \\ \hline 1 \end{array}$ $\dfrac{183}{7} = 26\dfrac{1}{7}$

57. $101 \cdot 2 = 202$ $202 + 1 = 203$ $101\dfrac{1}{2} = \dfrac{203}{2}$

61. $\begin{array}{r} 171 \\ 15\overline{)2565} \\ 15 \\ \hline 106 \\ 105 \\ \hline 15 \\ 15 \\ \hline 0 \end{array}$ $\dfrac{2565}{15} = 171$

65. $15 \cdot 4 + 6 = 60 + 6$
$= 66$

SECTION 2.3 (page 115)

1. Factorizations of 8:

$1 \cdot 8 \qquad 2 \cdot 4$

Factors of 8 are 1, 2, 4, 8.

5. Factorizations of 25:

$1 \cdot 25 \qquad 5 \cdot 5$

Factors of 25 are 1, 5, 25.

9. Factorizations of 40:

$1 \cdot 40 \qquad 2 \cdot 20 \qquad 4 \cdot 10 \qquad 5 \cdot 8$

Factors of 40 are 1, 2, 4, 5, 8, 10, 20, 40.

13. Because it can be divided by 3, 9 is composite.

17. Because it can be divided by 2 and 5, 10 is composite.

21. Because it can be divided by only itself and 1, 19 is prime.

25. Because it can be divided by 2 and 17, 34 is composite.

29. $\begin{array}{r} 2\overline{)6} \\ 3\overline{)3} \\ 1 \end{array}$ $6 = 2 \cdot 3$

33. $\begin{array}{r} 5\overline{)25} \\ 5\overline{)5} \\ 1 \end{array}$ $25 = 5^2$

37. $\begin{array}{r} 2\overline{)44} \\ 2\overline{)22} \\ 11\overline{)11} \\ 1 \end{array}$ $44 = 2^2 \cdot 11$

41. $\begin{array}{r} 3\overline{)75} \\ 5\overline{)25} \\ 5\overline{)5} \\ 1 \end{array}$ $75 = 3 \cdot 5^2$

45. $\begin{array}{r} 2\overline{)120} \\ 2\overline{)60} \\ 2\overline{)30} \\ 3\overline{)15} \\ 5\overline{)5} \\ 1 \end{array}$ $120 = 2^3 \cdot 3 \cdot 5$

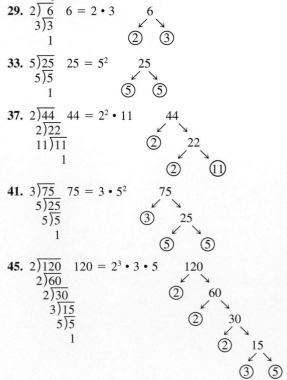

49.
```
  2)320    320 = 2⁶ • 5
   2)160
    2)80
     2)40
      2)20
       2)10
        5)5
          1
```

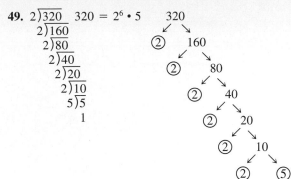

53. Answers will vary. A sample answer follows:
A composite number has a factor(s) other than itself or 1. Examples include 4, 6, 8, 9, 10.
A prime number is a whole number that has exactly two different factors, itself and 1. Examples include 2, 3, 5, 7, 11.

57. $5^3 = 5 • 5 • 5$
$\quad\quad = 125$

61. $2^2 • 3^3 = 2 • 2 • 3 • 3 • 3$
$\quad\quad\quad = 4 • 27$
$\quad\quad\quad = 108$

65. $3^5 • 2^2 = 3 • 3 • 3 • 3 • 3 • 2 • 2$
$\quad\quad\quad = 243 • 4$
$\quad\quad\quad = 972$

69.
```
  2)960    960 = 2⁶ • 3 • 5
   2)480
    2)240
     2)120
      2)60
       2)30
        3)15
         5)5
           1
```

73. $4 • 2 • 3 = 8 • 3$
$\quad\quad\quad\quad = 24$

77. $36 \div 4 = 9$

Section 2.4 (page 123)

1. $\dfrac{8}{16} = \dfrac{8 \div 8}{16 \div 8} = \dfrac{1}{2}$

5. $\dfrac{20}{32} = \dfrac{20 \div 4}{32 \div 4} = \dfrac{5}{8}$

9. $\dfrac{63}{70} = \dfrac{63 \div 7}{70 \div 7} = \dfrac{9}{10}$

13. $\dfrac{36}{63} = \dfrac{36 \div 9}{63 \div 9} = \dfrac{4}{7}$

17. $\dfrac{96}{132} = \dfrac{96 \div 12}{132 \div 12} = \dfrac{8}{11}$

21. $\dfrac{12}{18} = \dfrac{2 • 2 • \overset{1}{\cancel{3}}}{2 • 3 • \underset{1}{\cancel{3}}} = \dfrac{2}{3}$

25. $\dfrac{90}{180} = \dfrac{\overset{1}{\cancel{2}} • \overset{1}{\cancel{3}} • \overset{1}{\cancel{3}} • \overset{1}{\cancel{5}}}{2 • \underset{1}{\cancel{2}} • \underset{1}{\cancel{3}} • \underset{1}{\cancel{3}} • \underset{1}{\cancel{5}}} = \dfrac{1}{2}$

29. $\dfrac{77}{264} = \dfrac{7 • \overset{1}{\cancel{11}}}{2 • 2 • 2 • 3 • \underset{1}{\cancel{11}}} = \dfrac{7}{24}$

33.
$$12 • 12 = 144 \;\leftarrow$$
$$\dfrac{5}{12} \;\times\; \dfrac{12}{30} \qquad \text{Not equivalent}$$
$$5 • 30 = 150 \;\leftarrow$$

The fractions are not equivalent.

37.
$$16 • 35 = 560 \;\leftarrow$$
$$\dfrac{14}{16} \;\times\; \dfrac{35}{40} \qquad \text{Equivalent}$$
$$14 • 40 = 560 \;\leftarrow$$

The fractions are equivalent.

41.
$$30 • 65 = 1950 \;\leftarrow$$
$$\dfrac{25}{30} \;\times\; \dfrac{65}{78} \qquad \text{Equivalent}$$
$$25 • 78 = 1950 \;\leftarrow$$

The fractions are equivalent.

45. $\dfrac{224}{256} = \dfrac{\overset{1}{\cancel{2}} • \overset{1}{\cancel{2}} • \overset{1}{\cancel{2}} • \overset{1}{\cancel{2}} • \overset{1}{\cancel{2}} • 7}{\underset{1}{\cancel{2}} • \underset{1}{\cancel{2}} • \underset{1}{\cancel{2}} • \underset{1}{\cancel{2}} • \underset{1}{\cancel{2}} • 2 • 2 • 2} = \dfrac{7}{8}$

49. Factorizations of 15:
$$1 • 15 \qquad 3 • 5$$

Factors of 15 are 1, 3, 5, 15.

Section 2.5 (page 131)

1. $\dfrac{3}{8} \times \dfrac{1}{2} = \dfrac{3 \times 1}{8 \times 2} = \dfrac{3}{16}$

5. $\dfrac{3}{\underset{2}{\cancel{8}}} • \dfrac{\overset{3}{\cancel{12}}}{5} = \dfrac{3 • 3}{2 • 5} = \dfrac{9}{10}$

9. $\dfrac{\overset{1}{\cancel{3}}}{\underset{2}{\cancel{4}}} • \dfrac{5}{6} • \dfrac{\overset{1}{\cancel{2}}}{\underset{1}{\cancel{3}}} = \dfrac{1 • 5 • 1}{2 • 6 • 1} = \dfrac{5}{12}$

13. $\dfrac{\overset{\overset{1}{\cancel{3}}}{\cancel{21}}}{\underset{\underset{2}{\cancel{6}}}{\cancel{30}}} • \dfrac{\overset{1}{\cancel{5}}}{7} = \dfrac{1 • 1}{2 • 1} = \dfrac{1}{2}$

17. $\dfrac{\overset{1}{\cancel{16}}}{\underset{\underset{1}{\cancel{5}}}{\cancel{25}}} • \dfrac{\overset{7}{\cancel{35}}}{\underset{2}{\cancel{32}}} • \dfrac{\overset{3}{\cancel{15}}}{64} = \dfrac{1 • 7 • 3}{1 • 2 • 64} = \dfrac{21}{128}$

21. $\dfrac{4}{9} • 81 = \dfrac{4}{\underset{1}{\cancel{9}}} • \dfrac{\overset{9}{\cancel{81}}}{1} = \dfrac{4 • 9}{1 • 1} = \dfrac{36}{1} = 36$

SOLUTIONS

25. $42 \cdot \dfrac{7}{10} \cdot \dfrac{5}{7} = \dfrac{\overset{21}{\cancel{42}}}{1} \cdot \dfrac{\overset{1}{\cancel{7}}}{\underset{2}{\cancel{10}}} \cdot \dfrac{\overset{1}{\cancel{5}}}{\underset{1}{\cancel{7}}} = \dfrac{21 \cdot 1 \cdot 1}{1 \cdot 1} = \dfrac{21}{1} = 21$

29. $\dfrac{3}{5} \cdot 400 = \dfrac{3}{\underset{1}{\cancel{5}}} \cdot \dfrac{\overset{80}{\cancel{400}}}{1} = \dfrac{3 \cdot 80}{1 \cdot 1} = \dfrac{240}{1} = 240$

33. $\dfrac{28}{21} \cdot 640 \cdot \dfrac{15}{32} = \dfrac{\overset{4}{\cancel{28}}}{\underset{\underset{1}{3}}{\cancel{21}}} \cdot \dfrac{\overset{20}{\cancel{640}}}{1} \cdot \dfrac{\overset{5}{\cancel{15}}}{\underset{1}{\cancel{32}}} = \dfrac{4 \cdot 20 \cdot 5}{1 \cdot 1 \cdot 1}$

$= \dfrac{400}{1} = 400$

37. area = length • width

$= \dfrac{\overset{1}{\cancel{3}}}{4} \cdot \dfrac{7}{\underset{4}{\cancel{12}}}$

$= \dfrac{7}{16}$ square foot

41. area = length • width

$= \dfrac{\overset{1}{\cancel{7}}}{5} \cdot \dfrac{3}{\underset{2}{\cancel{14}}}$

$= \dfrac{3}{10}$ square inch

45.

	$\frac{2}{3}$ yard

2 yards

area = length • width

$= 2 \cdot \dfrac{2}{3}$

$= \dfrac{2}{1} \cdot \dfrac{2}{3}$

$= \dfrac{4}{3}$

$= 1\dfrac{1}{3}$ square yards

49. parking lot A   parking lot B

$\frac{3}{16}$ mile $\frac{1}{8}$ mile

$\frac{1}{4}$ mile $\frac{3}{8}$ mile

area = length • width   area = length • width

$= \dfrac{1}{4} \cdot \dfrac{3}{16}$ $= \dfrac{3}{8} \cdot \dfrac{1}{8}$

$= \dfrac{3}{64}$ square mile $= \dfrac{3}{64}$ square mile

Parking lot A and parking lot B are the same size.

SECTION 2.6 (page 137)

1. Multiply the length by the width.

$$\dfrac{\overset{1}{\cancel{3}}}{\underset{2}{\cancel{4}}} \cdot \dfrac{\overset{1}{\cancel{2}}}{\underset{1}{\cancel{3}}} = \dfrac{1}{2}$$

The area of the file cabinet top is $\frac{1}{2}$ square yard.

5. Multiply the fraction of New Englanders having over $700 million by the number of New Englanders.

$$\dfrac{2}{5} \cdot 25 = \dfrac{2}{\underset{1}{\cancel{5}}} \cdot \dfrac{\overset{5}{\cancel{25}}}{1} = 10$$

Ten New Englanders had over $700 million.

9. Multiply the number of freshmen by the fraction receiving scholarships.

$$\dfrac{5}{24} \cdot 1800 = \dfrac{5}{\underset{1}{\cancel{24}}} \cdot \dfrac{\overset{75}{\cancel{1800}}}{1} = \dfrac{375}{1} = 375$$

375 students received scholarships.

13. Add the income for all twelve months to find the income for the year.

$$3050 + 2875 + 3325 + 3020 + 2880$$
$$+ \, 3265 + 3160 + 2355 + 2780$$
$$+ \, 3675 + 3310 + 4305 = 38,000$$

The Gomes family had income of $38,000 for the year.

17. Multiply the total income by the fraction saved.

$$\dfrac{1}{16} \cdot 38,000 = \dfrac{1}{\underset{1}{\cancel{16}}} \cdot \dfrac{\overset{2375}{\cancel{38,000}}}{1} = \dfrac{2375}{1} = 2375$$

The Gomes Family saved $2375 for the year.

21. Multiply the amount of earnings for an 8-hour day by a fraction having a numerator of 3, the number of hours worked, and a denominator of 8.

$$112 \cdot \dfrac{3}{8} = \dfrac{\overset{14}{\cancel{112}}}{1} \cdot \dfrac{3}{\underset{1}{\cancel{8}}} = 42$$

Pamela earned $42 in 3 hours.

25. Multiply the remaining $\frac{1}{8}$ of the estate by the fraction going to the American Cancer Society.

$$\dfrac{1}{4} \cdot \dfrac{1}{8} = \dfrac{1}{32}$$

$\frac{1}{32}$ of the estate goes to the American Cancer Society.

SECTION 2.7 (page 145)

1. $\dfrac{1}{6} \div \dfrac{1}{3} = \dfrac{1}{\underset{1}{\cancel{6}}} \cdot \dfrac{\overset{1}{\cancel{3}}}{1} = \dfrac{1}{2}$

5. $\dfrac{3}{4} \div \dfrac{5}{3} = \dfrac{3}{4} \cdot \dfrac{3}{5} = \dfrac{9}{20}$

9. $\dfrac{\frac{7}{9}}{\frac{7}{36}} = \dfrac{7}{9} \div \dfrac{7}{36} = \dfrac{\overset{1}{\cancel{7}}}{\underset{1}{\cancel{9}}} \cdot \dfrac{\overset{4}{\cancel{36}}}{\underset{1}{\cancel{7}}} = \dfrac{4}{1} = 4$

13. $6 \div \dfrac{2}{3} = \dfrac{\overset{3}{\cancel{6}}}{1} \cdot \dfrac{3}{\underset{1}{\cancel{2}}} = \dfrac{9}{1} = 9$

17. $\dfrac{\frac{4}{7}}{8} = \dfrac{4}{7} \div \dfrac{8}{1} = \dfrac{\overset{1}{\cancel{4}}}{7} \cdot \dfrac{1}{\underset{2}{\cancel{8}}} = \dfrac{1}{14}$

21. Divide the number of cups of rice by $\frac{1}{4}$ cup, the size of the measuring cup.

$$4 \div \dfrac{1}{4} = \dfrac{4}{1} \cdot \dfrac{4}{1} = \dfrac{16}{1} = 16$$

Lisa needs 16 measuring cups.

25. Divide the total cords of wood by the amount per trip.

$$40 \div \dfrac{2}{3} = \dfrac{\overset{20}{\cancel{40}}}{1} \cdot \dfrac{3}{\underset{1}{\cancel{2}}} = \dfrac{60}{1} = 60$$

Pam has to make 60 trips.

29. Answers will vary. A sample answer follows:
Divide two fractions by inverting the second fraction (divisor) and multiplying.

33. First, divide the amount raised by the fraction of the total raised.

$$\$840{,}000 \div \dfrac{7}{8} = \dfrac{\overset{120,000}{\cancel{840,000}}}{1} \cdot \dfrac{8}{\underset{1}{\cancel{7}}} = \$960{,}000 \text{ needed}$$

Next, subtract the amount raised from the total funds needed.

$$\$960{,}000 - \$840{,}000 = \$120{,}000$$

The committee still needs $120,000.

37. $12\dfrac{1}{2}$ $12 \cdot 2 = 24$ $24 + 1 = 25$ $12\dfrac{1}{2} = \dfrac{25}{2}$

SECTION 2.8 (page 155)

1. *estimate* $2 \cdot 4 = 8$

exact

$2\dfrac{1}{4} \cdot 3\dfrac{1}{2} = \dfrac{9}{4} \cdot \dfrac{7}{2} = \dfrac{63}{8} = 7\dfrac{7}{8}$

5. *estimate* $3 \cdot 1 = 3$

exact

$3\dfrac{1}{9} \cdot 1\dfrac{2}{7} = \dfrac{\overset{4}{\cancel{28}}}{\underset{1}{\cancel{9}}} \cdot \dfrac{\overset{1}{\cancel{9}}}{\underset{1}{\cancel{7}}} = \dfrac{4}{1} = 4$

9. *estimate* $5 \cdot 2 \cdot 5 = 10 \cdot 5 = 50$

exact

$4\dfrac{1}{2} \cdot 2\dfrac{1}{2} \cdot 5 = \dfrac{9}{2} \cdot \dfrac{11}{\underset{1}{\cancel{2}}} \cdot \dfrac{\overset{1}{\cancel{5}}}{1} = 49\dfrac{1}{2}$

13. *estimate* $3 \div 3 = 1$

exact

$3\dfrac{1}{4} \div 2\dfrac{5}{8} = \dfrac{13}{4} \div \dfrac{21}{8} = \dfrac{13}{\underset{1}{\cancel{4}}} \cdot \dfrac{\overset{2}{\cancel{8}}}{21} = \dfrac{26}{21} = 1\dfrac{5}{21}$

17. *estimate* $6 \div 1 = 6$

exact

$6 \div 1\dfrac{1}{4} = \dfrac{6}{1} \div \dfrac{5}{4} = \dfrac{6}{1} \cdot \dfrac{4}{5} = \dfrac{24}{5} = 4\dfrac{4}{5}$

21. *estimate* $2 \div 6 = \dfrac{2}{6} = \dfrac{1}{3}$

exact

$1\dfrac{7}{8} \div 6\dfrac{1}{4} = \dfrac{15}{8} \div \dfrac{25}{4} = \dfrac{\overset{3}{\cancel{15}}}{\underset{2}{\cancel{8}}} \cdot \dfrac{\overset{1}{\cancel{4}}}{\underset{5}{\cancel{25}}} = \dfrac{3}{10}$

25. Multiply the number of wreaths by the ribbon per wreath.

estimate

$16 \cdot 2 = 32$ yards

exact

$16 \cdot 2\dfrac{1}{4} = \dfrac{\overset{4}{\cancel{16}}}{1} \cdot \dfrac{9}{\underset{1}{\cancel{4}}} = \dfrac{36}{1} = 36$

Shirley needs 36 yards of ribbon.

29. Multiply the ounces of chemical by the gallons of water.

estimate

$2 \cdot 13 = 26$ ounces

exact

$1\dfrac{3}{4} \cdot 12\dfrac{1}{2} = \dfrac{7}{4} \cdot \dfrac{25}{2} = \dfrac{175}{8} = 21\dfrac{7}{8}$

$21\dfrac{7}{8}$ ounces of chemical are needed.

33. Divide the total pounds of steel by the pounds of steel required for a fishing boat anchor.

estimate

$25{,}730 \div 10 = 2573$ anchors

exact

$25{,}730 \div 10\dfrac{3}{8} = 25{,}730 \div \dfrac{83}{8}$

$= \dfrac{\overset{310}{\cancel{25,730}}}{1} \cdot \dfrac{8}{\underset{1}{\cancel{83}}}$

$= \dfrac{2480}{1}$

$= 2480$

2480 anchors can be manufactured.

37. Multiply the number of rolls of film used at the wedding by the number of weddings and multiply the number of rolls of film used at the retirement party by the number of retirement parties. Add to get the total amount of film needed.

estimate

$\begin{array}{r} 13 \cdot 28 = \quad 364 \\ 7 \cdot 16 = \underline{+\ 112} \\ 476 \text{ rolls} \end{array}$

exact

$$12\frac{3}{4} \cdot 28 = \frac{51}{\overset{}{\underset{1}{\cancel{4}}}} \cdot \frac{\overset{7}{\cancel{28}}}{1} = \frac{357}{1} = 357$$

$$7\frac{1}{8} \cdot 16 = \frac{57}{\overset{}{\underset{1}{\cancel{8}}}} \cdot \frac{\overset{2}{\cancel{16}}}{1} = \frac{114}{1} = 114$$

$$\begin{array}{r} 357 \\ +\ 114 \\ \hline 471 \end{array}$$

471 rolls of film are needed.

41. $\dfrac{6}{8} = \dfrac{\overset{1}{\cancel{2}} \cdot 3}{\underset{1}{\cancel{2}} \cdot 2 \cdot 2} = \dfrac{3}{4}$

45. $\dfrac{56}{64} = \dfrac{\overset{1}{\cancel{2}} \cdot \overset{1}{\cancel{2}} \cdot \overset{1}{\cancel{2}} \cdot 7}{\underset{1}{\cancel{2}} \cdot \underset{1}{\cancel{2}} \cdot \underset{1}{\cancel{2}} \cdot 2 \cdot 2 \cdot 2} = \dfrac{7}{8}$

CHAPTER 3

SECTION 3.1 (page 175)

1. $\dfrac{3}{8} + \dfrac{2}{8} = \dfrac{3+2}{8} = \dfrac{5}{8}$

5. $\dfrac{1}{4} + \dfrac{1}{4} = \dfrac{1+1}{4} = \dfrac{2}{4} = \dfrac{1}{2}$

9. $\begin{array}{r}\dfrac{2}{9} \\ +\ \dfrac{1}{9} \\ \hline \dfrac{2+1}{9} = \dfrac{3}{9} = \dfrac{1}{3}\end{array}$

13. $\dfrac{3}{17} + \dfrac{2}{17} + \dfrac{5}{17} = \dfrac{3+2+5}{17} = \dfrac{10}{17}$

17. $\dfrac{2}{54} + \dfrac{8}{54} + \dfrac{12}{54} = \dfrac{2+8+12}{54} = \dfrac{22}{54} = \dfrac{11}{27}$

21. $\dfrac{11}{15} - \dfrac{4}{15} = \dfrac{11-4}{15} = \dfrac{7}{15}$

25. $\begin{array}{r}\dfrac{31}{21} \\ +\ \dfrac{7}{21} \\ \hline \dfrac{31-7}{21} = \dfrac{24}{21} = \dfrac{8}{7} = 1\dfrac{1}{7}\end{array}$

29. $\dfrac{103}{72} - \dfrac{7}{72} = \dfrac{103-7}{72} = \dfrac{96}{72} = 1\dfrac{1}{3}$

33. $\dfrac{746}{400} - \dfrac{506}{400} = \dfrac{746-506}{400} = \dfrac{240}{400} = \dfrac{3}{5}$

37. Add the distance down the ravine to the distance along the creek bed.

$$\frac{5}{12} + \frac{1}{12} = \frac{5+1}{12} = \frac{6}{12} = \frac{1}{2}$$

The search team traveled a total of $\frac{1}{2}$ mile.

41. Add the acres planted in the morning to the acres planted in the afternoon.

$$\frac{5}{12} + \frac{11}{12} = \frac{5+11}{12} = \frac{16}{12}$$

Then, subtract the acres destroyed by frost.

$$\frac{16}{12} - \frac{7}{12} = \frac{16-7}{12} = \frac{9}{12} = \frac{3}{4}$$

$\frac{3}{4}$ acre of seedlings remained.

45. $\begin{array}{l}2\overline{)100} \\ 2\overline{)50} \\ 5\overline{)25} \\ 5\overline{)5} \\ 1\end{array}$ $100 = 2 \cdot 2 \cdot 5 \cdot 5$

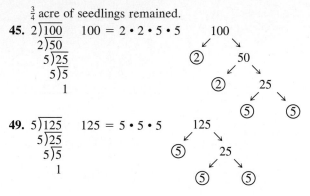

49. $\begin{array}{l}5\overline{)125} \\ 5\overline{)25} \\ 5\overline{)5} \\ 1\end{array}$ $125 = 5 \cdot 5 \cdot 5$

SECTION 3.2 (page 183)

1. Multiples of 10:

10, 20, 30, 40 . . .
└── First multiple divisible by 5.
    $(10 \div 5 = 2)$

The least common multiple of the numbers 5 and 10 is 10.

5. Multiples of 9:

9, 18, 27, 36, 45, 54 . . .
└── First multiple divisible by 4.
    $(36 \div 4 = 9)$

The least common multiple of 4 and 9 is 36.

9. Multiples of 10:

10, 20, 30, 40, 50 . . .
└── First multiple divisible by 4.
    $(20 \div 4 = 5)$

The least common multiple of 4 and 10 is 20.

13.

Prime	2	3
6 =	②	3
9 =		③ · ③
LCM =	②	③ · ③

LCM = $2 \cdot 3 \cdot 3 = 18$

17.

Prime	2	3	5
18 =	2 ·	③ · ③	
20 =	2 · 2 ·		⑤
24 =	② · ② · ② ·	3	
LCM =	② · ② · ②	③ · ③	⑤

LCM = $2 \cdot 2 \cdot 2 \cdot 3 \cdot 3 \cdot 5 = 360$

21.

Prime	2	3	5
12 =	(2 • 2)•	3	
15 =		3 •	(5)
18 =	2 •	(3 • 3)	
20 =	2 • 2 •		5
LCM =	(2 • 2)	(3 • 3)	(5)

LCM = 2 • 2 • 3 • 3 • 5 = 180

25. $\dfrac{1}{2} = \dfrac{1 \cdot 12}{2 \cdot 12} = \dfrac{12}{24}$ **29.** $\dfrac{3}{8} = \dfrac{3 \cdot 3}{8 \cdot 3} = \dfrac{9}{24}$

33. $\dfrac{9}{10} = \dfrac{9 \cdot 4}{10 \cdot 4} = \dfrac{36}{40}$ **37.** $\dfrac{5}{6} = \dfrac{5 \cdot 11}{6 \cdot 11} = \dfrac{55}{66}$

41. $\dfrac{9}{7} = \dfrac{9 \cdot 8}{7 \cdot 8} = \dfrac{72}{56}$ **45.** $\dfrac{8}{11} = \dfrac{8 \cdot 12}{11 \cdot 12} = \dfrac{96}{132}$

49. Answers will vary. A sample answer follows:
It depends on how large the numbers are. If the numbers are small, the method using multiples of the larger number seems best. If the numbers are larger, or there are more of them, then the factorization method will be better.

53.

Prime	2	3	7
1512 =	2 • 2 • 2 •	(3 • 3 • 3)•	7
392 =	(2 • 2 • 2)•		(7 • 7)
LCM =	(2 • 2 • 2)	(3 • 3 • 3)	(7 • 7)

LCM = 2 • 2 • 2 • 3 • 3 • 3 • 7 • 7 = 10,584

57. $\begin{array}{r} 1 \text{R7} \\ 8\overline{)15} \\ \underline{8} \\ 7 \end{array}$ $\dfrac{15}{8} = 1\dfrac{7}{8}$

SECTION 3.3 (page 191)

1. $\dfrac{3}{4} + \dfrac{1}{8} = \dfrac{6}{8} + \dfrac{1}{8} = \dfrac{6+1}{8} = \dfrac{7}{8}$

5. $\dfrac{9}{20} + \dfrac{3}{10} = \dfrac{9}{20} + \dfrac{6}{20} = \dfrac{9+6}{20} = \dfrac{15}{20} = \dfrac{3}{4}$

9. $\dfrac{2}{9} + \dfrac{5}{12} = \dfrac{8}{36} + \dfrac{15}{36} = \dfrac{8+15}{36} = \dfrac{23}{36}$

13. $\dfrac{1}{4} + \dfrac{2}{9} + \dfrac{1}{3} = \dfrac{9}{36} + \dfrac{8}{36} + \dfrac{12}{36} = \dfrac{9+8+12}{36} = \dfrac{29}{36}$

17. $\dfrac{4}{15} + \dfrac{1}{6} + \dfrac{1}{3} = \dfrac{8}{30} + \dfrac{5}{30} + \dfrac{10}{30} = \dfrac{8+5+10}{30} = \dfrac{23}{30}$

21.
$$\begin{aligned} \dfrac{5}{12} &= \dfrac{20}{48} \\ + \dfrac{1}{16} &= \dfrac{3}{48} \\ \hline &\dfrac{20+3}{48} = \dfrac{23}{48} \end{aligned}$$

25. $\dfrac{2}{3} - \dfrac{1}{6} = \dfrac{4}{6} - \dfrac{1}{6} = \dfrac{4-1}{6} = \dfrac{3}{6} = \dfrac{1}{2}$

29. $\dfrac{11}{12} - \dfrac{3}{4} = \dfrac{11}{12} - \dfrac{9}{12} = \dfrac{11-9}{12} = \dfrac{2}{12} = \dfrac{1}{6}$

33.
$$\begin{aligned} \dfrac{4}{5} &= \dfrac{12}{15} \\ - \dfrac{2}{3} &= \dfrac{10}{15} \\ \hline &\dfrac{12-10}{15} = \dfrac{2}{15} \end{aligned}$$

37. Add the cubic yards of corn, oats, and washed medium mesh gravel.

$$\dfrac{1}{3} + \dfrac{3}{8} + \dfrac{1}{4} = \dfrac{8}{24} + \dfrac{9}{24} + \dfrac{6}{24} = \dfrac{23}{24}$$

$\frac{23}{24}$ cubic yards of products were ordered.

41. Add the distances of each side of the field.

$$\dfrac{1}{6} + \dfrac{1}{4} + \dfrac{1}{16} + \dfrac{3}{8} = \dfrac{8}{48} + \dfrac{12}{48} + \dfrac{3}{48} + \dfrac{18}{48} = \dfrac{41}{48}$$

The total distance around the parcel of land is $\frac{41}{48}$ mile.

45. Answers will vary. A sample answer follows:
You cannot add or subtract until all the fractional pieces are the same size. For example, halves are larger than fourths, so you cannot add $\frac{1}{2} + \frac{1}{4}$ until you rewrite $\frac{1}{2}$ as $\frac{2}{4}$.

49. Change each fraction so that they have a common denominator.

$$\dfrac{1}{3} = \dfrac{8}{24} \quad \dfrac{1}{6} = \dfrac{4}{24} \quad \dfrac{1}{8} = \dfrac{3}{24} \quad \dfrac{1}{12} = \dfrac{2}{24} \quad \dfrac{7}{24} = \dfrac{7}{24}$$

The largest fraction is $\frac{8}{24}$. The greatest amount of time was spent in work and travel ($\frac{1}{3} = \frac{8}{24}$).
Multiply $\frac{1}{3}$ times 24 hours per day.

$$\dfrac{1}{3} \cdot 24 = \dfrac{1}{\underset{1}{\cancel{3}}} \cdot \dfrac{\overset{8}{\cancel{24}}}{1} = \dfrac{8}{1} = 8$$

8 hours were spent on work and travel.

53. $1\dfrac{3}{4} \cdot 2\dfrac{1}{2} = \dfrac{7}{4} \cdot \dfrac{5}{2} = \dfrac{35}{8} = 4\dfrac{3}{8}$

57. $1\dfrac{1}{2} \div 3\dfrac{3}{4} = \dfrac{\overset{1}{\cancel{3}}}{\underset{1}{\cancel{2}}} \cdot \dfrac{\overset{2}{\cancel{4}}}{\underset{5}{\cancel{15}}} = \dfrac{2}{5}$

SECTION 3.4 (page 199)

1.
$$\begin{array}{ll} \textit{estimate} & \textit{exact} \\ 6 \xleftarrow{\text{Rounds to}} & 5\dfrac{1}{2} = 5\dfrac{3}{6} \\ + 3 \longleftarrow & + 3\dfrac{1}{3} = 3\dfrac{2}{6} \\ \hline 9 & 8\dfrac{5}{6} \end{array}$$

5.
$$\begin{array}{ll} \textit{estimate} & \textit{exact} \\ 27 \xleftarrow{\text{Rounds to}} & 26\dfrac{5}{8} = 26\dfrac{15}{24} \\ + 9 \longleftarrow & + 9\dfrac{1}{12} = 9\dfrac{2}{24} \\ \hline 36 & 35\dfrac{17}{24} \end{array}$$

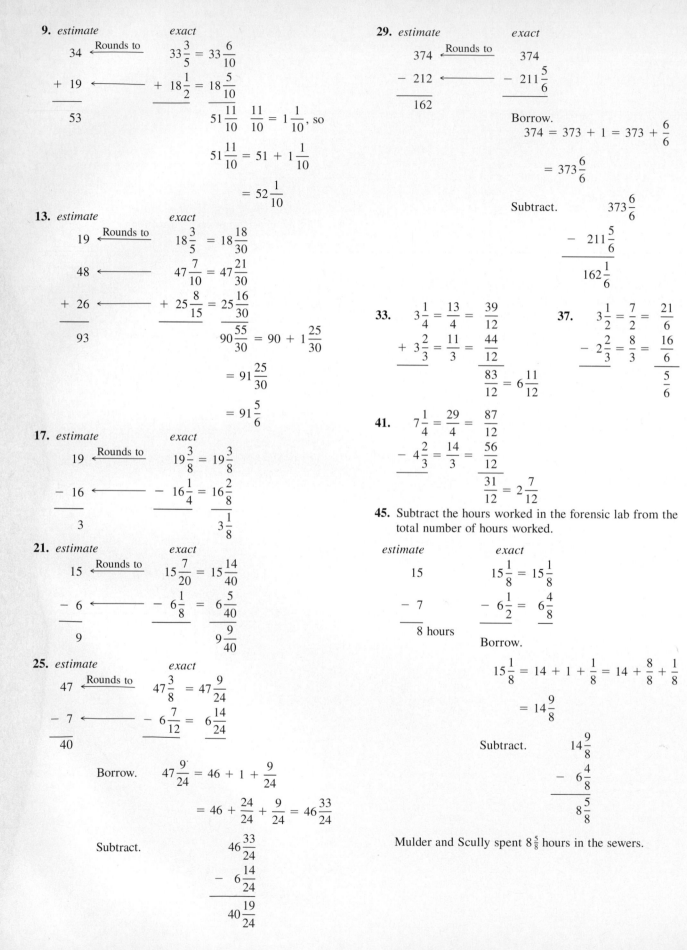

9. *estimate*   *exact*

$$34 \xleftarrow{\text{Rounds to}} 33\frac{3}{5} = 33\frac{6}{10}$$

$$+ \; 19 \longleftarrow + \; 18\frac{1}{2} = 18\frac{5}{10}$$

$$\overline{53} \qquad \overline{51\frac{11}{10}} \quad \frac{11}{10} = 1\frac{1}{10}, \text{ so}$$

$$51\frac{11}{10} = 51 + 1\frac{1}{10}$$

$$= 52\frac{1}{10}$$

13. *estimate*   *exact*

$$19 \xleftarrow{\text{Rounds to}} 18\frac{3}{5} = 18\frac{18}{30}$$

$$48 \longleftarrow 47\frac{7}{10} = 47\frac{21}{30}$$

$$+ \; 26 \longleftarrow + \; 25\frac{8}{15} = 25\frac{16}{30}$$

$$\overline{93} \qquad \overline{90\frac{55}{30}} = 90 + 1\frac{25}{30}$$

$$= 91\frac{25}{30}$$

$$= 91\frac{5}{6}$$

17. *estimate*   *exact*

$$19 \xleftarrow{\text{Rounds to}} 19\frac{3}{8} = 19\frac{3}{8}$$

$$- \; 16 \longleftarrow - \; 16\frac{1}{4} = 16\frac{2}{8}$$

$$\overline{3} \qquad \overline{3\frac{1}{8}}$$

21. *estimate*   *exact*

$$15 \xleftarrow{\text{Rounds to}} 15\frac{7}{20} = 15\frac{14}{40}$$

$$- \; 6 \longleftarrow - \; 6\frac{1}{8} = 6\frac{5}{40}$$

$$\overline{9} \qquad \overline{9\frac{9}{40}}$$

25. *estimate*   *exact*

$$47 \xleftarrow{\text{Rounds to}} 47\frac{3}{8} = 47\frac{9}{24}$$

$$- \; 7 \longleftarrow - \; 6\frac{7}{12} = 6\frac{14}{24}$$

$$\overline{40}$$

Borrow. $\quad 47\frac{9}{24} = 46 + 1 + \frac{9}{24}$

$$= 46 + \frac{24}{24} + \frac{9}{24} = 46\frac{33}{24}$$

Subtract. $\qquad\qquad 46\frac{33}{24}$

$$- \; 6\frac{14}{24}$$

$$\overline{40\frac{19}{24}}$$

29. *estimate*   *exact*

$$374 \xleftarrow{\text{Rounds to}} 374$$

$$- \; 212 \longleftarrow - \; 211\frac{5}{6}$$

$$\overline{162}$$

Borrow.

$$374 = 373 + 1 = 373 + \frac{6}{6}$$

$$= 373\frac{6}{6}$$

Subtract. $\qquad\qquad 373\frac{6}{6}$

$$- \; 211\frac{5}{6}$$

$$\overline{162\frac{1}{6}}$$

33.

$$3\frac{1}{4} = \frac{13}{4} = \frac{39}{12}$$

$$+ \; 3\frac{2}{3} = \frac{11}{3} = \frac{44}{12}$$

$$\overline{\frac{83}{12}} = 6\frac{11}{12}$$

37.

$$3\frac{1}{2} = \frac{7}{2} = \frac{21}{6}$$

$$- \; 2\frac{2}{3} = \frac{8}{3} = \frac{16}{6}$$

$$\overline{\frac{5}{6}}$$

41.

$$7\frac{1}{4} = \frac{29}{4} = \frac{87}{12}$$

$$- \; 4\frac{2}{3} = \frac{14}{3} = \frac{56}{12}$$

$$\overline{\frac{31}{12}} = 2\frac{7}{12}$$

45. Subtract the hours worked in the forensic lab from the total number of hours worked.

estimate   *exact*

$$15 \qquad\qquad 15\frac{1}{8} = 15\frac{1}{8}$$

$$- \; 7 \qquad\qquad - \; 6\frac{1}{2} = 6\frac{4}{8}$$

$$\overline{8 \text{ hours}}$$

Borrow.

$$15\frac{1}{8} = 14 + 1 + \frac{1}{8} = 14 + \frac{8}{8} + \frac{1}{8}$$

$$= 14\frac{9}{8}$$

Subtract. $\qquad\qquad 14\frac{9}{8}$

$$- \; 6\frac{4}{8}$$

$$\overline{8\frac{5}{8}}$$

Mulder and Scully spent $8\frac{5}{8}$ hours in the sewers.

49. Add the hours worked on Monday through Friday.

$$
\begin{array}{cc}
\textit{estimate} & \textit{exact} \\
5 & 4\frac{1}{2} = 4\frac{4}{8} \\
6 & 6\frac{3}{8} = 6\frac{3}{8} \\
5 & 5\frac{1}{4} = 5\frac{2}{8} \\
4 & 3\frac{3}{4} = 3\frac{6}{8} \\
\underline{+\ 7} & \underline{+\ 7\ =\ 7} \\
27 \text{ hours} & 25\frac{15}{8} = 25 + 1\frac{7}{8} \\
 & = 26\frac{7}{8}
\end{array}
$$

Mike worked $26\frac{7}{8}$ hours altogether.

53. Subtract the amount of peat moss the landscaper unloads at the first, second, and third stops from the original amount of peat moss.

$$
\begin{array}{cc}
\textit{estimate} & \textit{exact} \\
10 & 9\frac{5}{8} = 9\frac{5}{8} \\
\underline{-\ 2} & \underline{-\ 1\frac{1}{2} = 1\frac{4}{8}} \\
8 & \\
\underline{-\ 3} & 8\frac{1}{8} \quad \text{Amount left} \\
5 & \quad\quad \text{after first stop} \\
\underline{-\ 3} & \\
2 \text{ cubic yards} &
\end{array}
$$

$$
\begin{array}{c}
8\frac{1}{8} = 8\frac{1}{8} \\
\underline{-\ 2\frac{3}{4} = 2\frac{6}{8}}
\end{array}
$$

Borrow.

$$
8\frac{1}{8} = 7 + 1 + \frac{1}{8} = 7 + \frac{8}{8} + \frac{1}{8}
$$

$$
= 7\frac{9}{8}
$$

Subtract.

$$
\begin{array}{c}
7\frac{9}{8} \\
\underline{-\ 2\frac{6}{8}} \\
5\frac{3}{8} \quad \text{Amount left after} \\
\quad\quad \text{second stop}
\end{array}
$$

$$
\begin{array}{c}
5\frac{3}{8} \\
\underline{-\ 3} \\
2\frac{3}{8} \quad \text{Amount left} \\
\quad\quad \text{after third stop}
\end{array}
$$

$2\frac{3}{8}$ cubic yards of peat moss remain on the truck.

57. Add the weights of the four dinosaurs.

$$
\begin{array}{cc}
\textit{estimate} & \textit{exact} \\
3 & 3\frac{1}{4} = 3\frac{6}{24} \\
8 & 7\frac{1}{2} = 7\frac{12}{24} \\
2 & 2\frac{3}{8} = 2\frac{9}{24} \\
\underline{+\ 2} & \underline{+\ 1\frac{5}{6} = 1\frac{20}{24}} \\
15 \text{ tons} & 13\frac{47}{24} = 13 + 1\frac{23}{24} \\
 & = 14\frac{23}{24}
\end{array}
$$

$14\frac{23}{24}$ tons of dinosaurs were moved during the week.

61. First, add the length of the sections at each end.

$$
\begin{array}{c}
6\frac{1}{4} = 6\frac{2}{8} \\
\underline{+\ 1\frac{7}{8} = 1\frac{7}{8}} \\
7\frac{9}{8} = 7 + 1\frac{1}{8} \\
= 8\frac{1}{8}
\end{array}
$$

Next, subtract this total from the total length of the arrow.

$$
\begin{array}{c}
29\frac{1}{2} = 29\frac{4}{8} \\
\underline{-\ 8\frac{1}{8} = \ 8\frac{1}{8}} \\
21\frac{3}{8}
\end{array}
$$

The length of the section marked x is $21\frac{3}{8}$ inches.

65. $4 \cdot 1 + 8 \cdot 7 + 3 = 4 + 56 + 3$
$$= 60 + 3$$
$$= 63$$

69. $(16 - 6) \cdot 2^3 - (3 \cdot 9) = 10 \cdot 8 - 27$
$$= 80 - 27$$
$$= 53$$

SECTION 3.5 (page 209)

1., 5., and 9.

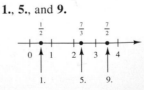

13. The least common multiple of 4 and 3 is 12.

$$
\frac{1}{4} = \frac{3}{12} \quad \text{and} \quad \frac{1}{3} = \frac{4}{12}; \quad \text{therefore,} \quad \frac{1}{4} < \frac{1}{3}
$$

17. The least common multiple of 8 and 12 is 24.

$$\frac{3}{8} = \frac{9}{24} \quad \text{and} \quad \frac{5}{12} = \frac{10}{24}; \quad \text{therefore,} \quad \frac{3}{8} < \frac{5}{12}$$

21. The least common multiple of 27 and 18 is 54.

$$\frac{19}{27} = \frac{38}{54} \quad \text{and} \quad \frac{13}{18} = \frac{39}{54}; \quad \text{therefore,} \quad \frac{19}{27} < \frac{13}{18}$$

25. $\left(\dfrac{3}{8}\right)^2 = \dfrac{3}{8} \cdot \dfrac{3}{8} = \dfrac{9}{64}$ **29.** $\left(\dfrac{2}{3}\right)^3 = \dfrac{2}{3} \cdot \dfrac{2}{3} \cdot \dfrac{2}{3} = \dfrac{8}{27}$

33. $\left(\dfrac{3}{2}\right)^4 = \dfrac{3}{2} \cdot \dfrac{3}{2} \cdot \dfrac{3}{2} \cdot \dfrac{3}{2} = \dfrac{81}{16} = 5\dfrac{1}{16}$

37. Answers will vary. A sample answer follows:
A number line is a horizontal line with a range of numbers placed on it. The lowest number is on the left with the highest on the right. It can be used to compare the size or value of numbers.

$$\overset{\displaystyle \leftarrow\ |\ |\ |\ |\ |\ |\ |\ |\ \rightarrow}{0\ \tfrac{1}{2}\ 1\ 1\tfrac{1}{2}\ 2\ 2\tfrac{1}{2}\ 3\ 3\tfrac{1}{2}\ 4}$$

41. $5 \cdot 2^2 - \dfrac{12}{3} = 5 \cdot 4 - \dfrac{12}{3}$
$$= 20 - 4$$
$$= 16$$

45. $\left(\dfrac{3}{4}\right)^2 \cdot \left(\dfrac{1}{3}\right) = \dfrac{\overset{3}{\cancel{9}}}{16} \cdot \dfrac{2}{\underset{1}{\cancel{3}}}$
$$= \dfrac{3}{16}$$

49. $6 \cdot \left(\dfrac{2}{3}\right)^2 \cdot \left(\dfrac{1}{2}\right)^3 = \overset{2}{\cancel{6}} \cdot \dfrac{4}{\underset{3}{\cancel{9}}} \cdot \dfrac{1}{8}$
$$= \dfrac{\overset{1}{\cancel{8}}}{3} \cdot \dfrac{1}{\underset{1}{\cancel{8}}}$$
$$= \dfrac{1}{3}$$

53. $\dfrac{1}{2} + \left(\dfrac{1}{2}\right)^2 - \dfrac{3}{8} = \dfrac{1}{2} + \dfrac{1}{4} - \dfrac{3}{8}$
$$= \dfrac{3}{4} - \dfrac{3}{8}$$
$$= \dfrac{3}{8}$$

57. $\dfrac{9}{8} \div \left(\dfrac{2}{3} + \dfrac{1}{12}\right) = \dfrac{9}{8} \div \dfrac{9}{12}$
$$= \dfrac{\overset{1}{\cancel{9}}}{\underset{2}{\cancel{8}}} \cdot \dfrac{\overset{3}{\cancel{12}}}{\underset{1}{\cancel{9}}}$$
$$= \dfrac{3}{2}$$
$$= 1\dfrac{1}{2}$$

61. $\dfrac{3}{8} \cdot \left(\dfrac{1}{4} + \dfrac{1}{2}\right) \cdot \dfrac{32}{3} = \dfrac{3}{8} \cdot \dfrac{\overset{1}{\cancel{3}}}{\underset{1}{\cancel{4}}} \cdot \dfrac{\overset{8}{\cancel{32}}}{\underset{1}{\cancel{3}}}$
$$= \dfrac{3}{\underset{1}{\cancel{3}}} \cdot \dfrac{1}{\cancel{8}}$$
$$= 3$$

65. $\left(\dfrac{7}{8} - \dfrac{1}{4}\right) - \left(\dfrac{3}{4}\right)^2 \cdot \dfrac{2}{3} = \dfrac{5}{8} - \left(\dfrac{3}{4}\right)^2 \cdot \dfrac{2}{3}$
$$= \dfrac{5}{8} - \dfrac{\overset{3}{\cancel{9}}}{\underset{8}{\cancel{16}}} \cdot \dfrac{\overset{1}{\cancel{2}}}{\underset{1}{\cancel{3}}}$$
$$= \dfrac{5}{8} - \dfrac{3}{8}$$
$$= \dfrac{2}{8}$$
$$= \dfrac{1}{4}$$

69. five thousand, seven hundred twenty-eight

CHAPTER 4

SECTION 4.1 (page 235)

1. 37.602
ones: 7
tenths: 6
tens: 3

5. 93.01472
thousandths: 4
ten-thousandths: 7
tenths: 0

9. 149.0832
hundreds: 1
hundredths: 8
ones: 9

13. 410.25
4 hundreds   2 tenths
1 ten   5 hundredths
0 ones

17. 5406.045
5 thousands   0 tenths
4 hundreds   4 hundredths
0 tens   5 thousandths
6 ones

21. $13.4 = 13\dfrac{4}{10} = 13\dfrac{2}{5}$ (lowest terms)

25. $0.66 = \dfrac{66}{100} = \dfrac{33}{50}$ (lowest terms)

29. $0.06 = \dfrac{6}{100} = \dfrac{3}{50}$ (lowest terms)

33. $5.002 = 5\dfrac{2}{1000} = 5\dfrac{1}{500}$ (lowest terms)

37. five *tenths*

41. one hundred five *thousandths*   (no "and")

45. one *and* seventy-five *thousandths*

49. 0.32

53. 0.0703

57. Use "and" only when there is a decimal point.

61. 1.006 which is part number 4-A.

65. Use the whole number place value chart on page 1 to find the names after hundred thousands. Then add "ths" to those names. So millions becomes million*ths* when used on the right side of the decimal point, and so on for ten-millionths, hundred-millionths, and billionths.

69. eight thousand six *and* five hundred thousand one *millionths*

73. ten   19,710   19,70**5**   5 or more, so add 1 to 0 in
      Tens →   tens place.
                 Change 5 to 0.

    hundred   19,700   19,**7**05   4 or less, leave 7 in hundreds
                    ↑   place unchanged.
           Hundreds   Change 5 to 0.

    thousand   20,000   1**9**,705   5 or more, so add 1 to 9.
                     ↑   Carry 1 to ten-thousands
        Thousands   place.
                       Change 7 and 5 to 0.

SECTION 4.2 (page 243)

1. ≈16.9   16.8|974   First digit cut is 5 or more.
        ↑   Round up by adding 1 tenth to the
  Tenths place part you are keeping.
                (Drop digits to right of tenths.)

5. ≈0.80   0.79|9   First digit cut is 5 or more,   0.79
      ↑   so round up by adding   + 0.01
           1 hundredth to the part   0.80
           you are keeping.
  Hundredths place

9. ≈794.0   793.9|88   First digit cut is 5 or more, 739.9
        ↑   so round up by adding + 0.1
             1 tenth to the part you 740.0
             are keeping.
    Tenths place

13. ≈49   48.|512   First digit cut is 5 or more, so round up by
      ↑   adding 1 one to the part you are keeping.
  Ones place

17. ≈82.0002   82.0001|51   First digit cut is 5 or more, so
              ↑   round up by adding 1
                 ten-thousandth to the part you
                 are keeping.
   Ten-thousandths place

21. $1.22 $1.22|25   First digit cut is 4 or less, so the part
        ↑   you are keeping stays the same.
  Cents place

25. ≈$17,250 $17,249.|70   First digit cut is 5 or $17,249
            ↑   more, so add $1 to + 1
  Dollars place the part you are $17,250
              keeping.

29. ≈$379 $378.|82   First digit cut is 5 or more $378
        ↑   so add $1 to the part you + 1
            are keeping. $379
    Dollars place

33. Guideline might be: Round amounts less than $1.00 to nearest cent.

37. ≈$1.00 $0.99|6   First digit cut is 5 or more so $0.09
        ↑   add $0.01 to the part you + 0.01
  Cents place   are keeping. $1.00

41.   estimate   exact
             1 1 2
      8000   7929
      6000   6076
   + 8000 + 8218
   22,000 22,223

SECTION 4.3 (page 247)

1.
  1 2
  5.69
  0.24
+ 11.79
 17.72

5.
  1 1
  8.763
 10.500
+ 339.250
 348.513

9.
 21 11
 14.230
  8.000
 74.630
 18.715
+ 0.286
115.861

13.
 232 21
 39.76005
182.00000
  4.79900
 98.31000
+ 5.99990
330.86895

17.   estimate   exact
             2 1
    40   37.25
    20   18.90
 + 8 + 7.50
   68   63.65

21.   estimate   exact
             112 11
    60   62.8173
   500 539.9900
  + 6 + 5.6290
  566 608.4363

25. Add the amounts of her paycheck and her refund check.
  estimate   exact
            1
  $300 $310.14
+ 1 + 0.95
 $301 $311.09

29. Add the cost of the muffins, croissants, and cookie.
  estimate   exact
           1 2
  $ 7 $ 7.42
  10 10.09
+ 1 + 0.69
 $18 $18.20

33. Add the hours worked on weekends (Saturday and Sunday).
  estimate   exact
    10   9.5
 + 5   + 4.8
 15 hours 14.3 hours

SOLUTIONS

37. Add the cost of the payroll and utilities.

estimate	*exact*
$5000	$4919.20
+ 700	+ 732.44
$5700	$5651.64

41. Add the business miles only.

35.4	Visit client
14.9	Visit client
+ 40.0	Business meeting
90.3	miles

45. Add the lengths of the four sides.

estimate	*exact*
20	19.75 inches
20	19.75 inches
6	6.30 inches
+ 6	+ 6.30 inches
52 inches	52.10 inches or 52.1 inches

49.

estimate	*exact*	6708
$7000 - 100 = 6900$		− 139
		6569

SECTION 4.4 (page 253)

1.
```
   73.5  ← Check:   19.2
 − 19.2          + 54.3
   54.3         → 73.5
```

5.
```
   58.254  ← Check:   19.700
 − 19.700          + 38.554
   38.554         → 58.254
```

9.
```
   15.700  ← Check:    2.852
 −  2.852          + 12.848
   12.848         → 15.700
```

13.
```
   0.400  ← Check:   0.291
 − 0.291          + 0.109
   0.109         → 0.400
```

17.
```
   15.000  ← Check:   8.339
 −  8.339          + 6.661
    6.661         → 15.000
```

21.

estimate	*exact*
$20	$19.74
− 7	− 6.58
$13	$13.16

25.

estimate	*exact*
2	2.000
− 2	− 1.981
0	0.019 inch

29. Estimate is $12 - 12 = 0$ so 0.275 is most reasonable answer.

33. Estimate is $457 - 455 = 2$ so 1.81 is most reasonable answer.

37. Subtract the hours so far from the hours agreed.

estimate	*exact*
40	42.50
− 20	− 16.35
20 hours	26.15 hours

Tom must work 26.15 hours.

41. Subtract the weight of the lighter box from the weight of the heavier box.

estimate	*exact*
11	10.50
− 10	− 9.85
1 ounce	0.65 ounce

The difference in weight is 0.65 ounce.

45. Subtract the grocery amount from the car payment.

```
  $190.78
 −  95.81
  $ 94.97  ← The difference in the amounts
```

49. Add the deposit to the balance, then subtract the checks and the service charge.

$129.86	Balance
+ 1749.82	Deposit
$1879.68	
− 1802.15	Checks
$77.53	
− 2.00	Service charge
$75.53	

Mitch has $75.53 in his account at the end of the month.

53. Subtract the two inside measurements from the total length.

3.00	Total length
− 0.91	Left-most measurement
2.09	
− 0.70	Middle measurement
1.39	

b is 1.39 cm.

57.

estimate	*exact*
80	83
× 30	× 28
2400	664
	166
	2324

SECTION 4.5 (page 259)

1.
```
   0.042  ← 3 decimal places
 ×  3.2   ← 1 decimal place
     84
    126
 0.1344   ← 4 decimal places in answer
```

5.
```
    23.4   ← 1 decimal place
 × 0.666  ← 3 decimal places
   1404
   1404
  1404
 15.5844  ← 4 decimal places in answer
```

9. $72 \times 0.6 = 43.2$

0 places + 1 place = 1 place

13. $0.72 \times 0.06 = 0.0432$

2 places + 2 places = 4 places

17. $(0.006)(0.0052)$
```
         0.0052  ← 4 decimal places
 ×       0.006   ← 3 decimal places
     0.0000312   ← 7 decimal places in answer
```
Write four zeros in order to get 7 decimal places.

21. When you multiply a number by 10, the decimal point in the answer is one place farther to the right.

$$5.96 \times 10 = 59.6 \qquad 3.2 \times 10 = 32$$

The decimal point moves two places to the right when multiplying by 100, three places for 1000.

25. *estimate*   *exact*

```
     40          37.1  ← 1 decimal place
  ×  40       ×   4 2  ← 0 decimal place
  ------       ------
  1600           74 2
              1484
              ----------
              1558.2  ← 1 decimal place in answer
```

29. *estimate*   *exact*

```
      3          2.809  ← 3 decimal places
   ×  7       ×  6.85   ← 2 decimal places
   -----       --------
     21           14045
               2 2472
              16 854
              -----------
              19.24165  ← 5 decimal places in answer
```

33. 60.5 inches is a reasonable height (about 5 feet).

37. 0.095 pounds is too small for a baby's weight (much less than 1 pound). Moving the decimal point around gives these possibilities: 0.95, 9.5, 95, or 950. The most reasonable weight for a baby is 9.5 pounds.

41. Multiply the cost of one meter of canvas by the number of meters needed.

```
      $4.09
   ×    0.6
   --------
    $2.454   Rounds to $2.45
```

Sid will spend ≈$2.45 on the canvas.

45. Multiply the cost of the home by 0.07.

```
    $125,300
   ×    0.07
   ---------
   $8771.00
```

Ms. Rolack's fee was $8771.

49. Multiply the number of sheets by the cost per sheet.

```
       5100
   × $0.015
   ---------
   $76.500   or   $76.50
```

The library will pay $76.50 for the paper.

53. First multiply to find the total gallons of fertilizer used on the corn. Then subtract the result from 600 gallons, the amount originally in the tank.

```
     158.2   acres
   ×   3.5   gallons for each acre
   -------
    7910
   4746
   -------
    553.70  gallons used

     600.0  gallons in tank
   − 553.7  gallons used
   -------
     46.3   gallons left
```

There are 46.3 gallons of fertilizer left in the tank.

57. Multiply to find the cost of the rope, then multiply to find the cost of the wire. Add the results to find Barry's total purchases. Subtract the purchases from $15 (three $5 bills is $15).

```
   Cost of rope        Cost of wire
      16.5                1.05
   ×  0.47             ×     3
   --------            --------
   $7.755               $3.15
   Rounds to $7.76
   Purchases           Change
      $7.76   rope       $15.00
   +  3.15   wire      − 10.91
   --------            --------
    $10.91               $4.09
```

Barry received $4.09 in change.

61. *estimate*   *exact*

```
         1000              905 R15
   20)20,000          21)19,020
                          18 9
                         ------
                           120
                           105
                          -----
                            15
```

SECTION 4.6 (page 269)

1.
```
      3.9
   7)27.3
     21
     ---
      6 3
      6 3
      ---
        0
```

5.
```
           400.2
   0.05)20.01 0
         20    ↓
         ------
         0 010
            10
            ---
             0
```

9.
```
         .06    Final answer is 0.06.
   1.8)0.1 08
```

13.
```
          25.3
   4.6)116.3 8
       92
       -----
       24 3
       23 0
       -----
        1 38
        1 38
        -----
           0
```

17. Enter on calculator: 240 ÷ 9.88 =
Round 24.29149798 to ≈24.291

21. When you divide a number by 10, the decimal point in the number is one place farther to the left.

$$3.77 \div 10 = 0.377 \qquad 406.5 \div 10 = 40.65$$

The decimal point moves two places to the left when dividing by 100, three places for 1000.

25. *estimate:*

```
       1
   50)50     The answer of 1.135 is reasonable.
```

29. *estimate:*

$$1\overline{)9}^{\,9}$$

The answer of 0.744 is unreasonable.
Correct answer:

$$
\begin{array}{r}
7.44 \\
1.25\,\overline{)9.30\,00} \\
8\,75 \\
\hline
550 \\
500 \\
\hline
500 \\
500 \\
\hline
0
\end{array}
$$

33. Divide the balance by the number of months.

$$
\begin{array}{r}
19.46 \\
21\,\overline{)408.66} \\
21 \\
\hline
198 \\
189 \\
\hline
9\,6 \\
8\,4 \\
\hline
1\,26 \\
1\,26 \\
\hline
0
\end{array}
$$

Aimee is paying $19.46 per month.

37. Divide the total earnings by the number of hours.

$$
\begin{array}{r}
5.89 \\
40\,\overline{)235.60} \\
200 \\
\hline
35\,6 \\
32\,0 \\
\hline
3\,60 \\
3\,60 \\
\hline
0
\end{array}
$$

Darren earns $5.89 per hour.

41. Find the average by adding the three lengths of U.S. athletes, then dividing by 3.

8.50 meters	8.303 Rounds to 8.30
8.24 meters	$3\overline{)24.910}$
+ 8.17 meters	24
24.91 meters	09

$$
\begin{array}{r}
8.303 \quad \text{Rounds to } 8.30\\
3\,\overline{)24.910} \\
24 \\
\hline
0\,9 \\
9 \\
\hline
01 \\
0 \\
\hline
10
\end{array}
$$

The average long jump length is ≈8.30 meters.

45. Add the first three lengths in the table.

$$
\begin{array}{r}
8.50 \text{ meters}\\
8.29 \text{ meters}\\
+\; 8.24 \text{ meters}\\
\hline
25.03 \text{ meters}
\end{array}
$$

The total length jumped was 25.03 meters.

49. $38.6 + 11.6 \cdot \underbrace{(13.4 - 10.4)}$ Work inside parentheses.

$38.6 + \underbrace{11.6 \cdot \qquad 3}$ Multiply next.

$38.6 + \qquad 34.8 \quad = 73.4$ Add last.

53. $33 - 3.2 \cdot \underbrace{(0.68 + 9)} - \overbrace{1.3^2}$ Parentheses first, then exponents.

$33 - \underbrace{3.2 \cdot \qquad 9.68} \quad - 1.69$ Multiply next.

$\underbrace{33 - \qquad 30.976} \qquad - 1.69$ Subtract last.

$\qquad 2.024 \qquad\quad - 1.69 = 0.334$

57. Divide by 4 to find the quarterly installment, then add the $2.75 service fee.

$$
\begin{array}{r}
234.5 \\
4\,\overline{)938.0} \\
8 \\
\hline
13 \\
12 \\
\hline
18 \\
16 \\
\hline
2\,0 \\
2\,0 \\
\hline
0
\end{array}
\qquad
\begin{array}{r}
\$234.50 \\
+\quad 2.75 \\
\hline
\$237.25
\end{array}
$$

Jenny's quarterly payment is $237.25.

61. The least common multiple of 6 and 9 is 18.

$$\frac{5}{6} = \frac{15}{18} \quad \text{and} \quad \frac{7}{9} = \frac{14}{18}; \quad \text{therefore} \quad \frac{5}{6} > \frac{7}{9}$$

Section 4.7 (page 277)

1.
$$
\begin{array}{r}
0.5 \\
2\,\overline{)1.0} \\
1\,0 \\
\hline
0
\end{array}
$$

5.
$$
\begin{array}{r}
0.3 \\
10\,\overline{)3.0} \\
3\,0 \\
\hline
0
\end{array}
$$

9.
$$
\begin{array}{r}
0.6 \\
5\,\overline{)3.0} \\
3\,0 \\
\hline
0
\end{array}
$$

13. $2\dfrac{1}{4} = \dfrac{9}{4}$

$$
\begin{array}{r}
2.25 \\
4\,\overline{)9.00} \\
8 \\
\hline
1\,0 \\
8 \\
\hline
20 \\
20 \\
\hline
0
\end{array}
$$

17. $3\dfrac{5}{8} = \dfrac{29}{8}$

$$
\begin{array}{r}
3.625 \\
8\,\overline{)29.000} \\
24 \\
\hline
5\,0 \\
4\,8 \\
\hline
20 \\
16 \\
\hline
40 \\
40 \\
\hline
0
\end{array}
$$

21. Enter on calculator: 5 $\boxed{\div}$ 6 $\boxed{=}$
Round 0.8333333 to ≈0.833

25. Divide numerator by denominator.

$$
\begin{array}{r}
0.5555 \quad \text{Rounds to } \approx 0.556\\
9\,\overline{)5.0000}
\end{array}
$$

29. $0.4 = \dfrac{4}{10}$ In lowest terms $\dfrac{4 \div 2}{10 \div 2} = \dfrac{2}{5}$

33. $0.35 = \dfrac{35}{100}$ In lowest terms $\dfrac{35 \div 5}{100 \div 5} = \dfrac{7}{20}$

37. $0.04 = \dfrac{4}{100}$ In lowest terms $\dfrac{4 \div 4}{100 \div 4} = \dfrac{1}{25}$

41.
$$5\overline{)\,1.0}^{\,0.2}$$
$$\underline{1\ 0}$$
$$0$$

45. Write two zeros to the right of 0.5 so it has the same number of decimal places as 0.505. Then you can compare the numbers $0.505 > 0.500$
There was too much calcium in each capsule.
Subtract to find the difference.

$$\begin{array}{r} 0.505 \\ -\ 0.500 \\ \hline 0.005 \end{array}\ \text{gram too much}$$

49. Write zeros so that all the numbers have four decimal places.
$1.0100 > 1.0020$ unacceptable
$0.9991 > 0.9980$ and $0.9991 < 1.0100$ acceptable
$1.0007 > 0.9980$ and $1.0007 < 1.0100$ acceptable
$0.9900 < 0.9980$ unacceptable
The lengths of 0.9991 cm and 1.0007 cm are acceptable.

53. $0.54 = 0.5400 = 5400$ ten-thousandths
$0.5455 = 5455$ ten-thousandths $\leftarrow$ 5455 is largest.
$0.5399 = 5399$ ten-thousandths $\leftarrow$ 5399 is smallest.
(smallest) $0.5399\quad 0.54\quad 0.5455$ (largest)

57. $0.628 = 0.62800 = 62800$ hundred-thousandths
$0.62812 = 62812$ hundred-thousandths
$0.609 = 0.60900 = 60900$ hundred-thousandths
$0.6009 = 0.60090 = 60090$ hundred-thousandths
(smallest) $0.6009\quad 0.609\quad 0.628\quad 0.62812$ (largest)

61.
$0.043 = 43$ thousandths
$0.051 = 51$ thousandths $\leftarrow$ Largest
$0.006 = 6$ thousandths $\leftarrow$ Smallest
$\dfrac{1}{20} = 0.05 = 0.050 = 50$ thousandths

(smallest) $0.006\quad 0.043\quad \dfrac{1}{20}\quad 0.051$ (largest)

65. $1\dfrac{7}{16} = \dfrac{23}{16}$
$$16\overline{)23.00}\,^{1.43}\ \text{Rounds to 1.4}$$
Length (a) is ≈ 1.4 inch

69. $8\overline{)3.00}\,^{0.37}$ Rounds to 0.4
Length (e) is ≈ 0.4 inches

73. $\dfrac{3}{11} \approx 0.273 \approx 273$ thousandths
$\dfrac{4}{15} \approx 0.267 \approx 267$ thousandths
$0.25 = 0.250 = 250$ thousandths $\leftarrow$ Smallest
$\dfrac{1}{3} \approx 0.333 \approx 333$ thousandths $\leftarrow$ Largest

(smallest) $0.25\quad \dfrac{4}{15}\quad \dfrac{3}{11}\quad \dfrac{1}{3}$ (largest)

77. $\dfrac{9}{12} = \dfrac{\cancel{3}\cdot 3}{2\cdot 2\cdot \cancel{3}} = \dfrac{3}{4}$

81. $\dfrac{96}{132} = \dfrac{\cancel{2}\cdot\cancel{2}\cdot 2\cdot 2\cdot 2\cdot\cancel{3}}{\cancel{2}\cdot\cancel{2}\cdot\cancel{3}\cdot 11} = \dfrac{8}{11}$

CHAPTER 5

SECTION 5.1 (page 303)

1. $\dfrac{8}{9}\ \substack{\leftarrow\ \text{Mentioned first}\\ \leftarrow\ \text{Mentioned second}}$

5. $\dfrac{30\ \text{minutes}}{90\ \text{minutes}} = \dfrac{30 \div 30}{90 \div 30} = \dfrac{1}{3}$

9. $\dfrac{6\ \text{hours}}{16\ \text{hours}} = \dfrac{6}{16} = \dfrac{6 \div 2}{16 \div 2} = \dfrac{3}{8}$

13. $\dfrac{15}{2\frac{1}{2}} = \dfrac{\frac{15}{1}}{\frac{5}{2}} = \dfrac{15}{1} \div \dfrac{5}{2} = \dfrac{\cancel{15}^{3}}{1}\cdot\dfrac{2}{\cancel{5}_{1}} = \dfrac{6}{1}$
(Do not write the ratio as 6.)

17. 4 feet $= 4 \cdot 12$ inches $= 48$ inches
$$\dfrac{4\ \text{feet}}{30\ \text{inches}} = \dfrac{48\ \text{inches}}{30\ \text{inches}} = \dfrac{48}{30} = \dfrac{48 \div 6}{30 \div 6} = \dfrac{8}{5}$$
(Do not write the ratio as $1\frac{3}{5}$.)

21. 2 days $= 2 \cdot 24$ hours $= 48$ hours
$$\dfrac{15\ \text{hours}}{2\ \text{days}} = \dfrac{15\ \text{hours}}{48\ \text{hours}} = \dfrac{15}{48} = \dfrac{15 \div 3}{48 \div 3} = \dfrac{5}{16}$$

25. $\dfrac{440\ \text{pounds}}{60\ \text{pounds}} = \dfrac{440}{60} = \dfrac{440 \div 20}{60 \div 20} = \dfrac{22}{3}$
The ratio of the tiger's weight to the weight of its meal is $\frac{22}{3}$.

29. $\dfrac{4}{400} = \dfrac{4 \div 4}{400 \div 4} = \dfrac{1}{100}$
The ratio of defective washers to the total number of washers is $\frac{1}{100}$.

33. $\dfrac{6\ \text{million}}{20\ \text{million}} = \dfrac{6}{20} = \dfrac{6 \div 2}{20 \div 2} = \dfrac{3}{10}$
The ratio of organ players to guitar players is $\frac{3}{10}$.

37. $\dfrac{\text{Taxes} \rightarrow \$400}{\text{Transportation} \rightarrow \$200} = \dfrac{400 \div 200}{200 \div 200} = \dfrac{2}{1}$

41. $\dfrac{7\ \text{feet}}{5\ \text{feet}} = \dfrac{7}{5}$
The ratio of the length of the longest side to the length of the shortest side is $\frac{7}{5}$.

45. $\dfrac{9\frac{1}{2}\ \text{inches}}{4\frac{1}{4}\ \text{inches}} = \dfrac{9\frac{1}{2}}{4\frac{1}{4}} = \dfrac{\frac{19}{2}}{\frac{17}{4}} = \dfrac{19}{2} \div \dfrac{17}{4}$

$$= \dfrac{19}{\cancel{2}_{1}}\cdot\dfrac{\cancel{4}^{2}}{17} = \dfrac{38}{17}$$

The ratio of the length of the longest side to the length of the shortest side is $\frac{38}{17}$.

49. $59\frac{1}{2}$ days $\div 7 = \dfrac{\overset{17}{\cancel{119}}}{2} \cdot \dfrac{1}{\underset{1}{\cancel{7}}} = \dfrac{17}{2}$ weeks

$$\frac{\frac{17}{2}\ \cancel{\text{weeks}}}{8\frac{3}{4}\ \cancel{\text{weeks}}} = \frac{\frac{17}{2}}{\frac{35}{4}} = \frac{17}{2} \div \frac{35}{4} = \frac{17}{\underset{1}{\cancel{2}}} \cdot \frac{\overset{2}{\cancel{4}}}{35} = \frac{34}{35}$$

The ratio of $59\frac{1}{2}$ days to $8\frac{3}{4}$ weeks is $\frac{34}{35}$.

53.
$$
\begin{array}{r}
0.0928 \quad \text{Rounds to} \approx 0.093 \\
7\overline{)0.6500} \\
\underline{63} \\
20 \\
\underline{14} \\
60 \\
\underline{56} \\
4
\end{array}
$$

57.
$$
\begin{array}{r}
9.4647 \quad \text{Rounds to} \approx 9.465 \\
0.71\overline{)6.720000} \\
\underline{6\ 39} \\
330 \\
\underline{284} \\
460 \\
\underline{426} \\
340 \\
\underline{284} \\
560 \\
\underline{497} \\
63
\end{array}
$$

SECTION 5.2 (page 311)

1. $\dfrac{10\text{ cups}}{6\text{ people}} = \dfrac{10\text{ cups} \div 2}{6\text{ people} \div 2} = \dfrac{5\text{ cups}}{3\text{ people}}$

5. $\dfrac{14\text{ people}}{28\text{ dresses}} = \dfrac{14\text{ people} \div 2}{28\text{ dresses} \div 2} = \dfrac{1\text{ person}}{2\text{ dresses}}$

9. $\dfrac{\$63}{6\text{ visits}} = \dfrac{\$63 \div 3}{6\text{ visits} \div 3} = \dfrac{\$21}{2\text{ visits}}$

13. $\dfrac{\$60}{5\text{ hours}} = \dfrac{\$12}{1\text{ hour}} = \$12/\text{hour}$ Divide: $5\overline{)\$60}^{\ 12}$

17. $\dfrac{7.5\text{ pounds}}{6\text{ people}} = \dfrac{1.25\text{ pounds}}{1\text{ person}} = 1.25\text{ pounds/person}$

Divide: $6\overline{)7.50}^{\ 1.25}$

21. Subtract to find miles traveled. Then divide miles by gallons to get miles per gallon.
27,758.2 − 27,432.3 = 325.9
and 325.9 ÷ 15.5 = 21.02 rounds to 21.0
Earl's car mileage was ≈21.0 miles per gallon.

25. 4 ounces: $\dfrac{\$0.89}{4\text{ ounces}} \approx \$0.223$ per ounce

8 ounces: $\dfrac{\$2.13}{8\text{ ounces}} \approx \$0.266$ per ounce

The best buy (lower cost per ounce) is 4 ounces for $0.89.

29. 12 ounces: $\dfrac{\$1.09}{12\text{ ounces}} \approx \$0.091$ per ounce

18 ounces: $\dfrac{\$1.41}{18\text{ ounces}} \approx \$0.078$ per ounce ← lowest cost per ounce

28 ounces: $\dfrac{\$2.29}{28\text{ ounces}} \approx \$0.082$ per ounce

40 ounces: $\dfrac{\$3.19}{40\text{ ounces}} \approx \$0.080$ per ounce

The best buy is 18 ounces for $1.41.

33. $\dfrac{10.5\text{ pounds}}{6\text{ weeks}} = \dfrac{1.75\text{ pounds}}{1\text{ week}} = 1.75\text{ pounds/week}$

Divide: $6\overline{)10.50}^{\ 1.75}$
Her rate of loss was 1.75 pounds/week.

37. $\dfrac{\$1725}{150\text{ shares}} = \dfrac{\$11.50}{1\text{ share}} = \$11.50/\text{share}$

Divide: $150\overline{)1725.00}^{\ 11.50}$
One share costs $11.50.

41. $\dfrac{\$51.75}{4.6\text{ yards}} = \dfrac{\$11.25}{1\text{ yard}} = \$11.25/\text{yard}$

Divide: $4.6\overline{)51.750}^{\ 11.25}$
One yard of fabric costs $11.25.

45. Brand G: $2.39 − $0.60 coupon = $1.79

$\dfrac{\$1.79}{10\text{ ounces}} = \$0.179$ per ounce

Brand K: $\dfrac{\$3.99}{20.3\text{ ounces}} \approx \$0.197$ per ounce

Brand P: $3.39 − $0.50 coupon = $2.89

$\dfrac{\$2.89}{16.5\text{ ounces}} \approx \$0.175$ per ounce

Brand P with the 50¢ coupon has the lowest cost per ounce and is the best buy.

49. $5\frac{2}{5} \cdot 20 = \dfrac{27}{\underset{1}{\cancel{5}}} \cdot \dfrac{\overset{4}{\cancel{20}}}{1} = \dfrac{108}{1} = 108$

SECTION 5.3 (page 317)

1. $\dfrac{\$9}{12\text{ cans}} = \dfrac{\$18}{24\text{ cans}}$

5. $\dfrac{120}{150} = \dfrac{8}{10}$ The common units (feet) cancel.

9. $\dfrac{1\frac{1}{2}}{4\frac{1}{2}} = \dfrac{6}{18}$

13. $\frac{5}{8}$ is already in lowest terms

$$\frac{25}{40} = \frac{25 \div 5}{40 \div 5} = \frac{5}{8} \quad \text{(lowest terms)}$$

$$\frac{5}{8} = \frac{5}{8} \text{ so the proportion is true}$$

17. $\frac{42}{15} = \frac{42 \div 3}{15 \div 3} = \frac{14}{5} \quad \text{(lowest terms)}$

$$\frac{28}{10} = \frac{28 \div 2}{10 \div 2} = \frac{14}{5} \quad \text{(lowest terms)}$$

$$\frac{14}{5} = \frac{14}{5} \text{ so the proportion is true}$$

21. $\frac{7}{6}$ is already in lowest terms

$$\frac{54}{48} = \frac{54 \div 6}{48 \div 6} = \frac{9}{8} \quad \text{(lowest terms)}$$

$\frac{7}{6}$ does not equal $\frac{9}{8}$ so the proportion is false.

25.

$$\frac{20}{28} = \frac{12}{16} \qquad \begin{array}{l} 28 \cdot 12 = 336 \\ 20 \cdot 16 = 320 \end{array} \Big] \text{Unequal}$$

Cross products are not equal, so proportion is false.

29.

$$\frac{3.5}{4} = \frac{7}{8} \qquad \begin{array}{l} 4 \cdot 7 = 28 \\ 3.5 \cdot 8 = 28 \end{array} \Big] \text{Equal}$$

Cross products are equal, so proportion is true.

33.

$$\frac{6}{3\frac{2}{3}} = \frac{18}{11} \qquad \begin{array}{l} 3\frac{2}{3} \cdot 18 = \frac{11}{\cancel{3}} \cdot \frac{\cancel{18}^{6}}{1} = 66 \\ 6 \cdot 11 = 66 \end{array} \Big] \text{Equal}$$

Cross products are equal, so proportion is true.

37. (Answers may vary: proportion may be set up with "at bats" in both numerators and "hits" in both denominators. Cross products will be the same.)

Jerome Walton   Mariano Duncan

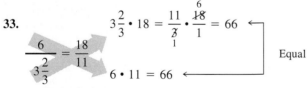

$$\frac{16 \text{ hits}}{50 \text{ at bats}} = \frac{128 \text{ hits}}{400 \text{ at bats}} \qquad \begin{array}{l} 50 \cdot 128 = 6400 \\ 16 \cdot 400 = 6400 \end{array}$$

Cross products are equal, so the proportion is true. Paul is correct.

41. $\frac{32}{40} = \frac{32 \div 8}{40 \div 8} = \frac{4}{5}$

45. $\frac{36}{63} = \frac{36 \div 9}{63 \div 9} = \frac{4}{7}$

SECTION 5.4 (page 325)

1. $\frac{1}{3} = \frac{x}{12}$ Check:

$3 \cdot x = 1 \cdot 12$

$3 \cdot x = 12$

$$\frac{\cancel{3}^{1} \cdot x}{\cancel{3}_{1}} = \frac{12}{3}$$

$x = 4$

$$\frac{1}{3} \diagdown \frac{4}{12} \qquad \begin{array}{l} 3 \cdot 4 = 12 \\ 1 \cdot 12 = 12 \end{array} \Big] \text{Equal}$$

Cross products are equal. Proportion is true.

5. $\frac{x}{11} = \frac{32}{4}$ Check:

$x \cdot 4 = 11 \cdot 32$

$x \cdot 4 = 352$

$$\frac{x \cdot \cancel{4}^{1}}{\cancel{4}_{1}} = \frac{352}{4}$$

$x = 88$

$$\frac{88}{11} \diagdown \frac{32}{4} \qquad \begin{array}{l} 11 \cdot 32 = 352 \\ 88 \cdot 4 = 352 \end{array}$$

Cross products are equal. Proportion is true.

9. $\frac{x}{25} = \frac{4}{20}$ Check:

$x \cdot 20 = 25 \cdot 4$

$x \cdot 20 = 100$

$$\frac{x \cdot \cancel{20}^{1}}{\cancel{20}_{1}} = \frac{100}{20}$$

$x = 5$

$$\frac{5}{25} \diagdown \frac{4}{20} \qquad \begin{array}{l} 25 \cdot 4 = 100 \\ 5 \cdot 20 = 100 \end{array}$$

Cross products are equal. Proportion is true.

13. $\frac{99}{55} = \frac{44}{x}$ Check:

$99 \cdot x = 55 \cdot 44$

$99 \cdot x = 2420$

$$\frac{\cancel{99}^{1} \cdot x}{\cancel{99}_{1}} = \frac{2420}{99}$$

$x \approx 24.44$
(rounded)

$$\frac{99}{55} \diagdown \frac{44}{24.44} \qquad \begin{array}{l} 55 \cdot 44 = 2420 \\ 99 \cdot 24.44 = 2419.56 \end{array}$$

Cross products are approximately equal (not exact due to rounding). Proportion is true.

17. $\frac{250}{24.8} = \frac{x}{1.75}$ Check:

$24.8 \cdot x = 250 \cdot 1.75$

$24.8 \cdot x = 437.5$

$$\frac{\cancel{24.8}^{1} \cdot x}{\cancel{24.8}_{1}} = \frac{437.5}{24.8}$$

$x \approx 17.64$
(rounded)

$$24.8 \cdot 17.64 = 437.472$$

$$\frac{250}{24.8} \diagdown \frac{17.64}{1.75}$$

$$250 \cdot 1.75 = 437.5$$

Cross products are approximately equal (not exact due to rounding). Proportion is true.

21. $\dfrac{15}{1\frac{2}{3}} = \dfrac{9}{x}$

$15 \cdot x = 1\frac{2}{3} \cdot 9$

$15 \cdot x = 15$

$\dfrac{\overset{1}{\cancel{15}} \cdot x}{\underset{1}{\cancel{15}}} = \dfrac{15}{15}$

$x = 1$

25. Change $\frac{1}{2}$ to a decimal by dividing: $1 \div 2 = 0.5$

$\dfrac{0.5}{x} = \dfrac{2}{0.8}$ Check:

$x \cdot 2 = 0.5 \cdot 0.8$ $\dfrac{0.5}{0.2} \bowtie \dfrac{2}{0.8}$ $0.2 \cdot 2 = 0.4$

$x \cdot \overset{1}{\cancel{2}} = 0.4$

$\dfrac{x \cdot \overset{1}{\cancel{2}}}{\underset{1}{\cancel{2}}} = \dfrac{0.4}{2}$ $0.5 \cdot 0.8 = 0.4$

$x = 0.2$ Cross products are equal. Proportion is true.

Change 0.8 to a fraction and write it in lowest terms:

$0.8 = \dfrac{8 \div 2}{10 \div 2} = \dfrac{4}{5}$

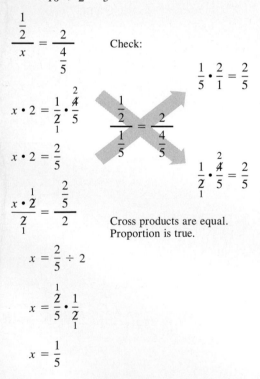

$\dfrac{\frac{1}{2}}{x} = \dfrac{2}{\frac{4}{5}}$ Check:

$\dfrac{1}{5} \cdot \dfrac{2}{1} = \dfrac{2}{5}$

$x \cdot 2 = \dfrac{1}{\cancel{2}} \cdot \dfrac{\overset{2}{\cancel{4}}}{5}$

$x \cdot 2 = \dfrac{2}{5}$

$\dfrac{x \cdot \overset{1}{\cancel{2}}}{\underset{1}{\cancel{2}}} = \dfrac{\frac{2}{5}}{2}$ Cross products are equal. Proportion is true.

$x = \dfrac{2}{5} \div 2$

$x = \dfrac{\overset{1}{\cancel{2}}}{5} \cdot \dfrac{1}{\underset{1}{\cancel{2}}}$

$x = \dfrac{1}{5}$

Notice that $\frac{1}{5}$ in decimal form is $1 \div 5 = 0.2$

29. 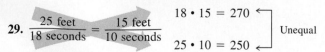 $\dfrac{25 \text{ feet}}{18 \text{ seconds}} \bowtie \dfrac{15 \text{ feet}}{10 \text{ seconds}}$ $\left.\begin{array}{l} 18 \cdot 15 = 270 \\ 25 \cdot 10 = 250 \end{array}\right\}$ Unequal

Cross products are not equal, so proportion is false.

SECTION 5.5 (page 329)

1. Let x represent the time to sketch 18 cartoon strips.

$\dfrac{5 \text{ hours}}{4 \text{ cartoon strips}} = \dfrac{x \text{ hours}}{18 \text{ cartoon strips}}$ Both rates compare hours to cartoon strips in the same order.

$\dfrac{5}{4} = \dfrac{x}{18}$

$4 \cdot x = 5 \cdot 18$

$4 \cdot x = 90$

$\dfrac{\overset{1}{\cancel{4}} \cdot x}{\underset{1}{\cancel{4}}} = \dfrac{90}{4}$

$x = 22.5$

It will take 22.5 hours to sketch 18 cartoon strips.

5. Let x represent the pounds of seed needed for 4900 square feet.

$\dfrac{5 \text{ pounds}}{3500 \text{ square feet}} = \dfrac{x \text{ pounds}}{4900 \text{ square feet}}$ Both rates compare pounds to square feet in the same order.

$\dfrac{5}{3500} = \dfrac{x}{4900}$

$3500 \cdot x = 5 \cdot 4900$

$3500 \cdot x = 24{,}500$

$\dfrac{\overset{1}{\cancel{3500}} \cdot x}{\underset{1}{\cancel{3500}}} = \dfrac{24{,}500}{3500}$

$x = 7$

7 pounds of grass seed are needed.

9. Let x represent the actual length of the kitchen.

$\dfrac{1 \text{ inch}}{4 \text{ feet}} = \dfrac{3.5 \text{ inches}}{x \text{ feet}}$ Both ratios compare inches to feet in the same order.

$1 \cdot x = 4 \cdot 3.5$

$1 \cdot x = 14$

$x = 14$

The kitchen is 14 feet long.

Let x represent the actual width of the kitchen.

$\dfrac{1 \text{ inch}}{4 \text{ feet}} = \dfrac{2.5 \text{ inches}}{x \text{ feet}}$ Both ratios compare inches to feet in the same order.

$1 \cdot x = 4 \cdot 2.5$

$1 \cdot x = 10$

$x = 10$

The kitchen is 10 feet wide.

13. Let x represent the runs given up in a 9-inning game.

$$\frac{78 \text{ runs}}{234 \text{ innings}} = \frac{x \text{ runs}}{9 \text{ innings}}$$

Both rates compare runs to innings in the same order.

$$\frac{78}{234} = \frac{x}{9}$$

$$234 \cdot x = 78 \cdot 9$$

$$234 \cdot x = 702$$

$$\frac{\overset{1}{\cancel{234}} \cdot x}{\underset{1}{\cancel{234}}} = \frac{702}{234}$$

$$x = 3$$

The Cardinal's pitcher will give up 3 runs in a 9-inning game.

17. Let x represent the number of people who choose vanilla ice cream.

$$\frac{4 \text{ choose vanilla}}{5 \text{ people}} = \frac{x \text{ choose vanilla}}{238 \text{ people}}$$

Both ratios compare people who choose vanilla to the total group of people.

$$5 \cdot x = 4 \cdot 238$$

$$5 \cdot x = 952$$

$$\frac{\overset{1}{\cancel{5}} \cdot x}{\underset{1}{\cancel{5}}} = \frac{952}{5}$$

$$x = 190.4 \text{ rounds to } 190$$

You would expect 190 people to choose vanilla ice cream. This is a reasonable answer.

Now flip one side of the proportion and solve.

$$\frac{4}{5} = \frac{238}{x}$$

$$4 \cdot x = 5 \cdot 238$$

$$4 \cdot x = 1190$$

$$\frac{\overset{1}{\cancel{4}} \cdot x}{\underset{1}{\cancel{4}}} = \frac{1190}{4}$$

$$x = 297.5 \text{ rounds to } 298$$

With an incorrect setup, 298 people choose vanilla ice cream. This is unreasonable because only 238 people attended the ice cream social.

21. Let x represent the number of stocks that went up.

$$\frac{5 \text{ stocks up}}{6 \text{ stocks down}} = \frac{x \text{ stocks up}}{750 \text{ stocks down}}$$

Both ratios compare stocks going up to stocks going down in the same order.

$$\frac{5}{6} = \frac{x}{750}$$

$$6 \cdot x = 5 \cdot 750$$

$$6 \cdot x = 3750$$

$$\frac{\overset{1}{\cancel{6}} \cdot x}{\underset{1}{\cancel{6}}} = \frac{3750}{6}$$

$$x = 625$$

625 stocks went up.

25. Let x represent the width of the wing.

$$\frac{8 \text{ length}}{1 \text{ width}} = \frac{32.5 \text{ meters length}}{x \text{ meters width}}$$

Both ratios compare length to width in the same order.

$$\frac{8}{1} = \frac{32.5}{x}$$

$$8 \cdot x = 1 \cdot 32.5$$

$$8 \cdot x = 32.5$$

$$\frac{\overset{1}{\cancel{8}} \cdot x}{\underset{1}{\cancel{8}}} = \frac{32.5}{8}$$

$$x = 4.0625 \text{ rounds to } 4.06$$

The wing must be ≈ 4.06 meters wide.

29. You cannot solve this problem using a proportion because the ratio of age to weight is not constant. As Jim's age increases from 25 to 50 years old, his weight may decrease, stay the same, or increase.

33. Let x represent the amount of margarine for 15 servings.

$$\frac{6 \text{ tablespoons}}{12 \text{ servings}} = \frac{x \text{ tablespoons}}{15 \text{ servings}}$$

Both rates compare tablespoons to servings in the same order

$$12 \cdot x = 6 \cdot 15$$

$$12 \cdot x = 90$$

$$\frac{\overset{1}{\cancel{12}} \cdot x}{\underset{1}{\cancel{12}}} = \frac{90}{12}$$

$$x = 7.5 \quad \text{or} \quad 7\frac{1}{2}$$

Use $7\frac{1}{2}$ tablespoons of margarine for 15 servings.

37. $0.06 \times 100 = 0.06 = 6$
Move the decimal point two places right.

41. $1.93 \div 100 = 01.93 = 0.0193$
Move the decimal point two places left.

CHAPTER 6
SECTION 6.1 (page 351)

1. $25\% = 025. = 0.25$ Percent sign is dropped.
Decimal is moved two places to the left.

5. $55\% = 055. = 0.55$

9. $7.8\% = 007.8 = 0.078$ 0 is attached so the decimal point can be moved two places to the left.

SOLUTIONS

13. $0.5\% = 0\underset{\smile\smile}{0}0.5 = 0.005$ Two zeros are attached.

17. $0.5 = 0.\underset{\frown}{50}\% = 50\%$ Percent sign is attached.
————————— Decimal point is moved two places to the right.

21. $0.03 = 0.0\underset{\smile}{3}\% = 3\%$

25. $0.629 = 0.6\underset{\smile}{2}9\% = 62.9\%$

29. $2.6 = 2.\underset{\smile}{60} = 260\%$ Add 1 zero to right so that decimal point can be moved 2 places to the right.

33. $4.162 = 4.1\underset{\smile}{62}\% = 416.2\%$

37. Answers will vary. A sample answer follows: No common denominators are needed with percents. The denominator is always 100 with percent which makes comparisons easier to understand.

41. 65% of the salespeople $= 0\underset{\smile}{65}. = 0.65$

45. Success rate is 2 times $= 2.\underset{\smile}{00}\% = 200\%$

49. Blood pressure is 153.6% of normal $= 1\underset{\smile}{53}.6 = 1.536$

53. There are 20 children in the preschool class. 100% of the children are served breakfast and lunch. The number served both meals is 20 children. Since 100% represents *all* of the children, all 20 children were served both meals.

57. John owes $285 for tuition. Financial aid will pay 50% of the cost. Financial aid will pay $142.50. Since 50% represents *half* of the cost of tuition, financial aid will pay $\frac{1}{2} \cdot \$285 = \$142.50$.

61. Since 100% means 100 parts out of 100 parts, 100% is all of the number.

65. 3 out of 10 shaded $= \dfrac{3}{10} = 0.3 = 30\%$ shaded

7 out of 10 unshaded $= \dfrac{7}{10} = 0.7 = 70\%$ unshaded

69. 55 out of 100 shaded $= \dfrac{55}{100} = 0.55 = 55\%$ shaded

45 out of 100 unshaded $= \dfrac{45}{100} = 0.45 = 45\%$ unshaded

73.
$$
\begin{array}{r}
0.75 \\
4\overline{)3.00} \\
2\,8 \\
\hline
20 \\
20 \\
\hline
0
\end{array}
\qquad \frac{3}{4} = 0.75
$$

77.
$$
\begin{array}{r}
0.8 \\
5\overline{)4.0} \\
4\,0 \\
\hline
0
\end{array}
\qquad \frac{4}{5} = 0.8
$$

SECTION 6.2 (page 361)

1. $20\% = \dfrac{20}{100} = \dfrac{1}{5}$ **5.** $55\% = \dfrac{55}{100} = \dfrac{11}{20}$

9. $6.25\% = \dfrac{6.25}{100} = \dfrac{6.25 \cdot 100}{100 \cdot 100} = \dfrac{625}{10,000} = \dfrac{1}{16}$

13. $6\dfrac{2}{3}\% = \dfrac{6\frac{2}{3}}{100} = \dfrac{\frac{20}{3}}{100} = \dfrac{20}{3} \div \dfrac{100}{1} = \dfrac{\overset{1}{\cancel{20}}}{3} \cdot \dfrac{1}{\underset{5}{\cancel{100}}}$

$\phantom{6\dfrac{2}{3}\%} = \dfrac{1}{15}$

17. $130\% = \dfrac{130}{100} = \dfrac{13}{10}$ or $1\dfrac{3}{10}$

21. $\dfrac{1}{4} = \dfrac{p}{100}$

$4 \cdot p = 1 \cdot 100$

$4 \cdot p = 100$

$\dfrac{\overset{1}{\cancel{4}} \cdot p}{\underset{1}{\cancel{4}}} = \dfrac{100}{4}$

$p = 25$

$\dfrac{1}{4} = 25\%$

25. $\dfrac{3}{5} = \dfrac{p}{100}$

$5 \cdot p = 3 \cdot 100$

$5 \cdot p = 300$

$\dfrac{\overset{1}{\cancel{5}} \cdot p}{\underset{1}{\cancel{5}}} = \dfrac{300}{5}$

$p = 60$

$\dfrac{3}{5} = 60\%$

29. $\dfrac{5}{8} = \dfrac{p}{100}$

$8 \cdot p = 5 \cdot 100$

$8 \cdot p = 500$

$\dfrac{\overset{1}{\cancel{8}} \cdot p}{\underset{1}{\cancel{8}}} = \dfrac{500}{8}$

$p = 62.5$

$\dfrac{5}{8} = 62.5\%$

33. $\dfrac{11}{25} = \dfrac{p}{100}$

$25 \cdot p = 11 \cdot 100$

$25 \cdot p = 1100$

$\dfrac{\overset{1}{\cancel{25}} \cdot p}{\underset{1}{\cancel{25}}} = \dfrac{1100}{25}$

$p = 44$

$\dfrac{11}{25} = 44\%$

37. $\dfrac{1}{20} = \dfrac{p}{100}$

$20 \cdot p = 1 \cdot 100$

$20 \cdot p = 100$

$\dfrac{\overset{1}{\cancel{20}} \cdot p}{\underset{1}{\cancel{20}}} = \dfrac{100}{20}$

$p = 5$

$\dfrac{1}{20} = 5\%$

41. $\dfrac{5}{9} = \dfrac{p}{100}$

$9 \cdot p = 5 \cdot 100$

$9 \cdot p = 500$

$\dfrac{\overset{1}{\cancel{9}} \cdot p}{\underset{1}{\cancel{9}}} = \dfrac{500}{9}$

$p \approx 55.6$

$\dfrac{5}{9} \approx 55.6\%$

45. Decimal: 0.75 (given)

Fraction: $0.75 = \dfrac{75}{100} = \dfrac{3}{4}$

Percent: $0.75 = 0.\underset{\smile}{75}\% = 75\%$

49. Decimal: 0.8 (given)

Fraction: $0.8 = \dfrac{8}{10} = \dfrac{4}{5}$

Percent: $0.8 = 0.\underset{\smile}{80}\% = 80\%$

53. Decimal: 0.25 (given)

Fraction: $0.25 = \dfrac{25}{100} = \dfrac{1}{4}$

Percent: $0.25 = 0.\underset{\smile}{25}\% = 25\%$

57. Fraction: $\dfrac{2}{3}$ (given)

Decimal: $\dfrac{2}{3} \approx 0.667$ (rounded)

Percent: $\dfrac{2}{3} \approx 0.667 = 0.667\%$

$\qquad \approx 66.7\%$ (rounded)

61. Fraction: $\dfrac{1}{100}$ (given)

Decimal: $\dfrac{1}{100} = 0.01$

Percent: $\dfrac{1}{100} = 0.01 = 0.01\% = 1\%$

65. Decimal: 2.5 (given)

Fraction: $2.5 = 2\dfrac{5}{10} = 2\dfrac{1}{2}$

Percent: $2.5 - 2.50\% - 250\%$

69. Answers will vary. A sample answer follows:

$$12.5\% = \frac{12.5}{100} = \frac{12.5 \cdot 10}{100 \cdot 10} = \frac{125}{1000} = \frac{1}{8}$$

$\qquad\qquad\qquad$ Multiply by 10 $\qquad$ Lowest terms

$\dfrac{4}{5} = \dfrac{p}{100}$ $5 \cdot p = 4 \cdot 100$ Cross products

$\qquad\qquad 5 \cdot p = 400$

$\dfrac{\overset{1}{\cancel{5}} \cdot p}{\underset{1}{\cancel{5}}} = \dfrac{\overset{80}{\cancel{400}}}{\underset{1}{\cancel{5}}}$ Divide each side by 5.

$\qquad\qquad p = 80 - 80\%$

73. Reduction is $150 out of $750.

Fraction: $\dfrac{150}{750} = \dfrac{1}{5}$

Decimal: $\dfrac{1}{5} = 0.20$

Percent: $0.20 = 20\%$

77. 64 out of 80 employees have cellular phones. Therefore, $80 - 64 = 16$ employees do not have cellular phones.

Fraction: $\dfrac{16}{80} = \dfrac{1}{5}$

Decimal: $\dfrac{1}{5} = 0.20$

Decimal: $0.20 = 20\%$

81. 1050 students out of 4200 students use public transportation.

Fraction: $\dfrac{1050}{4200} = \dfrac{1}{4}$

Decimal: $\dfrac{1}{4} = 0.25$

Percent: $0.25 = 25\%$

85. $\dfrac{8}{4} = \dfrac{x}{15}$

$4 \cdot x = 8 \cdot 15$

$4 \cdot x = 120$

$\dfrac{\overset{1}{\cancel{4}} \cdot x}{\underset{1}{\cancel{4}}} = \dfrac{120}{4}$

$x = 30$

89. $\dfrac{42}{30} = \dfrac{14}{b}$

$42 \cdot b = 30 \cdot 14$

$42 \cdot b = 420$

$\dfrac{\overset{1}{\cancel{42}} \cdot b}{\underset{1}{\cancel{42}}} = \dfrac{420}{42}$

$b = 10$

SECTION 6.3 (page 369)

1. $\dfrac{10}{b} = \dfrac{20}{100}$

$\dfrac{10}{b} = \dfrac{1}{5}$

$b \cdot 1 = 10 \cdot 5$

$b = 50$

5. $\dfrac{24}{b} = \dfrac{30}{100}$

$\dfrac{24}{b} = \dfrac{3}{10}$

$b \cdot 3 = 24 \cdot 10$

$b \cdot 3 = 240$

$\dfrac{b \cdot \overset{1}{\cancel{3}}}{\underset{1}{\cancel{3}}} = \dfrac{240}{3}$

$b = 80$

9. $\dfrac{55}{110} = \dfrac{p}{100}$

$\dfrac{1}{2} = \dfrac{p}{100}$

$2 \cdot p = 100$

$\dfrac{\overset{1}{\cancel{2}} \cdot p}{\underset{1}{\cancel{2}}} = \dfrac{100}{2}$

$p = 50$

The percent is 50%.

13. $\dfrac{1.5}{4.5} = \dfrac{p}{100}$

$\dfrac{1}{3} = \dfrac{p}{100}$

$3 \cdot p = 100$

$\dfrac{\overset{1}{\cancel{3}} \cdot p}{\underset{1}{\cancel{3}}} - \dfrac{100}{1}$

$p \approx 33.3$

The percent is $\approx 33.3\%$.

17. $\dfrac{a}{72} = \dfrac{30}{100}$

$\dfrac{a}{72} = \dfrac{3}{10}$

$a \cdot 10 = 72 \cdot 3$

$a \cdot 10 = 216$

$\dfrac{a \cdot \overset{1}{\cancel{10}}}{\underset{1}{\cancel{10}}} = \dfrac{216}{10}$

$a = 21.6$

21. Answers will vary. A sample answer follows:

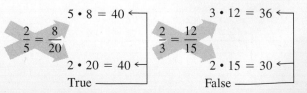

$\qquad$ True $\qquad\qquad\qquad\qquad$ False

SOLUTIONS

25. $\dfrac{20}{5000} = \dfrac{p}{100}$

$\dfrac{1}{250} = \dfrac{p}{100}$

$250 \cdot p = 100$

$\dfrac{\overset{1}{\cancel{250}} \cdot p}{\underset{1}{\cancel{250}}} = \dfrac{100}{250}$

$p = 0.4$

The percent is 0.4%.

33. $\dfrac{57}{100} = 0.57 = 57\%$

37. $\dfrac{2}{3} \approx 0.667 \approx 66.7\%$

29. $\dfrac{16.2}{6480} = \dfrac{p}{100}$

$6480 \cdot p = 16.2 \cdot 100$

$6480 \cdot p = 1620$

$\dfrac{\overset{1}{\cancel{6480}} \cdot p}{\underset{1}{\cancel{6480}}} = \dfrac{1620}{6480}$

$p = 0.25$

The percent is 0.25%.

Section 6.4 (page 373)

1. 20% of how many salespeople is 80 salespeople?
  p $\qquad$ b $\qquad$ a
$p = 20$ $b =$ unknown $a = 80$

5. What is 15% of $75?
  a $\quad$ p $\quad$ b $p = 15$ $b = 75$ $a =$ unknown

9. 34 trophies is 50% of 68 trophies.
  a $\qquad$ p $\qquad$ b
$p = 50$ $b = 68$ $a = 34$

13. $27.17 is 6.5% of what number?
  a $\qquad$ p $\qquad$ b
$p = 6.5$ $b =$ unknown $a = 27.17$

17. Answers will vary. A sample answer follows:
p (percent)—the ratio of the part to the whole. It appears with the word "percent" or "%" after it.
b (base)—the entire quantity or total. Often appears after the word "of."
a (amount)—the part being compared with the whole.

21. $p =$ unknown   (what percent)
$b = 15$ $\qquad$ (cost of compact disc)
$a = 0.90$ $\qquad$ (sales tax)

25. $p = 25$ $\qquad$ (25% of the bars)
$b =$ unknown   (total number of bars)
$a = 240$ $\qquad$ (candy bars sold)

29. $p = 32$ $\qquad$ (32% of the people)
$b = 272$ $\qquad$ (people tested)
$a =$ unknown   (number having high cholesterol)

33. $\dfrac{8}{x} = \dfrac{15}{30}$ (proportion)

$15 \cdot x = 8 \cdot 30$

$15 \cdot x = 240$

$\dfrac{\overset{1}{\cancel{15}} \cdot x}{\underset{1}{\cancel{15}}} = \dfrac{240}{15}$

$x = 16$ (missing number)

Section 6.5 (page 383)

1. $0.05 \cdot 1040 = 52$
$\qquad$ $a = 52$ bowlers

5. $0.04 \cdot 120 = 4.8$
$\qquad$ $a = 4.8$ feet

9. $0.225 \cdot 1100 = 247.5$
$\qquad$ $a = 247.5$ boxes

13. $2.5 \cdot 740 = 1850$
$\qquad$ $a = 1850$ sales

17. $0.009 \cdot 2400 = 21.6$
$\qquad$ $a = \$21.60$

21. $p = 20$ $a = 34$

$\dfrac{34}{b} = \dfrac{20}{100}$

$\dfrac{34}{b} = \dfrac{1}{5}$

$b = 170$

25. $p = 110$ $a = 748$

$\dfrac{748}{b} = \dfrac{110}{100}$

$\dfrac{748}{b} = \dfrac{11}{10}$

$11b = 7480$

$\dfrac{\overset{1}{\cancel{11}}b}{\underset{1}{\cancel{11}}} = \dfrac{7480}{11}$

$b = 680$ books

29. $b = 32$ $a = 16$

$\dfrac{16}{32} = \dfrac{p}{100}$

$32p = 1600$

$\dfrac{\overset{1}{\cancel{32}}p}{\underset{1}{\cancel{32}}} = \dfrac{1600}{32}$

$p = 50$

16 is 50% of 32.

33. $b = 200$ $a = 16$

$\dfrac{16}{200} = \dfrac{p}{100}$

$200p = 1600$

$\dfrac{\overset{1}{\cancel{200}}p}{\underset{1}{\cancel{200}}} = \dfrac{1600}{200}$

$p = 8$

16 is 8% of 200.

37. $b = 172$ $a = 32$

$$\frac{32}{172} = \frac{p}{100}$$

$$172p = 3200$$

$$\frac{\overset{1}{\cancel{172}}p}{\underset{1}{\cancel{172}}} = \frac{3200}{172}$$

$$p \approx 18.60 \text{ rounds to } 18.6$$

18.6 of 172 is 32.

41. 150% of \$30 cannot be less than \$30 because 150% is greater than 1 (100%). The answer must be greater than \$30.

25% of \$16 cannot be greater than \$16 because 25% is less than 1 (100%). The answer must be less than \$16.

45. $p = 38$ (% wearing seat belts)

$b = 2200$ (total number of drivers)

$a = \text{unknown}$ (number wearing seat belts)

$$\frac{a}{2200} = \frac{38}{100}$$

$$100a = 83600$$

$$\frac{\overset{1}{\cancel{100}}a}{\underset{1}{\cancel{100}}} = \frac{83600}{100}$$

$$a = 836$$

836 drivers were wearing seat belts.

49. First find the profit.

$p = 6.7$ (% of profit)

$b = 21,200$ (cost of van)

$a = \text{unknown}$ (profit)

$$\frac{a}{21,200} = \frac{6.7}{100}$$

$$100a = 142,040$$

$$\frac{\overset{1}{\cancel{100}}a}{\underset{1}{\cancel{100}}} = \frac{142,040}{100}$$

$$a = 1420.4$$

The profit on the van is \$1420.40. Therefore, the selling price of the van is \$21,200 + \$1420.40 = \$22,620.40.

53. **(a)** If 61% of Cuba's labor force is male, then 100% − 61% = 39% of Cuba's labor force is female.

(b) $p = 61\%$ (percent of male workers)

$b = 3.8$ million   (number of workers)

$a = \text{unknown}$ (number of male workers)

$$\frac{a}{3.8} = \frac{61}{100}$$

$$100a = 231.8$$

$$\frac{\overset{1}{\cancel{100}}a}{\underset{1}{\cancel{100}}} = \frac{231.8}{100}$$

$$a = 2.318 \text{ million} \approx 2.3 \text{ million}$$

About 2.3 million workers in Cuba are male.

57. $p = 12.7$ (percent of the total population)

$b = \text{unknown}$ (total American population, in millions)

$a = 31.5$ (Americans who are 65 or older, in millions)

$$\frac{31.5}{b} = \frac{12.7}{100}$$

$$12.7b = 3150$$

$$\frac{\overset{1}{\cancel{12.7}}b}{\underset{1}{\cancel{12.7}}} = \frac{3150}{12.7}$$

$$b \approx 248.03 \text{ rounds to } 248.0$$

There are 248.0 million Americans in the U.S.

61. The highest sales month was January when 15% of the total cans of chicken noodle soup were sold.

$p = 15$ (percent of cans sold)

$b = 350$ million   (number of cans sold annually)

$a = \text{unknown}$ (number of cans sold in January)

$$\frac{a}{350} = \frac{15}{100}$$

$$100a = 5250$$

$$\frac{\overset{1}{\cancel{100}}a}{\underset{1}{\cancel{100}}} = \frac{5250}{100}$$

$$a = 52.5 \text{ million cans}$$

There were 52.5 million cans of chicken noodle soup sold in January.

65. $p = 86$ (percent of products which failed)

$b = 15,401$ (number of new products)

$a = \text{unknown}$ (number of products which failed)

$$\frac{a}{15,401} = \frac{86}{100}$$

$$100a = 1,324,486$$

$$\frac{\overset{1}{\cancel{100}}a}{\underset{1}{\cancel{100}}} = \frac{1,324,486}{100}$$

$$a \approx 13,244.86 \text{ rounds to } 13,245$$

15,401 − 13,245 = 2156

2156 products reached their objectives.

69.
$$
\begin{array}{r}
2\,18.7 \leftarrow 1 \text{ decimal place} \\
\times\ 0.04\,2 \leftarrow 3 \text{ decimal places} \\
\hline
4\,37\,4 \\
8\,7\,48 \\
\hline
9.1\,85\,4 \leftarrow 4 \text{ decimal places}
\end{array}
$$

73.
$$
\begin{array}{r}
0.25 \\
688\overline{)172.00} \\
137\,6 \\
\hline
34\,40 \\
34\,40 \\
\hline
0
\end{array}
$$

SECTION 6.6 (page 393)

1. amount = percent • base
   $a = 0.35 • 660$ Write 35% as a decimal.
   $a = 231$
   35% of 660 programs is 231 programs.

5. amount = percent • base
   $a = 0.32 • 260$
   $a = 83.2$
   32% of 260 quarts is 83.2 quarts.

9. amount = percent • base
   $a = 0.124 • 8300$
   $a = 1029.2$
   12.4% of 8300 meters is 1029.2 meters.

13. amount = percent • base
   $142 = 0.50 • b$

   $$\frac{142}{0.50} = \frac{\overset{1}{\cancel{0.50}} • b}{\underset{1}{\cancel{0.50}}}$$

   $284 = b$

   142 employees is 50% of 284 employees.

17. amount = percent • base
   $476 = 0.70 • b$

   $$\frac{476}{0.70} = \frac{\overset{1}{\cancel{0.70}} • b}{\underset{1}{\cancel{0.70}}} • b$$

   $680 = b$

   476 circuits is 70% of 680 circuits.

21. amount = percent • base
   $3.75 = 0.0125 • b$

   $$\frac{3.75}{0.0125} = \frac{\overset{1}{\cancel{0.0125}} • b}{\underset{1}{\cancel{0.0125}}}$$

   $300 = b$

   $1\frac{1}{4}$% of 300 gallons is 3.75 gallons.

25. amount = percent • base
   $38 = p • 50$

   $$\frac{38}{50} = \frac{p • \overset{1}{\cancel{50}}}{\underset{1}{\cancel{50}}}$$

   $0.76 = p$ Percent = 76%

   38 styles is 76% of 50 styles.

29. amount = percent • base
   $2.4 = p • 160$

   $$\frac{2.4}{160} = \frac{p • \overset{1}{\cancel{160}}}{\underset{1}{\cancel{160}}}$$

   $0.015 = p$ Percent = 1.5%

   1.5% of 160 liters is 2.4 liters.

33. You must first change the fraction in the percent to a decimal, then divide the percent by 100 to change it to a decimal.

   $$2\tfrac{1}{2}\% = 2.5\% = 0.025$$
   Change $\frac{1}{2}$ $2\frac{1}{2}$% as a
   to decimal.   decimal

37. amount = percent • base
   $a = 0.84 • 1100$
   $a = 924$
   924 employers offer only one health plan.

41. amount = percent • base
   $338 = 0.186 • b$

   $$\frac{338}{0.186} = \frac{\overset{1}{\cancel{0.186}} • b}{\underset{1}{\cancel{0.186}}}$$

   $1817.2 \approx b$

   The total value of all mortgages made last year was about $1817.2 million.

45. amount = percent • base
   $a = 0.325 • 385,200$
   $a = 125,190$
   There was an increase in sales of $125,190. Therefore, the volume of sales this year is
   $385,200 + $125,190 = $510,390

49. 23% of 500 hinges is 115 hinges.
       p b a
   $p = 23$ $b = 500$ $a = 115$

53. What percent of $830 is $128.65?
       p b a
   $p = $ unknown $b = 830$ $a = 128.65$

SECTION 6.7 (page 401)

1. Sales tax = rate of tax • cost of item
          $= 0.06 • 100
          $= 6
   Sales tax is $6 and the total cost is $106 ($100 + $6).

5. Sales tax = rate of tax • cost of item
          $= 0.06 • 365
          $= 21.90
   Sales tax is $21.90 and the total cost is $386.90 ($365 + $21.90).

9. commission = rate of commission • sales
            $= 0.07 • 300
            $= 21
   The amount of commission is $21.

13. commission = rate of commission • sales
            $= 0.03 • 5783
            $= 173.49
   The amount of commission is $173.49.

17. discount = rate of discount • original price
          $= 0.15 • 100
          $= 15
   The amount of discount is $15 and the sale price is $85 ($100 − $15).

21. discount = rate of discount • original price
$$= 0.25 • \$17.50$$
$$= \$4.38 \text{ (rounded)}$$
The amount of discount is \$4.38 and the sale price is \$13.12 (\$17.50 − \$4.38).

25. Answers will vary. A sample answer follows:
On the basis of commission alone, I would choose Company A. Other considerations might be: reputation of the company; expense allowances; other fringe benefits; travel; promotion and training, to name a few.

29. discount = rate of discount • original price
$$= 0.40 • \$3850$$
$$= \$1540$$
The sale price of the diamond ring is \$2310 (\$3850 − \$1540).

33. amount of increase = \$1449 − \$1228 = \$221

$$\text{amount} = \text{percent} • \text{base}$$
$$221 = p • 1228$$

$$\frac{221}{1228} = \frac{p • \overset{1}{\cancel{1228}}}{\underset{1}{\cancel{1228}}}$$

$$0.180 \approx p \quad \text{Percent} \approx 18.0\%$$

Student tuition increased by about 18.0%.

37. discount = rate of discount • original price
$$= 0.45 • \$135$$
$$= \$60.75$$
The sale price of the ski parka is \$74.25 (\$135 − \$60.75).

41. discount = rate of discount • original price
$$= 0.18 • \$590$$
$$= \$106.20$$
The discount is \$106.20, and the sale price is \$483.80 (\$590 − \$106.20).

45. discount = rate of discount • original price
$$= 0.06 • \$18.50$$
$$= \$1.11$$
The sale price is \$17.39 (\$18.50 − \$1.11).
Sales tax = rate of tax • cost of item
$$= 0.06 • \$17.39$$
$$\approx \$1.04$$
The total cost of the dictionary is \$18.43 (\$17.39 + \$1.04).

49. discount = rate of discount • original price
$$= 0.18 • \$13,905$$
$$= \$2502.90$$
The sale price is \$11,402.10 (\$13,905 − \$2502.90).
Sales tax = rate of tax • cost of item
$$= 0.0475 • \$11,402.10$$
$$\approx \$541.60$$
The total cost of the boat is \$11,943.70 (\$11,402.10 + \$541.60).

53. amount = percent • base
$$a = 0.002 • 1920$$
$$a = 3.84$$
0.2% of \$1920 is \$3.84.

57. amount = percent • base

$$147.2 = p • 460$$

$$\frac{147.2}{460} = \frac{p • \overset{1}{\cancel{460}}}{\underset{1}{\cancel{460}}}$$

$$0.32 = p \quad \text{Percent} = 32\%$$

147.2 meters is 32% of 460 meters.

Section 6.8 (page 409)

1. $I = p • r • t$
$$= 200 • (0.04) • 1$$
$$= 8$$
The interest is \$8.

5. $I = p • r • t$
$$= 240 • (0.04) • 3$$
$$= 28.8$$
The interest is \$28.80.

9. $I = p • r • t$
$$= 9400 • (0.065) • 1.25$$
$$= 763.75$$
The interest is \$763.75.

13. $I = p • r • t$
$$= 750 • (0.05) • \frac{12}{12}$$
$$= 37.5 • 1$$
$$= 37.5$$
The interest is \$37.50.

17. $I = p • r • t$
$$= 1250 • (0.055) • \frac{3}{12}$$
$$= 68.75 • \frac{1}{4}$$
$$\approx 17.19$$
The interest is \$17.19.

21. $I = p • r • t$
$$= 300 • (0.04) • 1$$
$$= 12$$
The interest is \$12.
amount due = principal + interest
$$= \$300 + \$12$$
$$= \$312$$
The total amount due is \$312.

25. $I = p • r • t$
$$= 1500 • (0.10) • \frac{18}{12}$$
$$= 150 • \frac{3}{2}$$
$$= \$225$$
The interest is \$225.
amount due = principal + interest
$$= \$1500 + \$225$$
$$= \$1725$$
The total amount due is \$1725.

SOLUTIONS

29. $I = p \cdot r \cdot t$

$$= 17{,}800 \cdot (0.075) \cdot \frac{9}{12}$$

$$= 1335 \cdot \frac{3}{4}$$

$$= 1001.25$$

The interest is $1001.25.
amount due = principal + interest

$$= \$17{,}800 + \$1001.25$$

$$= \$18{,}801.25$$

The total amount due is $18,801.25.

33. $I = p \cdot r \cdot t$

$$= 4850 \cdot (0.06) \cdot 1$$

$$= 291$$

He will earn $291 in interest.

37. $I = p \cdot r \cdot t$

$$= 1000 \cdot (0.10) \cdot \frac{3}{12}$$

$$= 100 \cdot \frac{1}{4}$$

$$= 25$$

The interest is $25, so the total amount due will be $1025 ($1000 + $25).

41. $I = p \cdot r \cdot t$

$$= 8800 \cdot (0.0725) \cdot \frac{1}{4}$$

$$= 638 \cdot \frac{1}{4}$$

$$= 159.5$$

She will earn $159.50 in interest.

45. $\dfrac{2}{3} = \dfrac{2 \cdot 4}{3 \cdot 4} = \dfrac{8}{12}$ **49.** $\dfrac{5}{12} = \dfrac{5 \cdot 5}{12 \cdot 5} = \dfrac{25}{60}$

SECTION 6.9 (page 419)

1.

Year	Interest	Compound Amount
1	$1000 \cdot (0.05) \cdot 1 = \$50$	
	$\$1000 + \$50 =$	$1050
2	$1050 \cdot (0.05) \cdot 1 = \$52.50$	
	$\$1050 + \$52.50 =$	$1102.50

The compound amount is $1102.50.

5.

Year	Interest	Compound Amount
1	$3500 \cdot (0.07) \cdot 1 = \$245$	
	$\$3500 + \$245 =$	$3745
2	$3745 \cdot (0.07) \cdot 1 = \$262.15$	
	$\$3745 + \$262.15 =$	$4007.15
3	$4007.15 \cdot (0.07) \cdot 1 \approx \$280.50$	
	$\$4007.15 + \$280.50 =$	$4287.65
4	$4287.65 \cdot (0.07) \cdot 1 \approx \$300.14$	
	$\$4287.65 + \$300.14 =$	$4587.79

The compound amount is $4587.79.

9. $\$1200 \cdot \underbrace{1.03 \cdot 1.03 \cdot 1.03 \cdot 1.03 \cdot 1.03}_{\text{Year 1 Year 2 Year 3 Year 4 Year 5}}$

$$100\% + 3\% = 103\% = 1.03$$

$$\approx \$1391.13$$

The compound amount is $1391.13.

13. $\$9850 \cdot \underbrace{1.05 \cdot 1.05 \cdot 1.05 \cdot 1.05 \cdot 1.05}_{\text{Year 1 Year 2 Year 3 Year 4 Year 15}}$
$\cdot \underbrace{1.05 \cdot 1.05 \cdot 1.05 \cdot 1.05 \cdot 1.05}_{\text{Year 6 Year 7 Year 8 Year 9 Year 10}}$

$$100\% + 5\% = 105\% = 1.05$$

$$\approx \$16{,}044.61$$

The compound amount is $16,044.61.

17. 5% column, row 9 of the table gives 1.5513.

$$\$4000 \cdot 1.5513 = \$6205.20$$

The compound amount is $6205.20.
The interest is $2205.20 ($6205.20 − $4000).

21. Compound interest is interest paid on past interest as well as on the principal.
Many people describe compound interest as "interest on interest."

25. (a) Look in the table for 8% and 6 periods. Find the number 1.5869. Multiply this number and the principal of $7500.

$$\$7500 \cdot 1.5869 = \$11{,}901.75$$

The total amount that should be repaid is $11,901.75.

(b) Find the amount of interest by subtracting the $7500 loan from the total amount to be repaid.

$$\$11{,}901.75 - \$7500 = \$4401.75$$

The amount of interest earned is $4401.75.

29. $12.6 \div 8.4 \cdot 6.2 = 1.5 \cdot 6.2$

$$= 9.3$$

33. $5.34 - 2.6 \cdot 5.2 \div 2.6 = 5.34 - 13.52 \div 2.6$

$$= 5.34 - 5.2$$

$$= 0.14$$

CHAPTER 7

SECTION 7.1 (page 447)

1. 1 yard = 3 feet   **5.** 1 mile = 5280 feet

9. 1 minute = 60 seconds

13. $8 \text{ quarts} = \dfrac{\overset{2}{\cancel{8 \text{ quarts}}}}{1} \cdot \dfrac{1 \text{ gallon}}{\underset{1}{\cancel{4 \text{ quarts}}}} = 2 \text{ gallons}$

17. $12 \text{ feet} = \dfrac{\overset{4}{\cancel{12 \text{ feet}}}}{1} \cdot \dfrac{1 \text{ yard}}{\underset{1}{\cancel{3 \text{ feet}}}} = 4 \text{ yards}$

21. $5 \text{ quarts} = \dfrac{5 \text{ quarts}}{1} \cdot \dfrac{2 \text{ pints}}{1 \text{ quart}} = 10 \text{ pints}$

25. $3 \text{ inches} = \dfrac{\overset{1}{\cancel{3 \text{ inches}}}}{1} \cdot \dfrac{1 \text{ foot}}{\underset{4}{\cancel{12 \text{ inches}}}}$

$$= \frac{1}{4} \text{ foot or } 0.25 \text{ foot}$$

29. $5 \text{ cups} = \frac{5 \text{ cups}}{1} \cdot \frac{1 \text{ pint}}{2 \text{ cups}}$

$= \frac{5}{2} \text{ pints} = 2\frac{1}{2} \text{ pints or } 2.5 \text{ pints}$

33. $2\frac{1}{2} \text{ tons} = \frac{2\frac{1}{2} \text{ tons}}{1} \cdot \frac{2000 \text{ pounds}}{1 \text{ ton}}$

$= \frac{5}{\underset{1}{2}} \cdot \frac{\overset{1000}{2000}}{1} \text{ pounds} = 5000 \text{ pounds}$

37. $2\frac{3}{4} \text{ pounds} = \frac{2\frac{3}{4} \text{ pounds}}{1} \cdot \frac{16 \text{ ounces}}{1 \text{ pound}}$

$= \frac{11}{\underset{1}{4}} \cdot \frac{\overset{4}{16}}{1} \text{ ounces} = 44 \text{ ounces}$

The baby weighed 44 ounces.

41. $112 \text{ cups} = \frac{\overset{\overset{28}{56}}{112} \text{ cups}}{1} \cdot \frac{1 \text{ pint}}{\underset{1}{2} \text{ cups}} \cdot \frac{1 \text{ quart}}{\underset{1}{2} \text{ pints}} = 28 \text{ quarts}$

45. $1\frac{1}{2} \text{ tons} = \frac{1\frac{1}{2} \text{ tons}}{1} \cdot \frac{2000 \text{ pounds}}{1 \text{ ton}} \cdot \frac{16 \text{ ounces}}{1 \text{ pounds}}$

$= \frac{3}{\underset{1}{2}} \cdot \frac{\overset{1000}{2000}}{1} \cdot \frac{16}{1} \text{ ounces} = 48,000 \text{ ounces}$

49. $2\frac{3}{4} \text{ miles} = \frac{2\frac{3}{4} \text{ miles}}{1} \cdot \frac{5280 \text{ feet}}{1 \text{ mile}} \cdot \frac{12 \text{ inches}}{1 \text{ foot}}$

$= \frac{11}{\underset{1}{4}} \cdot \frac{5280}{1} \cdot \frac{\overset{3}{12}}{1} \text{ inches} = 174,240 \text{ inches}$

53. 24,000 ounces

$= \frac{24,000 \text{ ounces}}{1} \cdot \frac{1 \text{ pound}}{16 \text{ ounces}} \cdot \frac{1 \text{ ton}}{2000 \text{ pounds}}$

$= \frac{\overset{\overset{3}{12}}{24,000}}{1} \cdot \frac{1}{\underset{4}{16}} \cdot \frac{1}{\underset{1}{2000}} \text{ ton}$

$= \frac{3}{4} \text{ ton or } 0.75 \text{ ton}$

57. $2 \text{ weeks} = \frac{2 \text{ weeks}}{1} \cdot \frac{7 \text{ days}}{1 \text{ week}} = 14 \text{ days}$

2 weeks < 15 days

61. $32 \text{ days} = \frac{32 \text{ days}}{1} \cdot \frac{1 \text{ week}}{7 \text{ days}} = \frac{32}{7} \text{ weeks} = 4\frac{4}{7} \text{ weeks}$

32 days > 4 weeks

SECTION 7.2 (page 457)

1. *kilo* means 1000, so 1 km = 1000 m

5. *centi* means $\frac{1}{100}$ or 0.01, so 1 cm = $\frac{1}{100}$ m or 0.01 m

9. (Answers may vary.) The width of a person's thumb is about 20 mm.

13. Ming-Na swam in the 200 m backstroke race.

17. An aspirin tablet is 10 mm across.

21. Dave's truck is 5 m long.

25. $7 \text{ m} = \frac{7 \text{ m}}{1} \cdot \frac{100 \text{ cm}}{1 \text{ m}} = \frac{7 \cdot 100 \text{ cm}}{1} = 700 \text{ cm}$

or:
From m to cm is two places right.
7 m = 7.00 cm = 700 cm

29. $9.4 \text{ km} = \frac{9.4 \text{ km}}{1} \cdot \frac{1000 \text{ m}}{1 \text{ km}} = 9400 \text{ m}$

or:
From km to m is three places right.
9.4 km = 9.400 m = 9400 m

33. $400 \text{ mm} = \frac{\overset{40}{400} \text{ mm}}{1} \cdot \frac{1 \text{ cm}}{\underset{1}{10} \text{ mm}} = 40 \text{ cm}$

or:
From mm to cm is one place left.
400 mm = 400. cm = 40 cm

37. $82 \text{ cm} = \frac{82 \text{ cm}}{1} \cdot \frac{1 \text{ m}}{100 \text{ cm}} = \frac{82}{100} \text{ m} = 0.82 \text{ m}$

or:
From cm to m is two places left.
82 cm = 082. m = 0.82 m
82 cm is less than 1 m.
The difference in lengths is 0.18 m (1 m − 0.82 m) or 18 cm (100 cm − 82 cm).

41. From mm to km is six places left.
5.6 mm = 00.00005.6 km = 0.0000056 km
or, using unit fractions:

$\frac{5.6 \text{ mm}}{1} \cdot \frac{1 \text{ m}}{1000 \text{ mm}} \cdot \frac{1 \text{ km}}{1000 \text{ m}} = \frac{5.6}{1,000,000} \text{ km}$

$= 0.0000056 \text{ km}$

45. $0.08 = \frac{8}{100} = \frac{8 \div 4}{100 \div 4} = \frac{2}{25}$

SECTION 7.3 (page 465)

1. The glass held 250 mL of water. (Liquids are measured in mL or L.)

5. Our labrador dog grew up to weigh 40 kg. (Weight is measured in mg, g, or kg.)

9. Andre donated 500 mL of blood today. (Blood is a liquid, and liquids are measured in mL or L.)

13. The gas can for the lawn mower holds 4 L. (Gasoline is a liquid, and liquids are measured in mL or L.)

17. Unreasonable; 4.1 liters is about 4 quarts (1 gallon).

21. Reasonable; 15 milliliters is about 3 teaspoons.

25. Answers will vary. A sample answer follows:
Examples of metric capacity units: 2 liter bottles of soda and shampoo bottles marked in mL.
Examples of metric weight units: grams of fat listed on cereal boxes and vitamin doses in mg.

29. $15 \text{ L} = \frac{15 \text{ L}}{1} \cdot \frac{1000 \text{ mL}}{1 \text{ L}} = \frac{15 \cdot 1000 \text{ mL}}{1}$

$= 15,000 \text{ mL}$

or:
From L to mL is three places right.
15 L = 15.000 mL = 15,000 mL

33. $925 \text{ mL} = \dfrac{925 \text{ mL}}{1} \cdot \dfrac{1 \text{ L}}{1000 \text{ mL}} = \dfrac{925}{1000} \text{L} = 0.925 \text{ L}$

or:

From mL to L is three places left.
925 mL = 0 925. L = 0.925 L

37. $4.15 \text{ L} = \dfrac{4.15 \text{ L}}{1} \cdot \dfrac{1000 \text{ mL}}{1 \text{ L}} = 4150 \text{ mL}$

or:

From L to mL is three places right.
4.15 L = 4.150 mL = 4150 mL

41. $5.2 \text{ kg} = \dfrac{5.2 \text{ kg}}{1} \cdot \dfrac{1000 \text{ g}}{1 \text{ kg}} = 5200 \text{ g}$

or:

From kg to g is three places right.
5.2 kg = 5.200 g = 5200 g

45. $30{,}000 \text{ mg} = \dfrac{30{,}000 \text{ mg}}{1} \cdot \dfrac{1 \text{ g}}{1000 \text{ mg}}$

$= \dfrac{30{,}000}{1000} \text{ g} = 30 \text{ g}$

or:

From mg to g is three places left.
30,000 mg = 30000. g = 30 g

49. $60 \text{ mL} = \dfrac{60 \text{ mL}}{1} \cdot \dfrac{1 \text{ L}}{1000 \text{ mL}} = \dfrac{60}{1000} \text{L} = 0.06 \text{ L}$

or:

From mL to L is three places left.
60 mL = 0060. L = 0.06 L

53. $0.99 \text{ L} = \dfrac{0.99 \text{ L}}{1} = \dfrac{1000 \text{ mL}}{1 \text{ L}} = 990 \text{ mL}$

or:

From L to mL is three places right.
0.99 L = 0.990 mL = 990 mL

57. Buy a 60 mL jar of acrylic paint for art class. (Paint is a liquid and liquids are measured in mL or L.)

61. A single postage stamp weighs 90 mg. (Weight is measured in mg, g, or kg.)

65. $950 \text{ g} = \dfrac{950 \text{ g}}{1} \cdot \dfrac{1 \text{ kg}}{1000 \text{ g}} = 0.95 \text{ kg}$

or:

From g to kg is three places left.
950 g = 0950 kg = 0.95 kg
The premature infant weighed 0.95 kg.

69. $3 \text{ kg} = \dfrac{3 \text{ kg}}{1} \cdot \dfrac{1000 \text{ g}}{1 \text{ kg}} = 3000 \text{ g}$

or:

From kg to g is three places right.
3 kg = 3.000 g = 3000 g
The cat weighs 3000 g.

73. 1 kg = 1.000 g = 1000 g

or $\dfrac{1 \text{ kg}}{1} \cdot \dfrac{1000 \text{ g}}{1 \text{ kg}} = 1000 \text{ g}$

$\dfrac{1000 \text{ g}}{5 \text{ g}} = 200$; therefore, there are 200 nickels in 1 kg of nickels.

77. thousands: 7
hundredths: 1
thousandths: 8
hundreds: 2

SECTION 7.4 (page 471)

1. Write 2 kg 50 g in terms of kilograms (the unit in the price).

$\begin{array}{rl} 2 \text{ kg} \rightarrow & 2.00 \text{ kg} \\ 50 \text{ g} \rightarrow & + \; 0.05 \text{ kg} \\ \hline & 2.05 \text{ kg} \end{array}$

$\dfrac{\$0.65}{1 \text{ kg}} \cdot \dfrac{2.05 \text{ kg}}{1} = \$1.3325 \text{ rounds to } \$1.33$

Pam will pay $1.33 for the rice.

5. Write 5 L in terms of mL.

$5 \text{ L} = \dfrac{5 \text{ L}}{1} \cdot \dfrac{1000 \text{ mL}}{1 \text{ L}} = 5000 \text{ mL}$

or: 5 L = 5.000 mL = 5000 mL

$\dfrac{5000 \text{ mL}}{1} \cdot \dfrac{1 \text{ beat}}{70 \text{ mL}} = 71.42857 \text{ rounds to } 71 \text{ beats}$

It takes about 71 beats to pass all the blood through the heart.

9. Write 750 mL in terms of L (the unit called for in the answer). 750 mL = 0750. L = 0.75 L

$\dfrac{0.75 \text{ L}}{1 \text{ day}} \cdot \dfrac{30 \text{ days}}{1 \text{ month}} = 22.5 \text{ L per month}$

The caretaker should order 22.5 L of chlorine.

13. Given that the weight of one sheet of paper is 3000 mg, find the weight of one sheet in grams.

$$3000 \text{ mg} = 3000. \text{ g} = 3\text{g}$$

To find the net weight of the ream of paper, multiply 500 sheets by 3 g per sheet.

$$\dfrac{500 \text{ sheets}}{1} \cdot \dfrac{3 \text{ g}}{1 \text{ sheet}} = 1500 \text{ g}$$

The net weight of the paper is 1500 g.
Since the weight of the packaging is 50 g, the total weight is:

$$1500 \text{ g} + 50 \text{ g} = 1550 \text{ g}.$$

Then express the total weight in kilograms.

$$1550 \text{ g} = 1550. \text{ kg} = 1.55 \text{ kg}$$

17. $16 case:

$$\text{Capacity} = 12 \cdot 1 \text{ L} = 12 \text{ L}$$

$$\text{Cost per liter} = \dfrac{\$16}{12 \text{ L}} \approx \$1.33 \text{ per L}$$

$18 case:

$$400 \text{ mL} = 0.4 \text{ L}$$

$$\text{Capacity} = 36 \cdot 0.4 \text{ L} = 14.4 \text{ L}$$

$$\text{Cost per liter} = \frac{\$18}{14.4 \text{ L}} = \$1.25 \text{ per L}$$

The $18 case is the better buy.

21.   6.3 ← 1 decimal place
   $\times$ 0.9 1 ← 2 decimal places
   ———
   6 3
   5 6 7
   ———
   5.7 3 3 ← 3 decimal places in answer

Section 7.5 (page 477)

1. 1 meter $\approx$ 1.09 yards

$$\frac{20 \text{ meters}}{1} \cdot \frac{1.09 \text{ yards}}{1 \text{ meter}} = \frac{20 \cdot 1.09 \text{ yards}}{1} = 21.8 \text{ yards}$$

20 meters $\approx$ 21.8 yards

5. 1 foot $\approx$ 0.30 meters

$$\frac{16 \text{ feet}}{1} \cdot \frac{0.30 \text{ meters}}{1 \text{ foot}} = \frac{16 \cdot 0.30 \text{ meters}}{1} = 4.8 \text{ meters}$$

16 feet $\approx$ 4.8 meters

9. 1 pound $\approx$ 0.45 kilogram

$$\frac{248 \text{ pounds}}{1} \cdot \frac{0.45 \text{ kilogram}}{1 \text{ pound}} = \frac{248 \cdot 0.45 \text{ kilograms}}{1}$$
$$= 111.6 \text{ kilograms}$$

248 pounds $\approx$ 111.6 kilograms

13. 1 gallon $\approx$ 3.78 liters

$$\frac{16 \text{ gallons}}{1} \cdot \frac{3.78 \text{ liters}}{1 \text{ gallon}} = \frac{16 \cdot 3.78 \text{ liters}}{1} = 60.48 \text{ liters}$$

60.48 rounds to 60.5, so the gas tank holds $\approx$60.5 liters.

17. 40°C is the more reasonable temperature because normal body temperature is about 37°C.

21. A drop of 20 Celsius degrees is more than a drop of 20 Fahrenheit degrees. There are 180 degrees between freezing and boiling on the Fahrenheit scale, but only 100 degrees on the Celsius scale, so each Celsius degree is a greater change in temperature.

25. $C = \dfrac{5(F - 32)}{9}$

$$= \frac{5(104 - 32)}{9}$$

$$= \frac{5(72)}{9}$$

$$= \frac{5(\overset{8}{\cancel{72}})}{\underset{1}{\cancel{9}}}$$

$$= 40$$

Thus, 104°F = 40°C.

29. $F = \dfrac{9 \cdot C}{5} + 32$

$$= \frac{9 \cdot 35}{5} + 32$$

$$= \frac{9 \cdot \overset{7}{\cancel{35}}}{\underset{1}{\cancel{5}}} + 32$$

$$= 63 + 32$$

$$= 95$$

Thus, 35°C = 95°F.

33. Since the comfort range of the boots is from 24°C to 4°C, you would wear these boots in pleasant weather—above freezing, but not hot.

Change 24°C to Fahrenheit.   Change 4°C to Fahrenheit.

$$F = \frac{9 \cdot C}{5} + 32 \qquad\qquad F = \frac{9 \cdot C}{5} + 32$$

$$= \frac{9 \cdot 24}{5} + 32 \qquad\qquad = \frac{9 \cdot 4}{5} + 32$$

$$= \frac{216}{5} + 32 \qquad\qquad = \frac{36}{5} + 32$$

$$= 43.2 + 32 \qquad\qquad = 7.2 + 32$$

$$= 75.2 \qquad\qquad\qquad = 39.2$$

Thus, 24°C $\approx$ 75°F.   Thus, 4°C $\approx$ 39°F.

The boots are designed for Fahrenheit temperatures of $\approx$75°F to $\approx$39°F.

The range of metric temperatures in January would depend on where you live. In Minnesota it's 0°C to −40°C and in California it's 24°C to 0°C.

37. $\underbrace{2 \cdot 8} + \underbrace{2 \cdot 8}$ Multiply before adding.

   16 + 16 = 32

41. $\underbrace{(5^2)} + \underbrace{(4^2)}$ Do exponents before adding.

   25 + 16 = 41

Chapter 8

Section 8.1 (page 499)

1. This is line $\overleftrightarrow{CD}$ or $\overleftrightarrow{DC}$. A line is a straight row of points that goes on forever in both directions.

5. This is ray $\overrightarrow{PQ}$. A ray is a part of a line that has only one endpoint and goes on forever in one direction.

9. These are parallel lines. Parallel lines are lines in the same plane that never cross.

13. $\angle AOS$ or $\angle SOA$. The middle letter, O, identifies the vertex.

17. $\angle AQC$ or $\angle CQA$. The middle letter, Q, identifies the vertex.

21. This is an acute angle. Acute angles measure between 0° and 90°.

25. (a) If your car "did a 360," the car turned around in a complete circle.

   (b) If the governor's view on taxes "took a 180° turn," he or she took the opposite view. For example, he or she may have opposed taxes and now supports them.

29. False. $\angle USQ$ is a straight angle and so is $\angle PQR$, therefore each measures 180°. A true statement would be: $\angle USQ$ has the same measure as $\angle PQR$.

33. $(180 - 75) - 15 = 105 - 15 = 90$ Do what is inside parentheses first.

Section 8.2 (page 505)

1. $\angle EOD$ and $\angle COD$ are complementary angles because 75° + 15° = 90°

   $\angle AOB$ and $\angle BOC$ are complementary angles because 25° + 65° = 90°

SOLUTIONS

5. The complement of 40° is 50°, because
90° − 40° = 50°

9. The supplement of 130° is 50°, because
180° − 130° = 50°

13. If two angles are vertical angles, they are congruent.
∠SON ≅ ∠TOM and ∠TOS ≅ ∠MON

17. ∠EOF measures 37°. Since ∠EOF and ∠AOH are vertical angles, ∠EOF ≅ ∠AOH

21. ∠ABF ≅ ∠ECD. Both are 138°, since
180° − 42° = 138°. ∠ABC ≅ ∠BCD, both are 42°.

25. 16 − $\underbrace{(3 \cdot 3)}$ Work inside parentheses first.

16 −   9   = 7

29. $\underbrace{6 \cdot 8}$ − $\underbrace{5^2}$ Use exponent and multiply before subtracting.

48 − 25 = 23

SECTION 8.3 (page 513)

1. $P = 2 \cdot 8 \text{ yd} + 2 \cdot 6 \text{ yd} = 16 \text{ yd} + 12 \text{ yd} = 28 \text{ yd}$
$A = 8 \text{ yd} \cdot 6 \text{ yd} = 48 \text{ yd}^2$ (square units for area)

5. $P = 2 \cdot 10 \text{ ft} + 2 \cdot 1 \text{ ft} = 20 \text{ ft} + 2 \text{ ft} = 22 \text{ ft}$
$A = 10 \text{ ft} \cdot 1 \text{ ft} = 10 \text{ ft}^2$ (square units for area)

9. $P = 2 \cdot 76.1 \text{ ft} + 2 \cdot 22 \text{ ft} = 152.2 \text{ ft} + 44 \text{ ft}$
$= 196.2 \text{ ft}$
$A = 76.1 \text{ ft} \cdot 22 \text{ ft} = 1674.2 \text{ ft}^2$ (square units for area)

13.

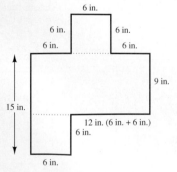

$P = 12 \text{ m} + 7 \text{ m} + 3 \text{ m} + 5 \text{ m} + 9 \text{ m} + 2 \text{ m}$
$= 38 \text{ m}$
$A = 7 \text{m} \cdot 3 \text{ m} + 9 \text{ m} \cdot 2 \text{ m} = 21 \text{ m}^2 + 18 \text{ m}^2$
$= 39 \text{ m}^2$ (square units for area)

17.

6 in.
6 in.   6 in.
6 in.   6 in.
                    9 in.
15 in.
          12 in. (6 in. + 6 in.)
          6 in.
6 in.

$P = 15 \text{ in.} + 6 \text{ in.} + 6 \text{ in.} + 6 \text{ in.} + 6 \text{ in.}$
$+ 9 \text{ in.} + 12 \text{ in.} + 6 \text{ in.} + 6 \text{ in.} = 78 \text{ in.}$
$A = 6 \text{ in.} \cdot 6 \text{ in.} + 18 \text{ in.} \cdot 9 \text{ in.} + 6 \text{ in.} \cdot 6 \text{ in.}$
$= 36 \text{ in.}^2 + 162 \text{ in.}^2 + 36 \text{ in.}^2$
$= 234 \text{ in.}^2$ (square units for area)

21. $P = 2 \cdot 4.4 \text{ m} + 2 \cdot 5.1 \text{ m}$
$= 8.8 \text{ m} + 10.2 \text{ m} = 19 \text{ m}$

$\text{Cost} = \dfrac{19 \text{ m}}{1} \cdot \dfrac{\$4.99}{1 \text{ m}} = \$94.81$

Tyra will have to spend $94.81 for the strip.

25. Length of inner rectangle = 172 ft − 12 ft − 12 ft
$= 148 \text{ ft}$
Width of inner rectangle = 124 ft − 12 ft − 12 ft
$= 100 \text{ ft}$

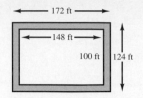

Shaded area = 172 ft · 124 ft − 148 ft · 100 ft
$= 21{,}328 \text{ ft}^2 − 14{,}800 \text{ ft}^2$
$= 6528 \text{ ft}^2$
The area of land that cannot be built on is 6528 ft².

29. $12 \text{ ft} = \dfrac{12 \text{ ft}}{1} \cdot \dfrac{12 \text{ in.}}{1 \text{ ft}} = 144 \text{ in.}$

SECTION 8.4 (page 521)

1. $P = 46 \text{ m} + 58 \text{ m} + 46 \text{ m} + 58 \text{ m} = 208 \text{ m}$
Do not include the height (43 m).

5. $P = 3 \text{ km} + 0.8 \text{ km} + 0.95 \text{ km} + 1.03 \text{ km}$
$= 5.78 \text{ km}$
Do not include the height (0.4 km).

9. $A = 5.5 \text{ ft} \cdot 3.5 \text{ ft} = 19.25 \text{ ft}^2$ (square units for area)

13. $A = \dfrac{1}{2} \cdot 45 \text{ ft} \cdot (80 \text{ ft} + 110 \text{ ft}) = \dfrac{1}{2} \cdot 45 \text{ ft} \cdot (\overset{95}{\cancel{190}} \text{ ft})$

$= 4275 \text{ ft}^2$

$\text{Cost} = \dfrac{4275 \text{ ft}^2}{1} \cdot \dfrac{\$0.22}{1 \text{ ft}^2} = \$940.50$

It will cost $940.50 to put sod on the yard.

17. $A = 1.3 \text{ m} \cdot 0.8 \text{ m} + 1.8 \text{ m} \cdot 1.1 \text{ m}$
$= 1.04 \text{ m}^2 + 1.98 \text{ m}^2$
$= 3.02 \text{ m}^2$ (square units for area)

21. $6\dfrac{1}{2} \cdot 8 = \dfrac{13}{2} \cdot \dfrac{\overset{4}{\cancel{8}}}{1} = \dfrac{52}{1} = 52$

SECTION 8.5 (page 527)

1. $P = 9 \text{ yd} + 7 \text{ yd} + 11 \text{ yd} = 27 \text{ yd}$

5. $A = \dfrac{1}{2} \cdot 60 \text{ m} \cdot 66 \text{ m} = 1980 \text{ m}^2$ (square units for area)
Do not use the sides labeled 72 m.

9. $A = 12 \text{ m} \cdot 12 \text{ m} + \dfrac{1}{2} \cdot 12 \text{ m} \cdot 9 \text{ m} = 144 \text{ m}^2$
$+ 54 \text{ m}^2 = 198 \text{ m}^2$ (square units for area)

13. *Step 1*   Add the two angles given. 90° + 58° = 148°
Step 2   Subtract the sum from 180°.
180° − 148° = 32°
The missing angle is 32°.

17. A triangle *cannot* have two right angles. Right angles are 90° so two right angles are 180°. This will not work since the sum of all *three* angles in a triangle equals 180°.

21. To find the amount of curbing needed to go around the space, find the perimeter of the triangle.

$P = 42 \text{ m} + 32 \text{ m} + 52.8 \text{ m} = 126.8 \text{ m}$

126.8 m of curbing will be needed.

To find the amount of sod needed to cover the space, find the area. It is a right triangle, so the perpendicular sides are the base and height.

$$A = \frac{1}{2} \cdot 32 \text{ m} \cdot 42 \text{ m} = 672 \text{ m}^2$$

672 m^2 of sod will be needed.

25. $\quad 3.14 \leftarrow$ 2 decimal places
$\underline{\times \quad 16} \leftarrow$ 0 decimal places
$\quad 18\ 84$
$\underline{31\ 4}$
$\quad 50.24 \leftarrow$ 2 decimal places in answer

SECTION 8.6 (page 535)

1. $d = 2 \cdot r = 2 \cdot 9 \text{ mm} = 18 \text{ mm}$

5. $C = 2 \cdot \pi \cdot r \approx 2 \cdot 3.14 \cdot 11 \text{ ft} \approx 69.08 \text{ ft}$ rounds to $\approx 69.1 \text{ ft}$
$A = \pi \cdot r^2 \approx 3.14 \cdot 11 \text{ ft} \cdot 11 \text{ ft} \approx 379.94 \text{ ft}^2$ rounds to $\approx 379.9 \text{ ft}^2$ (square units for area)

9. $C = \pi \cdot d \approx 3.14 \cdot 15 \text{ cm} \approx 47.1 \text{ cm}$
$r = \dfrac{d}{2} = \dfrac{15 \text{ cm}}{2} = 7.5 \text{ cm}$
$A \approx 3.14 \cdot 7.5 \text{ cm} \cdot 7.5 \text{ cm} \approx 176.625 \text{ cm}^2$ rounds to $\approx 176.6 \text{ cm}^2$ (square units for area)

13. $C \approx 3.14 \cdot 8.65 \text{ km} \approx 27.161 \text{ cm}$ rounds to $\approx 27.2 \text{ cm}$
$r = \dfrac{8.65 \text{ km}}{2} = 4.325 \text{ km}$
$A \approx 3.14 \cdot 4.325 \text{ km} \cdot 4.325 \text{ km}$ rounds to $\approx 58.7 \text{ km}^2$

17. $C \approx 3.14 \cdot 250 \text{ ft} \approx 785.0 \text{ ft}$
The circumference of the dome is about 785.0 ft.

21. First find the radius of the large pizza.
$$r = \frac{16 \text{ in.}}{2} = 8 \text{ in.}$$
Area of large pizza $\approx 3.14 \cdot 8 \text{ in.} \cdot 8 \text{ in.}$
$\approx 200.96 \text{ in.}^2$ rounds to 201.0 in.^2

25. Answers will vary. A sample answer follows:
π is the ratio of the circumference of a circle to its diameter. If you divide the circumference of any circle by its diameter, the answer is always a little more than 3. The approximate value is 3.14, which we call π (pi). Your test question could involve finding the circumference or the area of a circle.

29. Area of large circle $\approx 3.14 \cdot 12 \text{ cm} \cdot 12 \text{ cm}$
$\approx 452.16 \text{ cm}^2$
Area of small circle $\approx 3.14 \cdot 9 \text{ cm} \cdot 9 \text{ cm}$
$\approx 254.34 \text{ cm}^2$
Shaded area $\approx 452.16 \text{ cm}^2 - 254.34 \text{ cm}^2 \approx 197.8 \text{ cm}^2$

33. $\quad C = 2 \cdot \pi \cdot r$
$22 \text{ m} = 2 \cdot 3.14 \cdot r$
$\dfrac{22 \text{ m}}{6.28} = \dfrac{\overset{1}{\cancel{6.28}} \cdot r}{\underset{1}{\cancel{6.28}}}$
$3.5 \text{ m} \approx r$
The radius of the pool is about 3.5 m.

37. The *ra* prefix tells you that radius is a ray from the center of a circle. The *dia* prefix means the diameter goes through the circle, and *circum* means the circumference is the distance around.

41. $2\dfrac{3}{4} = \dfrac{11}{4}$
$\quad 2.75$
$4)\overline{11.00}$
$\quad \underline{8}$
$\quad 3\ 0$
$\quad \underline{2\ 8}$
$\quad\quad 20$
$\quad\quad \underline{20}$
$\quad\quad\ \ 0$

45. $10\dfrac{1}{2} = \dfrac{21}{2}$
$\quad 10.5$
$2)\overline{21.0}$
$\underline{20}$
$\ 1\ 0$
$\ \underline{1\ 0}$
$\quad\ 0$

SECTION 8.7 (page 545)

1. $V = l \cdot w \cdot h = 12.5 \text{ cm} \cdot 4 \text{ cm} \cdot 11 \text{ cm} = 550 \text{ cm}^3$ (cubic units for volume)

5. $V = \dfrac{2 \cdot \pi \cdot r^3}{3}$
$\approx \dfrac{2 \cdot 3.14 \cdot 12 \text{ in.} \cdot 12 \text{ in.} \cdot 12 \text{ in.}}{3}$
$\approx 3617.3 \text{ in.}^3$ (cubic units for volume)

9. $V = \dfrac{\pi \cdot r \cdot r \cdot h}{3} \approx \dfrac{3.14 \cdot 5 \text{ m} \cdot 5 \text{m} \cdot 16 \text{ m}}{3}$
$\approx 418.7 \text{ m}^3$ (cubic units for volume)

13. $V = l \cdot w \cdot h = 3 \text{ in.} \cdot 8 \text{ in.} \cdot \dfrac{3}{4} \text{ in.}$
$-\dfrac{\overset{6}{\cancel{24}}}{1} \text{ in.}^2 \cdot \dfrac{3}{\underset{1}{\cancel{4}}} \text{ in.}$
$= 18 \text{ in.}^3$
The volume of the box is 18 in.^{3}

17. $V = \pi \cdot r^2 \cdot h$
We are given a diameter of 5 ft, but we need to use the radius.
$$r = \frac{5 \text{ ft}}{2} = 2.5 \text{ ft}$$
$V \approx 3.14 \cdot 2.5 \text{ ft} \cdot 2.5 \text{ ft} \cdot 200 \text{ ft}$
$V \approx 3925 \text{ ft}^3$
The volume of the city sewer pipe is about 3925 ft^3.

21. First, the student used the diameter of 7 cm, but should have used the radius of 3.5 cm in the formula. Secondly, the units for volume are cm^3, not cm^2. The correct answer is $\approx 192.3 \text{ cm}^3$.

25. $8^2 = 8 \cdot 8 = 64$

29. $\sqrt{64} = 8$, since $8 \cdot 8 = 64$

SECTION 8.8 (page 551)

1. $\sqrt{16} = 4$, since $4 \cdot 4 = 16$

5. Using a calculator, $\sqrt{11} = 3.3166$ rounds to ≈ 3.317

9. Using a calculator, $\sqrt{73} = 8.5440$ rounds to ≈ 8.544

SOLUTIONS

13. Using a calculator, $\sqrt{190} = 13.7840$ rounds
to ≈ 13.784

17. $a^2 = 36$, $b^2 = 64$, $c^2 = 100$. Since $36 + 64 = 100$,
the Pythagorean Theorem is true.

21. hypotenuse $= \sqrt{(\text{leg})^2 + (\text{leg})^2}$
$= \sqrt{8^2 + 15^2}$
$= \sqrt{64 + 225}$
$= \sqrt{289}$
$= 17$

The hypotenuse is 17 in. long.

25. hypotenuse $= \sqrt{(\text{leg})^2 + (\text{leg})^2}$
$= \sqrt{8^2 + 3^2}$
$= \sqrt{64 + 9}$
$= \sqrt{73}$
≈ 8.544

The hypotenuse is about 8.544 in. long.

29. leg $= \sqrt{(\text{hypotenuse})^2 - (\text{leg})^2}$
$= \sqrt{22^2 - 17^2}$
$= \sqrt{484 - 289}$
$= \sqrt{195}$
≈ 13.964

The leg is about 13.964 cm long.

33. leg $= \sqrt{(\text{hypotenuse})^2 - (\text{leg})^2}$
$= \sqrt{11.5^2 - 8.2^2}$
$= \sqrt{132.25 - 67.24}$
$= \sqrt{65.01}$
≈ 8.063

The leg is about 8.063 cm long.

37. hypotenuse $= \sqrt{(\text{leg})^2 + (\text{leg})^2}$
$= \sqrt{4^2 + 7^2}$
$= \sqrt{16 + 49}$
$= \sqrt{65}$
≈ 8.1

The length of the loading ramp is about 8.1 ft.

41. leg $= \sqrt{(\text{hypotenuse})^2 - (\text{leg})^2}$
$= \sqrt{12^2 - 3^2}$
$= \sqrt{144 - 9}$
$= \sqrt{135}$
≈ 11.6

The ladder will reach about 11.6 ft high on the
building.

45. length $AC = \sqrt{12^2 + 9^2}$
$= \sqrt{144 + 81}$
$= \sqrt{225}$
$= 15$ ft

length $BC = 15$ ft $- 8.75$ ft $= 6.25$ ft

length $BD = \sqrt{6.25^2 - 5^2}$
$= \sqrt{39.0625 - 25}$
$= \sqrt{14.0625}$
$= 3.75$ ft

49. $\dfrac{x}{9.2} = \dfrac{15.6}{7.8}$

$x \cdot 7.8 = 9.2 \cdot 15.6$

$x \cdot 7.8 = 143.52$

$\dfrac{x \cdot \overset{1}{\cancel{7.8}}}{\cancel{7.8}} = \dfrac{143.52}{7.8}$

$x = 18.4$

Section 8.9 (page 559)

1. Similar because the triangles have the same shape.

5. Similar because the triangles have the same shape.

9. Corresponding angles: $\angle P$ and $\angle S$
$\angle N$ and $\angle R$
$\angle M$ and $\angle Q$

Corresponding sides:
$\overline{MP}$ and $\overline{QS}$ (the longest side in each triangle)
$\overline{MN}$ and $\overline{QR}$ (the shortest side in each triangle)
$\overline{NP}$ and $\overline{RS}$

13. $\dfrac{a}{10} = \dfrac{6}{12}$ $\qquad$ $\dfrac{b}{6} = \dfrac{6}{12}$

$a \cdot 12 = 10 \cdot 6$ $\qquad$ $b \cdot 12 = 6 \cdot 6$

$\dfrac{a \cdot \overset{1}{\cancel{12}}}{\cancel{12}} = \dfrac{60}{12}$ $\qquad$ $\dfrac{b \cdot \overset{1}{\cancel{12}}}{\cancel{12}} = \dfrac{36}{12}$

$a = 5$ mm $\qquad$ $b = 3$ mm

17. $\dfrac{x}{18.6} = \dfrac{28}{21}$

$x \cdot 21 = 18.6 \cdot 28$

$\dfrac{x \cdot \overset{1}{\cancel{21}}}{\cancel{21}} = \dfrac{520.8}{21}$

$x = 24.8$ m

perimeter $= 24.8$ m $+ 28$ m $+ 20$ m $= 72.8$ m

$\dfrac{y}{20} = \dfrac{21}{28}$

$y \cdot 28 = 20 \cdot 21$

$\dfrac{y \cdot \overset{1}{\cancel{28}}}{\cancel{28}} = \dfrac{420}{28}$

$y = 15$ m

perimeter $= 15$ m $+ 21$ m $+ 18.6$ m $= 54.6$ m

21. $\dfrac{x}{28} = \dfrac{60}{15}$

$x \cdot 15 = 28 \cdot 60$

$x \cdot 15 = 1680$

$\dfrac{x \cdot \overset{1}{\cancel{15}}}{\underset{1}{\cancel{15}}} = \dfrac{1680}{15}$

$x = 112 \text{ ft}$

The height of the wave was 112 ft.

25. Answers will vary. A sample answer follows:
One dictionary definition is, "Resembling, but not identical." Examples of similar objects are sets of different size pots or measuring cups; small and large size cans of beans; child's tennis shoe and adult tennis shoe.

29. $\dfrac{x}{120} = \dfrac{100}{100 + 140}$

$\dfrac{x}{120} = \dfrac{100}{240}$

$x \cdot 240 = 120 \cdot 100$

$\dfrac{x \cdot \overset{1}{\cancel{240}}}{\underset{1}{\cancel{240}}} = \dfrac{12000}{240}$

$x = 50 \text{ m}$

33. $13 - \underbrace{2 \cdot 5}_{} + 7$ Multiply first.

$\underbrace{13 - 10}_{} + 7$ Add and subtract from left to right.

$3 \quad + 7 = 10$

CHAPTER 9

SECTION 9.1 (page 587)

1. $+32$ Above zero is positive.

5. $+20$ Above flood stage is positive.

9. negative because of the "$-$" sign in front of the fraction.

13. negative because of the "$-$" sign in front of 6.3.

17.

21.

25. Since 9 is to the left of 14 on a number line, 9 is less than 14. Write it as $9 < 14$.

29. Since -6 is to the left of 3 on a number line, -6 is less than 3. Write it as $-6 < 3$.

33. Since -11 is to the left of -2 on a number line, -11 is less than -2. Write it as $-11 < -2$.

37. Since 2 is to the right of -1 on a number line, 2 is greater than -1. Write it as $2 > -1$.

41. $|5| = 5$ **45.** $|-1| = 1$

49. $|0| = 0$ **53.** $|-9.5| = 9.5$

57. $-|-10| = -10$ First, $|-10|$ is 10 but the negative sign *outside* the absolute value bars is not affected, so the final answer is negative.

61. $-|-7.6| = -7.6$ First, $|-7.6|$ is 7.6 but the negative sign *outside* the absolute value bars is not affected, so the final answer is negative.

65. The opposite of positive 2 is -2.

69. The opposite of -11 is 11 (positive 11).

73. The opposite of 0 is 0. Zero is neither positive nor negative.

77. The opposite of positive 5.2 is -5.2.

81. Answers will vary. A sample answer follows:
Opposite numbers are the same distance from 0 but on opposite sides of it.

-2 and $+2$ are opposites.

85. True. $|-5| = 5$, therefore $|-5| > 0$.

89. False. $-|-4| = -4$ and $-|-7| = -7$; -4 is to the right of -7 on the number line, therefore, $-4 > -7$ and $-|-4| < -|-7|$ is false.

93. $\dfrac{5}{6} - \dfrac{1}{4} = \dfrac{10}{12} - \dfrac{3}{12} = \dfrac{7}{12}$

SECTION 9.2 (page 599)

1. $-2 + 5 = 3$
See graph in the answer section.

5. $3 + (-4) = -1$
See graph in the answer section.

9. $|-1| = 1 \qquad |8| = 8 \qquad 8 - 1 = 7$
The positive number 8 has the larger absolute value, so the answer is positive.

$$-1 + 8 = 7$$

13. $|6| = 6 \qquad |-5| = 5 \qquad 6 - 5 = 1$
The positive number 6 has the larger absolute value, so the answer is positive.

$$6 + (-5) = 1$$

17. $|-10| = 10 \qquad |-10| = 10$
Add the absolute values.

$$10 + 10 = 20$$

Write a negative sign in front of the sum because both numbers are negative.

$$-10 + (-10) = -20$$

21. The overdrawn amount is negative ($-\$52.50$) and the deposit is positive.

$$-\$52.50 + \$50 = -\$2.50$$

25. $\left|-\dfrac{1}{2}\right| = \dfrac{1}{2} \qquad \left|\dfrac{3}{4}\right| = \dfrac{3}{4} \qquad \dfrac{3}{4} - \dfrac{1}{2} = \dfrac{3}{4} - \dfrac{2}{4} = \dfrac{1}{4}$

The positive number $\dfrac{3}{4}$ has the larger absolute value, so the answer is positive.

$$-\dfrac{1}{2} + \dfrac{3}{4} = \dfrac{1}{4}$$

SOLUTIONS

29. $\left|-\dfrac{7}{3}\right| = \dfrac{7}{3}$ $\left|-\dfrac{5}{9}\right| = \dfrac{5}{9}$ $\dfrac{7}{3} + \dfrac{5}{9} = \dfrac{21}{9} + \dfrac{5}{9} = \dfrac{26}{9}$

Write a negative sign in front of the sum, since both numbers are negative.

$$-\dfrac{7}{3} + \left(-\dfrac{5}{9}\right) = -\dfrac{26}{9}$$

33. The additive inverse of positive 3 is -3 (change sign).

37. The additive inverse of positive $\frac{1}{2}$ is $-\frac{1}{2}$ (change sign).

41. $19 - 5 = 19 + (-5)$ Change 5 to its opposite (-5) and add.

$\qquad\qquad = 14$

45. $7 - 19 = 7 + (-19)$ Change 19 to its opposite (-19) and add.

$\qquad\qquad = -12$

49. $-9 - 14 = -9 + (-14)$ Change 14 to its opposite (-14) and add.

$\qquad\qquad = -23$

53. $6 - (-14) = 6 + (+14)$ Change (-14) to its opposite $(+14)$ and add.

$\qquad\qquad = 20$

57. $-30 - 30 = -30 + (-30)$ Change 30 to its opposite (-30) and add.

$\qquad\qquad = -60$

61. $-\dfrac{7}{10} - \dfrac{4}{5} = -\dfrac{7}{10} + \left(-\dfrac{4}{5}\right)$ Change $\dfrac{4}{5}$ to its opposite $\left(-\dfrac{4}{5}\right)$ and add.

$\qquad\qquad = -\dfrac{7}{10} + \left(-\dfrac{8}{10}\right)$

$\qquad\qquad = -\dfrac{15}{10} = -\dfrac{3}{2}$

65. $-8.3 - (-9) = -8.3 + (+9)$ Change (-9) to its opposite $(+9)$ and add.

$\qquad\qquad = 0.7$

69. $-2 + (-11) - (-3) = -13 - (-3)$

$\qquad\qquad\qquad = -13 + 3$

$\qquad\qquad\qquad = -10$

73. $-12 - (-3) - (-2) = -12 + 3 - (-2)$

$\qquad\qquad\qquad = -9 - (-2)$

$\qquad\qquad\qquad = -9 + 2$

$\qquad\qquad\qquad = -7$

77. $\dfrac{1}{2} - \dfrac{2}{3} + \left(-\dfrac{5}{6}\right) = \dfrac{1}{2} + \left(-\dfrac{2}{3}\right) + \left(-\dfrac{5}{6}\right)$

$\qquad\qquad\qquad = \dfrac{3}{6} + \left(-\dfrac{4}{6}\right) + \left(-\dfrac{5}{6}\right)$

$\qquad\qquad\qquad = -\dfrac{1}{6} + \left(-\dfrac{5}{6}\right)$

$\qquad\qquad\qquad = -\dfrac{6}{6} = -1$

81. $-2 + (-11) + |-2| = -13 + |-2|$

$\qquad\qquad\qquad = -13 + 2$

$\qquad\qquad\qquad = -11$

85. $2\dfrac{1}{2} + 3\dfrac{1}{4} - \left(-1\dfrac{3}{8}\right) - 2\dfrac{3}{8}$ Write mixed numbers as improper fractions.

$\dfrac{5}{2} + \dfrac{13}{4} - \left(-\dfrac{11}{8}\right) - \dfrac{19}{8}$ Rewrite all fractions with lowest common denominator (8).

$\underbrace{\dfrac{20}{8} + \dfrac{26}{8}} - \left(-\dfrac{11}{8}\right) - \dfrac{19}{8}$ Add and subtract from left to right.

$\dfrac{46}{8} - \left(-\dfrac{11}{8}\right) - \dfrac{19}{8}$ Change subtraction to addition.

$\underbrace{\dfrac{46}{8} + \left(+\dfrac{11}{8}\right)} - \dfrac{19}{8}$

$\underbrace{\dfrac{57}{8} + \left(-\dfrac{19}{8}\right)}$

$\dfrac{38}{8} = \dfrac{19}{4}$ or $4\dfrac{3}{4}$

89.
```
      7 1.2 0  ← 2 decimal places
    × 2 1.2 5  ← 2 decimal places
      3 5 6 0 0
      1 4 2 4 0
      7 1 2 0
    1 4 2 4 0
  1 5 1 3.0 0 0 0  ← 4 decimal places in answer
                  1513.0000 or 1513
```

93. $\dfrac{7}{9} \div \dfrac{14}{27} = \dfrac{\overset{1}{7}}{\underset{1}{9}} \cdot \dfrac{\overset{3}{27}}{\underset{2}{14}} = \dfrac{3}{2}$ or $1\dfrac{1}{2}$

SECTION 9.3 (page 605)

1. $-5 \cdot 7 = -35$ (different signs, product is negative)

5. $3 \cdot (-6) = -18$ (different signs, product is negative)

9. $(-1)(40) = -40$ (different signs, product is negative)

13. $11 \cdot 7 = 77$ (same signs, product is positive)

17. $-13 \cdot (-1) = 13$ (same signs, product is positive)

21. $-\dfrac{1}{2} \cdot (-8) = -\dfrac{1}{\underset{1}{2}} \cdot \left(\dfrac{\overset{4}{-8}}{1}\right) = \dfrac{4}{1} = 4$ (same signs, product is positive)

25. $\left(\dfrac{\overset{1}{3}}{5}\right)\left(-\dfrac{1}{\underset{2}{6}}\right) = -\dfrac{1}{10}$ (different signs, product is negative)

29. $-\dfrac{\overset{1}{7}}{\underset{3}{15}} \cdot \dfrac{\overset{5}{25}}{\underset{2}{14}} = -\dfrac{5}{6}$ (different signs, product is negative)

33. $9 \cdot (-4.7) = -42.3$ (different signs, product is negative)

37. $-6.2 \cdot (5.1) = -31.62$ (different signs, product is negative)

41. $(-8.23)(-1) = 8.23$ (same signs, product is positive)

45. $\dfrac{-14}{7} = -2$ (different signs, quotient is negative)

49. $\dfrac{-28}{0}$ Division by 0 is undefined.

53. $\dfrac{-20}{-2} = 10$ (same signs, quotient is positive)

57. $\dfrac{-18}{18} = -1$ (different signs, quotient is negative)

61. $\dfrac{0}{-9} = 0$ (zero is neither positive nor negative)

65. $\dfrac{-\frac{5}{7}}{-\frac{15}{14}} = -\dfrac{\cancel{5}^{\,1}}{7} \cdot \left(-\dfrac{\cancel{14}^{\,2}}{\cancel{15}_{\,3}}\right) = \dfrac{2}{3}$ (same signs, quotient is positive)

69. $5 \div \left(-\dfrac{5}{8}\right) = \dfrac{\cancel{5}^{\,1}}{1} \cdot \left(-\dfrac{8}{\cancel{5}_{\,1}}\right) = -8$ (different signs, quotient is negative)

73. $\dfrac{-18.92}{-4} = 4.73$ (same signs, quotient is positive)

77. $\dfrac{45.58}{-8.6} = -5.3$ (different signs, quotient is negative)

81. $(-0.6)(-0.2)(-3) = (0.12)(-3) = -0.36$

85. Answers will vary. A sample answer follows:

(a) $6 \cdot (-1) = -6$; $2 \cdot (-1) = -2$;
$15 \cdot (-1) = -15$

(b) $-6 \cdot (-1) = 6$; $-2 \cdot (-1) = 2$;
$-15 \cdot (-1) = 15$

The result of multiplying any nonzero number times -1 is the number with the opposite sign.

89. $-36 \div (-2) \div (-3) \div (-3) \div (-1)$ Divide from left to right.
$\qquad 18 \qquad \div (-3) \div (-3) \div (-1)$
$\qquad\quad -6 \qquad\quad \div (-3) \div (-1)$
$\qquad\qquad\quad 2 \qquad\qquad\quad \div (-1) = -2$

93. $8 + 4 \cdot 2 \div 8$ Multiply first.
$\quad 8 + \quad 8 \ \div 8$ Divide next.
$\quad 8 + \qquad 1 \ \ = 9$ Add.

Section 9.4 (page 613)

1. $6 + \underbrace{3 \cdot (-4)}$ Multiply first.
$\underbrace{6 + (-12)}$ Add.
$\qquad -6$

5. $\underbrace{6^2} + \underbrace{4^2}$ Exponents first
$\underbrace{36 + 16}$ Add.
$\qquad 52$

9. $\underbrace{(-2)^5} + 2$ Exponents first
$\underbrace{-32 \ + 2}$ Add.
$\qquad -30$

13. $2 - (-5) + \underbrace{3^2}$ Exponents first
$2 - (-5) + 9$ Add and subtract from left to right.
$\underbrace{2 + (+5)} + 9$
$\underbrace{\quad 7 \quad + 9}$
$\qquad 16$

17. $3 + 5 \cdot \underbrace{(6 - 2)}$ Parentheses first
$3 + \underbrace{5 \cdot \quad 4}$ Multiply next
$\underbrace{3 + \quad 20}$ Add.
$\qquad 23$

21. $-6 + (-5) \cdot \underbrace{(9 - 14)}$ Parentheses first
$-6 + \underbrace{(-5) \cdot \quad (-5)}$ Multiply next.
$\underbrace{-6 + \qquad 25}$ Add.
$\qquad 19$

25. $9 \div \underbrace{(-3)^2} + (-1)$ Exponents first
$\underbrace{9 \div \quad 9} + (-1)$ Division next.
$\underbrace{\quad 1 \qquad + (-1)}$ Add.
$\qquad 0$

29. $\underbrace{(-2) \cdot (-7)} + \underbrace{3 \cdot 9}$ Multiply from left to right.
$\underbrace{\quad 14 \quad + \quad 27}$ Add.
$\qquad 41$

33. $\underbrace{2 \cdot 5} - \underbrace{3 \cdot 4} + \underbrace{5 \cdot 3}$ Multiply from left to right.
$\underbrace{10 \ - \ 12} + \ 15$ Add and subtract from left to right.
$\underbrace{\quad -2 \qquad + \ 15}$
$\qquad 13$

37. First do the work in the numerator.

$-1 + \underbrace{5^2} - (-3)$ Exponents first
$\underbrace{-1 + 25} - (-3)$ Add and subtract from left to right.
$\qquad 24 \quad - (-3)$
$\qquad\qquad \downarrow \quad\ \downarrow$
$\underbrace{\quad 24 \quad + \quad 3}$
Numerator $\to 27$

Now do the work in the denominator.

$-6 - \quad 9 \ + 12$ Add and subtract from left to right.
$\quad \downarrow \quad\ \downarrow$
$\underbrace{-6 + (-9)} + 12$
$\underbrace{\quad -15 \qquad + 12}$
Denominator $\to -3$

The last step is the division.

Numerator $\to \dfrac{27}{-3} \leftarrow$ Denominator $= -9$

41. First do the work in the numerator.

$2^3 \cdot \underbrace{(-2 - 5)} + 4 \cdot (-1)$ Parentheses first

$2^3 \cdot \quad (-7) \quad + 4 \cdot (-1)$ Exponents next

$\underbrace{8} \cdot \quad (-7) \quad + \underbrace{4 \cdot (-1)}$ Multiply from left to right.

$\underbrace{-56 \quad + \quad (-4)}$ Add.

Numerator $\rightarrow \quad -60$

Now do the work in the denominator.

$4 + 5 \cdot \underbrace{(-6 \cdot 2)} + \underbrace{(5 \cdot 11)}$ Parentheses first

$4 + \underbrace{5 \cdot \quad (-12)} + \quad 55$ Multiply next.

$\underbrace{4 + \quad (-60)} \quad + \quad 55$ Add from left to right.

$\underbrace{-56 \quad + \quad 55}$

Denominator $\rightarrow \quad -1$

The last step is the division.

$$\frac{\text{Numerator} \rightarrow -60}{\text{Denominator} \rightarrow -1} = 60$$

45. $\underbrace{(-0.3)^2} + \underbrace{(-0.5)^2} + 0.9$ Exponents first

$\underbrace{0.09 \quad + \quad 0.25} + 0.9$ Add from left to right.

$\underbrace{0.34 \quad\quad + 0.9}$

1.24

49. $\underbrace{(0.5)^2} \cdot (-8) - (0.31)$ Exponents first

$\underbrace{0.25 \cdot (-8)} - (0.31)$ Multiply next.

$\underbrace{-2 \quad + (-0.31)}$ Change subtraction to addition.

-2.31

53. $\left(-\dfrac{1}{2}\right)^2 - \underbrace{\left(\dfrac{3}{4} - \dfrac{7}{4}\right)}$ Parentheses first

$\underbrace{\left(-\dfrac{1}{2}\right)^2} - \left(-\dfrac{4}{4}\right)$ Exponents next

$\underbrace{\dfrac{1}{4} \quad + \quad \left(+\dfrac{4}{4}\right)}$ Change subtraction to addition.

$\dfrac{5}{4}$

57. $5^2 \cdot \underbrace{(9 - 11)} \cdot (-3) \cdot (-2)^3$ Parentheses first

$5^2 \cdot \quad (-2) \quad \cdot (-3) \cdot (-2)^3$ Exponents next

$\underbrace{25 \cdot \quad (-2)} \quad \cdot (-3) \cdot \underbrace{(-8)}$ Multiply from left to right.

$\underbrace{-50 \quad\quad \cdot (-3)} \cdot (-8)$

$\underbrace{150 \quad\quad\quad \cdot (-8)}$

-1200

61. $(-2)^2 = 4 \quad\quad (-2)^6 = 64$

$(-2)^3 = -8 \quad\quad (-2)^7 = -128$

$(-2)^4 = 16 \quad\quad (-2)^8 = 256$

$(-2)^5 = -32 \quad\quad (-2)^9 = -512$

There is a pattern. When a negative number is raised to an even power, the answer is positive. When a negative number is raised to an odd power, the answer is negative.

65. $-7 \cdot \left(6 - \underbrace{\dfrac{5}{\overset{}{\underset{1}{8}}} \cdot \overset{3}{24}} + \underbrace{\dfrac{1}{\overset{}{\underset{1}{3}}} \cdot \dfrac{8}{\overset{3}{3}}}\right)$ Work inside parentheses; do multiplications first.

$-7 \cdot (\underbrace{6 - \quad 15 \quad + \quad 8}\ )$ Add and subtract inside parentheses, from left to right.

$-7 \cdot \quad \underbrace{(-9 \quad + 8)}$

$\underbrace{-7 \cdot \quad\quad (-1)}$ Multiply.

7

69. $\dfrac{\overset{2}{\cancel{6}}}{\underset{1}{7}} \cdot \dfrac{\overset{2}{\cancel{14}}}{\underset{3}{\cancel{9}}} = \dfrac{4}{3}$ or $1\dfrac{1}{3}$

Section 9.5 (page 619)

1. $2 \underset{\downarrow}{r} + 4 \underset{\downarrow}{s}$

$\underbrace{2(2)} + \underbrace{4(6)}$ Multiply from left to right.

$\underbrace{4 \quad + \quad 24}$ Add.

28

5. $2 \underset{\downarrow}{r} + 4 \underset{\downarrow}{s}$

$\underbrace{2(-4)} + \underbrace{4(4)}$ Multiply from left to right.

$\underbrace{-8 \quad + \quad 16}$ Add.

8

9. $2 \underset{\downarrow}{r} + 4 \underset{\downarrow}{s}$

$\underbrace{2(0)} + \underbrace{4(-2)}$ Multiply from left to right.

$\underbrace{0 \quad + \quad (-8)}$ Add.

-8

13. $6 \underset{\downarrow}{k} + 2 \underset{\downarrow}{s}$

$\underbrace{6(1)} + \underbrace{2(-2)}$ Multiply from left to right.

$\underbrace{6 \quad + \quad (-4)}$ Add.

2

17. $-\underset{\downarrow}{m} - 3 \underset{\downarrow}{n}$

$-\left(\dfrac{1}{2}\right) - \underbrace{3\left(\dfrac{3}{8}\right)}$ Multiply first.

$-\dfrac{1}{2} - \dfrac{9}{8}$ Find common denominator.

$-\dfrac{4}{8} + \left(-\dfrac{9}{8}\right)$ Change subtraction to addition.

$-\dfrac{13}{8}$

21. $-4x - y$

$-4(5) - (-15)$ Multiply first.

$-20 - (-15)$

$-20 + (+15)$ Change subtraction to addition.

-5

25. $\dfrac{-3s - t - 4}{-s + 6 + t} = \dfrac{-3(-1) - (-13) - 4}{-(-1) + 6 + (-13)}$

$= \dfrac{3 + (+13) - 4}{1 + 6 + (-13)}$

$= \dfrac{16 - 4}{7 + (-13)}$

$= \dfrac{12}{-6}$

$= -2$

29. $P = 2l + 2w$

$P = 2(9) + 2(5)$

$P = 18 + 10$

$P = 28$

33. $A = \dfrac{1}{2}bh$

$= \dfrac{1}{2}(15)(3)$

$= \dfrac{1}{2}(45)$

$= \dfrac{45}{2}$ or $22\dfrac{1}{2}$

37. $d = rt$

$= (53)(6)$

$= 318$

41. The error was made when replacing x with -3; should be $-(-3)$, not -3.

$-x - 4y$ Replace x with (-3) and y with (-1).

$-(-3) - 4(-1)$

$(+3) - (-4)$ Opposite of (-3) is $+3$.

$3 + (+4)$ Change subtraction to addition.

7 Add.

45. $V = \dfrac{4\pi r^3}{3}$

$\approx \dfrac{4(3.14)(3)^3}{3}$

$\approx \dfrac{4(3.14)(27)}{3}$

$\approx \dfrac{12.56(27)}{3}$

$\approx \dfrac{339.12}{3} \approx 113.04$

Use this formula to find the volume of a sphere.

49. $-\dfrac{1}{9} \cdot 9 = -\dfrac{1}{\overset{1}{\cancel{9}}} \cdot \dfrac{\overset{1}{\cancel{9}}}{1} = \dfrac{-1}{1} = -1$

Section 9.6 (page 627)

1. $x + 7 = 11$

$4 + 7 = 11$

$11 = 11$ True

Yes, 4 is a solution of the equation.

5. $2z - 1 = -15$

$2(-8) - 1 = -15$

$-16 - 1 = -15$

$-17 = -15$ False

No, -8 is not a solution of the equation.

9. $k + 15 = 0$ Check:

$k + 15 - 15 = 0 - 15$ $-15 + 15 = 0$

$k = -15$ $0 = 0$ True

The solution is -15.

13. $8 = r - 2$ Check:

$8 + 2 = r - 2 + 2$ $8 = 10 - 2$

$10 = r$ $8 = 8$ True

The solution is 10.

17. $7 = r + 13$ Check:

$7 - 13 = r + 13 - 13$ $7 = -6 + 13$

$-6 = r$ $7 = 7$ True

The solution is -6.

21. $-12 + x = -1$ Check:

$-12 + 12 + x = -1 + 12$ $-12 + 11 = -1$

$x = 11$ $-1 = -1$ True

The solution is 11.

25. $d + \dfrac{2}{3} = 3$ Check:

$d + \dfrac{2}{3} - \dfrac{2}{3} = 3 - \dfrac{2}{3}$ $\dfrac{7}{3} + \dfrac{2}{3} = 3$

$d = \dfrac{7}{3}$ or $2\dfrac{1}{3}$ $\dfrac{9}{3} = 3$

$3 = 3$ True

The solution is $\frac{7}{3}$ or $2\frac{1}{3}$.

29. $\dfrac{1}{2} = k - 2$ Check:

$\dfrac{1}{2} + 2 = k - 2 + 2$ $\dfrac{1}{2} = \dfrac{5}{2} - 2$

$\dfrac{5}{2} = k$ $\dfrac{1}{2} = \dfrac{1}{2}$ True

The solution is $\frac{5}{2}$ or $2\frac{1}{2}$.

33. $x - 0.8 = 5.07$ Check:

$x - 0.8 + 0.8 = 5.07 + 0.8$ $5.87 - 0.8 = 5.07$

$x = 5.87$ $5.07 = 5.07$

True

The solution is 5.87.

SOLUTIONS

37. $6z = 12$ Check:

$$\frac{\overset{1}{\cancel{6}} \cdot z}{\underset{1}{\cancel{6}}} = \frac{12}{6} \qquad 6(2) = 12$$

$$12 = 12 \quad \text{True}$$

$$z = 2$$

The solution is 2.

41. $3y = 0$ Check:

$$\frac{\overset{1}{\cancel{3}} \cdot y}{\underset{1}{\cancel{3}}} = \frac{0}{3} \qquad 3(0) = 0$$

$$0 = 0 \quad \text{True}$$

$$y = 0$$

The solution is 0.

45. $-36 = -4p$ Check:

$$-36 = -4(9)$$

$$\frac{-36}{-4} = \frac{-\overset{1}{\cancel{4}} \cdot p}{\underset{1}{-\cancel{4}}} \quad -36 = -36 \quad \text{True}$$

$$9 = p$$

The solution is 9.

49. $-8.4p = -9.24$ Check:

$$-8.4(1.1) = -9.24$$

$$\frac{-\overset{1}{\cancel{8.4}} \cdot p}{\underset{1}{-\cancel{8.4}}} = \frac{-9.24}{-8.4} \quad -9.24 = -9.24 \quad \text{True}$$

$$p = 1.1$$

The solution is 1.1.

53. $11 = \dfrac{a}{6}$ Replace $\frac{a}{6}$ with $\frac{1}{6}a$ and multiply both sides by 6.

$$6 \cdot 11 = \frac{\overset{1}{\cancel{6}}}{1} \cdot \frac{1}{\underset{1}{\cancel{6}}} a$$

Check:

$$11 = \frac{66}{6}$$

$$11 = 11 \quad \text{True}$$

$$66 = a$$

The solution is 66.

57. $-\dfrac{2}{5}p = 8$

$$-\frac{\overset{1}{\cancel{5}}}{\underset{1}{\cancel{2}}} \cdot -\frac{\overset{1}{\cancel{2}}}{\underset{1}{\cancel{5}}}p = -\frac{5}{\underset{1}{\cancel{2}}} \cdot \overset{4}{\cancel{8}}$$

Check:

$$-\frac{2}{5}(-20) = 8$$

$$8 = 8 \quad \text{True}$$

$$p = -20$$

The solution is -20.

61. $6 = \dfrac{3}{8}x$ Check:

$$6 = \frac{3}{8}(16)$$

$$\frac{8}{\underset{1}{\cancel{3}}} \cdot \overset{2}{\cancel{6}} = \frac{\overset{1}{\cancel{8}}}{\underset{1}{\cancel{3}}} \cdot \frac{\overset{1}{\cancel{3}}}{\underset{1}{\cancel{8}}} x \qquad 6 = 6 \quad \text{True}$$

$$16 = x$$

The solution is 16.

65. $\dfrac{z}{-3.8} = 1.3$

Check:

$$\frac{\overset{1}{-\cancel{3.8}}}{1} \cdot \frac{1}{\underset{1}{-\cancel{3.8}}} z = (-3.8)(1.3) \qquad \frac{-4.94}{-3.8} = 1.3$$

$$1.3 = 1.3 \quad \text{True}$$

$$z = -4.94.$$

The solution is -4.94.

69. $x - 17 = 5 - 3$

$$x - 17 = 2$$

$$x - 17 + 17 = 2 + 17$$

$$x = 19 \qquad \text{The solution is 19.}$$

73. $\dfrac{7}{2}x = \dfrac{4}{3}$

$$\frac{\overset{1}{\cancel{2}}}{\underset{1}{\cancel{7}}} \cdot \frac{\overset{1}{\cancel{7}}}{\underset{1}{\cancel{2}}}x = \frac{2}{7} \cdot \frac{4}{3}$$

$$x = \frac{8}{21} \qquad \text{The solution is } \frac{8}{21}.$$

77. $\dfrac{1}{2} + \dfrac{\overset{1}{\cancel{3}}}{\underset{1}{\cancel{4}}} \cdot \dfrac{\overset{2}{\cancel{8}}}{\underset{3}{\cancel{9}}} - \dfrac{1}{6}$ Multiply first.

$$\underbrace{\frac{1}{2} + \frac{2}{3}} - \frac{1}{6}$$ Add and subtract from left to right.

$$\underbrace{\frac{7}{6} - \frac{1}{6}}$$

$$\frac{6}{6} = 1$$

SECTION 9.7 (page 635)

1. $7p + 5 = 12$

$$7p + 5 - 5 = 12 - 5 \quad \text{Subtract 5 from both sides.}$$

$$7p = 7$$

$$\frac{\overset{1}{\cancel{7}} \cdot p}{\underset{1}{\cancel{7}}} = \frac{7}{7} \qquad \text{Divide both sides by 7.}$$

$$p = 1$$

Check:

$$7(1) + 5 = 12$$

$$7 + 5 = 12$$

$$12 = 12 \quad \text{True}$$

The solution is 1.

5.
$$-3m + 1 = 1$$
$$-3m + 1 - 1 = 1 - 1 \quad \text{Subtract 1 from both sides.}$$
$$-3m = 0$$
$$\frac{\overset{1}{\cancel{-3}} \cdot m}{\underset{1}{\cancel{-3}}} = \frac{0}{-3} \quad \text{Divide both sides by } -3.$$
$$m = 0$$

Check:
$$-3(0) + 1 = 1$$
$$0 + 1 = 1$$
$$1 = 1 \quad \text{True}$$

The solution is 0.

9.
$$-5x - 4 = 16$$
$$-5x - 4 + 4 = 16 + 4 \quad \text{Add 4 to both sides.}$$
$$-5x = 20$$
$$\frac{\overset{1}{\cancel{-5}} \cdot x}{\underset{1}{\cancel{-5}}} = \frac{20}{-5} \quad \text{Divide both sides by } -5.$$
$$x = -4$$

Check:
$$-5(-4) - 4 = 16$$
$$20 - 4 = 16$$
$$16 = 16 \quad \text{True}$$

The solution is -4.

13. $6(x + 4) = 6 \cdot x + 6 \cdot 4 = 6x + 24$

17. $-3(m + 6) = -3 \cdot m + (-3) \cdot 6 = -3m - 18$

21. $-5(z - 9) = -5 \cdot z - (-5) \cdot 9 = -5z - (-45)$
$$= -5z + 45$$

25. $10x - 2x = (10 - 2)x = 8x$

29. $3y - y - 4y = (3 - 1 - 4)y = -2y$

33. $54 = 10m - m$ Combine like terms.
$$54 = 9m$$
$$\frac{54}{9} = \frac{\overset{1}{\cancel{9}} \cdot m}{\underset{1}{\cancel{9}}} \quad \text{Divide both sides by 9.}$$
$$6 = m$$

Check:
$$54 = 10(6) - 6$$
$$54 = 60 - 6$$
$$54 = 54 \quad \text{True}$$

The solution is 6.

37. $-12 = 6y - 18y$ Combine like terms.
$$-12 = -12y$$
$$\frac{-12}{-12} = \frac{\overset{1}{\cancel{-12}} \cdot y}{\underset{1}{\cancel{-12}}} \quad \text{Divide both sides by } -12.$$
$$1 = y$$

Check:
$$-12 = 6(1) - 18(1)$$
$$-12 = 6 - 18$$
$$-12 = -12 \quad \text{True}$$

The solution is 1.

41.
$$9 + 7z = 9z + 13$$
$$9 + 7z - 9z = 9z + 13 - 9z \quad \text{Subtract } 9z \text{ from both sides.}$$
$$9 - 2z = 13$$
$$9 - 2z - 9 = 13 - 9 \quad \text{Subtract 9 from both sides.}$$
$$-2z = 4$$
$$\frac{\overset{1}{\cancel{-2}} \cdot z}{\underset{1}{\cancel{-2}}} = \frac{4}{-2} \quad \text{Divide both sides by } -2.$$
$$z = -2$$

Check:
$$9 + 7(-2) = 9(-2) + 13$$
$$9 + (-14) = -18 + 13$$
$$-5 = -5 \quad \text{True}$$

The solution is -2.

45.
$$b + 3.05 = 2$$
$$b + 3.05 - 3.05 = 2 - 3.05 \quad \text{Subtract 3.05 from both sides.}$$
$$b = -1.05$$

Check:
$$-1.05 + 3.05 = 2$$
$$2 = 2 \quad \text{True}$$

The solution is -1.05.

49.
$$-10 = 2(y + 4) \quad \text{Use distributive property.}$$
$$-10 = 2y + 8$$
$$-10 - 8 = 2y + 8 - 8 \quad \text{Subtract 8 from both sides.}$$
$$-18 = 2y$$
$$\frac{-18}{2} = \frac{\overset{1}{\cancel{2}} \cdot y}{\underset{1}{\cancel{2}}} \quad \text{Divide both sides by 2.}$$
$$-9 = y$$

Check:
$$-10 = 2(-9 + 4)$$
$$-10 = 2(-5)$$
$$-10 = -10 \quad \text{True}$$

The solution is -9.

53.
$$6(x - 5) = -30 \quad \text{Use distributive property}$$
$$6x - 30 = -30$$
$$6x - 30 + 30 = -30 + 30 \quad \text{Add 30 to both sides.}$$
$$6x = 0$$
$$\frac{\overset{1}{\cancel{6}} \cdot x}{\underset{1}{\cancel{6}}} = \frac{0}{6} \quad \text{Divide both sides by 6.}$$
$$x = 0$$

Check:

$$6(0 - 5) = -30$$
$$6(-5) = -30$$
$$-30 = -30 \quad \text{True}$$

The solution is 0.

57. $\quad 30 - 40 = -2x + 7x - 4x \quad$ Combine like terms.

$$-10 = (-2 + 7 - 4)x$$
$$-10 = 1x$$
$$-10 = x$$

The solution is -10.

61. $\qquad \dfrac{y}{2} - 2 = \dfrac{y}{4} + 3$

$\dfrac{y}{2} - 2 + 2 = \dfrac{y}{4} + 3 + 2 \qquad$ Add 2 to both sides.

$\qquad \dfrac{y}{2} = \dfrac{y}{4} + 5 \qquad$ Write $\dfrac{y}{2}$ as $\dfrac{1}{2}y$ and $\dfrac{y}{4}$

$\qquad \dfrac{1}{2}y = \dfrac{1}{4}y + 5 \qquad$ as $\dfrac{1}{4}y$.

$\dfrac{1}{2}y - \dfrac{1}{4}y = \dfrac{1}{4}y - \dfrac{1}{4}y + 5 \qquad$ Subtract $\dfrac{1}{4}y$ from both

$\qquad\qquad\qquad\qquad$ sides.

$\qquad\qquad \dfrac{1}{4}y = 5$

$\dfrac{\overset{1}{\cancel{4}}}{1} \cdot \dfrac{1}{\underset{1}{\cancel{4}}}y = \dfrac{4}{1} \cdot 5 \qquad$ Multiply both sides by $\dfrac{4}{1}$.

$\qquad\qquad y = 20 \qquad$ The solution is 20.

65. Sales tax $= (0.06)(\$420) = \$25.20$

Section 9.8 (page 643)

1. $14 + x \quad$ or $\quad x + 14 \qquad$ **5.** $20 - x$

9. $x - 4 \qquad$ **13.** $2x$

17. $8 + 2x \quad$ or $\quad 2x + 8$

21. $2x + x \quad$ or $\quad x + 2x$

25. Let n represent the unknown number.

$$\underbrace{\text{four times a number}}_{4n} \underbrace{\text{decreased by 2}}_{-2} \underbrace{\text{result is}}_{=} \overset{\downarrow}{\underbrace{26}_{26}}$$

$$4n - 2 = 26$$
$$4n - 2 + 2 = 26 + 2$$
$$4n = 28$$
$$\dfrac{\overset{1}{\cancel{4}} \cdot n}{\underset{1}{\cancel{4}}} = \dfrac{28}{4}$$
$$n = 7$$

The number is 7.

Check: 4 times 7 is 28

        28 decreased by 2 is 26   True

29. Let n represent the unknown number.

$$\underbrace{\text{half a number}}_{\frac{1}{2}n} \underbrace{\text{is added to}}_{+} \underbrace{\text{twice the number}}_{2n} \underbrace{\text{answer is}}_{=} \overset{\downarrow}{\underbrace{50}_{50}}$$

$$2\dfrac{1}{2}n = 50$$
$$\dfrac{5}{2}n = 50$$
$$\dfrac{\overset{1}{\cancel{2}}}{\underset{1}{\cancel{5}}} \cdot \dfrac{\overset{1}{\cancel{5}}}{\underset{1}{\cancel{2}}}n = \dfrac{2}{\cancel{5}} \cdot \overset{10}{\cancel{50}}$$
$$n = 20$$

The number is 20.

Check: half of 20 is 10

        twice 20 is 40

        10 added to 40 is 50   True

33. Let d represent the number of days Kerwin rented the saw.

$$\underbrace{\text{one-time \$9 fee}}_{9} \underbrace{\text{added to}}_{+} \underbrace{\$16 \text{ per day}}_{16 \cdot d} \underbrace{\text{total is}}_{=} \overset{\downarrow}{\underbrace{\$89}_{89}}$$

$$9 + 16d = 89$$
$$9 + 16d - 9 = 89 - 9$$
$$16d = 80$$
$$\dfrac{\overset{1}{\cancel{16}} \cdot d}{\underset{1}{\cancel{16}}} = \dfrac{80}{16}$$
$$d = 5$$

Kerwin rented the saw for 5 days.

Check: \$16 per day for 5 days is \$80

        \$80 plus \$9 one-time fee is \$89   True

37. Let w represent the width of the rectangle. The length will be twice w, which is $2w$.

$$P = 2 \cdot \text{length} + 2 \cdot \text{width}$$
$$36 = 2 \cdot 2w + 2 \cdot w$$
$$36 = 4w + 2w$$
$$36 = 6w$$
$$\dfrac{36}{6} = \dfrac{\overset{1}{\cancel{6}} \cdot w}{\underset{1}{\cancel{6}}}$$
$$6 = w$$

Width $= w$ so width is 6 ft.

Length $= 2w$ so length is $2(6)$ or 12 ft.

Check: $36 = 2(12) + 2(6)$

        $36 = 24 + 12$

        $36 = 36 \qquad\qquad$ True

41. Let x represent the length of each equal part.
The length of the third part is $x + 25$

Total length is 706

$$x + x + x + 25 = 706$$
$$3x + 25 = 706$$
$$3x + 25 - 25 = 706 - 25$$
$$3x = 681$$

$$\frac{\overset{1}{\cancel{3}} \cdot x}{\underset{1}{\cancel{3}}} = \frac{681}{3}$$

$$x = 227$$

Two parts are each 227 m long.
The third part is $227 + 25$ or 252 m long.
Check: $227 + 227 + 252 = 706$ True

45. amount = percent • base

$$a = 0.80 \cdot 2900$$
$$a = 2320$$

80% of $2900 is $2320.

CHAPTER 10

SECTION 10.1 (page 667)

1. The total cost of adding the art studio is
$12,100 + $9800 + $2000 + $900 + $1800
+ $3000 + $2400 = $32,000

5. $\dfrac{\text{carpentry}}{\text{window coverings}} = \dfrac{12100}{900} = \dfrac{121}{9}$

9. $\dfrac{\text{business majors}}{\text{total students}} = \dfrac{4000}{12000} = \dfrac{1}{3}$

13. $\dfrac{\text{business majors}}{\text{science majors}} = \dfrac{4000}{1000} = \dfrac{4}{1}$

17. Total cost = $1,740,000
Cost of doors and thresholds = 10% of total cost
$$= 0.10 \cdot \$1,740,000$$
$$= \$174,000$$

21. Total employees = 8740
1–2 days of missed work = 22% of total employees
$$= 0.22 \cdot 8740$$
$$\approx 1923 \text{ employees}$$

25. Total employees = 8740
7 or more days of missed work
$$= 3\% \text{ of total employees}$$
$$= 0.03 \cdot 8740$$
$$\approx 262 \text{ employees}$$

29. 25% of total is rent.
number of degrees $= 25\% \cdot 360°$
$$= 0.25 \cdot 360°$$
$$= 90°$$

33. percent of total $= \dfrac{630}{4200} = 0.15 = 15\%$

37. (a) Total Sales
$$= \$12,500 + \$40,000 + \$60,000 + \$50,000$$
$$+ \$37,500 = \$200,000$$

(b) Adventure classes = $12,500
percent of total $= \dfrac{12,500}{200,000} = 0.0625 = 6.25\%$
number of degrees $= 0.0625 \cdot 360° = 22.5°$
Grocery and provision sales = $40,000
percent of total $= \dfrac{40,000}{200,000} = 0.2 = 20\%$
number of degrees $= 0.2 \cdot 360° = 72°$
Equipment rentals = $60,000
percent of total $= \dfrac{60,000}{200,000} = 0.3 = 30\%$
number of degrees $= 0.3 \cdot 360° = 108°$.
Rafting tours = $50,000
percent of total $= \dfrac{50,000}{200,000} = 0.25 = 25\%$
number of degrees $= 0.25 \cdot 360° = 90°$
Equipment sales = $37,500
percent of total $= \dfrac{37,500}{200,000} = 0.1875 = 18.75\%$
number of degrees $= 0.1875 \cdot 360° = 67.5°$

(c) See graph in the answer section.

41. $\dfrac{4}{5} = \dfrac{32}{40}$ and $\dfrac{7}{8} = \dfrac{35}{40}$; therefore, $\dfrac{4}{5} < \dfrac{7}{8}$

45. $25.9\% < 26.01\%$

SECTION 10.2 (page 677)

1. There were 8000 motorcycles stolen in 1994.

5. Motorcycles stolen in 1995 = 9500
Motorcycles stolen in 1993 = 7000

$$9500 - 7000 = 2500$$

There were 2500 more motorcycles stolen in 1995.

9. Unemployed workers in February of 1998 = 7000
Unemployed workers in February of 1997 = 5500

$$7000 - 5500 = 1500$$

There were 1500 more workers unemployed in
February of 1998.

13. 150,000 gallons of supreme unleaded gasoline were
sold in 1994.

17. 700,000 gallons of supreme unleaded gasoline were
sold in 1998. 150,000 gallons of supreme unleaded
gasoline were sold in 1994.

$$700,000 - 150,000 = 550,000$$

There was an increase of 550,000 gallons of supreme
unleaded gasoline sales.

21. Number of burglaries in March = 400
Number of burglaries in April   = 600

$$600 - 400 = 200$$

From March to April, the number of burglaries
increased by 200.

25. Chain store A sold 3,000,000 CDs in 1998.

29. Chain store B sold 3,500,000 CDs in 1997.

33. Answers will vary. A sample answer follows:
In a bar graph or line graph, a single bar or a single line must be used for each set of data. To show multiple sets of data, multiple sets of bars or lines must be used.

37. The total sales in 1996 were $25,000.

41. The decrease in sales from 1995 to 1996 may have resulted from poor service or greater competition. The increase in sales from 1996 to 1998 may have been a result of more advertising or better service.

45. Total budget = $59,703,876
Cost of salaries and benefits = 71% of total budget
$$= 0.71 \cdot \$59,703,876$$
$$\approx \$42,390,000$$

SECTION 10.3 (page 683)

1. The age group of 60–65 years has the greatest number of members, which is 16,000.

5. Total members in 41–60 years age group
$$= 10,000 + 15,000 + 11,000 + 14,000$$
$$= 50,000$$

9. There are 7 employees who earn $16–$20 thousand.

13. Class intervals are the result of combining data into groupings. Class frequency is the number of data items that fit in each class frequency. These are used to group data and to have multiple responses (frequency) in a class interval—this makes the data easier to interpret.

17.

Number of Sets	Tally	Number of Staff
140–149	⊔⊤⊤ I	6

21.

Number of Credits	Tally	Number of Employees
0–24	IIII	4

25.

Number of Credits	Tally	Number of Employees
100–124	⊔⊤⊤	5

29.

New Homes Sold	Tally	Number of Salespeople
1–5	⊔⊤⊤ ⊔⊤⊤ I	11

33.

New Homes Sold	Tally	Number of Salespeople
21–25	⊔⊤⊤ III	8

37. $(18 + 53 + 31 + 26) \div 5 = 128 \div 5 = 25.6$

41. $(4 \cdot 3) + (3 \cdot 9) + (2 \cdot 3) \div 15$
$$= 12 + 27 + 6 \div 15$$
$$= 12 + 27 + 0.4$$
$$= 39 + 0.4$$
$$= 39.4$$

SECTION 10.4 (page 691)

1. Mean
$$= \frac{3 + 5 + 7 + 12 + 18 + 27}{6} = \frac{72}{6} = 12 \text{ years}$$

5. Mean
$$= \frac{21,900 + 22,850 + 24,930 + 29,710 + 28,340 + 40,000}{6}$$
$$= \frac{167,730}{6}$$
$$= \$27,955$$

9.

Policy Amount	Number of Policies Sold	Product
$10,000	6	10,000 · 6 = 60,000
$20,000	24	20,000 · 24 = 480,000
$25,000	12	25,000 · 12 = 300,000
$30,000	8	30,000 · 8 = 240,000
$50,000	5	50,000 · 5 = 250,000
$100,000	3	100,000 · 3 = 300,000
$250,000	2	250,000 · 2 = 500,000
Totals	60	2,130,000

$$\text{Weighted mean} = \frac{2,130,000}{60} = \$35,500$$

13.

Value	Frequency	Product
12	4	12 · 4 = 48
13	2	13 · 2 = 26
15	5	15 · 5 = 75
19	3	19 · 3 = 57
22	1	22 · 1 = 22
23	5	23 · 5 = 115
Totals	20	343

$$\text{Weighted mean} = \frac{343}{20} = 17.15 \approx 17.2 \text{ (rounded)}$$

17. 328 420 483 549 592 715

Since there is an even number of customers the median is $(483 + 549) \div 2 = 516$ customers

21. 74, 68, 68, 68, 75, 75, 74, 74, 70

Because both 68 and 74 occur three times, each is a mode. This list is bimodal.

25. The median is a better average when the list contains one or more extreme values.
Find the mean and the median of the following home values. $82,000; $64,000; $91,000; $115,000; $982,000
mean home value = $266,800
median home value = $91,000

29. 23, 34, 40, 47, 48, 51, 56, 95, 96

Median is the middle number, which is 48.

Index

Video Correlation Guide

The purpose of this guide is to show those exercises from the text that are used in the Real to Reel videotape series that accompanies *Basic College Mathematics,* Fifth Edition.

Section	Exercises	Section	Exercises
1.1	13, 17, 21	6.1	7, 27, 47
1.2	37, 49, 85	6.2	9, 25, 77
1.3	9, 19, 35, 55, 69, 83	6.3	9, 19
1.4	17, 21, 41, 74, 87	6.4	15, 13, 31
1.5	13, 15, 17, 19, 25, 37,	6.5	13
	45, 75	6.6	3, 7, 13, 17
1.6	23, 41	6.7	31
1.7	37, 47, 63	6.8	17, 43
1.8	8, 55, 75	6.9	11
1.9	23	7.1	39
2.1	19	7.2	35
2.2	27	7.3	37
2.3	43	7.4	5
2.4	23	7.5	25
2.5	11, 43	8.1	19
2.6	5	8.2	7
2.7	7, 11, 17, 21	8.3	18
2.8	3, 17, 27	8.4	13
3.1	29	8.5	11
3.2	27	8.6	7
3.3	31	8.7	3, 7, 19
3.4	47	8.8	3, 19, 29, 41
3.5	55	8.9	9, 17
4.1	35	9.1	19
4.2	9	9.2	69
4.3	19	9.3	89
4.4	27	9.4	53
4.5	39	9.5	15
4.6	3, 17	9.6	25
4.7	13, 61	9.7	43
		9.8	33, 41
5.1	47	10.1	21
5.2	25	10.2	25
5.3	25	10.3	
5.4	21	10.4	17, 21
5.5	17		